S0-BLI-542

The chemistry of
the azido group

THE CHEMISTRY OF FUNCTIONAL GROUPS

A series of advanced treatises under the general editorship of Professor Saul Patai

The chemistry of alkenes (published in 2 volumes)
The chemistry of the carbonyl group (published in 2 volumes)
The chemistry of the ether linkage (published)
The chemistry of the amino group (published)
The chemistry of the nitro and nitroso group (published in two parts)
The chemistry of the carboxylic acids and esters (published)
The chemistry of the carbon–nitrogen double bond (published)
The chemistry of the amides (published)
The chemistry of the cyano group (published)
The chemistry of the hydroxyl group (published in 2 parts)
The chemistry of the azido group (published)

$—N_3$

The chemistry of
the azido group

Edited by

SAUL PATAI

The Hebrew University, Jerusalem

1971

INTERSCIENCE PUBLISHERS

a division of John Wiley & Sons

LONDON — NEW YORK — SYDNEY — TORONTO

Library of Congress Catalog Card No. 73-149579
ISBN 0 471 66925 3

Made and printed in Great Britain by
William Clowes & Sons, Limited, London, Beccles and Colchester

Contributing authors

R. A. Abramovitch	University of Alabama, Alabama 35486, U.S.A.
D. V. Banthorpe	University College London, London W.C.1.
M. E. C. Biffin	Defence Standards Laboratories, Ascot Vale, Victoria, Australia 3032.
J. E. Gurst	University of West Florida, Pensacola, Florida 32504, U.S.A.
E. P. Kyba	University of Alabama, Alabama 35486, U.S.A.
W. Lwowski	New Mexico State University, Las Cruces, New Mexico 88001, U.S.A.
J. Miller	University of Natal, Durban, S. Africa.
D. B. Paul	Defence Standards Laboratories, Ascot Vale, Victoria, Australia 3032.
C. A. Pryde	Bell Telephone Laboratories, New Jersey 07974, U.S.A.
A. Reiser	Research Laboratory, Kodak Ltd., Middlesex, England.
T. Sheradsky	The Hebrew University, Jerusalem, Israel.
G. Smolinsky	Bell Telephone Laboratories, New Jersey 07974, U.S.A.
A. Treinin	The Hebrew University, Jerusalem, Israel.
H. M. Wagner	Research Laboratory, Kodak Ltd., Middlesex, England.

Foreword

With this volume, the 'Chemistry of the Functional Groups' series is passing its half-way mark. It has been both praised and criticized as was to be expected but on the whole it seems that the series constitutes a serviceable and reasonably widely used contribution to the literature of organic chemistry.

The chemistry of the azido group, as far as I am aware, has not been treated as yet in a single volume, although, especially in the last two decades, many interesting studies have been carried out on this subject. I hope that the present volume will serve to stimulate interest and research on the azido group, which compared to the other main functional groups of organic compounds seems to be somewhat neglected.

Unfortunately, three of the originally planned chapters of the book did not materialize. These should have been chapters on the 'Biological Formation and Reactions of the Azido Groups', on the 'Mass Spectra of Azides' and on the 'Syntheses and Uses of Isotopically Labelled Azides.'

Jerusalem, April 1971

SAUL PATAI

The Chemistry of the Functional Groups

Preface to the series

The series 'The Chemistry of the Functional Groups' is planned to cover in each volume all aspects of the chemistry of one of the important functional groups in organic chemistry. The emphasis is laid on the functional group treated and on the effects which it exerts on the chemical and physical properties, primarily in the immediate vicinity of the group in question, and secondarily on the behaviour of the whole molecule. For instance, the volume *The Chemistry of the Ether Linkage* deals with reactions in which the C—O—C group is involved, as well as with the effects of the C—O—C group on the reactions of alkyl or aryl groups connected to the ether oxygen. It is the purpose of the volume to give a complete coverage of all properties and reactions of ethers in as far as these depend on the presence of the ether group, but the primary subject matter is not the whole molecule, but the C—O—C functional group.

A further restriction in the treatment of the various functional groups in these volumes is that material included in easily and generally available secondary or tertiary sources, such as Chemical Reviews, Quarterly Reviews, Organic Reactions, various 'Advances' and 'Progress' series as well as textbooks (i.e. in books which are usually found in the chemical libraries of universities and research institutes) should not, as a rule, be repeated in detail, unless it is necessary for the balanced treatment of the subject. Therefore each of the authors is asked *not* to give an encyclopaedic coverage of his subject, but to concentrate on the most important recent developments and mainly on material that has not been adequately covered by reviews or other secondary sources by the time of writing of the chapter, and to address himself to a reader who is assumed to be at a fairly advanced post-graduate level.

With these restrictions, it is realized that no plan can be devised for a volume that would give a *complete* coverage of the subject with *no* overlap between chapters, while at the same time preserving the readability of the text. The Editor set himself the goal of attaining *reasonable* coverage with *moderate* overlap, with a minimum of cross-references between the chapters of each volume. In this manner, sufficient freedom is given to each author to produce readable quasi-monographic chapters.

The general plan of each volume includes the following main sections:

(a) An introductory chapter dealing with the general and theoretical aspects of the group.

(b) One or more chapters dealing with the formation of the functional group in question, either from groups present in the molecule, or by introducing the new group directly or indirectly.

(c) Chapters describing the characterization and characteristics of the functional groups, i.e., a chapter dealing with qualitative and quantitative methods of determination including chemical and physical methods, ultraviolet, infrared, nuclear magnetic resonance, and mass spectra; a chapter dealing with activating and directive effects exerted by the group and/or a chapter on the basicity, acidity or complex-forming ability of the group (if applicable).

(d) Chapters on the reactions, transformations and rearrangements which the functional group can undergo, either alone or in conjunction with other reagents.

(e) Special topics which do not fit any of the above sections, such as photochemistry, radiation chemistry, biochemical formations and reactions. Depending on the nature of each functional group treated, these special topics may include short monographs on related functional groups on which no separate volume is planned (e.g. a chapter on 'Thioketones' is included in the volume *The Chemistry of the Carbonyl Group*, and a chapter on 'Ketenes' is included in the volume *The Chemistry of Alkenes*). In other cases, certain compounds, though containing only the functional group of the title, may have special features so as to be best treated in a separate chapter, as e.g., 'Polyethers' in *The Chemistry of The Ether Linkage*, or 'Tetraaminoethylenes' in *The Chemistry of the Amino Group*.

This plan entails that the breadth, depth and thought-provoking nature of each chapter will differ with the views and inclinations of the author and the presentation will necessarily be somewhat uneven. Moreover, a serious problem is caused by authors who deliver their manuscript late or not at all. In order to overcome this problem at least to some extent, it was decided to publish certain volumes in several parts, without giving consideration to the originally planned logical order of the chapters. If after the appearance of the originally planned parts of a volume it is found that either owing to non-delivery of chapters, or to new developments in the subject, sufficient material has accumulated for publication of an additional part, this will be done as soon as possible.

The overall plan of the volumes in the series 'The Chemistry of the Functional Groups' includes the titles listed below:

The Chemistry of the Alkenes (published in two volumes)
The Chemistry of the Carbonyl Group (published in two volumes)
The Chemistry of the Ether Linkage (published)
The Chemistry of the Amino Group (published)
The Chemistry of the Nitro and the Nitroso Group (published in two parts)
The Chemistry of Carboxylic Acids and Esters (published)
The Chemistry of the Carbon–Nitrogen Double Bond (published)
The Chemistry of the Cyano Group (published)
The Chemistry of the Amides (published)
The Chemistry of the Hydroxyl Group (published in two parts)
The Chemistry of the Azido Group (published)
The Chemistry of Carbonyl Halides (in press)
The Chemistry of the Carbon–Halogen Bond (in preparation)
The Chemistry of the Quinonoid Compounds (in preparation)
The Chemistry of the Carbon–Carbon Triple Bond
The Chemistry of Imidoates and Amidines
The Chemistry of the Thiol Group
The Chemistry of the Hydrazo, Azo and Azoxy Groups
The Chemistry of the SO, —SO_2, —SO_2H and —SO_3H Groups
The Chemistry of the —OCN, —NCO and —SCN Groups
The Chemistry of the —PO_3H_2 and Related Groups

Advice or criticism regarding the plan and execution of this series will be welcomed by the Editor.

The publication of this series would never have started, let alone continued, without the support of many persons. First and foremost among these is Dr. Arnold Weissberger, whose reassurance and trust encouraged me to tackle this task, and who continues to help and advise me. The efficient and patient cooperation of several staff-members of the Publisher also rendered me invaluable aid (but unfortunately their code of ethics does not allow me to thank them by name). Many of my friends and colleagues in Jerusalem helped me in the solution of various major and minor matters, and my thanks are due especially to Prof. Z. Rappoport and Dr. J. Zabicky. Carrying out such a long-range project would be quite impossible without the non-professional but none the less essential participation and partnership of my wife.

The Hebrew University, SAUL PATAI
Jerusalem, ISRAEL

Contents

1. General and theoretical aspects — 1
   A. Treinin

2. Introduction of the Azido Group — 57
   M. E. C. Biffin, J. Miller and D. B. Paul

3. Characterization and determination of organic azides — 191
   J. E. Gurst

4. The directing and activating effects of azido groups — 203
   M. E. C. Biffin, J. Miller, and D. B. Paul

5. Decomposition of organic azides — 221
   R. A. Abramovitch and E. P. Kyba

6. Azides as synthetic starting materials — 331
   T. Sheradsky

7. Rearrangements involving azide groups — 397
   D. V. Banthorpe

8. Photochemistry of the azido group — 441
   A. Reiser and H. M. Wagner

9. Acyl azides — 503
   W. Lwowski

10. The chemistry of vinyl azides — 555
   G. Smolinsky and C. A. Pryde

Author index — 587

Subject index — 617

CHAPTER 1

General and theoretical aspects

A. TREININ

The Hebrew University, Jerusalem, Israel

I. INTRODUCTION 2
A. Azides as Pseudohalides 2
B. Electronic Structure (Simplified Model) 4
1. σ bonds 5
2. π bonds 6
3. Lone pairs 7
C. Primitive Hückel-type Calculations 8
D. Bent Structure of the Azido Group 12
E. Some Data on Nitrogen Atoms and N—N Bonds . . . 13
II. SOME STRUCTURAL PROPERTIES OF THE AZIDO GROUP . . . 14
A. Geometry 14
B. Some Thermodynamic Data 16
C. Electrical Dipole Moments 18
D. Vibration Spectra 21
1. N_3^- and N_3 (azide radical) 21
2. Covalent azides 22
E. Force Constants 24
F. Internal Rotation and Hyperconjugation 25
G. Electronic Spectra 25
1. The azide ion 25
2. HN_3 and aliphatic azides 27
3. Aromatic azides 30
4. Rydberg transitions and ionization potentials . . . 30
H. Optical Rotatory Dispersion (o.r.d.) and Circular Dichroism (c.d.) 31
I. The Quadrupole Coupling and Nuclear Magnetic Resonance 33
III. THEORETICAL 34
A. General 34
B. Minimal Basis Set for N_3^- 37
C. Walsh's Correlation Diagram for N_3^- 38
D. Self-consistent Hückel Molecular Orbitals (HMO) . . . 40
1. The method of Wagner 41

2. The method of Mulliken–Wolfsberg–Helmholz . . . 44
E. Semi-empirical Self-consistent Field Methods 45
1. The Pariser–Parr–Pople method 45
2. All-valence electrons calculations 47
F. LCAO Self-consistent Field (Hartree–Fock) Molecular Orbitals 48
1. Slater-type atomic orbitals 48
2. Gaussian lobe functions representation 50
G. Other than LCAO Methods 50
H. Conclusion 52
IV. References 52

I. INTRODUCTION

The reviews by Evans, Gray and Yoffe[1] and by Gray[2] give very valuable surveys on the physical properties, thermodynamics and decomposition processes of inorganic azides. More recent are the chapters by Green[3] on bonding in nitrogen compounds, Yoffe[4] on inorganic azides, Mason[5] on hydrogen and metallic azides and Lappert and Pyszora[6] on the pseudohalides of group IIIB and IVB elements. Still, they are not up-to-date and most of the theoretical work on the electronic structure and properties of azides, carried out in the last ten years, has not yet been reviewed. As to organic azides, the situation is even less satisfactory: the structure and reactivity of organic azides were briefly reviewed by Lieber, Curtice and Rao[7], thus filling up some of the gap since the early review of Boyer and Canter[8].

The purpose of this chapter is to discuss the physical chemistry of the azido group. It consists of three parts. Section I gives a general discussion of the properties and structure of the azido group. It presents some very approximate calculations, which serve as introduction for the more refined theoretical treatment given in section III. (The reader who finds the last part difficult can still get some theoretical insight from section I.) Section II summarizes the basic experimental data concerning the azido group. The discussion of these data is based on the models and approximations outlined in section I.

A. Azides as Pseudohalides

The azide radical, N_3, belongs to a group of inorganic radicals which have certain properties in common with the halogen atoms. Other members of this group are CN, NCO, SCN, etc. For this reason the corresponding compounds are called *pseudohalides*. In common with the halogens they form either an anion X^- or a covalent bond R—X. Their acids HX are weaker than the hydrogen halides (the *dissociation constant* of HN_3 is $2 \cdot 04 \times 10^{-5}$ M at 25°C)[9]. The radicals are linear and more electronegative than carbon; hence they usually exert a negative *inductive effect* (−I) in organic compounds. Like the halogens they are

TABLE 1. Resemblance between N_3 and Br

Property	$X = N_3$	$X = Br$	Ref.
Normal boiling point (°C):[a]			
C_2H_5X	49	38	12
n-$C_7H_{15}X$	178	180	12
Dissociation constant at 26°C:[b]			
CH_2XCOOH	$0 \cdot 93 \times 10^{-3}$	$1 \cdot 38 \times 10^{-3}$	8
$CH_3CHXCOOH$	$0 \cdot 9 \times 10^{-3}$	$1 \cdot 1 \times 10^{-3}$	8
Dipole moment (D):[a]			
C_6H_5X	1·44	1·54	13, 14
Parachor (cm^3):[c]			
X	77	69	15
Molar refraction (cm^3):[d]			
X	9·4	8·9	13, 16
X^-	12·3	12·5	16, 17
Molar magnetic rotation:[d]			
C_6H_5X (at 15°C)	14·77	14·51	18
Effective radius (Å):[e]			
X^- in crystals	2·04	1·95	19
Charge-transfer-to-solvent spectrum:[f]			
λ_{max} of X^- (nm)	~200	~200	20
Electronegativity of X:[g]	2·71	2·8	2
Nucleophilic constant of X^-, E_n:[h]	1·58	1·51	

[a] More data on boiling points and dipole moments are available which show the resemblance between N_3 and Br (see the corresponding references).

[b] The large increase in dissociation constant of acetic acid ($K = 1 \cdot 8 \times 10^{-5}$) on substitution proves the −I inductive effect of X. The inductive effect of N_3 lies between that of Br and I[8].

[c] Comparison of *parachor* values amounts to a comparison of molecular volumes at constant surface tension. The parachor played an important role in establishing the linear structure of the azido group.

[d] The molar *refraction* [R] and *magnetic rotation* [M] were measured with the sodium D line; the value of [R] for the azido group was derived from alkyl azides. In phenyl derivatives it increases to about 10·2[13]. This results from conjugation of N_3 with the aromatic ring (*optical exaltation* effect); [M] is related to the *Faraday effect* (see reference 16 for a simple discussion, and reference 18a for a recent review.

[e] For discussion of the effective radius of N_3^- in crystals see section II.A.

[f] For discussion of the charge-transfer-to-solvent spectrum of N_3^- see section II.G.

[g] The electronegativity of N_3 is between that of Br and I (2·5). The Mulliken scale is used here[2].

[h] $E_n = E° + 2 \cdot 60$, where E_0 is the electrode potential; see J. O. Edwards, *J. Am. Chem. Soc.*, **76**, 1540 (1954).

also capable of *conjugation* through their non-bonding electrons and thus can act as electron-donors, i.e. they exert a positive *resonance effect* (+R). When this π-interaction is strong the pseudohalide group may become colinear with the atom to which it is bound[6]. This resonance effect is responsible for the effect of the azido group on a benzene ring, its *ortho–para* orienting character and its activating influence on electrophilic substitution.

A significant difference between halides and pseudohalides lies in the 'unsaturation' of the latter; they contain low-lying, unfilled π orbitals which can accept electrons. This puts them higher than halides in the *spectrochemical series*[10].

The reducing power of the halide and pseudohalide ions increases in the following order[11]: F^-, NCO^-, Cl^-, N_3^-, Br^-, SCN^-, I^-. In this and other respects the azido group shows close resemblance to bromine (Table 1).

The data recorded in Table 1 describe the behaviour of the azido group in its equilibrium nuclear configuration. However the main interest in this group lies in those properties which depend on changes in its geometry, its bending to form cyclic compounds[7] and its dissociation to $-N + N_2$. Obviously this has no parallel in the halogen series.

B. Electronic Structure (Simplified Model)

The general shape of *covalent azides* is shown in Figure 1.

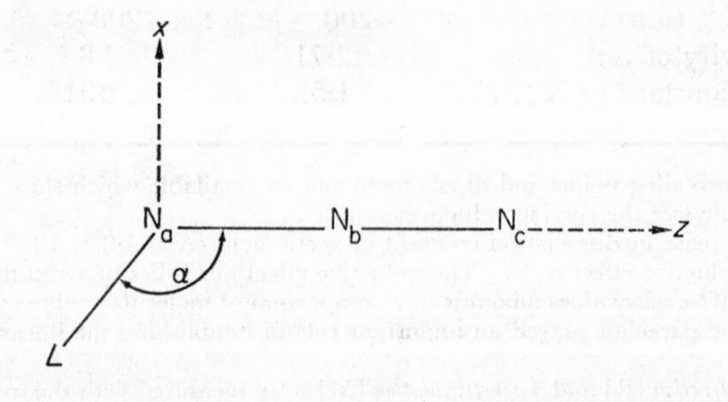

FIGURE 1. The geometry of covalent azides (L is the atom to which N_a is bound; all the nuclei lie in the xz plane).

As late as 1944 azides were considered to have the classical structure RN=N≡N, namely with pentavalent nitrogen[21]. This picture has been completely abandoned as it is now generally accepted that nitrogen obeys the octet rule. One can write two *canonical structures* in keeping with the octet rule and the 'adjacent charge rule'[22]:

$$\underset{\text{I}}{R-\ddot{N}=\overset{+}{N}=\overset{-}{\ddot{N}}:} \qquad \underset{\text{II}}{R-\overset{-}{\ddot{N}}-\overset{+}{N}\equiv N:}$$

Resonance of I and II, with equal contributions, leads to *bond order* 1·5 and 2·5 for the bonds N_a—N_b and N_b—N_c respectively, in agreement with the values derived from force constants (section II.E). The formal charges on N_a, N_b and N_c corresponding to the resonance hybrid I↔II are $-0{\cdot}5$, $+1$ and $-0{\cdot}5$ respectively. In both structures the central nitrogen atom is quadrivalent. This is in accord with its relatively small nuclear quadrupole coupling constant[23], which indicates that the surrounding valence-shell electrons are nearly spherically distributed[24] (section II.I). The non-integral bond orders indicate considerable delocalization of π electrons. The following simplified model[25,26] consists of both localized and delocalized *molecular orbitals*.

The shape of a molecule is essentially determined by the spatial distribution of the σ *bonds*, and these depend on the states of *hybridization* of the atoms. A simplified description of the valence states* of the three nitrogen atoms is given in Table 2 (only the *L*-shell orbitals are included).

TABLE 2. Valence states of nitrogen atoms in $RN_aN_bN_c$

	Lone pairs	σ Electrons	π Electrons
N_a	$(s\delta p)^2$	$p\delta s'$ $p\delta s''$	p_y
N_b		sp sp	$(p_y)^2$, p_x
N_c	$(s)^2$	p_z	p_y, p_x

I. σ bonds

The *sp hybridization* of the central N_b atom is responsible for the linear structure of the azido group. The σ orbitals of N_a are assumed to consist of three *non-equivalent* hybrids produced from s, p_z and p_x: one with more pronounced s character, $s\delta p$, is occupied by a *lone pair* of electrons, and the other two, $p\delta s'$ and $p\delta s''$, are involved in bonding N_a

* For a recent discussion of hybridization and valence states of nitrogen see reference 27.

to R and N_b, respectively. Notice that a pure sp^2 *hybridization* would lead to $\alpha = 120°$. When the angle α widens, the hybridization should also approach *sp*. (This has been considered to be the case of phenyl azide; section II.C.)

In this simplified picture N_c is assumed to be present in its ground state, using a p_z orbital for σ-bonding and an *s* orbital for the lone pair, but it is very probably also hybridized (see sections III.E.2 and III.F.1). A schematic picture of the σ-bonding in the azido group is shown in Figure 2. (For simplicity's sake, the hybrids were drawn as equivalent.)

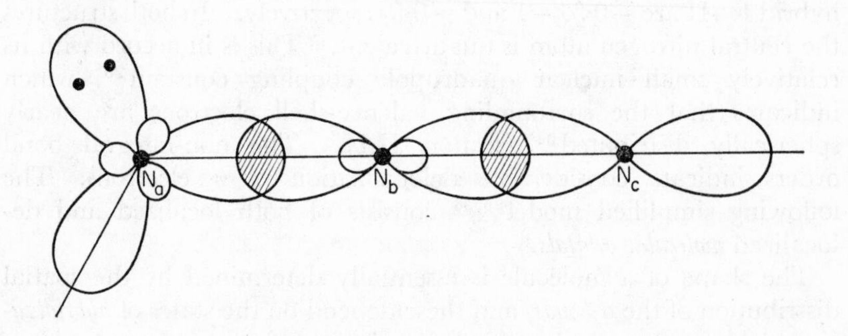

FIGURE 2. The σ-skeleton of azides.

The nature of the σ bonds may be summarized as follows:

$R—N_a: \sigma_R–p\delta s'$ (σ_R is a σ orbital of R);

$N_a—N_b: p\delta s''–sp$; $N_b—N_c: sp–p$ (or $sp–p\delta s$).

To each of these bonding molecular orbitals there is a corresponding *antibonding σ^* orbital* with relatively high energy.

2. π Bonds

(a) The p_x orbitals of N_b and N_c can form a *localized* orbital (π_x) which accommodates two electrons, and the corresponding antibonding π_x^* which is vacant. (b) the p_y orbitals of the three nitrogen atoms form three *delocalized* π orbitals, two accommodating four electrons and the third is a vacant antibonding orbital (π_y^*). Of the two filled π_y orbitals only one is bonding and the other is non-bonding (section I.C).

Figure 3 shows a schematic representation of the bonding π orbitals.

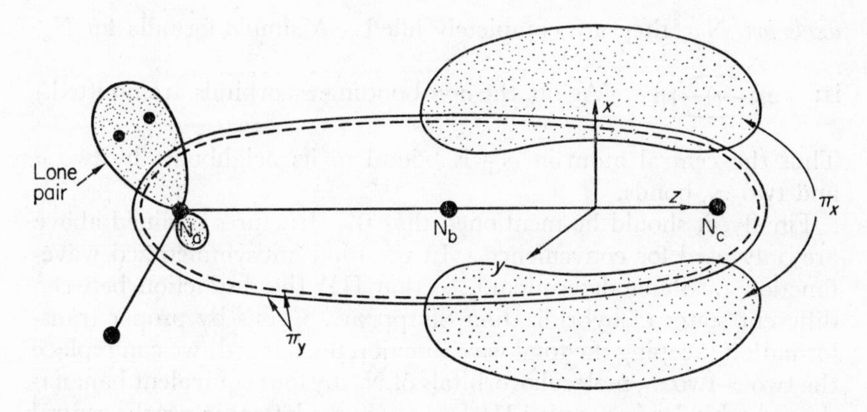

FIGURE 3. The bonding π orbitals of azides (the $(s\delta p)^2$ lone pair is also shown).

Here too the molecule is drawn in the plane of the paper. For the π_y orbital this is a nodal plane; the continuous and dashed curve, represent the two parts of the orbital, above and below the molecular plane, respectively.

3. Lone pairs

According to this picture there are two lone pairs in the valence shell of the azido group, one in the 2*s* orbital of N_c and the other in the $s\delta p$ hybrid of N_a. The latter is of special importance because its energy is higher and so can be more readily excited.

A simple formula which conveys much information on the bonding in the azido group is $\overset{\cdot\cdot}{\mathrm{N}}\!\!\frown$ N—N=N: where the arc represents the delocalized π bond. Thus the central N atom is bound to its neighbours by two σ, one localized π (π_L) and one delocalized $\pi(\pi_D)$ bonds. The 16 valence electrons of the azido group (5 from each nitrogen and 1 from R) are distributed as follows: 6 in the three σ bonds; 4 in the two lone pairs; 2 in π_L and 4 in the two π_D orbitals (bonding and non-bonding).

In the *azide radical* $N_aN_bN_c$, the valence states of N_a and N_c should be identical (approximately as that of N_c in Table 2). The p_x orbitals too can now be involved in three delocalized π orbitals, and together with those of p_y will form three *degenerate pairs* of π_D: bonding, non-bonding and antibonding, respectively. In the azide radical the bonding and non-bonding pairs are occupied by 7 electrons, but in the *azide ion*, N_3^-, they are completely filled. A simple formula for N_3^- is: $:\!N\!-\!N\!-\!N\!:$ (Again, the non-bonding π_D orbitals are omitted.)

Thus the central atom in N_3^- is bound to its neighbours by two σ and two π_D bonds.

Finally, it should be mentioned that the structures outlined above are only used for convenience. In the total antisymmetrized wavefunction (a Slater determinant, section III) the distinction between different states of hybridization disappears. Thus by proper transformation (keeping the total wavefunction unchanged) we can replace the two σ–two π_D molecular orbitals of N_3^- by four equivalent banana-shaped orbitals[3] (see section III.F.1). (In the latter picture the central nitrogen atom may be regarded as being in a hybridization state which resembles sp^3.) This transformation is similar to that employed for the nitrogen and acetylene molecules, where the three $\sigma\pi^2$ bonds can be replaced by three equivalent bonds[29].

C. Primitive Hückel-type Calculations

It is interesting to subject the azido group to a simple theoretical treatment based on the Hückel approximation*. More refined treatments will be discussed in section III.

In the azide ion, the π_x and π_y systems are identical; both can be written as linear combinations of three p orbitals, e.g.

$$\pi_y = \mathrm{a}p_y(\mathrm{a}) + \mathrm{b}p_y(\mathrm{b}) + \mathrm{c}p_y(\mathrm{c}) \tag{1}$$

(and the same for π_x). In the primitive treatment we assume that the *Coulomb integrals* of the three nitrogen atoms are identical (i.e. $\alpha_a = \alpha_b = \alpha_c$: this cannot be true; owing to different *formal charges* the Coulomb integral of N_b must be different from that of the marginal atoms). We thus set the following *secular equations* for each π system:

$$\begin{aligned}
\mathrm{a}(\alpha - E) + \mathrm{b}\beta \qquad\qquad\qquad &= 0 \\
\mathrm{a}\beta \qquad + \mathrm{b}(\alpha - E) + \mathrm{c}\beta &= 0 \\
\mathrm{b}\beta \qquad + \mathrm{c}(\alpha - E) &= 0
\end{aligned} \tag{2}$$

and the corresponding *secular determinant*

* The Hückel approximation has already been discussed in this series (see references 27 and 29). The notation used here is the usual: α and β designate the *Coulomb integral* and *resonance integral*, respectively.

$$\begin{vmatrix} \alpha - E & \beta & 0 \\ \beta & \alpha - E & \beta \\ 0 & \beta & \alpha - E \end{vmatrix} = 0 \quad (3)$$

Equations (2) and (3) and the normalization condition for the wavefunctions lead to energies and coefficients as shown in Table 3.

TABLE 3. Energies and coefficients of the π-molecular orbitals of N_3^- [a]

Orbital	Orbital energy	LCAO coefficients of molecular orbitals		
		a	b	c
$\pi_{x,y}$ (bonding, deloc.)	$\alpha + \sqrt{2}\beta$	1/2	$\sqrt{2}/2$	1/2
$\pi_{x,y}$ (non-bonding, deloc.)	α	$\sqrt{2}/2$	0	$-\sqrt{2}/2$
π^*_{xy} (antibonding, deloc.)	$\alpha - \sqrt{2}\beta$	1/2	$-\sqrt{2}/2$	1/2

[a] Based on the assumption of a single value for the Coulomb integral, α.

In the ground state of N_3^- the first two levels (each doubly degenerate) are occupied by 8 electrons, and the third level (also doubly degenerate) is vacant.

In covalent azides the π_y system of N_3 can be treated similarly, but here the foregoing approximation is even more primitive: owing to the asymmetry of the azido group (the distances N_a—N_b and N_b—N_c are different), the resonance integrals of the bonds N_a—N_b and N_b—N_c should be different and $\alpha_a \neq \alpha_b \neq \alpha_c$. If we still persist in using the same approximations, we can use Table 3 for the π_y system of N_3.

The π_x system of the azido group can be written as linear combination of the two p_x orbitals

$$\pi_x = \mathrm{b}p_x(\mathrm{b}) + \mathrm{c}p_x(\mathrm{c}) \quad (4)$$

The secular equations are now:

$$\begin{aligned} \mathrm{b}(\alpha - E) + \mathrm{c}\beta &= 0 \\ \mathrm{b}\beta + \mathrm{c}(\alpha - E) &= 0 \end{aligned} \quad (5)$$

and the secular determinant:

$$\begin{vmatrix} \alpha - E & \beta \\ \beta & \alpha - E \end{vmatrix} = 0 \quad (6)$$

The solutions for the π_x system together with those for π_y are summarized in Table 4. All the levels are non-degenerate.

TABLE 4. Energies and coefficients of the π-molecular orbitals of the azido group[a]

Orbital	Orbital energy	LCAO coefficients of molecular orbitals a	b	c
π_y (bonding, deloc.)	$\alpha + \sqrt{2}\beta$	$1/2$	$\sqrt{2}/2$	$1/2$
π_x (bonding, loc.)	$\alpha + \beta$	0	$\sqrt{2}/2$	$\sqrt{2}/2$
π_y (non-bonding, deloc.)	α	$\sqrt{2}/2$	0	$-\sqrt{2}/2$
π_x^* (antibonding, loc.)	$\alpha - \beta$	0	$\sqrt{2}/2$	$-\sqrt{2}/2$
π_y^* (antibonding, deloc.)	$\alpha - \sqrt{2}\beta$	$1/2$	$-\sqrt{2}/2$	$1/2$

[a] Based on the assumption of single values for α and β.

The *electronic charge densities* q_ν associated with any atom ν can be calculated by summing up the squares of the coefficients for this atom in the various molecular orbitals i, each multiplied by the number of electrons, n_i, in the corresponding orbital:

$$q_\nu = \sum_i n_i c_{\nu i}^2 \tag{7}$$

The *bond order* $P_{\mu\nu}$ associated with a bond between atoms μ and ν is evaluated by the expression

$$P_{\mu\nu} = \sum_i n_i c_{\nu i} c_{\mu i} \tag{8}$$

Using Table 4, we can draw Figure 4 to illustrate the formal

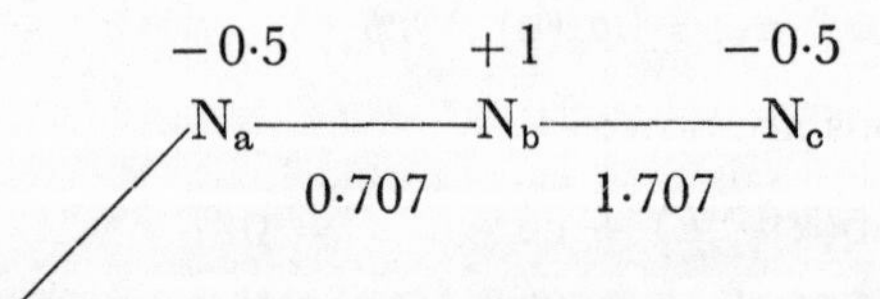

FIGURE 4. Primitive diagram of formal charges and bond orders of the π system in the azido group (based on single values for α and β).

charges and bond orders of the π system in the azido group. (The first three orbitals are doubly occupied and the rest are empty.) The formal charge is calculated as the number of electrons contributed to the π system by the atom ν (see Table 2) minus q_ν.

The overall bond orders are obtained by adding 1 to each π-bond order shown in Figure 4, to account for the single σ bond between two neighbouring atoms. This leads to: $P(N_aN_b) = 1{\cdot}707$ and $P(N_bN_c) = 2{\cdot}707$.

Our primitive calculations have led to the same distribution of formal charges as that of the resonance hybrid (section I.B).

Besides the π orbitals, it is important to discuss the $s\delta p$ lone pair (section I.B). Assuming that the orbital is one of three equivalent sp^2 hybrids, i.e.

$$s\delta p = \sqrt{\tfrac{1}{3}}\, s + \sqrt{\tfrac{2}{3}}\, p \tag{9}$$

($s\delta p$ etc. designate the wavefunctions), we obtain for the energy of $s\delta p$:

$$\varepsilon(s\delta p) = \langle s\delta p \,|\, \mathbf{H} \,|\, s\delta p\rangle = \tfrac{1}{3}\varepsilon_s + \tfrac{2}{3}\varepsilon_p \tag{10}$$

H is the Hamiltonian operator for the nitrogen atom (the usual notation for the matrix-element of the operator is used here); ε_s and ε_p are the energies of the atomic orbitals $2s$ and $2p$ of nitrogen, respectively.

Figure 5 shows schematic *energy level diagrams* for those molecular orbitals which are involved in the ultraviolet spectra of N_3^- and covalent azides.

Comparing the levels of N_3^- and RN_3 we can consider the effect of the off-axis R group as a *perturbation* on the azide ion, which reduces

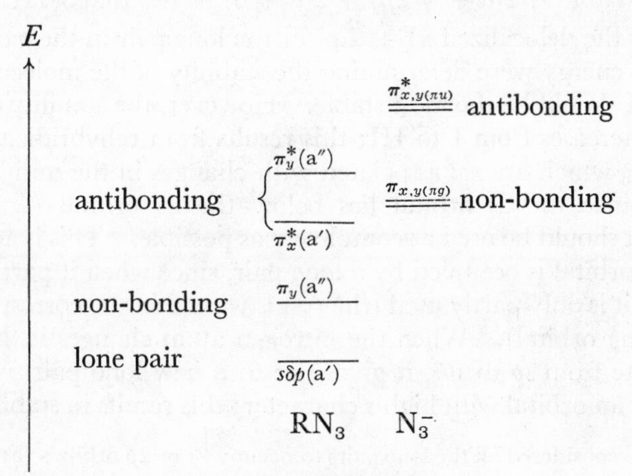

FIGURE 5. Energy levels of the highest-filled and lowest-unfilled orbitals of the azido group and of N_3^-.

the symmetry from $D_{\infty h}$ (for N_3^-) to C_s (for RN_3)[9]. Such perturbation should split the doubly degenerate π orbitals of N_3^- each into two orbitals, one symmetrical (a′) and the other antisymmetrical (a″) to reflexion in the plane of the molecule. Considering the *local symmetry* around the linear azido group, they are the localized π_x (and $s\delta p$) and the delocalized π_y orbitals, respectively (Figure 3).

D. Bent Structure of the Azido Group

Roberts[30] considered the following structures of RN_3*

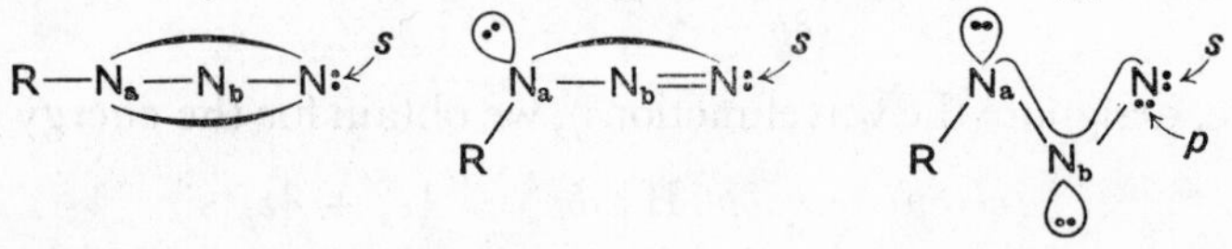

The delocalized π systems of I are similar to that of N_3^- (Table 3), while II and III have the non-degenerate delocalized orbitals shown in Table 4. II, which also contains a localized π orbital, is the real structure of RN_3 and so must be the most stable.

The total π energies of the three structures are:

$$\text{I: } 8\alpha + 5{\cdot}66\beta; \quad \text{II: } 6\alpha + 4{\cdot}83\beta; \quad \text{III: } 4\alpha + 2{\cdot}83\beta$$

The π energy of I is obtained from Table 3: $4(\alpha + \sqrt{2}\beta) + 4\alpha$; that of II from Table 4: $2(\alpha + \sqrt{2}\beta) + 2(\alpha + \beta) + 2\alpha$; that of III: $2(\alpha + \sqrt{2}\beta)$ (for the delocalized π) + 2α (for the lone pair in the p orbital).

If the π energy were determining the stability of the molecule then structure I should be the most stable. However, the stability of the σ skeleton increases from I to III; this results from rehybridizations of N_a and N_b which are *not* associated with changes in the number of σ bonds. Since the $2s$ orbital lies below the $2p$ orbital ($\varepsilon_p - \varepsilon_s$ = 10·2 eV) it should be used as completely as possible. This is achieved when the orbital is occupied by a lone pair, since when it participates in a bond it is only partly used (the rest is wasted in the corresponding antibonding orbital). When the nitrogen atom changes its hybridization state from sp to sp^2, it gives rise to a new lone pair, which is housed by an orbital with high s character; this results in stabilization

* Roberts considered all the lone pairs to occupy $2s$ or $2p$ orbitals (as in III). Here we give a more plausible picture, where (:) represents a lone pair in a $s\delta p$ hybrid.

of the σ skeleton. Denoting the corresponding energy change by Q, the energies associated with both the π systems and hybridization changes become:

$$\text{I}: 8\alpha + 5{\cdot}66\beta; \quad \text{II}: 6\alpha + 4{\cdot}83\beta + Q; \quad \text{III}: 4\alpha + 2{\cdot}83\beta + 2Q \quad (11)$$

Now, since II is the most stable structure we conclude that

$$8\alpha + 5{\cdot}66\beta > 6\alpha + 4{\cdot}83\beta + Q < 4\alpha + 2{\cdot}83\beta + 2Q$$

i.e.

$$2\alpha + 0{\cdot}83\beta > Q > 2\alpha + 2\beta \qquad (\alpha \text{ and } \beta \text{ are negative})$$

Assuming that Q lies in the middle, Roberts[30] obtained:

$$Q = 2\alpha + 1{\cdot}4\beta$$

Inserting this value of Q in equation (11) we get:

$$\text{I}: 8\alpha + 5{\cdot}66\beta; \quad \text{II}: 8\alpha + 6{\cdot}23\beta; \quad \text{III}: 8\alpha + 5{\cdot}63\beta$$

Thus only a loss of $\sim 0{\cdot}6\beta$ is involved in bending the linear configuration to an angle of $\sim 120°$. β for N—N bonds falls in the range of 10–30 kcal[30]; therefore the bending energy cannot be considerably higher than $\sim$20 kcal. This has an important bearing on the chemistry of azides[7].

E. *Some Data on Nitrogen Atoms and N—N Bonds*

Table 5 contains some useful data for the discussion of the azido group.

TABLE 5. Some data on N atoms and N—N bonds

Valence state[a]	energy, eV[b]	Valence state[a]	Energy, eV[b]
$N(s^2xyz)$	1·18	$N^+(t_1t_2t_3y)$	23·61
$N(s^2xy^2)$	2·92	$N^+(d_1d_2yx)$	24·25
$N(d_1^2d_2yx)$	7·70	$N^+(sxyz)$	26·18
$N(t_1^2t_2t_3y)$	9·19	$N^+(d_1d_2y^2)$	26·64
$N(d_1d_2y^2x)$	12·69	$N^+(sxy^2)$	28·57
$N(sxy^2z)$	14·23	$N^{2+}(d_1d_2y)$	53·54
		$N^{2+}(t_1t_2y)$	58·84
$N^+(s^2xy)$	15·02	$N^-(s^2xyz^2)$	0·32
$N^+(d_1^2d_2y)$	21·79	$N^-(d_1^2d_2y^2x)$	6·48
$N^+(t_1^2t_2t_3)$	23·20	$N^-(sx^2yz^2)$	12·65

[a] $s,x,y,z,d_1,d_2,t_1,t_2,t_3$ designate the orbital 2s, the three 2p orbitals (p_x, etc.), the two digonal and three trigonal hybrids, respectively.

[b] The reference state is the ground state of N. These data were taken from references 31, 32 and 33.

TABLE 5.—*continued*

Orbital electronegativity[c]	
p orbital in $N(s^2xyz)$	7·39
p orbital in $N(t_1^2t_2t_3y)$	7·95
trigonal hybrid in $N(t_1^2t_2t_3y)$	12·87

Bond	Bond energy[27] kcal/mole	Bond length[3] Å	Force constant[34] mdyne/Å
N—N	39·0	1·45	22·79
N=N	101·0	1·25	12·79
N≡N	225·8	1·10	3·59

[c] The orbital electronegativity (as defined by Mulliken) is half the sum of the ionization potential I and the electron affinity E. The electronegativities were taken from reference 27, which also includes information on I and E. (See also J. E. Huheey, *J. Phys. Chem.*, **70**, 2086 (1966).)

II. SOME STRUCTURAL PROPERTIES OF THE AZIDO GROUP

A. Geometry

Table 6 records *interatomic distances* and *RNN bond angles* (pertaining to the equilibrium configurations) of several azides. For some earlier data concerning inorganic azides see references 1 and 2.

TABLE 6. Interatomic distances and bond angles in $RN_aN_bN_c$

Azides	N_c—N_b Å	N_b—N_a Å	R—N_a Å	R—N_a—N_b Angle	Method	Ref.
N_3 (azide radical)	1·1815	1·1815	—	—	Ultraviolet spectroscopy	35
N_3^- (in $Ba(N_3)_2$)[a]	1·166 ± 0·002	1·166 ± 0·002	—	—	Neutron diffraction	36
HN_3	1·133 ± 0·002	1·237 ± 0·002	0·975 ± 0·015	114° 8′ ± 30′	Microwave spectroscopy	37
CH_3N_3	1·12 ± 0·01	1·24 ± 0·01	1·47 ± 0·02	120 ± 2°	Electron diffraction	38
C_3N_{12} (cyanuric triazide)	1·11	1·26	1·38	114°	X-ray diffraction	39

[a] N—N distances scattered around 1·17 ± 0·01 were reported for NaN_3, LiN_3 and $Sr(N_3)_2$ [G. E. Pringle and D. E. Noakes, *Acta Cryst.*, **24B**, 262 (1968)].

The N_3 group is linear. (A recent value for the NNN angle in $Ba(N_3)_2$ crystal is 179·7 ± 0·2[36].) This is in agreement with the correlation diagram of Walsh[40] which predicts that AB_2 or BAC molecules with 16 or less valence electrons should be linear in their ground states. (See section III.C.) The azide radical and N_3^- are symmetrical. The N—N distance in the radical is only 0·015 Å longer than in N_3^-. The two molecules differ in one non-bonding electron (sections I.C and I.D), and this is expected to have little effect on the bond properties.

As a first approximation, N_3^- can be considered as an ellipsoid of revolution with major and minor semi-axes, 2·54 and 1·76 Å, respectively[19]. For thermochemical calculations (e.g. crystal energies and hydration energies), N_3^- may be approximated as a sphere with effective radius 2·04 Å[19,41].

In covalent azides the N_3 group is asymmetric, the N_a—N_b bond being considerably longer. The geometry of HN_3 has been accurately determined by microwave spectroscopy[37]. The primary data consist of three moments of inertia of the molecule and its isotopic species. HN_3 is a planar asymmetric rotor. The positions of the atoms in the principal-axis system are shown in Figure 6; the co-ordinates x and z, respectively, are (in Å): H: +0·8151, −1·5922; N_a: −0·0605, −1·1636; N_b: −0·0184, +0·0730; N_c: +0·0202, +1·2052.

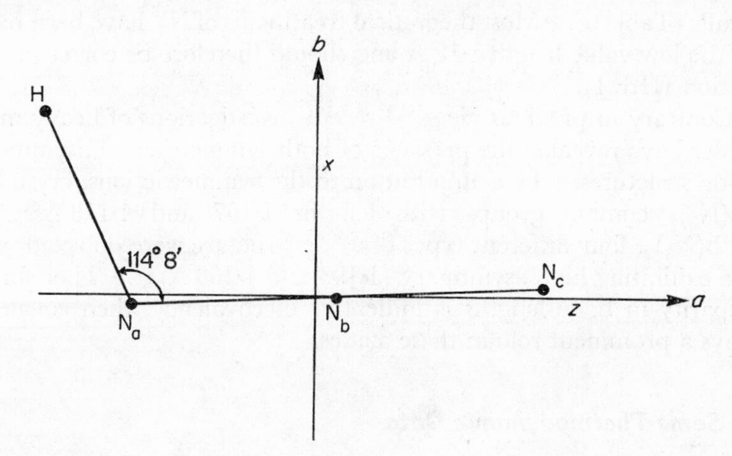

FIGURE 6. Positions of the atoms in the principal-axis system of HN_3.

Microwave spectroscopy has also supplied useful information on the centrifugal distortion constants (section II.E) and nuclear quadrupole coupling constants (section II.I) of HN_3, on the electric dipole-moments of HN_3 and CH_3N_3 (section II.C), and on the internal rotation potential barrier of CH_3N_3 (section II.F).

By plotting the lengths r of N—N bonds (Table 5) against the corresponding bond-orders P and interpolating, the bond-orders of N_a—N_b and N_b—N_c in HN_3 were determined as close to 2 and 2·8, respectively[3]. These are appreciably different from the values derived from force-constants (section II.E). However, P–r correlations should be handled carefully, since bond lengths largely depend on the states of hybridization of atoms. Thus an sp–sp^2 bond (as in N_b—N_a) should be shorter than an sp^3–sp^3 bond (closely represented by H_2N—NH_2). Orville-Thomas[26] has pointed out that no unique bond-length–bond-order curve exists for pairs of atoms, unless the states of hybridization of the two atoms remain the same for a series of compounds. He defined a *σ skeleton parameter* $\gamma = \sqrt{s_A s_B}$ (where s_A and s_B are the s character of the atomic orbitals joined to form the σ bond) and has shown that for bonds with the same γ there is a linear relation between length and order. There is insufficient data for plotting such P–r curves for the N—N bonds. Still, we can conclude that the N—N distance in N_3^- must be longer than that of N_b—N_c in RN_3; the σ-bonding in both cases is close to sp–p but the bond-order of N_b—N_c is higher (section I.B). This argument rules out some previous data for the length of N—N bonds in ionic azides[42] and is supported by recent results (Table 6). Most theoretical treatments of N_3^- have been based on the low value length 1·12 Å and should therefore be corrected (see section III.F.1).

Contrary to previous views[1,2] recent investigations of heavy metal azides have revealed the presence of both symmetric and asymmetric azide structures. Thus in addition to the symmetric ions, crystals of $Ba(N_3)_2$ contain groups with lengths 1·157 and 1·178 Å[36]. In α-$Pb(N_3)_2$, four different types of azide structure were detected, with one exhibiting high asymmetry (1·193 and 1·160 Å)[43]. Therefore, if disparity in bond-lengths is indicative of covalency, then covalency plays a prominent role in these azides.

B. Some Thermodynamic Data

The *heats of formation* of various azides were determined from their heats of combustion and decomposition (some of the data are collected in reference 1). The heats of formation of HN_3, cyclopentyl azide and cyclohexyl azide were compared with those calculated for the canonical structures RN=N=N (I) and RN—N≡N (II), in order to determine the *resonance energy* of the azido group[44,45]. However, the calculated values for II appear to be too high for effective resonance between I and II. The N≡N bond energy used for this calculation is that of molecular nitrogen, which is abnormally high[46] (see section III.G). Relative to structure I, resonance energies of 30, 41 and 45 were calculated for HN_3, $C_5H_9N_3$ and $C_6H_{11}N_3$, respectively[44,45]. The primitive Hückel-type of calculation (see section I.C) yields for structure I (which consists of two localized π bonds and a lone pair in a p orbital) a π energy of $6\alpha + 4\beta$, namely $0{\cdot}83\beta$ less than that of the resonance hybrid (section I.D). Comparing with the empirical resonance energy (average: 39 kcal) we can estimate: $\beta_{NN} \sim 47$ kcal, which is higher than the accepted values (10–30 kcal[7]).

The chemistry of covalent azides is primarily determined by the exceptional weakness of the $RN—N_2$ bond: they most readily decompose to yield N_2 and RN (*nitrenes*). Table 7 records the bond

TABLE 7. Bond dissociation energies of some azides (in kcal/mole)[a]

Compound	D(R—N$_3$)[b]	D(RN—N$_2$)	Method	Ref.
HN_3	90 ± 8	9 ± 2[c]	Thermal emission of N_3^- from heated filament in gaseous azide (magnetron cell)	47
CH_3N_3	88 ± 8			
NCN_3	96 ± 2	7 ± 2	Threshold energy of photodissociation in vacuum u.v.	48
$C_6H_5N_3$	84 ± 5		Computed from heats of formation of RN_3 and R	48
Cyclopentyl azide	74 ± 7			
Cyclohexyl azide	80 ± 7			

[a] Some earlier data are collected in reference 1.
[b] Apart from NCN_3 the data are based on the value 70 kcal for the electron affinity of N_3. The actual electron affinity of N_3 may be somewhat higher than the value used; see section II.G.
[c] From the appearance potential of N_2^+ produced by electron impact[48a].

dissociation energies of some azides and the methods used for their determination.

The C—N dissociation energies, computed from thermochemical data, show a considerable scattering, which is probably due to errors in ΔH_f of the organic azides[1]. D(H—N_3) and D(C—N_3) lie close to that of other single bonds between the same atoms but somewhat higher. [D(NH) = 84 ± 3 kcal/mole[49], D(C—N) = 73 kcal/mole[27]]. This may reflect the $p\delta s$ state of hybridization of the nitrogen atom (section I.B), with more *s*-character than that of N in amines or ammonia*. On the other hand the RN—N_2 bond is unusually weak, which again reflects the abnormal stability of molecular nitrogen. However the activation energy for dissociation is appreciably higher, since spin conservation rule requires the dissociation to excited nitrenes[1,2].

Table 8 records some thermodynamic data on the azide ion and the azide radical. The methods used for their determination are described in the corresponding references.

TABLE 8. Thermodynamic data on N_3^- and N_3 (per mole)

Molecule	ΔH_f° (gas) kcal	ΔH_f° (aq) kcal	ΔG_f° (aq) kcal
N_3^-	34·8[2]	65·5[50]	83·3[50]
N_3	105 ± 3[2]	—	—

Molecule	Hydration energy kcal	Entropy (gas) cal/deg	Electron affinity kcal
N_3^-	79[19]	50·7[51]	—
N_3	—	—	70 ± 5[2]

C. Electrical Dipole Moments

The most precise information on dipole moments can be derived from the effect of external electrical fields on the rotational spectra (*Stark effect*). For HN_3 and CH_3N_3 in their lowest vibrational states the following dipole components were determined: HN_3: μ_a = (0·847 ± 0·005) D[52]; CH_3N_3: $(\mu_a)^2$ = (3·53 ± 0·06) D^2, $(\mu_b)^2$ = (1·0 ± 1·0) D^2 [53]. (μ_a and μ_b are the components in the directions of the principal axes a and b; see for example, Figure 6).

The overall dipole moments of many covalent azides were obtained from measurements of the dielectric constants of their benzene solutions by the heterodyne beat method (Table 9).

* However the C—N distance in CH_3N_3 (Table 6) is the same as in CH_3NH_2[38].

TABLE 9. Electrical dipole moments of organic azides

Compound RN_3, R =	μ (Debye)	Ref.	Compound RN_3, R =	μ (Debye)	Ref.
C_2H_5—	2·13	13, 54	*o*-BrC_6H_4—	2·25	54
Cyclo-C_5H_9—	2·27	7	*m*-BrC_6H_4—	1·40	54
Cyclo-C_6H_{11}—	2·37	7	*p*-BrC_6H_4—	0·02	54
CH_2=CH—CH_2—	1·92	13	*o*-$NO_2C_6H_4$—	4·45	13, 54
$HOCH_2CH_2$—	2·48	13, 54	*m*-$NO_2C_6H_4$—	3·54	13, 54
CH_3COCH_2—	3·64	54	*p*-$NO_2C_6H_4$—	2·90	13, 54
—$COOCH_3$	1·73	13	2,4-$(NO_2)_2C_6H_3$—	2·66	13
—$CH_2COOC_2H_5$	2·79	13	C_6H_5CO—	2·60	13
C_6H_5—	1·44	13	1-naphthyl-	1·36	13
o-$CH_3C_6H_4$—	1·39	54	2-naphthyl-	1·60	13
m-$CH_3C_6H_4$—	1·75	54	1-Nitro-2-azido-naphthalene	4·44	13
p-$CH_3C_6H_4$—	1·90	54	1-Nitro-4-azido-naphthalene	3·12	13
o-ClC_6H_4—	2·37	54	8-Nitro-2-azido-naphthalene	4·59	13
m-ClC_6H_4—	1·45	54			
p-ClC_6H_4—	0·01	54			

The organic azides have moments close to the corresponding bromides and chlorides. (This is reflected in their close boiling points, Table 1.) The dipole moments of substituted phenyl azides show that the azido group has a dipole with negative pole directed away from the ring (i.e. negative group moment like that of the halogens). The results were interpreted as indicating that in aryl azides the C—N—N angles are not close to 120° (as in CH_3N_3 and HN_3; Table 6) but considerably wider. Thus in *p*-chloro- and *p*-bromophenyl azides the group moments nearly cancel each other (Table 9). A C—N—N angle close to 150° was calculated in order to account for the moments of *p*-tolyl, *p*-chloro and *p*-nitro azides[14]. Some π-bonding of the azido group to the ring, with negative charge flowing from the lone pair to the ring (section I.A), might explain this angle widening. As in the case of the halogen derivatives, because of this charge flow the resonance effect can also explain the lowering of moment when the azido group is attached to a benzene ring (compare ethyl and phenyl azides in Table 9). A similar resonance effect was also considered in the case of vinyl azide and methyl azidoformate[55]. However it must be realized that the behaviour of *para*- substituted azides can only prove that the dipole moment of phenyl azide is oriented nearly along the CN line, but this is not necessarily the orientation of the azido group. As to the reduction of moment on replacing the alkyl group by an aromatic ring, we must consider the *hybridization moment*, which reflects the effect of hybridization on the electronegativity of the hybrid orbital. On going from $C_2H_5N_3$ to $C_6H_5N_3$ the bond changes from C(sp^3)—N($\sim sp^2$) to C(sp^2)—N($\sim sp^2$) (assuming no change in the state of N) and so negative charge will flow from N to C (owing to the increase in the electronegativity of C).

Hybridization effects should also markedly affect the overall dipole moment of the azido group. Thus the N—N bonds should possess dipole moments since the two nitrogen atoms are in different hybridization states (section I.B); moreover, the dipoles of N_a—N_b and N_b—N_c may be different and so not necessarily cancel each other out as in the simple resonance picture (section I.B). Hybridization is also responsible for *atomic dipoles* of lone pairs. Thus the $s\delta p$ lone pair of N_a, directed at $\sim 120°$ to the azido group may contribute much to the overall dipole. (If the terminal nitrogen atom is also hybridized, we should also consider the atomic dipole of its lone pair.)

Favini[31] carried out calculations on the charge distribution and dipole moments of HN_3, $C_6H_5N_3$ and *para*-substituted phenyl azides. His separation of the total moment into a *σ moment* and a *π moment* is questionable (for a discussion of this point see reference 56), and the contributions of lone pairs and polarization effects were not analysed. Still it is interesting to note that he could account for the experimental dipole moments by assuming that in phenyl azide the CN_aN_b angle is 120°. The σ moment of the azido group was calculated as 1·35D, directed towards N_a, but this is outweighed by the π moment in the opposite direction. Altogether the overall moment of the azido group is rather low: almost all the moment of phenyl azide was attributed to the C—N bond (31,44). This is in accord with the simple resonance theory (section I.B).

The structure of phenyl azide should be determined in order to resolve the problem of bonding in aromatic azides.

The dipole moments of substituted phenyl azides were calculated by vector addition of the group moments, assuming that the substituted groups are colinear and lie in the plane of the ring[54]. Significant discrepancies from experimental values are displayed by *ortho* derivatives, where mutual *induction* between the close polar groups should occur (*ortho* effect). In nitro derivatives a more extended type of interaction was postulated to account for the discrepancy shown by the *para* compound[54].

D. Vibration Spectra

1. N_3^- and N_3 (azide radical)

These molecules possess three fundamental vibration frequencies: two *bond stretching* (symmetric ν_s and asymmetric ν_{as}) and one doubly degenerate *bond bending* vibration, ν_b (Figure 7). The symmetrical

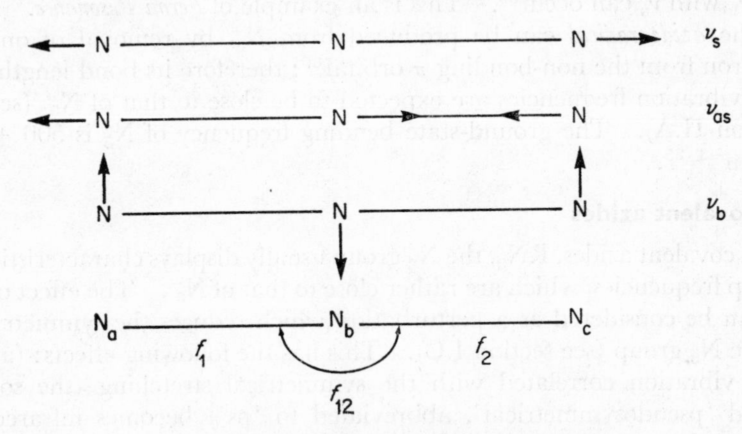

FIGURE 7. The fundamental vibration frequencies and bond-stretching force constants of molecules like N_3^- or N_3.

stretching ν_s is infrared inactive, but changes in the symmetry of N_3^- under the influence of neighbouring ions can somewhat relax this forbiddenness. Such effects can also remove the degeneracy of the bending vibration, which consequently splits into a doublet. ν_s is readily observed in Raman Spectra but it can also be analysed in the infrared as a component of *combination bands*[1,57,58,59].

A *binary combination band* is due to a transition in which two vibrations change by one quantum. In a *summation band* each vibration gains one quantum: $\nu_{sum} = \nu_i + \nu_k$, where ν_i and ν_k are the combining frequencies. If the initial state of the molecule is not the vibrationless ground state (e.g. ν_i is singly excited in the initial state) the two vibrations can combine to yield a difference band, $\nu_{diff} = \nu_k - \nu_i$, provided $\nu_k > \nu_i$. Normally difference bands are much weaker than summation bands owing to the Boltzmann factor $e^{-h\nu_i/kT}$. In general combination bands are weaker than the fundamental bands, since like *overtones* their occurrence is due to anharmonicity of the vibrations. Still they may be quite intense and in this way inactive vibrations when coupled with other vibrations can yield active combination bands. Thus although ν_s is forbidden, its combinations with ν_b and ν_{as} are significant bands in the infrared spectra of N_3^-*.

In ionic azides the vibration frequencies show little dependence on the nature of the cations, with $\nu_s \sim 1350 cm^{-1}$, $\nu_{as} \sim 2050 cm^{-1}$ and $\nu_b \sim 650 cm^{-1}$. The overtone $2\nu_b$ is only 80 cm^{-1} below the fundamental ν_s, and therefore—as observed in the Raman spectra—'mixing' of $2\nu_b$ with ν_s can occur[58]. This is an example of *Fermi Resonance.*

The *azide radical* can be produced from N_3^- by removal of one electron from the non-bonding π orbital[60]; therefore its bond lengths and vibration frequencies are expected to be close to that of N_3^- (see section II.A). The ground-state bending frequency of N_3 is 500 ± 50 cm^{-1} [35].

2. Covalent azides

In covalent azides, RN_3, the N_3 group usually displays characteristic group frequencies, which are rather close to that of N_3^-. The effect of R can be considered as a perturbation which reduces the symmetry of the N_3 group (see section I.C). This has the following effects: (a) The vibration correlated with the symmetrical stretching (the so-called 'pseudosymmetrical', abbreviated to 'ps') becomes infrared active; (b) the degeneracy of the bending vibration is often removed and the corresponding band splits into two components; (c) the lowering of symmetry (relaxation of selection rules) and increase in the number of molecular vibration modes, together give rise to higher probability of interaction between *accidentally degenerate* vibrational levels (Fermi resonance). Thus in many azides ν_{as} is close to the frequency of some combination tone or overtone and this often leads to Fermi resonance, which is reflected in the splitting of the ν_{as} band[61]. Many organic azides display two weak bands at about 3400 and 2500 cm^{-1} [62]. The former was ascribed to the combination band ν_{ps} + ν_{as}, and the latter to the overtone $2\nu_{ps}$. Interaction between ν_{ps} and the overtone of the bending vibration also occurs in some cases[63].

The infrared spectra of HN_3 [64] and CH_3N_3 [65] have been studied in detail. Information is now available on the frequencies of N_3 in many organic azides; some representative data are collected in Table 10.

The azido group is characterized by the strong asymmetric stretching band (ν_{as}) near 2100 cm^{-1} (Table 10). (In aliphatic azides the *integrated intensity* of the ν_{as} band is $\sim 4{\cdot}5 \times 10^4\ M^{-1}\ cm^{-2}$ [61].) The

* In crystalline azides combinations with *lattice modes* also occur.

TABLE 10. The azido group frequencies[a]

RN_3, R =	ν_{as}	ν_{ps}	ν_b	Combination band[b] or overtone	Ref.
HN_3 (gas)	2140	1274	522, 672	—	64
CH_3	2090	1270	—	3450, 2530	62
C_6H_5	2135[c]	1297	—	3320, 2550	62, 61
m-ClC_6H_4	2096	1288	—	—	66
o-ClC_6H_4	2088	1297	—	—	66
p-BrC_6H_4	2110	1287	—	—	66
o-$NO_2C_6H_4$	2130[c]	1295	—	—	61
C_6H_5CO	2141, 2179	1237, 1251	—	—	63
H_2NCO	2160	1217	—	—	63
$(CH_3)_2NCO$	2160, 2200	1225	—	—	63
CH_3OCH_2	2121, 2098	1232	—	2440, 2396	63
CH_3SCH_2	2113, 2091	1223	—	2420, 2370	63
$(CH_3)_2NCH_2$	2095	1250	—	2475, 2400	63
$(C_6H_5)_3C$	2110	1261	666	—	67
$(C_6H_5)_3Si$	2149	1308	660	—	67
$(C_6H_5)_3Ge$	2100	—	660	—	67
$(C_6H_5)_3Sn$	2093	—	658	—	67
$(C_6H_5)_3Pb$	2046	1261	655	—	67
Ferrocenyl	2113	1286	—	2111 etc.	61

[a] The samples were run as CCl_4 solutions, liquid films or Nujol mulls. For other bands of these and other compounds see references 61–67.
[b] The combination bands may involve frequencies of R or the bond R—N.
[c] These bands display splitting.

bending frequency around 650 cm^{-1} is generally weak. In most cases when ν_{as} splits the spectrum also shows a combination band or overtone with frequency close to ν_{as}[61].

Of special interest are the azides $(C_6H_5)_3XN_3$ where X is an element of group IVB. The Si compound has the highest ν_{as}; this was explained in terms of *p*–*d* π-bonding between Si and N, in agreement with the relatively high thermal stability of the compound. On the other hand, ν_{as} of $(C_6H_5)_3PbN_3$ is remarkably low (close to that of N_3^-), and this suggests a high degree of ionic character[67].

The variation in frequencies and integrated intensities of the N_3 bands were discussed[61] in relation to the nature of the radical R. The azide group was considered to act either as a donor or acceptor of electron (section I.A), depending on R. (However the variation in ν is small compared to the scatter of results in literature.) The conditions for Fermi interaction appear to be hindered by a barrier such as Fe atom in ferrocenyl azide.

E. Force Constants

In a linear triatomic molecule, the two parallel stretching vibrations can be characterized by three force constants (Figure 7): f_1 and f_2 are the *bond-stretching constants* and f_{12} is a bond–bond *interaction force constant* which accounts for the fact that vibration of one bond affects the other. In this case the stretching and bending vibrations belong to different symmetry species and therefore can be treated separately. The force constants can be determined from the values of the stretching frequencies of the molecule and its isotopic modifications. (The force constants remain the same in isotopic molecules.) Another relation between the three force constants can also be derived from the *centrifugal distortion* constants, which represent the extent of molecular distortion caused by rotation and reflected in its rotational spectrum[26]. The values of these constants for HN_3 and some of its isotopic modifications are given in reference 53.

The importance of bond stretching force constants is that they measure the strength of chemical bonds and therefore can be related to bond length r and bond order P. A useful relation between these properties is given by *Gordy's equation*[68]

$$f(\text{AB}) = 1{\cdot}67\,P(\text{AB})\left(\frac{X_A X_B}{r^2}\right)^{3/4} + 0{\cdot}3 \tag{12}$$

where X_A and X_B are Pauling's electronegativity values of atoms A and B, respectively; f is measured in millidynes/Å.

Table 11 records force constants and calculated bond orders in several azides. ($f(N_aN_b/N_bN_c)$ is the bond–bond interaction force constant.)

TABLE 11. Force constants f (in mdyne/Å) and bond order P in $RN_aN_bN_c$

Molecule	RN_a f	N_aN_b f	N_aN_b P	N_bN_c f	N_bN_c P	$f(N_aN_b/N_bN_c)$	Ref.
N_3^-	—	13·7	~2·0	13·7	~2·0	~1·7	1
HN_3	6·2	10·1	1·58	17·3	2·4	1·74	25
CH_3N_3	4·5	9·4	1·50	17·3	2·4	—	25

The bond properties of N_3 in HN_3 and CH_3N_3 appear to be similar. The non-integral bond orders and the considerable positive interaction force constant (~ 1·7) are indicative of charge delocalization. The force constant f(CN) = 4·5 mdyne/Å is typical for σ-CN bond (4·6 mdyne/Å for CH_3NH_2[25]).

Somewhat different force constants were assigned to N_3^- and HN_3 by D. R. Conant and J. C. Decius (*Spectrochim. Acta*, **23A**, 2931 (1967)).

F. Internal Rotation and Hyperconjugation

A vibrational degree of freedom may be replaced by *internal rotation* (torsion) around a σ bond. In this case the microwave spectrum of the molecule is modified by torsion–rotation interaction. By studying this effect on the rotational spectrum, the internal rotation potential barrier can be determined. The *hindering potential* of CH_3N_3 was found to be: $V_3 = 695 \pm 20$ cal/mole[53] (the subscript 3 stands for the 3-fold axis of the hindering potential). The potential is rather small but is not smaller than the value expected from a *hyperconjugation* effect[69].

There is no evidence for π bonding between methyl (or methylene) group and N_a in azides, i.e. for hyperconjugation. CH_3N_3 and CH_3NH_2 hardly differ in their CN distance (See section II.B) or force constant (section II.E). However Ladik and Messmer[70] presented some kinetic evidence for hyperconjugation in benzyl azides (its pronounced stability toward hydrolytic reactions) and carried out a Hückel-type calculation to evaluate the *hyperconjugation energy*. They obtained the following order for the interaction of CH_2 with various groups bound to it: $N_3 > CH_3 >$ phenyl $\gg$ Cl, with CH_2N_3 having hyperconjugation energy of 0·18β (β stands for the resonance integral between two carbon atoms in benzene).

As in other cases, here too the role of hyperconjugation is still uncertain.

G. Electronic Spectra

The absorption spectra of N_3^-, HN_3 and *n*-amylazide were studied in detail. Figure 8 shows their u.v. spectra above 180 nm (adapted from reference 9).

1. The azide ion

In molecular orbital language the electronic configuration of the first excited states of N_3^- is $\ldots(\pi_g)^3\pi_u$. (To simplify the discussion the orbitals below π_g were omitted; see Figure 5.) As in the case of CO_2, this configuration should give rise to three singlet states: $^1\Delta_u$, $^1\Sigma_u^-$ and $^1\Sigma_u^+$[71]. Transitions to these states should be responsible for the

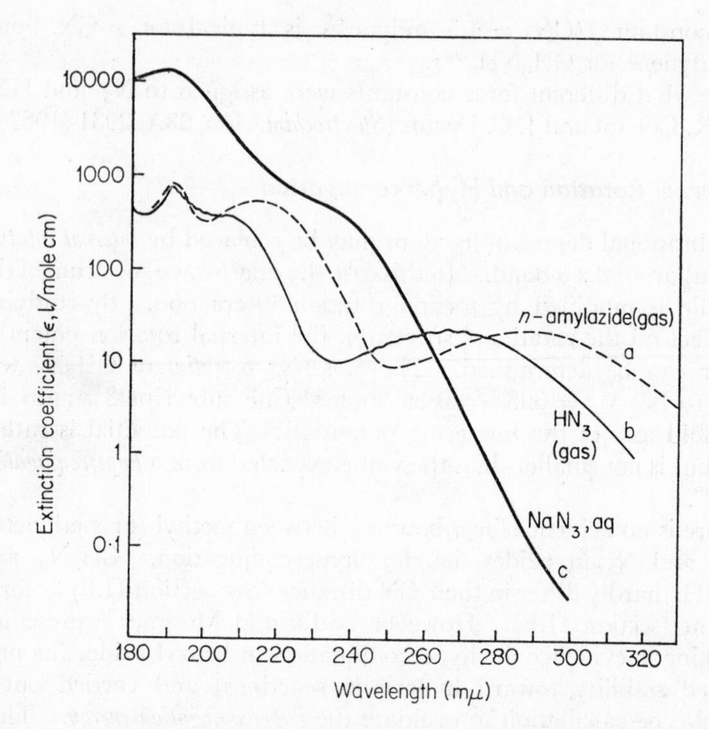

FIGURE 8. The electronic absorption spectra of *n*-amylazide gas (a), HN_3 gas (b) and NaN_3 in aqueous solution (c) (adapted from reference 9).

lowest singlet→singlet transitions. Of these only the $^1\Sigma_u^+ \leftarrow \Sigma_g^+$ transition is fully allowed. In addition, the relative low ionization potential of N_3^- should lead to an allowed low-energy electron transfer process with N_3^- acting as the donor. In solution the polarized medium around the ion acts as electron acceptor and so a *charge-transfer-to-solvent* (CTTS) transition occurs with $h\nu$ close to that of Br^-[41].

Much work has been conducted on the u.v. spectra of azide crystals but this has been reviewed at length[1,2]. In aqueous solution the spectrum of N_3^- reveals at least two bands (Figure 8): one appearing as a shoulder at about 230 nm with $\varepsilon_{max} \sim 400$ M^{-1} cm^{-1} and the other, much stronger, peaking at ~ 190 nm. In acetonitrile the spectrum reveals a third band which may be concealed under the 190 nm band[41]. By studying the effects of environmental changes and by comparing the spectrum with that of azide crystals and that of other isoelectronic molecules, the three bands were related to the following excited states (in the order of increasing energy): $^1\Delta_u(^1B_2)$, CTTS state and $^1\Sigma_u^+$[41].

The 230 nm band was correlated with the weak 225 nm band of crystalline alkali and alkali earth azides, which displays a vibrational structure[41,72]. The corresponding excited state being the lowest excited singlet, plays a major role in emission and photochemical processes. It is probably bent, as indicated by the large Stokes shift of the crystal fluorescence[72]. The assignment of this state to $^1\Delta_u(^1B_2)$ has been recently questioned and a $^1\Sigma_u^-$ state was proposed instead[9]. There is no convincing evidence for this or the other assignments, but the assignment of the strong 190 nm band to the forbidden $^1\Delta_u \leftarrow {}^1\Sigma_g^+$ transition[9] seems unlikely. The latter behaves, both physically and chemically, as a CTTS band[41,73].

The CTTS state gives rise to solvated electrons, a reaction which competes with fast internal conversion to the first excited singlet[73]. In *charge-transfer complexes* with an electron acceptor having higher electron-affinity than that of the solvation layer, e.g. Fe^{+3} or I_2, the transition energy of the electron transfer process shifts to longer wavelengths. From the transition energies of these and CTTS bands the electron affinity of N_3 was estimated as 81 kcal/mole[41] (for a lower value see Table 8).

2. HN_3 and aliphatic azides

Table 12 reports some properties of the bands displayed by gaseous HN_3 and amyl azide. The data were taken from reference 9.

TABLE 12. Electronic spectra of HN_3 and $C_5H_{11}N_3$

HN_3			$C_5H_{11}N_3$		
λ_{max} nm	ε_{max} M^{-1} cm^{-1}	Oscillator strength (approx.)	λ_{max} nm	ε_{max} M^{-1} cm^{-1}	Oscillator strength (approx.)
264	20	6×10^{-4}	287	22	6×10^{-4}
200	450	9×10^{-3}	214	530	—
190	740	$1{\cdot}5 \times 10^{-2}$	191	800	—
170	500	$1{\cdot}0 \times 10^{-2}$	176	12,000	0·2
156	2000	0·2	162	15,000	0·1
140	—	0·3			

Four Rydberg series below 160 nm (see section II. G. 4).

The gas-phase spectrum of HN_3 reveals a rich vibrational structure[9]

(Figure 8). Its analysis has led us to assign some vibrational frequencies to the first three excited states, which are appreciably lower than the corresponding frequencies in the ground state.

Many aliphatic azides display absorption bands near 285 and 215 nm with ε_{max} close to 25 and 500 M^{-1} cm^{-1}, respectively (Table 13).

TABLE 13. Electronic spectra of some aliphatic azides[a]

RN_3, R=	Solvent	λ_{max} (I) nm	ε_{max} (I) M^{-1} cm^{-1}	λ_{max} (II) nm	ε_{max} (II) M^{-1} cm^{-1}	Ref.
C_2H_5	Ethanol	287	20	222	150	74
n-Butyl	Methanol	287·5	23	216	462	28
n-$C_{10}H_{21}$	95% Ethanol	287	21	—	—	12
Cyclopentyl	95% Ethanol	288	24	—	—	12
Cyclohexyl	Methanol	287·8	25	216·6	410	28
2-Chloroethyl	Methanol	284	27	214·2	527	28
2-Hydroxyethyl	Methanol	285·3	19	214·2	527	28
—$CH_2COOC_2H_5$	Methanol	281·6	23	214·3	581	28
CH_3COCH_2—	Methanol	274	40	—	—	75
$CH_3CH_2OCH_2$—	Methanol	276	33	—	—	75
CH_3SCH_2—	Methanol	284	28	—	—	75
—CH_2SCH_2—	Methanol	284	55	—	—	75

[a] For more data see references 12, 28 and 75.

Closson and Gray[28] related the 285 and 215 nm bands of alkyl azides (bands I and II) to the single transition of N_3^- at 230 nm. If the latter is a $^1\Delta_u \leftarrow {}^1\Sigma_g^+$ transition (see foregoing discussion) then under an off-axis perturbation (section I.C) these bands should split into two components. In molecular orbital language, I and II are assigned to $\pi_y(a'') \rightarrow \pi_x^*(a')$ and $s\delta p(a') \rightarrow \pi_y^*(a'')$ transitions (see Figure 5). With an electron-releasing group attached to N_a the non-bonding π_y orbital is mainly localized on this atom, (i.e. becomes more negative than N_c), so in alkyl azides both transitions involve flow of charge from N_a to the other nitrogen atoms. Transition I is 'perpendicular' in the sense that the π_y and π_x orbitals are perpendicular to each other; this accounts for its low intensity. The higher intensity of band II may be ascribed to the partial *s* character of the $s\delta p$ orbital.

A primitive calculation of transition energies[28] is instructive. It ignores electronic repulsion terms on the basis of the assumption that in the final results (which express relations between energies of correlated transitions in RN_3 and N_3^-) these terms cancel out. (Thus, no physical significance should be given to various stages of the calculations.) According to the simple energy level diagram (Figure 5) and Tables 3 and 4:

$$RN_3: \begin{cases} h\nu_I = \varepsilon(\pi_x^*) - \varepsilon(\pi_y \text{ n.b.}) = -\beta & (13) \\ h\nu_{II} = \varepsilon(\pi_y^*) - \varepsilon(s\delta p) & (14) \end{cases}$$

$$N_3^- : h\nu_{230} = \varepsilon(\pi_u^*) - \varepsilon(\pi_g) = -\sqrt{2}\beta \qquad (15)$$

(ε here designates orbital energy).

Equations (13) and (15) lead to $h\nu_{230}(N_3^-)/h\nu_I = \sqrt{2}$ (instead of $285/230 \simeq 1{\cdot}2$). Equation (14) is handled by inserting $\varepsilon(s\delta p) = \frac{1}{3}\varepsilon_s + \frac{2}{3}\varepsilon_p$ (equation 10):

$$h\nu_{II} = \varepsilon(\pi_y^*) - (\tfrac{1}{3}\varepsilon_s + \tfrac{2}{3}\varepsilon_p) \qquad (16)$$

Closson and Gray[28] considered RN_3 as isolated RN and N_2 molecules, since the RN_a—N_b bond is considerably weaker than the N_b—N_c bond (section II.E). In this case π_y^* (and the bonding π_y) should be also localized, with $\varepsilon(\pi_y^*) = \varepsilon(\pi_x^*) = \alpha - \beta$. Using this, and the familiar approximation $\alpha = \varepsilon_p$, equation (16) reduces to

$$h\nu_{II} = -\beta + \tfrac{1}{3}(\varepsilon_p - \varepsilon_s) = \frac{1}{\sqrt{2}}\, h\nu_{230}(N_3^-) + \tfrac{1}{3}(\varepsilon_p - \varepsilon_s) \qquad (17)$$

The term $\frac{1}{3}(\varepsilon_p - \varepsilon_s)$ could be estimated from atomic spectral data as 14,000 cm^{-1}. Equation (17) leads to $h\nu_{II} \sim 44{,}000\ cm^{-1}$.

Instead of using the 'isolated molecules' approximation we could put $\varepsilon(\pi_y^*) = \alpha - \sqrt{2}\beta \sim \varepsilon_p - \sqrt{2}\beta$ (Table 4), and obtain

$$h\nu_{II} = h\nu_{230}\,(N_3^-) + \tfrac{1}{3}(\varepsilon_p - \varepsilon_s) \qquad (18)$$

i.e., $h\nu_{II} \sim 57{,}500\ cm^{-1}$. The average experimental value (46,000 cm^{-1}) lies closer to that of the 'isolated molecules' approximation.

The solvent sensitivity of bands I and II is low; the general trend of $h\nu$ is to increase with polarity of solvent[28]. The same applies to the 230 nm band of N_3^- [41]. This low sensitivity and the absence of Cotton effect in alkyl azides (see section II.H) was considered as evidence against the labelling of band I as an $n \rightarrow \pi^*$ transition. However, we can still keep this notation since the excitation involves a transfer of an electron from a non-bonding π to an antibonding π^*. Moreover, band II can also be considered as an $n \rightarrow \pi^*$ transition as it involves the non-bonding $s\delta p$ orbital. The low solvent sensitivity indicates that the interaction of the molecule with solvent hardly changes on excitation. This was explained by considering the dipole moments of ground and excited states.[28]

The spectra of N_3^-, HN_3 and vinyl azides were subjected to more refined molecular orbital calculations, but agreement was rather poor[9,32]. (Sections III, D.2, III, E.)

3. Aromatic azides

The spectra of phenyl azide and its derivatives are treated in references 12, 76–80. The spectra are ill-defined[76] owing to overlapping absorptions of the various chromophores. The absorptions of both N_3 and the benzene ring are highly intensified as a result of their interaction. The bands of the ring are considerably red-shifted; they appear at wavelengths close to those of the corresponding acetanilides. In general, the azido group resembles acylamino groups more than halogens in its interaction with the benzene ring[80].

The spectra of substituted phenyl azides have played an important role in determining the *Hammett substituent constants* of the azido group. However, no details were given on the resolution of the spectra to their components.

Some of the bands reported in the literature are collected in Table 14.

TABLE 14. Electronic spectra of some aromatic azides

Phenyl azide	Solvent	λ_{max}, nm	ε_{max}, M^{-1} cm^{-1}	Ref.
$C_6H_5N_3$	95% Ethanol	285; 277; 248	1550; 2270; 9940	12
o-amino-	Methanol	309; 252; 225	4800; 9210; 17,600	80
m-amino-	Methanol	302; 233	2660; 23,000	80
p-amino-	Iso-octane	307; 260	3100; 18,000	80
o-amino-HCl	Methanol	249	13,300	80
m-amino-HCl	Methanol	250	13,600	80
p-amino-HCl	Methanol	249	14,400	80
m-nitro-	Iso-octane	315; 242	1600; 20,500	80
p-nitro-	Iso-octane	299; 220	17,000; 14,000	80
m-bromo-	Iso-octane	293; 283; 251	1500; 2000; 12,000	80
p-bromo-	Iso-octane	297; 285; 255	2100; 3000; 17,800	80

4. Rydberg transitions and ionization potentials

Four Rydberg series have been mapped in the spectrum of HN_3 below 160 nm[9]. Three series originate in the highest-filled orbital $\pi_y(a'')$ ($a'' \rightarrow ns$, $a'' \rightarrow np$ and $a'' \rightarrow nd$) and the fourth ($a' \rightarrow ns$)

originates in the first inner orbital, $s\delta p(a')$ (Figure 5). From the analysis of these series the *ionization potentials* for these orbitals were found to be 11·5 and 12·6 eV, respectively. Similar values were obtained theoretically[9] (see section III.D.2). However, the mass spectrometric result is appreciably lower. Electron impact values for the vertical ionization potentials of HN_3 and CH_3N_3 are 10·3 ± 0·2 eV and 9·5 ± 0·1 eV, respectively[48a]. The reason for this discrepancy is not clear.

H. Optical Rotatory Dispersion (o.r.d.) and Circular Dichroism (c.d.)

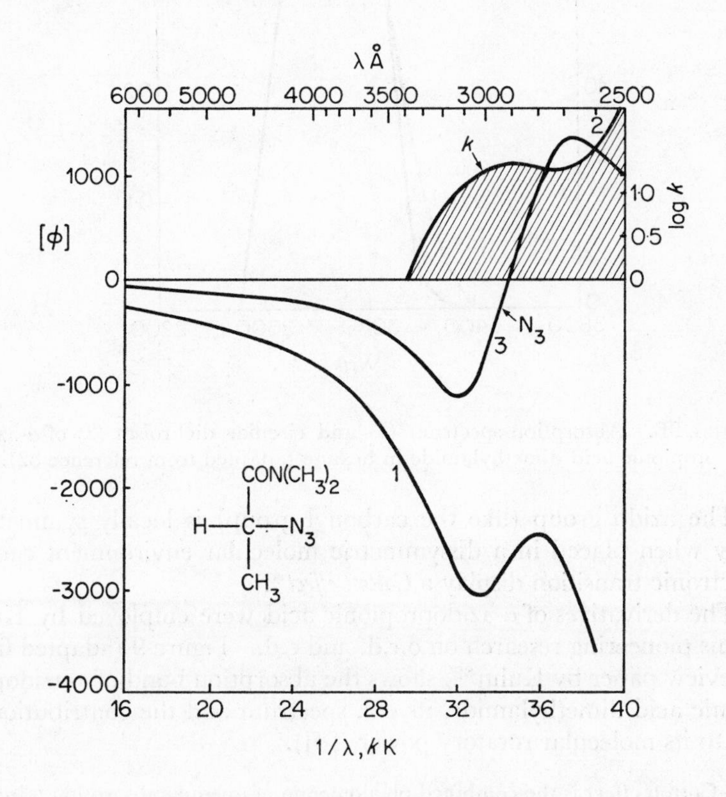

FIGURE 9A. Optical rotatory dispersion (1) and absorption spectrum (2) of α-azidopropionic acid dimethylamide in ether. Curve 3: the rotatory contribution of the azido group.

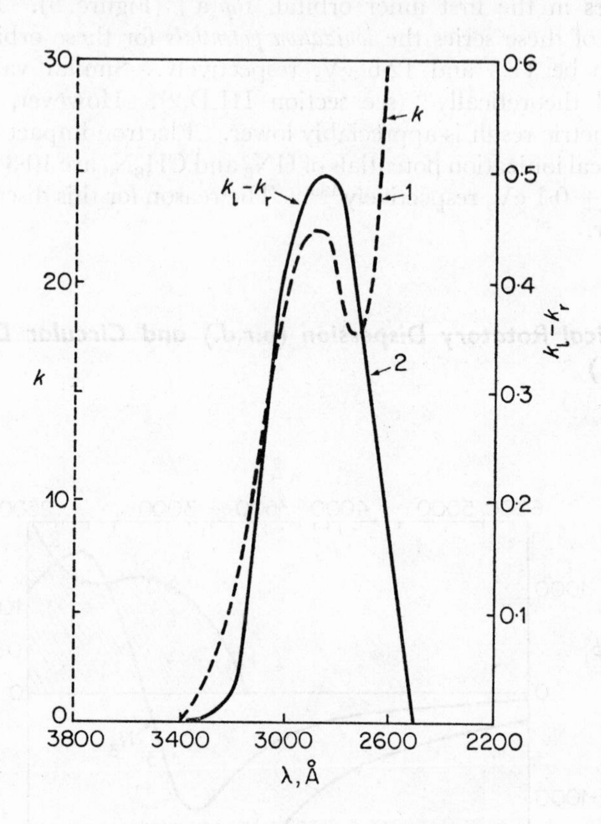

FIGURE 9B. Absorption spectrum (1) and circular dichroism (2) of α-azidopropionic acid dimethylamide in hexane (adapted from reference 82).

The azido group (like the carbonyl group) is locally symmetric; only when placed in a dissymmetric molecular environment can its electronic transition display a *Cotton effect**.

The derivatives of α-azidopropionic acid were employed by Kuhn in his pioneering research on o.r.d. and c.d. Figure 9 (adapted from a review paper by Kuhn[82]) shows the absorption band of α-azidopropionic acid dimethylamide, its c.d. spectrum and the contribution of N_3 to its molecular rotatory power ($[\phi]$).

* Cotton effect is the combined phenomenon of unequal absorption (circular dichroism) and unequal velocity of transmission of left and right circularly polarized light[81]. This series contains discussions on Cotton effect in references 56 and 81a.

Situated in the same environment the optical activity of N_3 will be smaller than that of the carbonyl group, mainly because its magnetic dipole transition moment is smaller[83] by a factor of ~5. This is because in the azido group the *n* and π^* orbitals are more separated in space than in >CO. Thus in all optically active alkyl azides which have been tested, the rotatory contribution of the 285 nm band is entirely negligible, while many analogous aldehydes do show Cotton effects[83a].

In more recent work the o.r.d. and c.d. of 34 steroidal azides were reported[83]. An *octant rule* similar to that of carbonyl was proposed, but the analysis of the results in terms of this rule was not successful. This failure was ascribed to free rotation of the chromophore in the steroidal azides, so that very many rotameric conformations are accessible.

I. The Quadrupole Coupling and Nuclear Magnetic Resonance

The nucleus of ^{14}N has a spin $I = 1$ and an *electric quadrupole moment* (which measures the deviation of nuclear charge from spherical symmetry), $Q = 0{\cdot}016 \times 10^{-24}$ cm^2 [84]. Its magnetic moment is relatively small ($\sim\frac{1}{7}$ that of ^1H), so that it is less sensitive to n.m.r. detection. The quadrupole moment couples the nuclear spin to the molecular rotation; therefore most ^{14}N n.m.r. spectra are strongly broadened and spin–spin splitting cannot be detected. This coupling involves the interaction of the quadrupole moment with the gradient of the electrical field centred on the nucleus. The quadrupole coupling energy is proportional to qQ, where q is the field gradient; q measures the departure from spherical symmetry of the charge distribution at the nucleus due to the electrons and other nuclei.

Nuclear quadrupole coupling brings about the splitting of the rotational lines in the microwave region into multiplets (hyperfine structure). From the analysis of the hyperfine structure, the *nuclear quadrupole coupling constants*, eqQ, can be determined (e is the electron charge). Some information concerning azides is included in Table 15. [In the case of HN_3, the constants recorded are the diagonal elements of the quadrupole coupling tensors along the principal inertial axis a (Figure 6); the axis a lies within a few degrees of the NNN line.] The magnitudes of the coupling constants indicate that in $HN_aN_bN_c$ the nucleus of N_a experiences a charge distribution which is less symmetrical than that around the other two nuclei, while the charge around N_b is nearly spherically distributed (section I.B).

In N_3^- the two terminal nuclei are equivalent and again the charge is more evenly distributed around the central atom (its coupling constant seems somewhat too high; another investigation is desirable). Of special interest is the fact that the field gradients at N_a and N_c are of different signs.

The quadrupole coupling constants for a molecule in a liquid are related to the width of the corresponding lines in its n.m.r. spectrum. The origin of the line-widths is mainly the spin–lattice relaxation, the mechanism of which involves quadrupole coupling. The larger the coupling constant the wider is the corresponding n.m.r. line. (For a study on quadrupole relaxation in nitrogen nuclear resonances, including N_3^-, see reference 85.)

Table 15 gives data on the n.m.r. spectra of several azides: the positions of the lines and their widths at half-height. The following regularities which emerge were helpful in constructing this table[87].

(a) The areas of the +277 and +128 signals of N_3^- are in the ratio 2:1, respectively.
(b) The chemical shift corresponding to the central atom, N_b, is the same for N_3^- and covalent azides.
(c) The chemical shift corresponding to the terminal atom, N_c, is the same for CH_3N_3 and $C_2H_5N_3$.
(d) The half-width increases with increase in the quadrupole coupling constant. (We assume that the coupling constants in alkyl azides follow the same sequence as in HN_3.)

Several theoretical studies have treated the chemical shifts[90,91] and quadrupole coupling constants[86,92] in N_3^-. The coupling constant can be calculated directly from the ground-state electronic charge distribution, and its evaluation is sensitive to the wavefunction employed. Thus the extent of agreement with experimental result may serve as a criterion for the suitability of the wavefunctions representing the valence-shell electrons. The calculation of chemical shifts requires in addition a knowledge of the average excitation energy, ΔE, for the magnetically allowed transitions. The assumptions concerning this parameter (see e.g. reference 90) make the calculation somewhat arbitrary.

III. THEORETICAL

A. General

The azide ion and to some less extent HN_3 have been subjected to numerous theoretical treatments, which altogether illustrate the various

TABLE 15. Quadrupole coupling constants and chemical shifts for $RN_aN_bN_c$

Azide	eqQ, Mc/s			Chemical shift, ppm[a]			Half width, c/s			Ref.
	N_a	N_b	N_c	N_a	N_b	N_c	N_a	N_b	N_c	
N_3^- [b]	+1·7	+1·0	+1·7	+277 (±1)	+128 (±0·5)	+277 (±1)	60 (±4)	20 (±2)	60 (±4)	86, 87
HN_3	+4·85 (±0·10)	<0·7[c]	−1·35 (±0·10)	—	—	—	—	—	—	23
CH_3N_3 [d]	—	—	—	+320 (±1)	+128 (±0·5)	+170·5 (±0·5)	101 (±4)	17 (±2)	19 (±2)	87
$C_2H_5N_3$	—	—	—	+305 (±1)	+129·5 (±0·5)	+167·5 (±0·5)	122 (±4)	22 (±2)	28 (±2)	87

[a] The shift is measured relative to the line of NO_3^-.
[b] See also references 88 and 89.
[c] The absolute value of the constant is smaller than 0·7.
[d] See also reference 88.

levels of approximation involved in molecular computations: (a) empirical methods such as simple molecular-orbital or valence-bond methods; (b) semi-empirical methods, e.g. those based on the Mulliken–Wolfsberg–Helmholz and the Pariser–Parr–Pople approximations; (c) non-empirical self-consistent field calculations.†

N_3^- and HN_3 are relatively small molecules (22 electrons); therefore satisfactory ground-state wavefunctions could be computed for these molecules on strict wave mechanical principles, using high-speed computers and refined numerical procedures. These and most of the empirical and semi-empirical calculations have been based on *linear combinations of atomic orbitals* (LCAO). (Some other approaches will be discussed in section III.G.)

The use of non-empirical LCAO methods for computing wavefunctions for small molecules, including N_3^- and HN_3, has been recently reviewed by Clark and Stewart[93]. An earlier valuable review is that of Nesbet[94] on 'approximate Hartree–Fock calculations on small molecules'. Some of the basic ideas of modern LCAO methods will be summarized here. (For a lucid discussion on this subject see reference 29.) A more advanced treatment has recently appeared in this series[94a].

The *N*-electron wavefunction of a singlet ground state is usually written as a *Slater determinant*

$$\Psi = (N!)^{-1/2} \det\{\chi_1\alpha(1)\chi_1\beta(2)\ldots\chi_{N/2}\alpha(N-1)\chi_{N/2}\beta(N)\} \quad (19)$$

where χ are the molecular orbitals, α and β the spin functions. If all the molecular orbitals χ are in their best possible forms (so as to give the lowest possible value for the energy $\langle\Psi|\mathbf{H}|\Psi\rangle$, where $\mathbf{H}$ is the complete Hamiltonian of the molecule), Ψ is described as a Hartree–Fock or *self-consistent field (SCF) wavefunction.*

The molecular orbitals χ are most commonly written as linear combinations of atomic orbitals and Ψ is then described as a LCAO wavefunction. The atomic orbitals are usually represented by *Slater-type atomic orbitals (STO)*

$$\psi = \Theta\Phi r^{n^*-1}e^{-\zeta r} \quad (20)$$

where the angular functions Θ and Φ are the same as for the hydrogen atom, n^* is the effective principal quantum number ($n^* = n$ for the first three rows in the periodic table), and ζ is the orbital exponent. The orbital exponents used for molecular computations are often the

† The terms 'empirical' and 'semi-empirical' describe calculations where all or some of the energy integrals are not evaluated mathematically but are treated as parameters which are determined experimentally.

best free atom values. (On the use of Gaussian functions see section III.F.2.)

The set of atomic orbitals used for LCAO (the basis set) can be either a *minimal basis set* or an *extended basis set.* In the first case only ground-state orbitals of the atoms are included in the combination, i.e. one atomic orbital is used for each independent occupied orbital in the component atom. SCF wavefunctions require the use of basis sets with more atomic orbitals (extended basis sets). Still, useful wavefunctions for ground states (but not excited states) can be obtained with minimal basis sets. Following the common usage these functions are also loosely described as SCF wavefunctions.

B. Minimal Basis Set for N_3^-

The ground-state orbitals of nitrogen atom are $1s$, $2s$ and three $2p$ orbitals. From symmetry of $N_aN_bN_c^-$ it is apparent that in the molecular orbital

$$\chi = a\psi_a + b\psi_b + c\psi_c \tag{21}$$

the atomic orbitals ψ_a and ψ_c must be equivalent and the coefficients must fulfil: $a = \pm c$. Hence it is convenient to consider the combinations $\psi_a + \psi_c$ and $\psi_a - \psi_c$ as group orbitals (denoted by ψ_g). Both the central atom and the marginal pair can contribute more than one orbital (or group orbital) to a particular molecular orbital.

For effective combination ψ_b and ψ_g must belong to the same irreducible representation (symmetry species). The azide ion has a $D_{\infty h}$ symmetry; Table 16 classifies its minimal basis set according to the symmetry properties of the orbitals.

TABLE 16. The symmetry properties of the minimal basis set of $N_aN_bN_c^-$ [a]

Symmetry species	ψ_b	ψ_g (group orbital)[b]
σ_g^+	$1s_b$, $2s_b$	$(1s_a + 1s_c)$, $(2s_a + 2s_c)$
		$(2p_{za} + 2p_{zc})$
σ_u^+	$2p_{zb}$	$(1s_a - 1s_c)$, $(2s_a - 2s_c)$
		$(2p_{za} - 2p_{zc})$
π_g	—	$(2p_{xa} - 2p_{xc})$
		$(2p_{ya} - 2p_{yc})$
π_u	$2p_{xb}$	$(2p_{xa} + 2p_{xc})$
	$2p_{yb}$	$(2p_{ya} + 2p_{yc})$

[a] The nuclei are considered to lie along the z axis. The $2p_z$ orbitals of the end atoms are taken with the same sign when directed in opposite directions. In the following discussion $2p_z$ will be designated as $2p_\sigma$, and $2p_x$ or $2p_y$ as $2p_\pi$.

[b] The group orbitals are not normalized. The normalization factor is $1/\sqrt{2}$.

From the fifteen orbitals recorded in Table 16 the following molecular orbitals can be constructed: five σ_g^+, four σ_u^+, one doubly degenerate π_g and two doubly degenerate π_u—altogether 15 molecular orbitals. Eleven of these can house the 22 electrons of N_3^- in its ground state; the rest can serve as *virtual orbitals* for excited states in rough calculations or to improve the ground-state wavefunction by 'mixing' it with that of excited states (*configuration interaction*). In order to obtain more reliable wavefunctions for excited states the basis set must be expanded.

The 1*s* orbitals are highly concentrated about their own nuclei and therefore they are little affected by molecule formation. Thus the three lowest orbitals are:

$$1\sigma_g^+ \sim 1s_b; \quad 1\sigma_u^+ \sim 1/\sqrt{2}(1s_a - 1s_c); \quad 2\sigma_g^+ \sim 1/\sqrt{2}(1s_a + 1s_c)$$

The energies of these orbitals are rather close, nearly the same as that of the 1*s* orbital of nitrogen atom.

From the foregoing discussion it appears that as a good approximation the valence shell of N_3^- consists of 12 molecular orbitals that can be constructed from three 2*s* and nine 2*p* orbitals of the nitrogen atoms; it accommodates 16 valence electrons.

The doubly degenerate orbital π_g involves only the end atoms (Table 16)

$$\pi_g = 1/\sqrt{2}(p_{\pi a} - p_{\pi c}) \tag{22}$$

Therefore it is non-bonding. The π_u orbital can be written as

$$\pi_u = \sin\theta . p_{\pi b} + 1/\sqrt{2}\cos\theta . (p_{\pi a} + p_{\pi c}) \tag{23}$$

where θ is a parameter to be determined by the variation method. This writing ensures that the function is normalized, if overlap integrals are ignored.

C. Walsh's Correlation Diagram for N_3^-

Figure 10 is a modified version of Walsh's diagram for triatomic molecules[40,71], showing in a schematic way how the orbital energies change with the interbond angle. The orbitals $1\sigma_g$, $2\sigma_g$ and $1\sigma_u$ are not shown in the diagram; their energies hardly change with angle (see section III.B). The principles underlying this diagram are the following:

(a) The correlations must follow symmetry rules which govern the resolution of species of the linear molecule into those of the bent molecule (see reference 71).

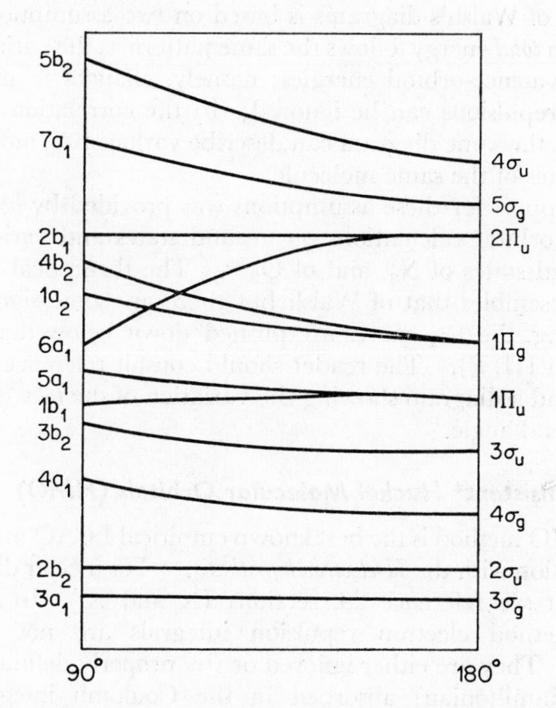

FIGURE 10. Walsh's diagram for ABA molecules (adapted from reference 71).

(b) The energy of a molecular orbital decreases as it changes from being built by a *p* orbital of the central atom to being built from its *s* orbital. In particular this is reflected in the $6a_1$–$2\pi_u$ orbital which changes from a pure $2s_b$ to a π_u orbital with no *s* character, as the angle changes from 90° to 180°. (At 90° there is no hybridization between *s* and *p* on the central atom.)

(c) If the orbital is bonding or antibonding between the end atoms, the energy tends to decrease or increase, respectively, as they come closer. This effect is less important than the hybridization effect.

Walsh's diagram shows that the energies of the first 11 orbitals either change little or steeply decrease as the angle changes from 90° to 180°. Therefore, triatomic non-hydrid molecules with 22 electrons (i.e. 16 valence electrons) or less should be linear. This is in accord with the linearity of N_3^- in its ground state. However, excitation of an electron from the highest filled $1\pi_g$ orbital to the vacant $2\pi_u$–$6a_1$ orbital should lead to a bent excited molecule after relaxation to equilibrium geometry (Figure 10. See section II.G).

The use of Walsh's diagrams is based on two assumptions: (a) the variation in *total* energy follows the same pattern as the variation in the sum over valence-orbital energies; namely, changes in nuclear and electronic repulsions can be ignored; (b) the correlation diagram is general i.e. the same diagram can describe various BA_2 molecules and various states of the same molecule.

Some support for these assumptions was provided by *ab initio* SCF molecular orbital calculations on ground states and various closed-shell excited states of N_3^- and of O_3[95]. The theoretical correlation diagram resembles that of Walsh but there are some significant departures, e.g. the $1\pi_u$ curves are pushed down below the $4\sigma_g$ curve (see section III, F). The reader should consult reference 95 for this diagram and a diagram showing the variation of the *total* energy with the interbond angle.

D. Self-consistent* Huckel Molecular Orbitals (HMO)

The HMO method is the best known empirical LCAO method, used in conjunction with the *Π-electron hypothesis*. (For a clear discussion of this subject see reference 29, sections IX and X.) In the simple Hückel method electron repulsion integrals are not considered explicitly. They are either ignored or (by properly defining the one-electron Hamiltonian) absorbed in the Coulomb integral α and resonance integral β, which are considered as empirical parameters†. The one-electron integrals α and β are not affected by replacing an antisymmetrized wavefunction by a simple-product wavefunction $\chi_1(1)\chi_1(2)\chi_2(3)\chi_2(4)\ldots$ Consequently the method does not distinguish between singlet and triplet states of the same configuration. There is some inconsistency here: in absence of spin correlation the one-electron operator **H** can mostly be a Hartree (not Hartree–Fock) Hamiltonian and so cannot be the same for all the electrons. The distinction between electrons disappears only as a result of antisymmetrization.

The naive Hückel approximation for the azido group, using single values for α and β, was described in section I.C, where its shortcomings were mentioned. In the case of the azido group $—N_aN_bN_c$ three Coulomb integrals and two resonance integrals should be considered in computing the delocalized π system; the HMO secular

* Not to be confused with 'self-consistent-field'.

† In this case the total energy of the π system is smaller than the sum of the orbital energies, because in the sum the repulsion between the π electrons is counted twice.

determinant (equation 3, section I.C) should be replaced by (equation 24)

$$\begin{vmatrix} \alpha_a - E & \beta_{ab} & 0 \\ \beta_{ab} & \alpha_b - E & \beta_{bc} \\ 0 & \beta_{bc} & \alpha_c - E \end{vmatrix} = 0 \tag{24}$$

where $\alpha_a = \langle p_{\pi a}|\mathbf{H}|p_{\pi a}\rangle$; $\beta_{ab} = \langle p_{\pi a}|\mathbf{H}|p_{\pi b}\rangle$, etc.

Improvement of the naive approach has been attempted by taking

$$\alpha_\mu = \alpha^0 + h\mu\beta^0;\ \beta_{\mu\nu} = k\mu\nu\beta^0 \tag{25}$$

where α^0 and β^0 refer to a particular bond taken as a standard (usually the C—C bond in benzene); $h\mu$ and $k\mu\nu$ are constants. In some works these constants were chosen in rather arbitrary way. For example, Ladik and Messmer took $h = 0{\cdot}58$ for the three N atoms in the azido group but considered the effect of bond distance on β_{NN} by assuming that it equals β_{CC} at the same interatomic distance[70]. More reasonable are the calculations based on the iterative method, which is the essence of all self-consistent methods.

The iterative method is incorporated into the HMO method as follows: starting with certain values of α_μ and $\beta_{\mu\nu}$, atomic charges q_μ and bond orders $P_{\mu\nu}$ are calculated. From these values new values of α_μ and $\beta_{\mu\nu}$ are obtained by using certain relationships (empirical or semi-empirical) between q_μ and α_μ, $P_{\mu\nu}$ and $\beta_{\mu\nu}$. The secular determinant (equation 24) is solved again with the new set of parameters and a new set of q_μ and $P_{\mu\nu}$ is derived. The process is repeated until two successive calculations give essentially the same result.

Several self-consistent HMO calculations were performed on N_3^- and on the azido group[34,96,97,98]. They vary in their degree of refinement and the expression used to relate charge and bond order with Coulomb and resonance integrals, respectively. The best work of this kind is that of Wagner and it may serve to illustrate the method.

I. The method of Wagner[98]

As a simple example consider the azide ion. Starting with the orbitals described by equations (23) and (22) (section III.B) we obtain the orbital energies

$$\varepsilon(\pi_u) = \langle\pi_u|\mathbf{H}|\pi_u\rangle = \sin^2\theta{\cdot}\alpha_b + \tfrac{1}{2}\cos^2\theta\,(\alpha_a + \alpha_c) + \sqrt{2}\sin\theta\cos\theta\,(\beta_{ab} + \beta_{bc}) \tag{26}$$

$$\varepsilon(\pi_g) = \langle\pi_g|\mathbf{H}|\pi_g\rangle = \tfrac{1}{2}\alpha_a + \tfrac{1}{2}\alpha_c \tag{27}$$

(Within the HMO approximation, $\beta_{ac} = 0$.) The total π-energy is considered to be the sum over the π orbitals, each accommodating four electrons (a procedure which is not correct; see previous section):

$$E_\pi = 4\varepsilon(\pi_u) + 4\varepsilon(\pi_g)$$
$$= q_a\alpha_a + q_b\alpha_b + q_c\alpha_c + 2P_{ab}\beta_{ab} + 2P_{bc}\beta_{bc} \quad (28)$$

where q_ν and $P_{\mu\nu}$ are the π-electron charge density and bond order, respectively (section I.C. μ and ν are running indexes). This is a general HMO expression for a molecule ABC. In the simple case considered we can put

$$\alpha_a = \alpha_c;\ \beta_{ab} = \beta_{bc}$$
$$q_a = q_c = 2\cos^2\theta + 2;\ q_b = 4\sin^2\theta \quad (29)$$
$$P_{ab} = P_{bc} = \sqrt{2}\sin 2\theta$$

Thus E_π is expressed in terms of one variable θ. For molecules with different end atoms (like the azido group in covalent azides), more parameters are required (see later).

Minimizing the energy with respect to θ gives

$$dE/d\theta = 2\xi_a dq_a/d\theta + \xi_b . dq_b/d\theta + 4\beta_{ab} . dP_{ab}/d\theta = 0 \quad (30)$$

where

$$\xi_\nu = \alpha_\nu + q_\nu . \partial\alpha_\nu/\partial q_\nu \quad (31)$$

From equations (29) and (30) we derive

$$\xi_b - \xi_a + 2\sqrt{2}\,\beta_{ab}\cot 2\theta = 0 \quad (32)$$

ξ_ν can be related to the corresponding valence state electronegativity X_ν by assuming: (a) Coulomb integrals are proportional to valence state electronegativities

$$\alpha_\nu = (\alpha_C^0/X_C^0)X_\nu \quad (33)$$

where α_C^0 and X_C^0 are the corresponding parameters for a carbon atom in benzene (arbitrarily chosen as a standard); (b) X_ν is linearly related to the formal charge on the atom, $Z_\nu - q_\nu$, where Z_ν is the number of electrons contributed by the atom ν to the π system (see section I.C):

$$X_\nu = X_\nu^0 + \delta_\nu(Z_\nu - q_\nu) \quad (34)$$

where X_ν^0 is the tabulated orbital electronegativity of the atom ν in some reference state with zero formal charge.

From equations (31), (33) and (34) we derive

$$\xi_\nu = (\alpha_C^0/X_C^0)\,(X_\nu^0 + \delta_\nu Z_\nu - 2\delta_\nu q_\nu) \quad (35)$$

Wagner tried various methods of determining δ, the change in electronegativity per unit charge, and found Pauling's method[22] to give the best results. Using equations (29), (32) and (35) and assuming that $\beta_{\mu\nu}$ is proportional to the overlap integral, $\beta_{\mu\nu} = (\beta_{CC}/S_{CC})S_{\mu\nu}$ (β_{CC} and S_{CC} correspond to the benzene bond), we can finally derive an equation relating θ to the overlap integral. This equation has the form:

$$A \cos 2\theta + BS_{NN} \cot 2\theta = C \tag{36}$$

where S_{NN} is $S_{ab} = S_{bc}$, and A,B,C are constants which depend on the particular choice of the various parameters*. A self-consistent solution to this equation can be found as follows: we start with some value of S, calculate θ and use θ to calculate charge densities and bond order (equation 29). Knowing the latter, a new value for S can be obtained and the whole process is repeated until two successive calculations yield almost the same result. Using the best values of θ and S, the best values of π-electron formal charges, bond orders and energy can be obtained (equations 28 and 29). The following results were obtained by Wagner:

$$\overset{-0{\cdot}806}{N_a}\underset{1{\cdot}387}{\text{———}}\overset{+0{\cdot}612}{N_b}\underset{1{\cdot}387}{\text{———}}\overset{-0{\cdot}806}{N_c}\ ,\ E_\pi = 32{\cdot}652\beta_{CC}$$

The calculations on unsymmetrical ABC molecules are more difficult. Wagner used '*unsymmetrical group orbitals*' for the end atoms: $\psi_{g_1} = \sin\gamma . p\pi_a + \cos\gamma . p\pi_c$; $\psi_{g_2} = \cos\gamma . p\pi_a - \sin\gamma . p\pi_c$. (When $p\pi_a$ and $p\pi_c$ are equivalent, as in N_3^-, $\gamma = \pi/4$.) ψ_{g_1} combines with $p\pi_b$ to form a nodeless wavefunction χ_1, and χ_2 is built-up solely from ψ_{g_2} with a node at the central atom (analogous to π_g, equation 22). Thus according to this method (*linear combination of group orbitals*), χ_2 is always non-bonding. Following the procedure outlined above we obtain an analogous equation for E, with q_ν and $P_{\mu\nu}$ now depending on θ and γ. Minimization with respect to these parameters leads to equations correlating θ, γ, S_{ab} and S_{bc}, from which their best values can be obtained by several iterations.

In the case of the azido group two different π systems should be considered, the delocalized π_D and the localized π_L (section I.B). The

* See reference 98 for the values of X_ν^0, δ_ν, α_C^0/β_{CC} and X_C^0/S_{CC}, used in Wagner's calculations.

full treatment requires three condition equations, from which the following results were derived[98]:

$$\underset{/}{\overset{-0{\cdot}733}{\text{N}}}\overset{}{\underset{0{\cdot}612}{\text{———}}}\overset{+0{\cdot}899}{\text{N}}\underset{1{\cdot}543}{\text{———}}\overset{-0{\cdot}166}{\text{N}}\;,\; E_\pi = 31{\cdot}965\beta_{CC}$$

Wagner claims that this structure corresponds to HN_3, but no consideration was given to the effect of the H atom. He carried out similar calculations on various pseudohalide anions and was able to show that the pK of the corresponding acids varies linearly with the difference in the π-energy between the ions and their acids. Linear correlations were also found between (a) calculated bond orders and *stretching force constants*; (b) calculated charge densities and ^{14}N *n.m.r. chemical shifts.*

2. The method of Mulliken–Wolfsberg–Helmholz[99, 100]

This method bears some resemblance to the 'charge-self-consistent' methods previously described, but it treats all the various types of valence orbitals (not only the π orbitals) and relates Coulomb and resonance integrals to ionization potentials. The secular determinant $|H_{ij} - \varepsilon S_{ij}| = 0$, obtained on application of the variation principle to LCAO wavefunction, is solved for the orbital energies (one determinant for each symmetry type of the orbitals), using a simple method to evaluate the Coulomb integrals H_{ii} and the resonance (or exchange) integrals H_{ij} $(i \neq j)$. H_{ii} is taken as the negative of the *valence-state ionization energy (VSIE)* of the corresponding atomic orbital. For H_{ij} the following approximation is used

$$H_{ij} = K\left(\frac{H_{ii} + H_{jj}}{2}\right)S_{ij} \tag{37}$$

where S_{ij} is the corresponding overlap integral and K is a constant for which Carroll and co-workers[101] set: $K = 2 - |S_{ij}|$.

Each VSIE is a function (empirically established) of charge and electronic configuration; the latter is obtained by *Mulliken's population analysis*[102] conducted on the molecular orbitals. Therefore an iterative method is required. Each secular determinant is solved in cycles until self-consistent charge distribution is obtained. Some relation between the output from one iteration and the input from the next is often necessary to ensure convergence[101].

The Mulliken–Wolfsberg–Helmholz (MWH) method appears to give a good description of the energy levels of small molecules, including the first unfilled orbitals. Spin correlation effects can be injected into the procedure when calculating transition energies. Therefore it is a helpful tool for spectroscopic calculations. As with other semi-empirical methods the success of this method is probably due to the empirical information absorbed in the calculation (e.g. concerning ionization potentials), which is closely related to the calculated data.

N_3^- and HN_3 were subjected to MWH computation[9], using the *double-zeta representation* of atomic orbitals (section III.F.1). The computed ionization potentials and oscillator strengths for HN_3 are in good agreement with experimental data, but the computed orbital energies are considerably different from that calculated by SCF method with minimal basis set[103]. In the case of N_3^- the method has failed and this was partly ascribed to solvent effects, since the calculations refer to the unknown free ion. The computed sequence of energy levels puts the orbital $1\pi_u$ above $4\sigma_g$, at variance with the order predicted by SCF methods (section III.F).

E. Semi-empirical Self-consistent Field Methods

In the framework of SCF methods the electronic repulsion terms are treated explicitly. Since these terms are considerably affected by spin correlation, the simple-product wavefunctions should be replaced by Slater determinants.

The number of the electronic repulsion integrals involved in SCF calculations increases roughly as the fourth power of the order of the basis set. Even with the fast computers now available the occurrence of these integrals in the energy expression (of other than the smallest molecules) prevents sufficient expansion of the basis set as necessary for obtaining good SCF wavefunctions.

I. The Pariser–Parr–Pople method

Pariser, Parr and Pople have simplified the evaluation of these integrals. Their method may be considered as a modified Hückel calculation. The electronic Hamiltonian of the π system is expressed as the sum of one-electron core Hamiltonians H^c plus the sum of two-electron repulsion energies between the π electrons. The expression for the π-energy includes three types of integrals: (1) one-electron Coulomb integrals: $H_{ii}^c = \langle\psi_i|\,H_c\,|\psi_i\rangle$, (2) one-electron resonance integrals: $H_{ij}^c = \langle\psi_i|\,H^c\,|\psi_j\rangle$; (3) two-electron repulsion integrals such as one-centre Coulomb integrals (ii, ii), two-centre Coulomb integrals (ii, jj), two-centre exchange integrals (ij, ij), and the difficult three- and four-centre integrals*.

The Pariser–Parr–Pople (PPP) method is based on three assumptions: (a) Whenever the factor $\psi_i(n)\psi_j(n)$ (where $i \neq j$, n is any electron index) appears in the integrand the integral vanishes (*zero differential overlap approximation*); (b) The resonance integrals between nearest neighbours are treated as empirical parameters; those between non-neighbours are neglected; (c) The one-centre Coulomb integrals (ii, ii) are taken to be equal to $I_i - E_i$, where I_i and E_i are the ionization potential and electron affinity, respectively, of the atomic orbital ψ_i, when the atom is in the appropriate valence state.

The integral H^c_{ii} (which represents the attraction between the core of the molecule and an electron occupying the orbital ψ_i) is usually calculated by the method introduced by Goeppert-Mayer and Sklar as the negative of the ionization potential of an electron occupying this orbital, corrected for the effect of molecular environment. This correction involves two-electron two-centre repulsion integrals and one-electron two-centre *penetration integrals* as (ii, Q); the latter represents the total energy of an electron occupying the atomic orbital ψ_i due to its interaction with the neutral atom Q. (The penetration integrals are often ignored.)

The empirical input in these calculations is orbital ionization potentials and electron affinities. Such data on nitrogen atoms can be inferred from Table 5.

Favini[32] applied the PPP method to vinyl azide. Here the delocalized π system consists of 6 electrons and 5 atomic $2p$ orbitals. The calculated data include charge densities, bond orders and dipole moments (obtained by vector addition of the σ and π moments). *Cis* and *trans* isomers were considered

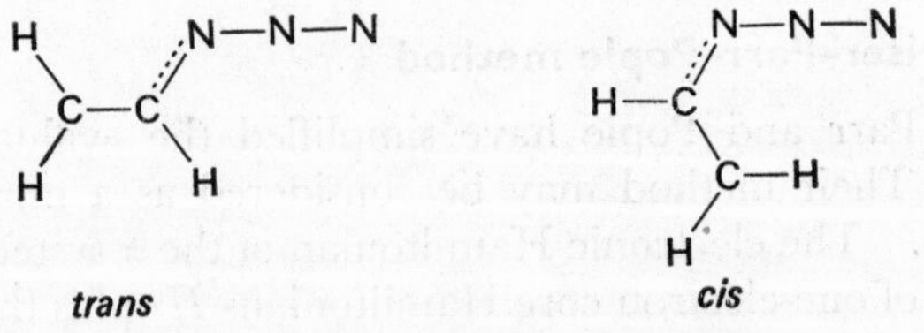

trans *cis*

* The notation (ij, ij) etc. is used for the integral

$$\int \psi_i(1)\, \psi_j(1)\, \frac{e^2}{r_{12}}\, \psi_i(2)\, \psi_j(2),\ d\tau_1 d\tau_2,$$

where arbitrary numerals 1 and 2 designate the two electrons.

Their computed π energies are rather close. Favini also carried out a configuration interaction calculation to determine the transition energies of vinyl azide. It displays a strong $\pi \rightarrow \pi^*$ band at 2380 Å. The theoretical values for this band are not far from the experimental, that of the *cis* being closer.*

2. All-valence electrons calculations

A '*complete neglect of differential overlap*' (CNDO) method was developed by Pople's group to include all valence electrons. Hendrickson and Kuznesof[91] applied this method to calculate excitation energies for N_3^-. Their ultimate goal was to calculate the ^{14}N *chemical shifts*; this requires the knowledge of the average excitation energy. All possible Slater determinants which result from the filled and virtual molecular orbitals were subjected to configuration interaction. According to their computation the lowest energy transition involves the Π_g state, at variance with other works[9,41].

In their treatment of N_3^- and HN_3 Yonezawa and co-workers[104] abandoned the zero differential overlap approximation and considered all valence electrons. The Mulliken approximation

$$(ij, kl) = \tfrac{1}{4}S_{ij}S_{kl}\,[(ii, kk) + (ii, ll) + (jj, kk) + (jj, ll)] \qquad (38)$$

was used to reduce three- or four-centre integrals to two- or one-centre integrals. For homonuclear molecules the two-centre integrals were further reduced:

$$(ii, jj) = [(ii, ii)^{-2} + R_{ij}{}^{2}]^{-1/2} \qquad (39)$$

where R_{ij} is the internuclear distance. The resonance integrals were evaluated by means of the MWH approximation (equation 37) with $K = 1{\cdot}1$.

The data calculated by Yonezawa and co-workers consist of orbital energies, charge distribution, atomic dipoles, proton affinities and excitation energies. (The latter are much higher than the experimental.) The energy of the highest occupied orbital of N_3^- (π_g) was calculated as $-4{\cdot}67$ eV, close to the experimental electron affinity of N_3 (section II.B). The computed atomic dipoles of the end nitrogen atoms are relatively large, 2·82D, suggesting that the lone pair orbitals are not pure *s* orbitals (as in Table 2, section I.B) but they are mixed with *p* orbitals. Thus N_3^- may act as strong π donor and also as σ donor, with the end atoms being the most reactive.

* A simple HMO-type of calculation was performed on phenyl azide[31].

F. LCAO Self-consistent Field (Hartree–Fock) Molecular Orbitals*

1. Slater-type atomic orbitals

The first SCF wavefunctions for the ground state of N_3^- were calculated by Clementi. Initially he used the minimal basis set (Table 16) with Slater-type atomic orbitals (STO)[105]. The orbital exponents ζ (equation 20) were chosen as the best free-atom values and the internuclear distance was taken as 1·12 Å. (As already discussed, section II.A, this distance is erroneous.) The total electronic energy of N_3^- was calculated as $-162{\cdot}5422H$ ($1\ H = 27{\cdot}209$ eV/atom).

In a later paper, Clementi and McLean[106] introduced three modifications: (a) The basis set was extended to include the $3d_\pi$ orbitals on the two end atoms and a $3d_\sigma$ orbital on the central atom; (b) Some of the orbital exponents were chosen as those optimized for N_2; (c) The basis set was further expanded by using a somewhat crude *double-zeta representation* of the $2p_\sigma$ orbital of the central atom. This representation replaces the single exponential factor $\exp(-\zeta r)$ (equation 20) by a linear combination of two exponential factors, $\exp(-\zeta_1 r)$ and $\exp(-\zeta_2 r)$, and the corresponding combination coefficients are included in the variation process†.

Table 17 records the orbital coefficients and energies computed with the expanded basis set. The total energy calculated is $-162{\cdot}7048\ H$, i.e. ~4·4 eV lower than that calculated with the minimal basis set. The table shows: (a) The electronic configuration of N_3^- is $(1\sigma_g)^2(1\sigma_u)^2(2\sigma_g)^2(3\sigma_g)^2(2\sigma_u)^2(1\pi_u)^4(4\sigma_g)^2(3\sigma_u)^2(1\pi_g)^4$ (the order follows that of increasing energy); (b) $1\sigma_g \sim 1s_b$, $1\sigma_u \sim 1/\sqrt{2}(1s_a - 1s_c)$ and $2\sigma_g \sim 1/\sqrt{2}(1s_a + 1s_c)$, i.e. they represent the inner shell orbitals of the atoms (see section III.B). This is also reflected in their energies; (c) the remaining σ orbitals and $1\pi_u$ orbital form the σ and π bonds, respectively‡, (d) $1\pi_g$ is a non-bonding orbital (see section III.B). The considerable positive value computed for its energy (Table 17) illustrates the shortcomings of this calculation. However, the order of orbital energies appears to be correct (see next section).

Bonaccorsi and co-workers[103] repeated these calculations of N_3^-,

* The orbitals discussed in this section are not real SCF wave functions, because of the limited basis sets employed; see section III.A.

† Optimum values of ζ_1 and ζ_2 are now available[107] for atomic orbitals with $Z \leq 36$. In this work[106], ζ_1 and ζ_2 were arbitrarily split around the best atomic exponent.

‡ The $2s$ orbitals of the end atoms also participate in bonding, at variance with the simple model outlined in section I.B (see also section III.E.2).

TABLE 17. LCAO coefficients and energies for SCF molecular orbitals in the azide ion, $N_aN_bN_c^-$

Atomic or group orbital	Orbital exponent	$1\sigma_g$	$1\sigma_u$	$2\sigma_g$	$2\sigma_u$	$3\sigma_g$	$3\sigma_u$	$4\sigma_g$	$1\pi_u$	$1\pi_g$
$1s_a + 1s_c$	6·6675	0·0027	—	0·7047	—	0·1045	—	0·1310	—	—
$1s_a - 1s_c$	6·6675	—	0·7047	—	−0·1380	—	−0·0976	—	—	—
$2s_a + 2s_c$	1·9170	−0·0103	—	0·0119	—	−0·2226	—	−0·7075	—	—
$2s_a - 2s_c$	1·9170	—	0·0187	—	0·4929	—	0·5522	—	—	—
$2p_{\sigma_a} + 2p_{\sigma_c}$	2·2524	−0·0055	—	0·0037	—	−0·1075	—	0·2981	—	—
$2p_{\sigma_a} - 2p_{\sigma_c}$	2·2524	—	0·0068	—	0·1656	—	−0·4100	—	—	—
$2p_{\pi_a} + 2p_{\pi_c}$	1·9090	—	—	—	—	—	—	—	0·3596	—
$2p_{\pi_a} - 2p_{\pi_c}$	1·9090	—	—	—	—	—	—	—	—	0·7171
$3d_{\pi_a} + 3d_{\pi_c}$	1·9090	—	—	—	—	—	—	—	0·0361	—
$3d_{\pi_a} - 3d_{\pi_c}$	1·9090	—	—	—	—	—	—	—	—	−0·0517
$1s_b$	6·670	0·9957	—	−0·0042	—	0·2050	—	−0·1047	—	—
$2s_b$	1·917	0·0266	—	−0·0078	—	−0·7032	—	0·5935	—	—
$2p_{\sigma_b}$	2·1	—	0·0920	—	0·5328	—	0·1229	—	—	—
$2p'_{\sigma_b}$	2·3	—	−0·0808	—	−0·9670	—	0·3186	—	—	—
$2p_{\pi_b}$	1·909	—	—	—	—	—	—	—	0·6503	—
$3d_{\sigma_b}$	2·2	0·0093	—	0·0005	—	−0·0891	—	0·0277	—	—
energy (H)		−15·3555	−15·0135	−15·0131	−0·8287	−1·0537	−0·1572	−0·2271	−0·2710	0·1239

using the minimal basis, and also applied the method to HN_3. They also computed the components of their dipole and quadrupole moments and have shown that average kinetic energy $\bar{T}$ and potential energy $\bar{V}$ are in keeping with the *virial theorem*: $-2\bar{T} = \bar{V}$. Using Mulliken's population analysis[102] they came to the conclusion that on protonation ($N_cN_bN_a \rightarrow N_cN_bN_aH$), there is a transfer of σ charge from the anion as a whole to the H atom, a transfer of π charge to N_a while the central atom N_b undergoes a minor population change. The *overlap population* of the bonds N_{bc} and N_{ab} increases and decreases, respectively, on protonation; this is in accord with the changes in force constants (section II.E).

By means of a unitary transformation the set of SCF molecular orbitals has been converted to localized orbitals[103]: the σ–π separation now disappears and the orbitals can be divided into lone-pair and bond orbitals (see also section I.B).

Reference 103 records maps of densities for some of these orbitals.

2. Gaussian lobe functions representation

The difficulty in calculating electron-repulsion integrals with Slater orbitals has led to development of this and similar methods, which employ basis sets consisting of Gaussian functions, exp $(-\zeta r^2)$, instead of the exponentials exp $(-\zeta r)$[93,94]. In the 'Gaussian lobe' method the angular functions $\Phi\Theta$ (equation 20) are also represented by Gaussian functions by suitable centring at proper points in space.

Peyerimhoff and Buenker[95] applied this method to calculate the correlation diagram of N_3^- (section III.C). The total energy of its ground state configuration was shown to have a minimum at 180°. The computed energies of the orbitals are in the same order as that obtained by Clementi and McLean[106], but there are two significant improvements: (a) the total energy is somewhat lower; (b) the energy of the non-bonding orbital $1\pi_g$ is negative.

G. Other than LCAO Methods

Three other approaches will be mentioned:

(a) A simple three-dimensional *free-electron model* (FEM) was applied to N_3^- for calculating its energy and charge distribution[108].

(b) Wulfman[109] replaced all the Coulomb potentials in the Hamiltonian of N_3^- by quadratic potentials of the harmonic oscillator treatment. Such potentials would be experienced by an electron when embedded in a uniformly charged sphere. Therefore this 'distributed atom' approach is more likely to be valid for a system of many moderately charged vibrating nuclei than for a system with few nuclei of large charge. The resulting Hamiltonian is factorable by the usual normal mode analysis and the Schrödinger equation can be thus solved exactly; the electronic wavefunction is expressed in terms of one-electron *single-centre Gaussian functions*. The quadratic potentials can then be modified to account for the behaviour of real molecules. In this way Wulfman was able to explain some characteristic features of Walsh's correlation diagram, in particular the linear and bent structures of N_3^- in its ground and excited states, respectively. However more accurate information on this subject was obtained from SCF calculations (section III.F).

(c) Singh[34] applied a *valence-bond method* to calculate the energy and interatomic distances of the hybrid structure arising from the two canonical structures of the azido group N_a—N_b—N_c (section I.B). The device of a 'phantom' electron[110] was employed. The energy was ultimately expressed in terms of σ-bond energy K, exchange integral J, repulsion energy between two lone pairs of electrons R, and promotion energy P. (The latter represents the energy required to change the interbond angle from 110°—the angle usually enclosed by two nitrogen bonds—to 180°.) J, K and R were related to the energies of the bonds C=C, C⋯C (benzene), N≡N, N=N and N—N. (The force constants of these bonds were used in the framework of the Morse equation to determine their energies at various interatomic distances.) P was treated as activation energy for inversion, as in the familiar case of ammonia. In estimating K and R two assumptions were made: (a) K is proportional to $S/1 + S$, where S is the overlap integral between the two orbitals involved in σ-bond formation; (b) R is proportional to the square of the overlap integral between the two lone-pair orbitals. The overlap integrals were calculated for the N=N and N≡N bonds at various bond distances and it was shown that the repulsion energy term in the triple bond is very small. (This can explain the relatively high stability of the N_2 molecule; see section II.B.) In calculating the overlap integrals for the σ bonds, changes in hybridization were taken into account: N_b was considered to hybridize diagonally and the states of hybridization of N_a and N_c were taken to be the same as in azomethane and N_2 molecule, respectively. The formal charges on the three nitrogen atoms were calculated by means of a simple molecular orbital treatment.

Assuming additivity of bond energies, the heat of formation of HN_3 (from atoms) was calculated from the bond energies of NH and the azido system as a function of the N—N distances; it was found to have a minimum value 319·1 kcal, with lengths of N_a—N_b and N_b—N_c bonds equal to 1·24 and 1·12 Å, respectively. (The experimental heat of formation is 318·3 kcal[2].)

The valence-bond method outlined here is largely empirical and loaded with many approximations. The calculation of total energy as a sum of contributions from the σ and π electrons is in principle wrong. (See for example reference 29, section IX.) Still some of its basic ideas may serve in more refined calculations.

H. Conclusion

The SCF methods described in section III.F appear to give an adequate picture of N_3^- in its ground state. The semi-empirical SCF method of Yonezawa and co-workers (section E.2) is a good approximation, giving results close to that of Clementi and McLean. However, no satisfactory calculation has been conducted on the excited states. Even the assignment of the first excited state is still unknown. More extended basis set and configuration interaction treatment should lead to better results.

IV. REFERENCES

1. B. L. Evans, P. Gray and A. D. Yoffe, *Chem. Rev.*, **59**, 515 (1959).
2. P. Gray, *Quart. Rev.*, **17**, 441 (1963).
3. M. Green in *Developments in Inorganic Nitrogen Chemistry*, Vol. I, (Ed. C. B. Colburn), Elsevier, 1966, p. 1.
4. A. D. Yoffe in *Developments in Inorganic Nitrogen Chemistry*, Vol. I (Ed. C. B. Colburn), Elsevier, 1966, p. 72.
5. K. G. Mason in *Mellor's Comprehensive Treatise on Inorganic and Theoretical Chemistry*, Vol. VIII, Supp. II, Part II, Longmans, 1967, pp. 1 and 16.
6. M. F. Lappert and H. Pyszora in *Advances in Inorganic and Radiochemistry*, Vol. 9 (Eds. H. J. Emeleus and A. G. Sharpe), Academic Press, 1966, p. 133.
7. E. Lieber, J. S. Curtice and C. N. R. Rao, *Chem. Ind.*, 586 (1966).
8. J. H. Boyer and F. C. Canter, *Chem. Rev.*, **54**, 1 (1954).
9. J. R. McDonald, J. W. Rabalais and S. P. McGlynn, *J. Chem. Phys.*, **52**, 1332 (1970).
10. C. S. G. Phillips and R. J. P. Williams, *Inorganic Chemistry*, Vols. 1 and 2, Oxford, Clarendon Press, 1965.
11. E. Gusarsky and A. Treinin, *J. Phys. Chem.*, **69**, 3176 (1965).
12. E. Lieber, C. N. R. Rao, T. S. Chao and W. H. Wahl, *J. Sci. Ind. Res.*, **16** B, 95 (1957).
13. Ya. K. Syrkin and E. A. Shott-L'vova, *Dokl. Akad. Nauk S.S.S.R.*, **87**, 639 (1952).
14. C. P. Smyth, *Dielectric Behaviour and Structure*, McGraw-Hill, 1955.

15. A. I. Vogel, *J. Chem. Soc.*, 1833 (1948).
16. S. Glasstone, *Textbook of Physical Chemistry*, 2nd ed. Van Nostrand, 1946.
17. A. Petrikalns and B. Ogrins, *Radiologica*, **3**, 201 (1938): *Chem. Abstr.*, **35**, 3146 (1) (1941).
18. W. H. Perkin, *J. Chem. Soc.*, **69**, 1025 (1896).
18a. P. N. Schatz and A. J. McCaffery, *Quart. Rev.*, **23**, 552 (1969).
19. P. Gray and T. C. Waddington, *Proc. Roy. Soc.* (*London*), **235A**, 481 (1956).
20. M. J. Blandamer and M. F. Fox, *Chem. Rev.*, **70**, 59 (1970).
21. R. J. Samuel, *J. Chem. Phys.*, **12**, 167, 180 (1944).
22. L. Pauling, *The Nature of the Chemical Bond*, 3rd ed. Cornell Univ. Press, Ithaca, N.Y., 1960.
23. R. A. Forman and D. R. Lide, *J. Chem. Phys.*, **39**, 1133 (1963).
24. W. J. Orville-Thomas, *Quart. Rev.*, **11**, 162 (1957).
25. W. J. Orville-Thomas, *Chem. Rev.*, **57**, 1179 (1957).
26. W. J. Orville-Thomas, *The Structure of Small Molecules*, Elsevier, 1966.
27. C. Sandorfy in *The Chemistry of the Carbon–Nitrogen Double Bond* (Ed. S. Patai), Interscience, London, 1970, p. 1.
28. W. D. Closson and H. B. Gray, *J. Am. Chem. Soc.*, **85**, 290 (1963).
29. C. A. Coulson and E. T. Stewart in *The Chemistry of Alkenes* (Ed. S. Patai), Interscience, London, 1964, p. 1.
30. J. D. Roberts, *Notes on Molecular Orbital Calculations*, W. A. Benjamin, New York, 1962, p. 131.
31. G. Favini, *Gazz. Chim. Ital.*, **91**, 270 (1961).
32. G. Favini, *Gazz. Chim. Ital.*, **92**, 468 (1962).
33. H. A. Skinner and H. O. Pritchard, *Trans. Faraday Soc.*, **49**, 1254 (1953).
34. K. Singh, *Proc. Roy. Soc.* (*London*), **225A**, 519 (1954).
35. A. E. Douglas and W. J. Jones, *Can. J. Phys.*, **43**, 2216 (1965).
36. C. S. Choi, *Acta Cryst.*, **25B**, 2638 (1969).
37. M. Winnewisser and R. L. Cook, *J. Chem. Phys.*, **41**, 999 (1964).
38. R. L. Livingston and C. N. R. Rao, *J. Phys. Chem.*, **64**, 756 (1960).
39. I. E. Knaggs, *Proc. Roy. Soc.* (*London*), **150A**, 576 (1935).
40. A. D. Walsh, *J. Chem. Soc.*, 2266 (1953).
41. I. Burak and A. Treinin, *J. Chem. Phys.*, **39**, 189 (1963).
42. *Interatomic Distances*, Special Publication, No. 11, The Chemical Society, London, 1958.
43. C. S. Choi and H. P. Boutin, *Acta Cryst.*, **25B**, 982 (1969).
44. V. N. Sidgwick, *The Chemical Elements and their Compounds*, Oxford, Clarendon Press, vol. I, 1950, p. 718.
45. T. F. Fagley and H. W. Myers, *J. Am. Chem. Soc.*, **76**, 6001 (1954).
46. L. Pauling, *Tetrahedron*, **17**, 229 (1962).
47. M. Chiang and R. Wheeler, *Can. J. Chem.*, **46**, 3785 (1968).
48. H. Okabe and A. Mele, *J. Chem. Phys.*, **51**, 2100 (1969).
48a. J. L. Franklin, V. H. Dibeler, R. M. Reese and M. Krauss, *J. Am. Chem. Soc.*, **80**, 298 (1958).
49. V. I. Vedeneyev, L. V. Gurvich, V. N. Kondratyev, V. A. Medvedev and Ye. L. Frankevich, *Bond Energies, Ionization Potentials and Electron Affinities*, Edward Arnold, London, 1966.
50. P. Gray and T. C. Waddington, *Proc. Roy. Soc.* (*London*), **235A**, 106 (1956).
51. A. P. Altshuller, *J. Chem. Phys.*, **28**, 1254 (1958).

52. E. Amble and B. P. Dailey, *J. Chem. Phys.*, **18**, 1422 (1950).
53. Landolt-Börnstein, *Molecular Constants from Microwave Spectroscopy*, Springer-Verlag, Berlin, Vol. 4 (new series), Group II.
54. H. O. Spauschus and J. M. Scott, *J. Am. Chem. Soc.*, **73**, 210 (1951).
55. V. N. Krishnamurthy and S. Soundararajan, *Indian J. Chem.*, **3**, 479 (1965).
56. G. Berthier and J. Serre, in *The Chemistry of the Carbonyl Group* (Ed. S. Patai), Interscience, London, 1966, p. 1.
57. H. A. Papazian, *J. Chem. Phys.*, **34**, 1614 (1961).
58. J. I. Bryant, *J. Chem. Phys.*, **45**, 689 (1966).
59. Z. Iqbal, C. W. Brown and S. S. Mitra, *J. Chem. Phys.*, **52**, 4867 (1970).
60. A. Treinin and E. Hayon, *J. Chem. Phys.*, **50**, 538 (1970).
61. Yu. N. Sheinker, L. B. Senyavina and V. N. Zheltova, *Dokl. Akad. Nauk SSSR*, **160**, 1339 (1965).
62. W. R. Carpenter, *Appl. Spectr.*, **17**, 70 (1963).
63. E. Lieber, C. N. R. Rao, A. E. Thomas, E. Oftedahl, R. Minnis and C. V. N. Nambury, *Spectrochim. Acta*, **19**, 1135 (1963).
64. D. A. Dows and G. C. Pimentel, *J. Chem. Phys.*, **23**, 1258 (1955).
65. F. A. Miller and D. Bassi, *Spectrochim Acta*, **19**, 565 (1963).
66. E. Lieber, C. N. R. Rao, T. S. Chao and C. W. W. Hoffman, *Anal. Chem.*, **29**, 916 (1957).
67. J. S. Thayer and R. West, *Inorg. Chem.*, **3**, 406 (1964).
68. W. Gordy, *J. Chem. Phys.*, **14**, 305 (1946).
69. C. A. Coulson, *Valence*, 2nd edition, Oxford University Press, 1961.
70. J. Ladik and A. Messmer, *Acta Chim. Hung.*, **34**, 7 (1962).
71. G. Herzberg, *Electronic Spectra of Polyatomic Molecules*, Van Nostrand, 1966.
72. S. K. Deb, *J. Chem. Phys.*, **35**, 2122 (1961).
73. I. Burak, D. Shapira and A. Treinin, *J. Phys. Chem.*, **74**, 568 (1970).
74. A. E. Gillam and E. S. Stern, *Electronic Absorption Spectroscopy*, Edward Arnold, London, 1954, p. 264.
75. E. Lieber and A. E. Thomas, *Appl. Spectr.*, **15**, 144 (1961).
76. Yu. N. Sheinker, *Dokl. Akad. Nauk SSSR*, **77**, 1043 (1951).
77. C. N. R. Rao, *J. Sci. Ind. Res.*, **17B**, 89 (1958).
78. C. N. R. Rao, *J. Sci. Ind. Res.*, **17B**, 504 (1958)
79. P. Grammaticakis, *Compt. Rend.*, **244**, 1517 (1957).
80. P. A. S. Smith, J. H. Hall and R. O. Kan, *J. Am. Chem. Soc.*, **84**, 485 (1962).
81. C. Djerassi, *Optical Rotatory Dispersion*, McGraw-Hill, 1960.
81a. R. Bonnett, in *The Chemistry of the Carbon–Nitrogen Double bond*, (Ed. S. Patai) Interscience, London, 1970, p. 181.
82. W. Kuhn, *Tetrahedron*, **13**, 1 (1961).
83. C. Djerassi, A. Moscowitz, K. Ponsold and G. Steiner, *J. Am. Chem. Soc.*, **89**, 347 (1967).
83a. P. A. Levene and A. Rothen, *J. Chem. Phys.*, **5**, 985 (1937).
84. C. C. Lin, *Phys. Rev.*, **119**, 1027 (1960).
85. D. Herbison-Evans and R. E. Richards, *Mol. Phys.*, **7**, 515 (1963/64).
86. C. W. Kern and M. Karplus, *J. Chem. Phys.*, **42**, 1062 (1965).
87. M. Witanowski, *J. Am. Chem. Soc.*, **90**, 5683 (1968).
88. T. Kanda, Y. Saito and K. Kawamura, *Bull. Chem. Soc. Japan*, **35**, 172 (1962).
89. R. A. Forman, *J. Chem. Phys.*, **39**, 2393 (1963).

90. T. K. Wu, *J. Chem. Phys.*, **49**, 1139 (1968).
91. D. N. Hendrickson and P. M. Kuznesof, *Theoret. Chim. Acta* (*Berlin*), **15**, 57 (1969).
92. W. D. White and R. S. Drago, *J. Chem. Phys.*, **52**, 4717 (1970).
93. R. G. Clark and E. T. Stewart, *Quart. Rev.*, **24**, 95 (1970).
94. R. K. Nesbet, *Advan. Quantum Chem.*, **3**, 1 (1967).
94a. E. Clementi, in *The Chemistry of the Cyano Group*, (Ed. Z. Rappoport), Interscience, London, 1970, p. 1.
95. S. D. Peyerimhoff and R. J. Buenker, *J. Chem. Phys.*, **47**, 1953 (1967).
96. A. Bonnemay and R. Daudel, *Compt. Rend.*, **230**, 2300 (1950).
97. A. Potier, *J. Chimie Physique*, **48**, 285 (1951).
98. E. L. Wagner, *J. Chem. Phys.*, **43**, 2728 (1965).
99. M. Wolfsberg and L. Helmholz, *J. Chem. Phys.*, **20**, 837 (1952).
100. C. J. Ballhausen and H. B. Gray, *Molecular Orbital Theory*, Benjamin, 1965.
101. D. G. Carroll, A. T. Armstrong and S. P. McGlynn, *J. Chem. Phys.*, **44**, 1865 (1966); D. G. Carroll and S. P. McGlynn, *J. Chem. Phys.*, **45**, 3827 (1966).
102. R. S. Mulliken, *J. Chem. Phys.*, **23**, 1833 (1955).
103. R. Bonaccorsi, C. Petrongolo, E. Scrocco and J. Tomasi, *J. Chem. Phys.*, **48**, 1500 (1968).
104. T. Yonezawa, H. Kato and H. Konishi, *Bull. Chem. Soc., Japan*, **40**, 1071 (1967).
105. E. Clementi, *J. Chem. Phys.*, **34**, 1468 (1961).
106. E. Clementi and A. D. McLean, *J. Chem. Phys.*, **39**, 323 (1963).
107. E. Clementi, *J. Chem. Phys.*, **40**, 1944 (1964); E. Clementi, R. Matcha and A. Veillard, *J. Chem. Phys.*, **47**, 1865 (1967).
108. H. Müller, *Z. Naturforsch.*, **19B**, 867 (1964).
109. C. E. Wulfman, *J. Chem. Phys.*, **33**, 1567 (1960).
110. L. Pauling and G. W. Wheland, *J. Chem. Phys.*, **1**, 362 (1933).

CHAPTER 2

Introduction of the Azido Group

M. E. C. BIFFIN*, J. MILLER† AND D. B. PAUL*

**Defence Standards Laboratories, Maribyrnong, Victoria, Australia*

†University of Natal, Durban, S. Africa

I. INTRODUCTION 58
II. TOXIC AND EXPLOSIVE PROPERTIES OF AZIDES 61
III. SYNTHESES INVOLVING NUCLEOPHILIC SUBSTITUTION REACTIONS OF AZIDE ION 63
A. Mechanisms of Nucleophilic Substitution by Azide Ion . 63
1. Introduction 63
2. Azide ion as a nucleophile 65
3. Role of solvent 70
4. General stereochemical considerations 72
B. Synthesis of Azides by Nucleophilic Substitution . 75
1. Aliphatic and alicyclic azides 75
2. Alkenyl azides 80
a. Vinyl azides 80
b. Allyl azides 83
3. Azides of the general structure $R—C(=X)—N_3$ 86
a. Acyl azides 86
b. Azidoformates 89
c. Carbamoyl azides 90
d. Azidoazomethines and related compounds . . 90
4. Syntheses from epoxides 93
5. Azidosteroids 95
6. Azidocarbohydrates 103
a. Bimolecular nucleophilic displacement at primary and secondary carbon atoms 104
b. Ring opening of sugar epoxides, epimines and episulphides 110
C. Synthesis of Aromatic Azides by Nucleophilic Substitution 113
1. Aryl azides 113

2. Heteroaromatic azides 115
3. Non-benzenoid aromatic azides 117
IV. Formation of Azides by Addition Reactions . . . 119
A. Introduction 119
B. Hydrazoic Acid Additions 120
1. General 120
2. Addition to olefins 120
a. Simple olefins 120
b. Conjugated olefins 122
3. Addition to alkynes 130
4. Addition to compounds with contiguous double bonds . 132
C. Addition of Trimethylsilyl Azide to Carbonyl Compounds 134
D. Mercuric Azide Additions 134
E. Electrophilic Addition of Halogen Azides 136
1. Addition to olefins 136
a. Simple olefins 136
b. Conjugated olefins 141
2. Addition to Alkynes 142
F. Free Radical Additions 144
1. Halogen azide additions 144
2. Additions in presence of redox systems 145
V. Synthesis of Azides from Diazotization and Related Reactions 147
A. Introduction 147
B. Reaction of Diazonium Compounds with Nucleophiles . 148
1. Ammonia and its derivatives 148
2. Hydrazine and its derivatives 152
3. Hydrazoic acid and its derivatives 156
C. Reaction of Hydrazoic Acid with Nitroso Compounds . 165
D. Diazo Transfer and Related Reactions 168
E. Nitrosation of Hydrazine Derivatives 171
VI. Miscellaneous Syntheses 176
A. Electrolytic Reactions 176
B. Oxidation Reactions 177
VII. References 177

I. INTRODUCTION

All organic azides are synthetic materials, there being no example in the literature to date of a naturally occurring compound in this class. The first organic azide, phenyl azide, was synthesized by Peter Griess[1] from benzenediazonium perbromide and ammonia (equation 1). The discovery by Curtius[2,3] in 1890 of the rearrangement of acyl azides to isocyanates (equation 2) stimulated interest in organic azides and as a consequence most of the general synthetic methods were soon elucidated.

Interest in the introduction of the azido group has continued

$$C_6H_5N_2^+Br_3^- + NH_3 \longrightarrow C_6H_5N_3 + 3\ HBr \qquad (1)$$

$$RCON_3 \longrightarrow [RCON:] + N_2 \longrightarrow RN{=}C{=}O \qquad (2)$$

with increasing awareness of the scope of reactions involving azide intermediates. For example, azides react with electrophiles and nucleophiles, undergo 1,3-dipolar addition reactions with olefins and reductive cleavage to amines[4] (equations 3–6). Controlled thermal or

$$O_2N\text{-}C_6H_3(NO_2)\text{-}N_3 \xrightarrow{HBr} O_2N\text{-}C_6H_3(NO_2)\text{-}NH_2 + Br_2 + N_2 \qquad (3)$$

$$RN_3 + PR_3 \longrightarrow R{-}N({}^+PR_3){-}N{=}N^- \xrightarrow{\Delta} R{-}N{=}PR_3 + N_2 \qquad (4)$$

$$RN_3 + \text{norbornene} \longrightarrow \text{triazoline (N, N, N-R)} \qquad (5)$$

$$RN_3 \xrightarrow{Ni/H_2} RNH_2 + N_2 \qquad (6)$$

photolytic decomposition of azides results in the generation of nitrenes by elimination of nitrogen. These may subsequently undergo rearrangement (equation 2), hydrogen abstraction, cyclization, dimerization or addition to C—H or C=C bonds[5–7] (equations 7–11).

$$RN_3 + \text{tetralin} \xrightarrow{180^\circ} RNH_2 + N_2 + \text{naphthalene} \qquad (7)$$

$$\text{2-azidobiphenyl} \longrightarrow [\text{2-biphenylylnitrene } (\ddot{N})] \longrightarrow \text{carbazole (N-H)} \qquad (8)$$

$$C_6H_5N_3 \xrightarrow{\Delta} C_6H_5N\!: + N_2 \longrightarrow C_6H_5N{=}NC_6H_5 \quad (9)$$

$$RN_3 + C_6H_6 \longrightarrow RNHC_6H_5 + N_2 \quad (10)$$

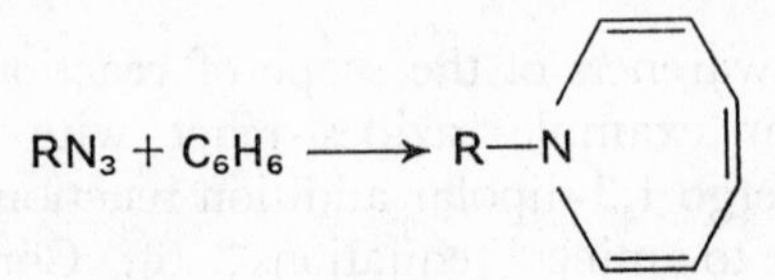

$$RN_3 + C_6H_6 \longrightarrow R{-}N\text{(azepine)} \quad (11)$$

Azides have found increasing application in bio-organic chemistry, particularly in the introduction of protective groups during peptide syntheses, the formation of the peptide link without racemization and stereospecific syntheses of amino derivatives of sugars and steroids (equations 12–15).

$$N_3CO_2R + H_2NCH_2CO_2H \longrightarrow RO_2CNHCH_2CO_2H + HN_3 \quad (12)$$

$$RO_2CNHCHR^1CON_3 + H_2NCHR^2CO_2Me \longrightarrow RO_2CNHCHR^1CONHCHR^2CO_2Me + HN_3 \quad (13)$$

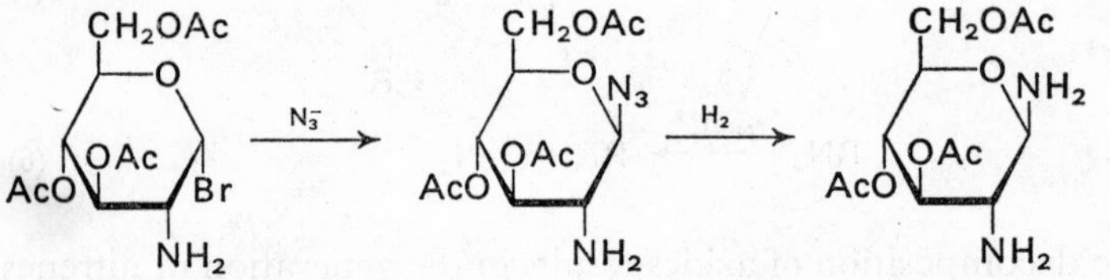

CH₂OAc, O, OAc, AcO, Br, NH₂ $\xrightarrow{N_3^-}$ CH₂OAc, O, N₃, OAc, AcO, NH₂ $\xrightarrow{H_2}$ CH₂OAc, O, NH₂, OAc, AcO, NH₂ (14)

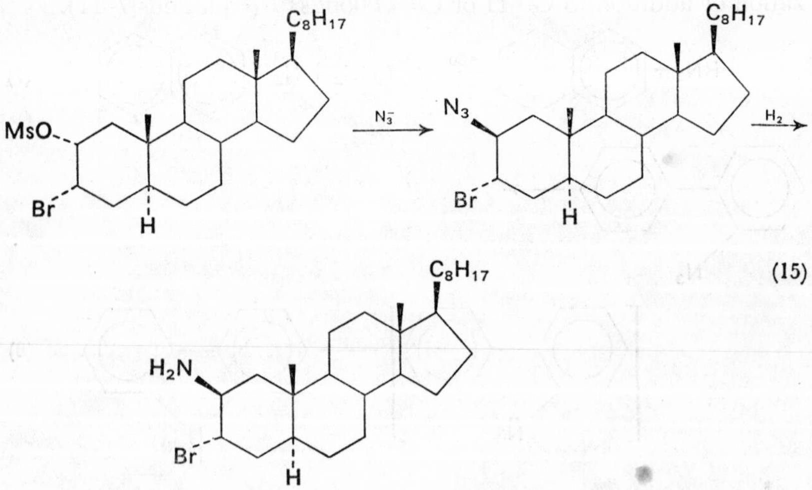

C₈H₁₇, MsO, Br, H $\xrightarrow{N_3^-}$ C₈H₁₇, N₃, Br, H $\xrightarrow{H_2}$ C₈H₁₇, H₂N, Br, H (15)

In recent reviews of the chemistry of organic azido compounds attention has been focused mainly on the reactions rather than the synthesis of azides[3,4,8–14]. The last comprehensive account of the introduction of the azido group was published by Boyer and Canter[15] in 1954 and for the purposes of this chapter we have concerned ourselves primarily with developments subsequent to this review. Reference to work cited in the review of Boyer and Canter is made only where it is pertinent to the discussion. Because of the very large number of azide syntheses described in the chemical literature no attempt has been made at an encyclopaedic coverage. Instead, each major synthetic method has been discussed from a mechanistic viewpoint with representative examples being used as illustration. Only reactions in which a carbon–azide bond is formed are considered; reactions in which organic azides are implicated as intermediates but are not isolated, e.g. the Schmidt reaction, were deemed to be outside the scope of this chapter.

II. TOXIC AND EXPLOSIVE PROPERTIES OF AZIDES

In view of the explosive and toxic nature of hydrazoic acid and its derivatives it is pertinent to discuss the hazards of handling azides in the context of this chapter. Heat, mechanical shock or exposure to some chemical reagents, e.g. concentrated sulphuric acid[16,17], will decompose organic azides. Molecular nitrogen is formed and the process is accompanied by the release of large amounts of energy. During controlled decompositions in solution this energy is absorbed by the solvent. In the absence of a solvent an explosion may occur unless this energy is dissipated over the organic fragment. The larger this fragment, the more effective the energy transfer process becomes; hence the generally higher thermal stability of aryl azides compared with alkyl or acyl azides. In this connexion it has been suggested[4] that a threshold value is given by the ratio (C+O)/N and that a violent decomposition may occur when this ratio is lower than approximately 3:1. Organic azides in consequence have been extensively investigated as both composite explosives and initiators. For example, triazidotrinitrobenzene (**1**) has some advantages over mercury fulminate as an initiator[18] and has been examined for development to practical application. In general, however, any consideration for the practical use of azides must include their high sensitivity to friction and impact. Thus cyanuric triazide (**2**) has attracted interest[19] since it is a stronger initiator than mercury fulminate[20], but its high sensitivity has caused many accidental explosions and therefore precluded its practical use[21]. Reports of the explosive character

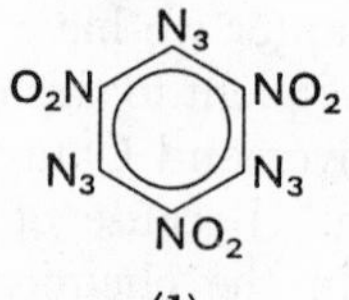

(1) 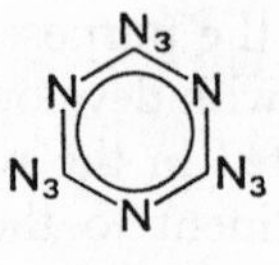 (2)

of azides have been numerous[22–28], *and it is therefore necessary to adopt suitable safety measures whenever handling azido compounds. Indeed some compounds, especially low molecular weight azides, are so treacherous that explosions may occur for unknown reasons during procedures which have previously proved uneventful. In particular, precautions are essential during distillation*[25] *and it is further recommended that low molecular weight organic azides should be stored in solutions of not more than* 10% *strength.*

Trimethylsilyl azide, which has been used as a reagent for azide syntheses, has the advantage of being non-toxic and possessing much higher thermal stability than most organic azides. The enhanced stability of this compound is attributed to $d\pi$–$p\pi$ bonding between silicon and nitrogen[29].

Hydrazoic acid and its alkali metal salts are often used in azide synthesis. *Pure hydrazoic acid is violently explosive* and the reagent is consequently used in dilute solution in which it is quite stable[2]. Solutions of hydrazoic acid in organic solvents may be conveniently prepared and find general application in azide synthesis[30–32]. Silver azide, which has occasionally been used for the preparation of organic azides, is impact sensitive and has been superseded by the alkali metal azides which are not considered explosive under most laboratory conditions.

Hydrazoic acid and its salts are very poisonous. The toxicity of hydrazoic acid is of the same order as hydrogen cyanide and concentrations in air of greater than 0·0005 mg/l produce marked symptoms of intoxication which are sometimes not apparent until the day following exposure[21]. The high vapour pressure of hydrazoic acid (b.p. 37°) accentuates this hazard and at least one serious case of poisoning has been documented[33]. The main physiological effect is a marked lowering in blood pressure with an accompanying rise in the rate of heart beat and respiration and it is thought that hydrazoic acid interferes with the oxidation–reduction processes of the human body[34]. Eye and nose irritation, headaches and unsteadiness may also occur[35–37]. Fortunately, recovery from the effects of low level exposure to hydrazoic acid is rapid and no evidence of cumulative damage has been reported[37].

III. SYNTHESES INVOLVING NUCLEOPHILIC SUBSTITUTION REACTIONS OF AZIDE ION

A. Mechanisms of Nucleophilic Substitution by Azide Ion

I. Introduction

In a discussion of the formation of the carbon–azide bond by nucleophilic substitution with azide ion at an electrophilic carbon centre, the consequences of mechanistic differences between reactions at saturated, unsaturated and aromatic centres must be considered.

Nucleophilic substitution (S_N) reactions of saturated aliphatic compounds may be either associative or dissociative and the majority lie between the limits set by S_N1 reactions, in which the rate-determining step is heterolysis of the bond to the leaving group, and typical S_N2 reactions with fully synchronous bond-formation and bond-rupture. S_N1-like reactions represent an intermediate case and are characterized by a greater extent of bond-rupture than bond-formation. Hence, in aliphatic S_N reactions the rate-limiting process involves some degree of prior or concurrent bond-rupture.

It is not possible in such reactions to have a transition state with a high degree of bond-formation and a low degree of bond-rupture. The total bond order of the entering and leaving groups must be less than or equal to unity, otherwise the total bond order of the carbon atom at the reaction centre will exceed 4 and it is generally accepted that this does not occur.

Most aromatic and heteroaromatic nucleophilic substitution reactions proceed by an addition–elimination S_N2 mechanism. In such reactions there are only three atoms bonded to the carbon at the reaction centre in the initial state (I.St.). Hence, by changing the hybridization at the reaction centre from sp^2p to sp^3, bond-formation can proceed in advance of bond-rupture and the reaction can pass through an intermediate complex (I.C.) with both the entering (nucleophile) and leaving groups fully bonded. The intermediates so formed are known as sigma complexes and are stable relative to the associated transition states (T.Sts.) along the reaction coordinate. In a number of cases sigma complexes are sufficiently stable to isolate[38].

Two sub-classes of this mechanism may be defined for the cases

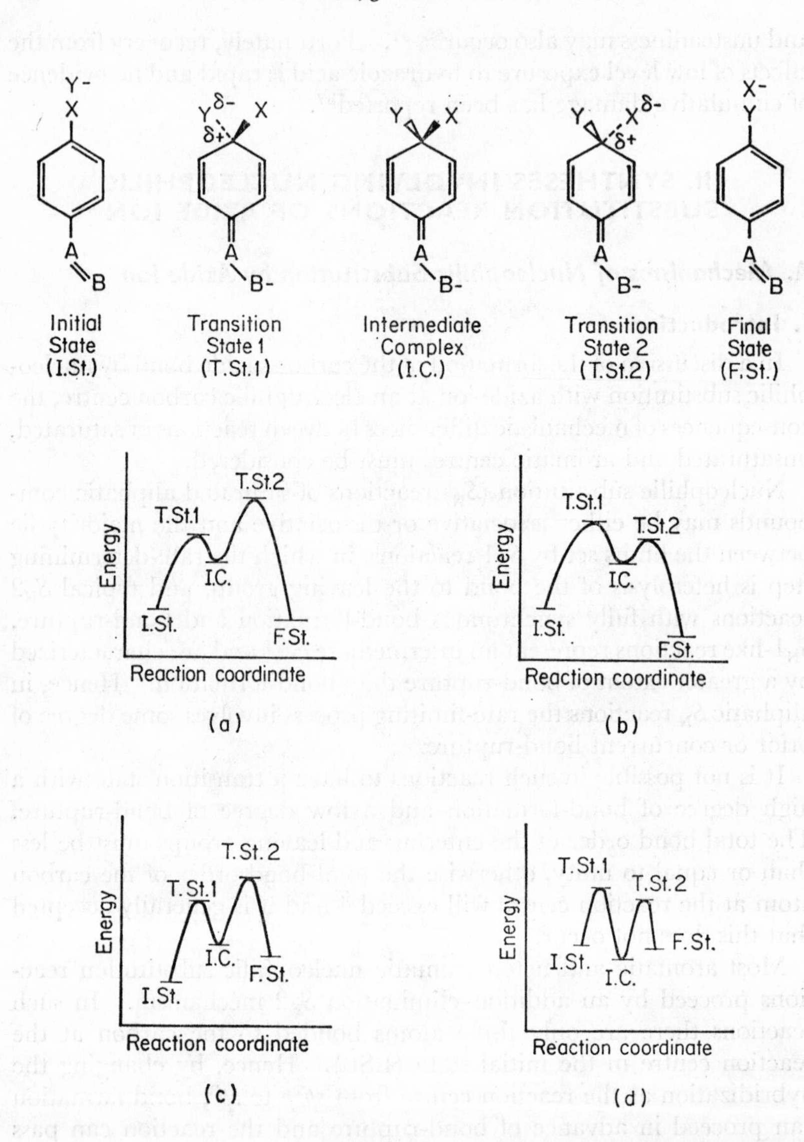

FIGURE 1. Potential energy (P.E.)-reaction coordinate curves for S_NAr reactions: (a) formation of T.St.2 rate-limiting; (b) formation of T.St.1 rate-limiting; (c) formation of T.St.2 rate-limiting and free energy of F.St. higher than I.St., e.g. I.St. = $\overline{N}_3$ + *p*-nitroanisole, F.St. = $\overline{O}$Me + *p*-azidonitrobenzene; (d) formation of relatively stable intermediate sigma complex.

where either the bond-forming step (formation of T.St.1) or the bond-breaking step (formation of T.St.2) is rate-limiting (Figure 1)[39]. In each of these sub-classes the sum of the bond orders of entering and leaving groups to the reaction centre approaches 2.

The case where the formation of T.St.2 is rate-limiting (Figure 1a) is similar in character to a saturated aliphatic S_N1 reaction and may be considered as an S_N1 reaction of the sigma complex. In the T.St. for the rate-limiting step, the bond from the nucleophile is fully formed and aliphatic in type (sp^3 sigma bond) and the order of the bond to the leaving group is still quite high. It is, however, more common for the bond-forming step to be rate-limiting (Figure 1b), with the bond from the nucleophile to the reaction centre in the T.St. substantially formed and the bond to the leaving group unbroken, although modified to an aliphatic sp^3 bond.

Among the S_N reactions of unsaturated compounds the most important in the present context are those of carbonyl compounds of the type —COX, in which X is a good leaving group. These would be expected to be intermediate in character between the S_N reactions of saturated aliphatic compounds and of aromatic compounds. In particular, the total bond order between the reaction centre and the entering and leaving groups should generally be greater than unity but less than for the case of aromatic S_N reactions.

2. Azide ion as a nucleophile

General concepts upon which the factors affecting nucleophilic strength may be specified and assessed have been independently proposed by Hudson[40,41] and Miller[42]. The effective strength of a nucleophile depends in part on its solvation and ionization energies, which are independent of the substrate. In addition, in the rate-limiting process the degree of bond-formation between the reagent and the substrate and the extent of rupture of the bond to the leaving group contribute to the total nucleophilic strength. Thus numerical values of the strength of nucleophiles as expressed by the log rate functions of Swain[43] and of Edwards[44] may vary quite widely with different substrates and reaction conditions. Nevertheless, azide ion can be qualitatively assessed as a moderately strong kinetic nucleophile. In terms of Pearson's HSAB principle[45], azide ion falls into the borderline category between hard and soft bases.

It is convenient to commence a discussion of the reactivity of azide ion in protic solvents with reference to aromatic S_N reactions since these have been investigated in greater depth than aliphatic S_N

reactions. The majority of reactions between aromatic substrates and azide ion are similar to those with other first row reagents in that the bond-forming step is rate-limiting, or so close to it as to give equivalent results. In azide ion reactions in which the energy of the second transition state is substantially higher than that of the first transition state, the free energy of the final state is also higher than the initial state. For example, azide ion may be displaced from *p*-nitroazidobenzene and 2,4-dinitroazidobenzene by methoxide ion in methanol[46,47], but these reactions are practically not reversible (Figure 1c). In more highly substituted systems such as picryl compounds, potential energy-reaction coordinate curves are of the type shown in Figure 1d and sigma complex formation by addition of azide ion to the substrate, or of a nucleophile to an aryl azide, can be observed[48].

The quite high values of solvation energy and electron affinity in protic solvents reflect weakness of azide ion as a nucleophile. However, the C—N_3 bond is quite strong and this indicates strength as a nucleophile. The resultant of these effects imparts moderate strength to azide ion when there is a substantial degree of bond-formation in the rate-limiting transition state. A quantative analysis of these factors has enabled some well known reagents to be placed in the order of reactivity $SMe^- > OMe^- > N_3^- > SCN^- > Hal^-$ for aromatic S_N reactions[42,49,50].

It is instructive to compare the nucleophilic strength of azide ion with that of amide and substituted amide ions. The significant difference is the low electron affinity of the amino group; amide ions are therefore much stronger nucleophiles than azide ion[42]. They are also very strong bases with reactivity such that they cannot be used in protic solvents which are stronger acids than the conjugate acids of amide ions, viz. amines.

In common S_NAr reactions of azide ion, where bond-formation is the rate-limiting process, the nature of the leaving group has little effect on reagent reactivity unless it is highly electronegative. For example, with azide ion, chlorides are little more reactive than bromides and iodides; displacement of the much more electronegative fluorine, however, proceeds some 10^3 times faster with activated aryl fluorides than with corresponding heavy halides[49–53].

Although the rates of azide exchange reactions have not been studied, it may be predicted that the enhancement of reactivity due to the electronegativity factor would be largely counteracted by the relatively greater stability of initial states in their reactions with azide ion. In fact, with some aromatic ethers such stabilization of the initial states prevents substitution with azide ion[46,47].

In S_N1 reactions of aliphatic substrates it is important to note that the electrophile is a carbonium ion and this is much less discriminating towards nucleophiles than are less reactive electrophilic species. In such reactions the nucleophilic strength of azide ion has no direct effect on rates, but is involved in determining product ratios in competitive reactions with other nucleophiles, including the solvent. Concentration factors are therefore important and salt effects of dissolved azides may also be significant[54].

Saturated aliphatic S_N2 reactions differ from aromatic S_N2 reactions in having substantially less bond-formation with the nucleophile in their transition states. Although the degree of deionization and desolvation may be less advanced in the formation of the transition state for aliphatic than for aromatic S_N2 reactions, the differences in ionization and solvation energies are proportionately less than the differences in bond energies. Hence the bond strength factor is less predominant and azide ion, which depends on the bond energy term for its relatively high place as a nucleophile in S_NAr reactions, appears as a weaker nucleophile in aliphatic S_N2 reactions. It is in fact a general phenomenon that first row nucleophiles which form strong bonds to carbon show up less well in aliphatic than in aromatic S_N reactions and, conversely, heavy nucleophiles which form weaker bonds to carbon show up relatively better. For example, in the reactions of nucleophiles with methyl bromide in water the order $I^- > SCN^- > N_3^-$, as shown in Swain's table of nucleophilic constants[43], is in the reverse order to that shown with aromatic substrates.

Changes in the leaving group are of little effect in differentiating between relative strengths of nucleophiles. The results of Parker and co-workers[53], for example, show that the N_3^-/SCN^- rate ratio in displacement reactions of methyl halides in methanol changes from 0·4 for displacement of chloride to 0·1 for iodide. The decrease in the ratio is due to a larger rate ratio MeI/MeCl for thiocyanate ($\simeq 600$) than for azide ion ($\simeq 150$). This result suggests that the degree of bond-rupture in the transition state of the reactions with thiocyanate ion is greater than for azide ion reactions. This is presumably accompanied by a larger degree of bond-formation in the transition state of thiocyanate ion reactions as is appropriate for this larger and more polarizable (softer) nucleophile. For the reactions with azide ion, the mobilities of the heavy halogens relative to Cl = 1 are Br = 135, I = 150 and this is consistent with the known patterns for aliphatic

S_N2 processes, although the ratios are a little larger than those typical with oxygen nucleophiles.

In the transition state for an aliphatic S_N2 reaction the bond between the reaction centre and the leaving group is partially broken. The inductive effect of strongly electronegative substituents is consequently offset by the energy required for this partial bond-rupture. Electronegativity therefore makes a lesser relative contribution to leaving group mobility in the aliphatic than in the aromatic S_N2 reactions where the formation of T.St.1 is rate-limiting. Groups such as fluorine and X^+ are generally less mobile in the aliphatic reactions but kinetic data concerning such relative mobility in reactions with azide ion are scarce. From the results of Hughes, Ingold and co-workers[55,56], however, it may be estimated that the SMe_2^+ group is about 5 times more mobile than chlorine. It is probable that the F/Cl ratio is about 10^{-1} to 10^{-2} but this has not been measured.

In unsaturated aliphatic systems the most important reactions are those of carbonyl compounds of the type —COX, in which X is a good leaving group such as halogen. In general, most displacement reactions of anionic nucleophiles on the carbonyl carbon atom of acyl halides involve an addition–elimination mechanism[57] (e.g. equation 16). In such reactions bond-formation is in advance of bond-rupture

$$X^- + R\overset{\overset{O}{\|}}{C}Y \longrightarrow R\overset{\overset{O^-}{|}}{\underset{\underset{X}{|}}{C}}Y \longrightarrow R\overset{\overset{O}{\|}}{C}X + Y^- \tag{16}$$

at the reaction centre and the total bond order of the entering and leaving groups is more than unity. Leaving group mobilities are particularly indicative of differences in total bond order. For example, in the reaction of benzoyl halides with hydroxide ion at 0·5° in 50% aqueous acetone[58], the rate ratio $k_F/k_{Cl} = 1{\cdot}4$. This halogen mobility pattern resembles that of aromatic S_N reactions rather than that of aliphatic S_N reactions but the differences are substantially smaller than in the aromatic series.

Thus one might expect nucleophilic displacements on —COX compounds with azide ion to be intermediate in character between saturated aliphatic S_N2 and S_NAr reactions. However, while there are numerous examples of the formation of azides by displacement of X from —COX compounds with azide ion, there appear to be no important mechanistic studies on such reactions.

Similar considerations apply to the reactions of azide ion with other species of the general type R—C(=Y)X (Y = S,NR), but again no supporting kinetic data are available and, in fact, compounds of the type R—C(=S)N_3 are unknown (section III.B.3).

Nucleophilic substitution reactions of unsaturated compounds containing the C=C double bond are well known and are distinguished by the relative location of the double bond and leaving group. Rappoport[59] has recently reviewed the wide and complex nature of nucleophilic vinylic substitutions. The susceptibility of simple vinyl compounds to nucleophilic attack is low and comparable to unactivated halobenzenes. As is the case with aromatic compounds, however, vinylic substrates may be activated by electron-withdrawing substituents conjugated with the reaction centre.

An addition–elimination mechanism (equation 17) appears to be operative in such reactions with azide ion, since it has been found that the k_{Br}/k_{Cl} rate ratios for a series of derivatives of $ArSO_2CH{=}CHHal$ are very similar[60]. This result indicates that C—N_3 bond formation precedes the C—Hal bond breaking step. The stereochemistry of the substitution depends on the specific addition–elimination pathway involved. In particular, the lifetime of a carbanionic intermediate is important in relation to the configuration of the final product and this is discussed in detail in section III.B.2.

(R¹)(X)C=C(Y)(R²) + Nu⁻ → (i) [R¹(X)C(Nu)····C(Y)R²]⁻ → −X⁻ → (R¹)(Nu)C=C(Y)R²

(ii) Nu—C(R¹)(X)—C⁻(Y)(R²) → −X⁻ → (R¹)(Nu)C=C(Y)R²

(iii) +H⁺ Nu—C(R¹)(X)—C(Y)(R²)—H → −HX → (R¹)(Nu)C=C(Y)R²

(17)

The relative nucleophilicities of anionic reagents towards standard vinylic substrates have been tabulated and from these it is apparent that azide ion is a moderately strong nucleophile in protic solvents, and even stronger in dipolar aprotic solvents[59,61]. The need for caution in construction of such nucleophilicity scales has been stressed, however, since trends appear to be dependent on the structure of the substrate and some anomalies in the effects of solvent have been noted.

Allyl compounds are highly reactive towards nucleophiles and examples of dissociative (S_N1) and associative (S_N2) reactions, and nucleophilic substitution with rearrangement are well documented[62,63]. Differentiation between these mechanisms when azide ion is the nucleophile is often extremely difficult due to the possibility of rearrangement of the allylic azide resulting from substitution, and for this reason the relative nucleophilic strength of azide ion in allylic S_N reactions has not been delineated.

3. Role of solvent

The role of solvent in substitution reactions may be twofold since it can act both as a reagent in solvolytic processes and as a reaction medium. S_N1 reactions in which the solvent competes with azide ion are well known and have been discussed previously. In this section consideration is given to the use of solvents as reaction media.

In recent years the marked effect of dipolar aprotic solvents on the rates of nucleophilic substitution reactions has become widely known[64,65]. Small anions, e.g. fluoride ion, are more solvated by protic than aprotic solvents and this is clearly indicated by the high kinetic nucleophilicity of fluoride ion in dimethylformamide or dimethyl sulphoxide. The levels of solvation of large anions, such as the intermediate complexes of S_NAr reactions, by both types of solvent are comparable, although the level in protic solvents is generally a little higher. Dipolar aprotic solvents have the negative end of their dipole more fully exposed than protic solvents and hence are more effective in the solvation of large cations. For this reason there is less ion-pair formation with caesium fluoride than with lithium fluoride in dipolar aprotic solvents. The solvation of neutral species is energetically unimportant compared with that of ions.

In common S_N1 reactions involving heterolysis to form ions, the differential solvation between protic and aprotic solvents for the anion X^- produced in the rate-limiting step is generally greater than for the cation R^+ (equation 18). Hence such reactions typically proceed faster in protic solvents in which X^- is better solvated.

$$Y^- + RX \xrightarrow{\text{r.d.s.}} R^+ + X^- + Y^- \xrightarrow{\text{fast}} RY + X^- \quad (18)$$

In S_N2 reactions and S_NAr reactions in which bond-formation is the rate-determining step, the differential solvation in the transition state is usually much smaller than the differences in levels of solvation of the anionic nucleophiles between protic and aprotic solvents. Therefore in dipolar aprotic solvents the activation energy $\Delta E^{\neq}$ is lowered and these reactions consequently proceed a good deal faster in such solvents (Figure 2). This is true of azide ion which is in general a stronger kinetic nucleophile in aprotic than in protic solvents.

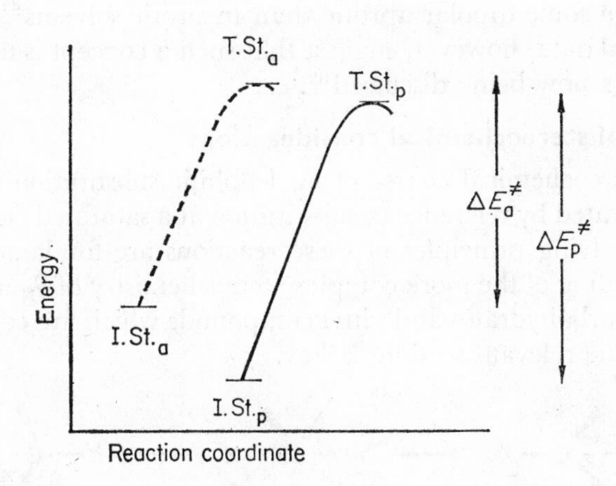

FIGURE 2. Effect of protic (p) and dipolar aprotic (a) solvents on the activation energy of S_N2 reactions and S_NAr reactions with the formation of T.St.1 rate-limiting. ---- dipolar aprotic solvent. —— protic solvent.

Some rate ratios (k_{DMF}/k_{MeOH}) for reaction in dimethylformamide compared with methanol are: azide ion with *p*-iodonitrobenzene[51,53] at 0°, $8{\cdot}7 \times 10^3$; azide ion with methyl bromide[53] at 0°, $1{\cdot}6 \times 10^4$. These may be compared with the solvent rate ratios for acetone: acetone–methanol (75:25 v/v): methanol, for reaction of *p*-nitrobenzyl bromide with azide ion at −20°, viz. $4{\cdot}5 \times 10^4$: 47:1, or approximately 10^3:1 for acetone: acetone–methanol[66]. Similarly the ratio acetone: acetone–methanol (75:25 v/v) for reaction of azide ion with *n*-butyl bromide[66] at −20° is $2{\cdot}35 \times 10^3$.

In addition to changes of rate in different solvents, solvent rate ratios have been shown to vary on changing the leaving group. In a simple approach the order of solvation energy (enthalpy) in both protic and aprotic solvents for S_N2 reactions is $N_3^- > \overset{\delta-}{N_3}\cdots R\cdots\overset{\delta-}{Cl} > \overset{\delta-}{N_3}\cdots R\cdots\overset{\delta-}{Br}$, and with the differences between protic and aprotic solvents decreasing in the same order. In such a pattern the solvent rate ratio k_{DMF}/k_{MeOH} should therefore be larger for displacement of bromine than of chlorine and this is apparent from the work of Parker and his colleagues[53] and of Miotti[66] which indicates that the solvent ratio for bromo compounds is about 3–5 times greater than that of the chloro compounds.

A more complex approach would require consideration of the possibility that some very large anionic transition states are better solvated in some dipolar aprotic than in protic solvents[51]. The experimental data, however, suggest that such a concept is not valid for the species now being discussed[67].

4. General stereochemical considerations

The stereochemical course of nucleophilic substitution reactions is best illustrated by reference to substitution at a saturated carbon atom. The underlying principles of these reactions are fundamental to an understanding of the more complex stereochemistry of S_N reactions on steroids, carbohydrates and vinyl compounds which are considered in detail in the relevant sections below.

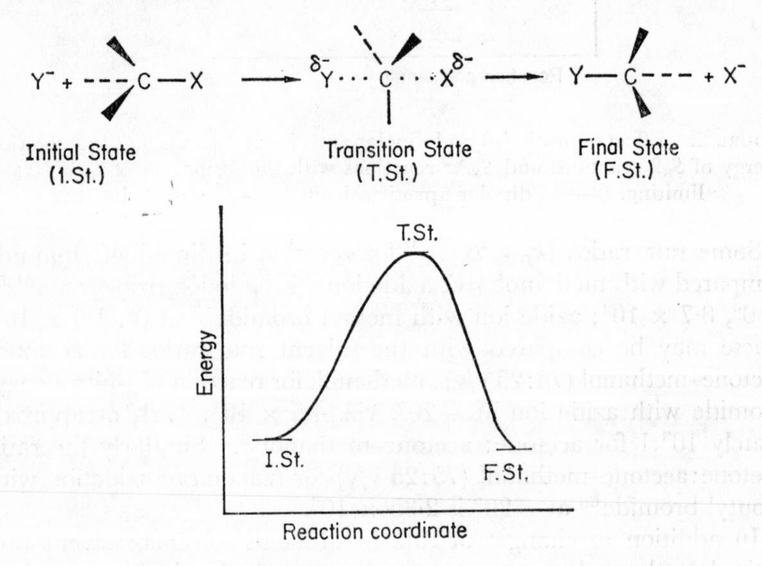

FIGURE 3. Walden Inversion during bimolecular substitution at a saturated carbon atom.

Synchronous S_N2 reactions are characterized by inversion of configuration at the reaction centre (Walden Inversion) for in the transition state the entering and leaving groups share a *p*-orbital of the carbon at the reaction centre and are thus at 180° to each other. The other attached groups in the transition state are bonded by sp^2 sigma bonds[68] (Figure 3). The pioneering studies of Hughes, Ingold and co-workers[69] conclusively demonstrated that Walden Inversion occurs in S_N2 reactions involving azide ion as the nucleophile.

In 1960 Hughes, Ingold and co-workers[55,56] indicated that this stereokinetic rule for S_N2 reactions, although generally accepted, had been established for the most part only for reactions of negative reagents with neutral substrates. Of the other possible S_N2 reaction types, that in which a negative reagent interacts with a positive substrate was considered the most likely to undergo substitution with retention of configuration (i.e. prove an exception to the stereokinetic rule). A study of some reactions of azide ion with alkyl halides and with dimethyl-1-phenylethylsulphonium chloride were, however, all shown to proceed with practically quantitative inversion of configuration, despite the countervailing electrostatic forces.

S_N2 reactions may, however, involve some complexities. An interesting example is the study by Sneen and co-workers[70–73] of the S_N2 solvolysis of 2-octyl brosylate in the presence of azide ion. In pure methanol or water, solvolysis with inversion occurs. In 75% aqueous dioxan, however, the inverted 2-octanol was obtained in only 77% optical purity. When azide ion was added optically impure 2-octyl azide (**3**) was obtained together with fully inverted 2-octanol (**4**). A sequence involving parallel solvolysis by direct reaction, and reaction through an ion-pair intermediate (**5**) was put forward to explain these results (equation 19). The formation of the alcohol (**6**) with reten-

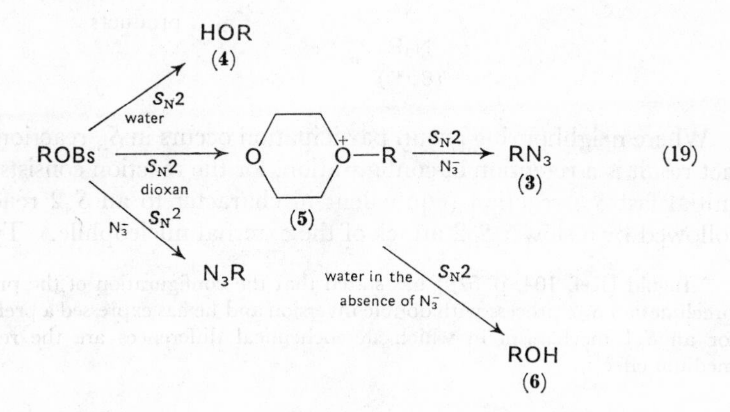

tion of configuration in aqueous dioxan proceeds by double inversion through the intermediate (5). The more nucleophilic azide ion competes for this intermediate so effectively that no product of solvolysis of the ion-pair is formed when azide ion is present.*

The stereochemical patterns in S_N1 reactions and S_N1-like reactions may vary widely. In the majority of S_N1 reactions the rate-limiting step involves the ionization of a neutral substrate. It is now generally agreed that ionization proceeds through the sequence: intimate ion-pair → solvent-separated ion-pair → dissociated (solvated) ions. The nucleophile may intercept the carbonium ion before the final stage and the stereochemistry of the products so obtained may range from substantial inversion to complete racemization. An example of substantial inversion, ascribed to early interception of the heterolysis, has been reported by Weiner and Sneen[72] for the solvolysis of 2-octyl methanesulphonate in 25% aqueous dioxan and its reaction with azide ion. From the kinetic data it is clear that azide ion is not involved in the rate-limiting step and yet it is pertinent that the 2-octyl azide produced is highly inverted (80% inversion). It has been suggested that formation of an ion-pair intermediate is rate-limiting (equation 20). Solvolysis also proceeds with inversion of configuration and has been explained in similar terms.

HOR
↑ OH_2
RX —rate-limiting→ R^+X^- —(dioxan)→ dioxan-$\overset{+}{O}$—R → products (20)
intimate ion-pair
↓ N_3^-
N_3R
(80%)

Where neighbouring group participation occurs in S_N reactions, the net result is a retention of configuration, for the reaction consists of an initial fast S_Ni reaction (equivalent in character to an S_N2 reaction) followed by a slower S_N2 attack of the external nucleophile. Typical

* Ingold (Ref. 104, p. 529) has stated that the configuration of the products precludes an S_N2 process with double inversion and he has expressed a preference for an S_N1 mechanism in which stereochemical differences are the result of medium effects.

neighbouring groups which participate in anchimeric reactions are nucleophilic species possessing unshared electrons or unsaturated π-systems. An example of the former[74] is the solvolysis with retention of configuration of the cyclohexyl derivative (7) (equation 21). Where the neighbouring group is a π-system, the neighbouring group effect involves the formation of a non-classical cation and such an example in the carbohydrate field is discussed in detail in section III.B.6. In either case an appropriate spatial relationship with the reaction centre is essential.

$$(7)\ \xrightarrow{}\ [\text{bridged } \overset{+}{N}N_2\ \ {}^{-}OTs]\ \xrightarrow{AcOH}\ \text{cis-}N{-}\overset{+}{N}{\equiv}N\ /\ OAc \qquad (21)$$

B. Synthesis of Azides by Nucleophilic Substitution

1. Aliphatic and alicyclic azides

The direct introduction of the azido group by nucleophilic displacement with azide ion constitutes the most convenient and general synthesis of aliphatic azides. Consequently the method has been extensively employed since its first application was reported in 1901 by Curtius[75].

Although some cases of S_N1 reactions occur, displacements are normally bimolecular. The associated Walden Inversion affords a configurational selectivity which is particularly valuable in steroid and carbohydrate metathesis. A wide range of leaving groups has been employed in the reaction, e.g. sulphate[76,77], nitro[78], nitrate[79,80], phenylazo[78] and iodoxy[81], but *p*-toluenesulphonyl, methanesulphonyl and halogeno derivatives are the most frequently used. Methyl and ethyl azides, however, cannot be obtained from the corresponding halides and are instead prepared from the alkyl sulphates[76,77,82]. The better leaving group mobility of *p*-toluenesulphonate than halogeno groups may be employed in the synthesis of haloazides[26] (equation 22).

Prior to the use of dipolar aprotic solvents for these reactions the general procedure involved interaction of the alkyl substrate with

$$ClCH_2CH_2Tos \xrightarrow{N_3^-} ClCH_2CH_2N_3 \qquad (22)$$

sodium azide in aqueous alcohol. Sealed tubes and a complicated work-up procedure including the separation of azeotropic mixtures were required[83,84] and in addition the facility of the reaction depended critically on the nature of the sodium azide employed[85]. Whereas early preparations of sodium azide from hydrazine, amyl nitrite and sodium methoxide gave a highly active reagent[86], commercial sodium azide prepared from sodamide and nitrous oxide is of variable activity and sometimes unreactive[87]. Treatment of the commercial product with hydrazine produces a reagent of greater and more uniform activity[88,89] and the use of such 'activated' sodium azide in azide synthesis has been widespread[90].

This situation existed until 1957 when Lieber, Chao and Rao demonstrated the advantages to be gained from using high boiling solvents such as the monoalkyl ethers of diethylene glycol[84]. This modification, which obviated the use of sealed tubes and the prior activation of sodium azide, enabled generally higher yields to be achieved and work-up procedures were simplified since azeotropic mixtures were not formed. Dipolar aprotic solvents are even more efficacious and dimethyl sulphoxide[91–93], dimethylformamide[94,95], dimethylacetamide[96], nitrobenzene[97] and hexamethylphosphoramide[98] are now used routinely as media for azide ion substitution reactions.

As expected, substitution by azide ion occurs more readily when the alkyl substrate bears electron-withdrawing groups. For example phenacyl bromide and its derivatives give high yields of azides when treated with sodium azide in the cold[99–101]. Secondary alkyl substrates undergo S_N2 reactions with azide ion[95,102,103] but with less facility than primary alkyl substrates in accordance with the normal polar influences and primary steric effects in S_N reactions[104]. Selective substitution is therefore possible and this has been effectively applied in carbohydrate and steroid synthesis (sections II.B.5,6). These effects are also exemplified in the alicyclic series where it has been reported that menthyl halides and 2-methylcyclohexyl halides afford unsatisfactory yields of azide[105]. The unsubstituted alicyclic azides, however, are obtained in good yields by the procedures outlined above[84,105–107] (Table 1).

Extension of the general synthetic method to the synthesis of vicinal[108–112] and geminal[112] diazides and α,ω-diazidoparaffins[28,102] is well documented. Some diazido compounds have received consideration as explosives and initiators[113].

Some syntheses of alkyl and alicyclic azides are considered to proceed

TABLE 1. Alkyl and alicyclic azides synthesized by nucleophilic substitution reactions

Substrate	Solvent	Product	Yield %	Reference
1-Iodopropane	Methyl carbitol	1-Azidopropane	64	84
1-Chlorobutane	DMF	1-Azidobutane	89	28
1-Bromobutane	Aq. methyl carbitol	1-Azidobutane	78	84
1-Bromobutane	Aq. methanol	1-Azidobutane	90	83
1-Iodopentane	Aq. carbitol	1-Azidopentane	84	84
1-Iodohexane	Aq. carbitol	1-Azidohexane	87	84
1-Iododecane	Aq. methyl carbitol	1-Azidodecane	89	84
1-Bromocyclopentane	Carbitol	1-Azidocyclopentane	82	84
1-Bromocyclopentane	Aq. ethanol	1-Azidocyclopentane	51	105
1-Iodocyclohexane	Carbitol	1-Azidocyclohexane	75	84
1-Iodocyclohexane	Aq. ethanol	1-Azidocyclohexane	68	105
1-Bromocycloheptane	Aq. ethanol	1-Azidocycloheptane	55	105
1,3-Dibromopropane	Aq. methanol	1,3-Diazidopropane	81	102
1,5-Dichloropentane	DMF	1,5-Diazidopentane	68	28
1,8-Dibromooctane	Aq. methanol	1,8-Diazidooctane	72	102
1,2-Dibromo-1-phenylethane	DMF	1-Azido-1-phenyl-2-bromoethane	—	94, 114
Phenacyl bromide	Aq. methanol–acetic acid	Phenacyl azide	93	99, 101
p-Bromophenacyl bromide	Aq. ethanol–acetic acid	*p*-Bromophenacyl azide	91	99
Ethyl 1-bromo-1-phenylacetate	Aq. ethanol	Ethyl 1-azido-1-phenylacetate	84	100
Ethyl 1-chloro-1-*p*-nitrobenzyl-acetate	Aq. ethanol	Ethyl 1-azido-1-*p*-nitrobenzyl-acetate	94	107
Chloromethyl methyl sulphide	Water	Azidomethyl methyl sulphide	52	115
Benzyl chloromethyl sulphide	Water	Azidomethyl benzyl sulphide	63	115
N,N-Dimethylamino-bromomethane	Dichloro-methane	*N,N*-Dimethyl-aminoazido-methane	95	116
Chloromethyl ethyl ether	Water	Azidomethyl ethyl ether	50	117
Bis(2-chloroethyl) ether	DMF	Bis(2-azidoethyl) ether	92	28
Bis(2-chloroethoxy) ethane	DMF	Bis(2-azidoethoxy) ethane	85	28
1-Chlorobutan-3-one	—	1-Azidobutan-3-one	—	118
Cyanogen chloride	Pentane	Cyanogen azide	—	119

by a unimolecular mechanism. The best known example involves the treatment of secondary and tertiary aryl carbinols with hydrazoic acid in the presence of trichloracetic acid and this constitutes a general route to the corresponding alkyl azides[120–123]. Tietz and McEwen[121] have shown that under such conditions benzhydrol derivatives (**8**) afford the azides (**9**) after 24 hours at room temperature, and Ege and Sherk[120] have similarly isolated 1,1-diphenylethyl azide (**10**) from the corresponding alcohol. The latter authors further observed that **10** was also produced from 1,1-diphenylethylene under the same conditions and invoked a carbonium ion as the common intermediate[120]. The formation of such intermediates from secondary and tertiary carbinols is in accordance with the well known unimolecular substitution reactions of such substrates (equation 23).

$$R^1,R^2\text{-}C_6H_3\text{-}CH(OH)\text{-}C_6H_5 \quad (\mathbf{8}) \qquad R^1,R^2\text{-}C_6H_3\text{-}CH(N_3)\text{-}C_6H_5 \quad (\mathbf{9})$$

$$(C_6H_5)_2C(Me)\text{-}OH \underset{-H^+}{\overset{H^+}{\rightleftharpoons}} (C_6H_5)_2C(Me)\text{-}\overset{+}{O}H_2 \underset{H_2O}{\overset{-H_2O}{\rightleftharpoons}} (C_6H_5)_2C^+(Me) \underset{+H^+}{\overset{-H^+}{\rightleftharpoons}} CH_2{=}C(C_6H_5)_2$$

$$(C_6H_5)_2C^+(Me) \underset{}{\overset{HN_3}{\rightleftharpoons}} (C_6H_5)_2C(Me)\text{-}\overset{+}{N_3}H \; (\mathbf{11}) \underset{H^+}{\overset{-H^+}{\rightleftharpoons}} (C_6H_5)_2C(Me)\text{-}N_3 \; (\mathbf{10}) \qquad (23)$$

An interesting application of this reaction is in the formation of 1-azido-1-ferrocenylethane (**12**) in 57% yield from the alcohol (**13**). The exceptionally stable carbonium ion (**14**) has been suggested as an intermediate in this reaction[123].

$$\text{FerrCH(N}_3\text{)Me} \; (\mathbf{12}) \qquad \text{FerrCH(OH)Me} \; (\mathbf{13}) \qquad \text{Ferr}\overset{+}{C}\text{HMe} \; (\mathbf{14})$$

The acid strength of the reaction medium is critical if azides are to be isolated from unimolecular reactions of carbinols in the presence of hydrazoic acid. Under strongly acidic conditions, deprotonation of the conjugate acid (**11**) is suppressed and its decomposition by loss of nitrogen may predominate, cf. Schmidt reaction[120,124–126]. Ege and Sherk have explained the instability of **11** in terms of the inhibition of resonance stabilization resulting on protonation of the azido group[120]. In this connexion it has been shown that with sulphuric acid as catalyst, substantial amounts of amines accompany azide formation[127–130] except at low temperatures[131,132]. Thus Arcus and Coombs obtained both azidofluorenes and phenanthridines from fluorenols in a sulphuric acid–hydrazoic acid medium (equation 24)[127]. Azidofluorenes were the exclusive products, however, when trichloracetic acid was used as catalyst[122].

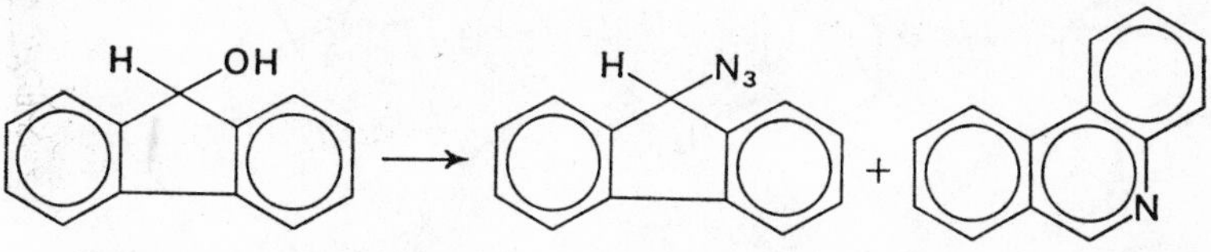

(24)

Other examples of azide syntheses have been reported where S_N1 mechanisms may be implicated but have not been substantiated by kinetic measurements. Bohme and Morf[115–117] have obtained the azides **15** and **16** simply by shaking the corresponding halo compounds and aqueous sodium azide at room temperature. It is well established that compounds such as **17**, which form a mesomeric cation such as the oxonium ion (**18**) by loss of chloride ion, undergo rapid unimolecular reactions[133].

$MeSCH_2N_3$ (**15**) $EtOCH_2N_3$ (**16**) $MeOCH_2Cl$ (**17**) $Me\overset{+}{O}{=}CH_2 \longleftrightarrow MeO{-}\overset{+}{C}H_2$ (**18**)

Kuczynski and Walkowicz[134] have described an azide synthesis in which the trishomocyclopropenyl cation[135] (**19**) is formed in the rate-limiting step of an S_N1 reaction of the *cis-p*-toluenesulphonate (**20**). In the presence of azide ion the product was a mixture of the *cis* and *trans* azides, **21** and **22**, and 3-azido-*cis*-2,2-dimethylbicyclo[3,1,0] hexane (**23**), which were identified after reduction to the amines. The *trans-p*-toluenesulphonate (**24**), however, afforded only the inverted azide (**21**) (equation 25), which seems surprising in view of the hindrance to backside attack by azide ion on **24** as suggested from stereo models.

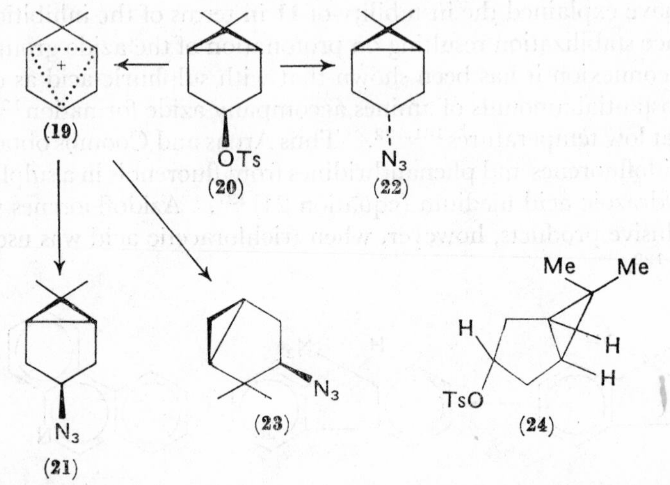

(19) (20) (22) (21) (23) (24)

2. Alkenyl azides

a. Vinyl azides. Owing to the general inertness of vinyl substrates towards nucleophilic substitution, vinyl azides are most conveniently obtained by dehydrohalogenation of azidohaloalkanes[26,114,136–138]. A general synthesis of this type has been described by Hassner and co-workers[136]: addition of iodine azide to olefins affords β-iodoazides (see section IV.E.1) and these compounds give vinyl azides on dehydro-iodination with potassium *t*-butoxide.

Replacement in vinylic derivatives is, however, facilitated by electron-withdrawing groups β to the site of substitution and addition–elimination reactions of this type have been reviewed by Rappoport[59] and by Patai and Rappoport[139]. Such reactions have been applied in the direct synthesis of vinyl azides. For example the synthesis in 81% yield of the β-azidovinyl ketone (**25**), of unspecified stereochemistry, from the chloro derivative and azide ion in dimethylformamide at 40° has been described by Maiorana[140].

$$N_3CH{=}C(Me)COC_6H_5$$

(**25**)

The mechanistic implications of the stereochemical course of such addition–elimination reactions with azide ion have been considered by several workers[59]. Meek and Fowler[141] have reported that the formation of vinyl azides from *cis*- and *trans*-1,2-di-*p*-toluenesulphonylethylene occurs with a high degree of retention of the configuration of the substrate. The azido product obtained from the *cis* isomer is comprised mostly (90%) of the *cis* compound (**26**) and the *trans* substrate yields exclusively the *trans* azide (**27**). The high degree of stereospecificity has been rationalized in terms of a kinetic preference for the product which is derived from an intermediate in which coplanarity of the leaving group and the developing π-orbital of the product is achieved with minimum eclipsing of bulky groups. This situation is illustrated for the *cis* substrate in equations (26) and (27).

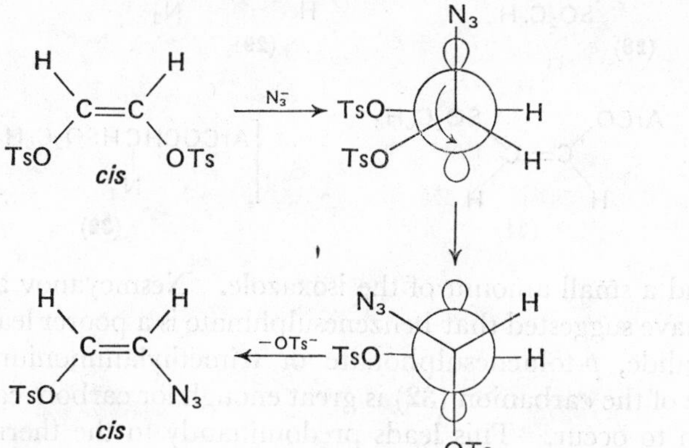

(26)

cis

(26)

Formation of the *trans* isomer from the *cis* substrate would entail an eclipsing of the bulky *p*-toluenesulphonate groups and hence is kinetically less favourable (equation 27).

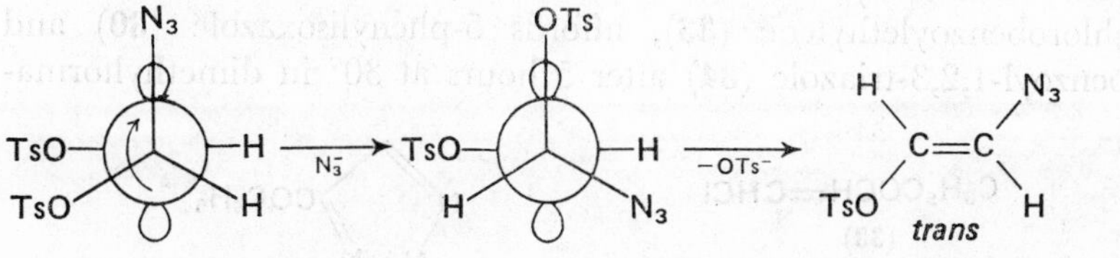

(27)

trans

(27)

Nesmeyanov and Rybinskaya have examined the products from the reaction of compounds of the general type *trans*-ArCOCH=CHX with azide ion in aqueous alcohol[142,143]. When X = Cl,Br,Me_3N^+ and NO_2, *trans* azides are isolated in high yield (> 85%), but with

$X = SO_3^-$ or CN, triazoles are formed. The formation of such cyclic products was thought to reflect the longer lifetime of the carbanionic intermediate.

Interesting behaviour, intermediate between these two extremes, was observed in the case when $X = C_6H_5SO_2$. On treatment with azide ion the *trans* compound (**28**) affords 88% of the *trans* azide (**29**), together with 12% of 5-phenylisoxazole (**30**) presumably derived from the *cis* azide. The *cis* substrate (**31**), however, also gives the *trans* azide

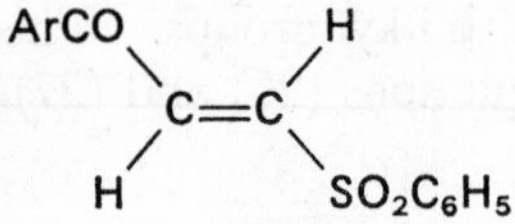

(ArCO)(H)C=C(H)($SO_2C_6H_5$) (**28**)

(ArCO)(H)C=C(H)(N_3) (**29**)

C_6H_5–isoxazole (5-phenylisoxazole) (**30**)

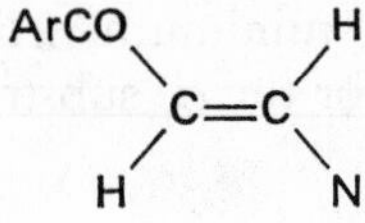

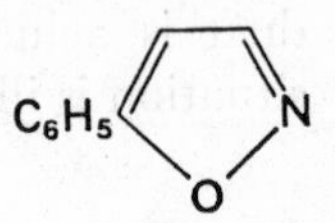

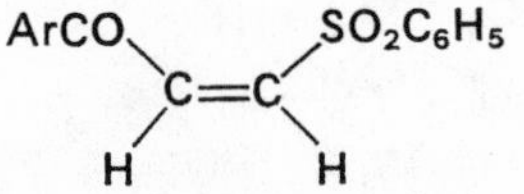

(ArCO)(H)C=C($SO_2C_6H_5$)(H) (**31**)

$[ArCOCHCH(N_3)SO_2C_6H_5]^-$ (**32**)

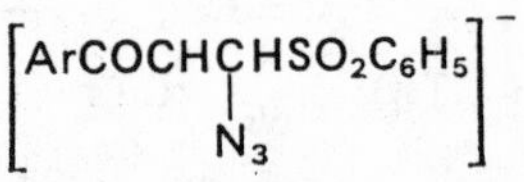

(**29**) and a small amount of the isoxazole. Nesmeyanov and Rybinskaya have suggested that benzenesulphinate is a poorer leaving group than halide, *p*-toluenesulphonate or trimethylammonium and the lifetime of the carbanion (**32**) is great enough for carbon–carbon bond rotation to occur. This leads predominantly to the thermodynamimically more stable *trans* product[143]. Although the lifetime of the carbanion exceeds the time needed for bond rotation, it is not great enough to allow cyclization to a triazole. The mechanism of such reactions, as indicated by product analysis, is dependent on the reaction conditions employed. For example, in contrast to the results of Nesmeyanov and Rybinskaya, it has been found that β-chlorobenzoylethylene (**33**), affords 5-phenylisoxazole (**30**) and 4-benzoyl-1,2,3-triazole (**34**) after 5 hours at 30° in dimethylformamide[140]. The dipolar aprotic solvent possibly stabilizes the intermediate carbanion and its lifetime is sufficient to enable cyclization to occur.

$C_6H_5COCH{=}CHCl$ (**33**)

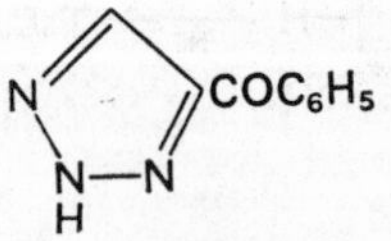

4-COC_6H_5-1,2,3-triazole (N–N(H)) (**34**)

The synthesis of azidoquinones from haloquinones is well established[144,145]. Replacement of more than one halo group may be effected, as for example in the synthesis of tetraazidobenzoquinones and 2,3-diazido-1,4-napthoquinone[146–148]. A related case is the formation of diazido-*N*-phenylmaleimide (**35**) from the corresponding dichloro derivative[149,150].

In the synthesis of the terminal vinyl azide (**36**) by reaction of 9-bromomethylenefluorene with azide ion in dimethylformamide[151] the halogen atom is activated by the pan-activating[152] hydrocarbon nucleus. Similar treatment of the dichloromethylene derivative (**37**) results in a more complex reaction. 9-Azido-9-fluorenecarbonitrile (**38**) is produced and Smolinsky and Pryde have rationalized the formation of this product by the sequence shown in equation (28).

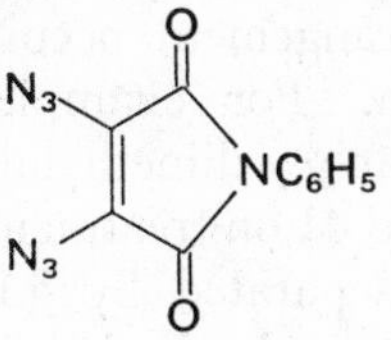

N_3 N_3 NC_6H_5 O O

(**35**)

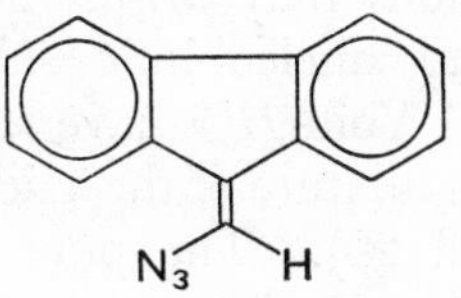

N_3 H

(**36**)

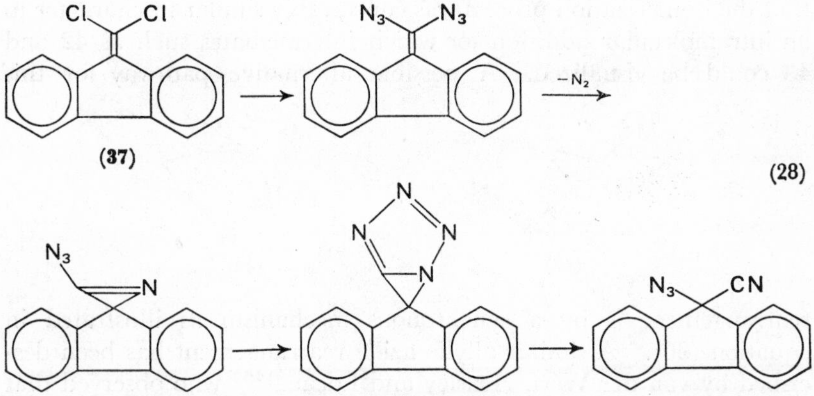

Cl Cl → N_3 N_3 $\xrightarrow{-N_2}$

(**37**)

(28)

N_3 N → N N N N → N_3 CN

(**38**)

b. Allyl azides. Several mechanisms have been delineated for reactions of allylic compounds with nucleophiles and this aspect of substitution processes has received extensive coverage in the literature[62,63]. Surprisingly, there are but few reports on the use of azide ion although from the data available it is apparent that with this nucleophile also a number of mechanisms require consideration.

Allyl chloride[153] and β-phenylallyl bromide[154] have been converted into allyl azide and β-phenylallyl azide, but to differentiate between possible mechanisms involved in these reactions labelling studies would be necessary. The primary and secondary allylic chlorides, geranyl chloride[155,156] and α-butylallyl chloride[157], appear to undergo normal S_N2 substitution by azide ion in aqueous alcohol to form geranyl azide and α-butylallyl azide respectively.

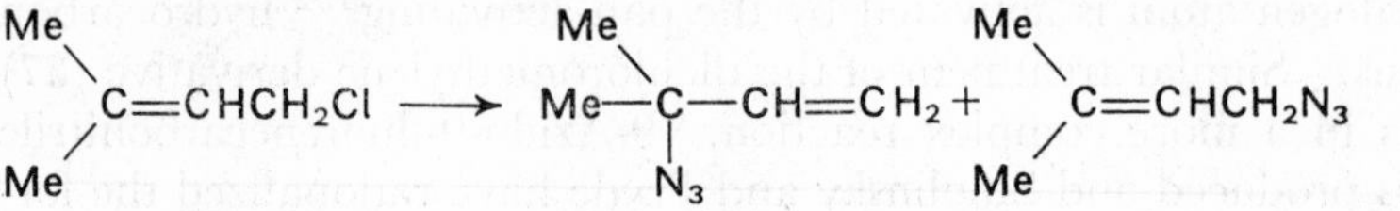

$$Me_2C{=}CHCH_2Cl \longrightarrow Me_2C(N_3){-}CH{=}CH_2 + Me_2C{=}CHCH_2N_3 \quad (29)$$

(39) **(40)** **(41)**

Complications may arise where rearrangement occurs in a step subsequent to nucleophilic substitution. For example Gagneux, Winstein and Young[158] have shown that γ,γ-dimethylallyl chloride (**39**) affords a mixture of the azides **40** and **41** on treatment with azide ion (equation 29). The pure azides, separated by g.l.c., rapidly equilibrated to the isomeric mixture. On the basis of the known ability of aryl and alkyl azides to add to olefins, these authors suggested that the isomerization process was conceivably similar in character to an intramolecular addition for which intermediates such as **42** and **43** could be visualized. A possible alternative pathway for this

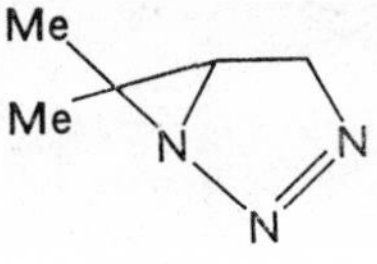

Me Me N N N

(42)

Me Me N N N

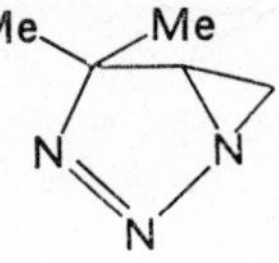

(43)

rearrangement is by a synchronous mechanism as illustrated in equation (30). A similar allylic azide rearrangement has been described by van der Werf, Heasley and Locatell[159] who observed that *cis*- and *trans*-1,4-diazido-2-butene spontaneously equilibrated at room temperature to give identical mixtures of 1,2- and 1,4-diazides.

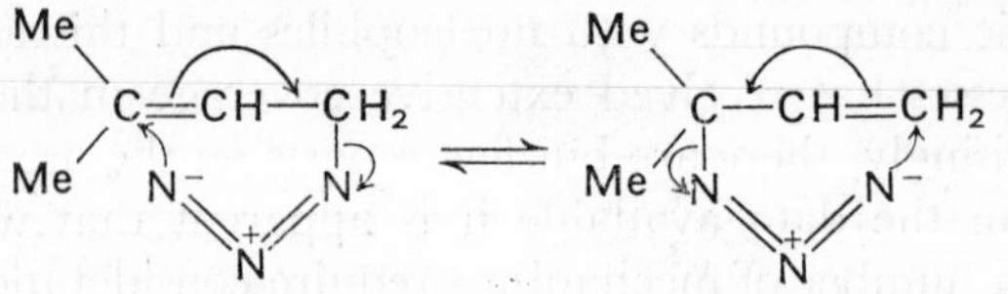

Me Me C=CH—CH₂ N⁻ N⁺ N ⇌ Me Me C—CH=CH₂ N N⁺ N⁻ (30)

Another mechanistic variation is the S_N2' reaction which has been discussed by Pegolotti and Young[160] in relation to the displacement reaction of azide ion with α-trifluoromethylallyl *p*-bromobenzenesulphonate (**44**). Second-order kinetics were observed and only one product, designated as the γ-trifluoromethylallyl azide (**45**), was isolated. Although these facts are consistent with an S_N2' reaction (equation 31), Pegolotti and Young appreciated that the possibility of

$$N_3^- \quad CH_2{=}CH{-}CH(CF_3)(OBs) \longrightarrow N_3CH_2CH{=}CHCF_3 \qquad (31)$$

(**44**) (**45**)

an S_N2 displacement followed by isomerization (equation 32) had not been adequately eliminated. In discussing this work it should be

$$N_3^- \quad CH(OBs)(CF_3){-}CH{=}CH_2 \longrightarrow CH(N_3)(CF_3){-}CH{=}CH_2 \xrightarrow{S_Ni'} F_3CCH{=}CHCH_2N_3 \qquad (32)$$

noted that an unambiguous structural identification of the azide product is not recorded in the paper and therefore this point might bear re-examination.

In 1937 Levene, Rothen and Kuna[157] reported that allyl halides undergo substitution by azide ion without inversion of configuration and quoted as an example the sequence in equation (33). In this

$$Br{-}C(H)(CH{=}CH_2)(C_4H_9) \xrightarrow[\text{alcohol}]{NaN_3} N_3{-}C(H)(CH{=}CH_2)(C_4H_9) \qquad (33)$$

work, absolute configurations were erroneously assumed on the basis of relative optical rotations. There appears no reason why Walden Inversion should not occur in these cases if a normal S_N2 mechanism is operative.

3. Azides of the general structure $R{-}C({=}X){-}N_3$

In this section azides of the general structure $R{-}C({=}X){-}N_3$ (where X is a heteroatom: O,N or S and R is alkyl or aryl, *O*-alkyl or amino) are considered. Compounds of this type may exist in tautomeric equilibrium with the cyclic form (equation 34), the best known example

$$R{-}C({=}X){-}N_3 \rightleftharpoons \text{(cyclic R-substituted five-membered ring: C, X, N, N, N)} \qquad (34)$$

being the azidoazomethine–tetrazole tautomerization, when X = NR′. Acyl azides (X = O) only exist in the open chain form and conversely only the cyclic tautomers of thiatriazoles (X = S) have been reported. Such cyclic compounds are not relevant within the context of this chapter and will only be discussed where pertinent to the synthesis of azides.

a. Acyl azides. Acyl and aroyl chlorides readily undergo nucleophilic substitution with azide ion to generate azides. Two synthetic procedures have been developed[85]. In the first of these, a concentrated aqueous solution ($\simeq 25\%$) of sodium azide is stirred into a solution of the acid chloride in an organic solvent which is miscible with water[161]. The most satisfactory solvent for the acid chloride is acetone, although the use of methanol, ethanol and dioxan has been described. This procedure gives a smooth, rapid reaction which is easy to control[85].

The use of aqueous solvents is obviously unsatisfactory for acyl halides which hydrolyse very rapidly, e.g. trifluoroacetyl bromide[162] and acetyl chloride, and in such cases it has been preferred to carry out a heterogeneous reaction using a suspension of powdered sodium azide[110b,163,164] in an inert solvent such as benzene, toluene, xylene, nitrobenzene or pyridine. Paraffin oil and perfluorokerosene have even been used for exceptionally reactive acid halides such as acryloyl chloride[165] and fluoroformylchloride (FCOCl)[166]. To achieve the best results by this method the sodium azide should be 'activated' (see section III.B.1). Notwithstanding this, the insolubility of sodium azide in organic solvents sometimes precludes satisfactory substitution[167–170]. Another shortcoming of the heterogeneous method is that heating is often required to initiate the reaction and decomposition of the azide product to an isocyanate by loss of nitrogen may ensue. The conversion of **46** into the acyl azide (**47**) using dimethyl sulphoxide as solvent has been reported[171]. However, the nucleophilic character of this solvent seriously limits its value in this connexion for it is well known to react vigorously at ambient temperature with many acid chlorides[172].

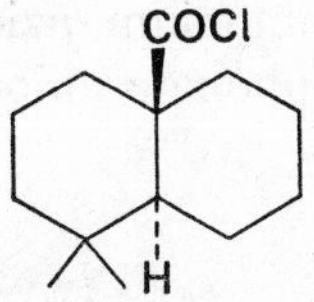

COCl (**46**) CON$_3$ (**47**) $BrCH_2CH_2COBr$ (**48**)

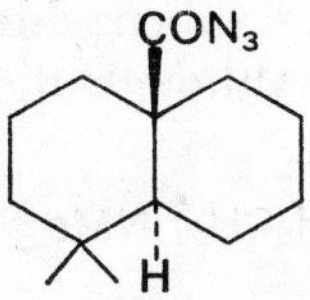

Horning and Muchowski[173] have recently presented a modification of the general procedures described above. In an attempt to replace selectively the acyl halogen in **48** with azide ion, these authors examined the reactions of dimethylformamide–acyl halide complexes (**49**) with nucleophiles (equation 35). The site of attack of a nucleophile on the ambident cation (**49**) is markedly influenced by the nature of the solvent and the temperature. Control reactions with aniline as the nucleophile enabled the optimum conditions for attack at the acyl carbon atom to be elucidated. With azide ion under these conditions, acyl azides were obtained in 60–100% yield (based on

$$RCOBr + Me_2NCHO \longrightarrow \left[R\overset{O}{\overset{\|}{C}}OCH{=}NMe_2 \right]^+ Br^- \xrightarrow{N_3^-} RCON_3 + Me_2NCHO \quad (35)$$

(**49**)

rearrangement to the isocyanate). The ambident cation may alternatively be prepared by interaction of carboxylate anions with the complex (**50**) which is derived from dimethylformamide and phosgene. Thus, addition of the carboxylic acid salt and sodium azide to the

$$[Me_2N{=}CHCl]^+ Cl^- + RCOO^- \xrightarrow{0°}$$

(**50**)

$$[Me_2N{=}CHOCOR]^+ Cl^- + Cl^- \xrightarrow{N_3^-} RCON_3 + Me_2NCHO \quad (36)$$

complex at 0° enabled the direct conversion of a carboxylic acid to its acyl azide (equation 36).

Other variations on the synthesis of acid azides have been described. Maffei and Bettinetti[174] for example, have employed thioacids as substrates (equation 37) and Shozda and Vernon[175] have reported the

$$C_6H_5COSH + HN_3 \xrightarrow[5\text{ hr}, 60°]{CHCl_3} C_6H_5CON_3 \quad (37)$$

synthesis of an azide using dibenzyldimethylammonium azidosulphonate (**51**) as reagent. Although the azidosulphonate anion is a

$$[(C_6H_5CH_2)_2NMe_2]^+ N_3SO_3^-$$

(**51**)

weak nucleophile and normally ineffective in substitution reactions (e.g. with alkyl halides) it interacts with acetyl chloride to give acetyl azide, together with methyl isocyanate. Preliminary decomposition of the azidosulphonate ion into azide ion does not appear to be involved, since it was found that treatment of the azidosulphonate with hexafluoroacetone followed by acetyl chloride did not give 2-azidohexafluoroisopropyl acetate (**52**). When an azide salt is used instead of azidosulphonate, however, the ester (**52**) forms in high yield (equation 38). It was suggested by Shozda and Vernon that direct reaction occurs between the acid chloride and azidosulphonate ion.

$$(F_3C)_2C{=}O \xrightarrow[MeCOCl]{N_3^-} N_3{-}C(CF_3)_2{-}O{-}C({=}O){-}CH_3 \quad (38)$$

(**52**)

$$(F_3C)_2C{=}O \xrightarrow[MeCOCl]{N_3SO_3^-} \not\rightarrow (\mathbf{52})$$

The preparation of acyl azides has received considerable attention due to the value of these compounds as synthetic intermediates. In the Curtius rearrangement for example, acyl azides are converted into isocyanates, urethans, ureas and amines and this aspect of the chemistry of acyl azides is considered in detail in a later chapter. The use of acyl azides in peptide synthesis has increased the scope of general chain-lengthening procedures[176] and this process has emerged in recent years as the only method in which no racemization of the activated peptide components takes place[177,178] (equation 39). The

$$\cdots\cdots CHR^1CONHCHR^2CONHNH_2 \xrightarrow{HONO} \cdots\cdots CHR^1CONHCHR^2CON_3 \xrightarrow{H_2NCHR^3CO_2H} \cdots\cdots CHR^1CONHCHR^2CONHCHR^3CO_2H \quad (39)$$

azidopeptide is commonly made by nitrosation of the corresponding hydrazide (see section V.D) and one particular aim in these syntheses has been to avoid the difficult hydrazinolysis of many long-chain peptide esters[176].

A method which avoids the use of hydrazides has been described by Weinstock[179] who demonstrated that mixed anhydrides interact with sodium azide in excellent yield. Thus the mixed anhydride (**53**) obtained from 2-phenylcyclopropane carboxylic acid and ethyl chloroformate is converted to the azide (**54**) after treatment for 30 minutes at 0° with sodium azide in aqueous acetone. It has been

C_6H_5–(cyclopropane)–C(=O)–O–C(=O)–OEt C_6H_5–(cyclopropane)–C(=O)N_3

(**53**)   (**54**)

suggested that this process could be of value in peptide chemistry, and in fact is a variation of the well established coupling reaction of amino esters with mixed anhydrides of acylamino acids with orthoformates[176].

b. Azidoformates. Interest in the synthesis of azidoformates stems from their increasing use as protecting agents for amino groups in peptide synthesis. Further, as they are rigid acyl azides, azidoformates are also valuable nitrene sources[6]. In general the procedures employed for the synthesis of azidoformates are the same as those described in the previous section.

The use of azidoformates as protecting groups deserves special comment. Increased importance has been attached to the *t*-butoxycarbonyl (BOC) function as a protecting group for amines due to the ease with which it may be introduced and removed[176,180–185]. Acylation may be carried out under mild conditions with *t*-butoxycarbonyl azide and subsequent removal of the BOC group may be achieved at room temperature with trifluoroacetic acid[182] or hydrogen bromide in glacial acetic acid or diethylphosphite[183]. In contrast, the commonly employed carbobenzoxy and trityl groups are stable under these conditions and thus selective removal of the BOC group has been accomplished[182,184]. *t*-Butoxycarbonyl azide has emerged as the agent of choice, since the corresponding chloroformate is unstable. The original synthesis of the azidoformate by Carpino[186,187] entailed nitrosation of the *t*-butyl carbamate (see section V.E). More recently Yajima and Kawatani[188] have synthesized the azide from phosgene and *t*-butanol as shown in equation (40) and similar

$$COCl_2 + t\text{-}BuOH \xrightarrow{\text{pyridine}} t\text{-}BuOCOCl \xrightarrow[EtN_3]{HN_3} t\text{-}BuOCON_3 \quad (40)$$

syntheses of other azidoformates have been reported[189–191]. The use of *p*-methoxybenzyloxycarbonyl azide as a protecting group has also been described[192].

c. Carbamoyl azides. These compounds[12] constitute another class of rigid azides which have been used to generate nitrenes[6]. A range of carbamoyl azides has been synthesized in good yield from the corresponding carbamoyl chlorides by nucleophilic substitution with azide ion[193–195] (equation 41). This procedure complements syntheses from isocyanates and hydrazoic acid (section IV.B.4) and from nitrosation of semicarbazides (section V.E).

$$R^1R^2NCOCl + NaN_3 \longrightarrow R^1R^2NCON_3 + NaCl \quad (41)$$

d. Azidoazomethines and related compounds. The proclivity of compounds of the general structure $R{-}C({=}NR^1){-}N_3$, e.g. azidoazomethines (R^1 = alkyl, aryl), azidooximes (R^1 = OH) and hydrazidic azides ($R^1 = NR^2R^3$) to exist in tautomeric equilibrium with the cyclic form (equation 42) has been extensively investigated during the last decade[196–207]. Spectroscopic methods have proved the most satisfactory for studying such equilibria. There is general agreement that

$$R{-}C({=}NR^1){-}N_3 \rightleftarrows \text{1-}R^1\text{-5-}R\text{-tetrazole (ring: } R^1{-}N{-}N{=}N{-}N{=}C(R){-}) \quad (42)$$

the nature of the substituents R and R^1 has a major effect in determining the relative proportions of the tautomers. When R and R^1 are electron-withdrawing groups, cyclization to the tetrazole form becomes less favourable[196–200]. An obvious prerequisite for cyclization is that the azido substituent should adopt the 'bent' form. An LCAO calculation for the azide system by Roberts[208] has indicated that very little energy is involved in bending the linear configuration to an angle of 120°, and such non-linear configurations are commonly invoked in cycloaddition reactions of azide ion and organic azides[9,209]. In considering the effect of substituents on the azidoazomethine–tetrazole equilibrium, Reynolds, van Allan and Tinker[196] have suggested that as the azide substituent bends, the canonical form (**55b**) becomes increasingly important and cyclization to **56** occurs when the terminal nitrogen of the azido group comes within bonding

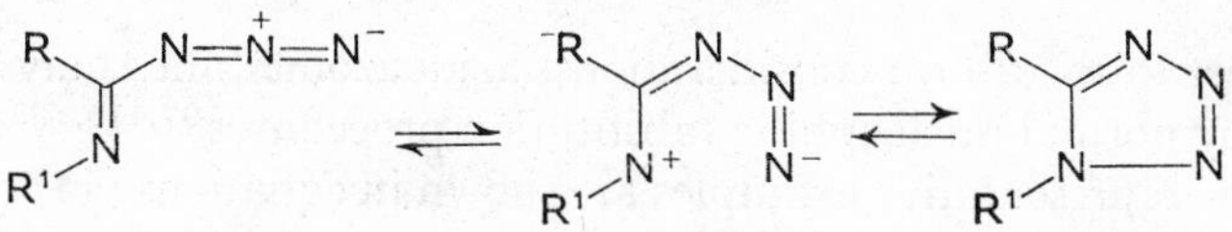

(**55a**) (**55b**) (**56**)

distance of the azomethine nitrogen. When R and R^1 are electron-withdrawing substituents, delocalization of the pair of electrons from the azomethine bond in **55a** into the azide side-chain is diminished and cyclization through a bent transition state similar to **55b** becomes energetically less favourable. Conversely, electron-donating groups R and R^1 favour cyclization.

This point has been well illustrated by Reynolds and co-workers in an investigation of the ring-chain tautomerism of a series of compounds of general structure **57** and **58**[196]. When X is an electron-withdrawing atom or group (O,SO_2 and NH) the open chain form prevails. The

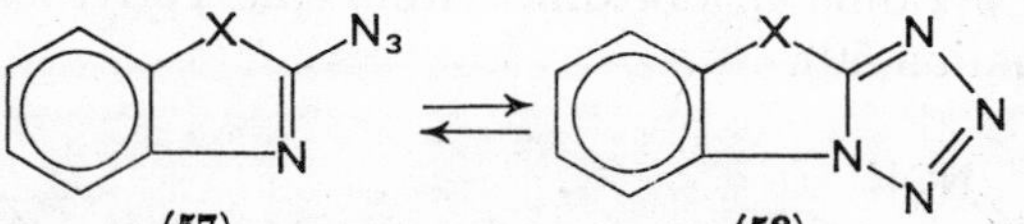

(**57**) (**58**)

equilibrium favours the tetrazole form, however, when X is less electronegative than nitrogen (S,Se). Parallel observations have been made by Boyer and Miller[197]. The dependence of the equilibrium on solvent and temperature was first recognized by Temple and Montgomery[207] and has since been confirmed by others[200–202]. In cases where the azidoazomethine moiety forms part of a heterocyclic ring, solvents which protonate the heterocyclic nitrogen atom will favour formation of the azido tautomer. As the ratio of the solubilities of the tautomers varies with the polarity of the medium, solvent is also important. These points are illustrated by the solvent dependence of the equilibrium shown in equation (43). In trifluoroacetic acid only the azide (**59**) is present, whereas in dimethyl sulphoxide the tetrazole (**60**) is the major component[207].

(**59**) ⇌ (**60**) (43)

Subject to the above considerations, azidoazomethines may be obtained by normal nucleophilic substitution procedures[97,113,199,210–214] and some representative examples are shown in equations (44) to (47).

$$F_2NCF{=}NF \xrightarrow{NaN_3} F_2NC(N_3){=}NF \quad \text{ref. 210} \qquad (44)$$

$$SF_5N{=}C(Cl)CF_3 \xrightarrow{N_3^-} SF_5N{=}C(N_3)CF_3 \quad \text{ref. 97} \qquad (45)$$

$$[(NH_2)_2CCl]^+X^- \xrightarrow{N_3^-} [(NH_2)_2CN_3]^+X^- \quad \text{ref. 211} \qquad (46)$$

$$[Cl_2C{=}NH_2]^+X^- \xrightarrow{N_3^-} [(N_3)_2C{=}NH_2]^+X^- \quad \text{ref. 211} \qquad (47)$$

The products from chlorooximes and azide ion were originally formulated as tetrazoles[215,216] but Eloy[199] has since established by infrared spectroscopy that these compounds exist as azides (equations 48,49) and this is consistent with the foregoing general considerations. These findings were corroborated by Chang and Matuszko[212] who synthesized pyridine azidooximes from the corresponding chlorooximes (equation 50).

$$C_6H_5C(=NOH)N_3 \xleftarrow{N_3^-} C_6H_5C(Cl){=}N{-}OH \not\rightarrow C_6H_5\text{-tetrazole (1H)} \qquad (48)$$

$$HON{=}C(Cl){-}C(Cl){=}NOH \longrightarrow HON{=}C(N_3){-}C(N_3){=}NOH \qquad (49)$$

$$\text{(pyridyl)}-\underset{\substack{|\\ Cl}}{C}{=}NOH \longrightarrow \text{(pyridyl)}-\underset{\substack{|\\ N_3}}{C}{=}NOH \qquad (50)$$

Hegarty, Aylward and Scott[217] have obtained hydrazidic azides (**61**) by treating hydrazidic bromides (**62**) with azide ion in aqueous dioxan (equation 51). As the reactions proceed considerably faster than normal S_N1 processes, these authors postulated the intermediacy of a 1,3-dipolar ion, formed by loss of HBr from the hydrazidic bromide.

$$\underset{(\mathbf{62})}{R-\underset{\substack{|\\ Br}}{C}{=}N-NHAr} \xrightarrow{-HBr} [R-\overset{+}{C}{=}N-\overset{-}{N}-Ar] \xrightarrow[\text{(ii) } H^+]{\text{(i) } N_3^-} \underset{(\mathbf{61})}{R-\underset{\substack{|\\ N_3}}{C}{=}NNHAr} \qquad (51)$$

4. Syntheses from epoxides

Epoxides undergo bimolecular nucleophilic displacement reactions with azide ion to produce azidoalcohols (Table 2). Azide ion preferentially attacks saturated unsymmetrical epoxides at the less substituted carbon atom in accordance with the normal pattern of polar

TABLE 2. Azides obtained by nucleophilic substitution reactions between azide ion and epoxides

Epoxide	Azide	Reference
1,2-Epoxycyclohexane	*trans*-2-Azidocyclohexanol	218
1,2-Epoxycyclopentane	*trans*-2-Azidocyclopentanol	218
2,3-Butylene oxide	3-Azido-2-butanol	218
Propylene oxide	1-Azido-2-propanol	218
Isobutylene oxide	1-Azido-2-methylpropan-2-ol	218
Epichlorhydrin	1,3-Diazidopropan-2-ol	218–220
Epichlorhydrin	1-Azido-3-chloropropan-2-ol	220
Butadiene monoxide	2-Azido-3-butene-1-ol 4-Azido-2-butene-1-ol	218
Styrene oxide	2-Azido-2-phenylethanol	218
1,2-Epoxy-1-phenylethylphosphonic acid diethyl ester	Diethyl 1-azido-2-hydroxy-1-phenylethylphosphonate	222
1,2-Epoxy-2-methyl-3-nonyne	1-Azido-2-methyl-3-nonyne-2-ol	223
1,1-Diphenylethylene oxide	2-Azido-1,1-diphenylethanol	151
1-Methyl-1-phenylethylene oxide	1-Azido-2-phenylpropan-2-ol	151

and primary steric effects in S_N2 displacement reactions of epoxides (equation 52). For example, 1-azido-2-propanol is formed readily at

$$R—CH\underset{O}{—}CH_2 \xrightarrow{N_3^-} R—CH(OH)—CH_2N_3 \quad (52)$$

room temperature from aqueous sodium azide and propylene oxide[218,219]. Selective ring opening of the epoxy group in epichlorhydrin to give 1-azido-3-chloro-2-propanol (**63**) has been accomplished by maintaining the solution at pH 6–9 during the reaction[220]. If the solution is unbuffered, 1,3-diazido-2-propanol (**64**) is formed[218,220], possibly as a result of the ring closure of **63** to the intermediate **65** at higher pH, followed by a second ring opening step with azide ion (equation 53). Similarly an epoxide intermediate may be invoked in

$$\underset{(63)}{N_3CH_2CH(OH)—CH_2Cl} \xleftarrow[pH6\text{-}9]{N_3^-} ClCH_2HC\underset{O}{—}CH_2 \xrightarrow{N_3^-} \underset{(64)}{N_3CH_2—CH(OH)—CH_2N_3}$$

$$(63) \longrightarrow \underset{(65)}{H_2C\underset{O}{—}CHCH_2N_3} \xrightarrow{N_3^-} (64) \quad (53)$$

the reaction of 2-chlorocyclohexanol with aqueous ethanolic sodium azide in the presence of sodium hydroxide. 2-Azidocyclohexanol is formed in 61% yield after 12 hours at 98° and it is significant that in the absence of sodium hydroxide no azide was produced even after 24 hours at the same temperature[105].

Epoxides which are fused to alicyclic systems undergo ring opening with azide ion in the usual *trans* diaxial manner[221]. This is demonstrated by the formation of the *trans*-2-azidoalcohols (**66**) and (**67**) from cyclohexene oxide and cyclopentene oxide respectively[218].

(**66**) (**67**)

A number of nucleophiles, including azide ion, form 'abnormal' products from unsaturated epoxides. For example, van der Werf and co-workers[218] established that 2-azido-2-phenylethanol (**68**) is the only isolable product from the reaction of styrene oxide and sodium azide in boiling aqueous dioxan. This attenuation of steric factors under typical S_N2 conditions has been rationalized in terms of the higher degree of resonance stabilization in the transition state (**69**) leading to the product (**68**). A similar rationalization may be made for the formation of 2-azido-3-butene-1-ol (**70**) from butadiene

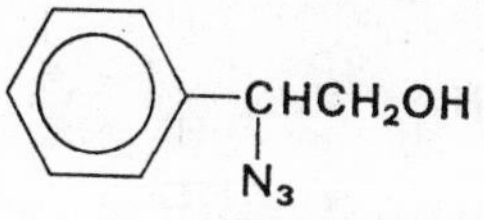

(68) 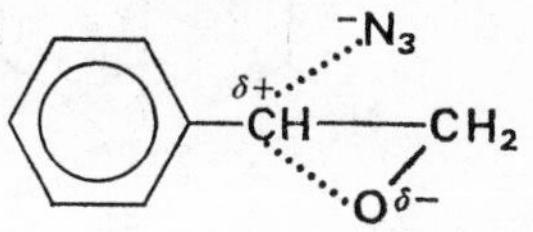 (69)

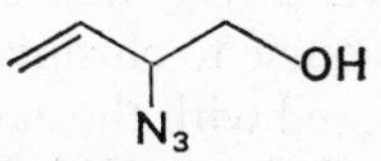

(70) 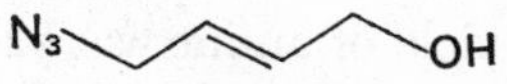 (71)

monoxide. 4-Azido-2-butene-1-ol (**71**), which is also formed in this reaction, probably arises from S_N2' attack of azide ion on the epoxide. The effect of vinyl and phenyl groups in promoting 'abnormal' attack is particularly marked when azide ion is the nucleophile[218]. This has been attributed, in part, to a combination of the low steric requirement of the linear azide ion and its polarizability which enhances the stability of transition states such as **69**.

5. Azidosteroids

The general procedures previously outlined in this section for the formation of the carbon to azide bond have been widely employed in the steroid field, particularly as a stage in the stereospecific synthesis of aminosteroids. Bimolecular nucleophilic displacement reactions of sterols substituted with *p*-toluenesulphonyl, methanesulphonyl or halogeno groups etc. with azide ion proceeds with Walden Inversion and enables the stereospecific introduction of the azido group, which may then be reduced to an amino group.

This sequence is well illustrated by the synthesis of 3α-aminocholestane (**72**) from cholestanyl-3β-*p*-toluenesulphonate[96]. Reaction of the equatorial *p*-toluenesulphonate with sodium azide in aqueous dimethylformamide or dimethylacetamide afforded the axial azide (**73**) which was reduced to the amine (**72**) with lithium aluminium hydride

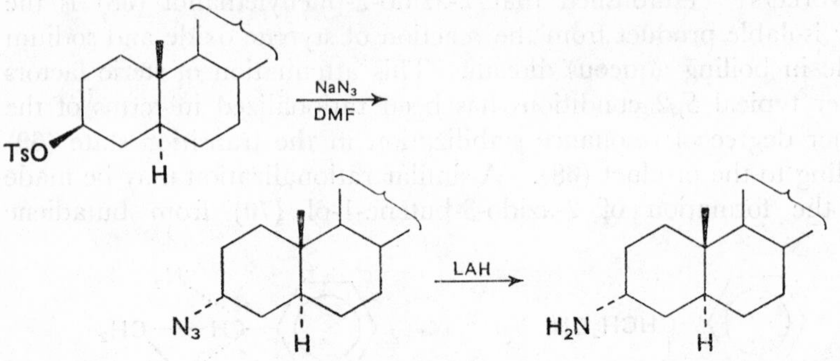

$$\text{TsO-steroid} \xrightarrow{\text{NaN}_3,\ \text{DMF}} \mathbf{(73)} \xrightarrow{\text{LAH}} \mathbf{(72)} \qquad (54)$$

(equation 54). No equatorial amine could be detected in the product, the only undesirable side reaction being the formation of some alkene by β-elimination of *p*-toluenesulphonic acid in the first step. The overall yield of amine was 62%, without recourse to chromatography. The obvious convenience of this route compared with the earlier, often non-stereospecific methods for synthesizing aminosteroids[224] has led to its further application by other workers in this field[225,226]. By appropriate selection of leaving groups in polysubstituted steroids, regiospecificity* may be realized as well as stereospecificity. Hence Ponsold and Groh[227] have utilized the better leaving group mobility of methanesulphonate compared with halogen to prepare 2α-bromo-3β-azidocholestane (**74**) from the methanesulphonate derivative (**75**) of 2α-bromo-3α-cholestanol, by reaction with sodium azide in dimethyl sulphoxide.

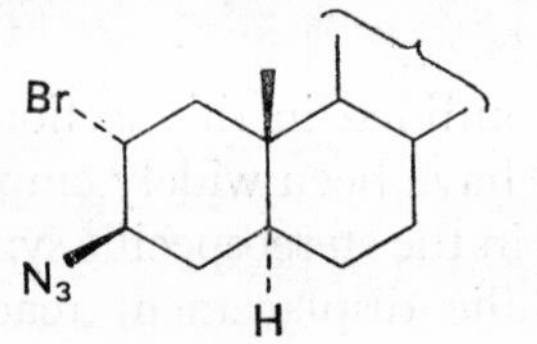

(74) 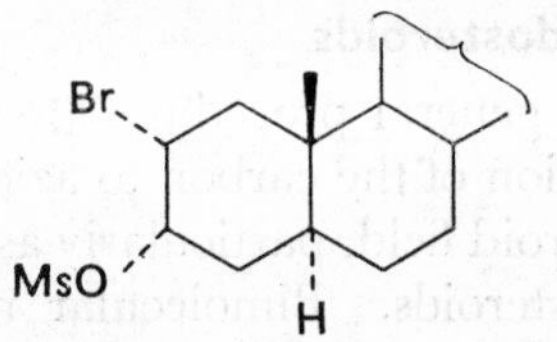 (75)

In an attempt to prepare 2β-azidocholestan-3-one (**76**) from the 2α-bromo derivative (**77**), Edwards and Purushothaman[228] observed an interesting shortcoming in this general procedure. Although 2-chlorocyclohexanone gave a high yield of the 2-azidoketone when treated with sodium azide in dimethylformamide, reaction of lithium azide and the bromo compound (**77**) in the same solvent produced

* For definitions of regiochemistry see section IV.E.1.

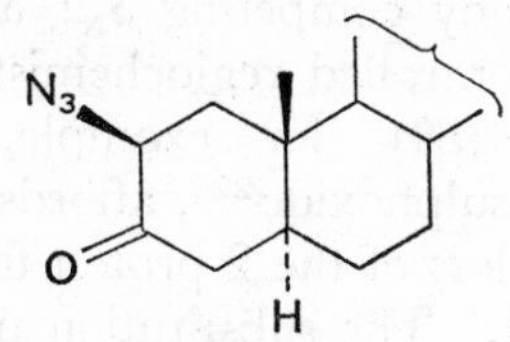

(76) 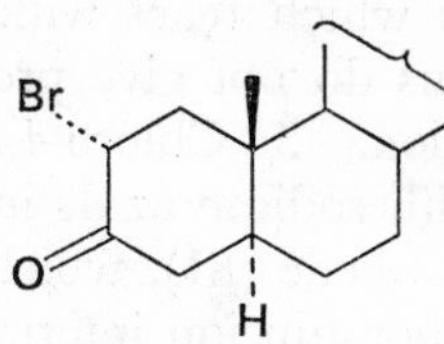 (77)

only the iminoketone (**78**). Infrared studies established that a steady state concentration of an azide, probably the β-epimer (**76**), was produced and this azide was continuously converted into the iminoketone.

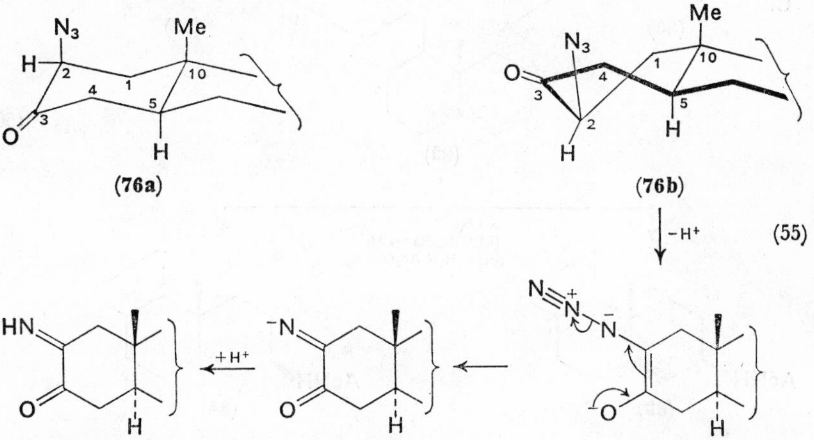

(76a) (76b) (55)

(78) (79)

Surprisingly, 2-azidocholestane-3-one was thermally stable in refluxing methanol, or dimethylformamide at 60°, or in these solvents containing lithium bromide. Clearly the lithium azide used as reagent catalyses the decomposition of the azidosteroid and Edwards and Purushothaman have suggested that the accelerated decomposition of the azide has a conformational driving force. A non-bonded interaction exists between the 10-methyl group and the 2-azido function in **76**, which probably adopts a half-boat ring A conformation (**76b**) to reduce compression in the chair form (**76a**). Although lithium azide is weakly basic, it assists abstraction of the 2-proton in **76b**, since the repulsive interaction between the 10-methyl group and the 2-azido group is relieved in the enolate anion (**79**). Rapid loss of nitrogen from this enolate would then give the iminoketone (equation 55).

Steroids which react with azide ion by competing S_N1 and S_N2 mechanisms do not give products of controlled regiochemistry and configuration. 3β-Chloro-4-cholestene (**80**), for example, when treated with sodium azide in dimethyl sulphoxide[229], affords mostly 2,4-cholestadiene (**81**), which arises by loss of the 2-proton from the allylic carbonium ion intermediate (**82**). The substitution product, isolated as the 3-acetamido derivative after reduction of the azide and

Cl (**80**) → (**82**) $\xrightarrow{-H^+}$ (**81**)

(**82**) → (i) N_3^- (ii) N_2H_4/Raney Ni (iii) C_5H_5N/Ac_2O → AcNH (**83**) + AcNH (**84**)

(56)

acylation of the resulting amine was a mixture of the 3α-and 3β-epimers (**83**) and (**84**) (equation 56).

Steroids which contain a homoallylic structural unit produce an even greater multiplicity of products by both S_N1 and S_N2 interactions with azide ion[91,93,230–235]. This point is best illustrated by reference to the elegant synthesis by Barton and Morgan[230–232] of the steroidal alkaloid conessine (**85**) (equation 57). Treatment of the di-*p*-toluene-sulphonate derivative (**86**) of pregn-5-ene-3β,20β-diol with azide ion afforded 3β, 20α-bisazidopregn-5-ene (**87**) which, after irradiation in cyclohexane, reduction and *N*-methylation gave a low yield (4·5%) of conessine. Substitution of a 3β-*p*-toluenesulphonate by azide ion without inversion at the 3-position was also demonstrated for cholesteryl *p*-toluenesulphonate (**88**), which was converted into 3β-azidocholest-5-ene (**89**) with lithium azide in methanol. This retention of configuration is associated with the *i*-steroid transformation[236], which is depicted as a unimolecular S_N1 process through the homoallylic bridged ion (**90**). The participation of the π-electrons of the 5,6-double bond is stereospecifically α and hence the configuration at the 3-position is maintained. No other azide products were reported from these reactions and this has stimulated comment from several groups of workers. Jones[91] observed that cholesteryl *p*-toluenesulphonate (**88**) afforded five products with sodium azide in dimethyl sulphoxide at 100° (Scheme 1). Solvolytic oxidation of the starting material and re-arrangement of the resulting cholest-5-ene-3-one (**91**) was invoked to rationalize the formation of **92**. The pattern of substitution products obtained[91] was that expected from reaction of a strong nucleophile with cholesteryl *p*-toluenesulphonate (see Ref. 237). Unimolecular heterolysis of the *p*-toluenesulphonate produced the non-classical carbonium ion (**90**) which reacted with azide ion to give **89** and **93**; 3α,5-cyclocholest-6-ene (**94**) was generated by loss of the 7-proton from **90**. A competing bimolecular reaction of azide ion at the 3-position of **88** gave the inverted α-epimer (**95**) in 16% yield.

The relative proportions of products from this reaction vary with different reaction conditions, particularly changes in solvent. This is illustrated in a further analysis of the products derived from cholesteryl 3β-*p*-toluenesulphonate by Freiberg[235] using methanol, dimethyl sulphoxide and *N*-methylacetamide (NMA) as solvents (Table 3).

TABLE 3. Reaction of cholesteryl *p*-toluenesulphonate with azide ion in various solvents

Product	MeOH[a]	DMSO[b]	NMA[b]
3α-Azidocholest-5-ene (**95**)	—	31	12
3β-Azidocholest-5-ene (**89**)	8 (57)[c]	13	19
3α,5-Cyclo-6β-azidocholestane (**93**)	35	16	31
3α,5-Cyclo-6α-azidocholestane (**97**)	21	6	10

[a] Reflux with 0·92 M LiN_3.
[b] At 85–90° with 1·5 M NaN_3.
[c] Barton and Morgan[230–232].

The results of Barton and Morgan, also included in Table 3, appear anomalous but it should be borne in mind that these latter workers purified their crude azide product by chromatography on alumina before conformational analysis was carried out. The facile rearrangement of 3α,5-cyclo-6β-azidocholestane (**93**) into 3β-azidocholest-5-ene (**89**) on prolonged contact with alumina has been well established[91] and it is possible that any 6β-azide (**93**) in the crude azide product isolated by Barton and Morgan may have rearranged to the 3β-isomer (**89**) during purification. Freiberg[235] employed Florisil columns during his separation process and no rearrangement occurred under these conditions. It is also significant that Goutarel and co-workers[238] isolated the 3β-azido-Δ^5-steroid (**89**) from 3α,5-cyclo-6β-cholestanol (**96**) and hydrazoic acid in the presence of Lewis acid catalysts. A three step mechanism was postulated (equation 58) in which ionization of the C—O linkage was followed by nucleophilic attack at the 6-position of the steroidal cation, and rearrangement

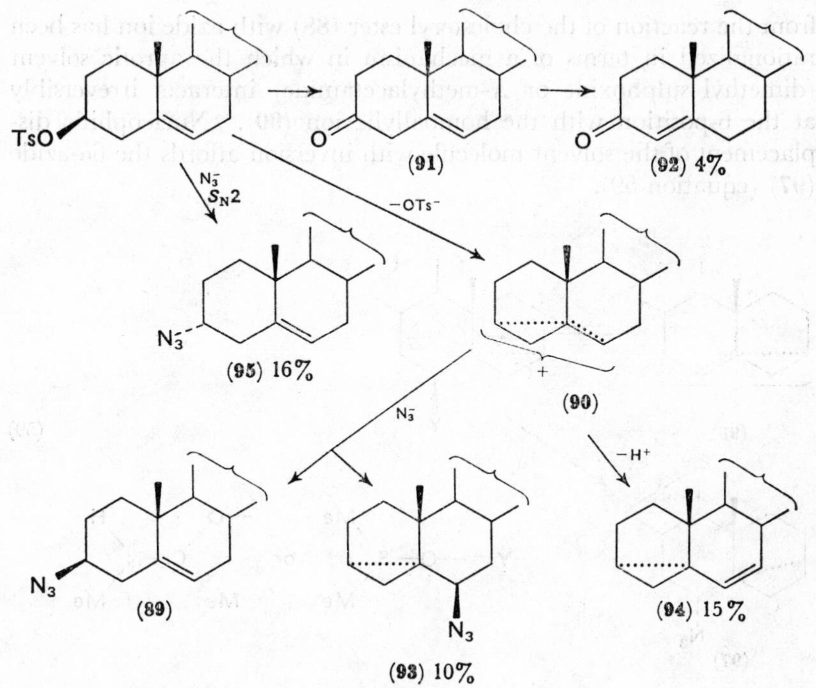

TsO — (91) — (92) 4%
N_3^- S_N2 — $-OTs^-$
N_3^- (95) 16% — (90)
N_3^- — $-H^+$
N_3 (89) — N_3 (93) 10% — (94) 15%

SCHEME 1

of the resulting 6β-azide (**93**) into 3β-azidocholest-5-ene (**89**). Homoallylic rearrangement of **93** into the 3β-azide (**89**) in the presence of boron trifluoride etherate was established separately.

The isolation by Freiberg[235] of 3α,5-cyclo-6α-azidocholestane (**97**)

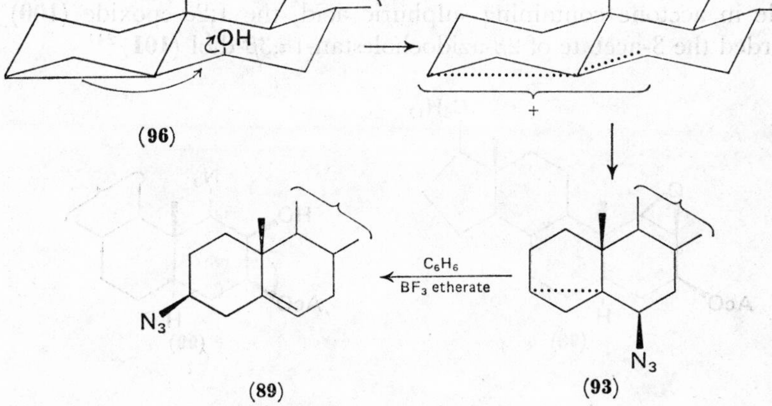

OH (96) → +
↓ (58)
N_3 (89) ← C_6H_6, BF_3 etherate — N_3 (93)

from the reaction of the cholesteryl ester (**88**) with azide ion has been rationalized in terms of a mechanism in which the aprotic solvent (dimethyl sulphoxide or *N*-methylacetamide) interacts irreversibly at the 6-position with the homoallylic ion (**90**). Nucleophilic displacement of the solvent molecule with inversion affords the 6α-azide (**97**) (equation 59).

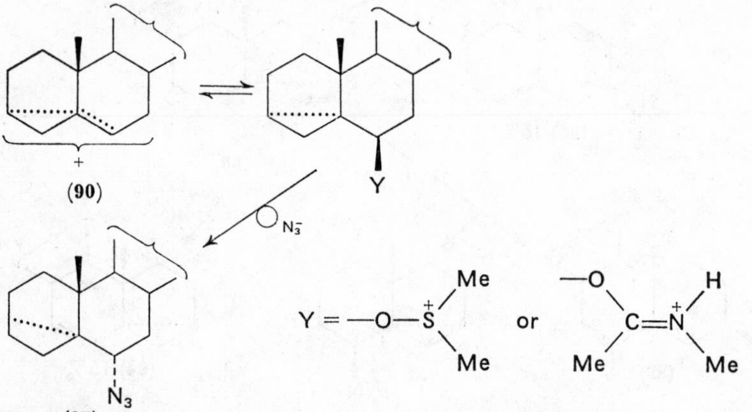

(90) (97) (59)

Y = —O—$\overset{+}{S}Me_2$ or —O(Me)C=$\overset{+}{N}$(H)Me

A considerable number of vicinal azidoalcohols have been synthesized by Ponsold and co-workers[92,239–241] by reactions of epoxysteroids with azide ion. In accordance with other nucleophilic ring opening reactions of steroidal epoxides[242], the diaxial product is mainly formed. Hence, whereas 1,2β-epoxy-3β-cholestanol acetate (**98**) gave 1α-azidocholestan-2β,3β-diol 3-acetate (**99**) with sodium azide in acetone containing sulphuric acid, the 1,2α-epoxide (**100**) afforded the 3-acetate of 2β-azidocholestan-1α,3β-diol (**101**)[241]

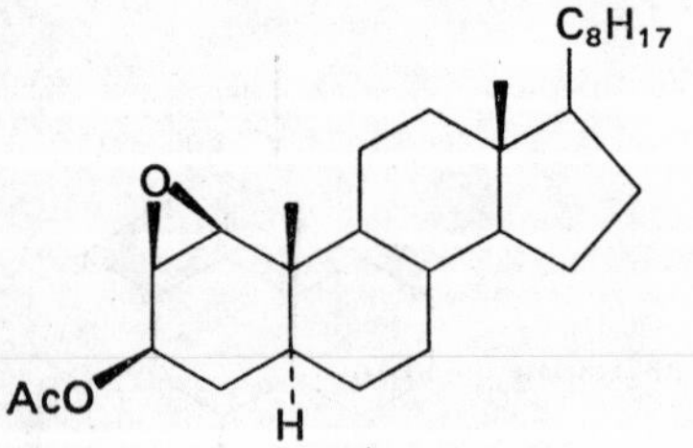

(98) 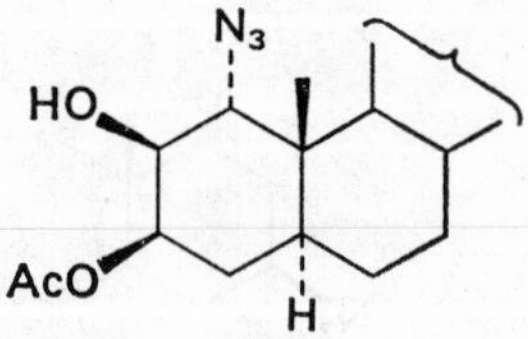 (99)

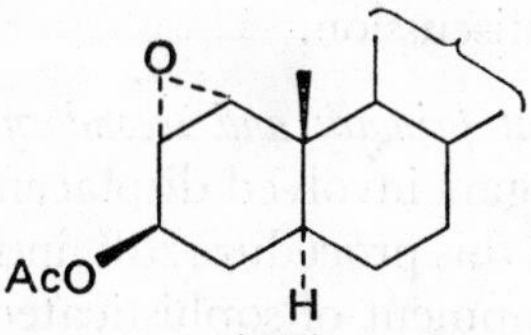

(100) 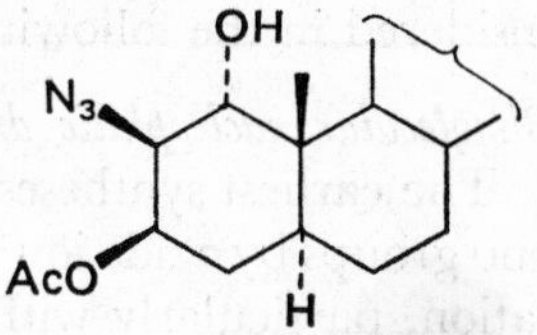 (101)

6. Azidocarbohydrates

Azido derivatives of carbohydrates are valuable intermediates in the synthesis of aminosugars. These latter compounds have elicited considerable interest in recent years since they are components of molecules such as streptomycin, carbomycin, erythromycin, streptothricin and neomycin which exert profound antibiotic activity on microorganisms[243]. These antibacterial agents are amino derivatives of oligosaccharides and it has been suggested that the amino functions are vital to their biological activity. When protonated, they may act as cationic centres which bind strongly to anionic sites in cell walls and enzymes[98]. Interest in the aminosugars also stems from their presence in biological tissues such as bacterial cell walls[244].

Aminosugars may be prepared conveniently by mild and stereospecific reduction of the corresponding azido derivatives[95,103,245–247]. Syntheses through azido intermediates have been generally employed in recent years since other methods are often complicated by undesirable side reactions[95,103].

The first azidosugar, tetra-*O*-acetyl-α-D-glucosyl azide (**102**; R = N_3), was synthesized by Bertho[248] in 1930 from the bromo derivative (**102**; R = Br) by treatment with sodium azide in acetonitrile at 30–40°. Similar bimolecular nucleophilic substitution reactions on

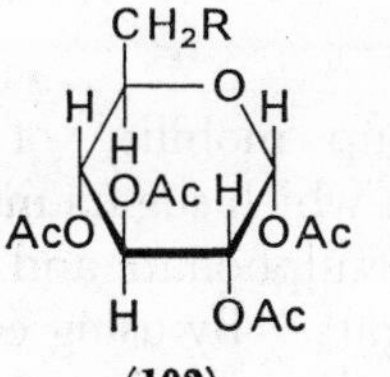

(102)

primary and secondary carbon atoms and also on epoxides, epimines and episulphides are now used routinely for the stereospecific introduction of the azido group in sugars. Only representative examples are considered in the following general discussion.

a. Bimolecular nucleophilic displacement at primary and secondary carbon atoms. The earliest syntheses of azidosugars involved displacement of halogeno groups by azide ion[248–251], and this procedure still finds some application, particularly with the development of sophisticated halogenation techniques such as that described by Hanessian and Plessas[252]. In this method, interaction of *N*,*N*-dimethylchloroformiminium chloride (**103**) with 1,2,3,4-di-*O*-isopropylidene-D-galactopyranose (**104**) under mild conditions led to the 6-chloro derivative (**105**), which was then converted into the azide as shown in equation (60).

$$(Me_2\overset{+}{N}{=}CHCl)Cl^- + (104) \longrightarrow \xrightarrow{} (105) \xrightarrow{N_3^-,\ DMF} \quad (60)$$

(**103**) (**104**) (**105**)

The high leaving group mobility of sulphonate (relative to halogen) and the ease with which sugars may be sulphonated has led to the wide use of *p*-toluenesulphonate and methanesulphonate esters in the synthesis of azidosugars. By using equimolar quantities of the reactants, specific sulphonation of primary hydroxyl groups in polyhydric sugars such as **106** may be achieved without recourse to the blocking of secondary hydroxyl functions[245,246], as illustrated in equation (61).

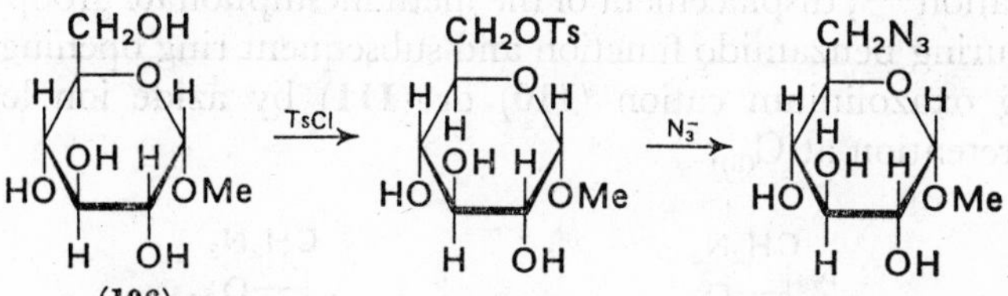

(106) (61)

The ease of bimolecular nucleophilic replacement of substituents generally follows the expected order primary > secondary. Primary substituents may often be replaced in polar solvents, but secondary substituents are only satisfactorily replaced in dipolar aprotic solvents since adverse steric and dipolar interactions hinder the approach of the azide ion to the secondary carbon atom[95,98,253,254]. Although dimethyl sulphoxide and dimethylformamide are acceptable solvents for such reactions, sulphonate esters may decompose at the temperatures required for replacement[98]. In a recent report the use of hexamethylphosphoric triamide has been advocated for in this solvent substitution may be effected at lower temperatures; decomposition is thus minimized and pure products are obtained in high yield[98]. Differential rates of substitution at primary and secondary carbon atoms may enable selective attack by azide ion at a primary carbon atom in the presence of secondary sulphonate groups[247a,254–256]. Hence, on brief treatment of 2-benzamido-2-deoxy-3,4,6-tri-*O*-methanesulphonyl-α-D-glucopyranoside (**107**) with azide ion in dimethylformamide at 85°, the 6-sulphonate group was selectively replaced to afford the 6-azido derivative (**108**) in 90% yield[256].

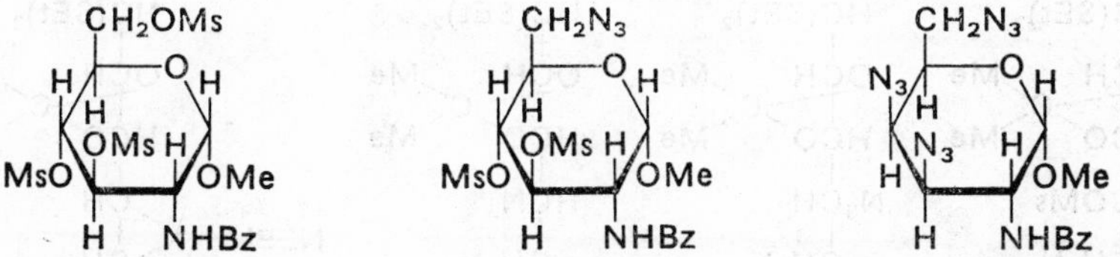

(**107**) (**108**) (**109**)

Displacement of secondary sulphonates usually proceeds with inversion and significant exceptions are stereochemically informative. When the reaction of **107** with azide ion in dimethylformamide was carried out over 24 hours at 85°, the triazide (**109**) was isolated. Substitution at $C_{(4)}$ proceeds with the expected Walden inversion, but at $C_{(3)}$ the azido function is introduced with retention of configuration. This result has been rationalized in terms of neighbouring group participation[256]; displacement of the methanesulphonate group by the neighbouring benzamido function and subsequent ring opening of the resulting oxazolinium cation (**110**) or (**111**) by azide ion leads to overall retention at $C_{(3)}$.

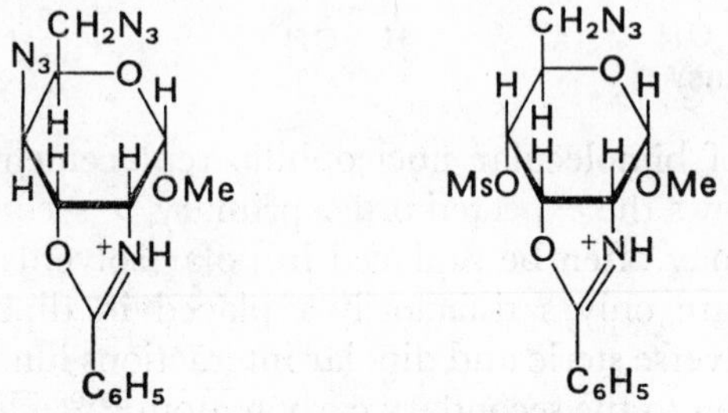

(**110**) (**111**)

Hanessian has described a case in which an azido substituent is involved in neighbouring group participation[257]. 5-Azido-5-deoxy-4-*O*-methanesulphonyl-D-arabinose-2,3-*O*-isopropylidene diethyl thioacetal (**112**), affords the L-xylose derivative (**113**) as the major product, together with a lesser amount of 4,5-diazido-4,5-dideoxy-2,3-*O*-isopropylidene-D-arabinose (**114**) on reaction with sodium azide in dimethylformamide. These products were considered to arise by way of an azidonium intermediate[74] (**115**); attack on the secondary carbon atom of **115** leads to the diazide (**114**) with overall retention, whereas the more facile substitution on the primary carbon affords the inverted product (**113**).

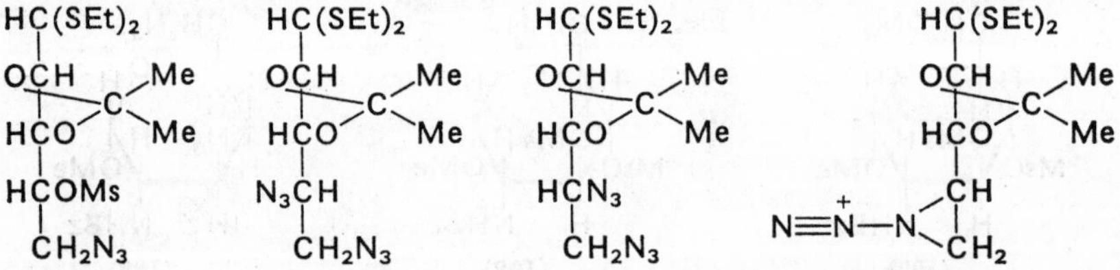

(**112**) (**113**) (**114**) (**115**)

Such neighbouring group participation is rare with azide ion, although it has been well established with other nucleophiles (e.g. I^-, SCN^-). Azide ion is a moderately powerful nucleophile and in dipolar aprotic solvents its strength as a nucleophile is enhanced. In consequence, normal bimolecular substitution with inversion generally proceeds to the exclusion of neighbouring group participation. Thus, whereas the 2,3-diacetate of methyl-α-D-glucopyranoside-4,6-dimethanesulphonate (**116**) afforded, in part, the 4,6-dithiocyanato derivative (**117**) on reaction with thiocyanate ion in dimethylformamide, through the intermediacy of the acetoxonium ion (**118**), no analogous product was obtained with azide ion[253,258]. A similar observation has been made by Baker and Haines with a mannitol derivative[259].

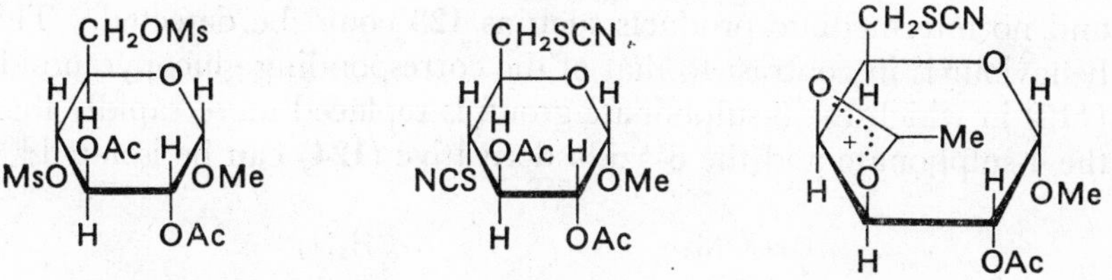

(**116**) (**117**) (**118**)

Since direct displacement with azide ion competes favourably with anchimeric reactions, less direct routes such as the double inversion procedure illustrated for conversion of **119** into the triazide **122** in equation (62), must often be used to achieve substitution with overall retention of configuration[256].

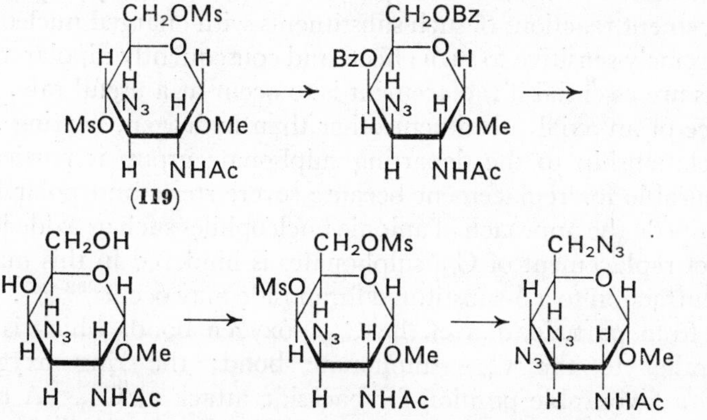

(**119**) (62)

(**120**) (**121**) (**122**)

A further interesting aspect of azidosugar synthesis is exemplified in equation (62). Replacement of the 6-sulphonates of galactopyranosides such as **121** is sluggish compared to the corresponding glucopyranosides. For galactopyranosides the terminal sulphonate ester is situated adjacent to a *cis*-axial group at $C_{(4)}$ and it has been suggested that for such conformations a combination of unfavourable steric, polar and electronic factors hinder the approach of the nucleophile to the primary carbon atom[253,256,260–262]. With azide ion, therefore, the initial attack occurs at the secondary carbon atom in **121**; with the configuration at $C_{(4)}$ thus changed, subsequent replacement at the primary carbon atom of the intermediate 6-sulphonate is extremely rapid. In support of this view, only the triazide **122** was isolated from the reaction of **121** with sodium azide in boiling dimethylformamide and no intermediate products such as **123** could be detected. This behaviour is in contrast to that of the corresponding glucopyranoside (**119**) in which the 6-sulphonate group is replaced more rapidly than the 4-sulphonate and the 6-azido derivative (**124**) can be isolated[256].

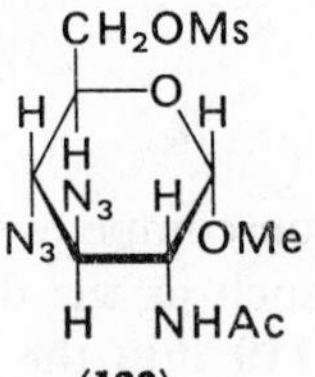

CH_2OMs ... N_3, H, N_3, H, OMe, H, NHAc

(**123**)

CH_2N_3 ... MsO, H, N_3, H, OMe, H, NHAc

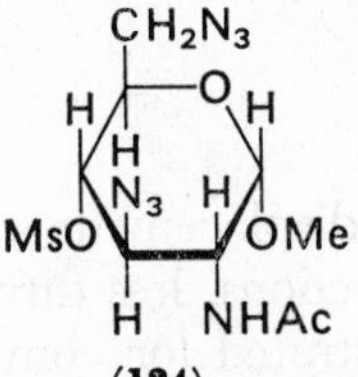

(**124**)

The adverse effect of polar, steric and electronic factors is more often encountered during attempts to replace secondary sulphonates. Displacement reactions of such substituents with charged nucleophiles are extremely sensitive to such effects and consequently dipolar aprotic solvents are essential if replacement is to occur at a useful rate. The presence of an axial substituent other than hydrogen, bearing a 1,3-*trans* relationship to the departing sulphonate group is particularly unfavourable for replacement because severe steric and polar repulsions impede the approach of anionic nucleophiles such as azide ion[253]. If direct replacement of $C_{(4)}$ sulphonates is hindered in this manner, ring contraction to a 5-substituted furanoside may occur[263–265]. This results from participation of the $C_{(5)}$—oxygen bond, which is *trans*-antiparallel to the $C_{(4)}$—sulphonate bond; the ring oxygen is thus in a favourable position for backside attack at $C_{(4)}$. A typical case has been described by Hanessian[263] who found that methyl 6-deoxy-2,3-*O*-isopropylidene-4-*O*-methanesulphonyl-α-L-mannopyranoside (**125**) was converted into the furanoside (**126**) by treatment with an excess of sodium azide in refluxing dimethylformamide.

It should be pointed out, however, that in some cases alternative chair conformations may be accessible in which the 1,3-*trans*-axial condition is removed and little hindrance to normal S_N2 replacement

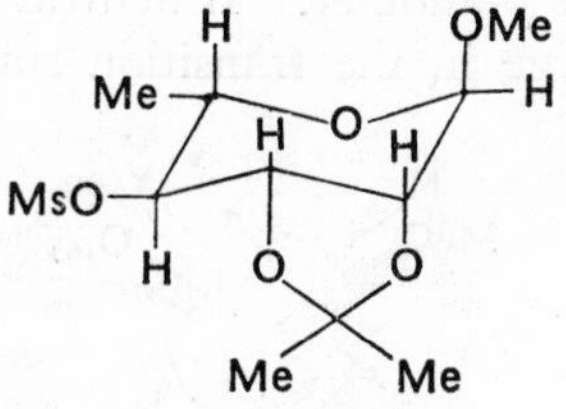

(125) 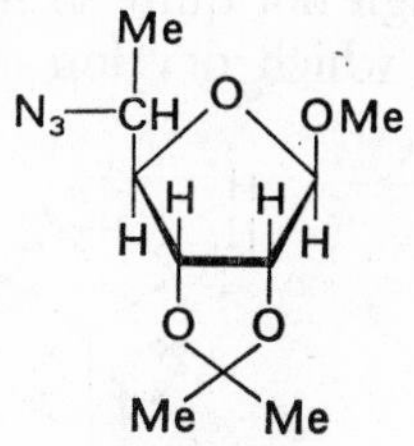 (126)

would be experienced. To account for the ease with which the secondary sulphonate group was replaced in the methanesulphonate derivative (**127**), for example, Ali and Richardson suggested a conformational change into the 1C form (**127b**), since in the more stable C1 conformation (**127a**) ring contraction would be preferred[98] (equation 63).

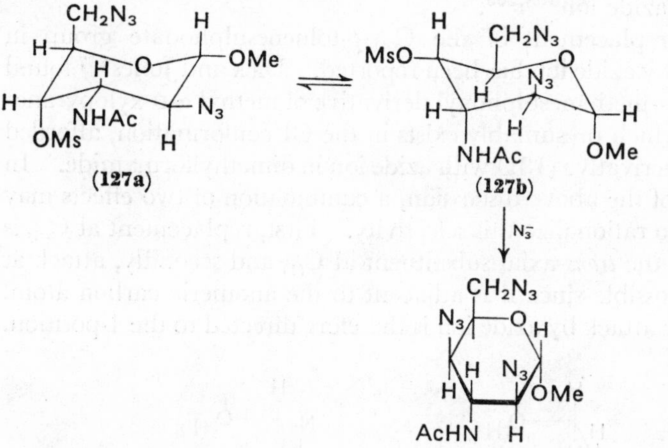

(127a) (127b) (63)

Replacement of secondary sulphonate esters which are adjacent to the anomeric carbon atom is not normally possible with azide ion owing to a combination of the electron-withdrawing effect of the acetal and unfavourable polar effects in the transition state[98]. The latter effect is best visualized by considering the Newman projections (**128**) and (**129**) along the $C_{(1)}$–$C_{(2)}$ bond for the S_N2 transition states of both α- and β-anomers. In the case of the α-anomer the developing dipoles of the transition state are almost parallel and opposed to the $C_{(1)}$–methoxyl and $C_{(1)}$-ring oxygen dipoles. The situation is similar, although not quite so marked, for the β-anomer. If neutral nucleophiles which develop a positive charge in the transition state (e.g.

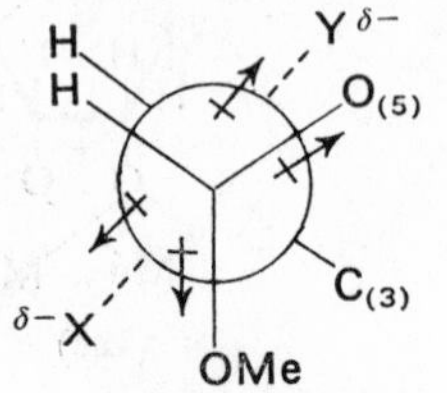

α-anomer
(128)

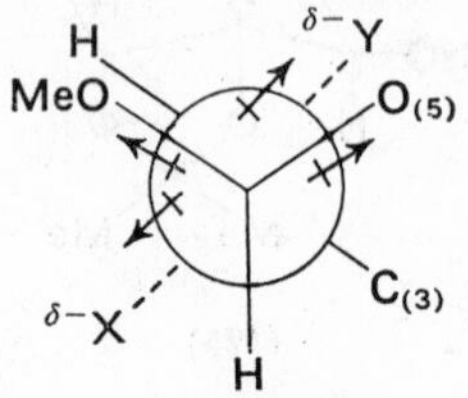

β-anomer
(129)

ammonia and hydrazine) are used, a reversal of polarity in one of the developing dipoles generates a dipolar attractive force[98]. This is generally true of substitution at all ring positions where dipolar interactions inhibit substitution by charged nucleophiles. Hence hydrazine or ammonia are often used to replace substituents which are not replaced by azide ion[103,266].

Selective replacement of the $C_{(4)}$–*p*-toluenesulphonate group in pyranosides by azide ion has been reported. Dick and Jones[267] found that the tri-*O*-methanesulphonyl derivative of methyl α-D-xylopyranoside (**130**), which presumably exists in the Cl conformation, afforded the 4-azido derivative (**131**) with azide ion in dimethylformamide. In the context of the above discussion, a combination of two effects may be invoked to rationalize this selectivity. First, replacement at $C_{(3)}$ is hindered by the *trans*-axial substituent at $C_{(1)}$ and secondly, attack at $C_{(2)}$ is not possible since it is adjacent to the anomeric carbon atom. Nucleophilic attack by azide ion is therefore directed to the 4-position.

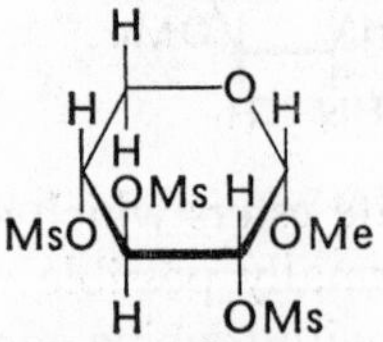

(130) 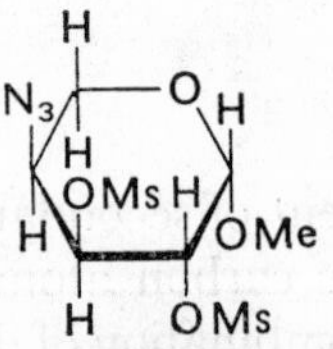 **(131)**

b. Ring opening of sugar epoxides, epimines and episulphides. Azidocarbohydrates possessing vicinal hydroxy, amino and thiol substituents are valuable intermediates in the synthesis of the respective amino- and diaminosugars, and may conveniently be prepared by ring opening reactions of epoxides, epimines and episulphides by azide ion. The first ring opening reaction of a sugar epoxide by azide ion was reported by Guthrie and Murphy[268] who converted methyl 2,3-anhydro-4,6-*O*-benzylidene-α-D-allopyranoside (**132**) into the 2-azidoaltroside (**133**) and the 3-azidoglucoside (**134**). Yields were low owing to decomposition which was presumably caused by liberated alkali. Addition of ammonium chloride, which removed hydroxide ions from the reaction,

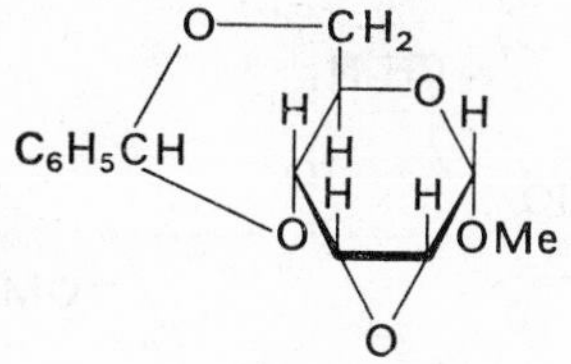

(**132**)

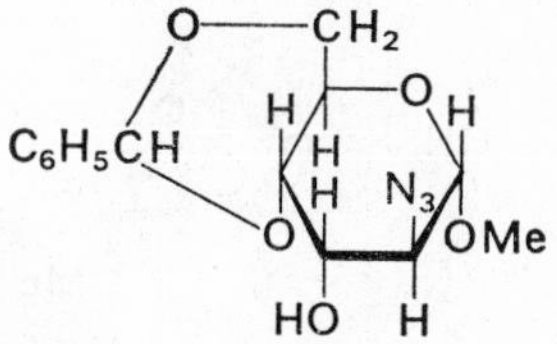

(**133**)

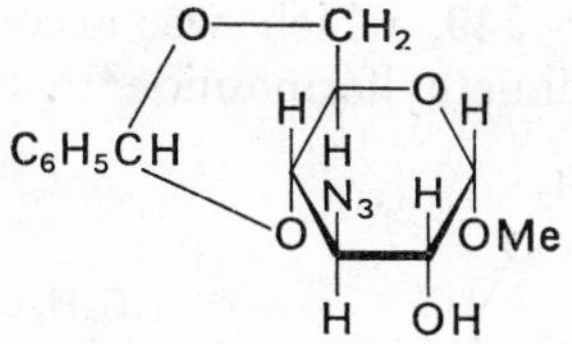

(**134**)

resulted in greatly increased yields and the proportion of **133** relative to **134**, 75% to 5%, is in accord with the normal predominance of diaxial ring opening of sugar epoxides with nucleophiles[221].

On the basis of the principle of diaxial scission of epoxides, sound configurational predictions may be made for systems such as **132** in which the conformation is locked by a fused ring. In less rigid systems, consideration must sometimes be given to reaction pathways involving all conformations of the ring. Hence Hanessian and Haskell[269] obtained methyl 3,6-diazido-3,6-dideoxy-α-D-idopyranoside (**135**) by reaction of methyl 2,3-anhydro-6-azido-6-deoxy-α-D-talopyranoside (**136**) with sodium azide in methyl cellosolve. A low yield of the galactoside (**137**) was also isolated. The predominant ido isomer (**135**) is thought to arise from diaxial ring opening of the less stable epoxide conformation (**136a**) since the alternative chair form (**136b**) would only give the ido isomer by the less favourable diequatorial ring opening process.

1,2-*trans*-Diaminosugars may similarly be synthesized by ring opening of epimines[270–274]. This reaction is exemplified by the

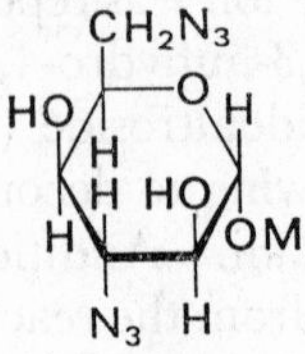

(135) 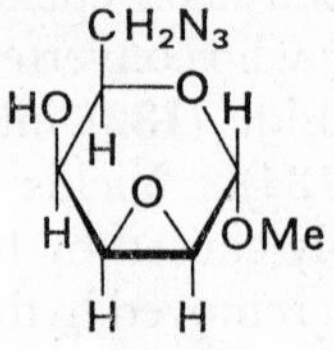 (136) 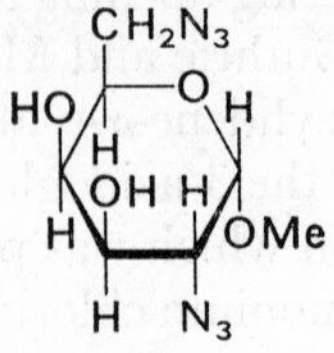 (137)

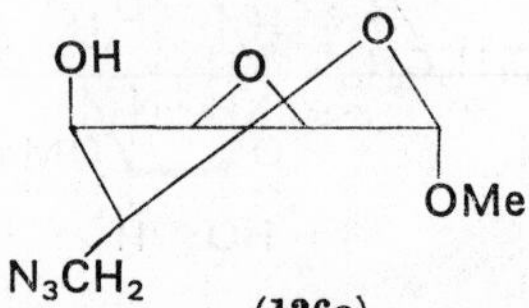

(136a) 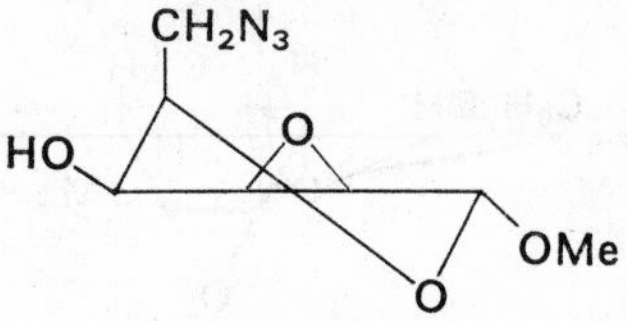 (136b)

conversion of **138** into **139**, which was accomplished by treatment with sodium azide in dimethylformamide[273].

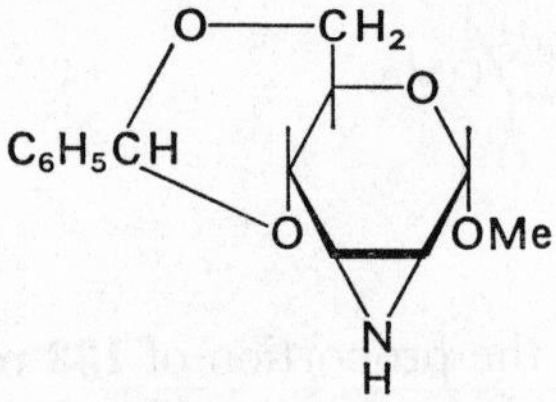

(138) 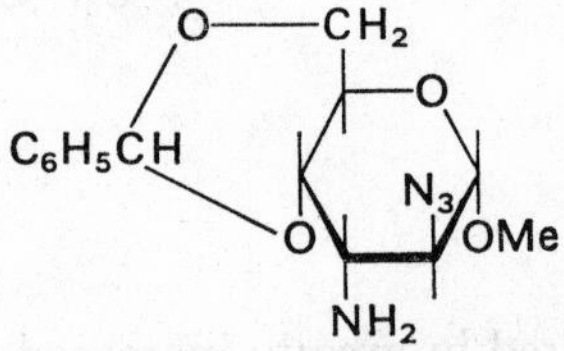 (139)

The intervention of episulphonium ion intermediates in the synthesis of azidosugars has been postulated by Christensen and Goodman[275] in order to rationalize the product (**140**), formed on interaction of azide ion with the furanoside (**141**). Direct replacement with inversion at $C_{(2)}$ did not occur; rather, a mixture of the diazides **140** and **142** was obtained by nucleophilic attack of azide ion on the episulphonium ion (**143**). A similar episulphonium intermediate (**144**) has been invoked[276] to rationalize the conversion of methyl 4,6-*O*-benzylidene-

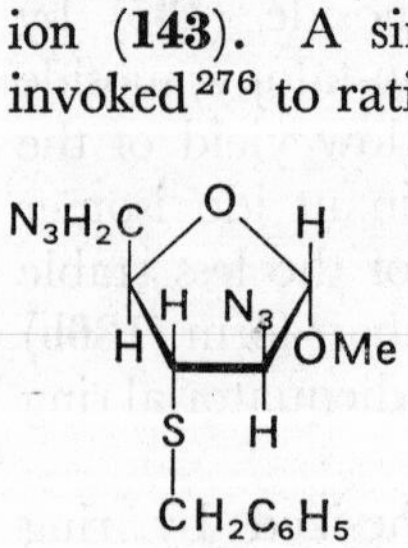

(140) 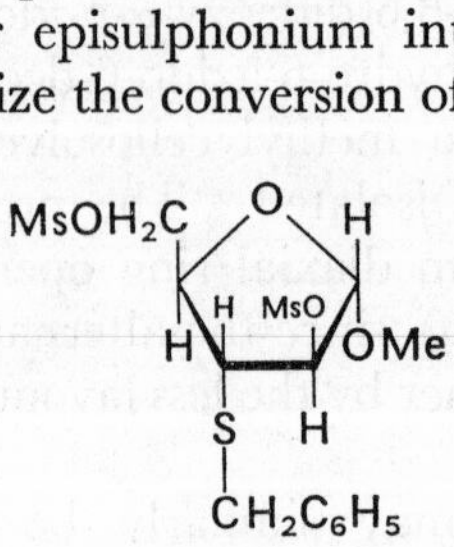 (141) 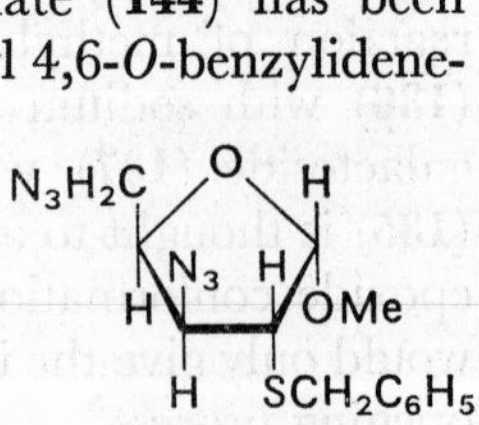 (142)

2-benzylthio-2-deoxy-3-*O*-*p*-toluenesulphonyl-α-D-altropyranoside (**145**) into the azide (**146**) in which the configuration at $C_{(3)}$ is retained.

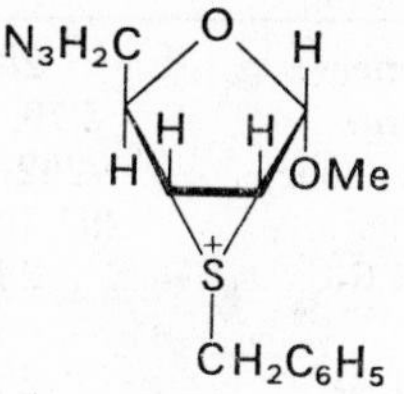

(**143**) 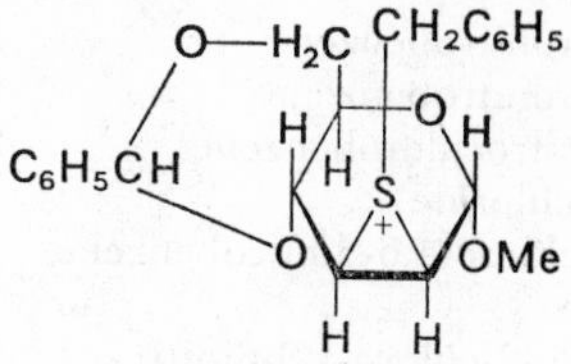 (**144**)

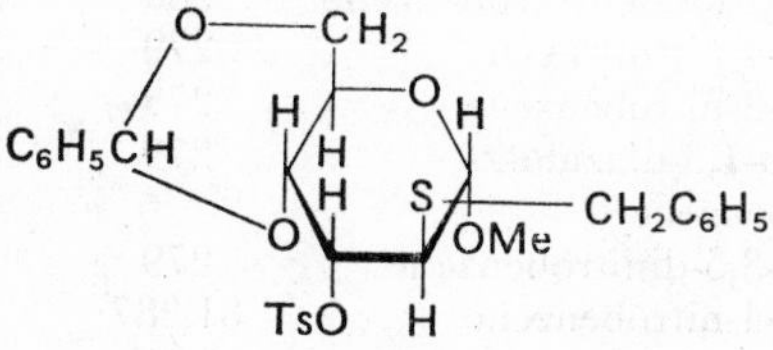

(**145**) 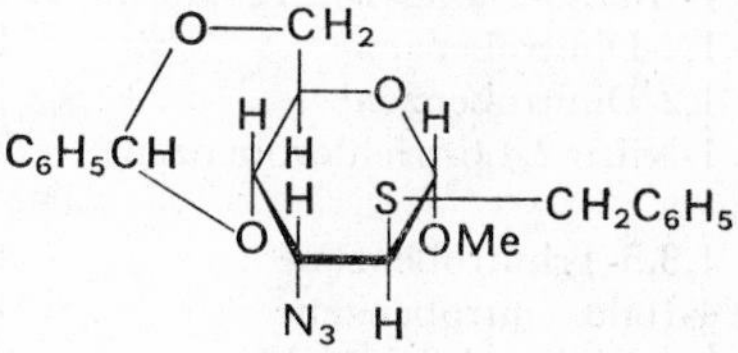 (**146**)

C. *Synthesis of Aromatic Azides by Nucleophilic Substitution*

I. Aryl azides

Aryl substrates containing suitable leaving and activating groups react readily with moderately strong nucleophiles such as azide ion and a number of aromatic azides have been prepared in this manner (Table 4). Such S_NAr reactions complement the synthesis of aromatic azides from diazonium compounds and are of particular value in forming azides of heteroaromatic systems in which diazotization procedures are unsatisfactory.

S_NAr reactions of azide ion proceed more readily in dipolar aprotic solvents, the use of which is thus preferred with substrates of low reactivity. The solvent differences are well illustrated by a comparison of rates and derived kinetic parameters for reaction of *p*-fluoro- and *p*-iodonitrobenzene in a range of solvents. This shows rate ratios of the order of 10^3–10^5 for reactions in dipolar aprotic solvents compared with methanol (section III.A.2).

As previously discussed in section III.A.1, azide ion undergoes addition–elimination S_NAr reactions with bond-formation being rate-limiting or sufficiently close to it to have equivalent results. At low temperatures the intermediate addition complex from the interaction of azide ion with 2,4,6-trinitroanisole has been detected by Caveng and Zollinger[48]. These workers observed characteristic p.m.r. resonances arising from the sigma complex (**147**) at −40° in acetonitrile and

TABLE 4. Synthesis of aryl azides by nucleophilic substitution

Substrate	Product	Reference
2-Chloronitrobenzene	2-Azidonitrobenzene	282
4-Chloronitrobenzene	4-Azidonitrobenzene	279, 282
2,4-Dinitrochlorobenzene	2,4-Dinitroazidobenzene	282,283
Picryl chloride	Picryl azide	282,284,285
1,2-Dichloro-4,6-dinitrobenzene	1-Azido-2-chloro-4,6-dinitrobenzene	286
5-Bromo-2-chloro-1,3-dinitrobenzene	2-Azido-5-bromo-1,3-dinitrobenzene	286
1-Amino-2-iodo-4-nitrobenzene	1-Azido-2-iodo-4-nitrobenzene	286
1,4-Dinitrobenzene	1-Azido-4-nitrobenzene	279
1,2-Dinitrobenzene	1-Azido-2-nitrobenzene	279
1-Nitro-2,4,6-trichlorobenzene	5-Chloro-1,3-diazido-2-nitrobenzene	279
1,3,5-Trinitrobenzene	1-Azido-3,5-dinitrobenzene	279
4-Iodoxynitrobenzene	1-Azido-4-nitrobenzene	81,287
2,4-Dinitro-1,3,5-trichlorobenzene	2,4-Dinitro-1,3,5-triazidobenzene	281

dimethylformamide. The activation energy for the formation of the complex was found to be 13·4 ± 1·0 kcal/mole; this figure is in agreement with that predicted from semi-empirical calculations[277] (13·5 kcal/mole), albeit in methanol as solvent. On treatment of picryl chloride with azide ion, the *gem*-diazido sigma complex (**148**) was observed rather than the monoazide (**149**). The former complex was also obtained directly from picryl azide and azide ion.

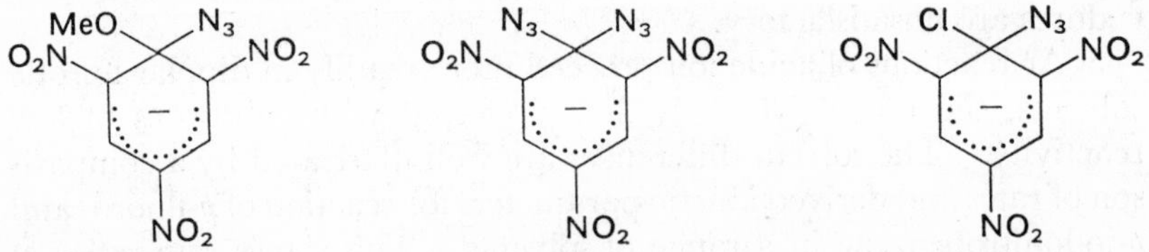

(**147**) (**148**) (**149**)

While S_N reactions of the anhydride group are common, S_N reactions of substrates in which the anhydride group activates an aromatic ring are rarely reported. It is of interest therefore that 3-azidophthalic acid and tetrazidophthalic acid have been obtained by reaction of azide ion with 3-nitro- and tetrachloro-phthalic anhydride respectively[278].

The displacement of the nitro group in pentachloronitrobenzene by azide ion has been reported, but unfortunately the yield of pentachloroazidobenzene was not quoted[279]. On the basis of the known activating power and leaving group mobility of chloro and nitro groups[152,280] one would anticipate that this is only a minor reaction compared with displacement of chlorine. Other experimental results also indicate the surprising nature of this report, for example the azidodechlorination occurring in reactions of 2,4,5- and 2,4,6-trichloronitrobenzene[279] and the formation of 1,3,5-triazido-2,4-dinitrobenzene from the corresponding trichloro compound[281]. The polysubstitution which occurs in these reactions is to be expected with excess of azide ion, since the azido group is weakly activating, though a little less so than chlorine[152] (see Chapter 4).

2. Heteroaromatic azides

As mentioned in the preceding section, synthesis of many classes of heteroaromatic azides may be achieved by nucleophilic displacement of a suitable leaving group by azide ion [e.g. 19,204,205, 288–295] and typical examples are shown in equations (64–68).

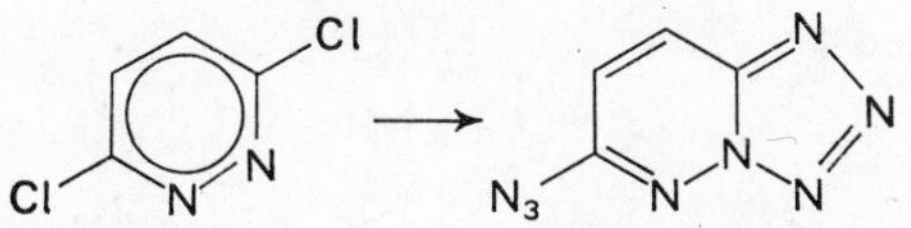

ref. 204 (64)

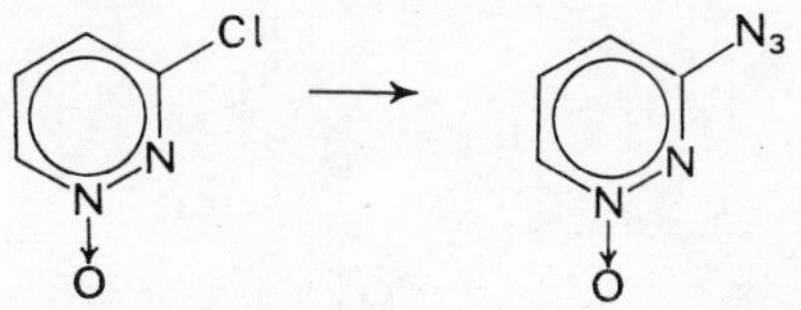

ref. 204 (65)

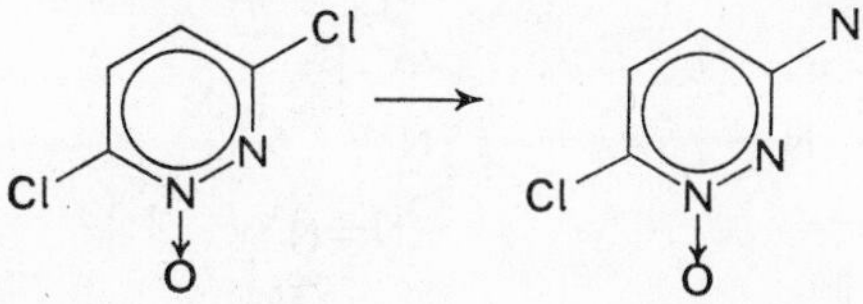

ref. 204 (66)

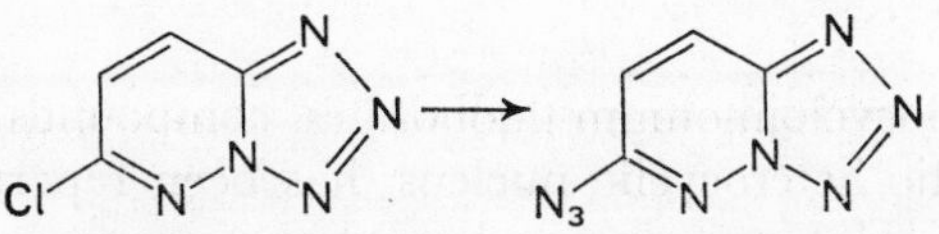

refs 205, 289 (67)

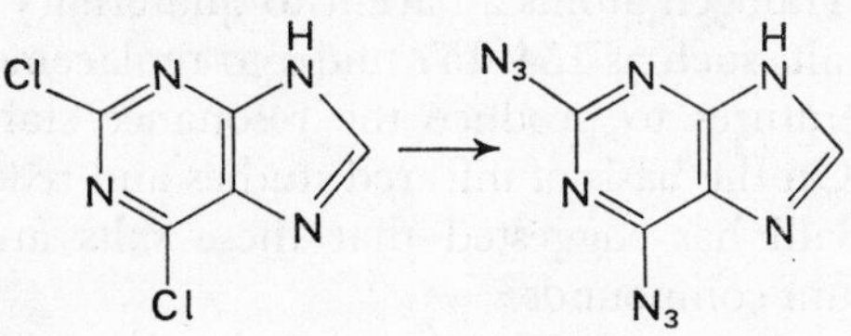

ref. 290 (68)

Compounds possessing the structural unit $N_3—\overset{|}{C}=N—$ may participate in the azidoazomethine–tetrazole equilibrium, which has been the subject of extensive investigation (section III.B.3), and the nature of this equilibrium process determines whether azides may be isolated. Interesting examples of azide syntheses which incorporate some features of this equilibrium have been reported by Stanovnik and his colleagues[205,288]. On treatment of the hydrazino derivative (**150**) with cyanogen bromide, a fused azolo ring is formed; concurrent opening of the tetrazole ring then affords the azide (**151**). It was suggested that destabilization of the tetrazole results from ring strain and that the bicyclic azide (**151**) is stabilized by electromeric electron-release from the azido group. Similarly, 6-azido-7-methyltetrazolo-[1,5-*b*]pyridazine (**152**) isomerizes into the thermodynamically more stable 6-azido-8-methyltetrazolo[1,5-*b*]pyridazine (**153**) on heating in dimethyl sulphoxide, the Arrhenius activation energy for this process being 20·5 kcal/mole.

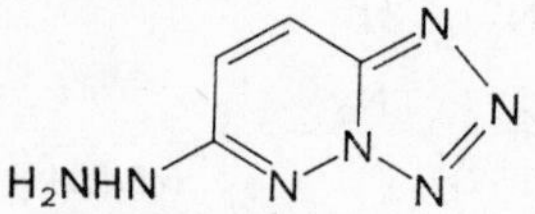

(**150**) 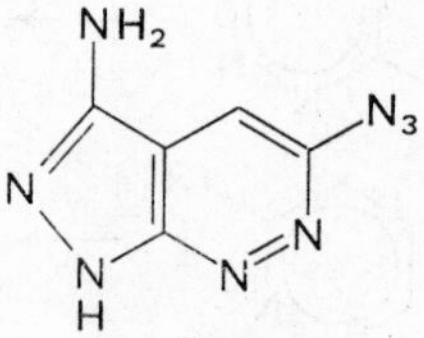 (**151**)

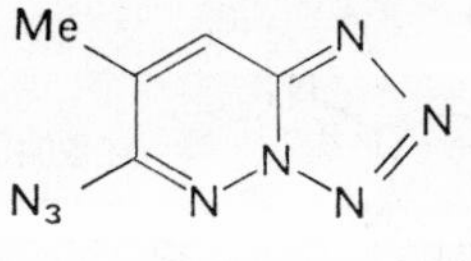

(**152**) 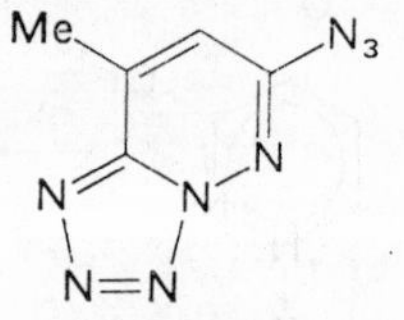 (**153**)

The synthesis of azidocycloimonium fluoborates, compounds which possess a quasi-aromatic heterocyclic nucleus, has been reported by Balli and Kersting[296]. Halogen atoms adjacent to quaternary nitrogens in heteroaromatic salts such as **154**–**157** undergo replacement by azide ion at low temperatures to produce the resonance stabilized azidinium salts (**158**). On the basis of infrared studies and reactivity towards nucleophiles, Balli has suggested that these salts are best considered as *N*-diazonium compounds[297].

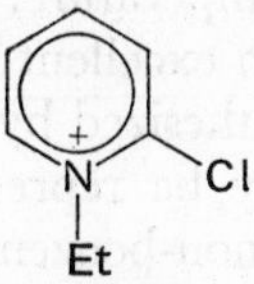

(154) 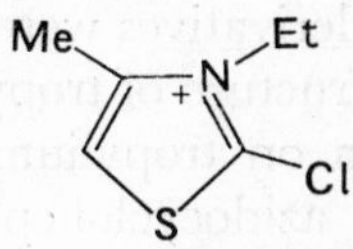 (155)

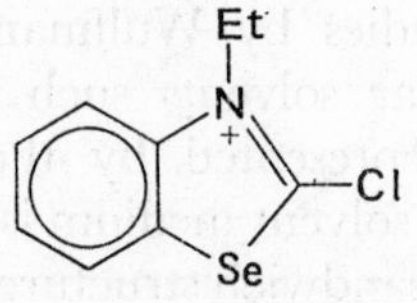

(156) 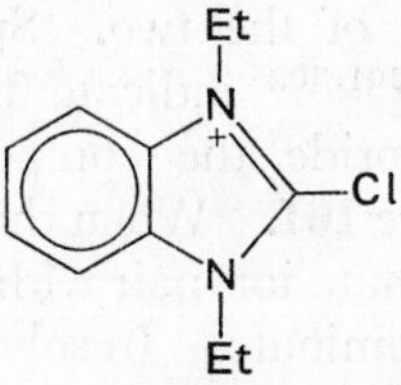 (157)

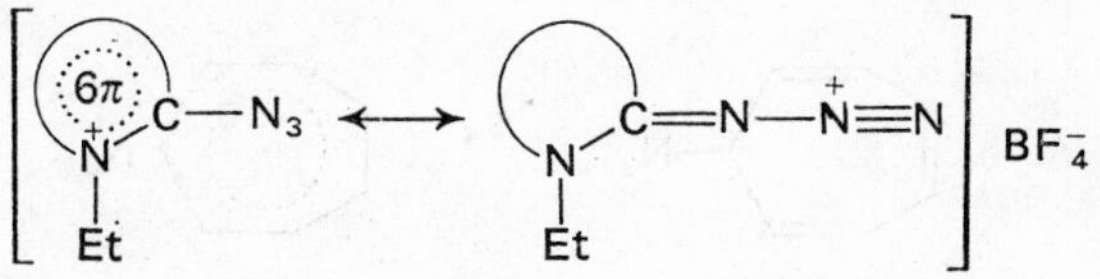

(158)

3. Non-benzenoid aromatic azides

Tropone may be represented structurally as a resonance hybrid (**159**) and the weight of evidence suggests **159b** as most nearly representing the true structure[298a,b]. The resemblance to heteroaromatic *N*-oxides is readily apparent. Correspondingly, just as suitable *N*-oxides, though readily susceptible to nucleophilic attack, are less reactive than quaternary heterocyclic systems, so suitable tropones are known to react with nucleophiles though with less ease than the corresponding tropylium compounds[298b]. Chlorotropones undergo substitution with azide ion but the facility of the reaction depends on the choice of solvent[299]. Thus in alcohol, 2-chlorotropone interacts only sluggishly with azide ion and the prolonged reaction time and elevated temperatures required lead to considerable decomposition of the product **160**. With dimethyl sulphoxide as solvent, however, the

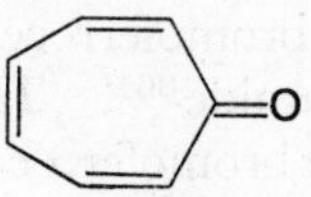

(159a) 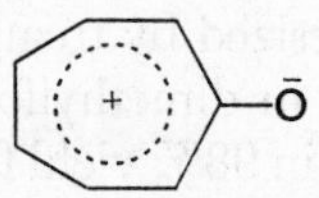 (159b) 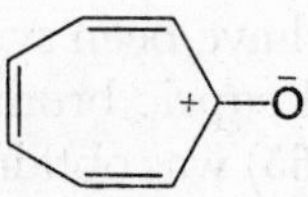 (159c) 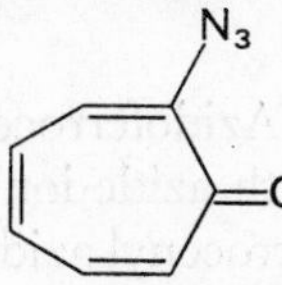 (160)

azide is formed quantitatively at ambient temperature. The 3- and 4-azido derivatives were similarly obtained in excellent yield[300].

The structure of tropyl azide, which is synthesized by the action of azide ion on tropylium perchlorate[301], may be represented as the covalent azidocycloheptatriene (**161**), the non-benzenoid aromatic tropylium salt with azide as the counter-ion (**162**), or an equilibrium mixture of the two. Spectroscopic studies by Wulfman and colleagues[301,302] indicate that in non-polar solvents such as carbon tetrachloride the compound is best represented by the covalent structure **161**. When the polarity of the solvent medium is increased, an intimate ion-pair with an open-faced sandwich structure is thought to predominate. In solvents of high ionizing power, such as aqueous ethanol, it has been shown that silver azide is precipitated on addition of silver nitrate.

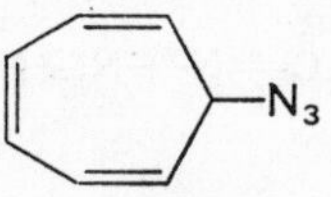

N_3 + $\bar{N}_3$

(**161**) 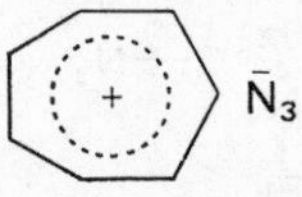   (**162**)

Interest in the chemistry of tropyl azide stems from the prediction[303] that a multiplicity of products might arise by unimolecular decomposition or self-reactions. Wulfman and Ward[303] have in fact presented spectroscopic evidence for the formation of the unique sandwich compound (**163**), resulting from the interaction of tropyl azide and the derived 1-aza-tricyclo [3,3,0,$0^{2,8}$] octa-3,6-diene (**164**).

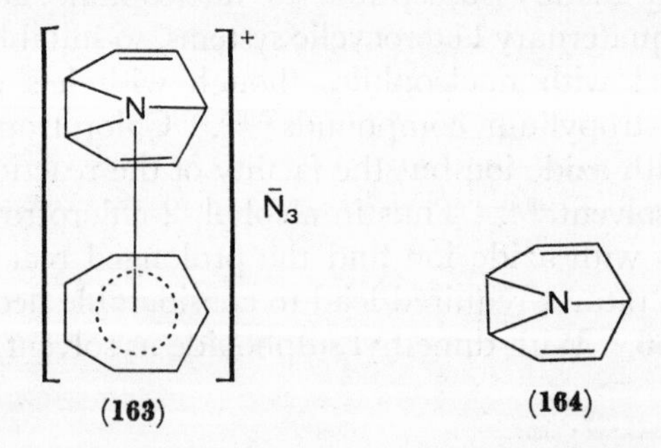

(**163**)   (**164**)

Azidoferrocenes have been synthesized by treating bromoferrocenes with azide ion and cupric bromide in dimethylformamide[304]. Thus ferrocenyl azide (**165**) was obtained in 98% yield from bromoferrocene and 1,1′-ferrocenylene diazide (**166**) was prepared from the corresponding dibromoferrocene. In contrast to ferrocene compounds, ferrocenium derivatives readily undergo nucleophilic substitution. It

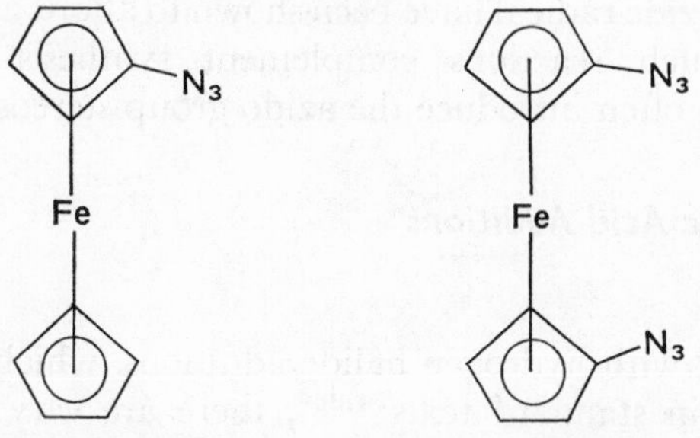

(165) (166)

seems probable therefore that the above reactions proceed by initial formation of a ferrocenium cation as shown below.

$$\text{FerrBr} + Cu^{2+} \longrightarrow [\text{FerrBr}]^{+} + Cu^{+}$$

$$[\text{FerrBr}]^{+} + N_3^{-} \longrightarrow [\text{FerrN}_3]^{+} + Br^{-}$$

$$[\text{FerrN}_3]^{+} + Cu^{+} \longrightarrow \text{FerrN}_3 + Cu^{2+}$$

IV. FORMATION OF AZIDES BY ADDITION REACTIONS

A. Introduction

The formation of organic azides by the addition of a preformed azide moiety to unsaturated systems seemed until recently to be unpromising as a synthetic procedure and consequently received scant attention. For example, simple alkenes were found to be inert towards hydrazoic acid in the conditions used by early workers[82,305], and additions to nitrile[306] and thione[307a–d] functions led to the formation of heterocyclic compounds rather than azides. Further, since formation of azides by nucleophilic substitution reactions had proved an adequate route to aliphatic azides, interest in addition reactions as preparative procedures languished.

Thus, although the reactions of quinones with hydrazoic acid (section IV.B.2) were established[308] in 1915, similar additions to a range of α,β-unsaturated carbonyl compounds were not demonstrated[17] until 1951. Since then, comprehensive studies have shown that other inorganic azides such as halogen and mercuric azides add to carbon–carbon double bonds, including those of simple alkenes, in mild conditions. More recently, trimethylsilyl azide and reagents which produce the azide radical have been shown to afford azides by addition processes. Such reactions complement synthesis by substitution reactions and often introduce the azido group stereospecifically.

B. *Hydrazoic Acid Additions*

1. General

In contrast with hydrogen halide additions, which have been well summarized in standard texts[54,309], there are very few mechanistic studies of hydrazoic acid additions. In some contexts it is convenient to regard the azido group and azide ion as pseudo-halogenoid and some resemblances between additions of hydrogen halide and of hydrazoic acid would be anticipated. This analogy has some value, but should not be carried too far for it should be borne in mind that hydrazoic acid is a much weaker acid than the hydrogen halides (except hydrogen fluoride which however is markedly different in other respects). The difference in acidity is probably greater in non-hydroxylic solvents in which addition reactions are often carried out. Further, hydrazoic acid can also react by 1,3-dipolar addition, a mode not available to the hydrogen halides.

2. Addition to olefins

a. Simple olefins. The addition of hydrazoic acid to simple alkenes and cycloalkenes was first reported in 1951 by Schaad[310] (equation 69), the reaction being carried out in an autoclave at 150° and catalysed by

$$CH_2{=}CHMe \xrightleftharpoons{H_3PO_4} Me{-}\overset{+}{C}H{-}Me \xrightleftharpoons{HN_3} Me{-}\underset{HN_3^+}{\underset{|}{C}H}{-}Me \xrightarrow{-H^+} Me{-}\underset{N_3}{\underset{|}{C}H}{-}Me \qquad (69)$$

phosphoric acid. The need for an acid catalyst suggests that hydrazoic acid is too weak to protonate an alkene (and thus act as an electrophilic reagent). It is not surprising, therefore, that the early attempts to carry out uncatalysed electrophilic additions were unsuccessful[305].

In these reactions it is likely that a carbonium ion, formed initially by protonation of the alkene by phosphoric acid, reacts with hydrazoic acid which acts as a nucleophile and the resulting triazenium ion subsequently loses a proton (cf. equation 69).

Hydrazoic acid adds readily to alkenes which are conjugated to powerful electron-withdrawing groups. Since such groups reduce the basicity of the carbon–carbon double bond, electrophilic attack by hydrazoic acid is more difficult, so that where reaction occurs it is likely to proceed by nucleophilic attack. The best known examples are the additions to α,β-unsaturated carbonyl compounds and these are discussed subsequently.

Formation of azides by nucleophilic attack on alkenes with non-conjugative electron-withdrawing groups is also known. It has been established for example, that *trans*-perfluoropropenyl azide (**167**) is formed by reaction of perfluoropropene (**168**) with trimethylammonium azide at 0°, and the intermediacy of a carbanion (**169**) has been postulated (Scheme 2)[311–313]. The formation of such carbanions as long-lived intermediates from monohydrofluorocarbons and base has been demonstrated by Andreades[314].

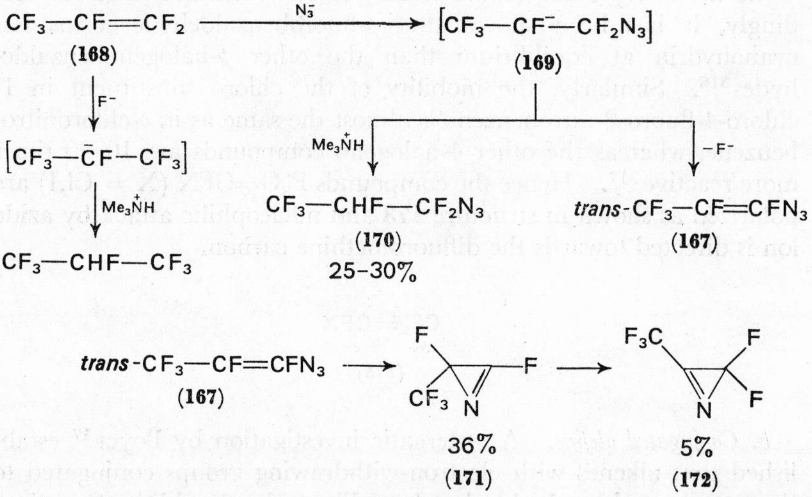

$$CF_3{-}CF{=}CF_2 \; (\mathbf{168}) \xrightarrow{N_3^-} [CF_3{-}\bar{C}F{-}CF_2N_3] \; (\mathbf{169})$$

$$(\mathbf{168}) \xrightarrow{F^-} [CF_3{-}\bar{C}F{-}CF_3] \xrightarrow{Me_3\overset{+}{N}H} CF_3{-}CHF{-}CF_3$$

$$(\mathbf{169}) \xrightarrow{Me_3\overset{+}{N}H} CF_3{-}CHF{-}CF_2N_3 \; (\mathbf{170}) \; 25\text{–}30\%$$

$$(\mathbf{169}) \xrightarrow{-F^-} trans\text{-}CF_3{-}CF{=}CFN_3 \; (\mathbf{167})$$

$$trans\text{-}CF_3{-}CF{=}CFN_3 \; (\mathbf{167}) \longrightarrow \mathbf{171}\;(36\%) \longrightarrow \mathbf{172}\;(5\%)$$

(**171**: azirine bearing F and CF_3 on the saturated carbon and F on C=N; **172**: azirine bearing F_3C on C=N and two F on the saturated carbon)

SCHEME 2

The azidoperfluoropropyl anion (**169**) undergoes further reaction by two pathways. In one, it reacts as a base to give the saturated azide (**170**) and the overall result is an Ad_N reaction. In the other it loses fluoride ion to form *trans*-propenyl azide (**167**), the overall result being nucleophilic substitution. By subsequent loss of nitrogen and cyclization, **167** forms an azirine (**171**) and an anionotropic shift converts this in part to the isomer (**172**). It is interesting that in an aprotic solvent, fluoride ion formed by elimination from **168** is a sufficiently powerful nucleophile to add to perfluoropropene[313].

The generality of the Ad_N reaction is evident from further work by Knunyants and Bykhovskaya[315] and representative examples are shown in equations (70–72). The orientation of products from

$$CF_2{=}CFCF_3 \longrightarrow N_3CF_2{-}CHFCF_3 \tag{70}$$

$$CF_2{=}CF_2 \longrightarrow N_3CF_2{-}CHF_2 \tag{71}$$

$$CF_2{=}CFHal \longrightarrow N_3CF_2{-}CHFHal \quad (Hal = Cl, I) \tag{72}$$

$CF_2 = CFHal$ is worthy of comment. Both the $-I$ and $+M$ effects of the halogens are in the order F > Cl > Br > I. Whereas for fluorine the two opposing effects are approximately equal, the $-I$ effect is greater in the case of the other halogens. As a result the total electronic effect of fluorine is about the same as that of hydrogen, while the heavy halogens are weakly electron-withdrawing. Accordingly, it has been shown that *p*-fluorobenzaldehyde forms less cyanohydrin at equilibrium than the other *p*-halogenobenzaldehydes[316]. Similarly, the mobility of the chloro substituent in 1-chloro-4-fluoro-2-nitrobenzene is almost the same as in *o*-chloronitrobenzene, whereas the other 4-halogeno compounds are 10–20 times more reactive[317]. Hence the compounds $F_2C{=}CFX$ (X = Cl,I) are polarized as shown in structure **173** and nucleophilic attack by azide ion is directed towards the difluoromethine carbon.

$$\underset{\delta+}{CF_2}{=}\underset{\delta-}{CFX}$$

(173)

b. Conjugated olefins. A systematic investigation by Boyer[17] established that alkenes with electron-withdrawing groups conjugated to the carbon–carbon double bond readily undergo addition reactions with hydrazoic acid. Carbonyl, carboxyl, alkoxycarbonyl, nitrile, azomethine and nitro groups were shown to activate olefins to such addition and typical examples are given in Table 5.

In view of the very low reactivity of simple alkenes towards hydrazoic acid, and since these electron-withdrawing groups render the homopolar double bond less basic, it is unlikely that the reactions are Ad_E in type. By comparison with hydrogen halide additions to such

TABLE 5. Synthesis of azides by hydrazoic acid additions to conjugated olefins

Compound	Product	Conditions	Yield (%)	Reference
$CH_2{=}CHCHO$	$N_3CH_2CH_2CHO$	A	71	17
		A	61·5	318
$CH_2{=}CHCOOMe$	$N_3CH_2CH_2COOMe$	B	35·4	17
$CH_2{=}CHCOOH$	$N_3CH_2CH_2COOH$	B	24	17
$(Me)_2C{=}CHCOMe$	$(Me)_2C(N_3)CH_2COMe$	C	38	17
		D	61	318
$CH_2{=}CHCOMe$	$N_3CH_2CH_2COMe$	A	70·6	118, 318
$CH_2{=}CHCOEt$	$N_3CH_2CH_2COEt$	A	75·3	318
$CH_2{=}C(Me)COMe$	$N_3CH_2CH(Me)COMe$	A	56·7	318
$MeCOCH{=}CH$— $C_6H_4NO_2$-*p*	$MeCOCH_2CH(N_3)$— $C_6H_4NO_2$-*p*	E	90	319
$C_6H_5CH{=}CHNO_2$	$C_6H_5CH(N_3)CH_2NO_2$	A	69	17
$CH_2{=}CHCN$	$N_3CH_2CH_2CN$	B	17	17
(pyridine ring, N) $CH{=}CH_2$	(pyridine ring, N) $CH_2CH_2N_3$	C	49·5	17
(maleimide ring: CO, NAr, CO)	(succinimide ring with N_3: CO, NAr, CO)	C	97	320
$Ar{=}C_6H_5$		F	80	149
p-$MeOC_6H_4$		F	80	149
p-ClC_6H_4		F	70	149

A: Ambient temperature in aqueous acetic acid for 1–4 hours
B: Ambient temperature in aqueous acetic acid for 1–3 days
C: Steam bath temperature in aqueous acetic acid for 1 day
D: Ambient temperature in trichloracetic acid–acetic acid–chloroform for 18 hours
E: Ambient temperature in trichloracetic acid–acetic acid–chloroform for 7 days
F: 70–80° in aqueous acetic acid for 30–45 minutes

substrates, an Ad_N mechanism appears more probable, although there are insufficient data to regard any mechanism as proven.

Although hydrazoic acid is a weak acid and is probably unable to protonate the hetero atom of the activating groups listed, hydrogen bonding may occur and this would result in mild catalysis of the Ad_N process. This is illustrated for α,β-unsaturated carbonyl compounds in equation (73). The examples in Table 5 include reactions in

$$R-\overset{|}{C}=\overset{|}{C}-\overset{|}{C}=O \ldots H-N_3 \;\underset{}{\overset{N_3^-}{\rightleftharpoons}}\; R-\underset{N_3}{\overset{|}{C}}-\overset{|}{C}=\overset{|}{C}-O^- \ldots H-N_3 \xrightarrow{HN_3} R-\underset{N_3}{\overset{|}{C}}-\underset{H}{\overset{|}{C}}-\overset{|}{C}=O \ldots H-N_3 \quad (73)$$

media containing trichloracetic acid, and these may be cases with genuine acid catalysis. The use of strongly acidic conditions is, however, unsuitable since the protonated azide may decompose by Schmidt type reactions[318,319] as shown in equation (74) (cf. section III.B.1).

$$-\ddot{N}=\overset{+}{N}=\ddot{N} \xrightarrow{H-X} -\underset{..}{\overset{H}{\overset{|}{N}}}-\overset{+}{N}\equiv N \;\; X^- \longrightarrow -\underset{..}{\overset{H}{\overset{|}{N}}}{}^+ + N_2 + X^- \longrightarrow \text{products} \quad (74)$$

It is noteworthy that compounds such as cinnamaldehyde and cinnamic acid derivatives, in which the conjugation is extended into a benzene ring, do not add hydrazoic acid even in abnormally vigorous conditions[17,321,322]. It has been suggested that the extension of conjugation causes the β-carbon atom to be less electrophilic and therefore of diminished reactivity[322].

The difference is interestingly exemplified in the differing modes of reaction of hydrazoic acid with some alkylidene (and arylidene) oxazolone derivatives[321,322]. The alkylidene compound **174** behaves as an α,β-unsaturated carbonyl compound with the hydrazoic acid adding to the exocyclic double bond. A subsequent ring opening with further addition leads to the formation of the diazide **175** (equation 75). In contrast the exocyclic double bond of the arylidene compound **176** is unreactive, and nucleophilic attack occurs at the electrophilic oximino carbon with ring opening. Recyclization leads to the formation of a tetrazole derivative (**177**) as shown in equation (76).

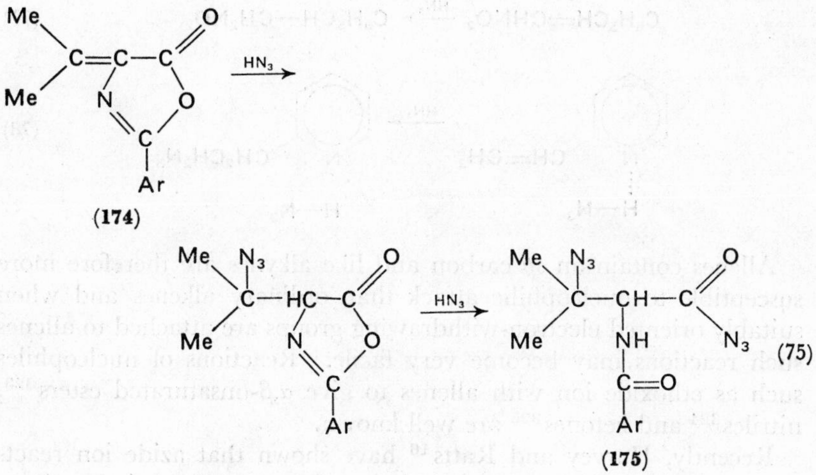

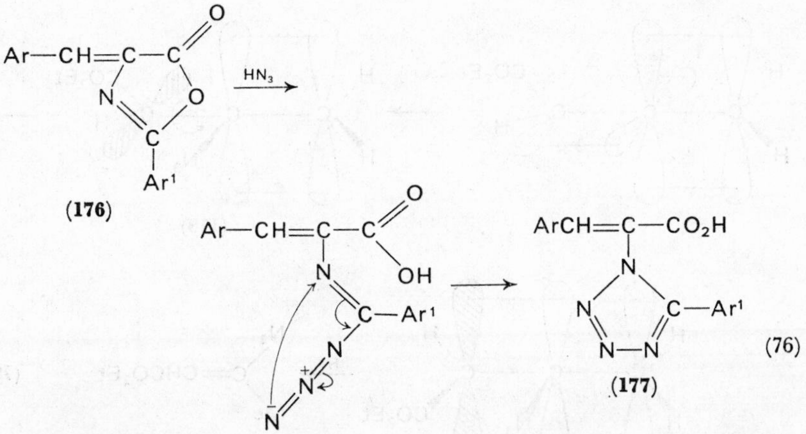

Me, Me C=C—C(=O) N O C Ar (**174**) $\xrightarrow{HN_3}$ Me, Me (N$_3$) C—HC—C(=O) N O C Ar $\xrightarrow{HN_3}$ Me, Me (N$_3$) C—CH—C(=O)N$_3$ NH C=O Ar (**175**) (75)

Ar—CH=C—C(=O) N O C Ar^1 (**176**) $\xrightarrow{HN_3}$ Ar—CH=C—C(=O)OH N=C—Ar^1 N=N$^+$=N$^-$ ⟶ ArCH=C—CO$_2$H, N (tetrazole) C—Ar^1 N—N (**177**) (76)

A combination of inductive and conjugative electron-withdrawal seems to be particularly effective in activating the α,β-double bond. Whereas cinnamaldehyde and cinnamic acid derivatives do not react with hydrazoic acid, nitrostyrene[17] and vinylpyridine[17,320] are reactive (equations 77, 78). The reactions of the vinylpyridines may involve hydrogen-bonding catalysis or even the formation of pyridinium species on protonation by hydrazoic acid.

$$C_6H_5CH{=}CHNO_2 \xrightarrow{HN_3} C_6H_5CH(N_3){-}CH_2NO_2 \qquad (77)$$

$$\text{2-}(CH{=}CH_2)\text{pyridine}\cdots H{-}N_3 \xrightarrow{HN_3} \text{2-}(CH_2CH_2N_3)\text{pyridine}\cdots H{-}N_3 \qquad (78)$$

Allenes contain an *sp* carbon and like alkynes are therefore more susceptible to nucleophilic attack than ordinary alkenes and when suitably oriented electron-withdrawing groups are attached to allenes such reactions may become very facile. Reactions of nucleophiles such as ethoxide ion with allenes to give α,β-unsaturated esters[323], nitriles[324] and ketones[325] are well known.

Recently, Harvey and Ratts[16] have shown that azide ion reacts similarly with allenic esters to give azidocrotonates and they have

(178)

$$\xrightarrow{H^+} (N_3)(Me)C{=}CHCO_2Et \qquad (79)$$

discussed the mechanism in some detail. It was suggested that an anionic intermediate (**178**) is formed by addition of azide ion (equation 79). However, this intermediate does not cyclize to a triazole because of a 90° rotation about the 2,3-bond which enables full π-orbital overlap. The *p*-orbital electrons of the carbanion are then orthogonal to the azide moiety and cyclization is precluded. The final addition of a proton takes place on the 1- rather than 3-carbon atom because conjugation then extends throughout the chain.

These authors contrasted this reaction with the behaviour of alkynes, which usually react with hydrazoic acid to form triazoles[306,326–329]. They suggested that an azidovinyl anion (**179**), formed by attack of azide ion on the alkyne, is locked in a configuration in which the non-bonded *p*-orbital electrons of the carbanion are in the same plane as the azido group (equation 80). This would constitute a favourable situation for ring closure to a triazole, provided the azido group is in the weakly-excited, bent configuration[9,208]. This argument implies that in the reaction of alkynes with hydrazoic acid, the rate-limiting

$$R{-}C{\equiv}C{-}R \xrightarrow{N_3^-} R{-}\overset{-}{C}{=}C(N_3){-}R \;(\mathbf{179}) \xrightarrow{H^+} \text{4,5-}R_2\text{-triazole (HN, N, N)} \qquad (80)$$

step is the addition of azide ion. While such a reaction may be possible in some cases, this mechanism also requires that cyclization proceeds in preference to protonation of the intermediate carbanion and this seems unlikely. A more probable mechanism is a synchronous 1,3-dipolar cycloaddition, and such reactions of covalent azides are well documented[4].

The addition of hydrazoic acid to quinones follows the general pattern for addition to α,β-unsaturated carbonyl compounds as previously discussed, with necessary modification to allow for the special interrelationships between quinonoid and benzenoid compounds. The product from reaction of *p*-benzoquinone and hydrazoic acid in benzene is 2-azido-1,4-benzohydroquinone, which arises by addition of the reagent followed by enolization[308,330]. When oxidation of the product is permitted, further addition to the resulting azidoquinone may occur[331,332]. Hence, with excess sodium azide in acetic acid, *p*-benzoquinone affords 2,5-diazido-1,4-benzohydroquinone[334] (equation 81).

Toluquinone and *ar*-tetrahydro-α-naphthoquinone undergo similar addition reactions to benzoquinone, but α- and β-naphthoquinone

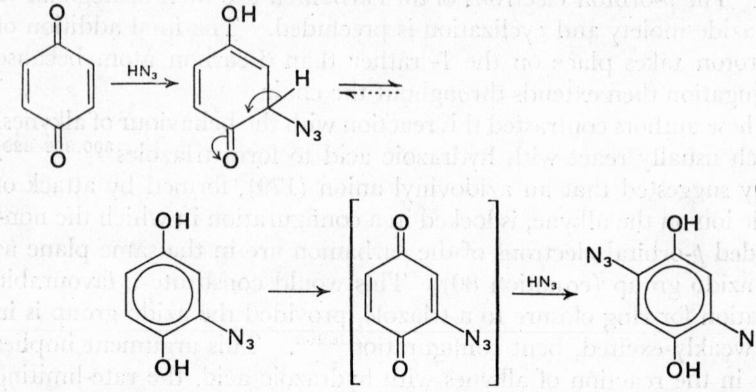

(81)

afford aminonaphthoquinones on treatment with hydrazoic acid[333]. The difference may be due to the lesser tendency to enolization in the bicyclic series and in consequence the elimination of nitrogen from

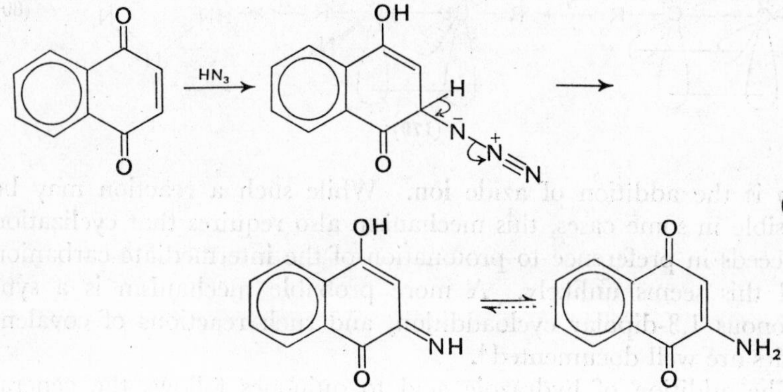

(82)

the azide adduct becomes preferable. A prototropic shift followed by rearrangement of the resulting quinone-imine would then afford the aminoquinone (equation 82).

Quinoneimides add hydrazoic acid in a similar manner to quinones[332,334]. The equilibrium in the amide–imide prototropic shift is more one-sided than that of the keto–enol system and thus, even in the naphthalene series, reaction with hydrazoic acid affords an azide (equation 83).

Because of the weakness of hydrazoic acid as a potential electrophilic reagent one would expect electrophilic addition of hydrazoic acid only in systems in which a powerful electron-releasing group is conjugated with the double bond. An interesting example of such an

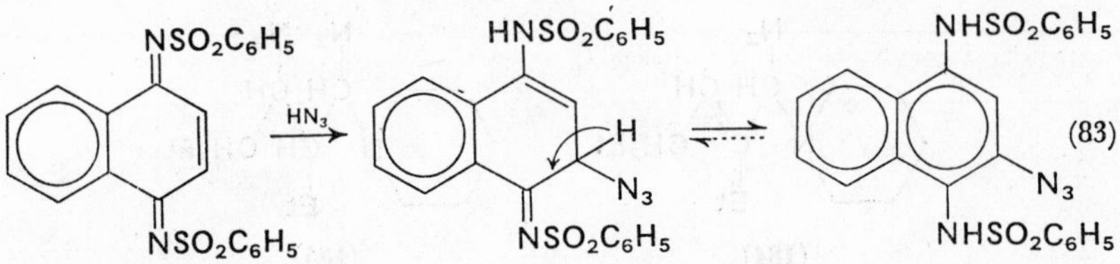

NSO$_2$C$_6$H$_5$ HNSO$_2$C$_6$H$_5$ NHSO$_2$C$_6$H$_5$

HN$_3$ H N$_3$ N$_3$ (83)

NSO$_2$C$_6$H$_5$ NSO$_2$C$_6$H$_5$ NHSO$_2$C$_6$H$_5$

Ad$_E$ reaction is the formation of 1-ferrocenylethyl azide by reaction of hydrazoic acid with vinylferrocene[123] (equation 84). Protonation of the β-sidechain carbon atom is considered to be rate-limiting and to afford the unusually stable carbonium ion intermediate (**180**).

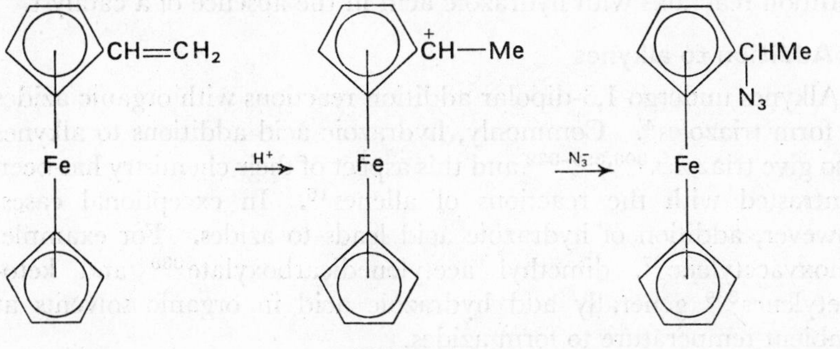

CH=CH$_2$ $\overset{+}{C}$H—Me CHMe N$_3$

Fe $\xrightarrow{H^+}$ Fe $\xrightarrow{N_3^-}$ Fe (84)

(180)

Unstable azido compounds which have been obtained by addition of hydrazoic acid to enamines such as **181** are also thought to arise from an Ad$_E$ process[335–337]. Although catalysis of an Ad$_N$ reaction by protonation of the heterocyclic nitrogen atom in **181** is possible, and in fact the salt (**182**) was isolated from the reaction, the position of the azido group in the products (**183**–**185**) indicates that an electrophilic addition on the unprotonated enamine is the rate-limiting step. Conjugation of the nitrogen lone-pair with the diene chain could provide sufficient activation for such an addition.

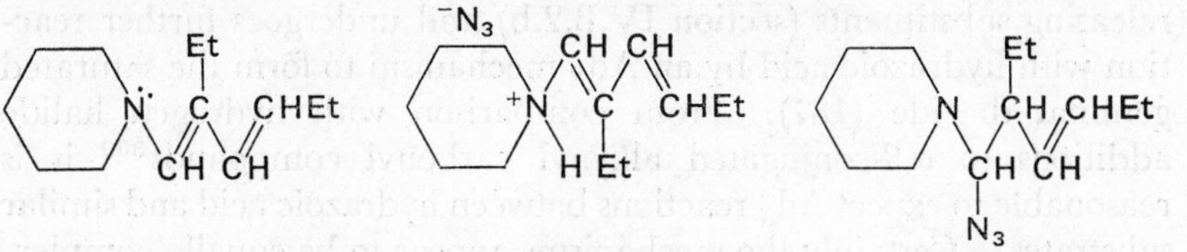

Et $^-$N$_3$ Et

N C CHEt CH CH N CH CHEt

CH CH $^+$N C CHEt CH CH

H Et N$_3$

(181) **(182)** **(183)**

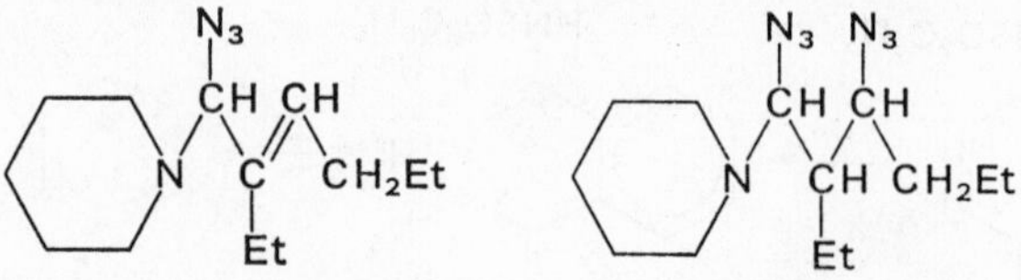

(184) (185)

The pattern of addition in polyenes in which the double bonds are neither contiguous nor conjugated does not require special comment. Compounds with conjugated double bonds are somewhat more susceptible to electrophilic addition than simple alkenes but nevertheless there appear to be no literature reports of such compounds undergoing addition reactions with hydrazoic acid in the absence of a catalyst.

3. Addition to alkynes

Alkynes undergo 1,3-dipolar addition reactions with organic azides to form triazoles[4]. Commonly, hydrazoic acid additions to alkynes also give triazoles[306,329–332] and this aspect of their chemistry has been contrasted with the reactions of allenes[16]. In exceptional cases, however, addition of hydrazoic acid leads to azides. For example, ethoxyacetylene[27], dimethyl acetylenedicarboxylate[338] and keto-acetylenes[339] generally add hydrazoic acid in organic solvents at ambient temperature to form azides.

Ethoxyacetylene possesses a strong electron-releasing substituent α- to the triple bond and as a result hydrazoic acid is able to react by an Ad_E mechanism (equation 85). The intermediate monoazide

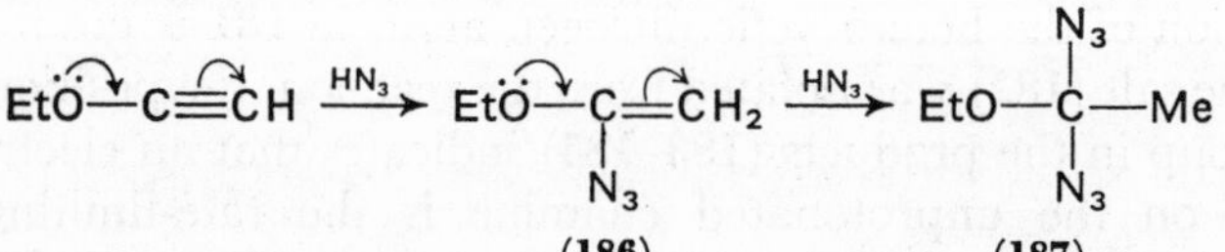

$$\text{EtO—C}\equiv\text{CH} \xrightarrow{HN_3} \text{EtO—C(N}_3\text{)=CH}_2 \xrightarrow{HN_3} \text{EtO—C(N}_3\text{)}_2\text{—Me} \qquad (85)$$

(186) (187)

(**186**) is another example of an alkene with conjugated electron-releasing substituents (section IV.B.2.b) and undergoes further reaction with hydrazoic acid by an Ad_E mechanism to form the saturated geminal diazide (**187**). From comparison with hydrogen halide additions to α,β-conjugated alkynyl carbonyl compounds[309] it is reasonable to expect Ad_N reactions between hydrazoic acid and similar substrates. Certainly the mechanisms appear to be equally complex. For example, in the reaction between hydrazoic acid and dimethyl acetylenedicarboxylate, Ostroverkhov and Shilov[338] have shown that in 90% acetic acid two moles of hydrazoic acid are required for each mole of substrate and in 90% methanol no simple kinetic relationship could be obtained. The additional hydrazoic acid molecules could be involved either in hydrogen bonding with the carbonyl groups of the substrate or in the dissociation of hydrazoic acid.

Electron-withdrawal by the carbonyl groups cannot be the only factor determining the nature of the product (**188**) since a triazole is

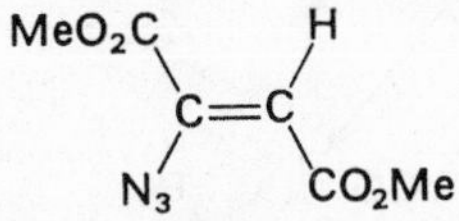

$$\underset{(188)}{(MeO_2C)(N_3)C{=}C(H)(CO_2Me)} \qquad \underset{(189)}{C_6H_5{-}C{\equiv}CCHO}$$

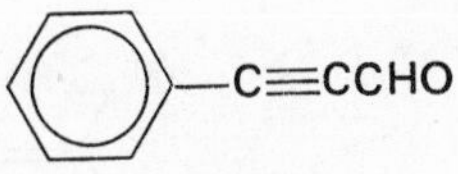

obtained from the corresponding diacid[330] (equation 86). For dimethyl acetylenedicarboxylate, the transition state for triazole formation would require the bulky carbomethoxy groups to adopt the sterically unfavourable *cis* configuration. In the case of nucleophilic addition to form methyl azidofumarate (**188**), however, the less hindered *trans* configuration of the ester groups is possible in the transition state. 1,3-Cycloaddition to acetylenedicarboxylic acid would involve a transition state with a favourable hydrogen bonding component which may offset steric compression (equation 86).

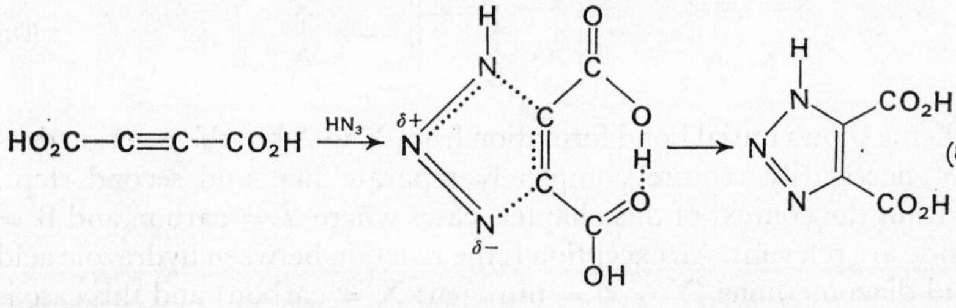

$$HO_2C{-}C{\equiv}C{-}CO_2H \xrightarrow{HN_3} [\text{transition state: } H{-}N{\cdots}N^{\delta+}{=}N^{\delta-}\text{, hydrogen-bonded } C(=O){-}O{-}H{\cdots}O{=}C(OH)] \longrightarrow \text{1H-1,2,3-triazole-4,5-}(CO_2H)_2 \qquad (86)$$

Addition of hydrazoic acid to carbonylacetylenes may also be envisaged as proceeding by an Ad_N reaction augmented by hydrogen bonding catalysis. This only applies for compounds in which the carbonyl group is of relatively high basicity. Thus propiolaldehyde (**189**) gives a triazole whereas arylketoacetylenes (**190**), which are stronger oxygen bases, can form azides[339]. In the latter cases the final product is determined by steric effects. When R is an aryl group the azido and ketoaryl substituents adopt a *cis* configuration about the double bond and further reaction affords an oxazole (**191**). If R is a hydrogen atom, both *cis* and *trans* configurations are possible and a mixture of the oxazole and *trans* azide (**192**) is formed (equation 87).

$$\text{ArCOC}{\equiv}\text{CR}\ (\mathbf{190}) \longrightarrow \text{ArCOC(H)}{=}\text{C(R)}{-}\bar{\text{N}}{-}\overset{+}{\text{N}}_2 \xrightarrow{\text{R = Ar}'} \text{oxazole } (\mathbf{191});\quad \xrightarrow{\text{R = H}} (\mathbf{191}) + \text{ArCO(H)C}{=}\text{C(R)N}_3\ (\mathbf{192}) \qquad (87)$$

4. Addition to compounds with contiguous double bonds

Compounds containing functional groups which can be generally represented as X=Y=Z̈ may react by electrophilic addition with suitable species A–B, where A is the electrophilic and B the nucleophilic centre. A general representation is shown in equation (88). This general

$$\text{X}{=}\text{Y}{=}\ddot{\text{Z}} + \text{A}{-}\text{B} \longrightarrow \left[\underset{\text{A}}{\text{X}}{-}\overset{\uparrow \bar{\text{B}}}{\text{Y}}{\equiv}\overset{+}{\text{Z}}\right] \longrightarrow \underset{\text{A}}{\text{X}}{-}\underset{\text{B}}{\text{Y}}{=}\text{Z} \qquad (88)$$

scheme shows initial bond formation from X to A but the reaction does not necessarily require completely separate first and second steps. Within the context of this chapter, cases where Y = carbon and B = azide are relevant. An exception is the reaction between hydrazoic acid and diazomethane (Y = Z = nitrogen; X = carbon) and this case is considered in section V.A.3.

The formation of carbamoyl azides from isocyanates is the best known application of this method of synthesis. The reaction was discovered in 1901 by Hantzsch and Vagt who prepared carbamoyl azide itself from isocyanic acid and hydrazoic acid[340]. Alkyl and aryl carbamoyl azides are usually obtained in good yields from low temperature reactions between hydrazoic acid and the corresponding isocyanate in non-aqueous media[12,194,307a,341–344] (equation 89)

$$RCNO + HN_3 \longrightarrow RNHCON_3 \quad (89)$$

Ketene also appears to react with hydrazoic acid according to the general mechanism outlined in equation (88). Carbamoyl azides are produced[345] and it is thought that the reaction proceeds by initial protonation and subsequent attack by azide ion to afford an acyl azide; further reaction of the derived isocyanate would then lead to the observed product (equation 90).

$$CR_2{=}C{=}O \xrightarrow{H^+} R_2CHC{\equiv}\overset{+}{O} \xrightarrow{HN_3}$$

$$R_2CHCON_3 \longrightarrow R_2CHNCO \longrightarrow R_2CHNHCON_3 \quad (90)$$

Additions of hydrazoic acid to carbon–sulphur double bonds may conceivably be of the general type outlined in equation (88) but do not constitute a synthetic route to azides. Recent corrections to the earlier literature relating to this field are, however, noteworthy. It was originally suggested that thiocarbamoyl azides[307] (**193**) were obtained from organic isothiocyanates and hydrazoic acid. The spectroscopic studies of Lieber and co-workers[346,347] have now established that the products are in fact thiatriazoles (**194**). The reactions of sodium azide with isothiocyanates[347], and carbon disulphide[348], which were also previously considered to furnish organic azides, have now been shown to produce the heterocyclic compounds **195** and **196** respectively.

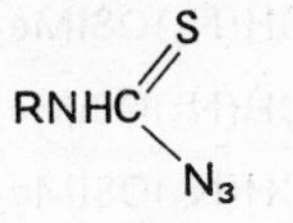

RNHC(=S)N$_3$

(193)

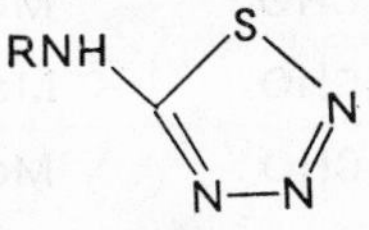

RNH–(thiatriazole: S, N, N–N)

(194)

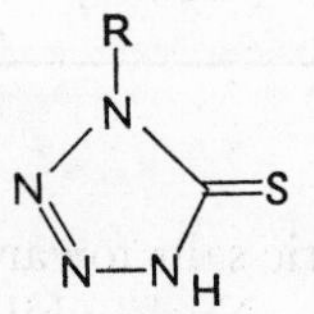

R–N, N, N–NH, C=S

(195)

S, N, N–NH, C=S

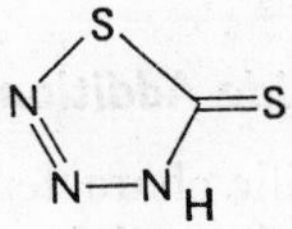

(196)

Numerous other systems (e.g. carbodiimides, nitrile oxides, carbon suboxide, etc.) could be of interest as substrates for such additions but these have not been examined experimentally.

C. Addition of Trimethylsilyl Azide to Carbonyl Compounds

Birkofer and co-workers[349] have recently reported that aldehydes react exothermally with trimethylsilyl azide, in the presence of the Lewis acid catalyst zinc chloride, to form α-trimethylsiloxyalkyl azides in good yield (equation 91). The main features of this reaction appear

$$\begin{array}{c} R{-}CH{=}O \\ + \\ ZnCl_2 \end{array} \rightleftharpoons \begin{array}{c} R{-}CH{=}\overset{+}{O}{-}\bar{Z}nCl_2 \\ \ddot{N}_3{-}SiR_3 \end{array} \longrightarrow \begin{array}{c} R{-}\underset{N_3}{\underset{|}{C}H}{-}OSiR_3 \\ + \\ ZnCl_2 \end{array} \quad (91)$$

to be the activation of the carbonyl group to nucleophilic attack by co-ordination of zinc chloride to the carbonyl oxygen, and the enhanced nucleophilic character of the azido group when bound to an element more electropositive than carbon. The α-trimethylsiloxyalkyl azides may be compared with the postulated intermediates of the Schmidt reaction of aldehydes. It was also claimed that ketones react with trimethylsilyl azide though less readily than aldehydes, as would be expected. Representative examples of the products obtained are listed in Table 6.

TABLE 6. Synthesis of α-trimethylsiloxyalkyl azides[349]

Aldehyde	Siloxyalkyl azide	Yield %
$Me(CH_2)_2CHO$	$Me(CH_2)_2CH(N_3)OSiMe_3$	73
$Me(CH_2)_3CHO$	$Me(CH_2)_3CH(N_3)OSiMe_3$	71
$Me(CH_2)_4CHO$	$Me(CH_2)_4CH(N_3)OSiMe_3$	73
$(Me)_2CHCHO$	$(Me)_2CHCH(N_3)OSiMe_3$	77
$(Me)_3CCHO$	$(Me)_3CCH(N_3)OSiMe_3$	78

D. Mercuric Azide Additions

The electrophilic character of mercuric salts towards π-bonded carbon compounds is well known[309,350]. Nucleophilic attack on the initially formed mercurinium cation leads to overall addition analogous to the common addition reactions of alkenes. When this second step is rate-limiting, the reactions may be classified as electrophile-catalysed nucleophilic additions. Synthetically a final reductive demercuration would normally be required. The formation of alcohols from alkenes is a typical application[351] (equation 92).

$$R^1R^2C{=}CR^3R^4 + HgX_2/H_2O \longrightarrow R^1R^2C(OH){-}CR^3R^4(HgX) \xrightarrow{NaBH_4} R^1R^2C(OH){-}CR^3R^4(H) \quad (92)$$

The method has recently been adapted for azide synthesis by Sokolov and Reutov who reacted mercuric nitrate and sodium azide in dimethylformamide with alkenes such as styrene and cyclohexene[352]. The azidomercurials were isolated as the mercurichlorides in 10–25% yield.

Heathcock[353] independently reported an essentially similar procedure using mercuric azide generated from sodium azide and mercuric acetate in 50% aqueous tetrahydrofuran. The products were reduced *in situ* with sodium borohydride to afford alkyl azides (equation 93). These results are summarized in Table 7.

TABLE 7. Preparation of azides by mercuration of alkenes[353]

Alkene	Product	Time hr	Temp °C	Yield %
1-Heptene	2-Heptyl azide	17	50	88
1-Octene	2-Octyl azide	24	30	55
3,3-Dimethyl-1-butene	3,3-Dimethyl-2-butyl azide	43	80	61
2-Methyl-1-heptene	2-Methyl-2-heptyl azide	40	90	50
Methylenecyclohexane	1-Methylcyclohexyl azide	68	90	60
Norbornene	*exo*-2-Norbornyl azide	16	50	75
Cyclohexene	Cyclohexyl azide	68	90	4
Methylcyclohexene	—	40	90	<1
Styrene	—	40	90	<1

Heathcock has observed that secondary and tertiary alkyl azides are obtained in good yields from terminal alkenes, but that non-terminal alkenes are relatively unreactive, except where reactivity is enhanced by steric strain as in norbornene. The order of reactivity in this reaction is similar to that in hydroxymercuration. A correlation between the reactivity sequence for azidomercuration and the known order of alkene-Ag^I stability constants[354] was also apparent and Heathcock[353] therefore proposed a mechanism which entails a rapid equilibrium involving the ion **197** which affords the organomercury derivative **198** in a rate-determining step with azide ion. In the last named process, azide ion competes favourably with water, which is a weaker nucleophile. Sneen and co-workers[355] have shown that the

$$\text{R}^1\text{R}^2\text{C}{=}\text{CR}^3\text{R}^4 + \text{Hg(N}_3)_2 \rightleftharpoons \left[\text{R}^1\text{R}^2\text{C}\overset{+}{\cdots}\text{CR}^3\text{R}^4 (\text{HgN}_3)\right] \text{N}_3^- \xrightarrow[\text{N}_3^-]{\text{r.d.s.}}$$

(**197**)

$$\text{R}^1\text{R}^2\text{C(N}_3)\text{–CR}^3\text{R}^4(\text{HgN}_3) \ (\mathbf{198}) \xrightarrow{\text{NaBH}_4} \text{R}^1\text{R}^2\text{C(N}_3)\text{–CR}^3\text{R}^4\text{H} \qquad (93)$$

rate ratio $k_{N_3^-}/k_{H_2O}$ increases greatly with increasing delocalization in the cation undergoing nucleophilic attack. This observation is consistent with the mechanism shown in equation (93). A rate-limiting nucleophilic attack by azide ion is also implied in the higher yields obtained from styrene and cyclohexene by Sokolov and Reutov[352] in a dipolar aprotic medium; Heathcock has reported very poor yields from the same substrates in a protic medium (Table 7). The low yield for styrene in both solvent systems reflects the extended delocalization in the arylcarbonium ion intermediate (cf. the lack of reactivity between cinnamic acid derivatives and hydrazoic acid).

E. Electrophilic Addition of Halogen Azides

I. Addition to olefins

a. Simple olefins. The facile addition of the pseudohalogen iodine isocyanate to alkenes[356] led Hassner and co-workers to investigate the analogous reaction between iodine azide and alkenes. Although iodine azide is unstable and explosive in the solid state, solutions of this reagent, generated *in situ* from iodine chloride and sodium azide in acetonitrile, may be handled with safety. Iodine azide adds stereospecifically to a variety of olefins in the cold to afford α,β-azidoalkyl iodides in good yield[136,357] (Table 8). Dehydroiodination of these products leads to vinyl azides[136], synthetically valuable compounds

TABLE 8. Products from iodine azide addition to simple olefins

Starting material		Product	Yield %
cis-MeCH=CHMe	(199)	MeCH(N$_3$)—CHIMe (*threo*)	90
trans-MeCH=CHMe	(200)	MeCH(N$_3$)—CHIMe (*erythro*)	88
cis-$C_6H_5CH{=}CHC_6H_5$	(201)	$C_6H_5CH(N_3)—CHIC_6H_5$ (*threo*)	80
trans-$C_6H_5CH{=}CHC_6H_5$	(202)	$C_6H_5CH(N_3)—CHIC_6H_5$ (*erythro*)	73
trans-$MeCH_2CH{=}CHCH_2Me$	(203)	$MeCH_2CH(N_3)—CHICH_2Me$ (*erythro*)	100
Cyclohexene	(204)	*trans*-1-Azido-2-iodocyclohexane	83
Cyclopentene	(205)	*trans*-1-Azido-2-iodocyclopentane	89
Cyclooctene	(206)	*trans*-1-Azido-2-iodocyclooctane	87
Cyclooctadiene	(207)	*trans*-3-Azido-4-iodocyclooct-1-ene	72
1,2-dihydronaphthalene	(208)	1-Azido-2-iodotetralin	90
Indene	(209)	1-Azido-2-iodoindane	76
2-Cholestene	(210)	2-β-Azido-3-α-iodocholestane	53
$C_6H_5CH{=}CH_2$	(211)	$C_6H_5CH(N_3)—CH_2I$	70
Methylenecyclohexane	(212)	1-Azido-1-iodomethylcyclohexane	85
$(C_6H_5)_2C{=}CH_2$	(213)	$(C_6H_5)_2C(N_3)—CH_2I$	75
$Me(CH_2)_3CH{=}CH_2$	(214)	$Me(CH_2)_3CH(N_3)—CH_2I$	100
$(Me)_3CCH_2CH{=}CH_2$	(215)	$(Me)_3CCH_2CH(N_3)—CH_2I$	79
$(Me)_3CCH{=}CH_2$	(216)	$(Me)_3CCH(I)CH_2N_3$	76
cis-$(Me)_2CHCH{=}CHMe$	(217)	$(Me)_2CHCH(I)CH(N_3)Me$ (*threo*)	95

TABLE 8—*continued*

Starting material		Product	Yield %
$(Me)_2CHCH{=}CH_2$	(**218**)	$(Me)_2CHCH(I)CH_2N_3$ + $(Me)_2CHCH(N_3)CH_2I$	92[a]
trans-$(Me)_3CCH{=}CHMe$	(**219**)	$(Me)_3CCH(I)CH(N_3)Me$ (*erythro*)	87
trans-$C_6H_5CH{=}CHMe$		$C_6H_5CH(N_3)CH(I)Me$ (*erythro*)	100

[a] Dehydroiodination of the product gave a mixture of the vinyl azides $(Me)_2CHCH{=}CHN_3$ (36%) and $(Me)_2CHC(N_3){=}CH_2$ (64%).

which are not readily accessible by other means. In contrast with the hydrazoic acid reactions, only very mild conditions are required for iodine azide additions to simple olefins. This probably indicates that the difference in bond strengths between H—N_3 and I—N_3 is greater than that between the C—H and C—I bond strengths.

The addition is stereospecific[357]; for example the adducts of the symmetrical *cis* olefins **199** and **201** have the *threo* structure (equation 94) and symmetrical *trans* olefins (**200, 202, 203**) give adducts with the *erythro* structure (equation 95). Both forms undergo *trans* elimination

$$\text{cis-RCH=CHR} \xrightarrow{IN_3} \text{threo-RCH(N}_3\text{)CH(I)R} \xrightarrow{-HI} \text{R(N}_3\text{)C=CHR} \quad (94)$$

$$\text{trans-RCH=CHR} \xrightarrow{IN_3} \text{erythro-RCH(N}_3\text{)CH(I)R} \xrightarrow{-HI} \text{R(N}_3\text{)C=CHR} \quad (95)$$

of hydriodic acid to produce vinyl azides. Addition of iodine azide to cyclic olefins such as cyclohexene also results in *trans* addition (Table 8, compounds **204**–**210**). As in isocyanate additions, this reaction proceeds by an electrophilic addition involving the intervention of a cyclic iodonium ion which may be described by the ground state resonance structures (**220a–d**). Similar species are commonly invoked in the mechanisms of *trans* additions to alkenes[309]. Considerable stability has been attributed to this iodonium ion for, even when the possibility of ring opening to a benzyl cation exists, little or no leakage to such a classical ion has been observed. Thus *cis*- and *trans*-stilbene give only *threo* and *erythro* adducts respectively.

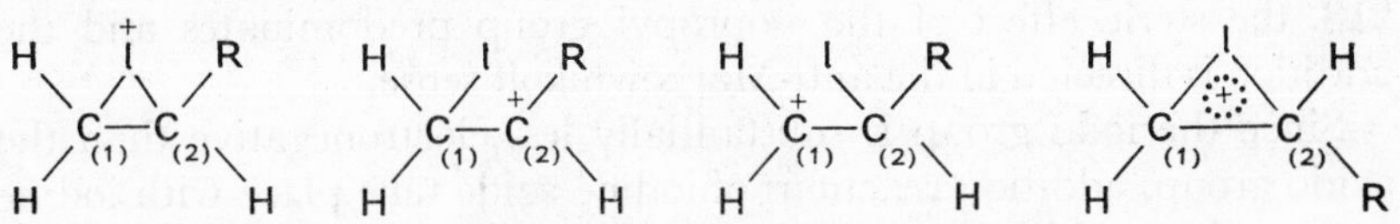

(a) (b) (c) (d)

(**220**)

It is axiomatic that an unsymmetrical iodonium ion is derived from an unsymmetrical olefin. If in the resonance structure **220d** R is an electron-donating group, e.g. alkyl or phenyl, then simple carbonium ion theory predicts that **220b** would contribute more to the ground state than **220c**. Ring opening of the iodonium ion by azide ion then occurs preferentially at $C_{(2)}$ and can lead to regiospecific products*. Thus, addition to terminal olefins affords an adduct in which the iodo group is attached to a primary carbon atom and the azido function occupies the internal position, as in **221** (cf. Table 8, compounds **211–213**).

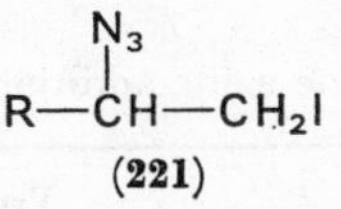

$$R\text{—}CH(N_3)\text{—}CH_2I$$

(**221**)

Cases exist where steric factors counteract these general polar directive effects (Table 8, compounds **214–219**). For example, whereas polar effects direct the addition in compounds **214** and **215**, for 3,3-dimethylbut-1-ene (**216**) adverse steric interactions between the approaching azide ion and the *t*-butyl group preclude the electronically controlled ring opening of the iodonium ion. Instead, anti-Markownikoff addition occurs with the formation of 1-azido-2-

* If the addition of an unsymmetrical reagent X–Y to an unsymmetrical olefin proceeds without skeletal rearrangement to give exclusively one of two or more possible isomers it is termed *regiospecific*. If, however, there is a significant preponderance of one isomer the reaction is said to be *regioselective* (See Ref. 358).

iodo-3,3-dimethylbutane[357]. An analogy is the bromination of 3-methylbut-l-ene in methanol which leads to 1-methoxy-2-bromo-3-methylbutane[359]. Intermediate situations with nearly balancing polar and steric factors are to be expected and in such cases regio-selective addition would result. 4-Methylbut-1-ene (**218**) falls into this category as neither the steric effect of the isopropyl group, nor electronic factors are sufficient to cause regiospecific addition. Thus both possible isomers are formed[136] (Table 8). When the electronic effects are nullified as in 5-methylpent-2-ene (**217**), the homologue of **218**, the steric effect of the isopropyl group predominates and the addition is directed in the anti-Markownikoff sense.

Since the iodo group is substantially less electronegative than the azido group, addition reactions of iodine azide take place with iodine as the electrophilic centre. Bromine is more electronegative than iodine, however, and consequently either heterolytic or homolytic cleavage of the Br—N_3 bond can occur depending on the polarity of the solvent[360]. Reactions involving the azide radical derived from homolytic cleavage of bromine azide are discussed in section IV.F.1. In mixed polar solvents such as nitromethane–methylene chloride, stereospecific ionic additions involving a bromonium ion intermediate occur in an analogous manner to those with iodine azide[360,361] (Table 9). With solvents such as methanol, acetone, acetic acid and acetonitrile, however, solvent participation can occur[362]. The intermediate bromonium ion is more reactive than the larger iodonium ion and therefore more susceptible to solvolysis, which affords products other than those of Br—N_3 addition.

TABLE 9. Bromine azide additions to simple olefins

Compound	Product	Yield %
$C_6H_5CH{=}CH_2$	$C_6H_5CH(N_3)CH_2Br$	95
2-Cholestene	2β-Azido-3α-bromocholestane	52
Cyclohexene	*trans*-1-Azido-2-bromocyclohexane	45
trans-MeCH=CHMe	MeCH(Br)CH(N_3)Me (*erythro*)	35
cis-MeCH=CHMe	MeCH(Br)CH(N_3)Me (*threo*)	35

In a recent extension of this work, Hassner and Boerwinkle[361] have indicated that chlorine azide is predominantly an azide radical source as would be expected from the electronegativity trend $I < N_3 \simeq Br < Cl$*. Solvents of low polarity, the presence of light and absence of oxygen enhance homolytic cleavage. Addition by Cl^+ attack on the olefinic linkage may therefore be induced in polar solvents in the presence of oxygen, which acts as a radical trap, and in this manner chlorine azide has been added to styrene to afford 1-azido-1-phenyl-2-chloroethane in high yield.

b. Conjugated olefins. Whereas both iodine azide and iodine isocyanate react similarly with alkenes, the latter being more reactive, only iodine azide adds readily to the carbon–carbon double bond of α,β-unsaturated carbonyl compounds. (Table 10). Hassner and Fowler have proposed that the orientation of this addition is explicable in terms of an Ad_E reaction involving an iodonium ion intermediate similar to that suggested for the addition to simple olefins[136,357].

TABLE 10. Addition of iodine azide to α,β-unsaturated carbonyl compounds

Substrate	Product	Yield %
trans-$C_6H_5CH{=}CHCOOMe$	$C_6H_5CH(N_3)CH(I)COOMe$	43
trans-$MeCH{=}CHCOOEt$	$MeCH(N_3)CH(I)COOEt$	81
$CH_2{=}CHCOOMe$	$N_3CH_2CH(I)COOMe$ + $ICH_2CH(N_3)COOMe$	86
$C_6H_5CH{=}CHCOC_6H_5$	$C_6H_5CH(N_3)CH(I)COC_6H_5$	100

The alternative explanation, that iodine azide additions to α,β-unsaturated carbonyl compounds are of the Ad_N type (equation 96) is,

$$RCH{=}CH{-}C({=}O)R^1 \xrightarrow{N_3{-}I} RCH(N_3){-}CH{=}C(O^-)R^1 \;\; I^+ \longrightarrow RCH(N_3)CH(I)COR^1 \qquad (96)$$

however, worthy of consideration. From studies of addition of pseudohalogens to simple olefins it is evident that iodine isocyanate is a more powerful electrophile than iodine azide[356,357]. However, only the latter reagent enters into addition reactions with α,β-unsaturated ketones and this is not consistent with an Ad_E mechanism. An Ad_N

* In the authors' opinion the sequence $I < Br < N_3 \simeq Cl$ more closely represents the electronegativity order.

mechanism could be invoked however, given that azide ion is a more reactive nucleophile than isocyanate ion. This seems probable, but reliable evidence is lacking.

Just as bromine chloride is a more reactive electrophile than iodine chloride[363], so one would expect bromine azide acting as an electrophile to be more reactive than iodine azide. In fact in solvents suitable for heterolytic reaction, bromine azide adds to chalcone *less* readily than iodine azide[357,360] and this further indicates an Ad_N mechanism. The evidence is strengthened by the occurrence of acid catalysis of the bromine azide addition.

Further investigation of such additions is clearly warranted and any mechanisms must be regarded as speculative at this stage.

2. Addition to alkynes

The addition of iodine azide to a number of substituted acetyllenes[357,364,365] has been shown to take place in the opposite regiochemical sense to their well known hydration in the presence of strong acid. Thus 1-phenylpropyne adds iodine azide to form *cis*- and *trans*-2-azido-1-iodo-1-phenylpropyne whereas the acid hydration gives propiophenone (equation 97). Addition of water or iodine azide to 2-bromo-1-phenylethyne on the other hand gives similarly oriented products (equation 98).

$$C_6H_5C(OH){=}CHMe \xleftarrow[H_2SO_4]{H_2O} C_6H_5C{\equiv}CMe \xrightarrow{IN_3} C_6H_5C(I){=}C(N_3)Me \quad (97)$$

$$C_6H_5C(OH){=}CHMe \rightleftharpoons C_6H_5COCH_2Me$$

$$C_6H_5C(OH){=}CHBr \xleftarrow[HgO]{H_2O} C_6H_5C{\equiv}CBr \xrightarrow{IN_3} C_6H_5C(N_3){=}C(I)Br \quad (98)$$

$$C_6H_5C(OH){=}CHBr \rightleftharpoons C_6H_5COCH_2Br$$

The additions of iodine azide (equations 99, 100) and the mercury catalysed hydration (equation 101) were rationalized on the basis of Ad_E reactions with cyclic intermediates having relatively little carbonium ion character. Orientation is then dependent on the inductive effects of the methyl and bromo substituents with little orientational effect of the phenyl group. In contrast, the orientation of the acid catalysed hydration is ascribed to the phenyl group as being the more important substituent in a reaction via a vinyl carbonium ion (equation 102). However, it has been mentioned that both *cis* and *trans* products are obtained from 1-phenylpropyne and it would appear that a multi-centre reaction is also implicated (equation 103).

$$C_6H_5C{\equiv}CMe \xrightarrow{IN_3} C_6H_5C\overset{+}{\underset{I}{\triangle}}CMe \longrightarrow C_6H_5C(I){=}C(N_3)Me \quad (99)$$

$$C_6H_5C{\equiv}CBr \xrightarrow{IN_3} C_6H_5C\overset{+}{\underset{I}{\triangle}}CBr \longrightarrow C_6H_5C(N_3){=}C(I)Br \quad (100)$$

$$C_6H_5C{\equiv}CBr \xrightarrow{HgX_2} C_6H_5C\overset{+}{\underset{HgX}{\triangle}}CBr \xrightarrow{H_2O} C_6H_5C(OH){=}CHBr \rightleftharpoons C_6H_5COCH_2Br \quad (101)$$

$$C_6H_5C{\equiv}CMe \xrightarrow{H^+} C_6H_5\overset{+}{C}{=}CHMe \xrightarrow{H_2O} C_6H_5C(OH){=}CHMe \rightleftharpoons C_6H_5COCH_2Me \quad (102)$$

$$C_6H_5C{\equiv}CMe + I{-}N{=}\overset{+}{N}{=}\overset{-}{N} \longrightarrow C_6H_5C(I){=}C(N_3)Me \quad (103)$$

Product orientation alone is insufficient for elucidation of the mechanisms of these addition reactions as the same products would be obtained by either electrophilic or nucleophilic addition. It is well known that the greater electronegativity of the *sp* carbon in alkynes leads to lower reactivity with electrophiles and greater reactivity with nucleophiles compared with the sp^2 carbon of alkenes. A number of nucleophilic additions to alkynes such as addition of alcohols and weak acids are known[309].

In view of the substantial nucleophilic strength of azide ion, an Ad_N mechanism for addition of iodine azide to alkynes cannot therefore be excluded. It is again relevant in this context that I—NCO does not add to diphenylacetylene whereas I—N_3 gives the addition product **222**. Since the isocyanato group is somewhat more electronegative than the azido group[366], electrophilic addition should proceed more readily with iodine isocyanate, whereas nucleophilic addition should proceed more readily with iodine azide. Reaction of diphenylacetylene with iodine azide alone therefore supports an Ad_N mechanism for this case. It is significant also that iodine isocyanate adds to phenylacetylene and it would be of interest to study the orientation of addition of iodine azide to this compound.

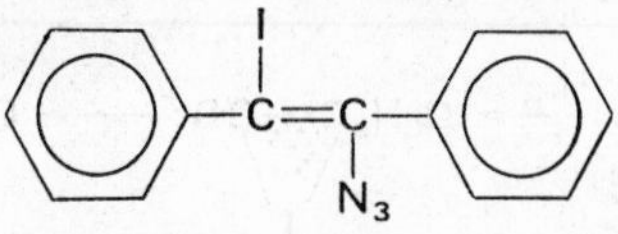

$$C_6H_5-C(I)=C(N_3)-C_6H_5$$

(222)

The present situation in regard to addition of halogen azide to alkynes is that the synthetic aspects are well established but that mechanistic evidence is sparse and equivocal. Clearly, some rate studies of these reactions would be of value in elucidating the mechanisms actually involved.

F. Free Radical Additions

I. Halogen azide additions

The possibility of both homolytic and heterolytic cleavage of halogen azides has previously been discussed (section IV.E.1). Such additions may be observed in solvents of low polarity, in the presence of light and the absence of radical trapping species such as oxygen[360,361].

Styrene, for example, is known to add both bromine and chlorine azide quantitatively and regiospecifically under such conditions to give the adduct (**223**; X=Br, Cl). This is consistent with an initial

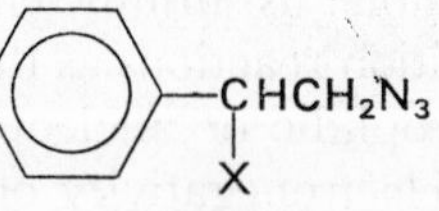

$$C_6H_5-CH(X)CH_2N_3$$

(223)

addition of the azide radical to form $C_6H_5—\dot{C}H—CH_2N_3$ which then adds halogen to give the adduct. Iodine azide, which has a lesser tendency to undergo homolytic cleavage, affords the iodo product (**223**; X=I) but in much lower yield.

This reaction has been utilized in the synthesis of azidosteroids[360,367]. Addition to 2-cholestene (**224**) occurs regioselectively and adducts **225** and **226** have been isolated in 37% and 27% yield respectively. By comparison, ionic addition leads to the *trans*-diaxial product[360] (**227**). Such differing orientations of the products obtained from homolytic and heterolytic addition may be synthetically valuable.

(**224**) $\xrightarrow{BrN_3,\ Ad_H}$ (**225**)

(**224**) $\xrightarrow{BrN_3,\ Ad_H}$ (**226**)

(**224**) $\xrightarrow{BrN_3,\ Ad_E}$ (**227**)

2. Additions in the presence of redox systems

Minisci, Galli and co-workers[368–375] have studied a variety of radical reactions which result in the formation of organic azides. These processes commonly involve the interaction of an organic peroxide, an alkene, and azide ion in the presence of a ferrous–ferric redox system. The initial step is the reduction of the peroxide by Fe^{2+} to form a free alkoxy radical (Fenton reaction, equation 104).

The alkoxy radicals are electrophilic in character and react readily with conjugated olefins such as butadiene or cyclopentadiene[369,370]

$$RO{-}OH + Fe^{2+} \longrightarrow RO^{\bullet} + Fe^{3+} + OH^{-} \quad (104)$$

$$Me_3COOH + Fe^{2+} \longrightarrow Me_3CO^{\bullet} + [FeOH]^{2+} \quad (105)$$

$$Me_3CO^{\bullet} + CH_2{=}CH{-}CH{=}CH_2 \longrightarrow Me_3COCH_2CH{=}CH\dot{C}H_2 \quad (106)$$

$$Me_3COCH_2CH{=}CH\dot{C}H_2 + (FeN_3)^{2+} \longrightarrow Me_3COCH_2CH{=}CHCH_2N_3 + Fe^{2+} \quad (107)$$

with the formation of resonance stabilized carbon radicals. These interact with an azide ion coordinated to ferric ion to afford an organo-azide (equations 105–107).

In a similar reaction[368,375] cyclohexanone peroxide forms ε-azidocaproic acid following ring fission of the original alkoxy radical to form an alkanoic acid radical (equation 108).

$$\text{1-hydroperoxy-1-hydroxycyclohexane} \xrightarrow{Fe^{2+}} \left[\text{1-hydroxycyclohexyloxy radical}\right] + (FeOH)^{2+} \longrightarrow \dot{C}H_2(CH_2)_4CO_2H \xrightarrow{[FeN_3]^{2+}} N_3(CH_2)_5CO_2H + Fe^{2+} \quad (108)$$

The electrophilic nature of alkoxy radicals is evident from their reactions with olefins which are conjugated to electron-withdrawing substituents[370,373]. With such substrates, both a diazide and the normal addition product are formed. The lower nucleophilicity of such olefins causes a commensurate reduction in their affinity for alkoxy radicals and formation of azido radicals from azidoferric ion and alkoxy radicals becomes a competitive process. Diazides are then formed as shown in equation (109).

$$RO^{\bullet} + C_6H_5CH{=}CH_2 \longrightarrow C_6H_5\dot{C}HCH_2OR \xrightarrow{[FeN_3]^{2+}} C_6H_5CH(N_3)CH_2OR + Fe^{2+}$$

$$RO^{\bullet} + [FeN_3]^{2+} \longrightarrow N_3^{\bullet} + Fe^{3+} + {}^{-}OR \quad (109)$$

$$N_3^{\bullet} + C_6H_5CH_2{=}CH_2 \longrightarrow C_6H_5\dot{C}HCH_2N_3 \xrightarrow{[FeN_3]^{2+}} C_6H_5CH(N_3)CH_2N_3 + Fe^{2+}$$

When hydrogen peroxide is employed instead of organic peroxides the diazide is formed exclusively, since the hydroxyl radical is more electrophilic than the alkoxy radical and reacts exclusively with azide ion rather than the olefin. In this manner diazides can even be obtained from non-conjugated olefins such as pent-2-ene, hex-1-ene, cyclohexene and steroids[370,372,373]. The procedure may be further extended to the synthesis of α-azidoesters and α-azidoketones by passing a stream of oxygen through the reaction mixture[374] (equation 110).

$$\dot{O}H + [FeN_3]^{2+} \longrightarrow \dot{N}_3 + [FeOH]^{2+}$$

$$\dot{N}_3 + C_6H_5CH{=}CH_2 \longrightarrow C_6H_5\dot{C}HCH_2N_3 \quad (110)$$

$$C_6H_5\dot{C}HCH_2N_3 + O_2 \longrightarrow C_6H_5CH(O{-}O^{\bullet})CH_2N_3 \xrightarrow{Fe^{2+}} C_6H_5COCH_2N_3 + [FeOH]^{2+}$$

In the presence of chloride ion, these reactions lead to the formation of chlorine atoms as well as azide radicals and the former react preferentially with organic (carbon) free radicals. This is demonstrated by the reaction of cyclohexanone peroxide with the ferrous–ferric system in presence of both azide and chloride ions to give exclusively ε-chlorocaproic acid (equation 111). This may be exploited syntheti-

$$\text{1-hydroperoxycyclohexanol} \xrightarrow{Fe^{2+}} [\text{1-hydroxycyclohexyloxyl radical}] \longrightarrow \dot{C}H_2(CH_2)_4CO_2H + (FeOH)^{2+} \xrightarrow{[FeCl]^{2+}} Cl(CH_2)_5CO_2H + Fe^{2+} \quad (111)$$

cally to accomplish azidochlorination of alkenes[371,373]. Thus cyclohexene reacts with sodium azide, ferric chloride, hydrogen peroxide and ferrous salts to give 1-azido-2-chlorocyclohexane (equation 112).

$$HO{-}OH \xrightarrow{Fe^{2+}} \cdot OH + (FeOH)^{2+} \xrightarrow{[FeN_3]^{2+}} \dot{N}_3 + 2(FeOH)^{2+}$$

$$\text{cyclohexene} \xrightarrow{\dot{N}_3} \text{2-azidocyclohexyl radical} \xrightarrow{[FeCl]^{2+}} \text{1-azido-2-chlorocyclohexane} + Fe^{2+} \quad (112)$$

Similar procedures have been used to introduce two functional groups into a steroid nucleus[373].

V. SYNTHESIS OF AZIDES FROM DIAZOTIZATION AND RELATED REACTIONS

A. Introduction

In the previous sections we have discussed reactions in which the carbon–azide bond is formed by substitution on carbon of a preformed azide moiety or by its addition to various multiple bonds. Processes in which the azide nitrogen atoms are introduced in a stepwise manner are now considered. These syntheses include the reactions of diazonium salts with nucleophiles such as ammonia, chloramine, hydroxylamine, hydrazine, sulphonamides and azide ion. Recent work on the reaction between diazonium salts and azide ion has shown that it should be included in this section rather than those concerned with reactions at carbon, since it has been established that an electrophilic nitrogen atom is the reactive centre in the diazonium compound. Also included are the nitrosation reactions of hydrazines, the reactions of hydrazoic acid with nitroso compounds and the topical and potentially valuable diazo transfer reactions.

B. Reaction of Diazonium Compounds with Nucleophiles

1. Ammonia and its derivatives

In his classical work on diazonium compounds, Griess[1] demonstrated that benzenediazonium tribromide reacts with ammonia to give phenyl azide in high yield (equation 113). It has since been estab-

$$(C_6H_5N_2)^+ \, Br_3^- + NH_3 \longrightarrow C_6H_5N_3 + 3HBr \qquad (113)$$

lished that this is a general reaction of diazonium perhalides. In particular, treatment of the plumbichlorides $(C_6H_5N_2^+)_2PbCl_6^{2-}$ with ammonia provides a very convenient synthetic route to aryl azides[376]. Some failures of the reaction have been reported, but these usually result from complications associated with the diazotization process (e.g. in 2-aminopyridine[377], 2-aminobenzaldehyde[378], etc.).

Boyer and Canter[15] suggested that the reaction proceeds through the intervention of a triazene, and further implied that this intermediate is subsequently oxidized to the azide by halogen or the anion (equation 114). In support of this suggestion it has been observed that in the absence of oxidant some diazonium compounds form triazenes with ammonia.

$$C_6H_5N_2^+ + 2NH_3 \longrightarrow C_6H_5N{=}NNH_2 + NH_4^+ \xrightarrow{\text{oxidation}} C_6H_5N_3 \qquad (114)$$

For example, the triazene (**228**) is obtained from diazotized anthraquinone and ammonium carbonate under these conditions[379] (equation 115). Further, the formation of aryl azides by oxidation of aryl triazenes with hypochlorite solution has been reported[380]. In a related process azides are formed by reaction of chloramine or chloramine-*T* with aromatic diazonium compounds[381], presumably through an unstable *N*-chlorotriazene (**229**) as illustrated in equation (116).

$$\text{1-diazonioanthraquinone} \xrightarrow{(NH_4)_2CO_3} \text{1-(N{=}N{-}NH_2)anthraquinone}\ \textbf{(228)} \quad (115)$$

$$C_6H_5N_2^+ + NH_2Cl \longrightarrow C_6H_5N{=}N{-}N(H)Cl\ \textbf{(229)} \xrightarrow{-HCl} C_6H_5N_3 \quad (116)$$

Clusius and co-workers[382,383] have established by ^{15}N-labelling experiments that in the reaction between benzenediazonium perbromide and ammonia, the terminal nitrogen of the azido group originates from a molecule of ammonia as shown in equation (117).

$$\begin{aligned} &C_6H_5N^aH_2{\cdot}HCl + HN^bO_2 \longrightarrow (C_6H_5N^a{=}N^b)^+\,Cl^- + 2\,H_2O \\ &(C_6H_5N^a{=}N^b)^+\,Br_3^- + 4\,N^cH_3 \longrightarrow C_6H_5N^aN^bN^c + 3\,N^cH_4Br \end{aligned} \quad (117)$$

The distribution of the isotopic label in the azide was deduced after degradation to aniline and ammonia (equation 118) and to *p*-chloroaniline and nitrogen (equation 119).

$$C_6H_5N_3 \xrightarrow{C_6H_5MgBr} C_6H_5N^a{=}N^bN^cHC_6H_5 \longrightarrow C_6H_5N^aH_2 + N^bH_3 + C_6H_5N^cH_2 \quad (118)$$

$$C_6H_5N_3 \xrightarrow{HCl} p\text{-}ClC_6H_4NH_2 + N_2 \quad (119)$$

These findings are consistent with the formation of a triazene intermediate.

Reaction of diazonium compounds with primary and secondary aliphatic amines usually results in the formation of substituted triazenes[1] (equation 120). However, an interesting exception with

$$C_6H_5\overset{+}{N}_2 + 2RNHR' \longrightarrow C_6H_5N{=}N{-}NRR' + RR'\overset{+}{N}H_2 \quad (120)$$

some value in azide synthesis has been reported[384]. Reaction of aryldiazonium salts with aziridine yields the corresponding aryl azide, presumably through the intermediacy of the triazene (**230**), which extrudes ethylene with concomitant azide formation (equation 121).

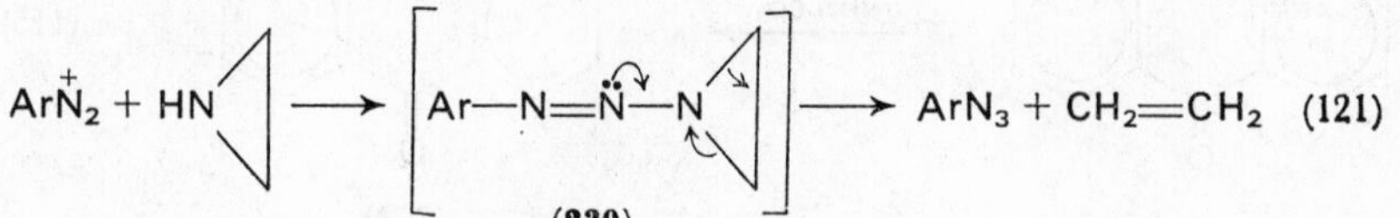

$$ArN_2^+ + HN\langle \longrightarrow [Ar{-}N{=}\ddot{N}{-}N\langle] \longrightarrow ArN_3 + CH_2{=}CH_2 \quad (121)$$

(**230**)

$Ar{=}C_6H_5$, p-MeC_6H_4, p-$MeOC_6H_4$, p-$NO_2C_6H_4$

The synthesis of phenyl azide from hydroxylamine and benzenediazonium chloride was first observed by Emil Fischer[385] and the reaction was further developed by Mai and co-workers[386,387]. As in the case of the ammonia reaction, a triazene intermediate is generally implicated (equation 122). In the anthraquinone series the triazene

$$C_6H_5\overset{+}{N}_2 + 2NH_2OH \longrightarrow [C_6H_5N{=}N{-}NHOH] \longrightarrow C_6H_5N{=}\overset{+}{N}{=}\overset{-}{N} + H_2O \quad (122)$$

(**231**) was actually isolated in the reaction of diazotized 1-aminoanthraquinone and hydroxylamine and was subsequently converted into 1-azidoanthraquinone (**232**) on treatment with alkali[388].

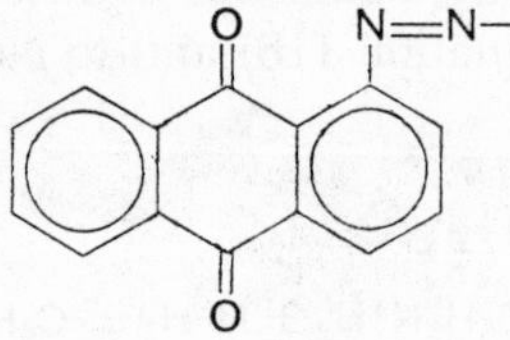

N=N—NHOH (**231**) N_3 (**232**)

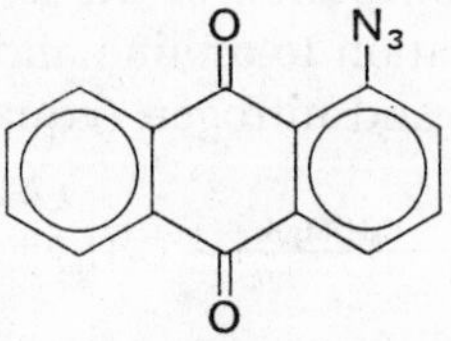

Diazotization of *o*-aminobenzaldoxime gave the bicyclic intermediate 1,2,3-benzotriazine-3-oxide (**233**) and further reaction of this compound with alkali yielded *o*-azidobenzaldehyde (**234**)[389–391]. This ring opening reaction probably leads initially to the triazene (**235**), which would arise from attack by hydroxide ion at the 4-position of the triazanaphthalene (**233**) followed by ring opening of the intermediate *N*-hydroxycarbinolamine (**236**).

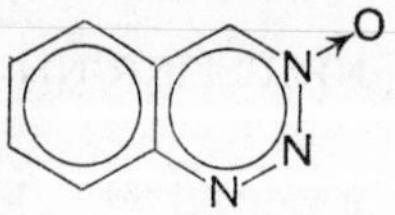

(**233**) CHO N_3 (**234**)

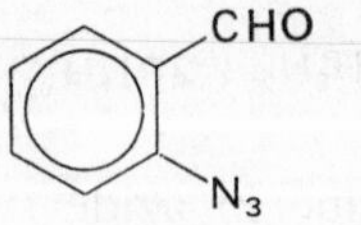

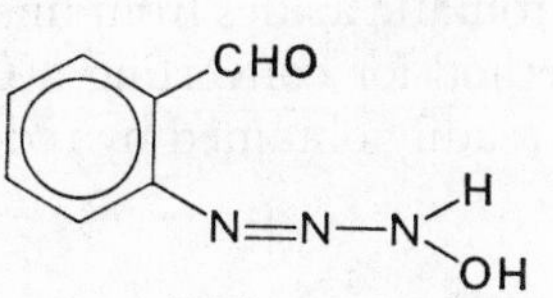

(235) 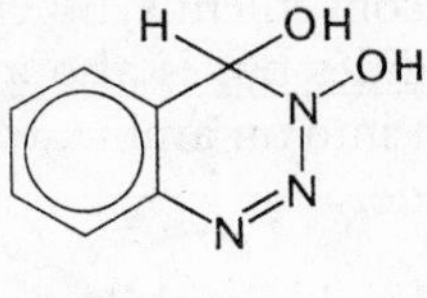 (236)

Methyl-, benzyl- and *p*-toluene-sulphonamide react with aromatic diazonium compounds to give isolable triazenes which decompose in the presence of alkali with the formation of the corresponding aromatic azide and the alkyl- or aryl-sulphinic acid salt[392–395]. This sequence (equation 123), known as the Dutt–Wormall reaction, is mechanistically similar to the Bamford–Stevens synthesis of aromatic diazo compounds from toluenesulphonylhydrazones and alkoxide ion[396,397] (equation 124).

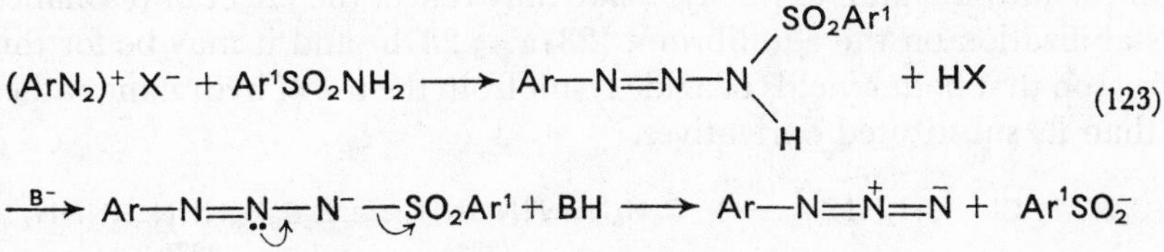

$$(ArN_2)^+X^- + Ar^1SO_2NH_2 \longrightarrow Ar{-}N{=}N{-}N(SO_2Ar^1)(H) + HX \qquad (123)$$

$$\xrightarrow{B^-} Ar{-}N{=}\ddot{N}{-}N^-{-}SO_2Ar^1 + BH \longrightarrow Ar{-}N{=}\overset{+}{N}{=}\overset{-}{N} + Ar^1SO_2^-$$

$$R^1RC{=}N{-}NH{-}SO_2R^2 \xrightarrow{B^-} R^1RC{=}\ddot{N}{-}\overset{-}{N}{-}SO_2R^2 + BH \longrightarrow R^1RC{=}\overset{+}{N}{=}\overset{-}{N} + R^2SO_2^- \qquad (124)$$

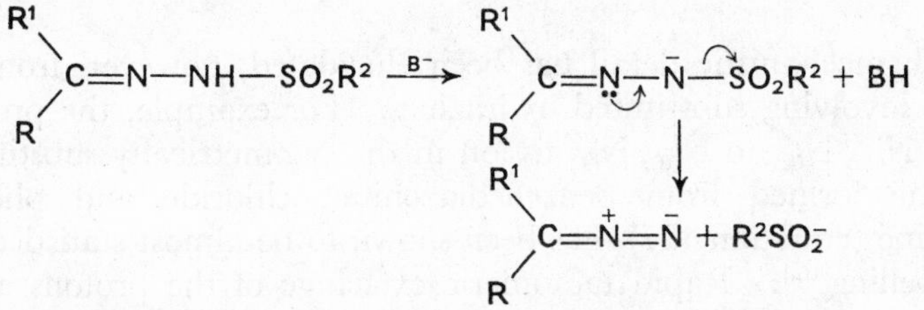

The reaction between sulphonyl azides and Grignard reagents also leads to stable triazenes[398,399] which are similarly degraded with alkali to aryl azides (equation 125). This reaction constitutes not

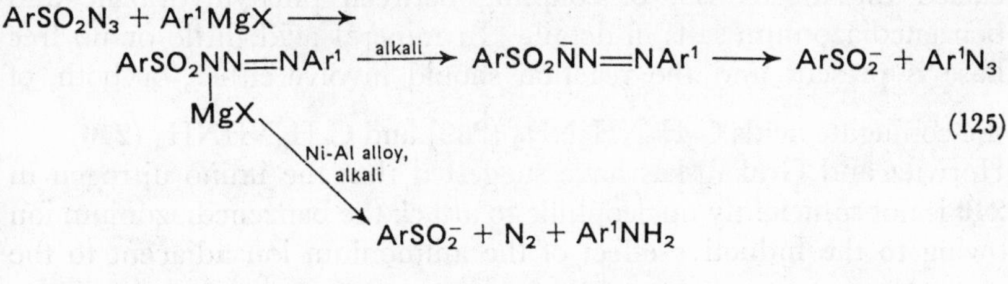

$$ArSO_2N_3 + Ar^1MgX \longrightarrow ArSO_2N(MgX)N{=}NAr^1 \xrightarrow{\text{alkali}} ArSO_2\overset{-}{N}N{=}NAr^1 \longrightarrow ArSO_2^- + Ar^1N_3 \qquad (125)$$

$$ArSO_2N(MgX)N{=}NAr^1 \xrightarrow{\text{Ni-Al alloy, alkali}} ArSO_2^- + N_2 + Ar^1NH_2$$

only a convenient synthetic route to aromatic azides from unactivated aryl halides but is also a valuable method for converting a Grignard reagent into an amine, which may be readily obtained by reduction of the azide.

2. Hydrazine and its derivatives

Benzenediazonium chloride reacts with hydrazine to form phenyl azide and ammonia in greater than 90% yield[385,400–402]. The unsymmetrical tetrazene (**237**) has been assumed as the intermediate in this reaction; the alternative decomposition pathway of this unstable intermediate, to give aniline and hydrazoic acid, accounts for less than 10% of the products (equation 126). The preferential decomposition of the intermediate to phenyl azide may reflect the effect of resonance stabilization on the equilibrium (**237a** ⇌ **237b**) and it may be for this reason that better yields of azide result from the use of hydrazine rather than its substituted derivatives.

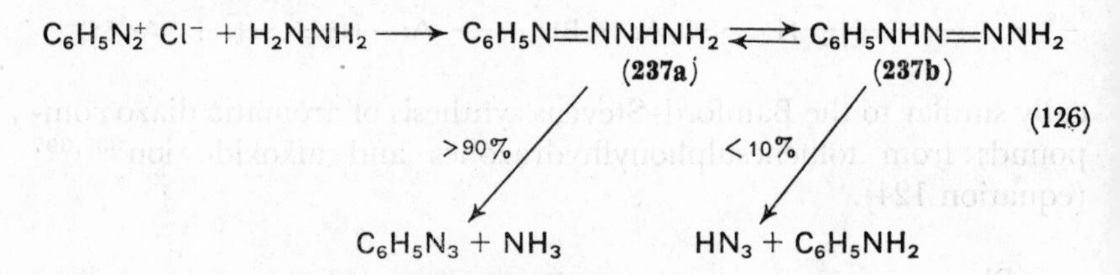

(126)

Much mechanistic detail has been elucidated, however, from reactions involving substituted hydrazines. For example, the proportion of $N_{(1)}$, $N_{(2)}$ to $N_{(3)}$, $N_{(4)}$ fission in the symmetrically substituted tetrazene formed from benzenediazonium chloride and phenylhydrazine (equation 127), has been shown to be almost statistical by ^{15}N-labelling[403]. Rapid tautomeric exchange of the protons along the nitrogen chain of the tetrazene was thereby established.

In reactions involving unsymmetrical diaryltetrazenes such as **238** four different products arise as a result of this statistical decomposition[404] (equation 128). Horwitz and Grakauskas[405] have discussed the mechanism of coupling between phenylhydrazine and benzenediazonium salts in detail. In mineral acid, little or no free base is present and the reaction should involve either, or both, of the conjugate acids $C_6H_5\overset{+}{N}H_2NH_2$ (**239**) and $C_6H_5NH\overset{+}{N}H_3$ (**240**). Horwitz and Grakauskas have suggested that the imino nitrogen in **240** is not sufficiently nucleophilic to attack the benzenediazonium ion owing to the inductive effect of the ammonium ion adjacent to the

$$C_6H_5NHNH_2 \xrightarrow{H^+} \begin{cases} C_6H_5\overset{+}{N}H_2NH_2 \text{ (239)} \xrightarrow{C_6H_5\overset{+}{N}_2} C_6H_5\overset{+}{N}H_2-NH\ddot{N}=NC_6H_5 \text{ (241)} \longrightarrow C_6H_5NH_2 + C_6H_5N_3 + H^+ \\ C_6H_5NH\overset{+}{N}H_3 \text{ (240)} \xrightarrow{C_6H_5\overset{+}{N}_2} C_6H_5N=NN(C_6H_5)NH_2 \text{ (242)} + 2H^+ \end{cases} \quad (127)$$

$$\underset{(238)}{Ar-N=NNH-NHAr^1} \longrightarrow \begin{cases} ArN_3 + Ar^1NH_2 \\ ArNH_2 + Ar^1N_3 \end{cases} \quad (128)$$

imino nitrogen atom. The positive charge on the imino nitrogen atom in **239**, however, has a comparatively small effect on the amino nitrogen since the phenyl group is an effective electron source. Consequently, coupling in an acidic medium involves nucleophilic attack by the amino nitrogen atom and the resulting 1,4-diaryltetrazene (**241**) subsequently decomposes to an azide (equation 127). In sodium acetate buffered solutions, however, the weakly basic phenylhydrazine ($K_b = 1{\cdot}6 \times 10^{-9}$) is present as the free base and coupling occurs predominantly at the imino nitrogen atom to give the stable 1,3-diphenyltetrazene[14] (**242**), although traces of phenyl azide and aniline in the products suggest some reaction at the amino nitrogen with transient formation of the unstable 1,4-diphenyltetrazene (**241**).

Horwitz and Grakauskas have presented experimental evidence in support of their suggestion that both conjugated acids should be considered in an acidic medium. Concurrent coupling at *both* the amino and imino nitrogen atoms of the conjugate acids **239** and **240** respectively was shown to occur with tetrazolediazonium chloride (**243**). Kuhn and Kainer[406] have suggested that in dilute aqueous solution this tetrazole has the resonance stabilized zwitterionic structure **244** and Horwitz and Grakauskas have postulated that both of the protonated forms **239** and **240** coordinate with this zwitterion. Proton transfer to the tetrazole ring and synchronous nucleophilic attack on the diazonium substituent, as in Scheme 3, results in the formation of both the 1,3- and 1,4-tetrazenes, (**245**) and (**246**), in approximately 35% and 20% yield respectively based on product analysis.

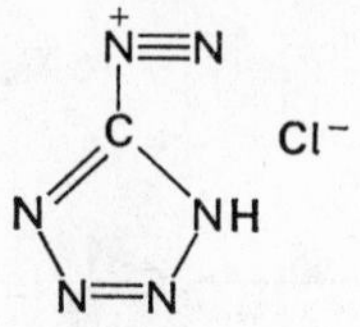

(**243**) 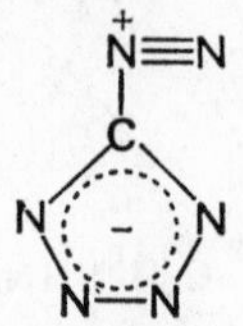 (**244**)

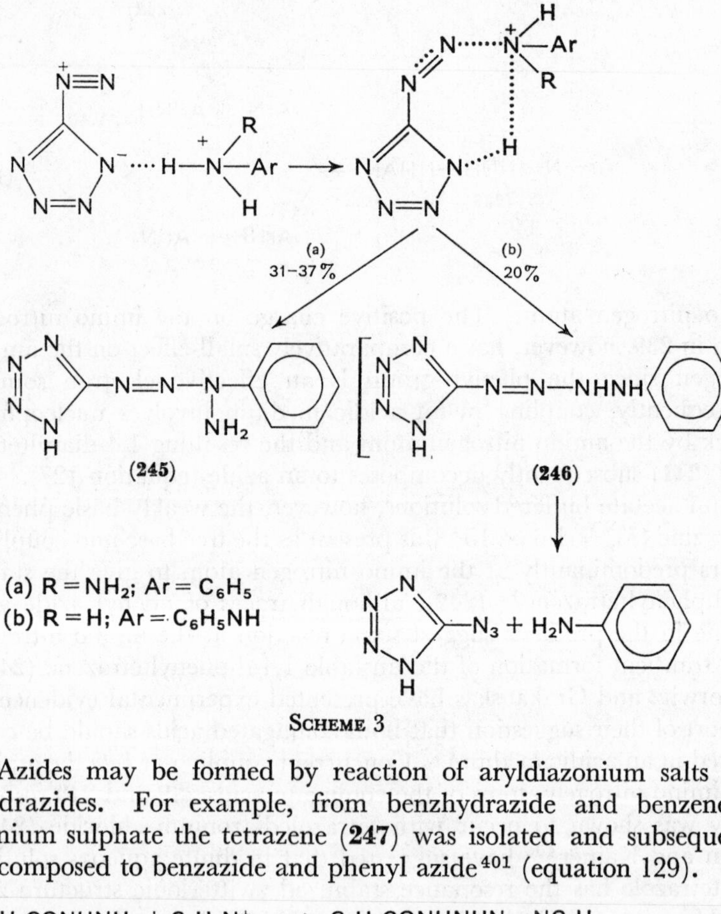

(a) $R = NH_2$; $Ar = C_6H_5$
(b) $R = H$; $Ar = C_6H_5NH$

(a) 31–37% (b) 20%

(**245**) (**246**)

SCHEME 3

Azides may be formed by reaction of aryldiazonium salts with hydrazides. For example, from benzhydrazide and benzenediazonium sulphate the tetrazene (**247**) was isolated and subsequently decomposed to benzazide and phenyl azide[401] (equation 129).

$$C_6H_5CONHNH_2 + C_6H_5N_2^+ \longrightarrow \underset{(\mathbf{247})}{C_6H_5CONHNHN{=}NC_6H_5} \qquad (129)$$

$$\longrightarrow C_6H_5CON_3 + C_6H_5NH_2 + C_6H_5N_3 + C_6H_5CONH_2$$

Similarly, treatment of the anthranilic acid derivative (**248**) with nitrous acid gave the cyclic product **249** which underwent hydrolysis in aqueous alkali to afford *o*-azidobenzoic acid[407]. 1,2-Diacetylhydra-

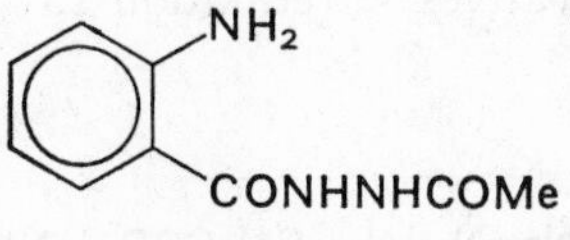

(**248**) 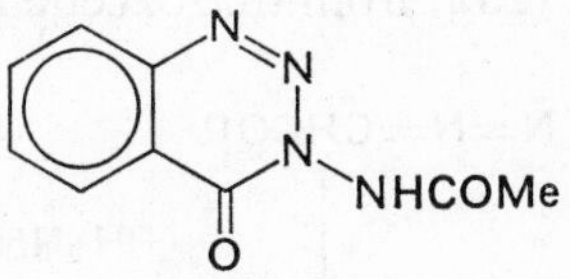 (**249**)

zine undergoes reaction with aryldiazonium salts in alkaline solution to form 1,6-bisaryl-3,4-diacetyl-1,5-hexazadiene (**250**)[408]. Hydrolysis of **250** at 5° in alcoholic potassium hydroxide affords an aryl azide and a 1-acetyl-3-aryltriazene (**251**) which may be further degraded to nitrogen and an aryl amine (equation 130). High yields of aryl azides are obtained from this reaction; for example 1,6-bis-(*p*-

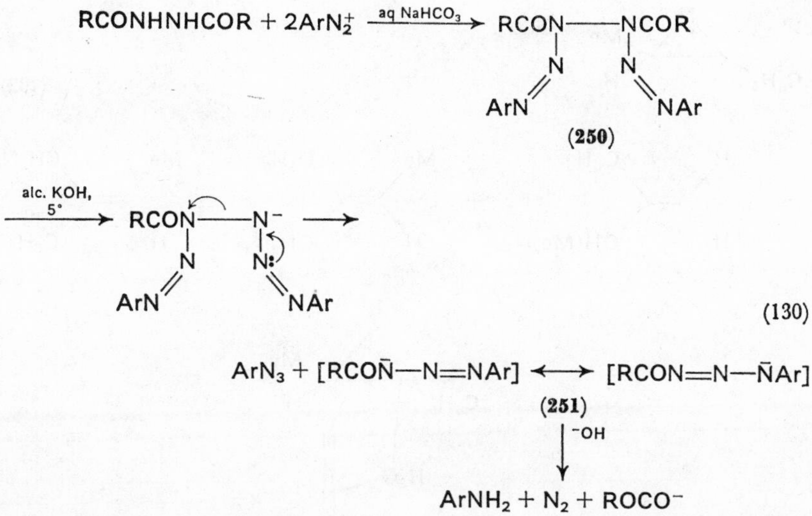

$$\mathrm{RCONHNHCOR} + 2\mathrm{ArN_2^+} \xrightarrow{\text{aq NaHCO}_3} \mathrm{RCON(N{=}NAr)}\!-\!\mathrm{N(N{=}NAr)COR} \quad (\mathbf{250})$$

$$\xrightarrow[5^\circ]{\text{alc. KOH,}} \mathrm{RCON(N{=}NAr)}\!-\!\mathrm{N^-}\!-\!\mathrm{N{=}NAr} \longrightarrow \qquad (130)$$

$$\mathrm{ArN_3} + [\mathrm{RCO\bar{N}{-}N{=}NAr}] \longleftrightarrow [\mathrm{RCON{=}N{-}\bar{N}Ar}] \quad (\mathbf{251})$$

$$\xrightarrow{^-\mathrm{OH}} \mathrm{ArNH_2} + \mathrm{N_2} + \mathrm{ROCO^-}$$

chlorophenyl)-3,4-diacetyl-1,5-hexazadiene gives a 94% yield of pure *p*-chlorophenyl azide on alkaline hydrolysis at 5°. However this procedure offers no advantage compared with the classical method of synthesis and has therefore found little practical application.

Syntheses of azides from diazonium compounds are generally

restricted to the aryl derivatives since aliphatic diazonium compounds spontaneously lose nitrogen to produce carbonium ions. Some aliphatic azides have been prepared from related reactions involving aliphatic *diazo* compounds. One such case is the formation of the azide (**252**) from diazoacetic acid derivatives[409] (equation 131).

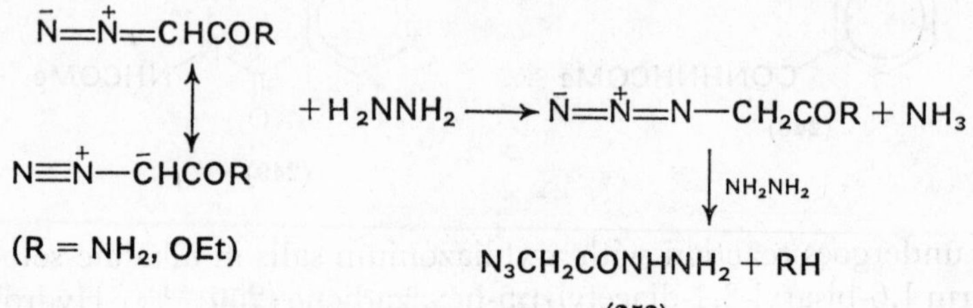

$$\bar{N}{=}\overset{+}{N}{=}CHCOR \longleftrightarrow N{\equiv}\overset{+}{N}{-}\bar{C}HCOR \quad (R = NH_2, OEt)$$

$$+ H_2NNH_2 \longrightarrow \bar{N}{=}\overset{+}{N}{=}N{-}CH_2COR + NH_3 \xrightarrow{NH_2NH_2} N_3CH_2CONHNH_2 + RH \quad (131)$$

(252)

The interaction of azirines and diazomethane at room temperature leads to a mixture of allylic azides[410] (equation 132). Although a triazoline intermediate such as **253** could be invoked, mechanistic speculation based on product analysis may be complicated by allylic isomerization (section III.B.2).

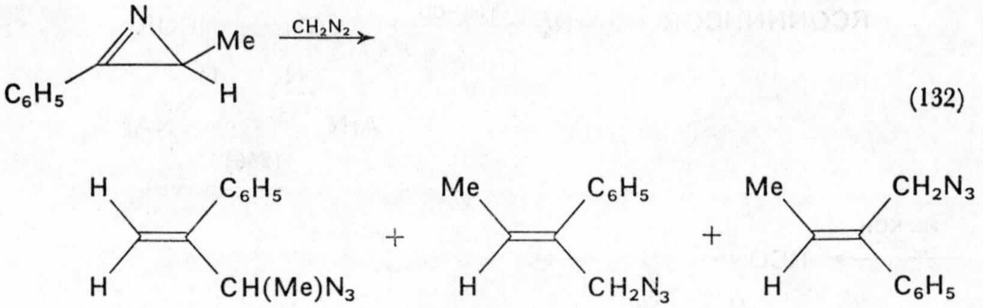

$$\text{2-}C_6H_5\text{-3-Me-3-H-azirine} \xrightarrow{CH_2N_2} H_2C{=}C(C_6H_5)CH(Me)N_3 + Me(H)C{=}C(C_6H_5)CH_2N_3 + Me(H)C{=}C(CH_2N_3)C_6H_5 \quad (132)$$

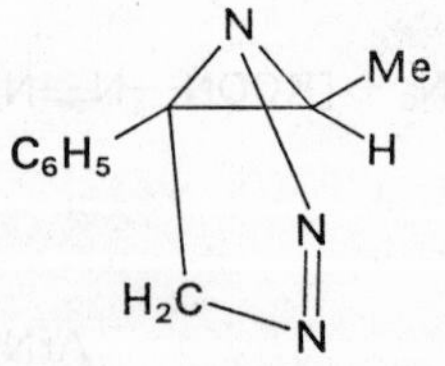

(253)

3. Hydrazoic acid and its derivatives

Aromatic azides may be synthesized from diazonium compounds using hydrazoic acid or azide ion as the nucleophile (equation 133).

$$Ar-\overset{+}{N}\equiv N \; \overset{-}{N}=\overset{+}{N}=\overset{-}{N} \longrightarrow$$

$$Ar-N=\ddot{N}-N=\overset{+}{N}=\overset{-}{N} \longleftrightarrow Ar-N=\ddot{N}-\overset{-}{N}-\overset{+}{N}\equiv N \qquad (133)$$

$$\downarrow$$

$$Ar-N=\overset{+}{N}=\overset{-}{N}+N_2$$

This reaction was first carried out by Noelting and Michel[400] and may be used in all cases in which a primary amine undergoes diazotization, except where the diazonium compound subsequently reacts before hydrazoic acid is introduced into the reaction medium (e.g. diazotization of *o*-aminodiphenylamine results in spontaneous cyclization of the diazonium compound with the formation of 1-phenylbenzotriazole[411]).

This procedure has found wide application in the synthesis of aromatic and heteroaromatic azides; the yields are usually high and often quantitative. General procedures have been developed by Smith and co-workers[412], the method of choice mainly being determined by the basicity of the amine involved or the solubility of its salts. Weakly basic amines, for example, are diazotized with amyl nitrite in an acetic acid–concentrated sulphuric acid mixture and aqueous sodium azide is subsequently added. Amines which form insoluble salts with common mineral acids are converted to the more soluble 2-hydroxyethanesulphonic acid salts prior to diazotization. This procedure has been applied in the diazotization of the *N*-(aminophenyl)phthalimides (**254**). Treatment of the resulting diazonium compounds (**255**) with hydrazoic acid and removal of the protecting phthalimido group affords the otherwise inaccessible azidoanilines[413] (equation 134). Some representative examples of azides recently synthesized by these methods, are shown in Table 11.

$$H_2N-C_6H_4-N(phthalimido) \longrightarrow N_2^+-C_6H_4-N(phthalimido)$$

(254) **(255)**

$$\xrightarrow{N_3^-} N_3-C_6H_4-N(\text{phthalimide}) \xrightarrow{H_2NNH_2} N_3-C_6H_4-NH_2 + \text{phthalhydrazide} \quad (134)$$

TABLE 11. Synthesis of aryl azides from diazonium compounds and hydrazoic acid

Amine	Azide	Yield %	Ref.
2-Azidoaniline	1,2-Diazidobenzene	100	414
2,2′-Diaminoazobenzene	2,2′-Diazidoazobenzene	93	415
1,4-Diaminobenzene	1,4-Diazidobenzene	59	416
2-(Aminophenyl)-phthalimide	2-Azidoaniline	57	413
5-Aminotropolone	5-Azidotropolone	88	417
3-Amino-2-piperidino-pyridine	3-Azido-2-piperidinopyridine	—	418
4-Aminofluorene	4-Azidofluorene	96·5	419
4-Aminofluorenone	4-Azidofluorenone	83·5	419
2-phenyl-1-naphthylamine	1-Azido-2-phenylnaphthalene	43	419
2-Aminocumene	2-Azidocumene	68	420
2-Amino-4′-hydroxy-biphenyl	2-Azido-4′-hydroxybiphenyl	57·5	419
2-Amino-2′-methoxy-biphenyl	2-Azido-2′-methoxybiphenyl	87	420
2,2′-Diaminobiphenyl	2,2′Diazidobiphenyl	83	411
2-Aminodiphenyl thioether	2-Azidodiphenyl thioether	64	411
2-Aminodiphenyl ether	2-Azidodiphenyl ether	92	411
2-Aminobenzophenone	2-Azidobenzophenone	12	411

The mechanisms of reactions between diazonium compounds and various nucleophiles, including azide ion, have been closely investigated by several groups of workers and have recently been discussed by Miller[421]. It is now generally agreed that the reaction between azide ion and diazonium compounds involves an initial attack of the nucleophile on the terminal nitrogen of the diazonium group. This has been clearly demonstrated by the kinetic investigation of Lewis and Johnson[422] who determined the rates of formation of *p*-substituted benzenediazonium ions from tetrazotized *p*-phenylenediamine and several nucleophiles (Cl^-, Br^-, SCN^-, N_3^-, OH^-) at 28·2°. Displacement of nitrogen was found to be bimolecular in each case. Nucleophilic reactivity of a reagent towards substitution on carbon may be approximated by its reactivity towards a standard substrate. This behaviour has been expressed by Swain and Scott[43b] in terms of the general equation (135) where *s* is the reaction constant and *n* is a measure of the nucleophilic character of the reagent. Lewis and Johnson derived a plot of log *k* vs. *n* (Figure 4) using the values of *n*

$$\log k = sn + \text{constant} \qquad (135)$$

determined experimentally by Swain and Scott. A straight line drawn

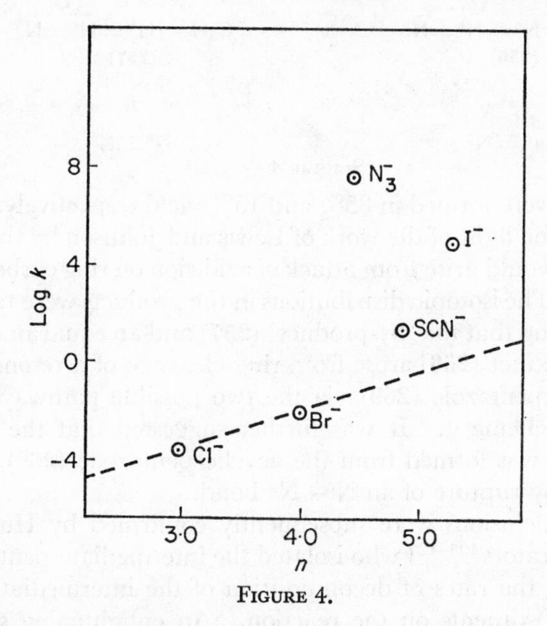

FIGURE 4.

through the points for Cl^- and Br^- has a slope (*s*) of 1·5, which is as large as any cited by Swain and Scott and it has been suggested that Cl^- and Br^- displace nitrogen by an activated S_N2 mechanism[421]. The point for azide ion, however, deviates by an enormous factor (2×10^9) which precludes a mechanism involving nucleophilic attack of azide ion on a carbon atom.

Comprehensive ^{15}N-labelling studies by Clusius, Huisgen and co-workers have led to the elucidation of the detailed mechanism which is shown in Scheme 4. A dual mechanism for this reaction was indicated by the initial investigations of Clusius and colleagues[423,424]. Interaction of benzenediazonium salts and azide ion variously labelled at N^a, N^b or N^c enabled the distinction between two products **256** and

$$[C_6H_5{-}N^a{\equiv}N^b]^+ + [N^c{=}N^d{=}N^c]^- \not\rightarrow C_6H_5{-}N^c{=}\overset{+}{N}{}^d{=}\overset{-}{N}{}^c \quad \textbf{(258)}$$

$$C_6H_5{-}\overset{-}{N}{}^a{-}N^b{=}N^c{-}\overset{+}{N}{}^d{\equiv}N^c \quad \textbf{(260)}$$

$$C_6H_5{-}N^a \text{ (cyclic: } N^b{\equiv}N^c \text{ (B)}, \; N^c{=}N^d \text{ (A))} \quad \textbf{(259)}$$

A; B

$$C_6H_5{-}N^a{=}\overset{+}{N}{}^b{=}\overset{-}{N}{}^c \quad \textbf{(256)} \quad + \quad N^d{\equiv}N^c$$

$$C_6H_5{-}N^a{=}\overset{+}{N}{}^c{=}\overset{-}{N}{}^d \quad \textbf{(257)} \quad + \quad N^b{\equiv}N^c$$

Scheme 4

257 which were formed in 85% and 15% yield respectively. It is significant in the light of the work of Lewis and Johnson[422] that product **258** which would arise from attack of azide ion on ring carbon, was not detected. The isotopic distributions in the products were rationalized by postulating that the 'by-product' (**257**) and an equal amount of the primary product (**256**) arose from ring cleavage of a resonance stabilized cyclic pentazole (**259**) via the two possible pathways A and B shown in Scheme 4. It was further suggested that the remaining 70% of **256** was formed from the acyclic pentazene (**260**) which decomposed by rupture of an N^c—N^d bond.

These conclusions were subsequently confirmed by Huisgen, Ugi and collaborators[425–427] who isolated the intermediate pentazoles and investigated the rates of decomposition of the intermediates and the effect of substituents on the reaction. An enlightening study[426] of the reaction between benzenediazonium ion, isotopically labelled on the terminal nitrogen, and lithium azide in monoglyme clearly indicated the two-path nature of the mechanism. At $-25°$, only unlabelled nitrogen was evolved from the system; on raising the temperature to 0–10°, however, a secondary reaction occurred which generated isotopically labelled nitrogen. Moreover, the phenyl azide produced in the overall reaction was also labelled (refer to Scheme 4 with the specific label on N^b). The percentage of nitrogen evolved in the two separate processes may be plotted against time. A typical diagram for the reaction of benzenediazonium chloride with lithium azide in methanol at −39·5° and −0·8° is shown in Figure 5. This graph and

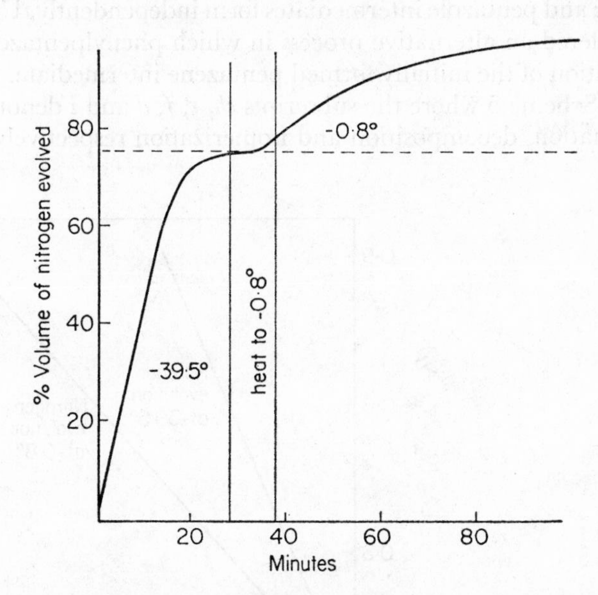

FIGURE 5. Plot of the volume of nitrogen evolved against time for the reaction of benzenediazonium chloride with lithium azide in methanol at −39·5° and −0·8° showing the two-path nature of the reaction.

the first-order plots derived from it (Figure 6) admirably demonstrate the primary and secondary reaction processes.

It is also noteworthy that crystalline arylpentazoles have been isolated from the reaction mixtures at low temperatures and found to decompose into aryl azides and nitrogen[427]. In particular, on decomposition of isotopically labelled *p*-ethoxyphenylpentazole (**261**) the ^{15}N was equally distributed between the *p*-ethoxyphenyl azide and nitrogen[428] (equation 136).

$$\mathrm{EtO{-}C_6H_4{-}\overset{+}{N}{\equiv}N^{*}} + \mathrm{\overset{-}{N}_3} \longrightarrow \mathrm{EtO{-}C_6H_4{-}N_5}\ (\textbf{261}) \longrightarrow \qquad (136)$$

where (**261**) is the pentazole ring N–N=N*–N=N (N* labelled).

$$\mathrm{EtO{-}C_6H_4{-}N{=}\overset{+}{N^{*}}{=}\overset{-}{N}} + \mathrm{EtO{-}C_6H_4{-}N{=}\overset{+}{N}{=}\overset{-}{N}} + \mathrm{N{\equiv}N} + \mathrm{N^{*}{\equiv}N}$$

In the above investigations of the main features of this reaction it was tacitly assumed, though not unequivocally established, that the pentazene and pentazole intermediates form independently. Ugi[429,430] has considered an alternative process in which phenylpentazole arises by cyclization of the initially formed pentazene intermediate. This is shown in Scheme 5 where the subscripts *ch*, *r*, *f*, *d* and *i* denote chain, ring, formation, decomposition and isomerization respectively.

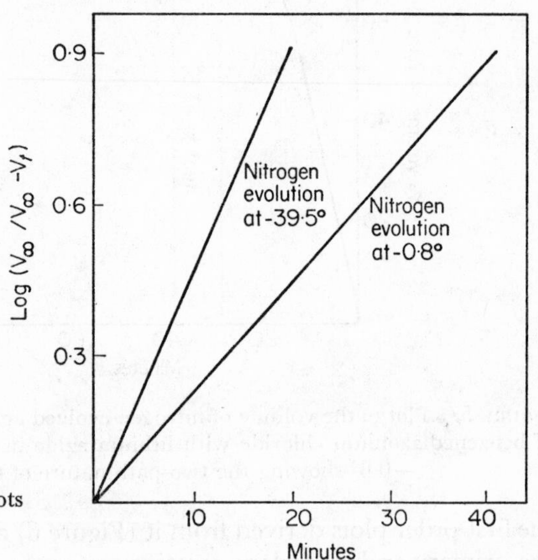

FIGURE 6. First-order plots derived from Figure 5.

According to this mechanism, the observed formation of phenyl azide with an isotopic distribution represented by C_6H_5—N^a=$\overset{+}{N}^c$=$\overset{-}{N}^d$ would result from the isomerization process (*i*). Decomposition of the pentazole would have the same effect, as discussed previously. Ugi has been able to differentiate between the mechanisms outlined in Schemes 4 and 5 by means of experiments in one-phase and two-phase media. In 80% aqueous methanol (phase I) at −35° it was found that lithium azide and benzenediazonium chloride interact to form the pentazene and pentazole in the proportion $Q_{ch/r} = 2{\cdot}03 \pm 0{\cdot}05$. Under these conditions the pentazene decomposes at a rate of $k_{ch.d}(\text{I}) = 5{\cdot}2 \times 10^{-3}\ \text{sec}^{-1}$. When the reaction was repeated in a two-phase system, where phase II (carbon tetrachloride: *n*-hexane, 22:78 wt.%) has almost the same density as the aqueous methanol of phase I and is kept in intimate contact with it by rapid stirring, the

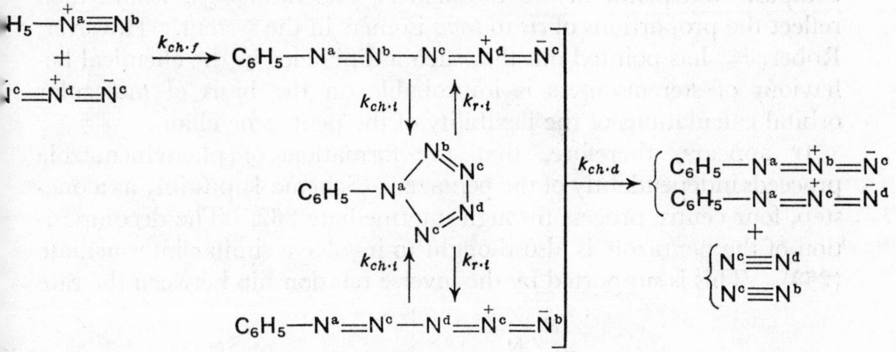

$$\mathrm{C_6H_5{-}\overset{+}{N}{}^a{\equiv}N^b + N^c{=}\overset{+}{N}{}^d{=}\overset{-}{N}{}^e \xrightarrow{k_{ch.f}} C_6H_5{-}N^a{=}N^b{-}N^c{=}\overset{+}{N}{}^d{=}\overset{-}{N}{}^e}$$

$$\mathrm{C_6H_5{-}N^a{=}N^b{-}N^c{=}\overset{+}{N}{}^d{=}\overset{-}{N}{}^e \underset{k_{r.l}}{\overset{k_{ch.l}}{\rightleftharpoons}} C_6H_5{-}N^a(N^bN^cN^dN^e) \underset{k_{r.l}}{\overset{k_{ch.l}}{\rightleftharpoons}} C_6H_5{-}N^a{=}N^e{-}N^d{=}\overset{+}{N}{}^c{=}\overset{-}{N}{}^b}$$

$$\xrightarrow{k_{ch.d}} \left\{\begin{matrix}\mathrm{C_6H_5{-}N^a{=}\overset{+}{N}{}^b{=}\overset{-}{N}{}^c} \\ \mathrm{C_6H_5{-}N^a{=}\overset{+}{N}{}^e{=}\overset{-}{N}{}^d}\end{matrix}\right. + \left\{\begin{matrix}\mathrm{N^c{\equiv}N^d} \\ \mathrm{N^e{\equiv}N^b}\end{matrix}\right.$$

SCHEME 5

ratio $Q_{ch/r(I+II)}$ is almost the same ($1{\cdot}96 \pm 0{\cdot}05$) as in phase I alone. On the other hand, the rate of decomposition of the pentazene is reduced by a factor of 2·5 ($k_{ch.d(I+II)} = 2{\cdot}3 \times 10^{-3}\ \text{sec}^{-1}$). In the mechanism proposed by Ugi (Scheme 5) the ratio $Q_{ch/r}$ is a function of the rate constant $k_{ch.d}(Q_{ch/r} \simeq k_{ch.d}/k_{ch/r})$. However, the experimental evidence shows that $k_{ch.d}$ and $Q_{ch/r}$ are independent of each other, and this can only be consistent with the mechanism in Scheme 5 if both $k_{ch.d}$ and $k_{ch/r}$ vary by exactly the same factor in different media, which does not seem likely. Ugi has also considered the possibility that the pentazene exists in both *cis* and *trans* configurations, one of

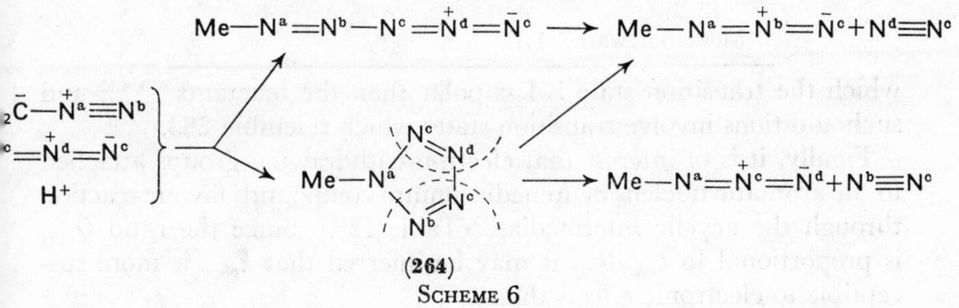

$$\left.\begin{matrix}\mathrm{H_2\overset{-}{C}{-}\overset{+}{N}{}^a{\equiv}N^b} \\ \mathrm{N^c{=}\overset{+}{N}{}^d{=}\overset{-}{N}{}^e} \\ \mathrm{H^+}\end{matrix}\right\} \rightarrow \mathrm{Me{-}N^a{=}N^b{-}N^c{=}\overset{+}{N}{}^d{=}\overset{-}{N}{}^e \rightarrow Me{-}N^a{=}\overset{+}{N}{}^b{=}\overset{-}{N}{}^c + N^d{\equiv}N^e}$$

$$\rightarrow \mathrm{Me{-}N^a(N^cN^dN^eN^b)} \rightarrow \mathrm{Me{-}N^a{=}\overset{+}{N}{}^c{=}\overset{-}{N}{}^d + N^b{\equiv}N^e}$$

(264)

SCHEME 6

which is rapidly converted into the pentazole whilst the other decomposes independently to the azide. The ratio $Q_{ch/r}$ would then reflect the proportions of *cis* to *trans* isomers in the system. However, Roberts[208] has pointed out that such a difference in the chemical behaviour of stereoisomers is improbable on the basis of molecular orbital calculations of the flexibility of the pentazene chain.

It appears, therefore, that the formation of phenylpentazole proceeds independently of the pentazene (Scheme 4) possibly as a one-step, four centre process through intermediate **262**. The decomposition of the pentazole is also thought to involve a similar intermediate (**263**). This is supported by the inverse relationship between the rate

C_6H_5—N (262) C_6H_5—N (263)

(262) **(263)**

of decomposition of the pentazole and the polarity of the medium (Table 12). A similar relationship exists in 1,3-dipolar additions in

TABLE 12. Rates of decomposition of phenylpentazole in various solvents

Solvent	Rate × 10^4 (sec^{-1})
n-Hexane	45·2
Carbon tetrachloride	34·0
Tetrahydrofuran	10·4
Methanol	9·8
Chloroform	8·9
Acetone	7·7
Acetonitrile	4·1
Formic acid	2·3
Carbon tetrachloride:acetonitrile (2:3)	8·0
Methanol:water (1:1)	5·7

which the transition state is less polar than the reactants[209,431] and such additions involve transition states which resemble **263**.

Finally, it is of interest that electron-withdrawing groups attached to the aromatic nucleus of arenediazonium compounds favour reaction through the acyclic intermediate (Table 13). Since the ratio $Q_{ch/r}$ is proportional to $k_{ch.f}/k_{r.f}$ it may be inferred that $k_{ch.f}$ is more susceptible to electronic effects than $k_{r.f}$.

The formation of methyl azide by interaction of hydrazoic acid and diazomethane at low temperatures in inert solvents was initially thought to involve an unstable cyclic pentazole intermediate[432] (**264**) as shown in Scheme 6.

Subsequent mechanistic studies by Clusius and Endtinger[433] using

TABLE 13. The formation and decomposition of substituted phenylpentazoles in methanol at 0°

R in $R—C_6H_4N_5$	$Q_{ch/r}$	Rate of decomp. × 10^4 (sec^{-1})
p-NO_2	6·1	59
p-$(Me)_2NH^+$	—	51
m-NO_2	4·6	36
m-Cl	3·3	23
p-Cl	3·5	12·1
H	3·2	8·4
m-Me	2·6	7·6
m-HO	2·0	7·1
p-Me	1·9	5·6
p-HO	1·9	3·2
p-EtO	1·9	3·0
p-$(Me)_2N$	1·2	1·7
o-NO_2	24	92
o-Cl	8	86
o-Me	3·4	31

variously ^{15}N-labelled hydrazoic acid and diazomethane established that the reaction did not proceed by either of the two pathways shown in Scheme 6. In fact, all of the evolved nitrogen originated from the diazo nitrogen atoms and Clusius and Endtinger have postulated the alternative mechanism for the reaction shown in Scheme 7.

$$\left.\begin{array}{l} H_2\overset{+}{C}—N^a{=}N^b \\ \bar{N}^c{=}\overset{+}{N}^d{=}N^cH \end{array}\right\} \longrightarrow \left[\begin{array}{c} H_2C—N^a{=}N^b—H \\ | \\ N^c{=}\overset{+}{N}^d{=}\bar{N}^c \end{array}\right] \longrightarrow \begin{array}{c} N^a{\equiv}N^b \\ + \\ Me—N^c{=}\overset{+}{N}^d{=}\bar{N}^c \end{array}$$

SCHEME 7

C. Reaction of Hydrazoic Acid with Nitroso Compounds

The synthesis of aryl azides in high yields and under mild conditions by the reaction of aromatic nitroso compounds with hydrazoic acid has been reported by Maffei and co-workers[434,435]. Their results are summarized in Table 14. No intermediates have yet been isolated from this reaction, although Maffei suggested the intervention of a diazohydroxide. If the quantity of hydrazoic acid employed is less

$$\text{Ar—N=O} + \text{HN}_3 \longrightarrow (\text{Ar—}\overset{+}{\text{N}}\equiv\text{N})\,\overset{-}{\text{O}}\text{H} + \text{N}_2 \xrightarrow{\text{HN}_3} \text{ArN}_3 + \text{N}_2 + \text{H}_2\text{O} \quad (137)$$

TABLE 14. Synthesis of azides from aromatic nitroso compounds and hydrazoic acid

Ar in ArNO	% Yield of ArN_3	Reference
o-$O_2NC_6H_4$	100	435
m-$O_2NC_6H_4$	100	435
p-$O_2NC_6H_4$	100	434
o-$HOOCC_6H_4$	100	435
p-$N_3C_6H_4$	100	435
o-ClC_6H_4	91	435
p-ClC_6H_4	92	435
o-BrC_6H_4	89	435
p-Me OC_6H_4	92	435
p-$H_2NC_6H_4$	75[a]	434
p-$(Me)_2NC_6H_4$	78[a]	434

[a] Isolated as its picrate.

than that represented in equation (137), the yield of azide is diminished accordingly and unreacted nitroso compound is isolated with the product. It appears that the intermediate formed from the nitroso compound and the first molecular equivalent of hydrazoic acid interacts more rapidly with hydrazoic acid than does the nitroso compound.

Geller and Samosvat[436] have carried out a detailed investigation of the mechanism of the reaction using the elegant ^{15}N-labelling technique of Clusius and collaborators (section V.B). Three reaction pathways were considered and these are outlined in Scheme 8.

When hydrazoic acid labelled on the two terminal nitrogen atoms, $H^+(^{15}\overset{-}{N} = \overset{+}{N} = {}^{15}\overset{-}{N})$, was used in the reaction the aryl azide produced had the isotopic structure $\text{Ar—N}={}^{15}\overset{+}{\text{N}}={}^{15}\overset{-}{\text{N}}$ as shown by reductive degradation of the azide chain. Only reaction pathway (*iii*) in Scheme 8 is consistent with this isotopic distribution in the product. Moreover, the ^{15}N content of the terminal nitrogen is lower than that of the central nitrogen in accordance with the formation of some aryl azide by decomposition of an aryl pentazole as shown in the secondary pathway of (*iii*) in Scheme 8. The proportion of the reaction proceeding through each pathway has been estimated (cf.

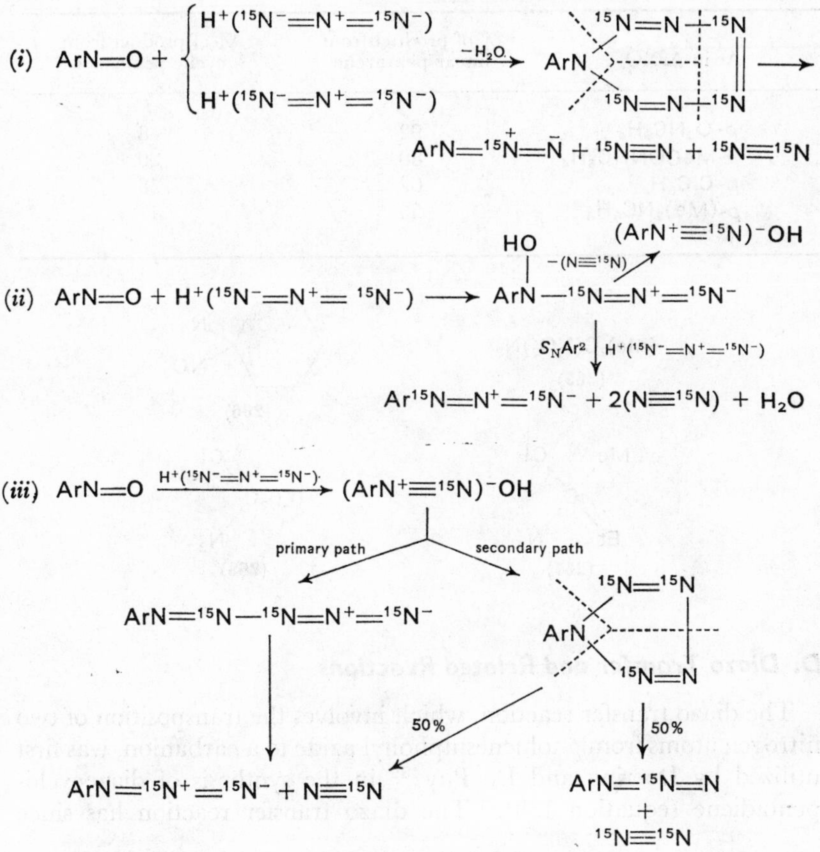

$$(i)\quad ArN{=}O + \begin{cases} H^+({}^{15}N^-{=}N^+{=}{}^{15}N^-) \\ H^+({}^{15}N^-{=}N^+{=}{}^{15}N^-) \end{cases} \xrightarrow{-H_2O} ArN\left\langle\begin{matrix}{}^{15}N{=}N{-}{}^{15}N \\ \| \\ {}^{15}N{=}N{-}{}^{15}N\end{matrix}\right. \longrightarrow ArN{=}{}^{15}\overset{+}{N}{=}\bar{N} + {}^{15}N{\equiv}N + {}^{15}N{\equiv}{}^{15}N$$

$$(ii)\quad ArN{=}O + H^+({}^{15}N^-{=}N^+{=}{}^{15}N^-) \longrightarrow \underset{HO}{|}ArN{-}{}^{15}N{=}N^+{=}{}^{15}N^- \xrightarrow{-(N\equiv{}^{15}N)} (ArN^+{\equiv}{}^{15}N)^-OH$$

$$\downarrow S_NAr2,\ H^+({}^{15}N^-{=}N^+{=}{}^{15}N^-)$$

$$Ar{}^{15}N{=}N^+{=}{}^{15}N^- + 2(N{\equiv}{}^{15}N) + H_2O$$

$$(iii)\quad ArN{=}O \xrightarrow{H^+({}^{15}N^-=N^+={}^{15}N^-)} (ArN^+{\equiv}{}^{15}N)^-OH$$

primary path:

$$ArN{=}{}^{15}N{-}{}^{15}N{=}N^+{=}{}^{15}N^- \longrightarrow ArN{=}{}^{15}N^+{=}{}^{15}N^- + N{\equiv}{}^{15}N$$

secondary path:

$$ArN\left\langle\begin{matrix}{}^{15}N{=}{}^{15}N \\ | \\ {}^{15}N{=}N\end{matrix}\right. \xrightarrow{50\%} ArN{=}{}^{15}N^+{=}{}^{15}N^- + N{\equiv}{}^{15}N$$

$$\xrightarrow{50\%} ArN{=}{}^{15}N{=}N + {}^{15}N{\equiv}{}^{15}N$$

SCHEME 8

section V.B) and it is apparent (Table 15) that electron withdrawing substituents on the aromatic nucleus favour reaction through the linear pentazene. This behaviour has been discussed previously in connexion with the formation of azides from hydrazoic acid and diazonium compounds (section V.B.3).

The reaction has been extended by Maffei and co-workers[437,438] to include the synthesis of the α-nitroazidoalkenes (**265**) and (**266**) from the corresponding pseudonitroles. Similarly the α-chloroazidoalkanes (**267**) and (**268**) have been isolated.

TABLE 15. Proportion of ArN_3 arising from linear and cyclic pentazene in the reaction of ArNO with HN_3

Ar in ArNO	% of product from linear pentazene	% of product from cyclic pentazene
p-$O_2NC_6H_4$	92	8
p-$MeCONHC_6H_4$	80	20
p-ClC_6H_4	62	38
p-$(Me)_2NC_6H_4$	53	47

$(Me)_2C(NO_2)N_3$
(265)

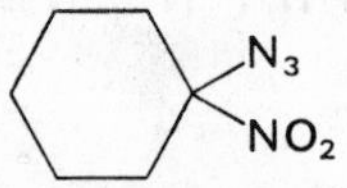

(266)

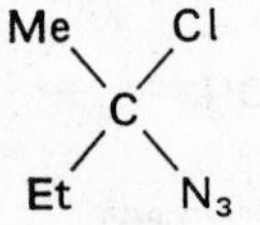

(267)

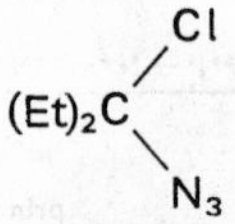

(268)

D. Diazo Transfer and Related Reactions

The diazo transfer reaction, which involves the transposition of two nitrogen atoms from p-toluenesulphonyl azide to a carbanion, was first utilized by Doering and De Puy[439] in the synthesis of diazocyclopentadiene (equation 138). The diazo transfer reaction has since

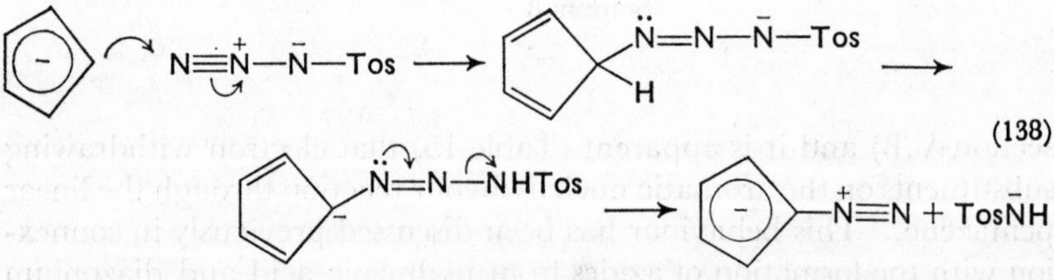

(138)

been used to advantage in the synthesis of diazoalkanes[440]. More recently it has been modified by Anselme and co-workers[441–443] to provide an interesting synthetic route to azides. It was found that the corresponding azide was formed in moderate yield (see Table 16) when p-toluenesulphonyl azide was reacted with the conjugate base of an

TABLE 16. Preparation of azides by the diazo transfer reaction

Amine	Azide[a] (% yield)	Recovered $TosN_3$ (%)	$TosNH_2$ (%)
Aniline	44	18·3	61
p-Toluidine	49(43)[b]	13·3	36
p-Chloroaniline	48	15·0	66
Benzylamine	26	15·0	36
Cyclohexylamine	(35)[b]	—	—

[a] Yields are based on the amount of unrecovered *p*-toluenesulphonyl azide.
[b] Figures in parentheses refer to reactions in which methyllithium was employed; all other figures refer to reactions using methylmagnesium chloride.

aromatic or aliphatic amine, generated by reaction with methylmagnesium chloride or methyllithium.

Although the mechanism of this reaction has yet to be fully established certain of its aspects have been defined. The intermediacy of the anion **269** is very probable, particularly since a solid which is presumably the intermediate (**269**; R = C_6H_5) has been isolated from the reaction of anilide ion with *p*-toluenesulphonyl azide. In support of the structural assignment this solid gave a mixture of phenyl azide (40%) and aniline when heated. The reaction sequence has been summarized as shown in equation (139).

$$R\bar{N}H + N_3Tos \rightleftharpoons RNH{-}N{=}N{-}\bar{N}Tos \quad (\mathbf{269})$$

$$RNHN{=}N\bar{N}Tos \rightarrow [\text{cyclic } RN{\cdots}H{\cdots}NTos \text{ via } N{=}N;\ \text{or } RN(H){-}N{=}N{-}\bar{N}Tos \text{ with } H_2O] \rightleftharpoons R\bar{N}{-}\ddot{N}{=}N{-}NHTos \rightarrow RN_3 + Tos\bar{N}H \quad (139)$$

Anselme and co-workers have suggested that the stability of the anion $R\bar{N}H$ relative to $Tos\bar{N}H$ may be of significance in the diazo transfer reaction. The poor conversion of *p*-nitroanilide ion into *p*-nitrophenyl azide was originally ascribed to a reaction between the organometallic reagent and the nitro group of the substrate. However, it may be due at least in part to the greater stability of the *p*-nitroanilide ion relative to $\text{Tos}\bar{\text{N}}\text{H}$. One might similarly rationalize the failure of this reaction when the conjugate bases of benzamide and *N*-benzyl-*p*-toluenesulphonamide are employed as substrates.

Nitrous oxide (**270**) is related electronically to the azido group and has been examined as a potential diazo transfer reagent. Meier[444]

$$\bar{N}{=}\overset{+}{N}{=}O \longleftrightarrow N{\equiv}\overset{+}{N}{-}\bar{O}$$

(270)

observed that with nitrous oxide, anilide ion gave a small amount of oil, thought to be phenyl azide, together with 57% of azobenzene. Anselme and Koga[445] reinvestigated this and related reactions at various pressures and obtained moderate yields of azides. Their

TABLE 17. Azides from reaction of N_2O and amine anions[a]

Azide	Pressure lb/in²	Time hr	% Yield (Gross)	% Yield (Net)[b]
$C_6H_5N_3$[c]	30–40	5	9 (10)	39
	80	70	18	32
	80	42	35	51
p-$MeC_6H_4N_3$	50–55	2·5	16	46
p-$MeOC_6H_4N_3$[c]	50–55	4·5	25 (19)	35
cyclo-$C_6H_{11}N_3$	30–50	5·0	10	15

[a] Anions prepared by the reaction of the amine with methyllithium.
[b] Based on the amount of unrecovered amine.
[c] Values in parentheses refer to initial runs carried out by bubbling N_2O through the reaction mixture.

findings are summarized in Table 17. No mechanistic details are as yet available but the sequence outlined in equation (140) would appear reasonable.

$$\bar{N}{=}\overset{+}{N}{=}O \longleftrightarrow N{\equiv}\overset{+}{N}{-}\bar{O} \xrightarrow{R\bar{N}H} RNHN{=}N{-}O^- \rightleftharpoons R\bar{N}{-}\bar{N}{=}N{-}OH \longrightarrow R\bar{N}{-}\overset{+}{N}{\equiv}N + \bar{O}H \quad (140)$$

There are some resemblances between this reaction and the formation of diazoalkanes from *N*-nitroso-*N*-*p*-toluenesulphonamides, for which the intermediate (**271**), presumably resulting from an anionotropic rearrangement, has been postulated[446] (equation 141). There

$$R^2CH{-}N(N{=}O){-}Tos \rightleftharpoons R^2CH{-}N{=}N{-}OTos \;(\mathbf{271}) \longrightarrow R^2CH{-}\overset{+}{N}{\equiv}N + Tos\bar{O} \;\dashrightarrow\; R^2C{=}\overset{+}{N}{=}\overset{-}{N} + TosOH \qquad (141)$$

is a closer analogy, however, with the conversion of *N*-nitrosohydrazides into azides which has been described by Ponzio and Canuto[447] and is discussed in section V.E.

A further similarity exists between the diazo transfer reaction and an azide synthesis involving the use of Grignard reagents and *p*-toluenesulphonyl azide, which has recently been described[398,399]. The reagents interact to form salts of *p*-toluenesulphonyltriazenes which undergo fragmentation in aqueous sodium pyrophosphate at 0–20°, sodium hydroxide, or on dry distillation, giving aryl or alkyl azides (equation 142). Moderate to good yields of aromatic azides can be

$$RCl \longrightarrow RMgX \xrightarrow{R^1SO_2N_3} (RNNNSO_2R^1)^- \, MgX^+ \longrightarrow RN_3 + R^1SO_2^- \, MgX^+ \qquad (142)$$

achieved from this process (which is essentially a triazo transfer), but aliphatic azides are obtained only in poor yields. Although nearly all simple aryl azides are as readily obtained by the conventional diazotization procedure, the principal utility of the synthesis is with more complex compounds where amino groups cannot readily be introduced (e.g. the synthesis of *o*-azidostyrene derivatives).

E. Nitrosation of Hydrazine Derivatives

Phenylhydrazine reacts typically with nitrous acid to form α-nitrosophenylhydrazine (**272**), which undergoes a facile dehydration to phenyl azide when treated with acid or alkali. When Fischer[448] discovered this reaction he postulated a cyclic structure (**273**) for phenyl azide, but this formulation was later abandoned in favour of the linear structure[14], which has now been firmly established by ^{15}N-labelling studies[449].

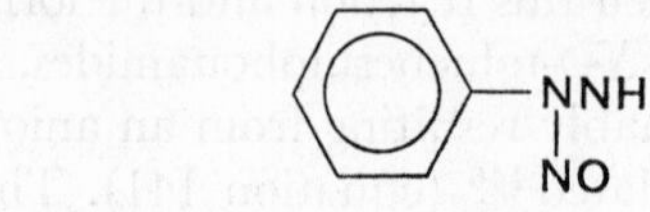

$C_6H_5N(NO)NH_2$ **(272)** $C_6H_5N_3$ (cyclic) **(273)**

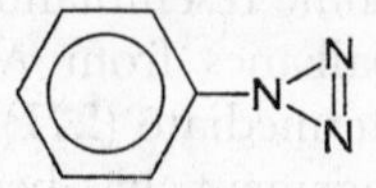

By use of tracer experiments, Clusius[450] showed that branched pathways are involved in this reaction; nitrosation of phenylhydrazine with ^{15}N-labelled nitrous acid produced two isotopic isomers of phenyl azide (**274**) and (**275**) (equation 143).

In a more detailed examination of the mechanism, Clusius and Schwarzenbach[451] found that the same mixture of isotopic isomers arises both from diazotization of phenylhydrazine with labelled nitrous acid and from dehydration of specifically labelled α-nitrosophenylhydrazine (**276**). It was concluded that α-nitrosophenylhydrazine is an intermediate in this reaction. Its dehydration through a 3-membered ring cation (conjugate acid of **277**) was proposed, as shown in Scheme 9.

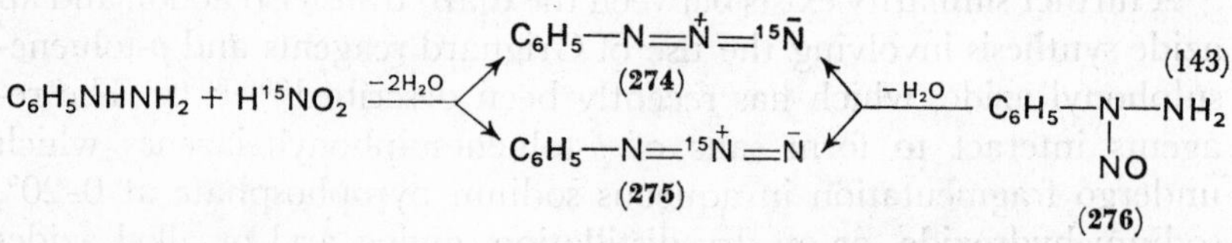

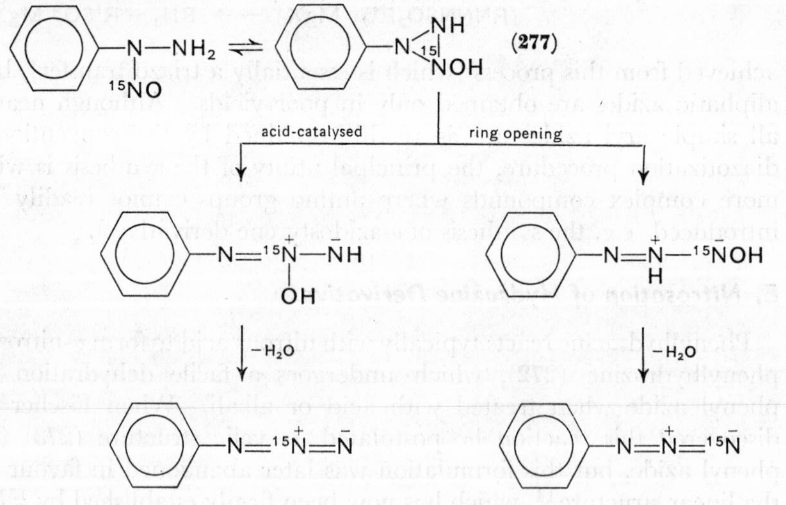

$$C_6H_5NHNH_2 + H^{15}NO_2 \xrightarrow{-2H_2O} C_6H_5{-}N{=}\overset{+}{N}{=}{}^{15}\overset{-}{N}\ \textbf{(274)} \;+\; C_6H_5{-}N{=}{}^{15}\overset{+}{N}{=}\overset{-}{N}\ \textbf{(275)} \xleftarrow{-H_2O} C_6H_5N(NO)NH_2\ \textbf{(276)} \qquad (143)$$

$$C_6H_5N(^{15}NO)NH_2 \rightleftharpoons C_6H_5N\langle NH{-}{}^{15}N(OH)\rangle\ \textbf{(277)}$$

acid-catalysed ring opening

$$C_6H_5{-}N{=}{}^{15}\overset{+}{N}(OH){-}NH \xrightarrow{-H_2O} C_6H_5{-}N{=}{}^{15}\overset{+}{N}{=}\overset{-}{N}$$

$$C_6H_5{-}N{=}\overset{+}{N}(H){-}{}^{15}\overset{-}{N}OH \xrightarrow{-H_2O} C_6H_5{-}N{=}\overset{+}{N}{=}{}^{15}\overset{-}{N}$$

Scheme 9

The acid catalysed conversion[452] of the nitroso derivative of benzylmethylhydrazine (**278**) into its isomer (**279**) is significant in the context of the above mechanism. In addition, the formation of aryl azides from nitrosohydrazides such as **280** by treatment with 20% sodium hydroxide solution[447] (equation 144) is consistent with the results of Clusius and co-workers.

$C_6H_5CH_2N(NO)NHMe$ **(278)** $\qquad$ $C_6H_5CH_2NH{-}N(Me)NO$ **(279)**

$$ArN(NO)NHCOR \xrightarrow{20\%\ NaOH} ArN_3 + RCO_2H \qquad (144)$$

(280)

In contrast to phenylhydrazine, 2,4-dinitrophenylhydrazine yields only the aryl azide of isotopic structure $ArN{=}\overset{+}{N}{=}{}^{15}\overset{-}{N}$ when treated with $H^{15}NO_2$[453]. This suggests that for arylhydrazines in which the aromatic nucleus is substituted with electron-withdrawing groups, nitrosation is directed to the β-nitrogen atom and the resulting β-nitrosohydrazine undergoes spontaneous decomposition to the terminally labelled aryl azide.

Synthesis of aromatic azides by nitrosation of hydrazine derivatives has usually been employed in preference to procedures involving diazotization of a primary amine only in cases where this latter procedure is of a difficult or uncertain nature. For this reason many heteroaromatic azides have been synthesized by nitrosation of their readily obtainable hydrazino derivatives (Table 18).

Alkyl azides are not generally obtained in appreciable yields by nitrosation of alkylhydrazines, although some examples have been reported (Table 18). Smith and co-workers[468] suggested that the intermediate α-nitrosoalkylhydrazines, which cannot give azides without rearrangement and dehydration (Scheme 9), decompose by alternative routes. Certainly there is at least one recorded example where loss of N_2O after rearrangement competes more favourably than dehydration[452]. However, the notorious instability of the lower alkyl azides might also be invoked to rationalize the poor yields in these reactions, even if dehydration were more facile than loss of N_2O. Indeed, Longo[469] has observed that substantial evolution of nitrogen accompanies azide formation from some nitrosohydrazine derivatives

TABLE 18. Preparation of azides by nitrosation of hydrazine derivatives

Hydrazine	Azide	Ref.
Phenylhydrazine	Phenyl azide	25,454
3-Hydrazinophenylsulphonic acid	3-Azidophenylsulphonic acid	455
2,4-Dinitrophenylhydrazine	2,4-Dinitrophenyl azide	456
2,3-Dichloro-5-nitrophenyl-hydrazine	2,3-Dichloro-5-nitrophenyl azide	457
5-Amino-2,4-disulphonamido-phenylhydrazine	5-Amino-2,4-disulphonamido-phenyl azide	458
Benzylhydrazine	Benzyl azide	459
Triphenylmethylhydrazine	Triphenylmethyl azide	460
Ethyl α-hydrazino-α-phenylacetate	Ethyl α-azido-α-phenylacetate	461
Ethyl α-hydrazino-β-phenyl-propionate	Ethyl α-azido-β-phenyl-propionate	461,462
Ethyl α-hydrazinohexanoate	Ethyl α-azidohexanoate	463
H_2NHN, N, N, N→O	N_3, N, N, N→O	204
H_2NHN, $NHNH_2$, N, N	N_3, N_3, N, N	464
N—N, H_2NHN, C_6H_5, O	N—N, N_3, C_6H_5, O	465
S, $H_2NN=$, N, Me	S, N_3, $\overset{+}{N}$, Me	466
NH_2, N, N, H_2NHN, N, N, N	NH_2, N, N, N_3, N, N, N	464

TABLE 18.—*continued*

Hydrazine	Azide	Ref.
2-hydrazino-tropone (O, $NHNH_2$)	2-azidotropone (O, N_3)	467

and it is also pertinent that higher molecular weight aliphatic compounds give reasonable yields of azides by this method.

A limited number of alkyl azides have been obtained by Smith, Clegg and Lakritz[468] by reaction of *N*-acyl-*N*-alkylhydrazines with nitrous acid (equation 145). In this manner, 2-hexahydrobenzylazide and 2-isobutyl azide were obtained from the corresponding

$$H_2NN(CONH_2)R \xrightarrow{HONO} RN_3 + CO_2 + N_2 \qquad (145)$$

alkyl semicarbazides and similarly *N*-*t*-butyl-*N*-benzoylhydrazine afforded a small quantity of *t*-butyl azide. However, as the starting materials are not readily accessible and the yields of azide are indifferent this synthesis does not provide a practical alternative to the standard procedure (section III.B).

Acyl azides may generally be prepared by nitrosation of acylhydrazines with nitrous acid[470–472], nitrosyl chloride[473] and organic nitrites[474]. The competitive production of amides which sometimes accompanies the formation of acyl azides is averted by use of nitrosyl chloride or nitrous acid esters in organic solvents[474]. This side reaction is also suppressed when the synthesis is carried out at high proton and nitrite concentrations[176], although no explanation of this observation has been forthcoming.

Preparation of acyl azides by the nitrosation method has been used to advantage in the synthesis of peptides[177,178,474–477] (cf. section III.B.3) and is particularly valuable since it is the only procedure for peptide chain lengthening which does not cause racemization of the peptide components[177,178]. The use of acyl azides as protecting agents[180,181,186,187,478] for labile amino groups in peptide synthesis has been discussed earlier. A sequential degradation of peptides involving the Curtius rearrangement of an acyl azide has also been reported[479].

Synthesis of azides by nitrosation of hydrazidines such as **281** has been achieved[217]. These compounds can undergo nitrosation on the imino nitrogen of the hydrazone group or on the hydrazino substituent. It has been shown that the latter possibility occurs when the hydrazone hydrogen atom is strongly hydrogen bonded to a nitro group in the adjacent aryl nucleus. In this case hydrazidic azides (**282**) are formed in good yield. In the absence of such hydrogen bonding, azides are not produced and the products, originally formulated as tetrazoles[480], are of uncertain structure.

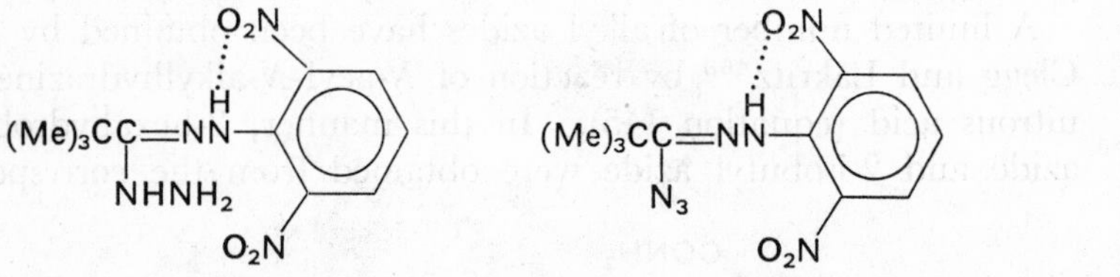

$$(Me)_3CC(NHNH_2){=}N{-}NH{-}C_6H_3(NO_2)_2$$
(281)

$$(Me)_3CC(N_3){=}N{-}NH{-}C_6H_3(NO_2)_2$$
(282)

Carbamoyl azides are produced by the action of nitrous acid on semicarbazides (equation 146) and this reaction has been used to prepare the parent compound and several *N*-substituted derivatives. A recent review on carbamoyl azides, including synthetic procedures, is available[12].

$$RNHCONHNH_2 + HONO \longrightarrow RNHCON_3 + 2\ H_2O \qquad (146)$$

VI. MISCELLANEOUS SYNTHESES

Some procedures for introducing the azido group cannot readily be incorporated into the above general classifications. Many of these reactions are mechanistically unclear and have not been widely investigated. For completeness these syntheses are now considered.

A. Electrolytic Reactions

An electro-oxidation procedure developed by Wright for the synthesis of azides from salts of carboxylic acids or nitroalkanes (equations 147, 148) has been briefly referred to by Smith in a recent review[481].

$$\text{cyclohexyl}^{-}\text{—}NO_2 + N_3^- \xrightarrow{\text{Pt anode } 0\cdot9\text{v}} \text{1-azido-1-nitrocyclohexane } (N_3, NO_2) + 2e \quad (147)$$

$$C_5H_{11}COO^- + N_3^- \xrightarrow{\text{Pt anode}} C_5H_{11}N_3 + CO_2 + 2e \quad (148)$$

B. Oxidation Reactions

Arylhydrazines undergo complex oxidation reactions from which a variety of products such as arylhydrocarbons, biaryls, azo compounds, tetrazenes and phenols may be obtained. Hypochlorite oxidation of aryl semicarbazides, however, affords aryl azides[454,482], possibly by oxidation to the azo compound (**283**) which would undergo Hofmann rearrangement to a triazene (**284**) with subsequent oxidation to afford the azide (equation 149).

$$\text{ArNHNHCONH}_2 \xrightarrow{\text{NaOCl}} \underset{(\mathbf{283})}{\text{ArN}{=}\text{NCONH}_2} \xrightarrow{\text{NaOCl}} \underset{(\mathbf{284})}{\text{ArN}{=}\text{NNH}_2} \xrightarrow{\text{NaOCl}} \text{ArN}_3 \quad (149)$$

Another oxidative procedure has recently been described by Koga and Anselme[483] who found that reaction of 1,1-dibenzylhydrazine (**285**) and of the symmetrical tetrazene (**286**) with lead tetraacetate in benzene gave benzyl azide in high yield (equation 150). The tetrazene intermediate which has been invoked is thought to arise from an *N*-nitrene.

$$2\,\underset{(\mathbf{285})}{(C_6H_5CH_2)_2NNH_2} \xrightarrow{Pb(OAc)_4} \underset{(\mathbf{286})}{(C_6H_5CH_2)_2NN{=}NN(CH_2C_6H_5)_2} \xrightarrow{Pb(OAc)_4}$$

$$Pb(OAc)_2 + MeCO_2H + C_6H_5\text{—}CH_2N(CH(C_6H_5)(OAc))\text{—}N{=}N\text{—}N(CH_2C_6H_5)_2 \longrightarrow \quad (150)$$

$$C_6H_5CH_2N_3 + (C_6H_5CH_2)_2NH + C_6H_5CHO$$

VII. REFERENCES

1. P. Griess, *Phil. Trans. Roy. Soc. London*, **13,** 377 (1964); *Ann. Chem.* **135**, 131 (1865); *Ann. Chem.* **137**, 39 (1966); *Ber.*, **2,** 370 (1869).
2. T. Curtius, *Ber.*, **23,** 3023 (1890); *J. Prakt. Chem.*, **50**, 275 (1894).

3. P. A. S. Smith in *Molecular Rearrangements* (Ed. P. de Mayo), Interscience, New York, 1963, p. 457.
4. P. A. S. Smith, *Open-Chain Nitrogen Compounds*, Vol. 2, Benjamin, New York, 1966, pp. 211–256.
5. R. A. Abramovitch and B. A. Davis, *Chem. Rev.*, **64,** 149 (1964).
6. L. Horner and A. Christmann, *Angew. Chem. Intern. Ed. Engl.*, **2,** 599 (1963).
7. G. L'Abbe, *Chem. Rev.*, **69,** 345 (1969).
8. C. Grundmann in *Houben-Weil, Die Methoden der Organischen Chemie*, Vol. 10, 1965, p. 781.
9. E. Lieber, J. S. Curtice and C. N. R. Rao, *Chem. Ind.* (*London*), 586 (1966).
10. A. Schonberg, *Preparative Organic Photochemistry*, Springer-Verlag, New York, 1968, 318.
11. R. O. Kan, *Organic Photochemistry*, McGraw-Hill, New York, 1966, p. 248.
12. E. Lieber, R. L. Minnis, Jr. and C. N. R. Rao, *Chem. Rev.*, **65,** 377 (1965).
13. C. G. Overberger, J.-P. Anselme and J. G. Lombardino, *Organic Compounds with Nitrogen–Nitrogen Bonds*, Ronald Press, New York, 1966, p. 99.
14. Sidgwick's *Organic Chemistry of Nitrogen*, 3rd ed. rewritten and revised by I. T. Millar and H. D. Springall, Oxford, London, 1966, p. 490.
15. J. H. Boyer and F. C. Canter, *Chem. Rev.*, **54,** 1 (1954).
16. G. R. Harvey and K. W. Ratts, *J. Org. Chem.*, **31,** 3907 (1966).
17. J. H. Boyer, *J. Am. Chem. Soc.*, **73,** 5248 (1951).
18. O. Tureck, *Chim. Ind.* (*Paris*), **26,** 781 (1931).
19. E. Ott and E. Ohse, *Ber.*, **54,** 179 (1921); *Chem. Zentr.*, **II**, 1118, 1194 (1922).
20. C. Taylor and W. H. Rinkenbach, *J. Franklin Inst.*, **19B,** 551 (1923).
21. T. Urbanski, *Chemistry and Technology of Explosives*, Vol. 3, Pergamon, London, 1967, 161.
22. C. Grundmann and H. Haldenwanger, *Angew. Chem.*, **62,** 4210 (1950).
23. C. L. Currie and B. de B. Darwent, *Can. J. Chem.*, **41,** 1048 (1963).
24. M. O. Forster and J. C. Withers, *J. Chem. Soc.*, **101,** 489 (1912).
25. R. O. Lindsay and C. F. H. Allen, *Org. Syn.*, Coll. Vol. III, 710 (1955).
26. R. H. Wiley and J. Moffat, *J. Org. Chem.*, **22,** 995 (1957).
27. Y. A. Sinnema and J. F. Arens, *Rec. Trav. Chim.*, **74,** 901 (1955).
28. J. A. Durden, Jr., H. A. Stansbury and W. H. Catlette, *J. Chem. Eng. Data*, **9,** 228 (1964).
29. L. Birkofer and A. Ritter, *Angew. Chem. Intern. Ed. Engl.*, **4,** 417 (1965).
30. J. von Braun, *Ann. Chem.*, **490,** 125 (1931).
31. W. S. Frost, J. C. Cothran and A. W. Browne, *J. Am. Chem. Soc.*, **55,** 3516 (1933).
32. L. F. Audrieth and C. F. Gibbs, *Inorg. Syn.*, **1,** 78 (1939).
33. R. Stern, *Klin. Wochschr.*, **6,** 304 (1927).
34. F. B. Shakhnovskaya, *Farmakol. i Toksikol.*, **13,** 3, 4 (1950).
35. T. Curtius, *Ber.*, **24,** 3341 (1891); *J. Prakt. Chem.*, **43,** 207 (1891); *Ber.*, **29,** 759 (1896); *J. Prakt. Chem.*, **48,** 273 (1893); *J. Prakt. Chem.*, **91,** 1, 415 (1915).
36. M. Oesterlin, *Angew. Chem.*, **45,** 536 (1932).
37. *Safety in Handling Hazardous Chemicals*, Matheson. Coleman and Bell Handbook, 1968, p. 7.

38. M. R. Crampton, *Advan. Phys. Org. Chem.*, **7**, 211 (1969).
39. J. Miller, *Aromatic Nucleophilic Substitution*, Elsevier, London, 1968, Chap. 1.
40. R. F. Hudson, *Chimia* (*Aarau*), **16**, 173 (1962).
41. R. F. Hudson and M. Green, *J. Chem. Soc.*, 1055 (1962).
42. J. Miller, *Aromatic Nucleophilic Substitution*, Elsevier, London, 1968, Chap. 6.
43a. C. G. Swain, *J. Am. Chem. Soc.*, **70**, 1119 (1948).
43b. C. G. Swain and C. B. Scott, *J. Am. Chem. Soc.*, **75**, 141 (1953).
43c. C. G. Swain and W. B. Langsdorf, *J. Am. Chem. Soc.*, **73**, 2813 (1951).
43d. C. G. Swain, R. B. Moseley and D. W. Bown, *J. Am. Chem. Soc.*, **77**, 3731 (1955).
44. J. O. Edwards, *J. Am. Chem. Soc.*, **76**, 1540 (1954); J. O. Edwards, *J. Am. Chem. Soc.*, **78**, 1819 (1956); J. O. Edwards and R. G. Pearson, *J. Am. Chem. Soc.*, **84**, 16 (1962).
45. R. G. Pearson in *Survey of Progress in Chemistry*, Vol. 5 (Ed. A. F. Scott), Academic Press, New York, 1969, p. 1.
46. J. Miller and K. W. Wong, unpublished results.
47. D. L. Hill, K. C. Ho and J. Miller, *J. Chem. Soc.*, **B**, 299 (1966).
48. P. Caveng and H. Zollinger, *Helv. Chim. Acta*, **50**, 861 (1967).
49. J. Miller, A. J. Parker and (in part) B. A. Bolto, *J. Am. Chem. Soc.*, **79**, 93 (1957).
50. K. C. Ho, J. Miller and K. W. Wong, *J. Chem. Soc.*, **B**, 310 (1966).
51. J. Miller and A. J. Parker, *J. Am. Chem. Soc.*, **83**, 117 (1961).
52. S. Patai and Y. Gotschal, *Israel J. Chem.*, **3**, 223 (1965–66).
53. B. O. Coniglio, D. E. Giles, W. R. McDonald and A. J. Parker, *J. Chem. Soc.*, *B*, 152 (1966).
54. E. S. Gould, *Mechanism and Structure in Organic Chemistry*, Holt, Rinehart and Winston, New York, 1959.
55. F. Hiron and E. D. Hughes, *J. Chem. Soc.*, 795 (1960).
56. S. H. Harvey, P. A. T. Hoye, E. D. Hughes and C. K. Ingold, *J. Chem. Soc.*, 800 (1960).
57. M. L. Bender, *Chem. Rev.*, **60**, 53 (1960).
58. C. G. Swain and C. B. Scott, *J. Am. Chem. Soc.*, **75**, 246 (1953).
59. Z. Rappoport, *Advan. Phys. Org. Chem.*, **7**, 1 (1969).
60. A. Campagni, G. Modena and P. E. Todesco, *Gazz. Chim. Ital.*, **90**, 694 (1960).
61. P. Beltrame, G. Favini, M. G. Cattania and F. Guella, *Gazz. Chim. Ital.*, **98**, 380 (1968).
62. P. B. D. de la Mare in *Molecular Rearrangements* (Ed. P. De Mayo), Interscience, New York, 1963, p. 27.
63. R. H. De Wolfe and W. G. Young, *Chem. Rev.*, **56**, 753 (1956).
64. A. J. Parker, *Quart. Rev.* (*London*), **16**, 163 (1962).
65. J. Miller, *Aromatic Nucleophilic Substitution*, Elsevier, London, 1968, Chap. 8.
66. U. Miotti, *Gazz. Chim. Ital.*, **97**, 254 (1967).
67. F. H. Kendall and J. Miller, unpublished work.
68. C. A. Bunton, *Nucleophilic Substitution at a Saturated Carbon Atom*, Elsevier, London, 1963, Chap. 3.
69. P. Brewster, E. D. Hughes, C. K. Ingold and P. A. D. S. Rao, *Nature*, **166**, 178 (1950); P. Brewster, F. Hiron, E. P. Hughes, C. K. Ingold and P. A. D. S. Rao, *Nature*, **166**, 179 (1950).

70. H. Wiener and R. A. Sneen, *J. Am. Chem. Soc.*, **84,** 3599 (1962).
71. H. Wiener and R. A. Sneen, *J. Am. Chem. Soc.*, **87,** 287 (1965).
72. H. Wiener and R. A. Sneen, *J. Am. Chem. Soc.*, **87,** 292 (1965).
73. R. A. Sneen and J. W. Larsen, *J. Am. Chem. Soc.*, **88,** 2594 (1966).
74. A. Streitwieser and S. Pulver, *J. Am. Chem. Soc.*, **86,** 1587 (1964).
75. T. Curtius and A. Darapsky, *J. Prakt. Chem.*, **63,** 428 (1901).
76. O. Dimroth and W. Wislicenus, *Ber.*, **38,** 1573 (1905).
77. H. Staudinger and E. Hauser, *Helv. Chim. Acta*, **4,** 872 (1921).
78. R. Meldola and H. Kuntzen, *J. Chem. Soc.*, **99,** 36 (1911).
79. M. O. Forster and F. M. van Gelderen, *J. Chem. Soc.*, **99,** 239 (1911).
80. M. O. Forster and F. M. van Gelderen, *J. Chem. Soc.*, **99,** 2059 (1911).
81. D. Vorlander and H. David, *Ber.*, **70,** 146 (1937).
82. E. Oliveri-Mandala and G. Caronna, *Gazz. Chim. Ital.*, **71,** 182 (1941).
83. J. H. Boyer and J. Hamer, *J. Am. Chem. Soc.*, **77,** 951 (1955).
84. E. Lieber, T. S. Chao and C. N. R. Rao, *J. Org. Chem.*, **22,** 238 (1957).
85. P. A. S. Smith, *Org. Reactions*, **3,** 374 (1946).
86. J. Thiele, *Ber.*, **41,** 2681 (1908).
87. C. Naegeli, L. Gruntuch-Jacobson and P. Lendorff, *Helv. Chim. Acta*, **12,** 227 (1929).
88. J. Nelles, *Ber.*, **65,** 1345 (1932).
89. M. S. Newman, *J. Am. Chem. Soc.*, **57,** 732 (1935).
90. K. Henkel and F. Weygand, *Ber.*, **76B,** 812 (1943).
91. D. N. Jones, *Chem. Ind.* (*London*), 179 (1962).
92. K. Ponsold, *Ber.*, **95,** 1727 (1962).
93. R. Goutarel, A. Cave, L. Tan and M. Leboeuf, *Bull. Soc. Chim. France*, 646 (1962).
94. G. Smolinsky, *J. Am. Chem. Soc.*, **83,** 4483 (1961).
95. E. J. Reist, R. R. Spencer, B. R. Baker and L. Goodman, *Chem. Ind.* (*London*), 1794 (1962).
96. A. K. Bose, J. F. Kistner and L. Farber, *J. Org. Chem.*, **27,** 2925 (1962).
97. A. L. Logothetis, *J. Org. Chem.*, **29,** 3049 (1964).
98. Y. Ali and A. C. Richardson, *J. Chem. Soc.*, **C,** 1764 (1968).
99. J. H. Boyer and D. Straw, *J. Am. Chem. Soc.*, **74,** 4506 (1952).
100. J. H. Boyer and D. Straw, *J. Am. Chem. Soc.*, **75,** 1642 (1953).
101. H. Bretschneider and H. Hormann, *Monatsh. Chem.*, **84,** 1021 (1953).
102. A. H. Sommers and J. D. Barnes, *J. Am. Chem. Soc.*, **79,** 3491 (1957).
103. M. L. Wolfrom, J. Bernsmann and D. Horton, *J. Org. Chem.*, **27,** 4505 (1962).
104. C. K. Ingold, *Structure and Mechanism in Organic Chemistry*, 2nd ed., Bell, London, 1969.
105. J. H. Boyer, F. C. Canter, J. Hamer and R. K. Putney, *J. Am. Chem. Soc.*, **78,** 325 (1956).
106. T. F. Fagley and H. W. Myers, *J. Am. Chem. Soc.*, **76,** 6001 (1954).
107. L. Horner and A. Gross, *Ann. Chem.*, **591,** 117 (1955).
108. M. O. Forster and R. Muller, *J. Chem. Soc.*, **97,** 126 (1910).
109. S. Gotsky, *Ber.*, **64,** 1555 (1931).
110a. G. Schroeter, *Ber.*, **42,** 2336 (1909).
110b. G. Schroeter, *Ber.*, **42,** 3356 (1909).
111. H. Lindemann and A. Muhlhaus, *Ann. Chem.*, **446,** 1 (1925).

112. M. O. Forster, H. E. Fierz and W. P. Joshua, *J. Chem. Soc.*, **93,** 1070 (1908).
113. C. J. Grundmann and W. J. Schnabel, U.S. Pat. 2,990, 412 (1961); *Chem. Abstr.*, **55,** 25256 (1961).
114. G. Smolinsky, *J. Org. Chem.*, **27,** 3557 (1962).
115. H. Bohme and D. Morf, *Chem. Ber.*, **90,** 446 (1957).
116. H. Bohme and D. Morf, *Chem. Ber.*, **91,** 660 (1958).
117. H. Bohme, D. Morf and E. Mundlos, *Chem. Ber.*, **89,** 2869 (1956).
118. A. S. R. Donald and R. E. Marks, *Chem. Ind. (London)*, 1340 (1965).
119. F. D. Marsh, Ger. Pat. 1,181,182 (1964); *Chem. Abstr.*, **62,** 9010 (1965).
120. S. N. Ege and K. W. Sherk, *J. Am. Chem. Soc.*, **75,** 354 (1953).
121. R. F. Tietz and W. E. McEwen, *J. Am. Chem. Soc.*, **77,** 4007 (1955).
122. C. L. Arcus, R. E. Marks and M. L. Coombs, *J. Chem. Soc.*, 4064 (1957).
123. G. R. Buell, W. E. McEwen and J. Kleinberg, *Tetrahedron Letters*, No. 5, 16 (1959).
124. L. P. Kuhn and J. Di Domenico, *J. Am. Chem. Soc.*, **72,** 5777 (1950).
125. C. Schuerch and E. H. Huntress, *J. Am. Chem. Soc.*, **71,** 704 (1949).
126. J. H. Boyer and F. C. Canter, *J. Am. Chem. Soc.*, **77,** 3287 (1955).
127. C. L. Arcus and R. J. Mesley, *J. Chem. Soc.*, 178 (1953).
128. C. L. Arcus and M. L. Coombs, *J. Chem. Soc.*, 4319 (1954).
129. C. L. Arcus and E. A. Lucken, *J. Chem. Soc.*, 1634 (1955).
130. Y. Yukawa and K. Tanaka, *Mem. Inst. Sci. Ind. Res., Osaka Univ.*, **14,** 199 (1957); *Chem. Abstr.*, **52,** 4513 (1958).
131. A. L. Logothetis, *J. Am. Chem. Soc.*, **87,** 749 (1965).
132. W. H. Saunders and J. C. Ware, *J. Am. Chem. Soc.*, **80,** 3328 (1958).
133. P. Ballinger, P. B. D. de la Mare, G. Kohnstam and B. M. Presst, *J. Chem. Soc.*, 3641 (1935).
134. H. Kuczynski and C. Walkowicz, *Roczniki Chem.*, **43,** 329 (1969); *Chem. Abstr.*, **70,** 114689 (1969).
135. S. Winstein and J. Sonnenberg, *J. Am. Chem. Soc.*, **83,** 3235, 3244 (1961).
136. A. Hassner and F. W. Fowler, *J. Org. Chem.*, **33,** 2686 (1968).
137. M. O. Forster and S. H. Newman, *J. Chem. Soc.*, **97,** 2570 (1910).
138. J. H. Boyer, W. E. Krueger and G. J. Mikol, *J. Am. Chem. Soc.*, **89,** 5504 (1967).
139. S. Patai and Z. Rappoport in *The Chemistry of Alkenes* (Ed. S. Patai), Interscience, London, 1964, p. 525.
140. S. Maiorana, *Ann. Chim. (Rome)*, **56,** 1531 (1966).
141. J. S. Meek and J. S. Fowler, *J. Org. Chem.*, **32,** 985 (1968).
142. A. N. Nesmeyanov and M. I. Rybinskaya, *Izv. Akad. Nauk SSSR, Otd. Khim. Nauk*, 816 (1962); *Chem. Abstr.*, **58,** 3408 (1963).
143. A. N. Nesmeyanov and M. I. Rybinskaya, *Dokl. Akad. Nauk SSSR*, **170,** 600 (1966); *Chem. Abstr.*, **66,** 55440 (1967).
144. H. W. Moore and H. R. Shelden, *Tetrahedron Letters*, 5431 (1968).
145. H. W. Moore, H. R. Shelden and W. Weyler, Jr., *Tetrahedron Letters*, 1243 (1969).
146. K. Fries, P. Ochwat and W. Pense, *Ber.*, **56,** 1291 (1923).
147. A. Korczynski and S. Namyslowski, *Bull. Soc. Chim. France*, **35,** 1186 (1924).
148. F. Sorm, *Chem. Obzor.*, **14,** 37 (1939); *Chem. Abstr.*, **33,** 7286 (1939).

149. W. I. Awad, S. M. A. R. Omran and F. Nagieb, *Tetrahedron*, **19,** 1591 (1963).
150. A. Mustafa, S. M. S. D. Zayed and S. Khattab, *J. Am. Chem. Soc.*, **78,** 145 (1956).
151. G. Smolinsky and C. A. Pryde, *J. Org. Chem.*, **33,** 2411 (1968).
152. J. Miller, *Aromatic Nucleophilic Substitution*, Elsevier, London, 1968, Chap. 4.
153. M. O. Forster and H. E. Fierz, *J. Chem. Soc.*, **93,** 1174 (1908).
154. A. G. Hortmann and J. E. Martinelli, *Tetrahedron Letters*, 6205 (1968).
155. M. O. Forster and D. Cardwell, *J. Chem. Soc.*, 1338 (1913).
156. D. A. Sutton, *J. Chem. Soc.*, 306 (1944).
157. P. A. Levene, A. Rothen and M. Kuna, *J. Biol. Chem.*, **120,** 777 (1937).
158. A. Gagneux, S. Winstein and W. G. Young, *J. Am. Chem. Soc.*, **82,** 5956 (1960).
159. C. A. van der Werf, V. Heasley and L. Locatell in *Abstracts of 144th National American Chemical Society*, Los Angeles, April 1963, p. 10.
160. J. A. Pegolotti and W. G. Young, *J. Am. Chem. Soc.*, **83,** 3258 (1961).
161. H. Lindemann and W. Schultheis, *Ann. Chem.*, **451,** 241 (1927).
162. N. N. Yarovenko, S. P. Motornyi, L. I. Kirenskaya and A. S. Vasil'eva, *Zh. Obshch. Khim.*, **27,** 2243 (1957); *Chem. Abstr.*, **52,** 8052 (1958).
163. M. O. Forster, *J. Chem. Soc.*, **95,** 433 (1909).
164. C. Naegeli and G. Stefanovitsch, *Helv. Chim. Acta*, **11,** 609 (1928).
165. T. Lieser, Ger. Pat., 860, 636 (1952); *Chem. Abstr.*, **48,** 10060 (1954).
166. R. J. Shozda, U.S. Pat., 3,418,341 (1968); *Chem. Abstr.*, **70,** 57199 (1969).
167. R. Graf, E. Lederer-Ponzer and L. Freiberg, *Ber.*, **69,** 21 (1931).
168. C. Naegeli and A. Tyabji, *Helv. Chim. Acta*, **16,** 349 (1933).
169. W. Steinkopf, H. F. Schmitt and H. Fiedler, *Ann. Chem.*, **527,** 237 (1937).
170. P. E. Spoerri and A. Erickson, *J. Am. Chem. Soc.*, **60,** 400 (1938).
171. R. F. C. Brown, *Australian J. Chem.*, **17,** 47 (1964).
172. C. Agami, *Bull. Soc. Chim. France,* 1021 (1965).
173. D. E. Horning and J. M. Muchowski, *Can. J. Chem.*, **45,** 1247 (1967).
174. S. Maffei and G. F. Bettinetti, *Ann. Chim.* (*Rome*), **49,** 1809 (1959).
175. R. J. Shozda and J. A. Vernon, *J. Org. Chem.*, **32,** 2876 (1967).
176. T. Wieland and H. Determann, *Angew. Chem. Intern. Ed. Engl.*, **2,** 358 (1963).
177. B. F. Erlanger, W. V. Curran and N. Kokowsky, *J. Am. Chem. Soc.*, **81,** 3051 (1959).
178. N. T. Smart, G. T. Young and M. W. Williams, *J. Chem. Soc.*, 3902 (1960).
179. J. Weinstock, *J. Org. Chem.*, **26,** 3511 (1961).
180. R. Schwyzer, P. Sieber and H. Kappeler, *Helv. Chim. Acta*, **42,** 2622 (1959).
181. E. Schnabel, *Ann. Chem.*, **702,** 188 (1967).
182. R. Schwyzer, W. Rittel, H. Kappeler and B. Iselin, *Angew Chem.*, **72,** 915 (1960).
183. G. W. Anderson and A. C. McGregor, *J. Am. Chem. Soc.*, **79,** 6180 (1957).
184. R. Schwyzer and W. Rittel, *Helv. Chim. Acta.*, **44,** 159 (1961).
185. L. A. Carpino, C. A. Giza and B. A. Carpino, *J. Am. Chem. Soc.*, **81,** 955 (1959).
186. L. Carpino and P. Crowley, *Org. Syn.*, **44,** 15 (1964).
187. L. A. Carpino, *J. Am. Chem. Soc.*, **79,** 98, 4427 (1957).
188. H. Yajima and H. Kawatani, *Chem. Pharm. Bull.* (*Tokyo*), **16,** 182 (1968).

189. G. Smolinsky and B. I. Feuer, *J. Am. Chem. Soc.*, **86,** 3085 (1964).
190. R. J. Cotter, U.S. Pat. 3,324,148 (1967); *Chem. Abstr.*, **68,** 59335 (1968).
191. D. S. Breslow, T. J. Prosser, A. F. Marcantonio and C. A. Genge, *J. Am. Chem. Soc.*, **89,** 2384 (1967).
192. F. Weygand and K. Hunger, *Chem. Ber.*, **95,** 1 (1962).
193. F. L. Scott and M. T. Scott, *J. Am. Chem. Soc.*, **79,** 6077 (1957).
194. R. Stolle and F. Henke-Stark, *J. Prakt. Chem.*, **124,** 261 (1930).
195. R. Stolle, N. Nieland and M. Merkle, *J. Prakt. Chem.*, **116,** 192 (1927); *J. Prakt. Chem.*, **117,** 185 (1297).
196. G. A. Reynolds, J. A. Van Allan and J. F. Tinker, *J. Org. Chem.*, **24,** 1205 (1959).
197. J. H. Boyer and E. J. Miller, *J. Am. Chem. Soc.*, **81,** 4671 (1959).
198. J. H. Boyer and H. W. Hyde, *J. Org. Chem.*, **25,** 458 (1960).
199. F. Eloy, *J. Org. Chem.*, **26,** 953 (1961).
200. C. Temple, C. L. Kussner and J. A. Montgomery, *J. Org. Chem.*, **31,** 2210 (1966).
201. L. F. Avramenko, T. A. Zakharova, V. Ya. Pochinok and Yi. S. Rozum, *Khim. Geterotsikl. Soedin.*, 423 (1968); *Chem. Abstr.*, **69,** 96591 (1968).
202. N. B. Smirnova, I. Ya. Postovskii, N. N. Vereshchagina, I. B. Lundina and I. I. Mudretsova, *Khim. Geterotsikl. Soedin.*, 167 (1968); *Chem. Abstr.*, **69,** 106652 (1968).
203. A. Alemagna, T. Bacchetti and P. Beltrame, *Tetrahedron*, **24,** 3209 (1968).
204. T. Itai and S. Kamiya, *Chem. Pharm. Bull.* (*Tokyo*), **11,** 348 (1963).
205. A. Kovacic, B. Stanovnik and M. Tisler, *J. Heterocyclic Chem.*, **5,** 351 (1968).
206. B. Stanovnik and M. Tisler, *Tetrahedron*, **23,** 387 (1967).
207. C. Temple and J. A. Montgomery, *J. Am. Chem. Soc.*, **86,** 2946 (1964).
208. J. D. Roberts, *Chem. Ber.*, **94,** 273 (1961).
209. R. Huisgen, *Angew. Chem. Intern. Ed. Engl.*, **2,** 633 (1963).
210. T. H. Brownlee, U.S. Pat. 3,405,144 (1968); *Chem. Abstr.*, **70,** 67583 (1969).
211. A. Schmidt, *Chem. Ber.*, **100,** 3725 (1967).
212. M. S. Chang and A. J. Matuszko, *J. Org. Chem.*, **28,** 2260 (1963).
213. R. Stolle and E. Helwerth, *Ber.*, **47,** 1132 (1914).
214. R. Stolle and A. Netz, *Ber.*, **55,** 1297 (1922).
215. M. O. Forster, *J. Chem. Soc.*, **95,** 184 (1909).
216. H. Wieland, *Ber.*, **42,** 4199 (1909).
217. A. F. Hegarty, J. B. Aylward and F. L. Scott, *J. Chem. Soc.*, **C,** 2587 (1967).
218. C. A. van der Werf, R. Y. Heisler and W. E. McEwen, *J. Am. Chem. Soc.*, **76,** 1231 (1954).
219. E. A. S. Cavell, R. E. Parker and A. W. Scaplehorn, *J. Chem. Soc.*, 4780 (1965).
220. J. D. Ingham, W. L. Petty and P. L. Nichols, Jr., *J. Org. Chem.*, **21,** 373 (1956).
221. F. H. Newth, *Quart. Rev.* (*London*), **13,** 30 (1959).
222. Brit. Pat. 1,087,066 (1967); *Chem. Abstr.*, **68,** 95956 (1968).
223. F. Ya. Perveer and A. P. Voitov, *Vestn. Leningr. Univ., Ser. Fiz. i Khim.*, **20,** 151 (1965); *Chem. Abstr.*, **63,** 1689 (1965).
224. C. W. Shoppee, D. E. Evans, H. C. Richards and G. H. R. Summers, *J. Chem. Soc.*, 1649 (1956); C. W. Shoppee, R. J. W. Cremlyn, D. E. Evans and G. H. R. Summers, *J. Chem. Soc.*, 4364 (1957); A. C. Cope, E. Ciganek,

L. J. Fleckenstein and M. A. P. Meisinger, *J. Am. Chem. Soc.*, **82,** 4651 (1960); R. D. Haworth, L. H. C. Lunts and J. McKenna, *J. Chem. Soc.*, 986 (1955); J. Sicher, F. Sipos and M. Tichy, *Collection Czech. Chem. Commun.*, **26,** 847 (1961).
225. A. Cave, P. Potier, A. Cave and J. Le Men, *Bull. Soc. Chim. France*, 2415 (1964).
226. P. D. Klimstra, U.S. Pat. 3,350,424 (1967); *Chem. Abstr.*, **68,** 114848 (1968).
227. K. Ponsold and H. Groh, *Chem. Ber.*, **98,** 1009 (1965).
228. O. E. Edwards and K. K. Purushothaman, *Can. J. Chem.*, **42,** 712 (1964).
229. K. Ponsold and W. Preibsch, *J. Prakt. Chem.*, **23,** 173 (1964).
230. D. H. R. Barton and L. R. Morgan, *Proc. Chem. Soc.*, 206 (1961).
231. D. H. R. Barton and L. R. Morgan, Jr., *J. Chem. Soc.*, 622 (1962).
232. D. H. R. Barton, *S. African Ind. Chemist*, **15,** 229 (1961).
233. A. Cave, F. X. Jarreau, Khuong-Huu-Qui, M. Leboeuf, N. Serban and R. Goutarel, *Bull. Soc. Chim. France*, 701 (1967).
234. L. Labler, J. Hora and V. Cerny, *Collection Czech. Chem. Commun.*, **28,** 2015 (1963).
235. L. A. Freiberg, *J. Org. Chem.*, **30,** 2476 (1965).
236. N. L. Wendler in *Molecular Rearrangements* (Ed P. de Mayo), Interscience, New York, 1963, p. 1019.
237. J. H. Pierce, H. C. Richards, C. W. Shoppee, K. J. Stephenson and G. H. R. Summers, *J. Chem. Soc.*, 694 (1955).
238. F. X. Jarreau, C. Monneret, Khuong-Huu-Qui and R. Goutarel, *Bull. Soc. Chim. France*, 2155 (1964).
239. K. Ponsold and G. Schubert, *J. Prakt. Chem.*, **311,** 445 (1969).
240. K. Ponsold, B. Schoenecker and I. Pfaff, *Chem. Ber.*, **100,** 2957 (1967).
241. K. Ponsold, *Chem. Ber.*, **96,** 1855 (1963).
242. D. H. R. Barton and R. Cookson, *Quart. Rev. (London)*, **10,** 67 (1965).
243. J. D. Dutcher, *Advan. Carbohydrate Chem.*, **18,** 259 (1963).
244a. N. Sharon in *The Amino Sugars*, Vol. IIA (Ed. E. A. Balazs and R. W. Jeanloz), Academic Press, New York, 1965, p. 1.
244b. N. Sharon, *Sci. Am.*, **220,** 92 (1969).
245. F. Cramer, H. Otterbach and H. Springmann, *Chem. Ber.*, **92,** 384 (1959).
246. F. D. Cramer in *Methods of Carbohydrate Chemistry*, Vol. 1 (Eds. R. L. Whistler and M. L. Wolfrom), Academic Press, New York, 1962, p. 242.
247a. W. Meyer zu Reckendorf, *Chem. Ber.*, **96,** 2017, 2019 (1963).
247b. W. Meyer zu Reckendorf, *Chem. Ber.*, **97,** 1275 (1964).
247c. W. Meyer zu Reckendorf, *Tetrahedron*, **19,** 2033 (1963).
247d. W. Meyer zu Reckendorf, *Angew. Chem. Intern. Ed. Engl.*, **5,** 967 (1966).
248. A. Bertho (with H. Nussel), *Ber.*, **63,** 836 (1930).
249. A. Bertho and D. Aures, *Ann. Chem.*, **592,** 54 (1955).
250. A. Bertho and A. Revesz, *Ann. Chem.*, **581,** 161 (1953).
251. F. Micheel and H. Wulff, *Chem. Ber.*, **89,** 1521 (1956).
252. S. Hanessian and N. R. Plessas, *Chem. Commun.*, 1152 (1967).
253. J. Hill, L. Hough and A. C. Richardson, *Carbohydrate Res.*, **8,** 7 (1968).
254. M. L. Wolfrom, Y.-L. Hung and D. Horton, *J. Org. Chem.*, **30,** 3394 (1965).
255. J. Cleophax, J. Hildesheim, R. E. Williams and S. D. Gero, *Bull. Soc. Chim. France*, 1415 (1968).

256. Y. Ali and A. C. Richardson, *J. Chem. Soc., C*, 320 (1969).
257. S. Hanessian, *Carbohydrate Res.*, **1,** 178 (1965).
258. J. Hill, L. Hough and A. C. Richardson, *Carbohydrate Res.*, **8,** 19 (1968).
259. B. R. Baker and A. H. Haines, *J. Org. Chem.*, **28,** 442 (1963).
260. S. Nadkarni and N. R. Williams, *J. Chem. Soc.*, 3496 (1965).
261. J. M. Sugihara and N. J. Teerlink, *J. Org. Chem.*, **29,** 550 (1964).
262. C. L. Stevens, P. Blumbergs, D. H. Otterbach and K. G. Taylor, *J. Org. Chem.*, **31,** 2822 (1966).
263. S. Hanessian, *Chem. Commun.*, 796 (1966).
264. C. L. Stevens, R. P. Glinski, K. G. Taylor, P. Blumbergs and F. Sirokman, *J. Am. Chem. Soc.*, **88,** 2073 (1966).
265. J. Jary, P. Novak, Z. Ksandr and Z. Samek, *Chem. Ind. (London)*, 1490 (1967).
266. D. Horton, M. L. Wolfrom and A. Thompson, *J. Org. Chem.*, **26,** 5069 (1961).
267. A. J. Dick and J. K. N. Jones, *Can. J. Chem.*, **43,** 977 (1965).
268a. R. D. Guthrie and D. Murphy, *Chem. Ind. (London)*, 1473 (1962).
268b. R. D. Guthrie and D. Murphy, *J. Chem. Soc.*, 5288 (1963).
269. S. Hanessian and T. H. Haskell, *J. Org. Chem.*, **30,** 1080 (1965).
270. R. D. Guthrie, D. Murphy, D. H. Buss, L. Hough and A. C. Richardson, *Proc. Chem. Soc.*, 84 (1963).
271. R. D. Guthrie and D. Murphy, *J. Chem. Soc.*, 3828 (1965).
272. W. Meyer zu Reckendorf, *Chem. Ber.*, **97,** 325 (1964).
273. Y. Ali and A. C. Richardson, *Chem. Commun.*, 554 (1967).
274. J. Cleophax, S. D. Gero and J. Hildesheim, *Chem. Commun.*, 94 (1968).
275. J. E. Christensen and L. Goodman, *J. Org. Chem.*, **28,** 2995 (1963).
276. J. E. Christensen and L. Goodman, *J. Am. Chem. Soc.*, **83,** 3827 (1961).
277. J. Miller, *J. Am. Chem. Soc.*, **85,** 1628 (1963).
278. S. Marburg and P. A. Grieco, *Tetrahedron Letters*, 1305 (1966).
279. P. A. Grieco and J. P. Mason, *J. Chem. Eng. Data*, **12,** 623 (1967).
280. J. Miller, *Aromatic Nucleophilic Substitution*, Elsevier, London, 1968, Chap. 5.
281. A. S. Bailey, *J. Chem. Soc.*, 4710 (1960).
282. K. Namba and T. Yamashita, *Kogyo Kayaku Kyokaishi*, **19,** 86 (1958); *Chem. Abstr.*, **53,** 11279 (1959).
283. G. Powell, *J. Am. Chem. Soc.*, **51,** 2436 (1929).
284. F. Moulin, *Helv. Chim. Acta.*, **35,** 167 (1952).
285. E. Schrader *Ber.*, **50,** 777 (1917).
286. D. S. Deorha, S. S. Joshi and V. K. Mahesh, *J. Indian Chem. Soc.*, **39,** 534 (1962).
287. D. Vorlander, *Rec. Trav. Chim.*, **48,** 912 (1929).
288. B. Stanovnik and M. Tisler, *Tetrahedron*, **25,** 3313 (1969).
289. I. B. Lundina, Yu. N. Sheinker and I. Ya. Postovskii, *Izv. Akad. Nauk SSSR, Ser. Khim.*, 66 (1967); *Chem. Abstr.*, **67,** 21884 (1967).
290. N. B. Smirnova and I. Ya. Postovskii, *Zh. Vses. Khim. Obshchestva im. D.I. Mendeleeva*, **9,** 711 (1964); *Chem. Abstr.*, **62,** 9130 (1965).
291. E. Schrader, *J. Prakt. Chem.*, **95,** 312 (1917).
292. F. R. Benson, L. W. Hartzel and E. A. Otten, *J. Am. Chem. Soc.*, **76,** 1858 (1954).
293. C. V. Hart, *J. Am. Chem. Soc.*, **50,** 1922 (1928).

294. R. Stolle and H. Storch, *J. Prakt. Chem.*, **135,** 128 (1932).
295. R. Stolle and F. Hanusch, *J. Prakt. Chem.*, **136,** 9, 120 (1933).
296. H. Balli and F. Kersting, *Ann. Chem.*, **647,** 1 (1961).
297, H. Balli, *Ann. Chem.*, **647**, 11 (1961).
298a. R. D. Brown, *J. Chem. Soc.*, 2670 (1951).
298b. J. Miller, *Aromatic Nucleophilic Substitution*, Elsevier, London, 1968, Chap. 7.
299. J. D. Hobson and J. R. Malpass, *J. Chem. Soc.*, **C,** 1645 (1967).
300. J. D. Hobson and J. R. Malpass, *J. Chem. Soc.*, **C,** 1499 (1969).
301. C. E. Wulfman, C. F. Yarnell and D. S. Wulfman, *Chem. Ind.* (*London*), 1440 (1960).
302. D. S. Wulfman, L. Durham and C. E. Wulfman, *Chem. Ind.* (*London*), 859 (1962).
303. D. S. Wulfman and J. J. Ward, *Chem. Commun.*, 276, 1967.
304. A. N. Nesmeyanov, V. N. Drozd and V. A. Sazonova, *Dokl. Akad. Nauk SSSR*, **150,** 321 (1963); *Chem. Abstr.*, **59,** 5196 (1963).
305. E. Oliveri-Mandala, *Mem. reale accad. naz. Lincei Classe sci. fis. mat. e nat.* (*vi*), **2,** 132 (1926); *Chem. Abstr.*, **21,** 3895 (1927).
306. O. Dimroth and G. Fester, *Ber.*, **43,** 2219 (1910).
307a. E. Oliveri-Mandala and F. Noto, *Gazz. Chim. Ital.*, **43,** I, 304 (1913).
307b. E. Oliveri-Mandala, *Gazz. Chim. Ital.*, **44,** I, 670 (1914).
307c. E. Oliveri-Mandala, *Gazz. Chim. Ital.*, **51,** II, 195 (1921).
307d. E. Oliveri-Mandala, *Gazz. Chim. Ital.*, **52,** II, 98 (1922).
308. E. Oliveri-Mandala, *Gazz. Chim. Ital.*, **45,** II, 120 (1915); E. Oliveri-Mandala and E. Calderaro, *Gazz. Chim. Ital.*, **45,** I, 307 (1915).
309. P. B. D. de la Mare and R. Bolton, *Electrophilic Additions to Unsaturated Systems*, Elsevier, London, 1966.
310. R. E. Schaad, U.S. Pat. 2,557,924 (1951); *Chem. Abstr.*, **46,** 1028 (1952).
311. I. L. Knunyants and E. G. Bykhovskaya, *Dokl. Akad. Nauk. SSSR*, **131,** 1338 (1960); *Chem. Abstr.*, **54,** 20840 (1960).
312. R. F. Banks and G. J. Moore, *J. Chem. Soc.*, 2304 (1966).
313. C. S. Cleaver and C. G. Krespan, *J. Am. Chem. Soc.*, **87,** 3716 (1965).
314. S. Andreades, *J. Am. Chem. Soc.*, **86,** 2003 (1964).
315. I. L. Knunyants and E. G. Bykhovskaya, *Zh. Vses. Khim. Obshchestva. im. D.I. Mendeleeva*, **7,** 585 (1962); *Chem. Abstr.*, **58,** 5509 (1963).
316. J. W. Baker and H. B. Hopkins, *J. Chem. Soc.*, 1089 (1949).
317. J. Miller and R. L. Heppolette, *J. Am. Chem. Soc.*, **75,** 4265 (1953).
318. A. J. Davies, A. S. R. Donald and R. E. Marks, *J. Chem. Soc.*, **C,** 2109 (1967).
319. A. S. R. Donald and R. E. Marks, *J. Chem. Soc.*, **C,** 1188 (1967).
320. R. D. Westland and W. E. McEwen, *J. Am. Chem. Soc.*, **74,** 6141 (1952).
321. W. I. Awad, A. F. M. Fahmy and A. M. A. Sammour, *J. Org. Chem.*, **30,** 2222 (1965).
322. W. I. Awad and A. F. M. Fahmy, *Can. J. Chem.*, **46,** 2207 (1968).
323. G. Eglinton, E. R. H. Jones, G. H. Mansfield and M. C. Whiting, *J. Chem. Soc.*, 3197 (1954).
324. P. Kurtz, H. Gold and H. Disselnkotter, *Ann. Chem.*, **624,** 1 (1959).
325. M. Gaudemar, *Ann. Chim.* (*Paris*), **1,** 161 (1956); M. Bertrans and J. Le Gras, *Compt. Rend.*, **260,** 6929 (1965).

326. L. W. Hartzel and F. R. Benson, *J. Am. Chem. Soc.*, **76,** 667 (1954).
327. F. Oliveri-Mandala and A. Coppola, *Atti reale accad. Lincei*, **19,** I, 563 (1910); *Chem. Abstr.*, **4,** 2455 (1910).
328. R. Huttel, *Ber.*, **74,** 1680 (1941).
329. J. C. Sheehan and C. A. Robinson, *J. Am. Chem. Soc.*, **73,** 1208 (1951).
330. R. Escales, *Chem.-Ztg.*, **29,** 31 (1905).
331. H. W. Moore, H. R. Shelden and D. F. Shellhamer, *J. Org. Chem.*, **34,** 1999 (1969).
332. R. Adams and W. Moje, *J. Am. Chem. Soc.*, **74,** 5560 (1952).
333. L. F. Fieser and J. L. Hartwell, *J. Am. Chem. Soc.*, **57,** 1482 (1935).
334. R. Adams and D. C. Blomstrom, *J. Am. Chem. Soc.*, **75,** 3405 (1953).
335. G. Opitz and W. Merz, *Ann. Chem.*, **652,** 158 (1962).
336. G. Opitz, A. Griesinger and H. W. Schubert, *Ann. Chem.*, **665,** 91 (1963).
337. I. Ugi in *Newer Methods of Preparative Organic Chemistry*, Vol. 4 (Ed. W. Foerst), Academic Press, New York, 1968, p. 1.
338. V. G. Ostroverkhov and E. A. Shilov, *Ukrain. Khim. Zhur.*, **23,** 615 (1957); *Chem. Abstr.*, **52,** 7828 (1958).
339. U. Tuerck and H. Behringer, *Chem. Ber.*, **98,** 3020 (1965).
340. A. Hantzsch and A. Vagt, *Ann. Chem.*, **314,** 361 (1901).
341. E. Oliveri-Mandala and F. Noto, *Gazz. Chim. Ital.*, **43,** I, 514 (1913).
342. K. H. Slotta and R. Tschesche, *Ber.*, **60,** 1021 (1926).
343. F. W. Brugman and J. F. Arens, *Rec. Trav. Chim.*, **74,** 209 (1955).
344. H. Saikachi and K. Takai, *Yakugaku Zasshi*, **89,** 34 (1969); *Chem. Abstr.*, **70,** 96507 (1969).
345. E. Oliveri-Mandala and E. Calderaro, *Gazz. Chim. Ital.*, **43,** I, 538 (1913).
346. E. Lieber, C. N. Pillai and R. D. Hites, *Can. J. Chem.*, **35,** 832 (1957).
347. E. Lieber and J. Ramachandran, *Can. J. Chem.*, **37,** 101 (1959).
348. E. Lieber, C. N. Pillai, J. Ramachandran and R. D. Hites, *J. Org. Chem.*, **22,** 1750 (1957); E. Lieber, E. Oftedahl and C. N. R. Rao, *J. Org. Chem.*, **28,** 194 (1963).
349. L. Birkofer, F. Muller and W. Kaiser, *Tetrahedron Letters*, 2781 (1967).
350. J. Chatt, *Chem Rev.*, **48,** 7 (1951).
351. H. C. Brown and P. Geoghan, *J. Am. Chem. Soc.*, **89,** 1522 (1967).
352. V. I. Sokolov and O. A. Reutov, *Izv. Akad. Nauk SSSR, Ser. Khim.*, 1632 (1967); *Chem. Abstr.*, **68,** 49704 (1968).
353. C. H. Heathcock, *Angew. Chem. Intern. Ed. Engl.*, **8,** 134 (1969).
354. S. Winstein and H. J. Lucas, *J. Am. Chem. Soc.*, **60,** 836 (1938).
355. R. A. Sneen, J. V. Carter and P. S. Kay, *J. Am. Chem. Soc.*, **88,** 2594 (1966).
356a. A. Hassner and C. H. Heathcock, *Tetrahedron Letters*, 1125 (1964).
356b. A. Hassner, M. Lorber and C. H. Heathcock, *J. Org. Chem.*, **32,** 540 (1967).
357a. A. Hassner and L. A. Levy, *J. Am. Chem. Soc.*, **87,** 4203 (1965).
357b. F. W. Fowler, A. Hassner and L. A. Levy, *J. Am. Chem. Soc.*, **89,** 2077 (1967).
358. A. Hassner, *J. Org. Chem.*, **33,** 2684 (1968).
359. W. H. Puterbaugh and M. S. Newman, *J. Am. Chem. Soc.*, **79,** 3469 (1957).
360. A. Hassner and F. Boerwinkle, *J. Am. Chem. Soc.*, **90,** 216 (1968).
361. A. Hassner and F. Boerwinkle, *Tetrahedron Letters*, 3309 (1969).
362. F. Boerwinkle and A. Hassner, *Tetrahedron Letters*, 3921 (1968).

363. J. Hine, *Physical Organic Chemistry*, 2nd ed., McGraw-Hill, New York, 1962.
364. A. Hassner, R. J. Isbister and A. Friederang, *Tetrahedron Letters*, 2939 (1969).
365. A. Hassner and R. J. Isbister, *J. Am. Chem. Soc.*, **91,** 6126 (1969).
366. A. F. Clifford, *J. Phys. Chem.*, **63,** 1227 (1959).
367. G. Drefahl, K. Ponsold and D. Eichhorn, *Chem. Ber.*, **101,** 1633 (1968).
368. F. Minisci, *Gazz. Chim. Ital.*, **89,** 626 (1959).
369. F. Minisci, R. Galli and U. Pallini, *Gazz. Chim. Ital.*, **91,** 1023 (1961).
370. F. Minisci and R. Galli, *Tetrahedron Letters*, 533 (1962).
371. F. Minisci and R. Galli, *Tetrahedron Letters*, 357 (1963).
372. F. Minisci and R. Galli, Fr. Pat. 1,350,360 (1964); *Chem. Abstr.*, **61,** 8239 (1964).
373. F. Minisci, R. Galli and M. Cecere, *Gazz. Chim. Ital.*, **94,** 67 (1964).
374. F. Minisci and R. Galli, *Chim. Ind. (Milan)*, **47,** 178 (1965).
375. F. Minisci, U.S. Pat. 3,026,334 (1962); *Chem. Abstr.*, **58,** 11222 (1963).
376. F. D. Chattaway, F. L. Garton and G. D. Parkes, *J. Chem. Soc.*, **125,** 1980 (1924).
377. J. H. Boyer, unpublished results cited in ref. 15.
378. E. Bamberger and E. Demuth, *Ber.*, **34,** 1309 (1901).
379. L. Wacker, *Ber.*, **35,** 3920 (1902).
380. O. Dimroth, *Ber.*, **40,** 2376 (1907).
381. M. O. Forster, *J. Chem. Soc.*, **107**, 260 (1915).
382. K. Clusius and H. Hurzeler, *Helv. Chim. Acta*, **37,** 383 (1954).
383. K. Clusius and M. Vecchi, *Ann. Chem.*, **607,** 16 (1957).
384. C. S. Rondesvedt, Jr. and S. J. Davis, *J. Org. Chem.*, **22,** 200 (1957).
385. E. Fischer, *Ber.*, **10,** 1334 (1877); *Ann. Chem.*, **190,** 67, 94, 96, 145 (1878).
386. J. Mai, *Ber.*, **24,** 3418 (1891); *Ber.*, **25,** 372, 1685 (1892); *Ber.*, **26,** 1271 (1893).
387. H. W. Bresler, W. H. Friedman and J. Mai, *Ber.*, **39,** 876 (1905).
388. L. Gatterman and R. Ebert, *Ber.*, **49,** 2117 (1916).
389. E. Bamberger and E. Demuth, *Ber.*, **34,** 2292 (1901).
390. J. Meisenheimer, O. Senn and P. Zimmerman, *Ber.*, **60,** 1736 (1927).
391. C. Sumuleanu, *Chem. Zentr.*, II, 31 (1903).
392. P. V. Dutt, H. R. Whitehead and A. Wormall, *J. Chem. Soc.*, **119,** 2088 (1921).
393. A. Key and P. V. Dutt, *J. Chem. Soc.*, 2035 (1928).
394. S. Skraup and K. Steinruck, Ger. Pat. 456, 857 (1923); *Brit. Chem. Abstr.*, **B364** (1930).
395. H. Bretschneidner and H. Rager, *Monatsh. Chem.*, **81,** 970 (1950).
396. W. R. Bamford and T. S. Stevens, *J. Chem. Soc.*, 4335 (1952).
397. D. G. Farnum, *J. Org. Chem.*, **28,** 870 (1963).
398. P. A. S. Smith, C. D. Rowe and L. B. Bruner, *J. Org. Chem.*, **34,** 3430 (1969).
399. S. Ito, *Bull. Chem. Soc. Japan*, **39,** 635 (1966).
400. E. Noelting and O. Michel, *Ber.*, **26,** 86 (1893).
401. T. Curtius, *Ber.*, **26,** 1263 (1893).
402. A. Wohl and H. Schiff, *Ber.*, **33,** 2741 (1900).
403. K. Clusius and H. Craubner, *Helv. Chim. Acta*, **38,** 1060 (1955).

404. P. Griess, *Ber.*, **9,** 1659 (1876).
405. J. P. Horwitz and V. A. Grakauskas, *J. Am. Chem. Soc.*, **80,** 926 (1958).
406. R. Kuhn and H. Kainer, *Angew. Chem.*, **65,** 442 (1953).
407. G. Heller, *J. Prakt. Chem.*, **111,** 42, 53 (1925).
408. J. P. Horwitz and V. A. Grakauskas, *J. Am. Chem. Soc.*, **79,** 1249 (1957).
409. T. Curtius, A. Darapsky and A. Bockmuhl, *Ber.*, **41,** 344 (1908).
410. V. Nair, *J. Org. Chem.*, **33,** 2121 (1968).
411. P. A. S. Smith, B. B. Brown, R. K. Putney and R. F. Reinisch, *J. Am. Chem. Soc.*, **75,** 6335 (1953).
412. P. A. S. Smith and B. B. Brown, *J. Am. Chem. Soc.*, **73,** 2438 (1951).
413. P. A. S. Smith, J. H. Hall and R. O. Kan, *J. Am. Chem. Soc.*, **84,** 485 (1962).
414. J. H. Hall, *J. Am. Chem. Soc.*, **87,** 1147 (1965).
415. R. A. Carboni and J. E. Castle, *J. Am. Chem. Soc.*, **84,** 2453 (1962).
416. D. L. Herring, U.S. Pat. 3,123,621 (1964); *Chem. Abstr.*, **60,** 13183 (1964).
417. T. Nozoe, H. Horino and T. Toda, *Tetrahedron Letters,* 5349 (1967).
418. O. Meth-Cohn, R. K. Smalley and H. Suschitzky, *J. Chem. Soc.*, 1666 (1963).
419. P. A. S. Smith, J. M. Clegg and J. H. Hall, *J. Org. Chem.*, **23,** 524 (1958).
420. G. Smolinsky, *J. Am. Chem. Soc.*, **83,** 2489 (1961).
421. J. Miller, *Aromatic Nucleophilic Substitution*, Elsevier, London, 1968, Chap. 2.
422. E. S. Lewis and M. D. Johnson, *J. Am. Chem. Soc.*, **82,** 5408 (1960).
423. K. Clusius and H. Hurzeler, *Helv. Chim. Acta,* **37,** 798 (1954).
424. K. Clusius and M. Vecchi, *Helv. Chim. Acta,* **39,** 1469 (1956).
425. I. Ugi, R. Huisgen, K. Clusius and M. Vecchi, *Angew. Chem.*, **68,** 753 (1956).
426. R. Huisgen and I. Ugi, *Angew. Chem.*, **68,** 705 (1956); *Chem. Ber.*, **90,** 2914 (1957).
427. I. Ugi, H. Perlinger and L. Behringer, *Chem. Ber.*, **91,** 2324 (1958).
428. I. Ugi, H. Perlinger and L. Behringer, *Chem. Ber.*, **92,** 1864 (1959).
429. I. Ugi, *Tetrahedron,* **19,** 1801 (1963).
430. I. Ugi in *Advances in Heterocyclic Chemistry,* Vol. 3, Academic Press, New York, 1964, p. 373.
431. R. Huisgen, *Proc. Chem. Soc.*, 357 (1961); *Naturw. Rundschau,* **14,** 43 (1961); *Angew. Chem.*, **75,** 742 (1963).
432. E. Oliveri-Mandala, *Gazz. Chim. Ital.*, **62,** 716 (1932).
433. K. Clusius and F. Endtinger, *Helv. Chim. Acta,* **41,** 1823 (1958).
434. S. Maffei and A. M. Rivolta, *Gazz. Chim. Ital.*, **84,** 750 (1954).
435. S. Maffei and L. Coda, *Gazz. Chim. Ital.*, **85,** 1300 (1955).
436. B. A. Geller and L. S. Samosvat, *Dokl. Akad. Nauk SSSR,* **141,** 847 (1961); *Chem. Abstr.*, **57,** 709 (1962).
437. S. Maffei and G. F. Bettinetti, *Ann. Chim. (Rome),* **47,** 1286 (1957).
438. S. Maffei, L. Coda and G. F. Bettinetti, *Ann. Chim. (Rome),* **46,** 604 (1956).
439. W. E. Doering and C. H. DePuy, *J. Am. Chem. Soc.*, **75,** 5955 (1953).
440. M. Regitz, *Chem. Ber.*, **99,** 3128 (1965) and references cited therein; M. Regitz, *Angew. Chem. Intern. Ed. Engl.*, **6,** 733 (1967).
441. W. Fischer and J.-P. Anselme, *J. Am. Chem. Soc.*, **89,** 5284 (1967).
442. J.-P. Anselme, W. Fischer and N. Koga, *Tetrahedron,* **25,** 89 (1969).
443. J.-P. Anselme and W. Fischer, *Tetrahedron,* **25,** 855 (1969).

444. R. Meier, *Chem. Ber.*, **86,** 1483 (1953); R. Meier and W. Frank, *Chem. Ber.*, **89,** 2747 (1956).
445. G. Koga and J.-P. Anselme, *Chem. Commun.*, 446 (1968).
446. E. H. White and C. A. Aufdermarsh, *J. Am. Chem. Soc.*, **83,** 1174, 1179 (1961).
447. G. Ponzio and G. Canuto, *Gazz. Chim. Ital.*, **45,** II, 29 (1915).
448. E. Fischer, *Ann. Chem.*, **232,** 236 (1886).
449. K. Clusius and H. R. Weisser, *Helv. Chim. Acta*, **35,** 1548 (1952).
450. K. Clusius, *Z. Electrochem.*, **58,** 586 (1954).
451. K. Clusius and K. Schwarzenbach, *Helv. Chim. Acta*, **42,** 739 (1959).
452. J. Thiele, *Ann. Chem.*, **376,** 264 (1910).
453. K. Clusius and K. Schwarzenbach, *Helv. Chim. Acta*, **41,** 1413 (1958).
454. A. Darapsky, *Ber.*, **40,** 3033 (1907).
455. H. Limpricht, *Ber.*, **21,** 3409 (1888).
456. A. Purgotti, *Gazz. Chim. Ital.*, **24,** I, 554 (1894).
457. E. Muller and W. Hoffmann, *J. Prakt. Chem.*, **111,** 293 (1925).
458. Fr. Pat. 1,476,505 (1967); *Chem. Abstr.*, **68,** 95867 (1968).
459. T. Curtius, *Ber.*, **33,** 2561 (1900).
460. F. D. Lewis and W. H. Saunders, Jr., *J. Am. Chem. Soc.*, **90,** 3828 (1968).
461. A. Darapsky and M. Prabhakar, *J. Prakt. Chem.*, **96,** 280 (1917).
462. A. Darapsky and H. Berger, *J. Prakt. Chem.*, **96,** 301 (1917).
463. A. Darapsky, J. Germscheid, C. Kreuter, F. Engelman, W. Engels and W. Trinius, *J. Prakt. Chem.*, **146,** 219 (1936).
464. W. D. Guither, D. G. Clark and R. N. Castle, *J. Heterocyclic Chem.*, **2,** 67 (1965).
465. R. Stolle and K. Feyrenbach, *J. Prakt. Chem.*, **122,** 289 (1929).
466. G. Henseke and H. Hanisch, *Ann. Chem.*, **643,** 184 (1961).
467. J. D. Hobson and J. R. Malpass, *Chem. Commun.*, 141 (1966).
468. P. A. S. Smith, J. M. Clegg and J. Lakritz, *J. Org. Chem.*, **23,** 1595 (1958).
469. G. Longo, *Gazz. Chim. Ital.*, **63,** 463 (1933).
470. H. Roesky and D. Glemser, *Chem. Ber.*, **97,** 1710 (1964).
471. A. Dornow and O. Hahmann, *Arch. Pharm.*, **290,** 61 (1957).
472. H. Neunhoffer, G. Cuny and W. K. Franke, *Ann. Chem.*, **713,** 96 (1968).
473. R. A. Clement, *J. Org. Chem.*, **27,** 1904 (1962).
474. J. Honzl and J. Rudinger, *Collection Czech. Chem. Commun.*, **26,** 2333 (1961).
475. F. Weygand and W. Steglich, *Chem. Ber.*, **92,** 313 (1959).
476. B. M. Iselin, *Arch. Biochem. Biophys.*, **78,** 532 (1958).
477. K. Hofmann, A. Lindenmann, M. Z. Magee and N. H. Khan, *J. Am. Chem. Soc.*, **74,** 470 (1952); M. Winitz and J. S. Fruton, *J. Am. Chem. Soc.*, **75,** 3041 (1953).
478. K. Inouye, M. Kanayama and H. Otsuka, *Nippon Kagaku Zasshi*, **85,** 599 (1964); *Chem. Abstr.*, **62,** 10333 (1965).
479. M. Bergmann and L. Zervas, *J. Biol. Chem.*, **113,** 341 (1936).
480. J. M. Burgess and M. S. Gibson, *Tetrahedron*, **18,** 1001 (1962).
481. C. M. Wright, quoted in ref. 4.
482. A. Darapsky, *J. Prakt. Chem.*, **76,** 433, 461 (1907).
483. G. Koga and J.-P. Anselme, *J. Am. Chem. Soc.*, **91,** 4323 (1969).

CHAPTER **3**

Characterization and determination of organic azides

JEROME E. GURST

Department of Chemistry, University of West Florida, Pensacola, Florida

I.	INTRODUCTION	191
II.	CHEMICAL MEANS OF ANALYSIS	192
	A. Qualitative	192
	B. Quantitative	193
III.	INFRARED SPECTROSCOPY	195
IV.	ULTRAVIOLET SPECTROSCOPY	197
V.	OPTICAL ROTATORY DISPERSION AND CIRCULAR DICHROISM	199
VI.	MASS SPECTROMETRY	199
VII.	REFERENCES	200

I. INTRODUCTION

The chemical literature has been surveyed through standard texts and through the indices of *Chemical Abstracts* from its inception, Vol. **1** (1906) to the present, Vol. **71**, (1969). The following reference words were used in the *Chemical Abstracts* search: acyl azides, alkyl azides, aryl azides, azides, azido group, sulphonyl azides. Unless specifically mentioned, organometallic azides, i.e. compounds in which the azido group is presumably bonded to or associated with the metal atom, have been ignored.

From this study, the overall impression arises that if the azide-containing compound is thermally stable and, of course, non-explosive, the proposed azide may be handled by any of the techniques available for the detection and characterization of organic compounds. The azido group is generally detected by infrared and ultraviolet spectrometry. Azides may be subjected to chromatographic techniques (which, of course, could purify but do not characterize azides). Mass spectrometry has been used with azides. Some effort has been expended to use the optical rotatory dispersion (o.r.d.) and circular dichroism (c.d.) measurements of the azido group as a means to determine the structure of the azide-containing molecule. Naturally, o.r.d./c.d. measurements would not lead one to assign an azido group to a molecule. Proton magnetic resonance (n.m.r.) measurements may be carried out; but again, this, in itself, does not lead to the assignment of an azide structure. Naturally, n.m.r. is useful in structure determination of azides, but not in any particular way. (Nitrogen n.m.r. spectra of ^{15}N-containing azides have not been reported.)

With this background this report will now review the chemical, infrared, ultraviolet, and mass spectrometric techniques as applied to the qualitative and quantitative measurements of organic azides. Additionally, a discussion of the o.r.d./c.d. properties of alkyl azides will be included in this chapter.

II. CHEMICAL MEANS OF ANALYSIS

A. Qualitative

Very little appears in the literature or in published reviews[1–5] which suggests qualitative tests for the azido group. A 1928 report[6] dealing with carboxylic acid azides suggests treatment with sodium hydroxide to displace the azide anion followed by its precipitation as silver azide after the addition of silver nitrate. An alternate suggestion in this paper is to treat the organic azide with hydrogen sulphide to yield an organic amine, nitrogen gas, and elemental sulphur.

The suggestion[7] that β-naphthoyl azide could be used as a reagent for the determination of amines could be used in reverse; i.e. treatment of a suspected acid azide with a known amine should yield a urethane and molecular nitrogen according to equation (1).

$$RCON_3 + R^1NH_2 \longrightarrow N_2 + RNH{-}\overset{\overset{\displaystyle O}{\|}}{C}{-}NHR^1 \qquad (1)$$

Sah showed that *p*-nitrobenzoyl azide[8] and *o*-nitrobenzoyl azide[9] could be used in similar fashion to obtain urethanes.

More recently published ideas include the formation of solid derivatives of alkyl azides by treatment with acetylene dicarboxylic acid[10] to form 1-alkyl-triazole-4,5-dicarboxylic acids. These frequently form hydrates with poor melting points and generally poor analytical properties. Alternatively, one can obtain bicyclic fused-ring triazolines by treatment with norbornadiene or dicyclopentadiene[10].

Fischer and Anselme[11] have reported the use of triphenylphosphine to form solid derivatives of aryl azides.

A more roundabout method has been used to form solid derivatives of sugar azides[12]. The azide is treated with methanolic hydrazine hydrate and Raney nickel, followed by direct benzoylation of the resulting amine. Lithium aluminium hydride is also used[12] to prepare a substrate which can be benzoylated.

The only other commonly reported technique is the acid-catalysed evolution of nitrogen gas. Typically, sulphuric acid is used as the catalyst[13]. This is not totally reliable, however. Coombs[14] reports that if carbon–nitrogen cleavage will yield a stable carbonium ion, then formation of HN_3 predominates over loss of N_2 and formation of imine. This is contradicted, however, by the report[10] that *t*-butyl azide in the presence of sulphuric acid yields acetone-*N*-methylimine which subsequently hydrolyses to give acetone, isolated as its 2,4-dinitrophenylhydrazone.

B. *Quantitative*

The difficulties of working with low molecular weight azides are characterized by the preparation of pivaloyl azide (trimethylacetyl azide)[15,16].

> 'The thermal instability of (pivaloyl azide) prevented elemental analysis . . . *Caution*: pivaloyl azide, like other acyl azides of low molecular weight, has been observed to explode without warning at room temperature. The vapors . . . are quite toxic, and (in some persons) even brief exposure causes rapid pulse, nausea, vertigo, and severe headache.'[16]

An earlier publication[17] also reports considerable difficulty in obtaining combustion analysis of azides.

Numerous techniques have been developed for the quantitative determination of the number of azido groups per molecule. The problem which must be solved is to release the covalently bound azido group. Most procedures yield one mole of nitrogen per azide leaving the third nitrogen covalently bonded to the original substrate. Nitrogen evolution is measured by the usual gasometric techniques.

Arsenite ion has been used[18–21] according to equation (2), but certain difficulties were noted[22] with simple alkyl azides. If acid treatment quantitatively liberates hydrazoic acid such as in acid azides,

$$ArN_3 + AsO_3^{-3} \xrightarrow{H_2O} ArNH_2 + AsO_4^{-3} + N_2 \tag{2}$$

some sugar azides and some alkyl azides as mentioned earlier[14], the hydrazoic acid thus liberated can be determined by ceric ion oxidation as in equation (3)[23]. An excess of ceric sulphate is added to the

$$2Ce^{+4} + 2HN_3 \longrightarrow 3N_2 + 2Ce^{+3} + 2H^+ \tag{3}$$

hydrazoic acid solution, the excess ceric ion is destroyed with excess potassium iodide, and the iodine is titrated against thiosulphate. Of course, this process works only if one can be assured that complete liberation of the azido group has occurred.

Water-soluble azides have been treated with hydriodic acid to obtain quantitative formation of molecular nitrogen and iodine, either of which can be measured[24]. Aryl azides have been handled in the same fashion[25]. These techniques suffer from the difficulty of working with iodine-free hydriodic acid and the difficulty of preventing air oxidation of hydriodic acid to yield excess iodine.

Sulphonyl azides have been analysed by two processes: release of nitrogen by reaction with triphenylphosphine and release of iodine by reaction with potassium iodide–acetic acid[26].

The best method for most azides seems to be a modification of the hydriodic acid reaction. In this analysis[22], the necessary hydriodic acid is generated *in situ* by using a 90% trichloroacetic acid (10% water) and sodium iodide system. The hydriodic acid is generated rapidly in an equilibrium situation which prevents build-up of HI and, hence, negligible formation of excess iodine by air oxidation. Excellent results are reported with a broad variety of substrates.

Acyl azides are easily handled by acetic acid-catalysed Curtius rearrangement which generates one mole of nitrogen per azide[27]. This can be done conveniently on a micro scale with no interference from such groups as nitro, nitroso, azoxy, azo, hydrazo, cyano, amido, imido, amino or ammonium. Diazo or *N*-nitroso groups do interfere.

Several other older and frequently more specific ideas are collected in standard texts[28].

III. INFRARED SPECTROSCOPY

At an early date it was reported[29] that the azido group (ionic) showed absorption in the infrared region at 4·6–4·8 μm (2120–2300 cm^{-1}). Since then organic azides have been the subject of extensive study. By a large margin, infrared absorptions in the region 2160–2090 cm^{-1} (asymmetric stretching), 1340–1180 cm^{-1} (symmetric stretching), and 680 cm^{-1} (bending) are most frequently cited for the characterization of azides[5].

Several extensive reports have appeared regarding the infrared spectra of organic azides[30–33]. All of these report approximately the same conclusion. The asymmetric stretching frequency in the region 2160–2090 cm^{-1} is very strong, only slightly influenced by substituents, and highly characteristic for the azido group.

The Russian workers[30] claim that substituents on the azido group which increase the asymmetry, presumably by changing the equivalence of the bonding, shift the bands in question to higher frequency. They also suggest that this asymmetric stretching band for aromatic sulphonyl azides can be correlated by a Hammett $\sigma_p \rho$ relationship where ρ has a small value. (They find the ultraviolet spectral parameters responding in similar fashion.)

The report[31] by Lieber and co-workers discusses such diverse classes of azides as acyl azides, carbamyl azides, vinyl azides, and α-azido ethers, thioethers and amines. None of these violates the limits of the asymmetric stretching frequency cited earlier.

Publications in 1951[34] and 1954[35] utilized this characteristic absorption at ca. 2100 cm^{-1} to demonstrate the structure of a β-keto azide and an α-hydroxy azide, respectively. The assumption that other functional groups in the molecule did not interfere was justified by a report[36] in which both alkyl and aryl azides were studied, followed by a paper[37] dealing with α-substitution. The clear result of all of this was the constancy of the azide asymmetric stretching frequency. Sheinker and co-workers[38,39] have made similar studies and report the interesting conclusion that the intensity of the band is more sensitive to structure than is the position of the band. The intensity is raised by electron donor groups and lowered by substitution of electron acceptor groups.

Three recent publications dealing with α-imino azides[40], α-azido chlorides[41] and vinyl azides[42] reflect the utility of infrared measurements with alkyl and aryl azides.

Sulphonyl azides have asymmetric stretching frequencies at the high end of the range mentioned earlier (2140 cm^{-1}). The success of an early preparation of methanesulphonyl azide[43] was characterized by i.r. absorption at 2137 cm^{-1} [44]. A 1964 study of infrared spectra in the series of compounds represented by Formula **1**, showed that the frequency of absorption (2140 cm^{-1}) was quite constant.

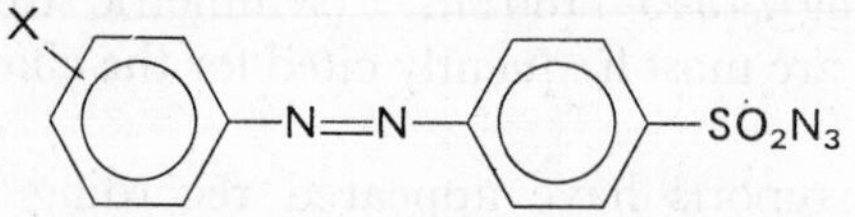

$$X\text{-}C_6H_4\text{-}N{=}N\text{-}C_6H_4\text{-}SO_2N_3$$

1 X = H; 2-OH; 4-OH; 2-OH, 5-Cl; 3-CH_3, 4-OH

Carboxylic acid azides have been analysed[45] and again, the characteristic asymmetric stretching frequency is 2140 cm^{-1}. Substitution alpha to the carbonyl group has little effect[46]. A review[47] of the infrared characteristics of carbamoyl azides has appeared which cites 2150 cm^{-1} as the characteristic frequency for the azido group.

It is not surprising that organometallic azides such as **2**[48] and **3**[49] still can be characterized by a band in the infrared spectrum at 2050–2100 cm^{-1}.

2 ϕ_3MN_3 M = carbon, germanium, tin, lead

3 $\phi_3M(N_3)_2$ M = arsenic, antimony

The azide–tetrazole equilibrium (equation 4) has been studied by infrared spectroscopy[50,51] and settled in favour of the azide form.

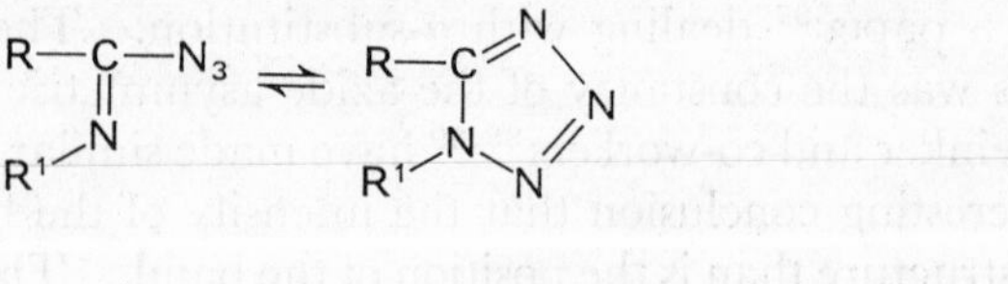

$$R\text{—}C(=N\text{—}R^1)\text{—}N_3 \rightleftharpoons R\text{—}C\!\!\begin{array}{c}\text{(tetrazole ring: }C{=}N\text{—}N{=}N\text{—}N(R^1)\text{)}\end{array} \tag{4}$$

The similar structural problem, structure **4** or **5**, has been resolved in favour of **5** due to the absence of the usual absorption for N_3 groups[52].

R—C(=S)—N_3

(4)

(5)

Infrared measurements of the complex between boron trifluoride (or trichloride) and carboxylic acid azides conclusively demonstrate that complexing occurred through the carbonyl oxygen rather than one of the nitrogens since the usual absorption is present[53].

One of the most interesting features of the i.r. spectral properties of azides is the frequent occurrence of Fermi resonance. If an overtone or combination band falls near a fundamental frequency, the band intensity of the fundamental may be anomalously enhanced or the band may be split. This coupling between a fundamental frequency and the combination or overtone band is called the Fermi resonance. Certain symmetry properties must be met[54].

Fermi resonance occurs only with aryl azides[31,55,56] or with acid azides[45,57]. It has been attributed[31] to a combination band from the azide symmetric stretch (ca. 1250 cm^{-1}) and an aromatic ring vibration (ca. 1000 cm^{-1}). Aryl acid azides are said[31] to have a similar vibration at ca. 900 cm^{-1}. Other possible combinations are the azide symmetric stretch (1250 cm^{-1}) and the C—N stretch (ca. 1150 cm^{-1}) or the C—N stretch and the aromatic vibrations mentioned earlier. Bhaskar[56] points out that in aromatic azides conjugation between the phenyl ring and the azide group is necessary, for benzyl azide shows a sharp singlet. Furthermore, he reports that the extent of interaction is dependent on the electronic properties of substituents on the aromatic ring. A maximum effect is found with electron releasing groups (*p*-dimethylamino) and a minimum with electron withdrawing groups (*p*-nitro).

The related phenomenon of Raman spectroscopy has been applied to azide chemistry, mainly in the early discussion of whether the azide structure was linear or cyclic[58–60].

IV. ULTRAVIOLET SPECTROSCOPY

Organic azides have been studied by ultraviolet absorption techniques since the early 1930's[61,62]. Alkyl azides are characterized by two relatively low intensity transitions[63]. These solvent insensitive bands centred at 287 nm (ε,25) and 216 nm (ε,500) have been described[63] as $\pi_y \rightarrow \pi_x^*$ and $s\text{–}p_x \rightarrow \pi_y^*$, respectively. Closson and Gray[63] also mention that substitution of electronegative groups in the alkyl portion would lower via induction the energy of the electrons on the nitrogen bonded to carbon. This would increase the energy (shift to shorter wavelength) of both of these transitions.

Precisely this effect was noted, first by Böhme[64,65] and then by Lieber and Thomas[37]. The German paper[64] reports absorption of α-azido ethers, thioethers, and amines in the range 275–290 nm with a blue shift due to the adjacent heteroatom. The more recent publication[37] cites absorption in the range 264–284 nm and a shift to shorter wavelength caused by sulphur, oxygen, or nitrogen when the azide is located alpha to thioether, ether, or amine functions. Nitrogen causes the greatest shift, sulphur the least. Böhme[65] also reports ultraviolet absorption for α-azido sulphones. The tetrazole-iminoazide equilibrium has been studied by ultraviolet as well as infrared spectroscopy[50].

As one would expect, the ultraviolet spectra of aromatic azides is decidedly more varied. Several papers have appeared reporting the general features of aryl azides[32,66–71]. Generally, it can be said that the intensity of the long wavelength band (ca. 285 nm) is substantially increased[67]. Reiser and co-workers present data for various aromatic and condensed aromatic azides and conclude that a red shift of the entire spectrum occurs with increased intensity. Due to the complex nature of polycyclic aromatic hydrocarbons it is difficult to say whether or how the azide transition energy is changed. It is noted[67] that *ortho* and *meta* substituted phenyl azides behave as if no interaction is present. For example, *ortho*- and *meta*- nitrophenyl azide each yield a spectrum which appears as a superposition of nitrobenzene and phenyl azide. However, the spectrum of *para*-nitrophenyl azide is substantially different and difficult to interpret.

No mention of the ultraviolet spectra of acyl azides was found. It was reported[67] that in $N_3CO_2CH_3$ the 285 nm band is not seen, perhaps due to interaction of the azido group with the carbonyl group.

Aroyl azides ($ArCON_3$) have been studied by some Japanese workers[72]. Benzoyl azide is reported to have a maximum at 245·5 nm while substitution in the *para* position causes a shift to longer wavelength. Methoxy changes the maximum to 283 nm while fluorine moves it to 251 and nitro to 260·5. Other substituents such as Cl, Br, I, methyl, and CON_3 caused the maximum to fall between that of the *p*-F and *p*-OMe derivatives.

The long wavelength band (285 nm) is not found in sulphonyl azides[30]. Perhaps, due to its low intensity it is hidden under the tail of other intense but shorter wavelength absorptions.

V. OPTICAL ROTATORY DISPERSION AND CIRCULAR DICHROISM

The low intensity long wavelength absorption of alkyl azides (λ_{max} 285 nm, $\varepsilon \sim 50$) is perfectly suited for measurements of optical rotatory dispersion and circular dichroism.

Indeed the rotatory dispersion properties of configurationally related alkyl azides were determined as early as 1937[73]. Further o.r.d./c.d. studies on azides apparently were discontinued until 1967 when Djerassi, Moscowitz, Ponsold and Steiner[74] reported the o.r.d./c.d. properties of 34 steroidal azides. An Octant Rule for the azide chromophore was presented. To date, one further application of the Rule has appeared. Paulsen[75] has studied the c.d. of sugar azides.

The Octant Rule for azides was established theoretically by recognizing the similarity between the $n \rightarrow \pi^*$ transition of ketones and the $s\text{–}p_x \rightarrow \pi_y^*$ transition of azides[63,74]. The geometry of the transitions appears similar; however, there are differences in the two situations. First, in the ketone, the atomic orbitals of both the ground and the excited state are associated with the oxygen atom. In azides different nitrogen atoms are involved in the two centres. This should result in a reduced magnetic dipole moment and would contribute to a decidedly lower amplitude for the o.r.d. curve. Secondly, conformational mobility from rotation about the C—N bond might contribute to low rotational strength.

Conformational mobility is a serious drawback to application of this Octant Rule. For a compound such as 2α-azidocholestane, the Rule is applied to each of six rotamers (generally depicted as Newman projections). By a combination of conformational analysis and this Octant Rule, one tries either to predict the sign of the rotation from the favoured structure or vice versa. In this particular case, no stereochemically significant conclusion was reached[74].

VI. MASS SPECTROMETRY

One communication[76] has appeared which addresses itself to the mass spectral properties of organic azides. The mass spectrum of phenyl azide[76] is reported to have a base peak at *m/e* 91 corresponding to loss of a nitrogen molecule (M-28). This is followed by the loss of HCN to yield the $C_5H_4^+$ ion. The authors report that the spectrum is the same as that from benzotriazole.

Other recent publications have utilized mass spectral measurements. Azidotropones are reported to yield a peak at M-28 rather than a molecular ion peak[77]. Vinyl azides generally do not show a molecular ion[78]. Various sugar azides were prepared and found to show no molecular ion, but peaks corresponding to loss of molecular nitrogen[79]. An α-azidolactam is reported to fragment to yield the M-28 ion as a result of the loss of N_2[80]. Moore and co-workers have employed mass spectrometry in their study of azidohydroquinones[81] and azidoquinones[82]. The spectra are not published but the descriptions suggest that molecular ions are not always found and that loss of molecular nitrogen is most prevalent.

VII. REFERENCES

1. I. T. Millar and H. D. Springall, *Sidgwick's Organic Chemistry of Nitrogen*, 3rd ed., Clarendon Press, Oxford, 1966, Chap. 14.
2. P. A. S. Smith, *The Chemistry of Open-Chain Organic Nitrogen Compounds*, Vol. 2, W. A. Benjamin, New York, 1966, Chap. 10.
3. C. G. Overberger, J.-P. Anselme and J. G. Lombardino, *Organic Compounds with Nitrogen–Nitrogen Bonds*, Ronald Press, New York, 1966, Chap. 8.
4. J. H. Boyer and F. C. Carter, *Chem. Rev.*, **54,** 1 (1954).
5. F. Lieber, J. S. Curtice and C. N. R. Rao, *Chem. Ind. (London)*, (14) 586 (1966).
6. C. V. Hart, *J. Am. Chem. Soc.*, **50,** 1922 (1928).
7. P. P. T. Sah, *J. Chinese Chem. Soc.*, **5,** 100 (1937).
8. P. P. T. Sah, *Rec. Trav. Chim. Pays-Bas.*, **59,** 231 (1940).
9. P. P. T. Sah and Wen-Hou Yin, *Rec. Trav. Chim. Pays-Bas.*, **59,** 238 (1940).
10. P. A. S. Smith, J. M. Clegg and J. Lakritz, *J. Org. Chem.*, **23,** 1595 (1958).
11. W. Fischer and J.-P. Anselme, *J. Am. Chem. Soc.*, **89,** 5284 (1967).
12. J. Cleophax, S. D. Gero and J. Hildesheim, *Chem. Commun.*, 94 (1968).
13. J. S. Fritz and G. S. Hammond, *Qualitative Organic Analysis*, J. Wiley, New York, 1957, pp. 110–111.
14. M. J. Coombs, *J. Chem. Soc.*, 4200 (1958).
15. A. Buhler and H. E. Fierz-David, *Helv. Chim. Acta.*, **26,** 2123 (1943).
16. G. T. Tisue, S. Linke and W. Lwowski, *J. Am. Chem. Soc.*, **89,** 6303 (1967).
17. P. A. S. Smith and J. H. Boyer, *J. Am. Chem. Soc.*, **73,** 2626 (1951).
18. A. Gutmann, *Ber.*, **45,** 821 (1912).
19. A. Gutmann, *Ber.*, **57,** 1956 (1924).
20. A. Gutmann, *Z. Anal. Chem.*, **66,** 224 (1925).
21. I. Ugi, H. Perlinger and L. Behringer, *Chem. Ber.*, **91**, 2330 (1958).
22. W. R. Carpenter, *Anal. Chem.*, **36,** 2352 (1964).
23. A. Messmer and S. Mlinko, *Acta. Chim. Sci. Hung.*, **29,** 119 (1961); *Chem. Abstr.*, **57,** 19531c (1962).
24. K. A. Hoffman, H. Hoch and H. Kirmreuther, *Justus Liebigs Ann. Chem.*, **380,** 131 (1911).

25. P. A. S. Smith, *J. Am. Chem. Soc.*, **73,** 2438 (1951).
26. J. E. Leffler and Y. Tsuno, *J. Org. Chem.*, **28,** 190 (1963).
27. W. I. Awad, Y. A. Gawargious and S. S. M. Hassan, *Talanta*, **14,** 1441 (1967).
28. N. D. Cheronis and T. S. Ma, *Organic Functional Group Analysis*, Interscience, New York, 1964, pp. 234–261.
29. A. D. Angstrom, *Z. Physik. Chem.*, **86,** 525 (1914).
30. V. A. Gal'perin and G. P. Balabanov, *Zh. Obshch. Khim.*, **38,** 926 (1968); *J. Gen. Chem. U.S.S.R.*, **38,** 889 (1968).
31. E. Lieber, C. N. R. Rao, A. E. Thomas, E. Oftedahl, R. Minnis and C. V. N. Namburg, *Spectrochim. Acta.*, **19,** 1135 (1963).
32. Gy. Varsányi, S. Holly and J. Szathmáry, *Periodica Polytech.*, **2,** 211 (1958); *Chem. Abstr.*, **54,** 20931c (1960).
33. E. Lieber, D. R. Levering and L. J. Patterson, *Anal. Chem.*, **23,** 1594 (1951).
34. J. H. Boyer, *J. Am. Chem. Soc.*, **73,** 5248 (1951).
35. C. A. VanderWerf, R. Y. Heisler and W. E. McEwen, *J. Am. Chem. Soc.*, **76,** 1231 (1954).
36. E. Lieber, C. N. R. Rao, T. S. Chao and C. W. W. Hoffman, *Anal. Chem.*, **29,** 916 (1957).
37. E. Lieber and A. F. Thomas, III, *Appl. Spectrosc.*, **15,** 144 (1961).
38. Yu. N. Sheinker and L. B. Senyavina, *Izv. Akad. Nauk. SSSR, Ser. Khim.*, 2113 (1964); *Chem. Abstr.*, **62,** 9942b (1965).
39. Yu. N. Sheinker, L. B. Senyavina and V. N. Zheltova, *Dokl. Akad. Nauk. SSSR*, **160,** 1339 (1965); *Proc. Acad. Sci., U.S.S.R.*, **160,** 215 (1965).
40. J.-P. Anselme, W. Fischer and N. Koga, *Tetrahedron*, **25,** 89 (1969).
41. A. Hassner, R. J. Isbister, R. B. Greenwald, J. T. Klug and E. C. Taylor, *Tetrahedron*, **25,** 1637 (1969).
42. G. Smolinsky, *Trans. N.Y. Acad. Sci.*, **30,** 511 (1968).
43. J. H. Boyer, C. H. Mack, N. Goebel and L. R. Morgan, Jr., *J. Org. Chem.*, **23,** 1051 (1958).
44. R. J. W. Cremlyn, *J. Chem. Soc.*, Suppl. II, 6235 (1964).
45. E. Lieber and E. Oftedahl, *J. Org. Chem.*, **24,** 1014 (1959).
46. H. Neunhoeffer, M. Neunhoeffer and W. Litzius, *Justus Liebigs Ann. Chem.*, **722,** 29, 38 (1969).
47. E. Lieber, R. L. Minnis, Jr. and C. N. R. Rao, *Chem. Rev.*, **65,** 377 (1965).
48. J. Jappy and P. N. Preston, *Inorg. Nucl. Chem. Lett.*, **4,** 503 (1968).
49. A. Schmidt, *Chem. Ber.*, **101,** 3976 (1968).
50. J. H. Boyer and E. J. Miller, Jr., *J. Am. Chem. Soc.*, **81,** 4671 (1959).
51. G. A. Reynolds, J. A. Van Allan and J. F. Tinker, *J. Org. Chem.*, **24,** 1205 (1959).
52. F. L. Scott, *Experentia*, **13,** 275 (1957).
53. E. Fahr and L. Neumann, *Justus Liebigs Ann. Chem.*, **715,** 15 (1968).
54. K. Nakanishi, *Infrared Absorption Spectroscopy*, Holden-Day, San Francisco, 1962, p. 19.
55. L. K. Dyall and J. E. Kemp, *Australian J. Chem.*, **20,** 1395 (1967).
56. K. R. Bhaskar, *Indian J. Chem.*, **5,** 416 (1967).
57. W. R. Carpenter, *Appl. Spectrosc.*, **17,** 70 (1963).
58. L. Kahovec, K. W. F. Kohlrausch, A. W. Reitz and J. Wanger, *Z. Phys. Chem.*, **B39,** 431 (1938).
59. E. H. Eyster and R. H. Gillette, *J. Chem. Phys.*, **8,** 369 (1940).

60. Yu. N. Sheinker and Ya. K. Syrkin, *Izv. Akad. Nauk. SSSR, Ser. Fiz.*, **14,** 478 (1950); *Chem. Abstr.*, **45,** 3246f (1951).
61. A. Hantzsch, *Ber.*, **66B,** 1349 (1933).
62. A. Hantzsch, *Ber.*, **67B,** 1674 (1934).
63. W. D. Closson and H. B. Gray, *J. Am. Chem. Soc.*, **85,** 290 (1963).
64. H. Böhme, D. Morf and F. Mundlos, *Chem. Ber.*, **89,** 2869 (1956).
65. H. Böhme and D. Morf, *Chem. Ber.*, **90,** 446 (1957).
66. P. Grammaticakis, *Compt. Rend*, **244,** 1517 (1957).
67. Yu. N. Sheinker, *Dokl. Akad. Nauk. SSSR*, **77,** 1043 (1951); *Chem. Abstr.*, **45,** 6927b (1951).
68. C. N. R. Rao, *Chem. Ind.* (*London*), 666 (1956).
69. P. A. S. Smith, J. H. Hall and R. O. Kan, *J. Am. Chem. Soc.*, **84,** 480 (1962).
70. C. N. R. Rao and C. W. W. Hoffman, *Sci. Cult.* (*Calcutta*), **22,** 463 (1957); *Chem. Abstr.* **51,** 11853h (1957).
71. A. Reiser, G. Bowes and R. J. Horne, *Trans. Faraday Soc.*, **62,** 3162 (1966).
72. S. Munekata and S. Kikuchi, *Seisan-Kenkyu*, **20,** 519 (1968); *Chem. Abstr.*, **70,** 67362h (1969).
73. P. A. Levene and A. Rothen, *J. Chem. Phys.*, **5,** 985 (1937).
74. C. Djerassi, A. Moscowitz, K. Ponsold and G. Steiner, *J. Am. Chem. Soc.*, **89,** 347 (1967).
75. H. Paulsen, *Chem. Ber.*, **101,** 1571 (1968).
76. W. D. Crow and C. Wentrup, *Tetrahedron Letters.*, 4379 (1967).
77. J. D. Hobson and J. R. Malpass, *J. Chem. Soc.* (*C*), 1499 (1969).
78. A. Hassner and F. W. Fowler, *J. Org. Chem.*, **33,** 2686 (1968).
79. J. Cleophax, J. Hildesheim, R. E. Williams and S. D. Gero, *Bull. Soc. Chim. Fr.*, 1415 (1968).
80. A. K. Bose, V. Sudarsanam, B. Anjaneyulu and M. S. Manhas, *Tetrahedron*, **25,** 1191 (1969).
81. H. W. Moore and H. R. Sheldon, *J. Org. Chem.*, **33,** 4918 (1968).
82. H. W. Moore, H. R. Sheldon and D. F. Shellhamer, *J. Org. Chem.*, **34,** 1999 (1969).

CHAPTER 4

The directing and activating effects of the azido group

M.E.C. Biffin,* J. Miller† and D. B. Paul*

* *Defence Standards Laboratories, Maribyrnong, Victoria, Australia*

† *University of Natal, Durban, S. Africa*

I. Introduction 203
II. The Directing and Activating Effects of the Azido Group in Substitution Reactions 204
A. Polarization and Polarizability Effects 204
B. Electrophilic Substitution 209
C. Nucleophilic Substitution 212
III. The Directing and Activating Effects of the Azido Group in Addition and Elimination Reactions 216
A. Addition Reactions 216
B. Elimination Reactions 217
IV. References 219

I. INTRODUCTION

In contrast to most common functional groups, the azido group has been the subject of comparatively few investigations of specific relevance to its directing and activating effects. The reason may perhaps be due to the fact that the azido substituent is often the most reactive centre in a molecule owing to its general lability in many chemical environments. Thus investigations of the chemical effects of its electronic properties have been restricted. The reactive nature of the azido group is, however, of obvious value in organic synthesis and indeed the main interest in the chemistry of the organic azides has been focused on this aspect.

Despite the brevity of this chapter, sufficient data are available to illustrate the influence of the azido group on both electrophilic and nucleophilic substitution reactions, particularly aromatic substitutions, and its role in neighbouring group participation. Some evidence of relevance to the directing and activating effects of the azido group in addition and elimination reactions is also available.

In this chapter we have limited our discussion to include only those processes in which the azido group maintains its integrity. Hence reactions in which the azido group may strictly be deemed to be activating, but which result in the formation of a product other than an organic azide, are not considered.

II. THE DIRECTING AND ACTIVATING EFFECTS OF THE AZIDO GROUP IN SUBSTITUTION REACTIONS

A. *Polarization and Polarizability Effects**

Determination of the dissociation constants of azidoaliphatic acids has conclusively shown that the azido group is acid strengthening. In Table 1 the acid dissociation constants of α-azidoacetic and α-azidopropionic acids obtained by Philip[1] are compared with the K_a values for the corresponding haloacetic acids[2–4]. From these figures, the substituent constant σ′ for the azido group may be evaluated as

TABLE 1. Ionization constants[a] of substituted acetic and propionic acids[1–4].

Acid	$10^3 K_a$
α-Chloroacetic acid	1·55
α-Bromoacetic acid	1·38
α-Azidoacetic acid	0·93
α-Iodoacetic acid	0·75
Acetic acid	0.018
α-Chloropropionic acid	1·5
α-Bromopropionic acid	1·1
α-Azidopropionic acid	0·9
α-Iodopropionic acid	0·62
Propionic acid	0·014

[a] Measured at 26°C.

* The symbols K (conjugative effect) I (inductive effect) M (mesomeric effect) and E (electromeric effect) are employed throughout this chapter and follow the notation employed by C. K. Ingold in *Structure and Mechanism in Organic Chemistry*, 2nd Ed., Bell, London, 1969.

+0·41 by using a procedure based on the Hammett–Burkhardt relationship[5]. Hence in the ground state the azido group exerts an electron-withdrawing effect by induction ($-I_s$), the magnitude of which falls between those of the bromo and iodo groups and is closely comparable to that of the trifluoromethyl substituent (Table 2).

TABLE 2. Substituent constant values[4,5]

Substituent	σ'	Substituent	σ'
t-Butyl	−0·07	Trifluoromethyl	+0·41
Methyl	−0·05	Bromo	+0·45
Hydrogen	0	Chloro	+0·47
Methoxy	+0·23	Fluoro	+0·50
Ethoxycarbonyl	+0·30	Cyano	+0·59
Iodo	+0·38	Nitro	+0·63
Azido	+0·41	Trimethylammonio	+0·86

In the aromatic series it has also been shown that the electron-withdrawing effect of the azido substituent is similar to those of the bromo and iodo groups. Smith, Hall and Kan[6] have recorded the ionization constants of the *ortho*-, *meta*- and *para*-azidobenzoic acids and azidoanilines (Table 3) and from the Hammett–Burkhardt relationship these authors derived values for the constant σ_m of the azido

TABLE 3. Ionization constants of azidobenzoic acids and azidoanilines[6]

	$10^6 K_a$	$10^{11} K_b$	Hammett substituent constant
Benzoic acid[a]	5·76	—	—
o-Azidobenzoic acid[a]	29·5	—	—
m-Azidobenzoic acid[a]	13·5	—	σ_m = 0·37
p-Azidobenzoic acid[a]	6·92	—	σ_p = 0·08
Aniline[b]	—	25·7	—
o-Azidoaniline[b]	—	2·30	—
m-Azidoaniline[b]	—	3·31	σ_m = 0·33
p-Azidoaniline[b]	—	12·6	σ_p = 0·11

[a] Apparent K_a measured at 25°C in 45·4% aqueous methanol.
[b] Measured at 25°C in 45·4% aqueous methanol.

substituent of 0·37 from the benzoic acid series and 0·33 from the anilines. These are of the same order as the currently accepted σ_m values of the halogens (Table 4) and slightly greater than that of the

TABLE 4. Hammett and electrophilic substituent constants of some common functional groups[5,7–9]

Substituent	σ_m	σ_p	σ_p^+
Dimethylamino	−0·21	−0·83	−1·7
Hydroxy	0·121	−0·37	−1·0
Methoxy	0·115	−0·27	−0·78
Azido	0·37	0·08	−0·54
Methyl	−0·07	−0·17	−0·31
Trifluoromethyl	0·42	0·54	0·61
Phenyl	0·06	−0·01	−0·18
Fluoro	0·337	0·062	−0·07
Chloro	0·37	0·23	0·11
Bromo	0·39	0·23	0·15
Iodo	0·35	0·18	0·135
Ethoxycarbonyl	0·37	0·45	0·48
Cyano	0·56	0·66	0·66
Nitro	0·71	0·78	0·79
Diazonio	1·76	1·91	—
Acetamido	—	−0·015	−0·6
Benzamido	0·217	0·078	—
Amino	−0·16	−0·66	−1·3
Trimethylammonio	1·01	0·88	0·408

benzamido group (0·217). The lower, yet still positive values calculated for σ_p indicate that the azido group is inductively electron-withdrawing, but mesomerically electron-releasing in the *para* position ($-I_s$, $+M$), with the inductive effect in ascendance ($-I_s > +M$). The extent of the conjugative effect is comparable with those of the fluoro and benzamido substituents which have σ_p values of 0·062 and 0·078 respectively (Table 4). The differences in the orders of electron-withdrawing ability that are apparent from the aliphatic series and the *meta*-substituted aromatic compounds (compare Tables 2 and 4) may be rationalized on the ground that the $+M$ effect also has some influence on positions *meta* to the azido group in the aromatic nucleus[5]. The conjugative effect is obviously more significant from the *para*-position and this is reflected in the lower σ_p values.

Estimations of the Hammett σ constants for substituted aromatic compounds by application of the Doub and Vandenbelt[10] empirical relationship between (σ_p–σ_m) and ultraviolet wavelength shifts have been described. In particular the values for the azido group obtained by Smith and co-workers have been confirmed by this procedure[6,11].

Further evidence that the polarization effect of the azido group affords a net electron-withdrawal is provided by the magnitude and direction of the dipole moments of organic azides[12–16]. In phenyl azide for example, the dipole is directed with the negative pole away from the ring and the value of the dipole moment (1·44 D) is approximately the same as that of the C—$N_{(1)}$ contribution. This low figure is consistent with nearly equivalent and opposing dipoles in the major resonance contributors **1** and **2**. The dipole moments of aliphatic azides appear to be generally larger than those of aryl azides (see

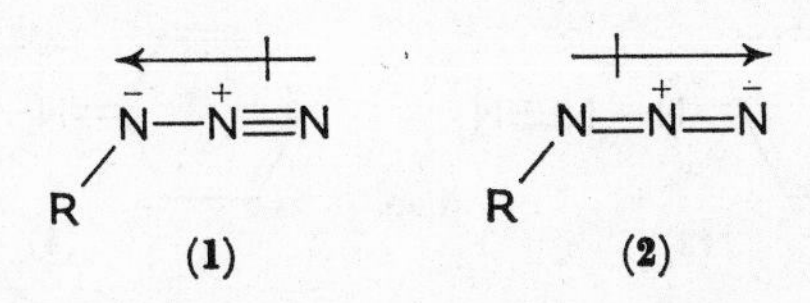

Table 9 in Chapter 1) and this is presumably due to the absence of attenuating conjugative components in the polarization of the aliphatic derivatives. The low value of the dipole moment for *p*-chloroazidobenzene is consistent with the similar dipole contributions of the azido and the halogeno groups.

The polarizability effects of the azido group impart strong *ortho–para* activation towards electrophilic substitution. This is exemplified in the large Brown σ_p^+ constant (−0·54) obtained from competitive bromination studies on mixtures of phenyl azide with benzene, toluene or anisole[6]. The conjugative influence (+K effect) of the azido group is thus greater than that of fluoro (σ_p^+ = −0·073), intermediate between methyl (σ_p^+ = −0·31) and methoxy (σ_p^+ = −0·78), and only slightly less than that of acetamido (σ_p^+ = −0·6) (Table 4). Additional evidence for the +K effect of the azido group, albeit qualitative, is provided by the infrared carbonyl stretching frequency of *p*-azidoacetophenone[6], which occurs at 1680 cm^{-1}. This frequency is lower than that of acetophenone and substantially higher than that of *p*-aminoacetophenone in which the *p*-amino substituent exerts a powerful +K effect (σ_p^+ NH_2 = −1·3). Again there is a close similarity with the acetyl carbonyl stretching frequency of *p*-acetamidoacetophenone which is only slightly lower than that of *p*-azidoacetophenone.

Information concerning the directing and activating effects of the azido group on nucleophilic substitution is scarce but from the existing evidence it has been suggested that in S_NAr reactions the azido substituent is activating to *para*-substitution[17]. Nevertheless, the electron-withdrawing influence is relatively weak and comparable with the effects of the p-CO_2^- and p-SO_3^- groups.

In summary, it is apparent that the polarizability of the azido group is such that the substituent may be either electron-donating or electron-attracting according to the demands of the approaching reagent. In electrophilic aromatic substitution the conjugative effect overrides the opposing electron-withdrawal by induction ($+K \gg -I_d$) and the importance of structures such as **3** and **4** is implied. Similarly, for nucleophilic substitution the activating effect may be considered as

(3) **(4)**

$-I_d$, $-K$, with activation by conjugation occurring as shown in **5**. Internal conjugation in the azido group places a fractional negative

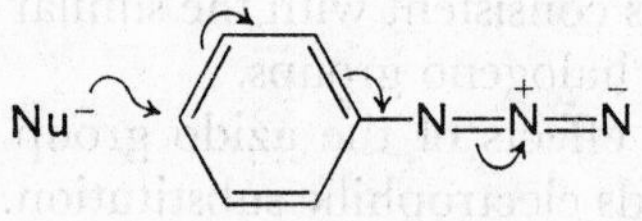

Nu⁻ N=N=N

(5)

charge on the α-nitrogen atom as in **6**, and it is for this reason that only a small activating effect has been observed in S_NAr reactions.

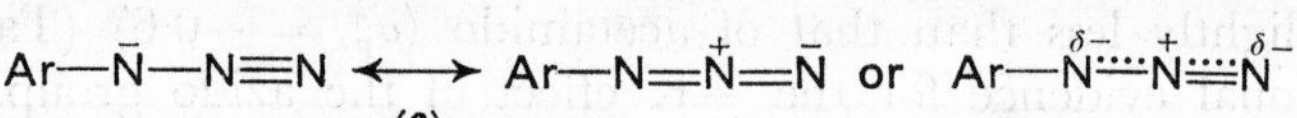

$$\mathrm{Ar{-}\bar{N}{-}N{\equiv}N \longleftrightarrow Ar{-}N{=}\overset{+}{N}{=}\bar{N}} \text{ or } \mathrm{Ar{-}\overset{\delta-}{N}{\cdots}\overset{+}{N}{\cdots}\overset{\delta-}{N}}$$

(6)

Substituents which are capable of activating both electrophilic and nucleophilic substitution have been termed pan-activating[18] and the azido group falls into this category, together with such structural units as aromatic N-oxide, arylazo, azoxy, thiocyanato, nitroso and phenyl.

Finally, the azido group would be expected to activate homolytic aromatic substitution by virtue of its ability to provide additional stabilization to free radical intermediates. No evidence in support of this prediction is available however and investigation in this area seems warranted.

B. *Electrophilic Substitution*

Under conditions of strong electron demand the conjugative polarizability (+K) effect in aromatic azides is the controlling factor. The azido group thus activates electrophilic substitution reactions and directs the incoming group to the *ortho* and *para* positions.

This activating influence by the azido group was predicted on electronic grounds by Miller and Parker[17] in 1958. Both the susceptibility of aromatic azides to electrophilic substitution and the *ortho*–*para* directing effect could perhaps be inferred from earlier qualitative investigations of the nitration of aromatic azides which are summarized in Table 5. These conclusions have since been placed on a

TABLE 5. Electrophilic substitution of aryl azides

Substrate	Reagent	Product	Yield (%)	Reference
Phenyl azide	HNO_3 (d = 1·44)	*p*-Nitrophenyl azide	30	19
Phenyl azide	HNO_3-*c*.H_2SO_4	*p*-Nitrophenyl azide	37	19
Phenyl azide	HNO_3-*c*.H_2SO_4	*p*-Nitrophenyl azide	44	20
Phenyl azide	HNO_3 (d = 1·5)	*p*-Nitrophenyl azide, 2,4-dinitrophenyl azide	ca.45 'small'	21
Phenyl azide	Nitrating conditions	*p*-Nitrophenyl azide, 2,4-dinitrophenyl azide, picryl azide	—	22
Phenyl azide	Br_2–CCl_4	*p*-Bromophenyl azide	54	6
Phenyl azide	Br_2–Fe–CCl_4	*p*-Bromophenyl azide	45	6
o-Nitrophenyl azide	HNO_3 (d = 1·5)	2,4-Dinitrophenyl azide	—	23
o-Nitrophenyl azide	HNO_3-*c*.H_2SO_4	Picryl azide	—	24
p-Nitrophenyl azide	HNO_3 (d = 1·5)	2,4-Dinitrophenyl azide	—	23
p-Nitrophenyl azide	HNO_3-*c*.H_2SO_4	Picryl azide	—	24
m-Nitrophenyl azide	Fuming HNO_3	3,4-Dinitrophenyl azide	94	25
m-Nitrophenyl azide	HNO_3-*c*.H_2SO_4	2,4,5-Trinitrophenyl azide	76	25
1,5-Diazido-2,4-dinitrobenzene	Fuming HNO_3	1,3-Diazido-2,4,6-trinitrobenzene	76	25
1,3,5-Triazido-2,4-dinitrobenzene	HNO_3-*c*.H_2SO_4	1,3,5-Triazido-2,4,6-trinitrobenzene	62	25
1-Azidonaphthalene	HNO_3 (d = 1·4)	1-Azido-4-nitro-napthalene	—	26
2-Azidonaphthalene	HNO_3 (d = 1·4)	2-Azido-1-nitro-napthalene	—	26
2-Azidonaphthalene	Br_2–CCl_4	2-Azido-1-bromo-naphthalene	57	6
2-Azidoanthraquinone	Fuming HNO_3	2-Azido-1-nitro-anthraquinone	—	27
2,6-Diazidoanthra-quinone	Fuming HNO_3	2,6-Diazido-1,5-dinitroanthraquinone	—	27

quantitative basis by Smith, Hall and Kan[6] who obtained a value of σ_p^+ for the azido group from a study of the competitive bromination of phenyl azide in the presence of other aromatic substrates. Bromination of an equimolar mixture of anisole and phenyl azide using molecular bromine and sodium acetate in glacial acetic acid afforded *p*-bromoanisole and *p*-bromophenyl azide in the ratio of 13:1 and similarly 12 times as much *p*-bromophenyl azide as *p*-bromotoluene was formed in a competitive reaction with toluene. From these results the σ_p^+ value of $-0{\cdot}54$ was calculated for the azido substituent which thus activates electrophilic substitution to approximately the same degree as the methylthio and acetamido groups (Table 4).

Systematic surveys in this area have been limited, however, due to the sensitivity of the azido group in strongly acidic media[28]. Under these conditions aromatic azides generally form ring-substituted anilines[6,19,29]. This probably occurs by *nucleophilic* attack and a subsequent prototropic shift. It is presumed that the initial process involves protonation of the azido group to produce the conjugate acid (**7**). Subsequent loss of nitrogen leads to an intermediate delocalized cation (**8**). This may react with the conjugate base of the acid, or the solvent (as nucleophiles), and by subsequent prototropic shift afford a substituted aniline. This is illustrated in equation (1), although it is possible that the addition of a second proton may occur earlier than illustrated. For brominations with molecular bromine in acetic acid the extent of the process illustrated in equation (1) has been minimized by employing sodium acetate to remove hydrogen bromide generated by electrophilic bromination[6].

It appears that in nitration reactions the nitronium ion is sufficiently powerful as an electrophile to enable ring-substitution of the aryl azide to compete adequately with the decomposition pathway, and modest yields of nitroaryl azides may be achieved[19–21]. It is significant that high yields of di- and tri-nitroazidobenzenes have been obtained when nitroaromatic azides are employed as starting materials[23–25] and this presumably reflects the lower basicity of such substrates and the consequent attenuation of the competitive decomposition through the conjugate acid. This situation is somewhat similar to the case of dimethylaniline and *p*-nitrodimethylaniline which are both nitrated at comparable rates despite the powerful electron-withdrawing effect of the nitro group in the latter. It is thought that the lower basicity of *p*-nitrodimethylaniline results in relatively higher concentrations of the unprotonated nitroamine in the nitrating mixture and that this offsets the deactivating effect of the nitro group on the rate of nitration[30].

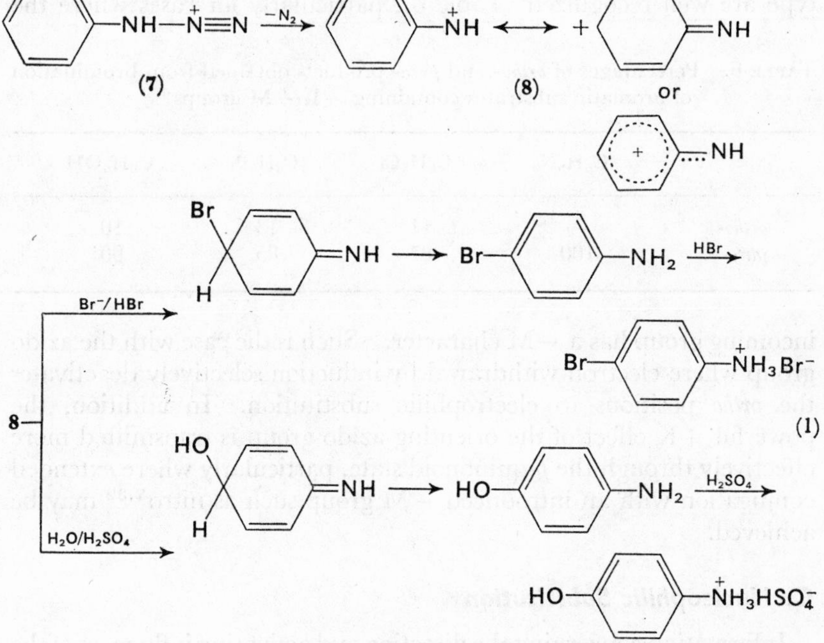

A further example of the diminished significance of the decomposition pathway for nitroaromatic azides is implicit in the observation of Bailey and White[24] that whereas phenyl azide is an unsuitable starting material in the synthesis of picryl azide, both *o*- and *p*-nitrophenyl azide react smoothly in a mixture of nitric and sulphuric acids at low temperature to yield the *sym*-trinitro derivative.

It is apparent from Table 5 that in the benzene series the incoming electrophile is directed preferentially to the *para* position by the azido group, except in those cases where this position is already substituted. In the naphthalene series *ortho*-substitution is only preferred for cases such as 2-azidonaphthalene[6,26] where substitution involving a *p*-quinonoid type intermediate such as **9** would result in the loss of resonance energy in both rings. The very high *para* to *ortho* ratios in

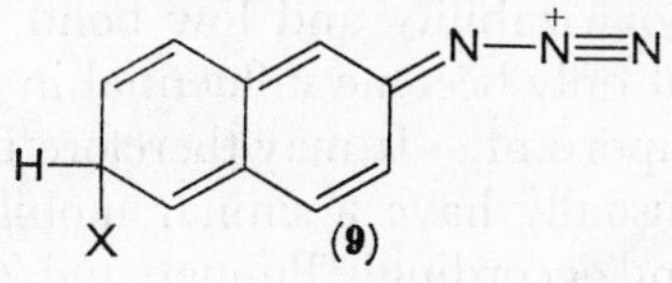

(9)

electrophilic substitutions influenced by substituents of the $-$I $+$K type are well recognized (Table 6), particularly for cases where the

TABLE 6. Percentages of *ortho*- and *para*- products obtained from bromination of aromatic substrates containing $-$I, $+$M groups[6,31]

	$C_6H_5N_3$	C_6H_5Cl	C_6H_5Br	C_6H_5OH
ortho-	—	11	13	10
para-	100	87	85	90

incoming group has a $-$M character. Such is the case with the azido group where electron withdrawal by induction selectively deactivates the *ortho* positions to electrophilic substitution. In addition, the powerful $+$K effect of the orienting azido group is transmitted more effectively through the *p*-quinonoid state, particularly where extended conjugation with an introduced $-$M group such as nitro[7,32] may be achieved.

C. *Nucleophilic Substitution*

Information concerning the directing and activating influences of the azido group on nucleophilic substitution reactions is somewhat fragmentary and is related mainly to aromatic substitutions. In this latter connexion it is pertinent to compare the ease of displacement of the azido group with the mobilities of other common functional groups, since in many reactions between aromatic azides and nucleophiles the azido group is itself displaced[22,26,33–37] and is subject to the directing and activating effects of substituents with lower mobilities.

Miller[38] has suggested that in activated (addition–elimination) aromatic S_N2 reactions, the electron-attracting power of the replaceable group usually exerts the major influence on mobility (for cases where the formation of T.St.1 is rate-limiting). Thus one would anticipate the order of replacement $X^+ > X^0 > X^-$ and within each polar category it would be expected that electronegativity determines the relative mobilities (e.g. F > OR > NR_2; and F > Cl > Br > I). In addition, high polarizability and low bond dissociation energy enhance mobility but only become influential in cases where electronegativities are unimportant. It may therefore be predicted that the azido group would usually have a similar mobility to the halogens, excluding fluorine, and accordingly Bunnett and Zahler[39] have placed a number of common leaving groups in the sequence F > NO_2 > Cl,Br,I > N_3 > OSO_2R > OAr > OR > SR,SAr,SO_2R > NR_2 (R = alkyl) with the qualification that changes in reagent may vary this order.

The mobilities of some common functional groups X relative to chlorine (Cl = 1) for the reaction of 1-X-4-nitrobenzenes with methoxide ion in methanol at 50° have been determined by Miller and co-workers (Table 7). These data indicate that the leaving group

TABLE 7. Leaving group mobilities in reactions of 1-X-4-nitrobenzenes with OMe^- in MeOH at 50°[38]

Leaving group (X)	Mobility relative to Cl = 1	$\Delta E^{\neq}$ (kcal $mole^{-1}$)	$\log_{10}B$
SMe_2^+	$2{\cdot}09 \times 10^5$	24·5	16·8
NMe_3^+	$2{\cdot}36 \times 10^3$	20·0	11·8
NO_2	$1{\cdot}83 \times 10^2$	22·4	12·6
N_3	1·05	25·4	12·1
NMe_2	≪1	—	—
F	$3{\cdot}12 \times 10^2$	21·2	11·7
Cl	1	24·0	11·2

mobility of the azido group (1·05) is almost equivalent to that of the chloro group[38].

Selective nucleophilic replacement of the chloro group is therefore possible in systems such as 1-azido-4-chloro-3-nitrobenzene since the nitro group exerts a substantially greater activating effect at the *ortho* and *para*- than at the *meta*- positions, and this has been used to advantage to determine the activating effect of the azido group in nucleophilic substitution reactions. Miller and Parker[17] have investigated the kinetics of the reaction of methoxide ion on a series of 1-chloro-2-nitro-4-X-benzenes (Table 8) and from their results a value for σ_p^- of 0·116 was derived for the azido group. Although this value implies some activation, the effect is very weak and comparable to that of p-CO_2^-. It is also noteworthy that no reduction in $\Delta E^{\neq}$, compared with the case where X = H, has been observed.

The weak activation by the azido group is not unexpected since internal conjugation places a fractional negative charge on the α-nitrogen atom, and leads to an N_α—N_β bond-order between 1 and 2, as in **6**. The conjugative polarization of the azido group tends towards the dipolar structure (**3**) and it is generally considered that only weak

TABLE 8. Substituent constants[a] and reaction parameters for reactions of 1-Cl-2-NO_2-4X-benzenes with OMe^- in MeOH at 50°[2,18]

X	Rate relative to H = 1	σ_p^-	$\Delta E^{\neq}$ (kcal mole^{-1})
H	1	0	23·6
CO_2^-	3·37	0·135	21·1
$CONH_2$	$2{\cdot}62 \times 10^2$	0·627	21·3
SO_2Me	$1{\cdot}28 \times 10^4$	1·049	18·6
N_3	2·84	0·116	24·2
$C_6H_5N{=}N{-}$	$4{\cdot}16 \times 10^2$	0·672	20·3
$C_6H_5N(O){=}N{-}$[b]	$2{\cdot}52 \times 10^2$	0·616	—
$C_6H_5N{=}N(O){-}$	$1{\cdot}02 \times 10^3$	0·772	19·5
NO_2	$1{\cdot}14 \times 10^5$	1·270	17·4
NO	$6{\cdot}27 \times 10^5$	1·486	16·2

[a] $\rho = 3{\cdot}90$.
[b] From reaction with OH^- in dioxan–water (75:25 v/v).

$-$E effects apply in such cases[32]. Conversely, powerful +E effects are the norm for +M substituents, and the results of electrophilic substitution with aromatic azides are in accord with this pattern. Additional confirmatory evidence concerning the magnitude of the activating effect of the azido group in S_NAr reactions is lacking, however, and further studies in this area would be of value.

The activating effect on nucleophilic substitution of substituents such as $-$COX, which incorporate both +M and $-$M components has been examined by Fuller and Miller[40] and in particular the influence of the azido moiety in the azidocarbonyl group has been compared with the related cases where X = O^-, NH_2, OMe, Me and C_6H_5. The electron-withdrawing power of groups $-$COX where X is a +M substituent depends on the extent of internal conjugation[32], this effect being particularly noticeable for the carboxylate substituent (structure **10**) which is only mildly activating in S_NAr reactions ($\sigma_p^- = 0{\cdot}135$). On theoretical grounds one would expect the activating power of $-$COX groups in S_NAr reactions to be in the order $-$CHO > $-$ COR > $-$COOR > $-CON_3$ > $-CONH_2$ > $-CONR_2$ > $-CO_2^-$. The σ_p^- values for the various $-$COX groups determined from reactions of the 4-iodo-3-nitro-COX-benzenes with azide ion indicated the validity of this suggested sequence. The activating effect of the azidocarbonyl group ($\sigma_p^- = 0{\cdot}780$) is of the same order as that of $-CO_2Me$ ($\sigma_p^- = 0{\cdot}819$) and substantially greater than that of $-CONH_2$ ($\sigma_p^- = 0{\cdot}627$).

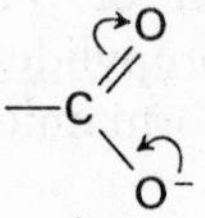

(10)

Information concerning the directing effect of a neighbouring azido group and its influence in rates of solvolytic reactions in the alicyclic series has been provided by Streitwieser and Pulver[41] from a study of the acetolysis of *trans*-2-azidocyclohexyl *p*-toluenesulphonate (**11**). These authors determined the rate of acetolysis of **11** at 100° and compared this result with the acetolysis rate for cyclohexyl *p*-toluenesulphonate[42,43] extrapolated to the same temperature. It was concluded that cyclohexyl *p*-toluenesulphonate is 280 to 380 times more reactive than the azido derivative (**11**). Using an approximate value of σ^* (0·968) from the dissociation constant of azidoacetic acid in water, and a ρ value ($-3{\cdot}15$) from the known correlation between the acetolysis rates of sulphonates and σ^*, a value of $2{\cdot}5 \times 10^3$ was estimated for the expected relative rate. However, Streitwieser and Pulver pointed out that this value may be in error by as much as a factor of ten because of the approximate nature of the derived σ^* for the azido group and concluded that little, if any, anchimeric acceleration is associated with the participation of the neighbouring azido group.

A specific directing effect of the azido group operates in this reaction since the acetolysis of **11** is accompanied by complete retention of configuration to afford *trans*-2-azidocyclohexyl acetate (**13**). It is thought that an intermediate azidonium ion, probably with structure **12**, is produced and that its participation at the rate-limiting transition state is small. Hence the azido group displays pseudohalogenoid character in that it forms a bridged ion which controls the stereochemistry of a reaction without greatly affecting the rate.

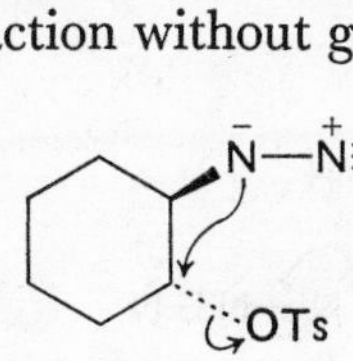

(11) 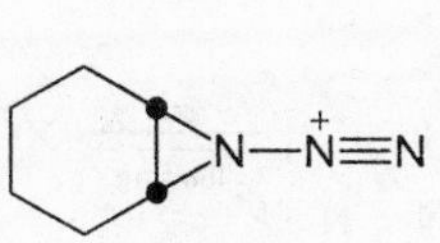 (12) 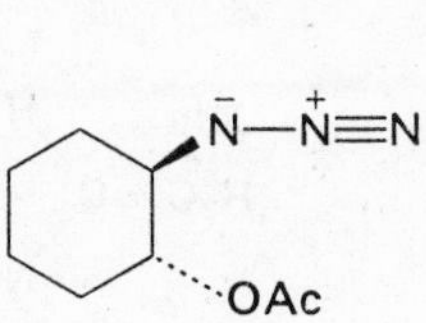 (13)

Azidonium ions similar to **12** have recently been formulated as reaction intermediates by other authors[44,45]. For example, Hanessian has invoked the intermediate **15** to rationalize the partial retention of configuration in the reaction of 5-azido-5-deoxy-4-*O*-methanesulphonyl-D-arabinose-2,3-*O*-isopropylidene diethyl dithioacetal (**14**) with sodium azide in dimethylformamide (equation 2). Neighbouring

$$\text{HC(SEt)}_2\text{–OCH–HCO(CMe}_2\text{)–HCOMs–CH}_2\text{N}_3\ (\mathbf{14}) \longrightarrow \text{HC(SEt)}_2\text{–OCH–HCO(CMe}_2\text{)–CH(N}\equiv\overset{+}{\text{N}}\text{–N)CH}_2\ (\mathbf{15}) \xrightarrow{N_3^-} \text{HC(SEt)}_2\text{–OCH–HCO(CMe}_2\text{)–HCN}_3\text{–CH}_2\text{N}_3 \quad (2)$$

group participation by the azido group is the exception rather than the rule, however, in carbohydrate chemistry[46] (cf. chapter 2).

III. THE DIRECTING AND ACTIVATING EFFECTS OF THE AZIDO GROUP IN ADDITION AND ELIMINATION REACTIONS

A. *Addition Reactions*

Although there are virtually no experimental data relating to the influence of the azido group on addition reactions it is pertinent to consider the probable consequences particularly in the case of vinyl azides where both the inductive and conjugative effects of the azido group would be of significance.

It might be anticipated that when the azido substituent is conjugated to the olefinic bond, electrophilic addition (Ad_E) reactions would be activated by the undeniably powerful $+K$ effect and directed as shown in equation (3). An example, albeit an indirect one, where this effect may be inferred occurs in the reaction of ethoxyacetylene with

$$X^+ + H_2C{=}C(H){-}N{=}\overset{+}{N}{=}\overset{-}{N} \underset{\text{limiting}}{\overset{\text{rate}}{\rightleftharpoons}} XCH_2{-}C(H){=}N{-}\overset{+}{N}{\equiv}N \;(+\,Y^-) \xrightarrow{\text{fast}} XCH_2C(H)(Y){-}N_3 \quad (3)$$

hydrazoic acid[47]. When these reagents are mixed in benzene a spontaneous addition occurs with the formation of the diazide (**16**). No vinylic azide (**17**) could be isolated and it seems probable that **17** is formed initially but undergoes addition even more rapidly than the alkyne because of the activation by both the azido and ethoxy groups.

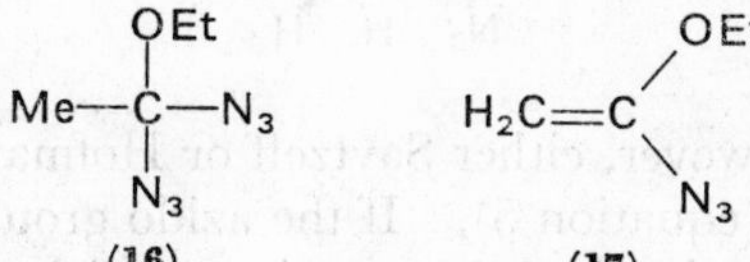

$$\mathrm{Me{-}C(OEt)(N_3){-}N_3}\quad\textbf{(16)}\qquad\qquad \mathrm{H_2C{=}C(OEt)N_3}\quad\textbf{(17)}$$

Inductive electron-withdrawal by the azido group, perhaps reinforced by the weak conjugative (−K) effect, should activate nucleophilic addition (Ad_N) reactions of olefins conjugated to the azido substituent. The only data available in this area concerns the addition in a dipolar aprotic medium of azide ion to picryl azide (**18**), in which the nucleophile is directed to the $C_{(1)}$ position of the aromatic ring with the formation of the benzenide anion (**19**)[48]. Although activation is predominantly due to the nitro groups, the inductive

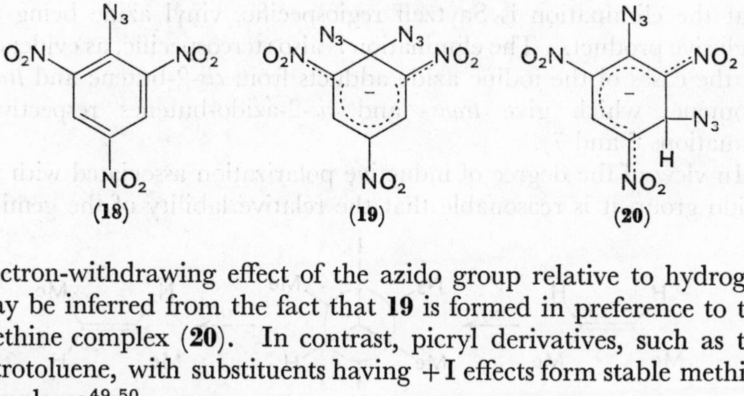

(18) **(19)** **(20)**

electron-withdrawing effect of the azido group relative to hydrogen may be inferred from the fact that **19** is formed in preference to the methine complex (**20**). In contrast, picryl derivatives, such as trinitrotoluene, with substituents having +I effects form stable methine complexes[49,50].

B. *Elimination Reactions*

The synthesis of vinyl azides by regiospecific elimination of hydrogen iodide from vicinal iodoazides has been developed by Hassner and co-workers[51,52] and constitutes the only noteworthy example of the directing effect of the azido group on elimination reactions.

Iodine azide adds to terminal olefins[51] as shown in equation (4) (cf. Chapter 2) and elimination of hydrogen iodide from the β-iodo-azide using potassium *t*-butoxide in ether affords the vinyl azide. In the case of β-iodoazides derived from addition of iodine azide to

$$RCH{=}CH_2 \xrightarrow{IN_3} R(N_3)CH{-}CH_2I \xrightarrow{-HI} R{-}C(N_3){=}CH_2 \qquad (4)$$

internal olefins, however, either Saytzeff or Hofmann elimination, or both, are possible (equation 5). If the azido group has no directing influence on the elimination, then a mixture of the vinylic and allylic azides would result. However, Hassner and Fowler[51] have established

$$RCH{=}CHMe \xrightarrow{IN_3} RCH(N_3){-}CHIMe \; \begin{cases} \xrightarrow{\text{Saytzeff}} R(N_3)C{=}CHMe \\ \xrightarrow{\text{Hofmann}} RCH(N_3){-}CH{=}CH_2 \end{cases} \qquad (5)$$

that the elimination is Saytzeff regiospecific, vinyl azide being the exclusive product. The elimination is also stereospecific, as evidenced by the cases of the iodine azide adducts from *cis*-2-butene and *trans*-2-butene, which give *trans*- and *cis*-2-azido-butenes respectively (equations 6 and 7).

In view of the degree of inductive polarization associated with the azido group it is reasonable that the relative lability of the geminal

$$\underset{cis}{MeCH{=}CHMe} \xrightarrow{IN_3} \underset{threo}{MeCH(N_3){-}CHIMe} \xrightarrow{-HI} \underset{trans}{Me(N_3)C{=}CHMe} \qquad (6)$$

$$\underset{trans}{MeCH{=}CHMe} \xrightarrow{IN_3} \underset{erythro}{MeCH(N_3){-}CHIMe} \xrightarrow{-HI} \underset{cis}{Me(N_3)C{=}CHMe} \qquad (7)$$

proton should be enhanced in such β-iodoazides; when the opportunity exists for *trans* elimination, even though this may entail the eclipsing of bulky substituents as in equation (7), stereospecific elimination to the vinyl azide occurs. In the case of the iodine azide adducts derived from simple cyclic olefins, however, the stereoelectronic preference for *trans* diaxial elimination predominates and the azido group has no directing control over the elimination which consequently proceeds exclusively to the allylic azide (equation 8).

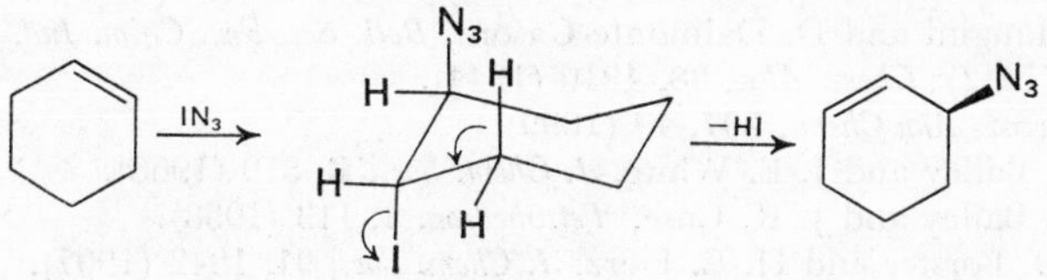

IN_3 → ... $-HI$ → ... N_3 (8)

With larger ring compounds such as cyclooctene iodoazides, production of vinyl azides is possible by *trans* elimination of hydrogen iodide (equations 9 and 10).

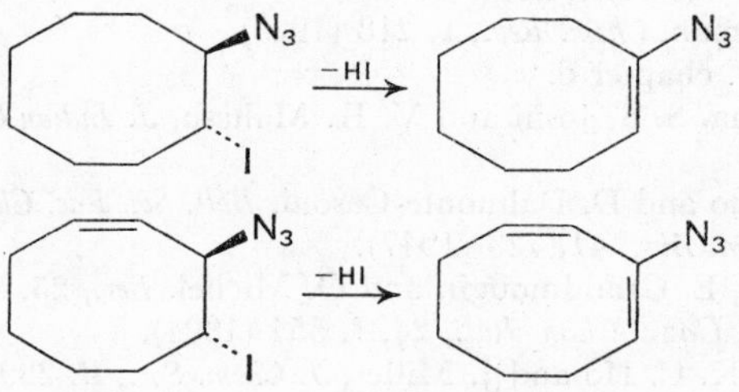

N_3, I $-HI$ → N_3 (9)

N_3, I $-HI$ → N_3 (10)

IV. REFERENCES

1. J. C. Philip, *J. Chem. Soc.*, **95**, 925 (1908).
2. J. H. Boyer and F. C. Canter, *Chem. Rev.*, **54**, 1 (1954).
3. E. Hannerz, *Ber.*, **59**, 1367 (1926).
4. R. O. C. Norman and R. Taylor, *Electrophilic Substitution in Benzenoid Compounds*, Elsevier, Amsterdam, 1965, chapter 1.
5. C. K. Ingold, *Structure and Mechanism in Organic Chemistry*, Bell, London, 1969, chapter 16.
6. P. A. S. Smith, J. H. Hall and R. O. Kan, *J. Am. Chem. Soc.*, **84**, 485 (1962).
7. Ref. 4, chapter 11.
8. H. H. Jaffe, *Chem Rev.*, **53**, 191 (1953).
9. H. C. Brown and Y. Okamoto, *J. Am. Chem. Soc.*, **80**, 4979 (1958).
10. L. Doub and J. M. Vandenbelt, *J. Am. Chem. Soc.*, **69**, 2714 (1947).
11. P. Grammaticakis, *Compt. Rend.*, **244**, 1517 (1957).
12. E. Bergmann and W. Schutz, *Z. Physik. Chem.*, **19B**, 389 (1932).

13. L. E. Sutton, *Nature*, **128**, 639 (1934).
14. G. Favini, *Gazz. Chim. Ital.*, **91**, 270 (1961).
15. E. Lieber, J. S. Curtice and C. N. R. Rao, *Chem. Ind.* (*London*), 586 (1966).
16. A. L. McClellan, *Tables of Experimental Dipole Moments*, Freeman and Co. San Francisco, 1963.
17. J. Miller and A. J. Parker, *Australian J. Chem.*, **11**, 302 (1958).
18. J. Miller, *Nucleophilic Aromatic Substitution*, Elsevier, Amsterdam, 1968, chapter 4.
19. C. Culmann and K. Gasiorowski, *J. Prakt. Chem.*, (2), **40**, 97 (1899).
20. A. Michael, F. Luehn and H. H. Higbee, *Am. Chem. J.*, **20**, 377 (1898).
21. W. A. Tilden and J. H. Millar, *J. Chem. Soc.*, **63**, 256 (1893).
22. A. Mangini and D. Dalmonte-Casoni, *Boll. Sci. Fac. Chim. Ind. Bologna*, **3**, 173 (1942); *Chem. Abs.*, **38**, 4916 (1944).
23. P. Drost, *Ann Chem.*, **307**, 49 (1899).
24. A. S. Bailey and J. E. White, *J. Chem. Soc.*, **B**, 819 (1966).
25. A. S. Bailey and J. R. Case, *Tetrahedron*, **3**, 113 (1958).
26. M. O. Forster and H. E. Fierz, *J. Chem. Soc.*, **91**, 1942 (1907).
27. *Ger. Pat.* 337, 734; *Chem. Abs.*, **17**, 1804 (1923).
28. P. A. S. Smith, *Open-Chain Nitrogen Compounds*, Vol. 2, Benjamin, New York, 1966, chapter 10.
29. P. A. S. Smith and B. B. Brown, *J. Am. Chem. Soc.*, **73**, 2438 (1951).
30. J. Glazer, E. D. Hughes, C. K. Ingold, A. T. James, G. T. Jones and E. Roberts, *J. Chem. Soc.*, 2657 (1950).
31. A. F. Holleman, *Chem. Rev.*, **1**, 218 (1925).
32. Reference 5, chapter 6.
33. D. S. Deorha, S. S. Joshi and V. K. Mahesh, *J. Indian Chem. Soc.*, **39**, 534 (1962).
34. R. Andrisano and D. Dalmonte-Casoni, *Boll. Sci. Fac. Chim. Ind. Bologna*, 1 (1943); *Chem. Abs.*, **41**, 723 (1947).
35. E. Noelting, E. Grandmougin and O. Michel, *Ber.*, **25**, 3328 (1892).
36. A. Purgotti, *Gazz. Chim. Ital.*, **24**, I, 554 (1894).
37. D. L. Hill, K. C. Ho and J. Miller, *J. Chem. Soc.*, **B**, 299 (1966).
38. Reference 18, chapter 5.
39. J. F. Bunnett and R. E. Zahler, *Chem. Rev.*, **49**, 273 (1951); J. F. Bunnett, *Quart. Rev.*, **12**, 1 (1958).
40. M. W. Fuller and J. Miller, unpublished work.
41. A. Streitwieser and S. Pulver, *J. Am. Chem. Soc.*, **86**, 1587 (1964).
42. H. C. Brown and G. Ham, *J. Am. Chem. Soc.*, **78**, 2753 (1956).
43. J. D. Roberts and V. C. Chambers, *J. Am. Chem. Soc.*, **73**, 5034 (1951).
44. S. Hanessian, *Carbohydrate Res.*, **1**, 178 (1965).
45. K. Kischa and E. Zbiral, *Tetrahedron*, **26**, 1417 (1970); E. Zbiral, G. Nestler and K. Kischa, *Tetrahedron*, **26**, 1427 (1970).
46. H. Kuzuhara, H. Ohrui and S. Emoto, *Tetrahedron Letters*, 1185 (1970).
47. Y. A. Sinnema and J. F. Arens, *Rec. Trav. Chim.*, **74**, 901 (1955).
48. P. Caveng and H. Zollinger, *Helv. Chim. Acta.*, **50**, 861 (1967).
49. E. Buncel, A. R. Norris and W. Proudlock, *Can. J. Chem.*, **46**, 2759 (1968).
50. A. R. Norris, *Can. J. Chem.*, **47**, 2895 (1969).
51. A. Hassner and F. W. Fowler, *J. Org. Chem.*, **33**, 2686 (1968).
52. F. W. Fowler, A. Hassner and L. A. Levy, *J. Am. Chem. Soc.*, **89**, 2077 (1967).

CHAPTER 5

Decomposition of organic azides

R. A. Abramovitch and E. P. Kyba

University of Alabama, University, Alabama 35486

I. Introduction 222

II. Acid-Catalysed Decompositions 222
A. Protonic Acids 223
1. Alkyl azides 223
2. Aryl azides 234
3. Sulphonyl azides 242
B. Lewis Acids 242
1. Alkyl azides 242
2. Aryl azides 243
3. Sulphonyl azides 245

III. Thermally Induced Decompositions 245
A. Alkyl Azides 245
B. Aryl Azides 256
1. Introduction 256
2. Mechanism of decomposition of aryl azides . . . 257
3. Decomposition in the presence of unsaturated compounds 260
4. Assisted intramolecular cyclization 261
5. Decomposition in the presence of aliphatic substrates . 265
6. Decomposition in the presence of aromatic substrates . 267
7. High temperature pyrolyses 271
8. Decomposition in the presence of nucleophiles . . 274
9. Decomposition in the presence of radicals . . . 275
10. Decomposition of azidoquinones and azidotropones . 277
C. Sulphonyl Azides 279
1. Mechanism of the decomposition 279
2. Decomposition in the presence of aliphatic substrates . 284
3. Decomposition in the presence of aromatic substrates . 285
4. Decomposition in the presence of nucleophiles . . 287

5. Decomposition in the presence of unsaturated compounds 289
6. Decomposition in the presence of radicals 291
7. Decomposition in the presence of organic anions . 292

IV. DECOMPOSITION WITH TRIVALENT PHOSPHORUS COMPOUNDS . 294
A. Alkyl Azides 294
B. Aryl Azides 295
C. Sulphonyl Azides 296

V. PHOTOLYTIC DECOMPOSITION 297
A. Alkyl Azides 297
B. Aryl Azides 308
C. Sulphonyl Azides 314

VI. TRANSITION METAL-CATALYSED DECOMPOSITIONS 317
A. Alkyl Azides 317
B. Aryl Azides 317
C. Sulphonyl Azides 320

VII. REFERENCES 322

I. INTRODUCTION

The decomposition of organic azides has been studied from a number of points of view, the objectives being mainly synthetic or mechanistic. The nature of the intermediates formed have received much attention. In this chapter we shall consider the decomposition of alkyl, aryl and sulphonyl azides under various conditions. Acyl and vinyl azides are considered elsewhere in this volume.

This present treatment is subdivided into the five major modes of decomposition: acid-catalysed, thermal, trivalent phosphorus compound-catalysed, photolytic and transition metal-catalysed decompositions. Each section is further subdivided into alkyl, aryl and sulphonyl azide decomposition. The nature of the products formed and how these are affected by the nature of the reactive intermediates is discussed.

II. ACID-CATALYSED DECOMPOSITIONS

A considerable amount of work has been carried out in this area and several reviews have given fairly extensive coverage of some aspects of this topic[1–4]. The discussion below is divided into protonic and

Lewis acid-catalysed decompositions. Reactions in which the azide (derived from an alcohol or carbonyl compound) is not isolated, but reacts further to give various products (Schmidt reaction), are included.

The protonic acid-catalysed decomposition of azides is conceived of as involving an initial protonation of the α-nitrogen atom, subsequent to which nitrogen may be eliminated either in a non-concerted or a concerted process.

$$R-\bar{N}-\overset{+}{N}\equiv N \underset{}{\overset{H^+}{\rightleftharpoons}} R-\overset{H}{\overset{|}{\overset{+}{N}}}-\overset{+}{N}\equiv N \xrightarrow[\text{(ii) } -H^+]{\text{(i) } -N_2} \text{products} \quad (1)$$

A. Protonic Acids

I. Alkyl azides

Curtius and co-workers[5–7] studied the acid-catalysed decomposition of alkyl azides such as benzyl azide. This was decomposed in either warm 1:1 (v/v) sulphuric acid–water or with concentrated hydrochloric acid to give a mixture of products corresponding to hydrogen migration [benzaldimine (**1**)], phenyl migration [formaldehyde anil (**2**)], the azide reduction product [benzylamine (**3**)], and the solvolysis product [benzyl alcohol (**4**)]. The first two were obtained as the

$$PhCH_2N_3 \xrightarrow{H^+} \underset{(\mathbf{1})}{PhCH=NH} + \underset{(\mathbf{2})}{PhN=CH_2} + \underset{(\mathbf{3})}{PhCH_2NH_2} + \underset{(\mathbf{4})}{PhCH_2OH} \quad (2)$$

hydrolysis products, benzaldehyde and aniline, respectively; the latter two were formed to a lesser extent.

When treated with fuming sulphuric acid followed by hydrolysis, ethyl azide produced methylamine (methyl migration, 14%) and acetaldehyde (hydrogen migration, 86%)[8]. Similar behaviour was exhibited by *n*-butyl azide in concentrated sulphuric acid, but it was reported that *n*-hexyl and *n*-dodecyl azides gave products corresponding only to hydrogen migration[9].

Pritzkow and Mahler[10] studied the four isomeric azido *n*-heptanes and obtained directly information concerning the migratory aptitudes of different alkyl groups. Contrary to earlier observations[9], they found that *n*-hexyl migration did occur (10–30%) to give *n*-hexylamine after hydrolysis. The secondary heptyl azides also gave rise to alkyl migration in which the longer chain group migrated to a greater extent (1·5–2·5 times) than the short chain. Perhaps the discrepancy in results might be due to the difference in reaction conditions used: Pritzkow and Mahler[10] used glacial acetic acid–perchloric acid, whereas Boyer and co-workers[9] used concentrated sulphuric acid. Second-order kinetics (first-order in each of azide and perchloric acid) were found for the four azido-*n*-heptanes only for the first 25% of the reaction[10]. This is not surprising since the amounts of reagents used (10 ml 2N perchloric acid in acetic acid and 10 ml 2M azide in acetic acid) were such that the acidity changed appreciably during the reaction, and the concentration of perchloric acid rather than the activity was used in the calculations. The rates of evolution of nitrogen and the Arrhenius activation energies (24·9–27·4 kcal/mole) varied little among the four isomers.

Boyer and co-workers[9] studied the sulphuric acid-catalysed decomposition of three secondary cyclic azides: cyclopentyl, cyclohexyl and cycloheptyl, and found that ring expansion, i.e. alkyl migration, competes quite favourably with hydrogen migration. For example, cyclopentyl azide in chloroform, gave cyclopentanone (as its 2,4-dinitrophenylhydrazone, 2·2%) and piperideine (**5**) (79·8%), on

H, N_3 —($CHCl_3$, H_2SO_4)→ N (**5**) + =NH (3)

reaction with sulphuric acid. The other two azides also gave appreciable amounts of ring-expanded products, but polymerization was so rapid that the products were not characterized.

Other work in which migratory aptitudes have been compared has involved the formation of the azide (or its conjugate acid) *in situ* from the corresponding alcohol or olefin, and its decomposition without isolation (the Schmidt reaction). For example, 1-substituted cyclohexanols underwent predominantly ring expansion when the 1-substituent was methyl, ethyl or cyclohexyl[11]. A series of tertiary carbinols were studied in which it was found that the migration tendency was in the order Ph ≈ *i*-Pr ≈ C_6H_{11} ≫ Et ≈ Me[12,13]. The products of the Schmidt reaction on menthol (**6**, equation 4) indicate the intermediacy of a secondary carbonium ion **7**, which can rearrange to more stable tertiary carbonium ions **8** and **9** before reacting with hydrogen azide[14]. The ring expansion product, 4-methyl-7-isopropyl-3,4,5,6-tetrahydro-2*H*-azepine (**10**) (6·7%) was derived from **8** while **9** yielded methyl 4-methylcyclohexyl ketone (**11**) (methyl migration, 23%) and cyclohexylamine (**12**) (methylcyclohexyl migration, 14%) after a hydrolytic workup. This illustrates the danger of utilizing the Schmidt reaction for alcohols or olefins as a proof of

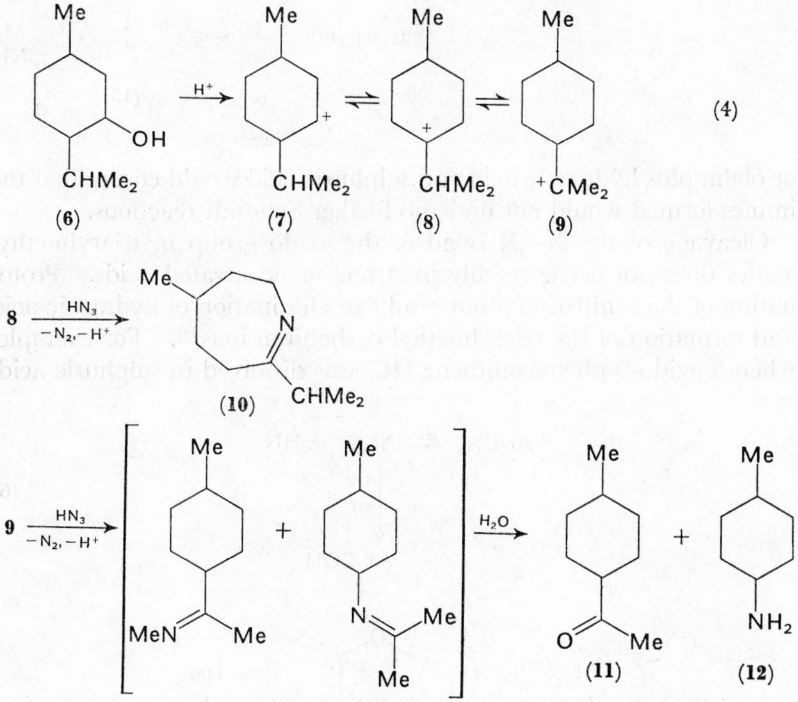

(4)

(6) (7) (8) (9)

(10)

(11) (12)

structure where rearrangements of intermediate carbonium ions are possible.

It was shown that ring expansion could occur in the Schmidt reaction of norborneol (**13**) or norbornene (**14**) to give the trimer of 3-azabicyclo(3,2,1)oct-2-ene, which could be reduced to 3-azabicyclo(3,2,1)octane (**15**)[15].

There is a lack of reliable quantitative data concerning migratory aptitudes in acid-catalysed decomposition of alkyl azides. Of particular interest would be the acid-catalysed decomposition of tertiary aliphatic azides, coupled with modern quantitative techniques, such as gas liquid chromatography, in which a minimum of manipulation of products would be involved. The use of azide rather than carbinol

(13) (OH) or **(14)** $\xrightarrow[H_2SO_4]{HN_3/CHCl_3}$ $(C_7H_{11}N)_3$ $\xrightarrow{[H]}$ **(15)** (NH) (5)

or olefin plus hydrazoic acid and sulphuric acid would ensure that the imines formed would not undergo further Schmidt reactions.

Cleavage of the N—N bond of the azido group in triarylmethyl azides does not occur readily in strong, concentrated acids. Protonation of the α-nitrogen atom results in elimination of hydrazoic acid and formation of the triarylmethyl carbonium ions[16]. For example, when 9-azido-9-phenylxanthene (**16**) was dissolved in sulphuric acid,

$$Ar_3CN_3 \overset{H^+}{\rightleftharpoons} Ar_3C^+ + HN_3 \xrightarrow[-H^+]{H_2O} Ar_3COH \qquad (6)$$

(16) (Ph, N_3) $\overset{H_2SO_4}{\rightleftharpoons}$ **(17)** (Ph) + HN_3 → **(18)** (NH); **(17)** $\xrightarrow{H_2O}$ **(19)** (Ph, OH) (7) the yellow colour and green fluorescence characteristic of the 9-phenylxanthenyl cation (**17**) was observed[16]. The presence of free hydrazoic acid was demonstrated by the addition of fluorenone to the reaction mixture, which gave phenanthridone (**18**) (*vide infra*). Dilution of this mixture with water gave 9-hydroxyphenylxanthene (**19**). The azide (**16**) was largely recovered when the reaction mixture was poured into water (without the addition of fluorenone) even after 24 hours.

The Schmidt reaction of diarylmethylcarbinols or 1,1-diarylethylenes and benzhydrols, or acid-catalysed decomposition of 1,1-diarylethyl azides or benzhydryl azides has received considerable attention, and reliable data are available. It was found that a Hammett relationship was followed in the migratory aptitudes of substituted aryl groups. For the Schmidt reaction of unsymmetrical 1,1-diarylethylenes, the Hammett relationship was[17]

$$\log(\text{migratory aptitude}) = -2{\cdot}11\sigma + 0{\cdot}293$$

and for the acid-catalysed decomposition of benzhydryl azides[18]

$$\log(\text{migratory aptitude}) = -2{\cdot}03\sigma + 0{\cdot}237.$$

A later paper[19] modified the latter *rho* value to $-2{\cdot}26$.

The question of whether migration of the aryl group is concerted with elimination of nitrogen (equation 8) or whether the protonated nitrene (nitrenium ion) (**20**) is a discrete intermediate was considered by Gudmunsen and McEwen[19] who used the rate law

$$-d(\text{azide})/dt = k_2(\text{azide})h_0,$$

where h_0 is the negative antilogarithm of the acidity function H_0. By considering the relationship between the rate of evolution of nitrogen, the product ratio (i.e. substituted phenyl compared with phenyl migration), and k_2, they concluded that the nitrenium ion is a discrete

$$\text{Ph—C(Ar)(H)—}\overset{\text{H}}{\text{N}}\text{—}\overset{+}{\text{N}}\equiv\text{N} \xrightarrow{-\text{N}_2} \text{PhCH=}\overset{+}{\text{N}}\text{HAr} \xrightarrow{-\text{H}^+} \text{PhCH=NAr}$$

$$\text{Ph—C(H)(Ar)—}\overset{\text{H}}{\text{N}}\text{—}\overset{+}{\text{N}}\equiv\text{N} \xrightarrow{-\text{N}_2} \text{Ph—C(H)(Ar)—}\overset{+}{\text{N}}\text{H} \;(\mathbf{20}) \rightarrow \text{PhCH=}\overset{+}{\text{N}}\text{HAr} \quad (8)$$

intermediate. However, exactly the opposite conclusion was reached[2] by considering the rate law to be of the form

$$d(N_2)/dt = k_N K_b H_0 c$$

where k_N was the rate constant for the evolution of nitrogen, K_b, the basicity constant for the substituted benzhydryl azides, and c, the stoichiometric concentration of azide. Thus, by considering the effect of substitution on the basicity of the azido group it was concluded that migration and elimination of nitrogen are synchronous. Whether migration and elimination of nitrogen are concerted when the groups attached to the carbon bearing the azide are aliphatic has not been established. The activation energies are somewhat higher (25–27 kcal/mole) for the decomposition of the azidoheptanes[10] than for benzhydryl azides[19] (20–23 kcal/mole), perhaps supporting the stepwise formulation given for the acid-catalysed decomposition of the former[10].

The migratory aptitudes of substituted phenyl groups have been determined for the Schmidt reaction of unsymmetrical diarylethylenes[17] and are, in order of decreasing mobility (X in XC_6H_4): *p*-MeO > 3,4-$(Me)_2$ > *p*-Me > *p*-Et > *m*-Me > *p*-Ph > *p*-F > H > *p*-Cl > *p*-Br. The migratory aptitudes relative to phenyl ranged from 6·12 for *p*-MeO to 0·54 for *p*-Br. This is a very narrow range of reactivities for a process involving participation (cf. carbonium ions), which is indicative of the intervention of a very reactive species. A similar order exists for benzhydryl azides[18], although not all the above substituents were studied.

It has been found that both ruthenocenylphenylcarbinyl azide[20], and ferrocenylphenylcarbinyl azide[21,22] undergo phenyl migration to give ruthenocenyl and ferrocenyl carboxaldehyde (after hydrolysis), but no products due to the metallocenyl group migration. The ferrocenyl derivative **21** was studied more fully, and it was found that, in addition to the rearrangement product **22** (6·7%), considerable amounts of 1,2-diferrocenyl-1,2-diphenylethane (**23**) (31%, two diastereoisomers) were formed, along with several other minor products.

$$\mathrm{Fc{-}C(H)(Ph){-}N_3} \xrightarrow{H^+} \mathrm{FcCH{=}NPh} + \mathrm{Fc{-}CH(Ph){-}CH(Ph){-}Fc} \quad (9)$$

(**21**) (**22**) (**23**)

Fc = ferrocenyl

It was suggested that the complete lack of ferrocenyl migration was due to protonation of the metal atom, thereby greatly reducing its electron-releasing properties[22].

The Schmidt reaction of substituted fluorenols, **24** and the acid-catalysed decomposition of the corresponding azides **25** have been studied extensively[23–28]. In particular, Arcus and Coombs[24] found that a substituent's electron-releasing ability markedly affected the product distribution. When R was electron-withdrawing, the ring

HN_3/H_2SO_4, $-H_2O$; H^+, $-N_2$, $-H^+$

(24) **(25)** (10)

(26) + **(27)**

not bearing the substituent migrated preferentially (with R = 3-NO_2, ratio **26**: **27** = 1:16), whereas an electron-releasing substituent caused the substituent-bearing ring to migrate more readily (with R = 1,2-benzo, ratio **26**:**27** = 2·9:1)[26]. This is only a qualitative relationship, however, and not all the results are fully understood. For example, with R = 2-Me, the ratio of **26**:**27** was 53:47, whereas only **26** was formed with R = 3-Me. The kinetics of the decomposition of 9-azidofluorene were studied in acetic acid–sulphuric acid media in the range $H_0 = -2{\cdot}87$ to $-4{\cdot}85$[27]. The reaction was found to be first-order in azide, and went via the monoprotonated species present in small equilibrium amounts. With $H_0^{25°} = -3{\cdot}21$, the energy of activation was 24·0 kcal/mole using the initial rate technique, or 22·8 kcal/mole, from pseudo first-order rate constants measured over times for half reaction.

Treatment of *N*-benzhydrylidene azidobenzhydrylamine (**28**) with trifluoroacetic acid at room temperature for 2 hr yielded 79% benzanilide (**33**) and 65% benzophenone imine (**34**)[29]. Similar results could be obtained using concentrated sulphuric acid at room temperature, or aluminium chloride and gaseous hydrogen chloride at

$-30°$. It was suggested that the imino nitrogen was suitably located to provide anchimeric assistance for the expulsion of nitrogen to give the diaziridinium ion (**30**). Since *N*-phenyl-*N'*-benzhydrylidene benzamidine (**31**) was stable to the reaction conditions, it was

$$Ph_2C(N{=}CPh_2)(N_3)\ \textbf{(28)} \xrightarrow{H^+} Ph_2C(N{=}CPh_2)(NH{-}N_2^+)\ \textbf{(29)} \xrightarrow{-N_2} \quad (11)$$

$$Ph_2C(\overset{+}{N}{=}CPh_2)(NH) \longleftrightarrow Ph_2C(N{-}\overset{+}{C}Ph_2)(NH)\ \textbf{(30)}$$

suggested that **31** was not an intermediate in the rearrangement, but rather that migration of phenyl was assisted by a *gegen ion* attack

$$Ph{-}C(N{=}CPh_2){=}N{-}Ph\ \textbf{(31)}$$

at benzhydryl carbon in **30** to give **32**, hydrolysis of which would give the observed products. The possible incursion of non-classical ions was also suggested[29].

$$X^- + Ph_2C(\overset{+}{N}{=}CPh_2)(NH) \longrightarrow Ph{-}C(N{=}CPh_2)(X){-}NHPh\ \textbf{(32)} \xrightarrow{H_2O} PhCONHPh\ \textbf{(33)} + Ph_2C{=}NH\ \textbf{(34)} \quad (12)$$

The Schmidt reaction of carbonyl compounds[30] follows a different course to that of alcohols and olefins. This reaction has been reviewed[31], and the work of Arcus and colleagues[32–35] and of Pritzkow and Schuberth[36] are representative of more recent investigations. Smith[37] postulated that the conjugate acid (**36**) of the carbonyl compound (**35**) adds to hydrazoic acid to give the protonated α-hydroxy azide (**37**), which then eliminates water to give **38**. Elimination of nitrogen and migration of one of the groups to electron-deficient

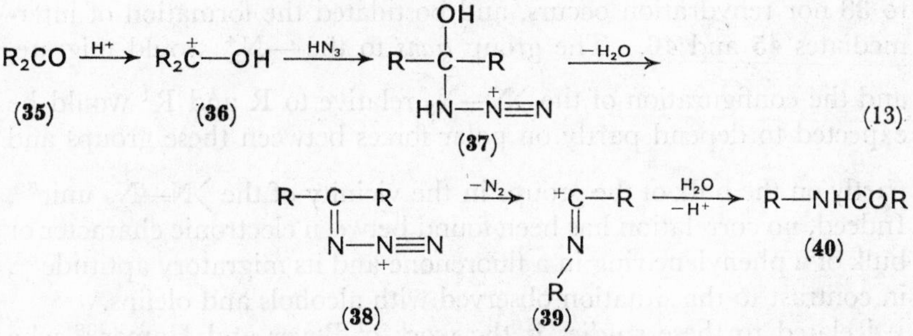

$$R_2CO \xrightarrow{H^+} R_2\overset{+}{C}{-}OH \xrightarrow{HN_3} R{-}C(OH)(R){-}HN{-}\overset{+}{N}{\equiv}N \xrightarrow{-H_2O} \quad (13)$$

(35) (36) (37)

$$R{-}C(R){=}N{-}\overset{+}{N}{\equiv}N \xrightarrow{-N_2} {}^{+}C(R){=}N{-}R \xrightarrow[-H^+]{H_2O} R{-}NHCOR$$

(38) (39) (40)

nitrogen would give **39** which, upon hydration and loss of a proton, would give the amide (**40**). Arcus, Coombs and Evans[32] reasoned, however, that if dehydration of **37** did occur, it would be unlikely for rehydration to occur under the reaction conditions (concentrated sulphuric acid). They assumed that hydration occurred on workup. To test this hypothesis they attempted to trap **43** from fluorenone (**41**) by adding anhydrous methanol at the completion of the reaction and keeping it completely free of water. Only amide (**42**) was formed and no methoxy-derivative (**44**) could be detected, even though it had been established that **44** was stable to the reaction and workup conditions. Arcus and co-workers concluded that neither dehydration

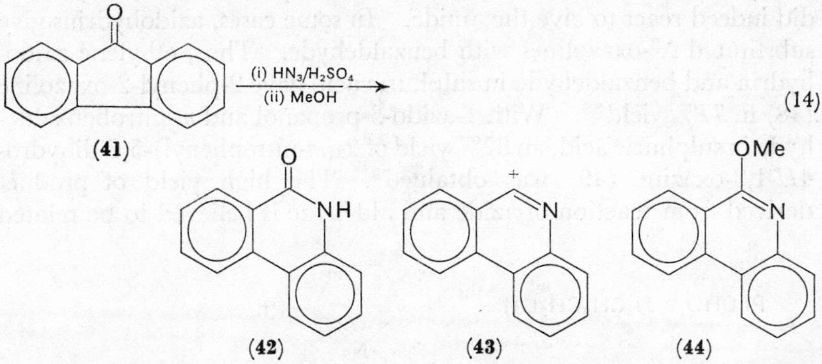

$$\textbf{(41)} \xrightarrow[\text{(ii) MeOH}]{\text{(i) } HN_3/H_2SO_4} \quad (14)$$

(42) (43) (44)

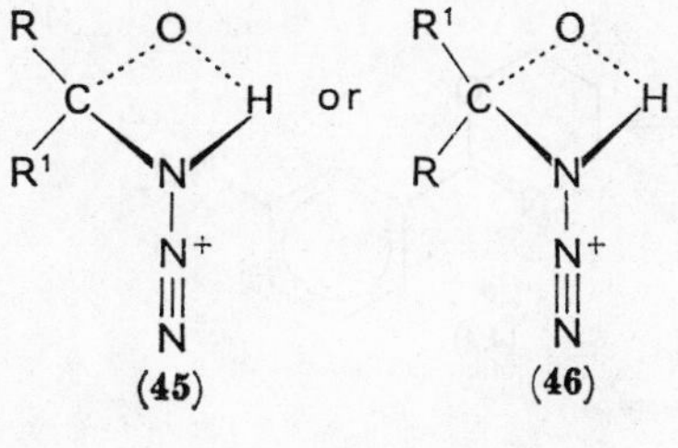

(45) or (46)

to **38** nor rehydration occurs, and postulated the formation of intermediates **45** and **46.** The group *trans* to the $—N_2^+$ would migrate, and the configuration of the $\rangle N—\overset{+}{N}_2$ relative to R and R^1 would be expected to depend partly on polar forces between these groups and partly on the bulk of the groups in the vicinity of the $\rangle N—\overset{+}{N}_2$ unit[32]. Indeed, no correlation has been found between electronic character or bulk of a phenylene ring in a fluorenone and its migratory aptitude[32], in contrast to the situation observed with alcohols and olefins.

Related to these studies is the work of Boyer and Hamer[38] who treated benzaldehyde with 2-phenylethyl azide in the presence of sulphuric acid and obtained *N*-(2-phenylethyl) benzamide (**47**) in

$$PhCHO \underset{}{\overset{H^+}{\rightleftharpoons}} Ph\overset{+}{C}HOH \underset{}{\overset{PhCH_2CH_2N_3}{\rightleftharpoons}} Ph—\underset{}{\overset{OH}{\overset{|}{C}H}}—\underset{\underset{CH_2CH_2Ph}{|}}{N}—\overset{+}{N}{\equiv}N \xrightarrow{-N_2} \tag{15}$$

$$Ph\overset{OH}{\overset{|}{C}H}—\overset{+}{N}CH_2CH_2Ph \xrightarrow{-H^+} Ph—\overset{OH}{\overset{|}{C}}{=}NCH_2CH_2Ph \longrightarrow \underset{(\mathbf{47})}{PhCONHCH_2CH_2Ph}$$

10% yield. No amide formation was observed with *n*-butyl azide or benzyl azide, but in a later paper[39] it was reported that *n*-butyl azide did indeed react to give the amide. In some cases, azidohydrins give substituted Δ^2-oxazolines with benzaldehyde. Thus, ethylene azidohydrin and benzaldehyde in sulphuric acid gave 2-phenyl-2-oxazoline (**48**) in 77% yield[38]. With 1-azido-3-propanol and *m*-nitrobenzaldehyde in sulphuric acid, an 82% yield of 2-(*m*-nitrophenyl)-5,6-dihydro-4*H*-1,3-oxazine (**49**) was obtained[9]. The high yield of product derived from reaction of azide and aldehyde is believed to be related

$$PhCHO + N_3CH_2CH_2OH \xrightarrow[C_6H_6]{H_2SO_4} \underset{(\mathbf{48})}{\text{2-phenyl-2-oxazoline}}$$

$$m\text{-}O_2NC_6H_4CHO + N_3(CH_2)_3OH \xrightarrow[CHCl_3]{H_2SO_4} \underset{(\mathbf{49})}{\text{2-}(m\text{-nitrophenyl})\text{-5,6-dihydro-4}H\text{-1,3-oxazine}} \tag{16}$$

to the fact that the azidohydrin can undergo intramolecular hydrogen bonding. It is thought that this might retard nitrogen elimination from the protonated azide, thereby increasing the opportunity for reaction with the aldehyde[9].

Two reports[40,41] have appeared concerning an extension of the Schmidt reaction of aryl alkyl ketones in which the ketone was treated with an alkyl azide to give benzaldehyde, an aliphatic aldehyde and an amine. The yields of benzaldehyde were claimed to range from

$$PhCOCH_2R \xrightarrow{H^+} Ph{-}\overset{OH}{\underset{+}{C}}{-}CH_2R \xrightarrow{R^1N_3} Ph{-}\overset{OH}{\underset{CH_2R}{C}}{-}N(R^1){-}N^+{\equiv}N \xrightarrow[-N_2]{-H_2O}$$

$$R{-}CH{=}\underset{Ph}{C}{-}\overset{+}{N}{-}R^1 \longleftrightarrow R\overset{+}{C}H{-}\underset{Ph}{C}{=}N{-}R^1 \xrightarrow[-H^+]{2H_2O}$$

$$R{-}\overset{OH}{CH}{-}\overset{OH}{\underset{Ph}{C}}{-}NHR^1 \longrightarrow PhCHO + RCHO + R^1NH_2 \qquad (17)$$

62–85% with either acetophenone or propiophenone and *n*-butyl, *n*-hexyl, cyclohexyl and *n*-octyl azides. Considerable doubt has recently been cast on these results[42].

The Schmidt reactions discussed above involve the reaction of a carbonium ion with an azide. There has been a report of the reaction of carbonium ions, generated either from triethyloxonium tetra-

$$CH_3CH_2CH_2CH_2N_3 + CH_3^+ \longrightarrow CH_3CH_2CH_2CH_2{-}\overset{\overset{+}{N}\equiv N}{N}{-}CH_3 \xrightarrow{-N_2}$$

$$\underset{\mathbf{(50)}}{CH_3CH_2CH_2CH_2{-}\overset{+}{N}{-}CH_3} \xrightarrow[\sim 80\%]{-H^+} \underset{\mathbf{(51)}}{CH_3CH_2CH_2CH{=}NCH_3}$$

$$\mathbf{(50)} \xrightarrow[\sim 10\%]{-H^+} \underset{\mathbf{(52)}}{CH_3CH_2CH_2CH_2{-}N{=}CH_2} \qquad (18)$$

$$\mathbf{(50)} \xrightarrow{\sim 5\text{–}10\%} \underset{\mathbf{(53)}}{CH_3CH_2CH_2{-}\overset{+}{N}(=CH_2)CH_3} \xrightarrow[-H^+]{+H_2O} \underset{\mathbf{(54)}}{CH_3CH_2CH_2NHCH_3} + CH_2O$$

fluoroborate or from alkyl halides and silver perchlorate, with alkyl azides[43]. For example, *n*-butyl azide was decomposed by the methyl cation (from methyl iodide and silver perchlorate) to give three products (equation 18). Two were the imines **51** and **52**, which could result either from removal of a proton from one of the carbon atoms adjacent to the positive nitrogen in **50**, or by hydrogen migration to positive nitrogen followed by loss of a proton from nitrogen. The third product arose from alkyl migration to electron-deficient nitrogen, which, upon hydrolysis and loss of a proton, gave *N*-methyl-*n*-propylamine (**54**) and formaldehyde.

2. Aryl azides

The protonic acid-catalysed decomposition of aryl azides was first observed by Griess[44,45], who reported that the decomposition of phenyl azide in hydrochloric acid gave a mixture of *o*- and *p*-chloroanilines, whereas with 'strong' sulphuric acid *p*-aminophenol was formed[46]. The hydroxylation either *ortho* or *para* to the original azido group has proved to be quite a general reaction using a concentrated sulphuric acid to water ratio of 2:1. Thus, 4-amino-*m*-cresol was formed from *o*-tolyl azide[47] and 4-amino-2-nitrophenol was obtained from *m*-nitroazidobenzene[48]. Brass and co-workers found that 2- and 4-azidophenanthraquinone (**55**) gave the ring-hydroxylated products [2- and 4-aminophenol derivatives (**56**), respectively][49,50], but 3-azidophenanthraquinone was much more stable to acid-catalysed decomposition and yielded only 3-aminophenanthraquinone[51].

N_3 (**55**) —aq. H_2SO_4→ H_2N, OH (**56**) (19)

Friedlander[47,48] assumed that the phenols arose by way of a rearrangement of a hydroxylamine which was formed by the reaction of water with a nitrene. Bamberger[52] compared the behaviour of arylhydroxylamines and aryl azides under similar conditions and concluded that hydroxylamines are not intermediates in the acid-catalysed decompositions of aryl azides. He proposed instead that with phenyl azide, for example, the initially formed nitrene was attacked at the *para* position by water, followed by tautomerization to give an iminoquinol (**57**) which would further tautomerize to the *p*-aminophenol. It is unlikely in such strongly acidic media that a

$$PhN_3 \xrightarrow{-N_2} C_6H_5\text{–}N \xrightarrow{H_2O} \text{(57)} \longrightarrow HO\text{–}C_6H_4\text{–}NH_2 \quad (20)$$

free nitrene would be involved. A much more probable sequence of events would involve the formation of the conjugate acid (**58**) of the azide, which could then lose nitrogen to give a resonance-stabilized nitrenium ion (**59**)[3,53]. Reaction of **59** with a nucleophile at either the activated *ortho* or *para* positions would lead eventually to the ring-substituted amine. Alternatively, attack by the nucleophile and elimination of nitrogen could be synchronous. Unfortunately, as

$$C_6H_5N_3 \underset{}{\overset{H^+}{\rightleftharpoons}} C_6H_5\text{–}NH\text{–}\overset{+}{N_2}\ \text{(58)} \xrightarrow[\text{(ii) } -H^+]{\text{(i) } H_2O,\ \text{concerted},\ -N_2} \text{iminoquinol} \longrightarrow HO\text{–}C_6H_4\text{–}NH_2$$

$$\text{(58)} \xrightarrow{-N_2} \text{(59)} \xrightarrow[\text{(ii) } -H^+]{\text{(i) } H_2O} \text{iminoquinol} \quad (21)$$

Smith and Brown[54] have pointed out, the interpretation of many of the decompositions in acidic media are subject to question due to the failure to differentiate clearly between acid-catalysed and thermally-induced reactions. Since it is known that thermal decomposition of phenyl azides occurs at appreciable rates at temperatures in the range of 140–170°[4], it is probable that the reactions investigated by

Bamberger[52] were acid-catalysed, since many were carried out, at or near, room temperature or on a water-bath. For example, 2,4-xylyl azide was decomposed in sulphuric acid–ethanol medium at a temperature $\leq 35°$[55], or in a 1:2 mixture of sulphuric acid and water at 65°[56]. *p*-Tolyl azide was decomposed in concentrated sulphuric acid at $-20°$[57]. On the other hand, it was also decomposed under conditions in which some thermolysis may well have been occurring, i.e. in a 1:3 (v/v) mixture of boiling sulphuric acid–water[58]. More quantitative data are needed in this area, and kinetic data along with activation parameters would aid in clarifying the mode of decomposition. This type of data has been obtained for the acid-catalysed decomposition of a similar system, the arylhydroxylamines, and has helped in establishing that this is a nucleophilic intermolecular rearrangement[59].

Bamberger and Brun[58] reported the rearrangement of *para*-substituted aryl azides in acid. When *p*-tolyl azide was boiled in a mixture of sulphuric acid and water (1:3, v/v) for two hours, toluhydroquinone (**60**) was obtained in 21% yield along with *p*-toluidine, ammonia and the usual resinous material. The toluhydroquinone

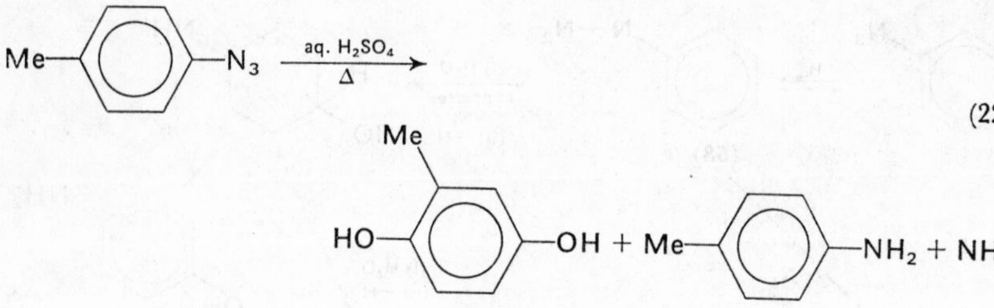

Me–C₆H₄–N₃ →(aq. H_2SO_4, Δ) (**60**) + Me–C₆H₄–NH_2 + NH_3 (22)

possibly arose via the hydrolysis and rearrangement of the intermediate imine (**61**). Under less vigorous conditions (sulphuric acid at $-20°$),

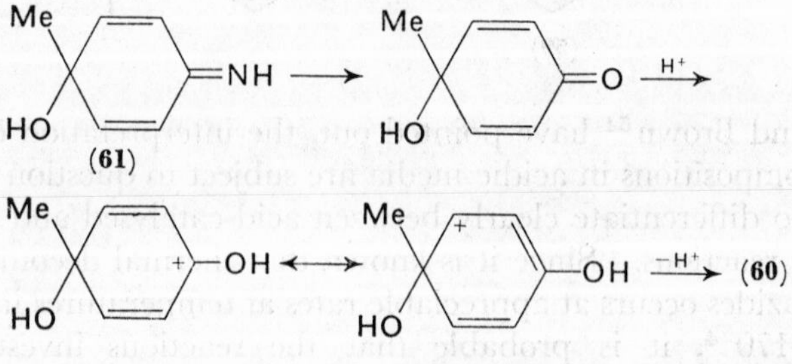

(**61**) =NH → =O →(H^+) ... –OH ⟷ ... –OH →($-H^+$) (**60**) (23)

Bamberger[57] was able to isolate a low molecular weight polymer having a composition corresponding to a quinonoid monomeric structure **62** from *p*-tolyl azide.

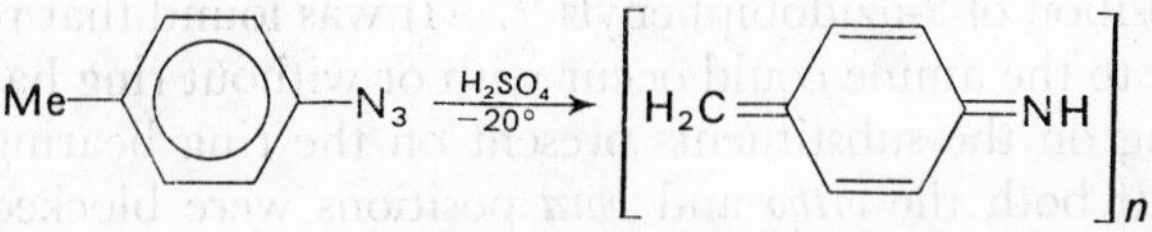

$$Me-C_6H_4-N_3 \xrightarrow[-20^\circ]{H_2SO_4} \left[H_2C{=}C_6H_4{=}NH \right]_n \qquad (24)$$

(62)

More recently, Sherk, Houpt and Brown[8] decomposed phenyl azide in fuming sulphuric acid and isolated phenylhydroxylamine *O,m*-disulphonic acid (**63**) which, upon hydrolysis, rearranged to 4-aminophenol-2-sulphonic acid (**64**). The same product **64** (38%) was obtained[52] in addition to *p*-aminophenol (11%) on heating the azide in a 1:3 (v/v) mixture of sulphuric acid and water.

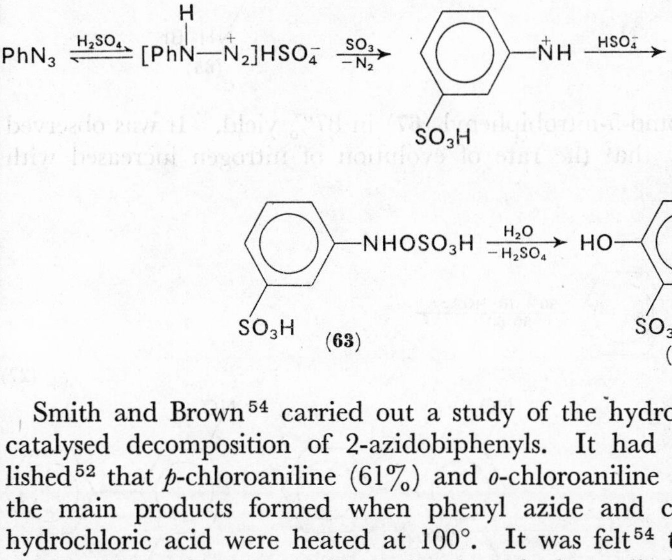

$$PhN_3 \overset{H_2SO_4}{\rightleftharpoons} [Ph\overset{+}{N}H-N_2]HSO_4^- \xrightarrow[-N_2]{SO_3} m\text{-}HO_3S-C_6H_4-\overset{+}{N}H \xrightarrow{HSO_4^-} \qquad (25)$$

$$m\text{-}HO_3S-C_6H_4-NHOSO_3H \xrightarrow[-H_2SO_4]{H_2O} HO-C_6H_3(SO_3H)-NH_2$$

(63) **(64)**

Smith and Brown[54] carried out a study of the hydrogen halide-catalysed decomposition of 2-azidobiphenyls. It had been established[52] that *p*-chloroaniline (61%) and *o*-chloroaniline (18%) were the main products formed when phenyl azide and concentrated hydrochloric acid were heated at 100°. It was felt[54] that a clear differentiation between acid-catalysed and thermally-induced reactions had not been established. Their choice of 2-azidobiphenyls was predicated on the possibility of competitive reactions of ring halogenation accompanying reduction of the azide, and cyclization. It was found that the reactivity of 2-azidobiphenyl with hydrogen halides was in the order of their reducing abilities and acid strength: HI > HBr > HCl > HF. The azide was almost completely inert to hydrogen fluoride, but decomposed instantaneously with hydrogen iodide, with the liberation of free iodine.

Although the possibility existed for cyclization to the carbazole, in practice this was not detected in the hydrogen bromide catalysed decomposition of 2-azidobiphenyls[54]. It was found that reduction of the azide to the amine could occur with or without ring halogenation, depending on the substituents present on the ring bearing the azide group. If both the *ortho* and *para* positions were blocked, no halogenation occurred; otherwise, halogenation always accompanied reduction. For example, 2-azidobiphenyl gave 2-amino-5-bromobiphenyl (**65**) in 82% yield, and 2-azido-5-nitrobiphenyl (**66**) gave 2-

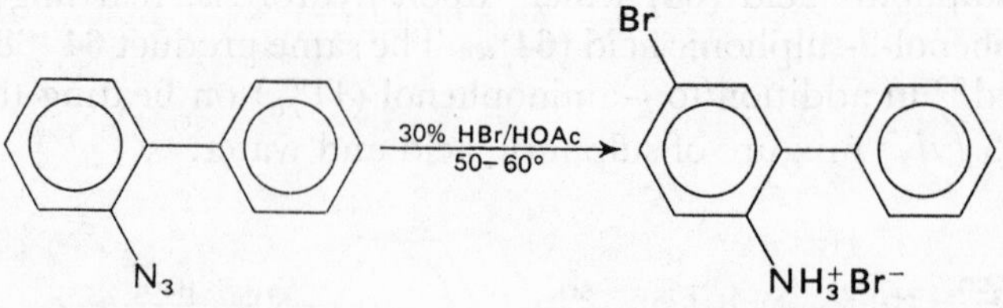

Br

N_3 —30% HBr/HOAc, 50–60°→ $NH_3^+Br^-$ (26)

(**65**)

amino-3-bromo-5-nitrobiphenyl (**67**) in 87% yield. It was observed qualitatively that the rate of evolution of nitrogen increased with

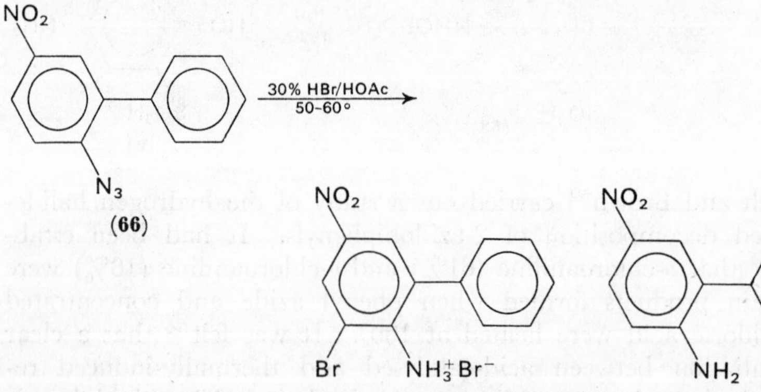

NO_2

N_3 —30% HBr/HOAc, 50–60°→ (27)

(**66**)

NO_2 Br $NH_3^+Br^-$ (**67**)

NO_2 NH_2 (**68**)

increasing acid strength of the medium, but was retarded by base-weakening substituents on the azide. For example, 2-azido-3,5-dinitrobiphenyl was recovered from the reaction mixture in 90% yield, even though the temperature was raised to 70°. In accord with these observations, Smith and Brown[54] proposed the following mechanism (equation 28). In support of molecular bromine being

$$\begin{aligned}
&ArN_3 + H^+ \rightleftharpoons (ArNH{-}N_2)^+ \\
&(ArNH{-}N_2)^+ + Br^- \longrightarrow ArNHBr + N_2 \\
&ArNHBr + HBr \longrightarrow Br_2 + ArNH_2\ (\overset{HBr}{\rightleftharpoons} ArNH_3^+Br^-) \\
&ArNH_2 + Br_2 \longrightarrow \text{brominated products}
\end{aligned} \tag{28}$$

the brominating agent, it was found that in the presence of phenol, **66** was reduced to the unhalogenated derivative **68** in 80% yield. Whether hydrogen halide catalysed azide decompositions follow a different course to those with aqueous sulphuric acid is not certain, but it has been proposed (*vide supra*) that in aqueous sulphuric acid, ring substitution is due to a nucleophilic attack, whereas Smith and Brown supported an electrophilic attack for the case of hydrogen bromide.

Recently, a series of papers has been published concerning the Schmidt reaction of *p*-quinones[61], and the acid-catalysed decomposition of azidoquinones[62–66]. The Schmidt reaction of various *p*-quinone derivatives gave rise to substituted 2,5*H*-2,5-azepindiones in yields ranging from 75–85%[61]. For example, 3,4-benzo-6-methyl-2,5-azepindione (**73**) was obtained from the reaction of hydrazoic acid with 2,3-benzo-5-methyl-1,4-quinone (**69**) in concentrated sulphuric

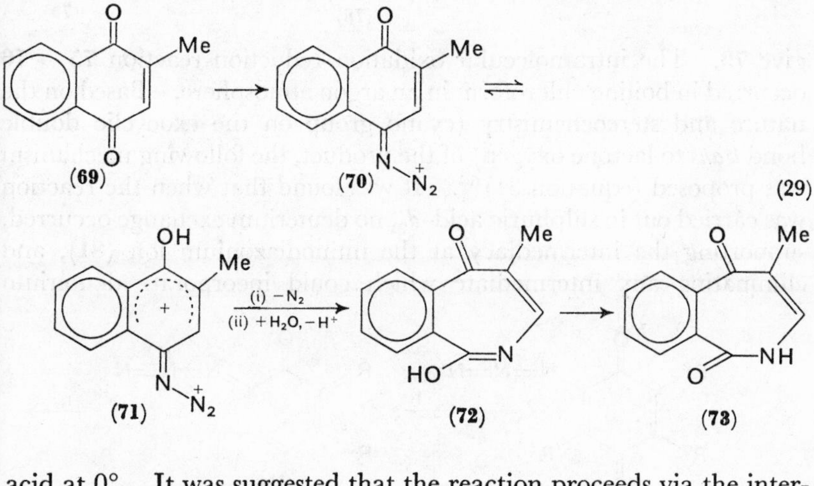

(69) (70) (29)

(i) $-N_2$
(ii) $+H_2O, -H^+$

(71) (72) (73)

acid at 0°. It was suggested that the reaction proceeds via the intermediates **70–72**[61].

The reaction of thymoquinone (**74**) to yield the γ-alkylidene-$\Delta^{\alpha,\beta}$-butenolide (**79**) illustrates the interesting chemistry that 1,4-quinones exhibit upon reaction with sodium azide in trichloroacetic acid[62]. Intermediates **75, 76** and **78** were synthesized and shown to

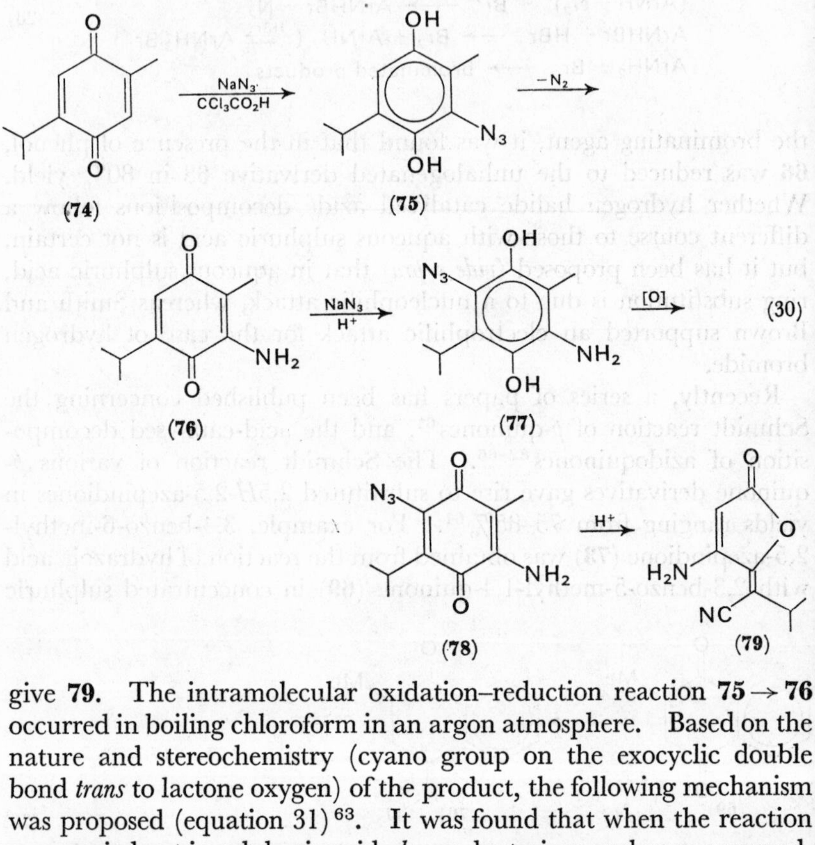

NaN_3, CCl_3CO_2H; $-N_2$; NaN_3, H^+; [O]; H^+ (30)

(**74**) (**75**) (**76**) (**77**) (**78**) (**79**)

give **79**. The intramolecular oxidation–reduction reaction **75** → **76** occurred in boiling chloroform in an argon atmosphere. Based on the nature and stereochemistry (cyano group on the exocyclic double bond *trans* to lactone oxygen) of the product, the following mechanism was proposed (equation 31)[63]. It was found that when the reaction was carried out in sulphuric acid-d_2, no deuterium exchange occurred, supporting the intermediacy at the iminodiazonium ion (**81**), and eliminating any intermediate which could incorporate deuterium

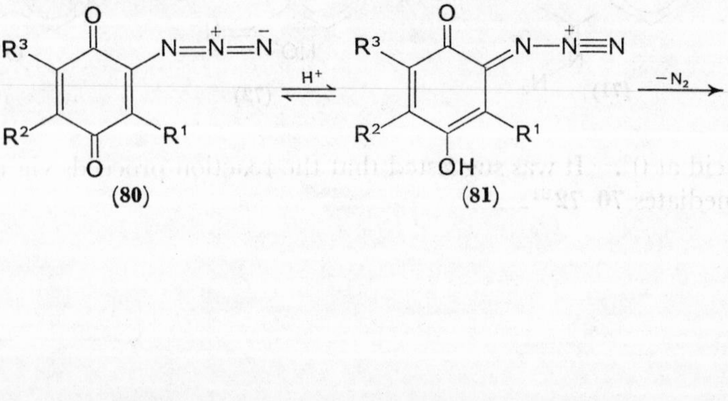

H^+; $-N_2$

(**80**) (**81**)

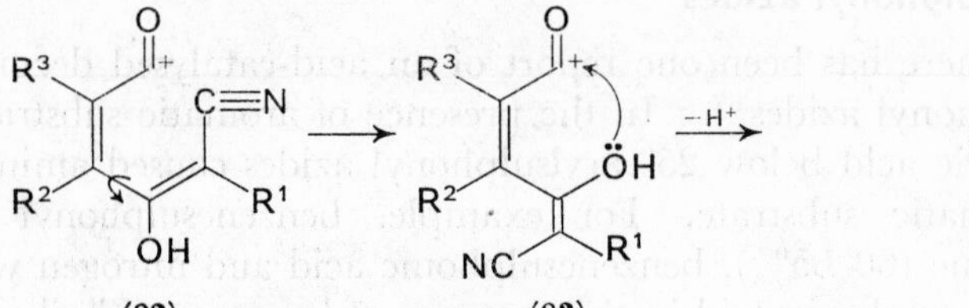

(31)

(82) (83)

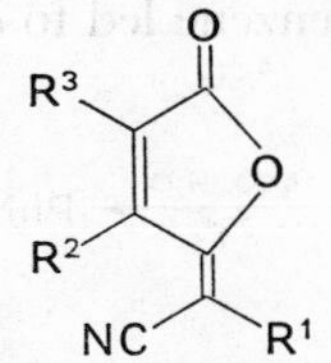

(84) 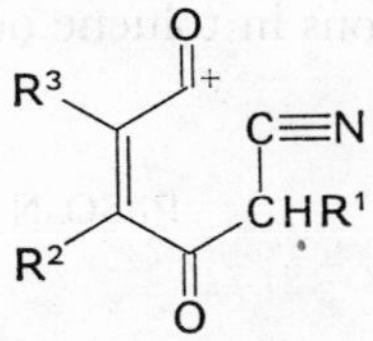 (85)

(e.g. a ring protonated species)[66]. Spectroscopic evidence for the existence of **81** was also obtained. For example, treatment of 2-azido-3,6-dimethyl-1,4-benzoquinone (**80**) ($R^1 = R^3 = Me, R^2 = H$) with concentrated sulphuric acid at 5–10° resulted in an absorption at 567 nm (**80** absorbed at 495 nm). The half-life of this species at 25·5° was 35 sec and it disappeared in a first-order process. It was felt that the iminodiazonium ion (**81**) was responsible for this long-wave absorption, although the protonated azide was a possible intermediate. Since the latter would lead to an intermediate such as **85**, whose subsequent reactions would not lead to the observed stereospecificity (cyano group *trans* to the lactone oxygen), it was rejected.

It was shown that for a variety of substituents the γ-alkylidene-(or arylidene)-$\Delta^{\alpha,\beta}$-butenolides (**84**) could be obtained in good yields (60–90%, mostly over 80%)[62,66]. It has been pointed out that the reaction conditions are important[62]; for example, if instead of trichloroacetic acid, sulphuric acid were used, the ring-expanded azepindione would be produced. A good yield (82%) of the ring-expanded product (**87**) was obtained by acid-catalysed decomposition of the benzo-1,2-quinone (**86**)[65].

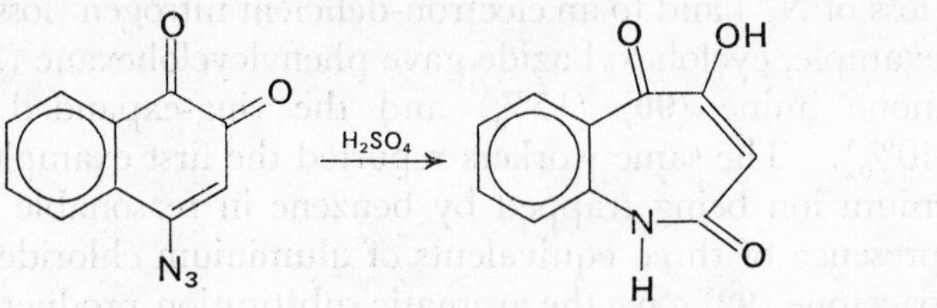

$\xrightarrow{H_2SO_4}$ (32)

(86) (87)

3. Sulphonyl azides

There has been one report of an acid-catalysed decomposition of sulphonyl azides[67]. In the presence of aromatic substrates and sulphuric acid below 25° arylsulphonyl azides caused amination of the aromatic substrate. For example, benzenesulphonyl azide gave aniline (60–65%), benzenesulphonic acid and nitrogen when treated with sulphuric acid in the presence of benzene. Similar decompositions in toluene or chlorobenzene led to *o*- and *p*- substituted anilines

$$PhSO_2N_3 + C_6H_6 \xrightarrow[<25°]{H_2SO_4/H_2O} PhNH_2 + PhSO_3H + N_2 \qquad (33)$$

(35–60%). The relative amounts of the *o*- and *p*-isomers were not reported[67].

B. *Lewis Acids*

I. Alkyl azides

Treatment of methyl azide with antimony pentachloride gave an adduct (**88**) with an intact azido group, as shown by its infrared spectrum[68]. The adduct could be decomposed to methylene-

$$\underset{}{MeN_3} \xrightarrow{SbCl_5} \underset{\textbf{(88)}}{MeN_3SbCl_5} \xrightarrow[CH_2Cl_2,\ \text{reflux}]{\text{dry HCl}} \underset{\textbf{(89)}}{CH_2{=}\overset{+}{N}H_2SbCl_6^-} + N_2 \qquad (34)$$

imonium hexachloroantimonate (**89**) with dry hydrogen chloride on heating.

Decomposition of alkyl azides with aluminium chloride in benzene at 50° gave products which corresponded to the formation of a carbonium ion (loss of N_3^-) and to an electron-deficient nitrogen (loss of N_2)[69,70]. For example, cyclohexyl azide gave phenylcyclohexane (30%), cyclohexanone imine (**90**) (15%) and the ring-expanded imine (**91**) (30–40%). The same workers reported the first example of an alkyl nitrenium ion being trapped by benzene in reasonable yield[71]. In the presence of three equivalents of aluminium chloride in benzene, azidoacetone (**92**) gave the aromatic substitution product (**93**) in 35%

$$C_6H_{11}N_3 \xrightarrow[C_6H_6,\ 50°,\ -N_2]{AlCl_3} C_6H_{11}{-}C_6H_5 + \text{(90)} + \text{(91)} \qquad (35)$$

(90) cyclohexylidene=NH; (91) N (azepine ring)

$$\underset{(92)}{MeCOCH_2N_3} \xrightarrow[C_6H_6]{3AlCl_3} \underset{(93)}{MeCOCH_2NHC_6H_5} \qquad (36)$$

yield. An aluminium chloride complex of a nitrenium ion could well be involved.

2. Aryl azides

Borsche[72] reported that phenyl azide decomposed to intractable tars in nitrobenzene in the presence of aluminium chloride. In carbon disulphide, two products were obtained in addition to tars: phenylisothiocyanate (**95**) (30%) and 1-phenyl-2-phenylimino-5-thio-3,4-dithiazolidine (**96**) (6%). It was felt that phenyl azide reacted under the influence of aluminium chloride to give the addition product **94**, which could then extrude sulphur to give **95**. This, in the presence of **94**, could form the 5-membered heterocycle (**96**). The reaction of

$$PhN_3 \xrightarrow[CS_2,\ -N_2]{AlCl_3} \underset{(94)}{PhN(S)C{=}S} \xrightarrow{-S} \underset{(95)}{PhNCS}$$

$$\mathbf{94} + \mathbf{95} \longrightarrow \underset{(96)}{S{=}C(S{-}S)C({=}NPh)N(Ph)} \qquad (37)$$

phenyl azide with aluminium chloride and aromatic substrates gave diarylamines in yields of 30–40%[73]. For example, phenyl azide and aluminium chloride in toluene gave 4-methyldiphenylamine (**97**) (35%) and aniline (22%). When phenyl azide was decomposed with aluminium chloride in acetyl chloride, acetic anhydride, or better still,

$$PhN_3 + PhMe \xrightarrow{AlCl_3} \underset{(97)}{p\text{-}MeC_6H_4NHPh} + PhNH_2 \qquad (38)$$

a mixture of the chloride and anhydride, *p*-chloroacetanilide (**98**) was formed[73].

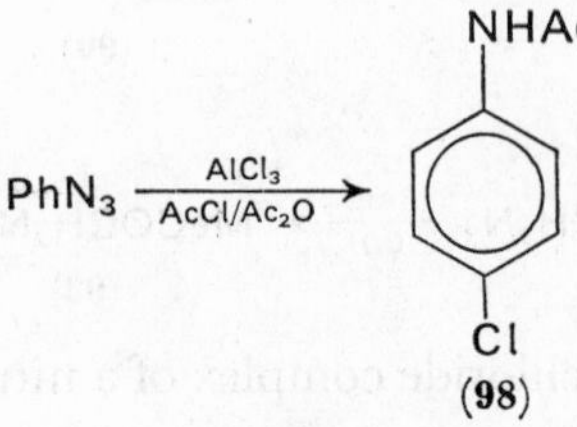

$$PhN_3 \xrightarrow[AcCl/Ac_2O]{AlCl_3} p\text{-}ClC_6H_4NHAc \quad \mathbf{(98)} \tag{39}$$

Hoegerle and Butler[74] decomposed phenyl azide in the presence of alkyl aluminiums to give, in addition to aniline, *N*- and ring-alkylated products and an olefin. For example, phenyl azide reacted with triethylaluminium at −70° to form a hydrocarbon-soluble complex which decomposed slowly on warming, evolving nitrogen to give aniline (63·5%), *N*-ethylaniline (**99**) (16·3%), *o*-ethylaniline (**100**) (9·0%), *p*-ethylaniline (**101**) (11·2%), and a trace of *N*,*o*-diethylaniline (**102**).

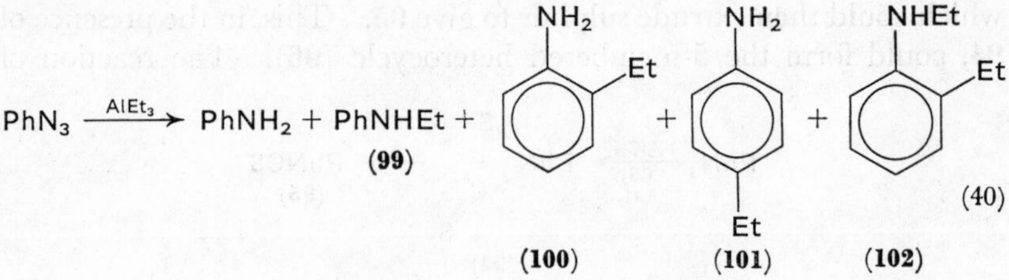

$$PhN_3 \xrightarrow{AlEt_3} PhNH_2 + \underset{\mathbf{(99)}}{PhNHEt} + \underset{\mathbf{(100)}}{o\text{-}EtC_6H_4NH_2} + \underset{\mathbf{(101)}}{p\text{-}EtC_6H_4NH_2} + \underset{\mathbf{(102)}}{o\text{-}EtC_6H_4NHEt} \tag{40}$$

These products could be formed from a phenylnitrenium–aluminiumalkyl complex ($Ph\overset{+}{N}\overset{-}{Al}Et_3$).

From the kinetics of the aluminium bromide-catalysed decomposition of phenyl azide in toluene at 0° in the presence of traces of hydrogen bromide, it was concluded[75] that two paths were available for the azide decomposition, as shown below.

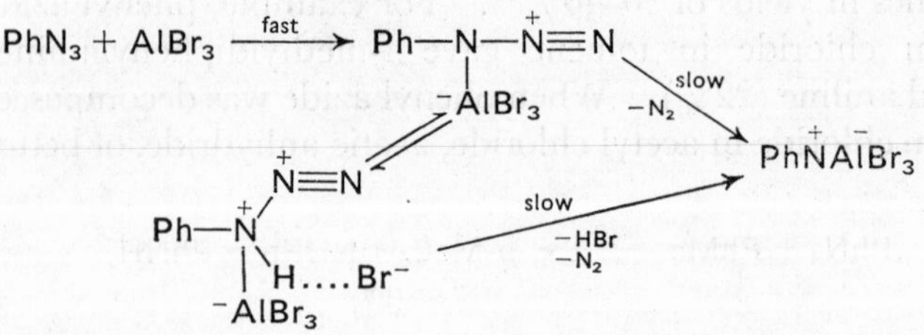

$$PhN_3 + AlBr_3 \xrightarrow{fast} Ph{-}N(\overset{-}{Al}Br_3){-}\overset{+}{N}{\equiv}N \xrightarrow[-N_2]{slow} Ph\overset{+}{N}\overset{-}{Al}Br_3 \tag{41}$$

$$Ph{-}\overset{+}{N}(\overset{-}{Al}Br_3)(\overset{+}{N}{\equiv}N)H\cdots Br^- \xrightarrow[-HBr,\ -N_2]{slow} Ph\overset{+}{N}\overset{-}{Al}Br_3$$

3. Sulphonyl azides

Kreher and Jager[70] reported that arylphenylsulphones and anilines were formed from arylsulphonyl azides in the presence of aluminium chloride and benzene. The yields and the specific azides used were not given.

III. THERMALLY INDUCED DECOMPOSITIONS

A. Alkyl Azides

Some aspects of this topic have been reviewed[76–78]. The thermal decomposition of alkyl azides was first demonstrated by Senior[79], working with Stieglitz. He showed that benzophenone anils, nitrogen and tars were obtained by heating triarylmethyl azides at temperatures of about 200°. The effect of substituents on the nature of the products was considered also, but the analytical techniques available were quite crude. The work was repeated and extended more recently, and will be discussed later.

The gas phase thermolysis (245°) of methyl azide was first studied by Ramsperger[80], who found it to be a homogeneous first-order reaction, but did not determine the nature of the products formed. Leermakers[81] showed the presence of hexamethylenetetramine, hydrazoic acid, ethylene, ethane, ammonia and nitrogen (in high conversion experiments at 200–240° and pressures of 0·08–46·6 cm). In order to account for the initial to final pressure ratio (1·66) at constant volume, he proposed that methyl azide decomposed according to two primary modes: 75% to give methylnitrene and nitrogen and 25% to give methylene and hydrazoic acid. The reactive inter-

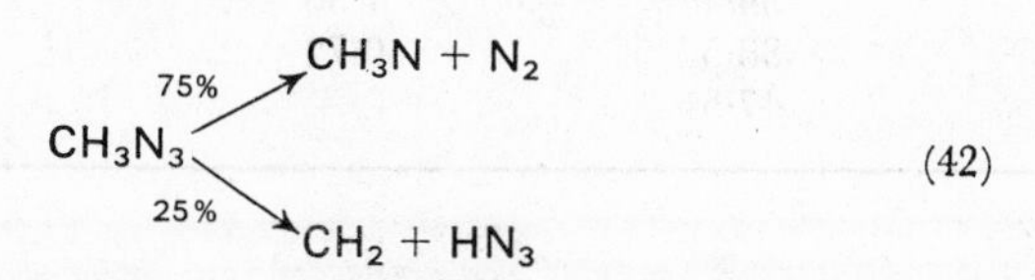

(42)

mediates generated initially then reacted further to give the final products. Similar decomposition modes were postulated for ethyl azide, via ethylnitrene (80%) and methylcarbene (20%)[82]. It was pointed out in both papers[81,82], however, that the reaction scheme merely agreed with the initial to final pressure data and product distribution, and that no claim was made that all of the reactions occurred as written. An attempt to stabilize methylnitrene by decomposing 1·0 mole of methyl azide at 900° in a flow system in which the decomposition products were frozen out at −196° gave nitrogen (0·97 mole), hydrogen (0·25 mole), ammonia (0·29 mole), hydrogen cyanide (0·23 mole), methane (trace) and polymer[83]. The difference between these products and those observed by Leermakers[81] is probably due to the use of a flow system and high temperatures in the latter work. For example, it is possible that at 900° methylnitrene is sufficiently vibrationally excited to give hydrogen cyanide and hydrogen, whereas at 200–240° these products were not observed.

Very recently, the gas phase pyrolysis (155–200°) of methyl azide at low conversions (< 1%) was studied[84]. Nitrogen was the major non-condensable gas, in addition to small amounts of hydrogen (6% at lowest initial pressure of azide to less than 1% at the highest pressure) and methane (< 2%). Ethane, ethylene, ammonia and hydrazoic acid were not detected, although the ethane, ethylene and hydrazoic acid determinations were subject to some uncertainty. Two solid white products were also obtained which were not characterized. The results showed that the thermolysis of methyl azide is of first-order, homogeneous and free from chains, in agreement with previous work[80,81]. The Arrhenius activation parameters (see Table 1)

TABLE 1. Arrhenius activation parameters for gas phase pyrolysis of alkyl azides

	E_a (kcal/mole)	$A \times 10^{-14}$ (sec^{-1})	$k_{200°} \times 10^5$ (sec^{-1})
MeN_3 [81]	47·5[a], 43·5, 40·8[b]	28·0[a], 30·2, 2·85[b]	2·3, 3·0[b]
EtN_3 [86]	40·1	3·30	9·6
n-PrN_3 [86]	39·4	1·50	10·2
i-PrN_3 [86]	38·5	0·72	16·6
$C_6H_{11}N_3$ [87]	47·5[c]		

[a] Reference 80.
[b] Reference 84.
[c] Ethyl benzoate solution.

obtained in this study[84] were significantly different from those obtained earlier[81].

It was felt that the most important step in the decomposition was the formation of the nitrene and nitrogen. In contrast to a previous proposal[81], C—N bond cleavage to give methylene and hydrazoic acid was not thought to be a significant process. Since added olefins did not reduce the rate of nitrogen evolution, but did inhibit the formation of polymer, it was suggested that methylnitrene molecules did not react with methyl azide to give nitrogen and azomethane, but rather dimerized to the latter which then gave polymers.

Since no ethane could be detected, it was felt that methane had to be formed directly, and not by way of methyl radicals. It was proposed that methane arose from two methylnitrene molecules reacting at a surface to give methane and diazomethane. There was, however,

$$2CH_3N \xrightarrow{\text{wall}} CH_4 + CH_2N_2$$

no effect observed on the $[CH_4]/[N_2]$ ratio when the surface to volume ratio of the reaction vessel was increased by a factor of 18·7, as might have been expected from such a reaction at the glass surface.

Finally, arguments were presented that the nitrene was formed in the $X^3\Sigma^-$ state,

$$CH_3N_3 \longrightarrow CH_3N(X\,^3\Sigma^-) + N_2(X\,^1\Sigma_g^+) \tag{43}$$

rather than

$$CH_3N_3 \longrightarrow CH_3N(a\,^1\Delta) + N_2(X\,^1\Sigma_g^+) \tag{44}$$

contrary to the rules of spin conservation. The authors felt that the most compelling reason for preferring a reaction which forms triplet $MeN(X^3\Sigma^-)$ rather than singlet $MeN(a^1\Delta)$ was the observed difference in behaviour of the thermally and photochemically[97] generated species. Thus, the former did not appear to react with methyl azide, while the latter did. It was assumed that photolysis of methyl azide gave a singlet nitrene (this is not necessarily so; a nitrene may not be formed in photolysis, see section V.A.). If the triplet nitrene were indeed formed and did not undergo reaction with methyl azide, it would appear to contradict the observation of Reiser and co-workers (*vide infra*)[85], who found that low concentrations of triplet arylnitrenes react with unchanged aryl azides at diffusion-controlled rates.

The gas phase thermolysis of ethyl, *n*-propyl and *i*-propyl azides has been studied[86] and the results reaffirmed the previous conclusion[80–82] that it is a homogeneous first-order process. Radical inhibitors such as nitrous oxide had no effect on the rate of decomposition. The problem of whether a radical chain mechanism was operative in the explosive decomposition of ethyl azide was considered. It was shown that the activation energies of the explosive and non-explosive decompositions were essentially the same, from which it was concluded that the same mechanism probably prevails in both processes. In this context, it is interesting to note the observation that hydrogen cyanide occurred only in traces in the slow decompositions of ethyl azide, but in much greater quantities in explosions. In such rapid decompositions, local temperatures would be much higher than the initial temperature (ca. 200°), and enough energy might be available for hydrogen cyanide to be formed from ethylnitrene. Arrhenius parameters for the thermolysis of some alkyl azides are presented in Table 1. The activation energy for the decomposition of cyclohexyl azide seems abnormally high, although it was obtained in ethyl benzoate solution, whereas the others were from gas phase pyrolyses. The trend in activation energies from methyl to *i*-propyl azide might indicate an inductive stabilization of the developing electron-deficient nitrene.

Pritzkow and Timm[88] studied the gas phase pyrolysis of eight alkyl azides, in which quantitative product determinations were carried out. The decompositions were effected in a flow system using argon as the carrier gas at 350–410°. The major products obtained were those of rearrangement in every case but one. Hydrogen migration predominated, with H/R migratory ratios ranging from 2·2 (H/*n*-

$$R-CH_2-N_3 \xrightarrow[-N_2]{\Delta} RCH{=}NH + CH_2{=}NR \qquad (45)$$

heptyl) to 9·2 (H/*i*-propyl). Only in the case of ethyl azide, did a non-rearrangement process occur to any significant extent. It was found that in addition to rearrangement products, aziridine (**103**) was formed in 35–46% yield, in agreement with previous observations[82,86].

$$CH_3CH_2N_3 \xrightarrow[-N_2]{\Delta} CH_3CH{=}NH + CH_2{=}NCH_3 + \underset{\mathbf{(103)}}{H_2C{-}CH_2 \text{ (bridged by N–H)}} \qquad (46)$$

The formation of **103** may be thought of as an intramolecular C—H insertion by the nitrene (**104**), but in comparison with all other alkyl

$$\underset{(\mathbf{104})}{\overset{\displaystyle CH_2—CH_2}{\underset{H\qquad N}{|\qquad\;\;|}}} \longrightarrow \mathbf{103} \tag{47}$$

azides studied thus far, this behaviour is anomolous and deserves further attention.

With the exception of the above report, there exists very little evidence for the intermediacy of discrete nitrene intermediates in alkyl azide thermolyses. A unimolecular process may arise either from a rearrangement concerted with elimination of nitrogen, or from a rate-determining elimination of nitrogen followed by a fast rearrangement (equation 48). Several tertiary aliphatic azides were synthesized to determine whether processes other than rearrangement could be

$$R_3C—N_3 \xrightarrow[-N_2]{\text{concerted}} R_2C{=}N—R \qquad R_3C—N_3 \xrightarrow[-N_2]{\text{stepwise}} R_3C—N \xrightarrow{\text{fast}} R_2C{=}N—R \tag{48}$$

observed on thermolysis and photolysis. In particular, the question of whether an alkylnitrene could be trapped by aromatic substitution was probed. Equation (49) illustrates the results of the thermolysis of 1-biphenyl-2-yl-1-methylethyl azide (**105**)[89], with percentage yields in parentheses, and equation (50), the results from the thermolysis of biphenyl-2-yldiphenylmethyl azide (**112**)[90]. Both **105** and **112** were decomposed in the absence of solvent. Products **111** and **118** show that an alkylnitrene is indeed formed and can undergo aromatic substitution. In the case of azide (**105**) this process competes favourably with rearrangement to imines (**109**) and (**110**). (A discussion of migratory aptitudes of aryl and alkyl groups is presented in the section on photochemical decomposition of alkyl azides.) Azide (**112**) was studied so that elimination of hydrazoic acid could be circumvented, but **113** and **114** were formed.

That the absence of hydrogens at the α-carbon atom in azides (**105**) and (**112**) does increase the lifetime of the nitrene sufficiently to make possible non-rearrangement processes may be seen from the

$$(105) \xrightarrow{-HN_3} \underset{(106)\ (62 \cdot 0\%)}{Me(Ar)C{=}CH_2} + (107)\ (1 \cdot 7\%)$$

$$(105) \xrightarrow{-N_2} (108) \longrightarrow \underset{(109)\ (7 \cdot 4\%)}{ArN{=}CMe_2} + \underset{(110)\ (7 \cdot 6\%)}{Me(Ar)C{=}NMe} + (111)\ (6 \cdot 5\%) \tag{49}$$

Ar = *o*-Biphenylyl

$$(112) \xrightarrow{-HN_3} (113)\ (5 \cdot 5\%) + \underset{(114)\ (16 \cdot 2\%)}{ArPh_2CH}$$

$$(112) \xrightarrow{-N_2} (115) \longrightarrow \underset{(116)\ (20 \cdot 2\%)}{ArN{=}CPh_2} + \underset{(117)\ (36 \cdot 0\%)}{ArPhC{=}NPh} + (118)\ (1 \cdot 7\%) \tag{50}$$

results of the thermolysis of biphenyl-2-ylmethyl azide (**119**) at 248–250° in diphenyl ether[91]. The aromatic substitution product **122** was not observed, rather, the formaldehyde anil (**120**) and imine (**121**) were found. Imine (**121**) corresponds formally to the coupling

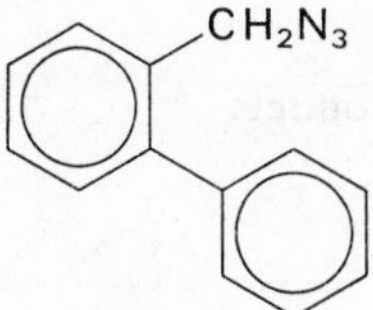

$$\mathbf{(119)}\ \xrightarrow[Ph_2O]{\Delta} ArN{=}CH_2\ \mathbf{(120)}\ (19\%) + ArCH_2N{=}CHAr\ \mathbf{(121)}\ (65\%) \tag{51}$$

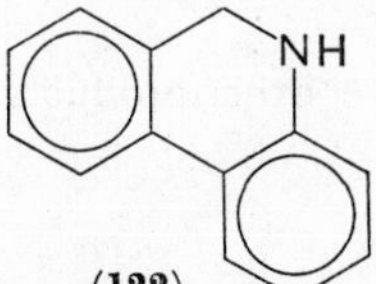

(122) ArCH (123) ArCH$_2$N (124) ArCH=NH (125)

Ar = *o*-Biphenylyl

of *o*-biphenylylcarbene (**123**) with *o*-biphenylylmethylnitrene (**124**), a most unlikely process. In light of the observation that no product of hydrogen migration (**125**) was detected, and that hydrazoic acid was formed in the reaction, a plausible sequence of events would be the formation of **125**, followed by a 1,3-dipolar addition[4] of azide (**119**) to give the tetrazoline (**126**) which, on elimination of hydrazoic acid, would give **121**.

$$\mathbf{119} + \mathbf{125} \longrightarrow \mathbf{(126)} \longrightarrow \mathbf{121} + HN_3 \tag{52}$$

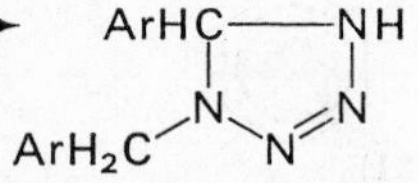

(126): ArHC—NH, ArH$_2$C—N, N=N ring

Ar = *o*-Biphenylyl

The formation of small amounts of the hydrogen abstraction products, amines (**128**) and (**130**) from the thermolysis in the absence of solvent of azides (**127**) and (**129**) also lends some support to the view that a free alkylnitrene is formed[90]. However, routes to the amines other than hydrogen abstraction may be envisaged. For example, 1,3-dipolar additions (of the type described for azide **119**) to olefins and imines formed in the thermolysis of the azide, followed decomposition of the adducts and hydrolysis could give **128** and **130**.

The results described above indicate that for tertiary aliphatic azides (as for primary aliphatic azides[82,83]), two primary modes of decomposition exist, one involving C—N bond fission, and the other, N—N bond fission. The C—N bond cleavage usually predominates

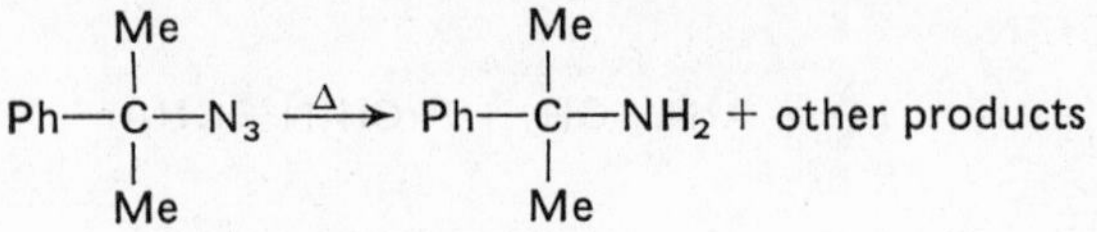

$$\mathrm{Ph{-}C(Me)_2{-}N_3} \xrightarrow{\Delta} \mathrm{Ph{-}C(Me)_2{-}NH_2} + \text{other products}$$

(127) (128)

(53)

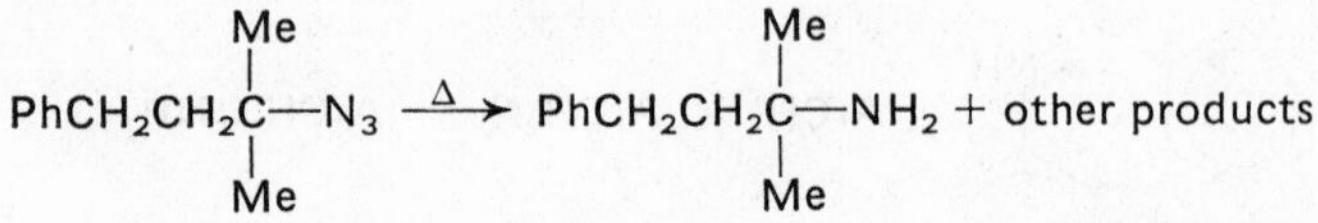

$$\mathrm{PhCH_2CH_2C(Me)_2{-}N_3} \xrightarrow{\Delta} \mathrm{PhCH_2CH_2C(Me)_2{-}NH_2} + \text{other products}$$

(129) (130)

with tertiary azides, except in the case of the triarylmethyl azide (**112**). The formation of the triarylmethane derivative **114** from **112** and of 2-methyl-4-phenylbutane in the thermolysis of azide (**129**)[90] suggests that a homolytic C—N bond fission is occurring as well in these cases. Thus, when the carbonium ion is generated from the triarylmethanol (**131**), a quantitative yield of cyclized product (**113**) is obtained[92].

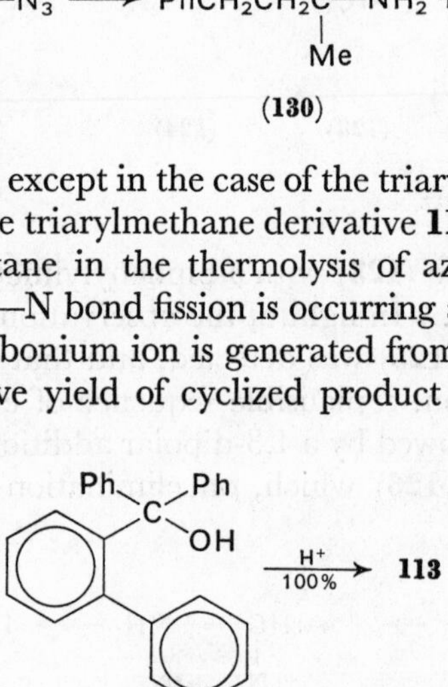

$$\mathbf{131} \xrightarrow[100\%]{H^+} \mathbf{113} \qquad (54)$$

(131)

Since the yield of **114** is about three times that of **113** in the azide pyrolysis, it appears that a different reactive intermediate is involved. The occurrence of both **113** and **114** may be rationalized on the basis of the formation of the tertiary radical **132** and azide radical[89].

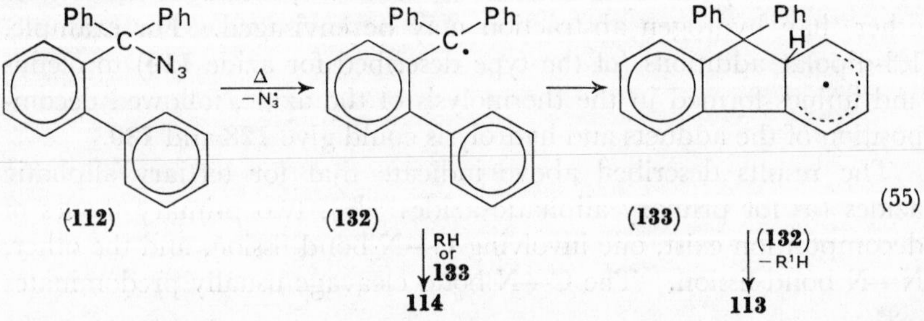

$$\mathbf{112} \xrightarrow[-N_3^{\cdot}]{\Delta} \mathbf{132} \longrightarrow \mathbf{133} \qquad (55)$$

(112) (132) (133)

$$\mathbf{132} \xrightarrow{RH \text{ or } \mathbf{133}} \mathbf{114} \qquad \mathbf{133} \xrightarrow[-R'H]{(\mathbf{132})} \mathbf{113}$$

Thermolysis of α-azidocarbonyl compounds (**134**) at 200° to give imines (**135**) and (**136**) may proceed via a nitrene intermediate[93–95]. It was found that the acyl group never migrated, and when R^1 = H,

$$R{-}\overset{O}{\overset{\|}{C}}{-}\underset{R^2}{\overset{R^1}{C}}{-}N_3 \xrightarrow{200^\circ} R{-}\overset{O}{\overset{\|}{C}}{-}\underset{R^2}{\overset{R^1}{C}}{-}\ddot{N} \longrightarrow \quad (56)$$

(**134**)

$$R{-}\overset{O}{\overset{\|}{C}}{-}\underset{R^2}{C}{=}N{-}R^1 + R{-}\overset{O}{\overset{\|}{C}}{-}\underset{R^1}{C}{=}N{-}R^2$$

(**135**) (**136**)

R^2 = Me or Ph, only H migrated, whereas if R^1 = Me, R^2 = Ph, Ph migration predominated. Either the migratory aptitude of hydrogen is enhanced or those of phenyl and methyl are depressed in the presence of the α-carbonyl group, since it has been found (*vide supra*) that alkyl group migration competes quite well with hydrogen migration[86].

When *N*-benzhydrylidene azidobenzhydrylamine (**137**) was heated at 200° in decalin a 20% yield of *N*-benzhydryl benzophenone imine (**141**) was obtained together with much tar[96]. It was suggested that the reaction proceeded via the dimerization of the nitrene (**138**) to give **139**, which then decomposed to give radical (**140**). (Thus

$$Ph_2C(N_3)(N{=}CPh_2) \xrightarrow[\text{decalin}]{-N_3^\bullet} Ph_2\dot{C}{-}N{=}CPh_2 \xrightarrow{RH} Ph_2CHN{=}CPh_2 \quad (57)$$

(**137**) (**140**) (**141**)

$$\textbf{137} \xrightarrow{-N_2} Ph_2C(\ddot{N})(N{=}CPh_2)\ (\textbf{138}) \longrightarrow Ph_2C(N{=}CPh_2){-}N{=}N{-}CPh_2(N{=}CPh_2)\ (\textbf{139}) \xrightarrow{-N_2} \textbf{140}$$

far, there is only one tentative report of an alkylnitrene dimerization[97].) Abstraction of hydrogen from solvent would give **141**. It was considered less likely[96] that **140** would be produced directly by loss of azide radical. In view of the findings on the ease of C—N bond fission in tertiary alkyl azides[89,90], the latter pathway becomes more plausible and indeed more likely. Thermolysis of **137** in naphthalene followed a different course, with *N*-phenyl-*N'*-benzhydrylidene benzamidine (**142**) being isolated in 23% yield, in addition to tars[96]. No

$$\mathbf{137} \xrightarrow[C_{10}H_8]{-N_2} Ph_2C(N{=}CPh_2)(\ddot{N}) \longrightarrow Ph{-}C(N{=}CPh_2){=}N{-}Ph \quad (\mathbf{142}) \tag{58}$$

evidence is available as to whether the rearrangement is concerted with nitrogen evolution or not.

The thermolysis of triarylmethyl azides to give benzophenone anils (**144**) and (**145**) has been carefully studied by Saunders and coworkers[98–100]. Table 2 gives the kinetic data for the thermolysis of

TABLE 2. Kinetic data for the thermolysis of p-$XC_6H_4(C_6H_5)_2CN_3$ in dibutyl carbitol and migration aptitudes in absence of solvent[98]

X	Migration aptitude[c,d]	Relative rate[e]	$\Delta H^\ddagger$ (kcal/mole)	$\Delta S^\ddagger$ (cal/deg mole)
H[a]	—	0·94	31·3	−11·0
H[b]	1·0	1·13	34·3	−4·7
H	—	1·00	32·0	−9·8
Cl	0·39	1·11	33·8	−5·6
NO_2	0·20	1·07	34·3	−4·6
Me	1·8	1·08	29·0	−16·1
OMe	2·5	1·53	28·9	−16·0
NMe_2	6·7	2·50	25·4	−22·5

[a] Solvent–nitrobenzene.
[b] Solvent–hexadecane.
[c] p-XC_6H_4:C_6H_5, corrected for statistical preference.
[d] Decompositions carried out in the absence of solvent.
[e] Rate of decomposition of Ph_3CN_3(X=H) in dibutyl carbitol defined as 1·00.

diphenyl(4-X-phenyl)methyl azides (**143**). The strong variation in the enthalpy and entropy of activation with a *para*-substituent X in the thermal decomposition was in the direction expected for rearrangement to an electron-deficient nitrogen intermediate, but was felt to be too large to be explained in terms of just an inductive effect (path *a*), and more in agreement with a concerted process (path *b*).

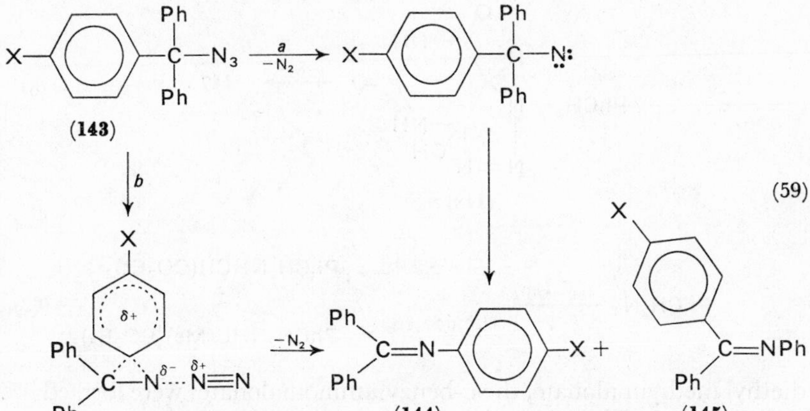

X—C_6H_4—C(Ph)(Ph)—N_3 $\xrightarrow[-N_2]{a}$ X—C_6H_4—C(Ph)(Ph)—N:

(**143**)

b ↓

(Ph)(Ph)C—N$^{\delta-}$···$N^{\delta+}$≡N (bridged $\delta+$ aryl) $\xrightarrow{-N_2}$ Ph(Ph)C=N—C_6H_4—X + (X—C_6H_4)(Ph)C=NPh

(**144**) (**145**)

(59)

The thermal migratory aptitudes varied from Ar/Ph = 6·7 for X = p-NMe_2 to 0·20 for X = p-NO_2. It has been pointed out, however, that these rate differences are small compared with the corresponding migratory aptitudes observed in assisted migrations involving carbonium ion intermediates, and that if the effect of substituents on $\Delta H^{\ddagger}$ only is considered, the inductive order is indeed followed[3]. A study of the effects of *meta*-substituents would be of value since these would not be expected to facilitate or retard aryl participation by direct conjugation[3].

When benzyl azide was decomposed in the presence of barbituric acid (**146**) at 165°, a quantitative yield of benzylaminobarbituric acid (**147**) resulted[101]. This was thought to be a C—H insertion reaction

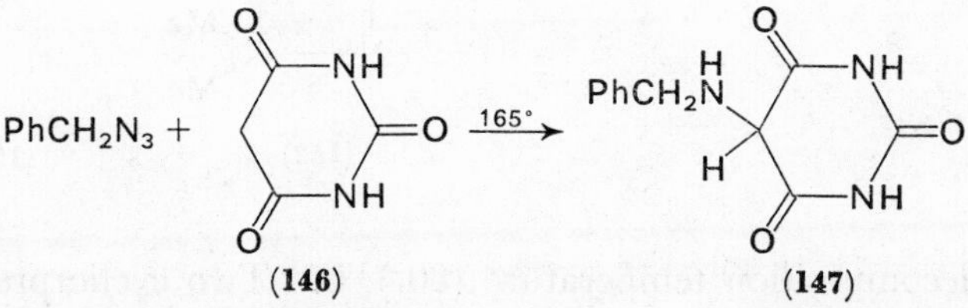

$PhCH_2N_3$ + (**146**) $\xrightarrow{165°}$ $PhCH_2NH$—(**147**) (60)

of the nitrene[101]. Since intramolecular H-migration would be expected to be much faster than an intermolecular insertion reaction, this might more probably be a 1,3-dipolar addition to the tautomeric enol form to give a triazoline (**148**), which could then lose nitrogen to give **147**. Other apparent C—H insertions have been reported. When benzyl azide was heated with either diethyl malonate or with

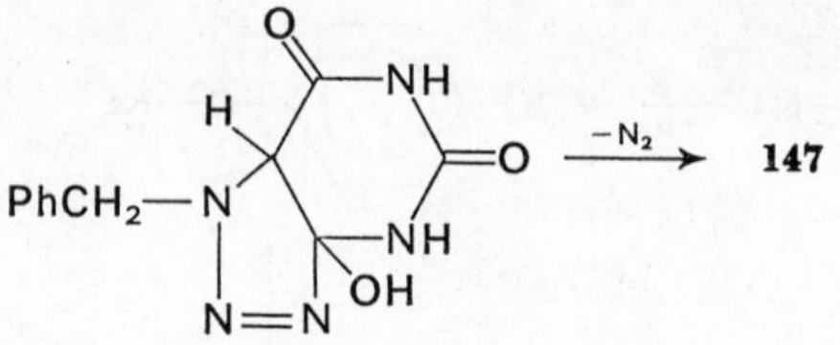

$$\text{(148)} \xrightarrow{-N_2} \mathbf{147} \qquad (61)$$

(148)

$$PhCH_2N_3 \xrightarrow{160-170^\circ} \begin{cases} \xrightarrow{CH_2(CO_2Et)_2} PhCH_2NHCH(CO_2Et)_2 \\ \xrightarrow{MeCH(CO_2Et)_2} PhCH_2NHC(Me)(CO_2Et)_2 \end{cases} \qquad (62)$$

diethyl methylmalonate, the α-benzylaminomalonates were formed[102]. Before any conclusion can be reached concerning the mechanism of these decompositions, it will be necessary to determine whether the rates are dependent upon the concentration of barbituric acid or of the malonates.

The thermal decomposition of olefinic azides such as 5-azido-5-methylhex-1-ene (**149**) proceeds by way of the isolable triazoline (**150**), and a nitrene mechanism was ruled out, as was indicated by

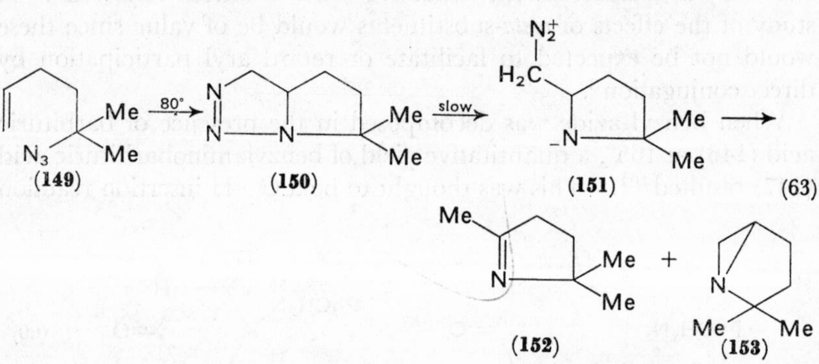

$$\text{(149)} \xrightarrow{80^\circ} \text{(150)} \xrightarrow{\text{slow}} \text{(151)} \longrightarrow \text{(152)} + \text{(153)} \qquad (63)$$

(149) (150) (151) (152) (153)

the low decomposition temperature (80°)[103]. Two cyclic products, **152** and **153**, were obtained.

B. Aryl Azides

1. Introduction

The thermal decomposition of aryl azides can lead to a much wider variety of products than alkyl azides. One of the major reasons for this is that rearrangement does not occur as readily as with alkyl azides (though the rearrangements are much more complex) and consequently the nitrene obtained on thermolysis of aryl azides is longer-lived. Thus, the probability of intermolecular reactions is much higher.

2. Mechanism of decomposition of aryl azides

Addition of phenyl azide to an approximately equivalent amount of aniline at 150° gave a compound $C_{12}H_{12}N_2$ (6–10%) that Wolff named dibenzamil[104]. Several incorrect structures were proposed initially for this compound until Huisgen and co-workers[105] obtained improved yields (up to 54%) by adding phenyl azide to a two hundred-fold excess of aniline at 165°, and suggested the structure 2-anilino-7*H*-azepine, which was later modified to 2-anilino-3*H*-azepine (**154**)[106]. The intermediacy of the 1*H*-azirene derivative (**156**) was eliminated by

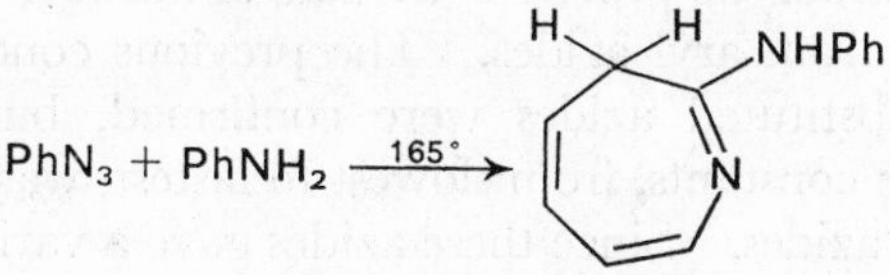

$$PhN_3 + PhNH_2 \xrightarrow{165°} \textbf{(154)} \qquad (64)$$

^{14}C labelling (equation 65)[107]. If **156** were an intermediate, the labelled carbon would be found in both the 2 and 7 positions of **154**.

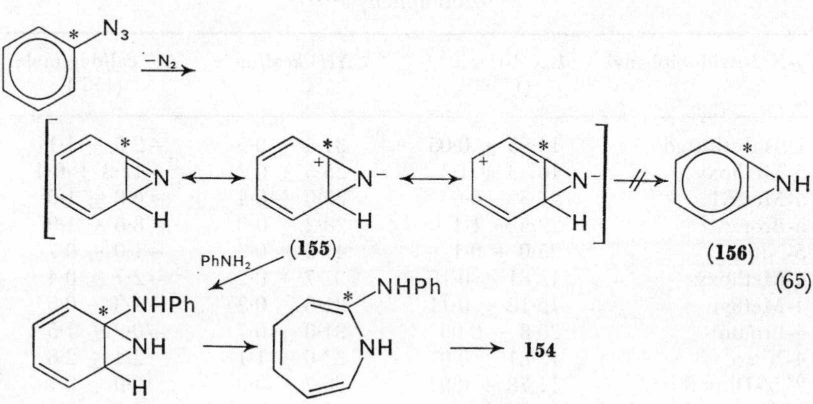

(65)

In fact, 100% of the label was found at C-2, thus eliminating **156** as a possible intermediate. Finally, it was found that substituents *meta* to the azido group had virtually no effect on the rates of decomposition[108]. This indicated that the decomposition of an aryl azide led to a nitrene intermediate, and was not a concerted process, since the latter would be sensitive to electronic effects of substituents. It was suggested that the equilibrium **157** ⇌ **155** lay well to the side of

$$PhN_3 \xrightarrow[-N_2]{\Delta} Ph\ddot{N}: \rightleftharpoons \mathbf{155} \quad (\mathbf{157}) \tag{66}$$

157 so that only strong nucleophiles could trap **155**[3]. (The existence of such an equilibrium has recently received more support[109].) There was only a very small change in the rate of evolution of nitrogen on changing the solvent from aniline to nitrobenzene, hence solvent does not participate in the decomposition, again supporting the formation of a discrete nitrene intermediate[108].

Smith and Hall[110] carried out extensive studies which showed that the primary thermolytic process is the loss of nitrogen with formation of aryl nitrenes from aryl azides. The previous conclusions[108] concerning *meta*-substituted azides were confirmed, but an eight-fold variation in rate constants, from slowest to fastest, was found for *para*-substituted aryl azides. Since these azides gave a variety of products, 2-azidobiphenyls were investigated because these compounds generally give high yields of carbazoles. Table 3 gives the kinetic data and

TABLE 3. Rate constants and activation parameters for thermolysis of 2-azidobiphenyls[110]

γ-X-2-Azidobiphenyl	$k \times 10^3$ min^{-1} (155·3°)	$\Delta H^{\ddagger}$ kcal/mole	$\Delta S^{\ddagger}$ cal/deg mole (156°)
Unsubstituted	14·55 ± 0·05	31·4 ± 0·5	−2·5 ± 1·1
5-Methoxy	107·3 ± 1·3	25·5 ± 0·2	−12·3 ± 0·4
5-Methyl	27·3 ± 0·6	29·0 ± 0·4	−6·8 ± 1·0
5-Bromo	29·5 ± 1·1	28·2 ± 0·8	−8·6 ± 1·9
5-Nitro	25·0 ± 0·1	41·6 ± 0·3	−1·0 ± 0·7
4-Methoxy	15·81 ± 0·18	31·7 ± 0·2	−2·7 ± 0·4
4-Methyl	15·18 ± 0·11	31·8 ± 0·2	−1·3 ± 0·6
4-Bromo	20·8 ± 0·05	31·8 ± 0·7	−0·9 ± 1·5
4-Nitro	12·81 ± 0·07	33·6 ± 1·1	2·4 ± 2·6
2′,5′-Dimethoxy	11·78 ± 0·04	36·7 ± 0·3	9·0 ± 0·7
2′,3′-Benzo	12·72[a,b]	36·0 ± 2·5	8·0 ± 8
3′,4′-Benzo	18[a,b]	29 ± 6	

[a] 156·7°.
[b] One determination only.

activation parameters for the thermolysis of substituted 2-azidobiphenyls. The rates were, in all cases, strictly first-order, and the yields of carbazoles were almost quantitative. The effect of a 4-substituent (*meta* to the azido group) had a negligible effect on the rate of decomposition, whereas a 5-substituent (*para* to the azido group), where direct conjugation between the substituent and azido group was possible, caused a ten-fold variation in rates. The rate constant for decomposition of 2-azidobiphenyl was the same in decalin as in nitrobenzene, but was about doubled in ethylene glycol and tripled in benzyl alcohol[110]. It was felt that this acceleration might be due to acid catalysis by hydroxylic solvents. In general, this work showed that thermolysis of aryl azides is a unimolecular process leading to an arylnitrene, and that inert solvents play no part in the decomposition.

Walker and Waters[87] considered the effects of various solvents on the rates of decomposition of phenyl, *p*-methoxyphenyl and cyclohexyl azides and on the nature of the products obtained. Some of their results are summarized in Table 4. The decompositions could

TABLE 4. Rate constants and activation parameters for azide thermolysis[87]

Azide	$k \times 10^5$ sec^{-1} (132·6°)		E_{Arr} kcal/mole		$\Delta S^{\ddagger}$ cal/deg/mole (132·6°)	
	Ethyl benzoate	Indene	Ethyl benzoate	Indene	Ethyl benzoate	Indene
$C_6H_5N_3$	0·78	150	(39·0)[b]	23·6	(18·7)[b]	−4·6
p-$MeOC_6H_4N_3$	9·1	166	38·5	23·2	19·6	−4·9
$C_6H_{11}N_3$	3·8[a]	8·7	47·5	34·6	32·2[a]	17·4

[a] 194·4°.
[b] Decomposition in decalin solution, calculated from results of Smith and Hall[110].

be classified into slow, intermediate and fast rates. The slow rates were observed in inert solvents such as hydrocarbons (saturated and some aromatic), benzophenone, ethyl benzoate and tetrachloroethylene. The intermediate rates were found in hydroxylic solvents (benzyl alcohol and diphenylmethanol), and the fast rates in indene and styrene. Product studies showed that amines and azo compounds resulted from slow decompositions in inert solvents, a complex mixture of products, but no azo compounds in hydroxylic solvents, and a single compound involving the solvent was characteristic of decomposition in indene. Since the rates of disappearance of azide and evolution of nitrogen were equal it was suggested that the decomposition involved a concerted elimination of nitrogen to give the aziridine (**158**), but it is possible that the triazoline intermediate was formed and decomposed in a fast step. The energies of activation for

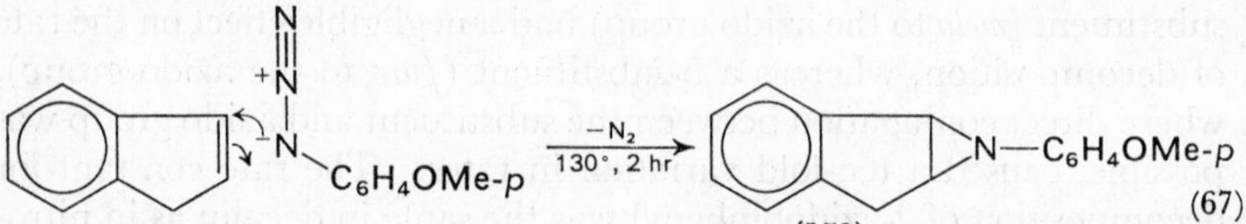

$$\xrightarrow[130°,\ 2\ hr]{-N_2} N-C_6H_4OMe\text{-}p \quad (67)$$

(**158**)

decomposition in indene were lower by 15–20 kcal/mole than in inert solvents[87].

3. Decomposition in the presence of unsaturated compounds

In general, aryl azides undergo 1,3-dipolar addition reactions with strained olefins to give 1,2,3-Δ^2-triazolines, rather than direct decomposition of the azido group. This subject has been reviewed recently[4], and since the azido group is not decomposed directly, it will be described only in a cursory fashion. Unactivated olefins react quite slowly with aryl azides, whereas strained bicyclic systems are very reactive. For example, the addition of phenyl azide to dicyclopentadiene (**159**) occurs exclusively at the double bond of the norbornene nucleus to give **160**[111,112]. At more elevated temperatures (>100°) the Δ^2-triazolines decompose, evolving nitrogen and giving aziridines

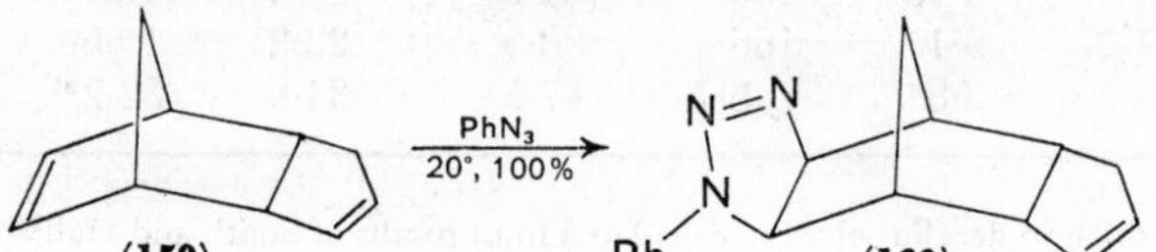

$$\xrightarrow[20°,\ 100\%]{PhN_3} \quad (68)$$

(**159**) (**160**)

and imines[4]. For example, the adduct (**161**) from phenyl azide and norbornene gave the aziridine (**162**) (55%) and anil (**163**) (27%) at 116°[113]. *p*-Methoxyphenyl azide reacted with indene to give the aziridine (**158**) (45%) as the only product isolated[87] (*vide supra*),

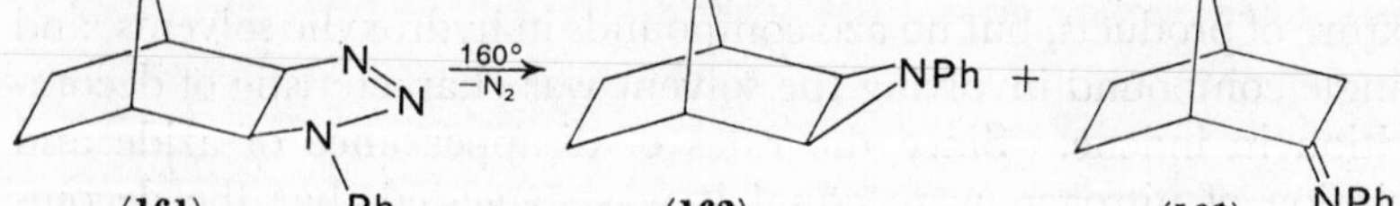

$$\xrightarrow[-N_2]{160°} NPh + \quad (69)$$

(**161**) (**162**) (**163**)

whereas phenyl azide gave the anil (**164**) (79%) as the sole product[114]. Thus, if aryl azides are heated (>100°) in the presence of olefins,

$$\text{indene} + PhN_3 \xrightarrow[-N_2]{91^\circ} \text{(164)} \qquad (70)$$

(164)

products are obtained which may correspond to those expected from a nitrene. In cases where substrate participation is suspected, it is essential to establish that the rate of decomposition is independent of its concentration before a conclusion is reached that a nitrene intermediate is formed or not[78]. Aromatic azides also undergo 1,3-dipolar additions with enamines, aldimines, ketimines, vinyl ethers and α,β-unsaturated esters and nitriles[4].

4. Assisted intramolecular cyclization

There are many cases where a suitable *ortho* substituent assists in the elimination of nitrogen and no nitrene is formed. It is characteristic of these reactions that they have a lower activation energy than those in which nitrenes are formed, and consequently the decompositions occur at significantly lower temperatures. For example, *o*-nitrophenyl azide (**165**) decomposes at 65–90° to benzofuroxan (**166**)[115,116], whereas unassisted decompositions occur at 140–170°[4]. Dyall and Kemp[117,118] carried out a systematic study of neighbouring-group

$$\text{(165)} \xrightarrow[-N_2]{\Delta} \text{(166)} \qquad (71)$$

(165) **(166)**

participation in the pyrolysis of aryl azides, and the results are presented in Table 5. It was found that phenylazo, nitro, acetyl and benzoyl groups lend anchimeric assistance in the displacement of nitrogen from the corresponding 2-substituted phenyl azides. The energies of activation were in the range of 22·4 to 27·2 kcal/mole, whereas for unassisted reactions the values ranged from 32·5 to 40·6 kcal/mole[118]. An *ortho*-hydroxymethyl substituent was found not to participate as a neighbouring group. In agreement with previous

TABLE 5. Kinetic data and activation parameters for pyrolysis of substituted phenyl azides in decalin[118]

Substituent in phenyl azide	k_{rel}[a] (161°)	E_a kcal/mole	ΔS_a e.u.
2-N=N—Ph	6680	22·4 ± 0·1	−8·3 ± 0·2
2-Ac	254	25·8 ± 0·3	−7·1 ± 0·8
2-Bz	45·1	27·2 ± 0·2	−7·4 ± 0·5
2-NO_2	537 (294)[b]	26·2 ± 0·3	−4·9 ± 0·8
2-NO_2-4-Me	323	27·3 ± 0·2	−3·1 ± 0·5
2-NO_2-3-Me	47·6	29·0 ± 0·2	−3·2 ± 0·5
2-NO_2-6-Me[b]	7·0	28·8 ± 0·8	−6 ± 2
2-CH_2OH	0·82		
4-NO_2	1·4	40·6 ± 1·2	16 ± 2·8

[a] Relative to PhN_3.
[b] Solvent, di-*n*-butyl phthalate.

suggestions[114,119–121] it was proposed that a concerted π-bond reorganization leading to the new heterocyclic ring provides the driving force in assisted reactions[118]. Using this concept, it was reasoned that the failure of the phenyl ring in 2-azidobiphenyls to

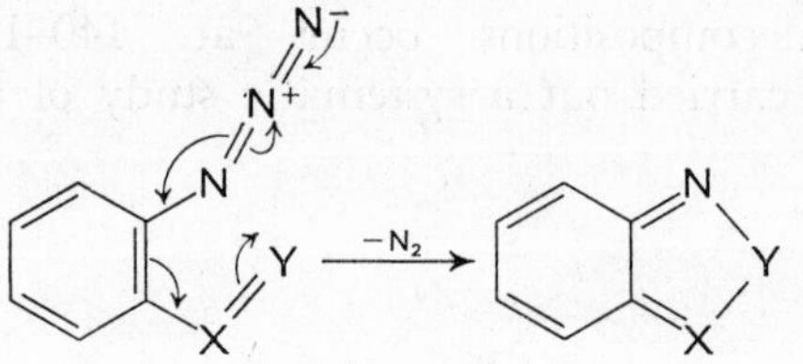

(72)

assist in elimination of nitrogen is due to the fact that the very unstable tautomer **167** would have to be formed, in which the aromaticity of

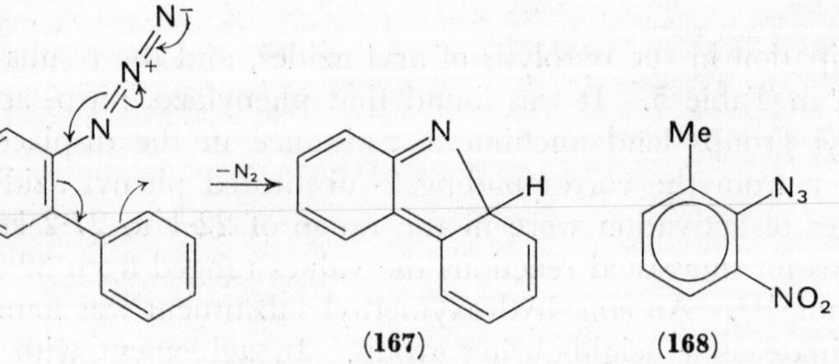

(73)

(167) **(168)**

both phenyl rings would be lost. Substituents *ortho* to the assisting group[121] or to the azide[118] interfere with anchimeric assistance. For example, 6-methyl-2-nitrophenyl azide (**168**) showed virtually no evidence of assisted nitrogen elimination, indicating the need for coplanarity between the azido and assisting group[118].

Thermolysis of *o*-azidobenzophenone (**169**) gave 3-phenylanthranil (**170**)[122], and a kinetic study of the thermal decomposition of *o*-azidobenzophenones of the type *o*-$N_3C_6H_4COC_6H_4R$-*p*, to give derivatives of **170** showed that the rate of nitrogen loss was accelerated by electron-withdrawing, and retarded by electron-donating groups R[123]. The enthalpies and entropies of activation varied over a wide range: $\Delta H^{\ddagger}$, from 20·5 to 27·0 kcal/mole, and $\Delta S^{\ddagger}$, from −5·9 to −21·4 cal/deg mole. The data are inconsistent with either a nitrene or a concerted nitrogen-elimination process. A mechanism involving an intramolecular 1,3-dipolar addition appears more likely[78,123].

(**169**) → $\xrightarrow{-N_2}$ (**170**) (74)

The first elimination of nitrogen in the double cyclization of 2,2′-diazidoazobenzene (**171**), which occurs at 58°[124,125] involves a concerted mechanism[3]. The second step proceeds at 170° to give

(**171**) $\xrightarrow[-N_2]{58°}$ $\xrightarrow[-N_2]{170°}$ (**172**) (75)

dibenzo-1,3*a*,4,6*a*-tetraazapentalene (**172**) and probably involves a nitrene. The last step is similar to the formation of pyrido[1,2-*b*]-

indazole which had been reported earlier[126]. It was found that when 2-(2′-azidophenyl)-5-methylbenzotriazole (**173**) was pyrolysed at 170°, the nitrene cyclized more readily onto the nitrogen *para* to the methyl rather than *meta* to it (**175**:**174** = 1·9)[127].

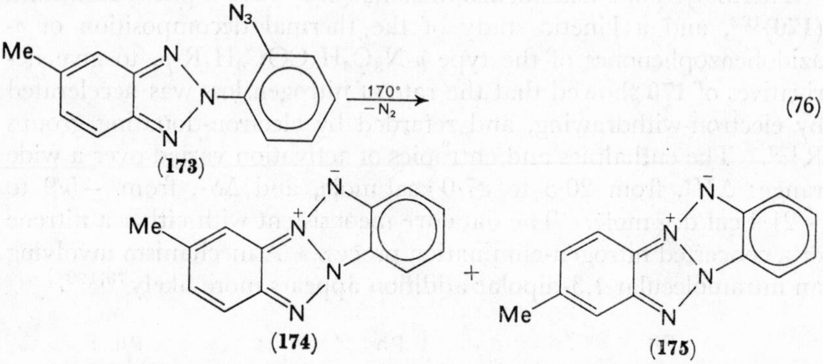

(76)

(**173**) (**174**) (**175**)

As mentioned above, aryl azides normally react with olefins to form 1,2,3-Δ^2-triazolines or aziridines with concerted nitrogen evolution. Pyrolysis of *o*-azidostyrenes leads to good yields of the 2-substituted indoles **177** instead[128]. For example, with R = *n*-Pr in **176**, the

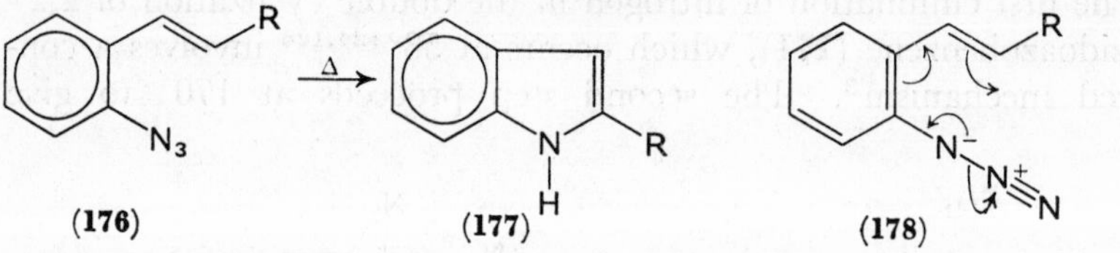

(77)

(**176**) (**177**) (**178**)

corresponding indole (**177**) was obtained in 81% yield. The reaction probably involves an assisted elimination (**178**) of nitrogen of the type described by Dyall and Kemp[118].

Thermal decomposition of *o*-azidobenzylidene amines leads to the corresponding indazole derivatives in good yields (75–97%)[129]. An interesting case was the two-stage thermolysis of **179**, which, at 120–130° gave **180** (79%), and at 150° gave a 90% yield of **181**. Krbechek and Takimoto assumed that a nitrene was involved and commented on the high degree of stereospecificity for 5-membered ring formation. No evidence was presented, however, that a nitrene was actually involved.

$$\text{(179)} \xrightarrow[-N_2]{120\text{–}130^\circ} \text{(180)};\quad \text{(179)} \xrightarrow[-2N_2]{150^\circ} \text{(181)};\quad \text{(180)} \xrightarrow[-N_2]{155^\circ} \text{(181)} \qquad (78)$$

5. Decomposition in the presence of aliphatic substrates

When thermolysis of an aryl azide leads to an arylnitrene, the latter is generated in the singlet (**182**) state, which can be in equilibrium with the triplet (**183**). Electron spin resonance experiments have

$$ArN_3 \xrightarrow{-N_2} Ar\ddot{N}\colon \rightleftharpoons Ar\ddot{N}\cdot \qquad (79)$$
$$\text{(182)} \qquad \text{(183)}$$

shown that the ground state of alkyl-, aryl- and sulphonylnitrenes is the triplet[130–133]. Intermediate **182** would be expected to behave as an electrophilic species, whereas **183** would exhibit the properties of a diradical[3]. The nature of the products obtained upon decomposition of an azide may then be used to infer the electronic state of the nitrene produced. For example, it is generally agreed that hydrogen abstraction in aliphatic solvents is characteristic of the triplet diradical, regardless of the group attached to the electron-deficient nitrogen atom[4,78,134–136] whereas stereospecific C—H bond insertion reactions are due to the singlet nitrene[134,137]. In fact, one of the most common products of aryl azide thermolysis is the corresponding aniline. Where readily abstractable hydrogens are available, this may become the dominant reaction[87,138]. On the other hand, insertions into aliphatic C—H bonds are normally low yield processes. For example, when phenyl azide was pyrolysed in *n*-pentane, aniline was obtained in 30% yield and the mixture of solvent insertion products (**184**) was formed in 10% yield[139]. Thermolysis of 2-azido-4,6-

$$PhN_3 \xrightarrow[n\text{-}C_5H_{12}]{\Delta} PhNH_2 + \underset{\text{(184)}}{PhNHC_5H_{11}} \qquad (80)$$

dimethylpyrimidine (**186**) (which is in equilibrium with the tetrazole **185**) in cyclohexane gave a 48% yield of hydrogen abstraction product (**187**) and 11% insertion product (**188**)[137]. On the other hand, when 2-pyridyl azide (also in equilibrium with the corresponding tetrazolo

(185) ⇌ **(186)** $\xrightarrow[185°]{C_6H_{12}}$ **(187)** + **(188)** (81)

compound) was heated in cyclohexane at 195°, only non-volatile materials were obtained, in addition 1% of hydrogen abstraction product[140]. Thermolysis of **185** in isobutane at 175° gave **187** (48%) and the insertion products **189** (5·4%) and **190** (1·6%). Recently, some evidence has been presented in favour of a triplet phenylnitrene

185 ⇌ **186** $\xrightarrow[175°]{(CH_3)_3CH}$ **187** + **(189)** + **(190)** (82)

from the pyrolysis of phenyl azide in mixtures of cyclohexane–neopentane, giving rise to both hydrogen abstraction and insertion products[141]. It was found that there was no change in the aniline: *N*-cyclohexylaniline ratio with increasing concentration of the neopentane inert solvent. As the concentration of substrate (cyclohexane) is decreased, the probability of intersystem crossing should increase. If the singlet nitrene (**191**) were responsible for the formation of *N*-cyclohexylaniline and the triplet that of aniline, then dilution should have led to the formation of more aniline at the expense of the

$$PhN_3 \xrightarrow[-N_2]{160°} Ph\ddot{N}: \longrightarrow Ph\dot{N}\cdot \xrightarrow{C_6H_{12}} Ph\dot{N}H \xrightarrow{RH} PhNH_2 \quad (83)$$
(191) (192)

$Ph\dot{N}H \xrightarrow{C_6H_{11}\cdot} PhNHC_6H_{11}$

'insertion' product. Since this did not occur, it was argued that both products must arise from the same intermediate, the triplet nitrene (**192**).

Aliphatic C—H insertions become dominant only in intramolecular cyclizations[142–146]. For example, **194**, obtained in 50–60% yield from azide (**193**), was 100% optically active when produced by vapour phase pyrolysis, and about 60% optically active from solution phase decomposition[142,146]. This evidence lends support to the idea that

(193) $\xrightarrow{\Delta}$ (194) (195) (84)

stereospecific insertion is due to the singlet species, whereas the non-stereospecific cyclization involves the abstraction of one hydrogen to give **195**, followed by radical coupling. It is expected that singlet→triplet conversion would occur more readily in solution than in the vapour phase, due to the greater number of collisions suffered in solution. Other yields of intramolecular C—H insertion products have ranged as high as 86%[144,145].

6. Decomposition in the presence of aromatic substrates

Bertho[138] first studied the decomposition of phenyl azide in benzene and *p*-xylene. He found that when the decomposition was carried out in benzene under pressure at 150–160° for 7–8 hours, azobenzene (11%) and aniline (18%) were obtained. When the reaction was carried in *p*-xylene under the same conditions, aniline (85%), a small amount of azobenzene, and *p*,*p*′-ditolylethane were obtained.

$$PhN_3 \xrightarrow[p\text{-}MeC_6H_4Me]{150\text{–}160°} PhNH_2 + PhN{=}NPh + Me\text{–}C_6H_4\text{–}CH_2CH_2\text{–}C_6H_4\text{–}Me \quad (85)$$

When *o*- or *p*-halogenophenyl azides (**196**) were heated in boiling chlorobenzene, toluene or xylene, to which a few drops of aniline had been added, a sublimate was produced consisting mainly of anilinium halide (**197**), a little halogenoanilinium halide (**198**) and ammonium halide[147]. In the absence of aniline the sublimate was formed more

$$X\text{-}C_6H_4N_3 \ (\mathbf{196}) \xrightarrow[PhNH_2]{\Delta} C_6H_5NH_3^+X^- \ (\mathbf{197}) + X\text{-}C_6H_4NH_3^+X^- \ (\mathbf{198}) + NH_4X + \text{tars} \qquad (86)$$

slowly and in smaller quantity, and consisted mainly of **198**, together with some ammonium halide. With *p*-chlorophenyl azide in the absence of aniline, the amount of chloride liberated corresponded to about 7% of the starting azide, whereas in the presence of excess aniline, it was about 70% of the azide used. Evidence was presented which indicated that the source of the halide ion was the reaction of the halogenoaniline (**199**) (formed by hydrogen abstraction) and 2-anilino-3*H*-halogenoazepine (**200**). It could be determined whether

$$X\text{-}C_6H_4NH_2 \ (\mathbf{199}) + X\text{-}\mathbf{200} \longrightarrow X\text{-}C_6H_4NH_3^+X^- + \text{tars} \qquad (87)$$

the halogen removal was the result of a substitution or an elimination reaction[146].

The thermal decomposition of ferrocenyl azide in benzene gave ferrocene (13·6%), phenylferrocene (1·9%), azoferrocene (17·8%), and ferrocenylamine (16·9%)[148]. Thermolysis of this azide in cyclohexane gave ferrocene (7·3%), and azoferrocene (20·5%). No product of aliphatic C—H insertion nor of aromatic substitution by the nitrene was observed[148]. (For more results and a comparison with photolysis, see section V.B.).

Aromatic azides readily undergo intramolecular—but not intermolecular—aromatic substitution[110,149]. An interesting comparison of intramolecular aromatic substitution and aliphatic C—H insertion may be made. 2-Azido-2′-methylbiphenyl (**201**) gives the carbazole (**202**) exclusively in 91% yield[91], and only when aromatic substitution becomes much more difficult, as in the case of 2,4,6-trimethyl-2′-azidobiphenyl (**203**) does C—H insertion become dominant[150]. When 2-azidodiphenylmethane (**204**) was pyrolysed in trichloroben-

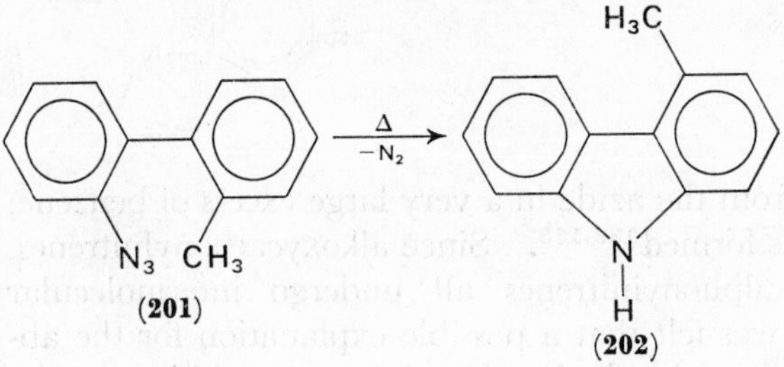

(**201**) $\xrightarrow[-N_2]{\Delta}$ (**202**) (88)

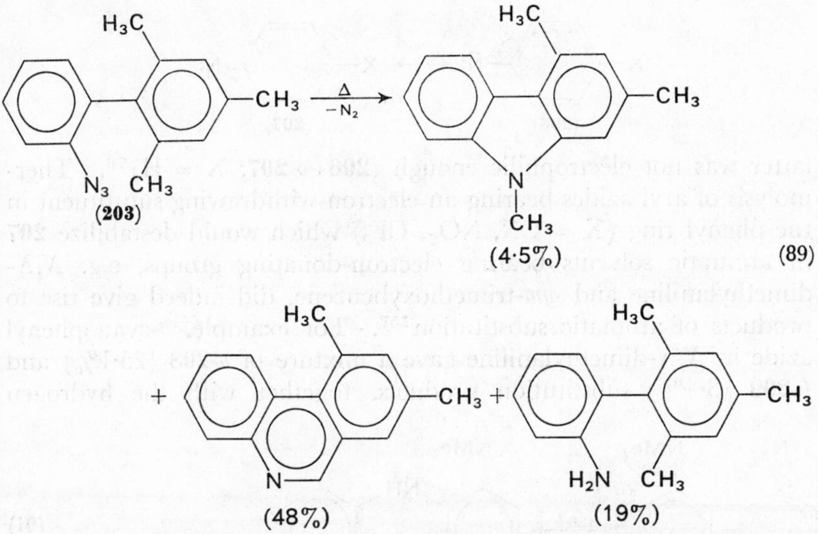

(**203**) $\xrightarrow[-N_2]{\Delta}$ (4·5%) + (48%) + (19%) (89)

zene, azepino[2,1-*a*]-11*H*-indole (**205**) was obtained in 66% yield[151]. This lends support to the idea that aromatic substitution is due to the addition of a singlet nitrene to the aromatic π-electron system followed by electrocyclic ring expansion[152–155].

That no intermolecular aromatic substitution was observed has been attributed to rapid intersystem crossing of the thermally generated singlet nitrene to the triplet. Even when phenylnitrene was

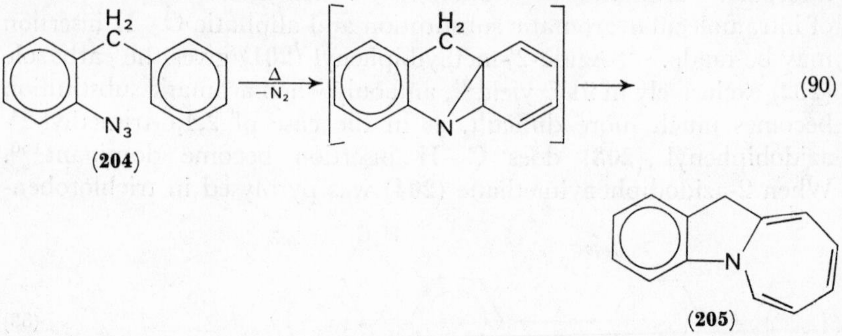

(90)

(204) (205)

generated thermally from the azide in a very large excess of benzene, no diphenylamine was formed[138,156]. Since alkoxycarbonylnitrenes, cyanonitrenes and sulphonylnitrenes all undergo intermolecular aromatic addition, it was felt that a possible explanation for the absence of a similar reaction with singlet phenylnitrene could be that the

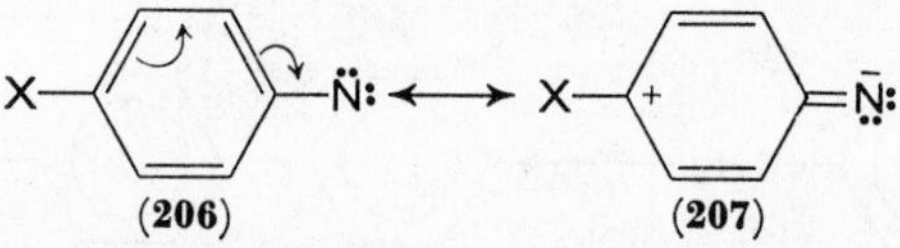

(206) (207)

latter was not electrophilic enough (**206**↔**207**, X = H)[78]. Thermolysis of aryl azides bearing an electron-withdrawing substituent in the phenyl ring (X = CN, NO_2, CF_3) which would destabilize **207** in aromatic solvents bearing electron-donating groups, e.g. *N,N*-dimethylaniline and *sym*-trimethoxybenzene, did indeed give rise to products of aromatic substitution[157]. For example, *p*-cyanophenyl azide in *N,N*-dimethylaniline gave a mixture of *o*-**208** (25·1%) and *p*-**209** (3·4%) substitution products, together with the hydrogen

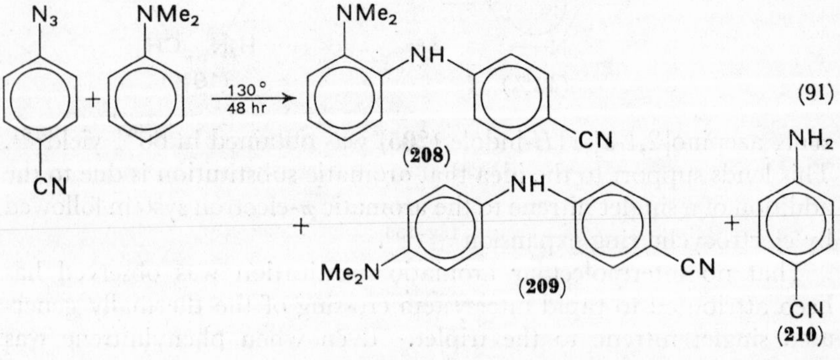

(91)

(208) (209) (210)

abstraction product (**210**) (20·3%)[157]. The results of a study of the thermolysis of *p*-X-phenyl azides in various solvents are given in Table 6[157].

TABLE 6. Products of thermolysis of aryl azides in aromatic solvents[157]

p-$XC_6H_5N_3$	Solvent	Temp (°C)	Time (hr)	% Diphenyl-amine	% Azo-compound	% Ani-line
CN	C_6H_6	140	45	—	25·2	4·9
	C_6H_5OMe	155	15·5	—	2·4	18·1
	p-$(MeO)_2C_6H_4$	130	50	—	3·4	41·0
	1,3,5-$(MeO)_3C_6H_3$	130	50	19·2	2·0	13·6
	$C_6H_5NMe_2$	130	48	25·1[a], 3·4[b]	—	20·3
	1,3,5-$Me_3C_6H_3$[c]	165	12	13·2	—	16·4
NO_2	$C_6H_5NMe_2$[d]	130	50	13·5[a]	1·0	18·3
	1,3,5-$(MeO^f)_3C_6H_3$[e]	130	50	18·9	—	16·8
CF_3	$C_6H_5NMe_2$	150	16	13·4[a]	—	trace

[a] *o*-Isomer.
[b] *p*-Isomer.
[c] Also obtained 3,3′,5,5′-tetramethylbibenzyl (23%).
[d] Also obtained 4,4′-methylenebis(*N*,*N*-dimethylaniline) (23·7%).
[e] Also obtained 2,4,6-trimethoxy-4′-nitrobiphenyl (3·4%).
[f] Also obtained 4,4′-methylenebis(*N*,*N*-dimethylaniline) (9·0%).

The kinetics of the decomposition of *p*-cyanophenyl azide in chlorobenzene at 132° with or without *N*,*N*-dimethylaniline were studied[157]. It was shown that the rate of decomposition of this azide was unaffected by the presence of the amine, even when in five-molar excess, and remained zero order in amine. This eliminated the possibility of a bimolecular decomposition involving a nucleophilic attack by the substrate on the azide.

Huisgen and von Fraunberg[140] have reported a number of aromatic substitutions by 2-pyridyl- and 2,4-dimethyl-2-pyrimidyl azides with reactive nuclei e.g. naphthalene, anthracene, anisole and others. It is not yet known whether these are direct substitutions, or additions followed by ring-opening of an aziridine intermediate. Since the pyridyl- and pyrimidylnitrenes can also be looked upon as arylnitrenes bearing a strongly electron-attracting substituent, this work supports the suggestion made concerning the electrophilicity requirement of arylnitrenes for them to undergo aromatic substitution[78].

7. High temperature pyrolyses

Recently, a considerable amount of work has been carried out on the high temperature (300–900°) vapour phase pyrolysis of aryl azides[158–165]. At these elevated temperatures, rearrangement reactions are important and most of the products can be rationalized on the

$$PhN_3 \xrightarrow{-N_2} \mathbf{191} \rightleftharpoons \cdots \rightleftharpoons \cdots \rightleftharpoons \cdots \quad (92)$$

(191) ⇅ (192)

basis of the following equilibrium (equation 92)[78]. Most of this work is beyond the scope of this treatment, but several examples will be given. The pyrolysis of phenyl azide at 300–670° led either to azobenzene and aniline or to a cyanocyclopentadiene (**212**), depending on the rate of introduction of phenyl azide into the pyrolysis furnace[158].

$$PhN_3 \xrightarrow[-N_2]{300-670°} \mathbf{191} \xrightleftharpoons{\Delta} [\mathbf{211}]^* \rightleftharpoons \cdots \rightarrow \mathbf{212}\ (-CN) \quad (93)$$

$$\mathbf{192} \longrightarrow PhN{=}NPh + PhNH_2$$

It might have been expected that high concentrations of reactant (high rate of introduction) would lead to increased yields of azobenzene and aniline. In fact, the opposite was found. It was felt that at high rates of introduction, the singlet nitrene (**191**) might be unable to diffuse from the hot zone of the furnace and might absorb additional energy to become a vibrationally excited nitrene (**211**) (a 'hot' nitrene). It is this species that would then insert intramolecularly into the benzene ring, eventually leading to ring contraction. At low rates of introduction, the nitrene (**191**) would be better able to diffuse from the hot surfaces of the furnace, undergo intersystem crossing to the triplet (**192**) and then undergo dimerization or hydrogen abstraction.

Nitrene–carbene interconversions have been observed at elevated temperatures. For example, 2,6-dimethylphenyl azide was pyrolysed at 900°/0·02 mm to give an 8% yield of 2-vinyl-6-methyl-pyridine (**213**)[164]. As with the nitrene–carbene interconversion, nitrogen

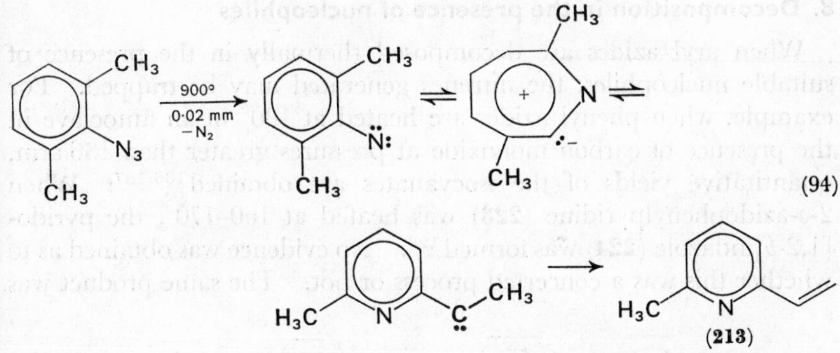

(94)

(213)

scrambling was demonstrated when substituted 2-pyridyl azides (in equilibrium with the tautomeric tetrazolo compound) were pyrolysed at 380°/0·05 mm. For example, starting from either 4- (**214**) or 5-methyltetrazolo[1,5-*a*]pyridine (**215**), the same products **218**–**222** in virtually identical yields, were obtained (equation 95)[165].

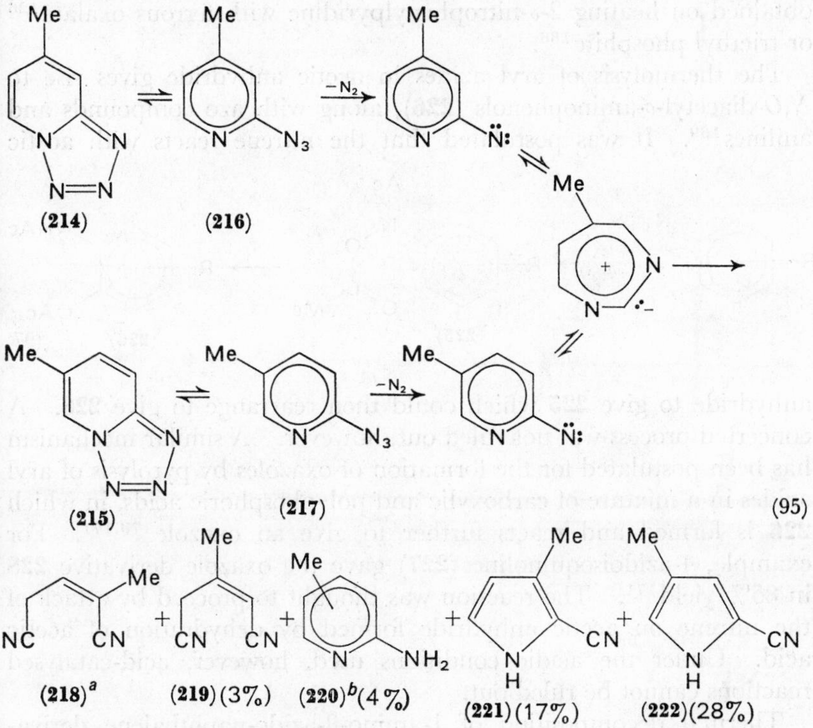

(214) (216)

(215) (217) (95)

(218)[a] (219) (3%) (220)[b] (4%) (221) (17%) (222) (28%)

[a] Yield from **216**, 1·6%; from **217**, 6%.
[b] 1:1 Mixture of 4- and 5-methyl-2-aminopyridine.

8. Decomposition in the presence of nucleophiles

When aryl azides are decomposed thermally in the presence of suitable nucleophiles, the nitrenes generated may be trapped. For example, when phenyl azides are heated at 180° in an autoclave in the presence of carbon monoxide at pressures greater than 136 atm, quantitative yields of the isocyanates are obtained[166,167]. When 2-*o*-azidophenylpyridine (**223**) was heated at 160–170°, the pyrido-[1,2-*b*]indazole (**224**) was formed[126]. No evidence was obtained as to whether this was a concerted process or not. The same product was

(**223**) $\xrightarrow[-N_2]{\Delta}$ (**224**) (96)

obtained on heating 2-*o*-nitrophenylpyridine with ferrous oxalate[126] or triethyl phosphite[168].

The thermolysis of aryl azides in acetic anhydride gives rise to *N*,*O*-diacetyl-*o*-aminophenols (**226**), along with azo compounds and anilines[169]. It was postulated that the nitrene reacts with acetic

$R-C_6H_4-N_3 \xrightarrow[Ac_2O,\ -N_2]{\Delta}$ (**225**) $\longrightarrow$ (**226**) (97)

anhydride to give **225** which could then rearrange to give **226**. A concerted process was not ruled out, however. A similar mechanism has been postulated for the formation of oxazoles by pyrolysis of aryl azides in a mixture of carboxylic and polyphosphoric acids, in which **226** is formed and reacts further to give an oxazole[170,171]. For example, 4-azidoisoquinoline (**227**) gave the oxazole derivative **228** in 85% yield[171]. The reaction was thought to proceed by attack of the nitrene on acetic anhydride formed by dehydration of acetic acid. Under the acidic conditions used, however, acid-catalysed reactions cannot be ruled out.

Thermal decomposition of 1-amino-8-azido-naphthalene derivatives (**229**) gives rise to oxazoles (**230**) and perimidines (**231**)[172].

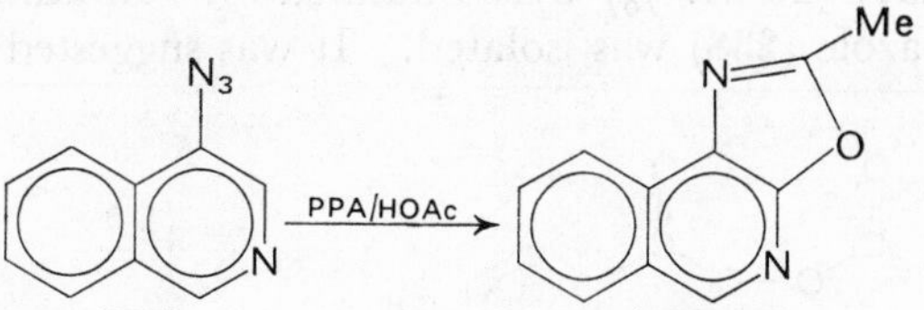

(227) $\xrightarrow{\text{PPA/HOAc}}$ (228) (98)

The former are thought to arise via an intramolecular acid-catalysed reaction (amide N—H), and the latter via a concerted reaction of

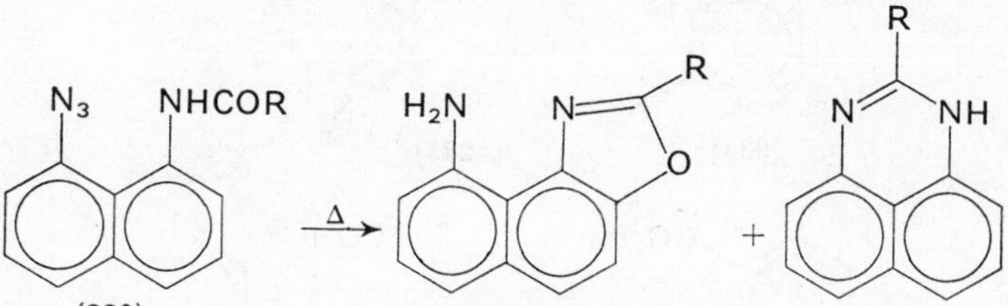

(229) $\xrightarrow{\Delta}$ (230) + (231) (99)

the azide and carbonyl group. If the nitrogen is tertiary, e.g. —NMeAc or —NAc_2, benzo[*c*, *d*]indazoles result[172].

9. Decomposition in the presence of radicals

There have been very few reports of azide decompositions induced by radicals. It has been found that the decomposition of phenyl azide was accelerated by thiyl radicals, particularly when it was carried out in thiols as solvents in the presence of free radical sources, such as tetraphenylhydrazine, hexaphenylethane and triphenylmethyl hydroperoxide[173]. When the decomposition was carried out in thiophenol, the major product was aniline, along with a small amount of *o*-aminodiphenylsulphide (**232**)[173].

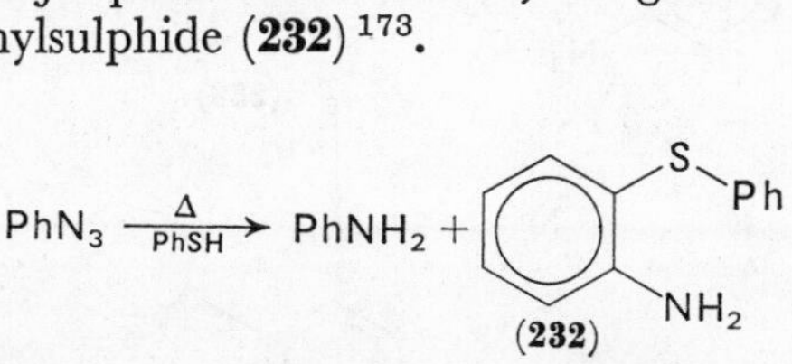

$$PhN_3 \xrightarrow[\text{PhSH}]{\Delta} PhNH_2 + \text{(232)} \quad (100)$$

Thermolysis of bis(*o*-azidobenzoyl)peroxide (**233**) at 80° in benzene in the absence of air led to a number of products in which the azido group was left intact: *o*-azidobenzoic acid (**234**) (102%), *o*-azidobiphenyl (**235**) (18%), phenyl *o*-azidobenzoate (**236**) (20%), and a mixture of unidentified, apparently aliphatic, esters of *o*-azidobenzoic acid (**237**) (20 wt. %) were obtained[174]. In addition, a 10% yield of carbazole (**238**) was isolated. It was suggested that the source of

(233) $\xrightarrow[80^\circ]{C_6H_6,\ \Delta}$ (234) + (235) + (236) + (237) + (238) (101)

235 and **238** was intermediate **239**. Carbazole (**238**) would come from **240**, which could result from a radical-induced intramolecular reductive decomposition of the azido group in **239**.

233 $\xrightarrow{-CO_2}$ $\xrightarrow{PhH}$ (239) $\xrightarrow{-[H]}$ **235**; (239) $\xrightarrow{-N_2}$ (240) $\longrightarrow$ **238** (102)

Thermolysis of benzoyl peroxide in a degassed solution of phenyl azide in benzene at 80° led to no decomposition of the azide over two half-lives of the peroxide[174]. On the other hand, the decomposition of phenyl azide in carbon tetrachloride at 80° was quite rapid in the presence of benzoyl peroxide, and yielded *p*-chlorophenylcarbonimidoyl dichloride (**241**) (67%)[174]. Evidence was presented which suggested that the radical responsible for inducing the decomposition of phenyl azide was trichloromethyl, and the mechanism proposed for the formation of **241** is outlined in equation (103).

$$(PhC(=O)-O)_2 \xrightarrow{CCl_4} CCl_3\cdot \xrightarrow{PhN_3} PhNN{=}\dot{N}CCl_3 \xrightarrow{-N_2}$$
$$Ph\dot{N}CCl_3 \xrightarrow{CCl_4} p\text{-}ClC_6H_5\dot{N}CCl_3 \longrightarrow \quad (103)$$
$$p\text{-}ClC_6H_4N(H)CCl_3 \xrightarrow{-HCl} \underset{(\mathbf{241})}{p\text{-}ClC_6H_4N{=}CCl_2}$$

Aryl azides undergo induced decomposition when heated at 50–80° in *i*-propyl alcohol in the presence of diethyl peroxydicarbonate (**243**)[175]. For example, 2,4,6-tribromoazidobenzene (**242**) and **243** (ca. 1:1 mole ratio) in *i*-propyl alcohol gave a 35% yield of 2,4,6-tribromoaniline (**244**). It was felt that the radical responsible for the induced decomposition was the 2-hydroxy-*i*-propyl radical.

$$\underset{(\mathbf{242})}{2,4,6\text{-}Br_3C_6H_2N_3} + \underset{(\mathbf{243})}{(EtOC(=O)-O)_2} \xrightarrow[Me_2CHOH]{\Delta} \underset{(\mathbf{244})}{2,4,6\text{-}Br_3C_6H_2NH_2} \quad (104)$$

Decomposition of 4-azido-2-picoline in the presence of hydrogen peroxide gave 4,4′-azoxy-2,2′-picoline[60].

10. Decomposition of azidoquinones and azidotropones

The thermolysis of azidoquinones probably involves a concerted ring-opening and elimination of nitrogen. The cyclized product **247** may well arise from a nitrene (**246**) formed in the pyrolysis of the azidoquinone (**245**)[176].

Thermolysis of 2,5-diazido-3,6-di-*t*-butyl-1,4-benzoquinone in boiling benzene is a convenient route to *t*-butylcyanoketene[176a].

Thermolysis of 2-azidotropone (**248**) in boiling cyclohexane gave *o*-cyanophenol, whereas in protic solvents (methanol, water, aniline) the ring-opened products **250** were obtained[177]. Both may be

(245) (70%) (ca. 5%) (246) (247) (20%) (105)

rationalized on the basis of a concerted ring-opening and nitrogen loss to give the ketene (**249**). Pyrolysis of 3- and 4-azidotropones gave

(248) (249) (250) (106)

X = MeO, HO, PhNH

mainly intractable tars in addition to small amounts of hydrogen abstraction products[178].

C. *Sulphonyl Azides*

I. Mechanism of the decomposition

Certain aspects of this topic have been reviewed[179,180]. The uncatalysed thermal decomposition of sulphonyl azides is believed to give sulphonylnitrene intermediates and nitrogen. It has been observed that arylsulphonyl azide decomposition is independent of

$$RSO_2N_3 \xrightarrow{\Delta} RSO_2N + N_2 \qquad (107)$$

solvent[181] and follows first-order kinetics in chlorobenzene, nitrobenzene and *p*-xylene[182,183]. More recently, it was found that the thermolysis of *p*-toluenesulphonyl azide was first-order in a variety of solvents with rates of decomposition varying little ($\pm 15\%$ relative to decomposition in diphenyl ether) from solvent to solvent[134]. These results are shown in Table 7. The greatest rate observed was in

TABLE 7. Decomposition of *p*-toluenesulphonyl azide[a] in various solvents[134]

Solvent	Temp °C	$k_1 \times 10^4$ sec^{-1}	Rel. rate	Gas evolved % of theory
Diphenyl ether	155	3·43	1	100
Tetradecane	155	3·80	1·11	—
Nitrobenzene	155	3·97	1·15	119[b]
1-Octanol	155	3·63	1·06	96
n-Hexanoic acid	155	2·97	0·86	114[c]
Dimethyl terephthalate	155	3·23	0·95	104
Diphenyl ether	145	1·44	1	—
1,4-Dichlorobutane	145	1·70	1·18	140[d]
Diphenyl ether	130	0·330	—	—

[a] Approximately 0·14M in azide.
[b] Contained NO and other gases.
[c] Contained CO_2.
[d] Contained a chlorobutene.

nitrobenzene and the smallest in dimethyl terephthalate. The average half-life of *p*-toluenesulphonyl azide at 155° was 33 min, with $\Delta H^{\ddagger} = 35{\cdot}1$ kcal/mole and $\Delta S^{\ddagger} = 7$ e.u.[134]. Pyrolysis of benzenesulphonyl azide in cyclohexane at 135–150° was a first-order reaction with an activation energy of 33 kcal/mole and $\Delta S^{\ddagger} = 5{\cdot}2$ e.u.[184]. On the other hand, the activation energy for the thermolysis of benzenesulphonyl azide in naphthalene at 110–135° was found to be 36·4 kcal/mole[185]. The rate constants for the decomposition of *p*-substituted derivatives have been reported to correlate well in a Hammett plot[186], but other groups found that there is a negligible substituent effect ($\rho = -0{\cdot}1$) using a modified Hammett equation[183,185]. Table 8 gives the rate constants for decomposition of substituted aryl-

TABLE 8. Decomposition of *p*-R-substituted benzenesulphonyl azides in naphthalene at 120°[185]

R	MeO	Me	H	Cl	Br	NO_2
$k_d \times 10^5\ sec^{-1}$	1·31	1·12	1·07	1·15	1·36	1·60

sulphonyl azides in naphthalene solution. In general, the rate-determining step in the thermal decomposition of arylsulphonyl azides is the loss of nitrogen to give the electron-deficient nitrene intermediate, with little interaction between the electrons on the nitrogen atom and the aromatic nucleus. This is further supported by e.s.r. measurements of the triplet species, generated photochemically in a glass, in which the zero-field parameters D and E indicated the lack of important delocalization of the unpaired electron on nitrogen. For example, for triplet *p*-fluorobenzenesulphonylnitrene, $D = 1{\cdot}555$ cm^{-1} and $E < 0{\cdot}005\ cm^{-1}$ [133].

Aliphatic mono- and disubstituted sulphonyl azides also decompose thermally in diphenyl ether in a first-order process, but are somewhat more stable than the aromatic sulphonyl azides[134] (compare rates in Table 7 with those in Table 9). Thermolysis of 1-pentane and 2-

TABLE 9. Decomposition of various sulphonyl azides[a] in diphenyl ether[134]

Sulphonyl azide	Temp °C	$k_1 \times 10^4\ sec^{-1}$	Gas evolved % of theory
1-Pentane-	166	4·46	102
1,4-Butanedi-	163	5·02	99
1,6-Hexanedi-	163	5·02	98
1,9-Nonanedi-	170	4·25	—
	160	2·25	—
	150	0·884	—
1,10-Decanedi-	163	4·45	100
1,4-Dimethylcyclohexane-α,α', di-	163	4·82	98
m-Xylene-α,α'-di-	163	6·09	101
p-Xylene-α,α'-di-	163	5·78	96

[a] Solutions approximately 1M in azide.

propanesulphonyl azides in mineral oil did not give good first-order plots[134,187]. In the former case, evolved gases were identified as nitrogen, sulphur dioxide and a small amount of *n*-pentane. Also, the solvent incorporated some azide, as was indicated by the infrared absorption band at 2100 cm^{-1} after heating the sulphonyl azide for 60 half-lives. The amount of sulphur dioxide evolved with 1-pentanesulphonyl azide was 16–20%. With added radical inhibitors such as hydroquinone, however, this could be reduced to about 3% and good first-order kinetics were observed. In comparison with alkane derivatives, *p*-toluenesulphonyl azide in tetradecane gave a 1·3% yield of sulphur dioxide[134,187], and mesitylenesulphonyl azide in dodecane at 150° gave a 22% yield of sulphur dioxide[188]. In all cases, essentially quantitative yields of gas were obtained from alkanesulphonyl azides[134]. In order to account for these observations, Breslow and coworkers[134] postulated the following mechanism (equations 108–111) for the process in which SO_2 is formed. Since each azide evolves either N_2 or SO_2, a quantitative yield of gas is to be expected. In

$$C_5H_{11}SO_2N_3 \longrightarrow C_5H_{11}SO_2\ddot{N}: + N_2 \qquad (108)$$

$$R^\bullet + C_5H_{11}SO_2N_3 \longrightarrow C_5H_{11}SO_2^\bullet + RN_3 \qquad (109)$$

$$C_5H_{11}SO_2^\bullet \longrightarrow C_5H_{11}^\bullet + SO_2 \qquad (110)$$

$$C_5H_{11}^\bullet + RH \longrightarrow C_5H_{12} + R^\bullet \qquad (111)$$

support of step (109), both Leffler and Tsuno[182], and Breslow and co-workers[134] showed that added radical sources such as *t*-butylhydroperoxide catalyse the cleavage of the sulphur–nitrogen bond.

A careful study of the evolution of sulphur dioxide showed that this reaction was completed by the time half the sulphonyl azide had decomposed. The source of the radical $R^\bullet$ was not established, although peroxidic impurities in the solvents and impurities in the azides were ruled out. Another possibility is that initially produced singlet nitrene undergoes intersystem crossing to give the triplet, which could then abstract a hydrogen atom from solvent to give $R^\bullet$. This, at first glance, seems unlikely since the triplet should be expected to be formed continuously during the run and not stop when only half the azide has been decomposed, *unless* an inhibitor of the singlet to triplet conversion process were formed, or a radical trap were generated in the reaction[179]. This could, for instance, be the sulphonamide or the amine formed by Curtius rearrangement (*vide infra*). Some of the SO_2 probably comes from the Curtius rearrangement, particularly since radical inhibitors do not completely eliminate SO_2 evolution. Another source of radicals might be the homolysis of the S—N bond, analogous to the C—N fission which occurs with alkyl azides (*vide supra*). The same problem exists for this suggestion as for the triplet nitrene. There must be a mechanism by which SO_2 evolution is complete in about one half-life of azide.

In agreement with equation (109), recent work has shown that the thermolysis (150°) under nitrogen of mesitylene-2-sulphonyl azide (**251**) in *n*-dodecane gave rise to dodecyl azides (2·3%)[188]. This was thought to occur by the triplet nitrene (**253**) abstracting a hydrogen from the solvent to produce $C_{12}H_{25}^{\bullet}$ which could then attack azide (**251**) to produce an alkyl azide and **254**[180]. The same products

$$\text{Me}_3\text{C}_6\text{H}_2\text{–SO}_2\text{N}_3\ (\mathbf{251}) \xrightarrow[-N_2]{\Delta} \text{Me}_3\text{C}_6\text{H}_2\text{–SO}_2\ddot{\text{N}}\text{:}\ (\mathbf{252}) \rightleftharpoons \text{Me}_3\text{C}_6\text{H}_2\text{–SO}_2\ddot{\text{N}}^{\bullet}_{\bullet}\ (\mathbf{253})$$

$$\mathbf{253} + n\text{-}C_{12}H_{26} \longrightarrow \text{Me}_3\text{C}_6\text{H}_2\text{–SO}_2\dot{\ddot{\text{N}}}\text{H} + C_{12}H_{25}^{\bullet} \qquad (112)$$

$$C_{12}H_{25}^{\bullet} + \mathbf{251} \longrightarrow C_{12}H_{25}N_3 + \text{Me}_3\text{C}_6\text{H}_2\text{–SO}_2^{\bullet}\ (\mathbf{254})$$

could have been obtained by homolysis of the S—N bond to give **254** and azide radical. Radical (**254**) could lose SO_2 to give the aryl radical (**255**) which could then abstract hydrogen from solvent to give mesitylene or combine with other radicals in solution ($C_{12}H_{25}^{\bullet}$, $N_3^{\bullet}$, **255**) to give various products, most of which would be difficult to separate from solvent. There is some evidence for the formation of aryl radicals on thermolysis of arylsulphonyl azides. Thus diphenyl

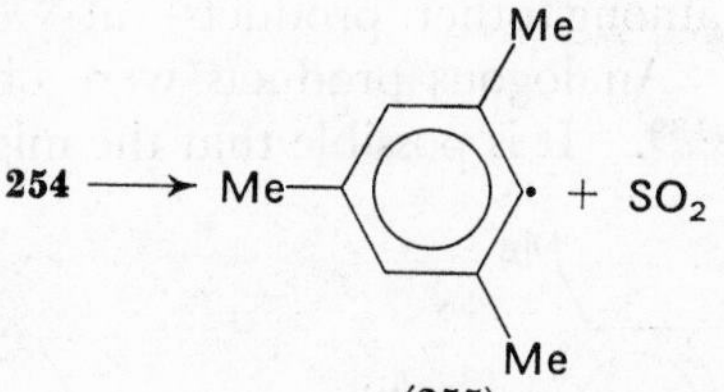

$$\mathbf{254} \longrightarrow \text{Me}-\text{C}_6\text{H}_2(\text{Me})_2\cdot + SO_2 \qquad (113)$$

(255)

sulphone-2-sulphonyl azide (**256**) in *n*-dodecane at 150° gave diphenyl sulphone (**257**) (27%) together with the hydrogen abstraction product (**258**); in Freon E-4 at 150° some diphenylene sulphone (**259**) (1·3%) was obtained[189]. Thermolysis of ferrocenylsulphonyl azide in cyclohexane at 150° under pressure gave ferrocene (20%)[148]. The detec-

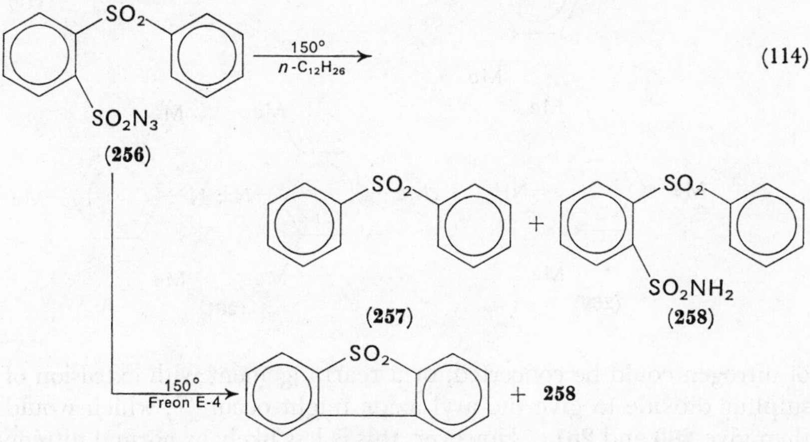

(114)

SO_2, SO_2N_3 **(256)** $\xrightarrow[n\text{-}C_{12}H_{26}]{150°}$ **(257)** + SO_2NH_2 **(258)**

(256) $\xrightarrow[\text{Freon E-4}]{150°}$ **(259)** + **258**

tion of *n*-pentane from the thermolysis of *n*-pentanesulphonyl azide speaks for the occurrence of alkyl radicals[134].

Until recently, Curtius' classification of sulphonyl azides as 'rigid' or 'starre' azides, that is, incapable of rearrangement, had found few exceptions. The vapour phase pyrolysis of benzenesulphonyl azide at 625° gave a 17·5% yield of azobenzene[190], and traces of azo compound could be obtained in boiling cyclohexanone[184]. Surprisingly, no C—H insertion product was obtained in the latter case, and only about one third of the azide was accounted for[184]. Two products were obtained from the decomposition of mesitylene-2-sulphonyl azide (**251**) which can best be explained by a Curtius-type rearrangement. 2,4,6-Trimethylaniline (**260**) and the corresponding azobenzene (**261**) were formed (among other products) in yields of 20·7 and 0·4% respectively[188]. Analogous products were obtained with durene-3-sulphonyl azide[189]. It is possible that the migration and elimination

251 $\xrightarrow{-N_2}$ Me–C$_6$H$_2$(Me)$_2$–$SO_2\ddot{N}$: ⟶ Me–C$_6$H$_2$(Me)$_2$–N=SO_2 $\xrightarrow{-SO_2}$ Me–C$_6$H$_2$(Me)$_2$–N ⟶ (115)

Me–C$_6$H$_2$(Me)$_2$–NH_2 + Me–C$_6$H$_2$(Me)$_2$–N=N–C$_6$H$_2$(Me)$_2$–Me

(**260**) (**261**)

of nitrogen could be concerted, or a rearrangement with extrusion of sulphur dioxide to give the aryl azide might occur[191], which would then give **260** and **261**. However, this is less likely as normal nitrene products were also obtained.

2. Decomposition in the presence of aliphatic substrates

Thermolysis of alkanesulphonyl azides in cyclohexane led to unsubstituted sulphonamides (H abstraction) in yields of 0–3%, but the yields of C—H insertion products ranged above 50%[134]. For example, 2-propanesulphonyl azide decomposed in cyclohexane to give *N*-cyclohexyl-2-propanesulphonamide in 60% yield, and a maximum of 3% of 2-propanesulphonamide[134]. On the other hand, aryl sulphonyl azides gave higher yields of hydrogen abstraction products, generally in the range of 5–20%. Good yields of aliphatic C—H insertion products were also obtained[134,179,189]. Two examples of intramolecular C—H insertion have been reported. Mesitylene-2-sulphonyl azide (**251**)[188] and durene-3-sulphonyl azide (**263**)[189] gave the corresponding sultams (**262**) and (**264**) in 2 and 15% yields, respectively, on thermolysis in dodecane at 150°. A high yield (48%) of hydrogen abstraction product (**266**), in addition to the insertion product (**267**) (24%), was produced from the thermolysis of ferrocenylsulphonyl azide (**265**) in cyclohexane (equation 117)[148,179].

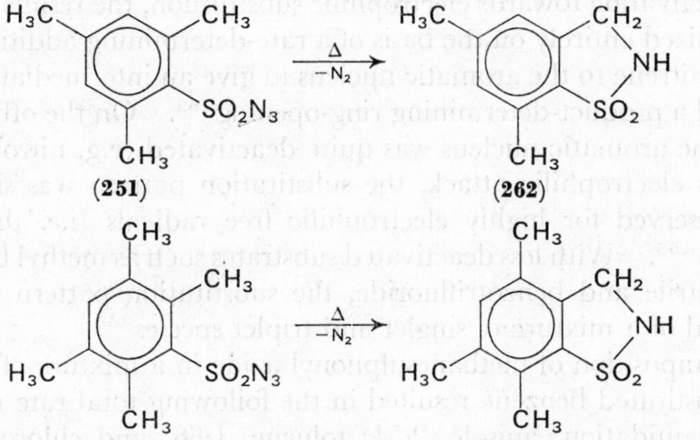

$$(\mathbf{251}) \xrightarrow[-N_2]{\Delta} (\mathbf{262})$$

$$(\mathbf{263}) \xrightarrow[-N_2]{\Delta} (\mathbf{264}) \qquad (116)$$

$$\underset{(\mathbf{265})}{Fc{-}SO_2N_3} \xrightarrow[C_6H_{12}]{\Delta} \underset{(\mathbf{266})}{FcSO_2NH_2} + \underset{(\mathbf{267})}{FcSO_2NHC_6H_{11}} \qquad (117)$$

3. Decomposition in the presence of aromatic substrates

Decomposition of sulphonyl azides in aromatic solvents may lead to aromatic 'substitution', which is thought to involve the addition of the singlet nitrene to the aromatic nucleus, followed by further rearrangement to give products[152,153]. In aromatic solvents, the yields of unsubstituted sulphonamides (hydrogen abstraction) are better than in aliphatic hydrocarbons[152,192]. Due to the absence of biaryls in these decompositions, it was postulated that this reaction is due to a singlet nitrene which added to the aromatic nucleus to give **268** and then decomposed to amide and **269** in a concerted or almost concerted abstraction of two hydrogen atoms[152]. Thus, no aryl radicals would be formed. No evidence was obtained for the intervention of benzynes (**269**) in these reactions but tars are always formed. When the thermolysis of methanesulphonyl azide was carried out in various

$$MeSO_2N_3 + X\text{-}C_6H_5 \xrightarrow[-N_2]{\Delta} X\text{-}(\mathbf{268})\text{-}NSO_2Me \longrightarrow X\text{-}(\mathbf{269}) + MeSO_2NH_2 \quad (118)$$

substituted benzenes, in which the substituents were activating or not too deactivating towards electrophilic substitution, the results could be rationalized entirely on the basis of a rate-determining addition of the singlet nitrene to the aromatic nucleus to give an intermediate such as **268** and a product-determining ring-opening[152]. On the other hand, when the aromatic nucleus was quite deactivated (e.g. nitrobenzene) towards electrophilic attack, the substitution pattern was similar to that observed for highly electrophilic free radicals (i.e. the triplet nitrene)[155]. With less deactivated substrates such as methyl benzoate, benzonitrile and benzotrifluoride, the substitution pattern was that expected of a mixture of singlet and triplet species[155].

Decomposition of methanesulphonyl azide in a mixture of benzene and substituted benzene resulted in the following total rate ratios for sulphonamidation; anisole, 2·54; toluene, 1·86; and chlorobenzene, 0·44[152]. For benzenesulphonyl azide the values were 0·96, 1·00 and 0·69 respectively[192]. In a similar experiment, ring substitution in *p*-xylene by *p*-toluenesulphonylnitrene was shown to take place 2·2 times as rapidly as in benzene[134]. When *p*-toluenesulphonyl azide was decomposed in an equimolar mixture of benzene and cyclohexane at 165°, it was found that the benzene 'double bond' is about eight times more reactive than a C—H bond in cyclohexane[134].

Decomposition of suitable *o*-substituted benzenesulphonyl azides can result in intramolecular aromatic substitution. For example, biphenyl-2-sulphonyl azide (**270**) gave the cyclized product (**271**) in 38–70% yield on thermolysis in *n*-dodecane or in the absence of solvent at 150°[189].

$$\underset{(\mathbf{270})}{\text{biphenyl-}SO_2N_3} \xrightarrow[-N_2]{150^\circ} \underset{(\mathbf{271})}{O_2S\text{—}NH} \quad (119)$$

A great deal of interesting work was carried out by Curtius and co-workers[193,196], who first studied the thermolysis of sulphonyl azides. Some of the work certainly deserves reinvestigation using modern analytical tools. For example, the report[195] that **272** and **273** were obtained from decomposition of sulphonyl azide in *p*-xylene must be viewed with scepticism, but it would certainly be interesting to determine the correct structures of the products formed.

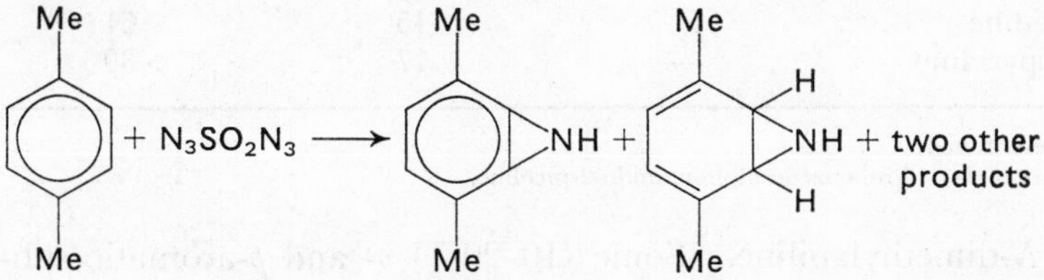

$$p\text{-xylene} + N_3SO_2N_3 \longrightarrow \mathbf{272} + \mathbf{273} + \text{two other products} \quad (120)$$

(272) **(273)**

4. Decomposition in the presence of nucleophiles

Thermolysis of sulphonyl azides in the presence of pyridine results in the formation of pyridinium ylids[197,198]. It is not known whether the reaction is concerted or involves a free nitrene, but when benzenesulphonyl azide was decomposed in the presence of 2,6-lutidine, at least part of the azide decomposed to the free nitrene, as evidenced by the formation of 3-benzenesulphonamido-2,6-lutidine (**275**), in

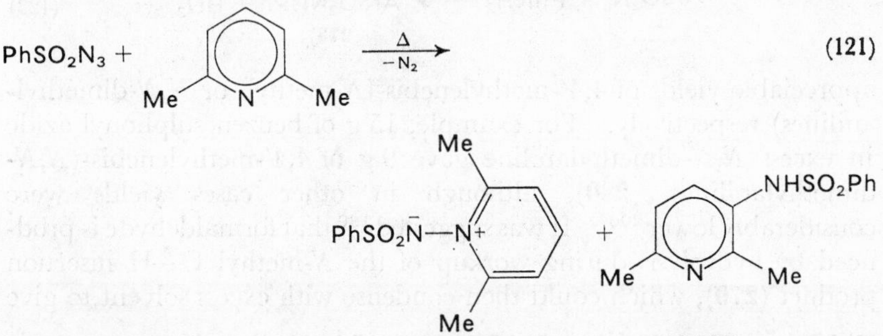

$$PhSO_2N_3 + \text{2,6-lutidine} \xrightarrow[-N_2]{\Delta} PhSO_2\bar{N}-N^+(\text{2,6-lutidine}) + \text{3-}NHSO_2Ph\text{-2,6-lutidine} \quad (121)$$

(274) **(275)**

addition to the ylid (**274**)[199]. Table 10 presents the results of the thermolysis of benzenesulphonyl azide in the presence of various pyridine derivatives[199].

Curtius and co-workers[196] studied the thermolysis of a number of aromatic sulphonyl azides in the presence of aniline, *N*-methylaniline,

TABLE 10. Yields of products from the thermolysis of benzenesulphonyl azide with pyridine derivatives[199]

Substrate	3-Substitution product %	Pyridinium ylid %	Benzenesulphonamide %
Pyridine	—	30	—[a]
2-Picoline	8·8[b]	37	45
2,6-Lutidine	13	18	57
sym-Collidine	15	15	61
4-Cyanopyridine	—	17	> 30

[a] Not determined.
[b] Mixture of 3- and 5-benzenesulphonamido-2-picoline.

and *N,N*-dimethylaniline. Some (10–20%) *o*- and *p*-aromatic substitution products (**276**) along with the hydrogen abstraction product (**277**) and nitrogen were obtained. With aniline, the major product was usually the aniline (**278**), formed by a nucleophilic displacement

$$ArSO_2N_3 + PhNR_2 \xrightarrow[-N_2]{\Delta} \underset{(\mathbf{276})}{ArSO_2NHC_6H_4NR_2} + \underset{(\mathbf{277})}{ArSO_2NH_2} \quad (122)$$

of the azido group, to give hydrazoic acid. Decomposition of most arylsulphonyl azides in *N*-methylaniline or *N,N*-dimethylaniline gave

$$ArSO_2N_3 + PhNH_2 \xrightarrow{\Delta} \underset{(\mathbf{278})}{ArSO_2NHPh} + HN_3 \quad (123)$$

appreciable yields of 4,4′-methylenebis-(*N*-methyl- or *N,N*-dimethylanilines) respectively. For example, 15 g of benzenesulphonyl azide in excess *N,N*-dimethylaniline gave 9 g of 4,4′-methylenebis-(*N,N*-dimethylaniline) (**280**), although in other cases yields were considerably lower[200]. It was suggested[180] that formaldehyde is produced by hydrolysis during workup of the *N*-methyl C—H insertion product (**279**), which could then condense with excess solvent to give

$$PhSO_2N_3 + Me_2NPh \longrightarrow \underset{(\mathbf{279})}{PhSO_2NHCH_2N(Me)Ph}$$

$$\xrightarrow{H_2O} PhSO_2NH_2 + HCHO + PhNHMe \quad (124)$$

$$Me_2NPh + HCHO \longrightarrow \underset{(\mathbf{280})}{Me_2N{-}C_6H_4{-}CH_2{-}C_6H_4{-}NMe_2}$$

280. No hydrazine derivatives were observed in any of the thermolyses of sulphonyl azides in anilines studied by Curtius and coworkers[196].

Thermolysis of sulphonyl azides in dimethyl sulphoxide leads to *N*-sulphonylsulphoximines. For example, *p*-methoxybenzenesulphonyl azide gave a 31% yield of the sulphoximine (**281**)[181]. No evidence is

$$p\text{-}MeOC_6H_4SO_2N_3 \xrightarrow[Me_2SO,\ -N_2]{\Delta} p\text{-}MeOC_6H_4SO_2N{=}\overset{\overset{\displaystyle O}{\|}}{S}Me_2 \qquad (125)$$

(**281**)

available concerning the effect of dimethyl sulphoxide on the rate of decomposition of sulphonyl azides.

5. Decomposition in the presence of unsaturated compounds

Sulphonyl azides react with strained olefins, at temperatures below those required for the formation of the corresponding nitrenes, to give aziridines and anils. These reactions have been reviewed recently[4,180]. Benzenesulphonyl azide reacts with norbornene in acetonitrile solution at 55–60° to give the aziridine (**282**) (64%) in 2–3 hr[201,202]. It was found, however, that when benzene was the solvent, a quantitative yield of **282** was obtained at room temperature in about 1·5 hr[203].

$$\text{norbornene} + PhSO_2N_3 \xrightarrow[55\text{–}60°]{MeCN} \qquad (126)$$

NSO₂Ph (**282**) N=N, N, SO₂Ph (**283**)

Although an unstable triazoline intermediate was a possibility, it was thought that a concerted mechanism involving a species such as **283** was more likely[202]. When benzenesulphonyl azide was heated at 100° for 1 hr with maleic anhydride, *N*-phenylmaleimide, divinyl sulphone, mesityl oxide, cyclohexene, cyclopentene, styrene, vinyl acetate or *p*-quinone, no gas evolution was observed, nor was there any change in the concentration of sulphonyl azide (as indicated by infrared measurement)[201].

Sulphonyl azides react slowly (1–2 weeks at 70–80°) with some acetylene derivatives to yield the corresponding 1,2,3-triazoles (**284**)[204].

$$ArSO_2N_3 + R{-}C{\equiv}C{-}R^1 \xrightarrow[70\text{-}80^\circ,\ 50\text{-}75\%]{1\text{-}2\ \text{weeks}} \text{1-}ArSO_2\text{-4-R-5-R}^1\text{-1,2,3-triazole}\ (\mathbf{284}) \qquad (127)$$

Enamines react readily with sulphonyl azides to give various products, depending on the structure of the enamine. For example, with enamine (**285**) and *p*-toluenesulphonyl azide, the sulphonamidine (**287**) was obtained in good yield, presumably via the triazoline intermediate (**286**)[205]. The reaction of 2-morpholino-1-benzoylpropene (**288**) with *p*-nitrobenzenesulphonyl azide illustrates two other ways

$$CH_3CH{=}CHNHC_3H_7\ (\mathbf{285}) \xrightarrow{TsN_3} CH_3{-}CH{-}CH(NHC_3H_7)\ [\text{triazoline ring } N{=}N{-}N(Ts)]\ (\mathbf{286}) \xrightarrow{-N_2} CH_3CH_2C({=}NTs)NHC_3H_7\ (\mathbf{287}) \qquad (128)$$

which the unisolable triazoline intermediate (**289**) may react (equation 129)[206].

$$PhCO{-}CH{=}C(Me)(N\text{-morpholino})\ (\mathbf{288}) + p\text{-}O_2NC_6H_4SO_2N_3 \longrightarrow \text{triazoline: } PhCO{-}HC{-}N{=}N{-}N(SO_2C_6H_4NO_2\text{-}p){-}C(Me)(N\text{-morpholino})\ (\mathbf{289}) \qquad (129)$$

$$\mathbf{289} \longrightarrow PhCOCH{=}N_2 + Me{-}C({=}NSO_2C_6H_4NO_2\text{-}p){-}N(CH_2CH_2)_2O$$

$$\mathbf{289} \longrightarrow \text{(PhCO, Me-triazole, N-H)} + p\text{-}O_2NC_6H_4SO_2{-}N(CH_2CH_2)_2O \quad (129)$$

Vinyl ethers also react readily with sulphonyl azides to give imino-esters. For example, from dihydropyran and *p*-toluenesulphonyl azide, the arylsulphonylimine of δ-valerolactone (**290**) was obtained in quantitative yield[207].

$$\text{dihydropyran} + TsN_3 \xrightarrow[12\ \text{hr}\ 96\%]{80°} \text{(C=NTs lactone imine)}\ (\mathbf{290}) \quad (130)$$

The reaction of benzene- and toluenesulphonyl azides with tetra-methyllallene has been studied recently[208]. A 26% yield of *N*-(1,2,3-trimethyl-2-butylidene)benzenesulphonamide (**291**) was obtained with benzenesulphonyl azide.

$$Me_2C{=}C{=}CMe_2 + PhSO_2N_3 \xrightarrow[-N_2]{\Delta} Me_2C{=}C(Me){-}C(Me){=}N{-}SO_2Ph\ (\mathbf{291}) \quad (131)$$

6. Decomposition in the presence of radicals

It has been reported that decomposition of benzenesulphonyl azide in thiols is induced by thiyl radicals, particularly in the presence of free radical sources[173]. The major product was benzenesulphonamide. Benzenesulphonyl azide and *t*-butyl hydroperoxide exhibit mutually induced decomposition in chlorobenzene at 126·7°[182]. The rate of decomposition of the azide at the beginning and end of the reaction was that expected for the uncatalysed decomposition, but in the intervening period there was a very rapid nitrogen (and oxygen) evolution. This induced decomposition could be inhibited completely by addition of iodine or by using *p*-xylene as the solvent. The reaction proved to be quite complex and no specific mechanism was proposed.

When *p*-toluenesulphonyl azide was heated at 50–80° in isopropyl alcohol in the presence of diethyl peroxydicarbonate (20·3 azide: 1·4 peroxide), *p*-toluenesulphonamide and acetone were obtained in 75 and 81% yields respectively[175]. It was thought that the 2-hydroxy-*i*-propyl radical (**292**) added to the azide to give **293** (equation 132) and that an intramolecular reduction and elimination of nitrogen occurred via a cyclic intermediate (**294**) to give the radical (**295**) and acetone. Hydrogen abstraction by (**295**) would then give the sulphonamide.

$$(EtO{-}\overset{O}{\overset{\|}{C}}{-}O)_2 \longrightarrow EtO{-}\overset{O}{\overset{\|}{C}}{-}O^{\bullet} \longrightarrow EtO^{\bullet} + CO_2$$

$$EtO^{\bullet} + Me_2CHOH \longrightarrow EtOH + \underset{(\mathbf{292})}{Me_2\dot{C}OH} \qquad (132)$$

$$ArSO_2{-}N{=}\overset{+}{N}{=}\overset{-}{N} + Me_2\dot{C}OH \longrightarrow \underset{(\mathbf{293})}{ArSO_2{-}N{=}N{-}N^{\bullet}{-}C(Me)_2{-}O{-}H} \longleftrightarrow \underset{(\mathbf{294})}{\text{cyclic } ArSO_2{-}\dot{N}{-}N{=}N{-}CMe_2{-}O{-}H}$$

$$\mathbf{294} \xrightarrow{-N_2,\ -Me_2CO} \underset{(\mathbf{295})}{ArSO_2NH^{\bullet}} \xrightarrow{RH} ArSO_2NH_2$$

7. Decomposition in the presence of organic anions

Reactions of this type were first studied by Curtius and co-workers[196]. For example, diethyl malonate and *p*-toluenesulphonyl azide reacted in the presence of cold sodium ethoxide to give the sodium salt of the 5-hydroxytriazole (**296**)[209]. Acidification gave the 5-triazolone (**297**) which isomerized to the diazo compound (**298**) (overall yield 90%). (The same product could be obtained in 85% yield without base by heating the azide and malonate at 100° and a pressure of 20 mm)[209]. This reaction provided the basis for the so-called diazo-transfer reaction, an extremely useful method of synthesizing diazo compounds, which has been reviewed[210]. The reaction has been formulated[210] as shown in equation (134), and has been extended to the synthesis of azides by a diazo transfer to amine anions[211,212]. *p*-Toluenesulphonyl azide reacts with hydrazone

$$TsN_3 + CH_2(CO_2Et)_2 \xrightarrow{NaOEt} Ts{-}N(N{=}N)C(ONa){=}C{-}CO_2Et \quad \mathbf{(296)}$$

$$\xrightarrow{H^+} \mathbf{(297)} \longrightarrow N_2C(CO_2Et)(CONHTs) \quad \mathbf{(298)} \qquad (133)$$

$$EtO_2C{-}\bar{C}H{-}COOEt\ Na^+ \xrightarrow{TsN_3} EtO_2CCO_2Et + TsNHNa,\quad EtO_2C{-}C(=N_2){-}CNHTs(=O) \qquad (134)$$

$$-EtOH,\quad H^+,\quad EtO_2C{-}C(N_2){-}C(O)\cdots O{-}S(O)(N){-}C_6H_4Me\text{-}p$$

anions to give diazo compounds. For example, benzophenone hydrazone anion (**299**) in tetrahydrofuran reacted immediately with this azide to give diphenyl diazomethane (**300**) (50%), *p*-toluenesulphonamide anion (**301**) and nitrogen[213].

$$\underset{\mathbf{(299)}}{Ph_2C{=}N{-}\bar{N}H} + TsN_3 \longrightarrow \underset{\mathbf{(300)}}{Ph_2C{=}\overset{+}{N}{=}\bar{N}} + \underset{\mathbf{(301)}}{Ts\bar{N}H} + N_2 \qquad (135)$$

Recently, it has been found that when benzenesulphonyl azide and quinaldine are heated at 110–115° for 40 hr, a 70% yield of the diazo-transfer product (**302**) is obtained[199]. This type of reaction was not

$$\text{2-methylquinoline} + PhSO_2N_3 \xrightarrow{\Delta} \textbf{(302)} + PhSO_2NH_2 \qquad (136)$$

observed with 2-picoline or *sym*-collidine (see Table 10), but has been extended to 6-methylphenanthridine and 1-methylisoquinoline[199].

IV. DECOMPOSITION WITH TRIVALENT PHOSPHORUS COMPOUNDS

These reactions were first studied by Staudinger and co-workers[214,215] who showed that the initial product is a very unstable 1:1 adduct, which in some cases could be isolated by low temperature workup[215]. In almost all cases, the reaction occurs readily at room temperature.

A. Alkyl Azides

The decomposition of methyl and ethyl azide with triphenylphosphine has been studied. When ethyl azide in petroleum ether was treated with a suspension of triphenylphosphine at −20°, little nitrogen evolution occurred, but the primary adduct (**303**) was not isolated[215]. When the mixture was warmed to room temperature,

$$EtN_3 + Ph_3P \xrightarrow{-20^\circ} \underset{\textbf{(303)}}{EtN_3.PPh_3} \xrightarrow[-N_2]{R.T.} \underset{\textbf{(304)}}{EtN{=}PPh_3} \qquad (137)$$

nitrogen was evolved and the phosphimine (**304**) obtained. Some difficulty was encountered in purifying **304** sufficiently for analysis[215]. The primary adduct formed from *n*-butyl azide and triphenylphosphine was isolated but no analysis was given[216]. It was found that benzyl azide reacted with triphenylphosphine about one tenth as fast as phenyl azide in xylene[216].

Triphenylmethyl azide[217] and 9-phenyl-9-fluorenyl azide[218] give quite stable 1:1 adducts with triphenyl phosphine, which decompose only at elevated temperatures in neutral solvents at rates which appear to be similar to those of the azides in the absence of the phosphine. The major product from the triphenylmethyl azide adduct is benzophenone anil[218].

B. Aryl Azides

When phenyl azide was treated with phenyldiethylphosphine in ether at $-80°$, the primary adduct (**305**) was isolated at low temperatures[215]. In solution at 0° nitrogen was evolved and the phosphimine (**306**) was produced. The kinetics of the reaction of

$$PhN_3 + Et_2PhP \xrightarrow[-80°]{Et_2O} \underset{(\mathbf{305})}{PhN{=}N{-}N{=}PPhEt_2} \xrightarrow[-N_2]{0°} \underset{(\mathbf{306})}{PhN{=}PPhEt_2} \quad (138)$$

substituted phenyl azides with triphenylphosphine were studied by Horner and Gross[216], who found that the rate was accelerated by electron-withdrawing substituents and retarded by electron-donating groups. The decomposition followed a rough Hammett relationship with $\rho = +1{\cdot}36$. The *p*-methoxy substituent gave a point significantly off the Hammett plot line[216]. More recently, a very careful study of the Staudinger reaction of aryl azides showed that adherence to second-order kinetics is only approximate. Deviations increased when electron-releasing substituents were present in the triphenylphosphine and electron-withdrawing substituents in the phenyl azide[219]. Evidence was presented for a reaction proceeding through two isomeric transition states and an intermediate complex, all having the empirical formula $Ar_3P\ Ar^1N_3$. The first transition state involved the attack of the phosphine at the terminal nitrogen of the azide to give the 1:1 adduct (**307**), whereas in the second transition state (**308**) the phosphorus was at least partially bonded to the α-nitrogen atom[219].

$$PhN_3 + Ph_3P \longrightarrow \underset{(\mathbf{307})}{PhN{=}N{-}N{=}PPh_3}$$

$$\longrightarrow \underset{(\mathbf{308})}{\left[\begin{array}{c} N{=}N \\ \vdots \quad \vdots \\ Ph{-}N{-}PPh_3 \end{array}\right]} \xrightarrow{-N_2} PhN{=}PPh_3 \quad (139)$$

Although the apparent k_2 was complex, its Arrhenius plot showed no significant deviation from linearity. From this were calculated $\Delta H^\ddagger = 9{\cdot}6 \pm 0{\cdot}9$ kcal/mole and $\Delta S^\ddagger = -33 \pm 3$ cal/deg. mole for the reaction of phenyl azide and triphenylphosphine[219].

C. Sulphonyl Azides

Trialkylphosphites and thiophosphites react with a variety of sulphonyl azides at room temperature or above to form phosphinimine derivatives (**309**) and nitrogen[220]. It was shown that in benzene at low temperature, a 1:1 adduct, probably the triazine (**310**), could be

$$RSO_2N_3 + (R^1O)_3P \xrightarrow{-N_2} \underset{(\mathbf{309})}{RSO_2N{=}P(OR^1)_3} \qquad (140)$$

isolated, which on warming decomposed to the phosphinimine (**311**) in a clean first-order reaction[182]. In chloroform and similar solvents,

$$PhSO_2N_3 + Ph_3P \longrightarrow \underset{(\mathbf{310})}{PhSO_2N{=}N{-}N{=}PPh_3} \xrightarrow[-N_2]{\Delta} \underset{(\mathbf{311})}{PhSO_2N{=}PPh_3} \qquad (141)$$

the reaction was more complicated. The evolution of nitrogen was not quantitative, benzenesulphonamide and triphenylphosphine oxide were obtained and no **311** was formed[182]. The reaction was investigated in more detail, and diphenyl sulphide (**315**) and disulphide (**314**) were isolated when the reaction was carried out in acetonitrile or chloroform[221]. It was postulated (equation 142) that the phosphine was deoxygenating the sulphonyl azide to give the sulphinyl azide (**312**), and then the sulphenyl azide (**313**), which would then decompose to give the disulphide (**314**) and nitrogen[221]. Diphenyl sulphide (**315**) could then arise by reaction of **314** with the phosphine. The formation and thermal decomposition of benzene-

$$\begin{aligned}
&PhSO_2N_3 + Ph_3P \longrightarrow Ph_3PO + \underset{(\mathbf{312})}{PhSON_3}\\
&\mathbf{312} + Ph_3P \longrightarrow Ph_3PO + \underset{(\mathbf{313})}{PhSN_3}\\
&\mathbf{313} \longrightarrow \underset{(\mathbf{314})}{PhSSPh} + N_2\\
&\mathbf{314} + Ph_3P \longrightarrow \underset{(\mathbf{315})}{PhSPh} + Ph_3PS
\end{aligned} \qquad (142)$$

sulphinyl azide (**312**) has been described recently[222]. This gave **314** and triphenylphosphine oxide in yields of 80 and 75%, respectively, on treatment with Ph_3P[222].

V. PHOTOLYTIC DECOMPOSITION

This is a convenient method of introducing the energy required to effect decomposition and has been used, for example, in electron spin resonance studies, where it was necessary to decompose the azide in a solid matrix at low temperatures[130–133].

A. Alkyl Azides

The vapour phase photolysis of methyl azide at low conversions gave nitrogen as the predominant gaseous product[97]. In addition, hydrogen (6–11%), methane (1%), ethane (1%), hydrogen cyanide (approximately equal to hydrogen yield), and a polymeric material of empirical formula CH_3N were obtained. The polymer contained some hexamethylenetetramine (3%). The quantum yields for N_2 production at 30° ranged from 1·7 to 2·3 depending on the wavelength used for irradiation, but were independent of intensity, temperature and pressure. Free radical scavengers inhibited the production of N_2, and an inert gas (CO_2) also inhibited N_2 production but to a much lesser extent. Based on the above observations, it was postulated that the only primary photolytic process of any import was the cleavage of the N—N bond to give nitrogen and methylnitrene, probably in a vibrationally excited state, since the excitation energy (91–112 kcal/mole) far exceeded the energy of the N—N bond (ca. 40 kcal/mole)[97]. A short chain, carried mainly by the CH_3N radical, was thought to be probable[97].

On the other hand, a study[223] of the solution phase photolysis of hydrazoic acid, methyl and ethyl azides showed that for hydrazoic acid and methyl azide, the quantum yields were concentration dependent, and reached a maximum value of 2·1 ± 0·2 at azide concentrations somewhat over 1M. At first glance, these results might appear to contradict those obtained in the gaseous phase[97]. It is probably very significant, however, that in most cases hydroxylic solvents were used. In particular, the concentration dependence studies were carried out in methanol. It is known that alcohols play important roles in determining the nature of the products of sulphonyl azides (*vide infra*). It is possible that the solvent could play a role in stabilizing the excited azide sufficiently (perhaps it is the hydrogen-bonded azide which undergoes excitation) so that its reaction with unexcited azide could become important and the quantum yield for nitrogen evolution consequently would be dependent on the azide concentration. The fact that the molar extinction coefficients of hydrazoic acid and methyl azide are concentration dependent[223] suggests that some interaction is occurring between the azide and methanol or between azide molecules themselves. It was concluded[223] that the unimolecular decomposition of hydrazoic acid and methyl azide to the corresponding nitrene was an insignificant process. The possibility of solvent association with azide affecting the course of the reaction was not considered as likely as that of the azide merely being excited and existing in this state long enough to encounter an unexcited azide molecule[223]. Since it was found that the quantum yield of nitrogen for ethyl azide photolysis in methanol was unity and independent of azide concentration, it was felt that a nitrene was formed in this case. It would seem then, that in this case association with solvent does not affect the lifetime of the excited azide sufficiently for bimolecular processes to be significant. This might be due to an inductive stabilization of the incipient nitrene (**316**) or to a stabilization of a developing partial positive charge on the α-carbon if hydrogen migrates as nitrogen departs (**317**), i.e. no free nitrene is formed. The concerted pro-

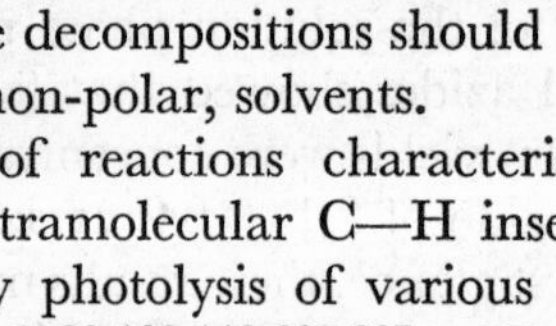

$$CH_3 \rightarrow CH_2 \rightarrow \bar{N}-\overset{+}{N}\equiv N \quad \textbf{(316)} \qquad \overset{H_3C\;\;H}{\underset{H}{C^{\delta+}}}\cdots N^{\delta-}\cdots N^{\delta+}\equiv N \quad \textbf{(317)}$$

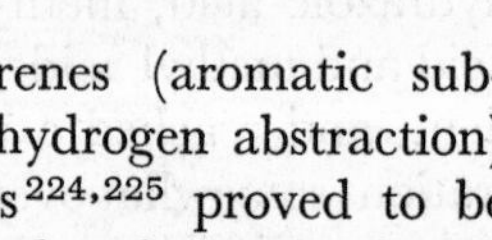

cess might explain the fact that no 'nitrene-adducts' were found[223] when the photolysis was carried out in methanol. In order to understand the processes involved in the photolysis of simple alkyl azides better, these decompositions should be carried out in non-hydroxylic, preferably non-polar, solvents.

Reports of reactions characteristic of nitrenes (aromatic substitution, intramolecular C—H insertion and hydrogen abstraction) obtained by photolysis of various alkyl azides[224,225] proved to be irreproducible[88,132,142,226,227]. Thus, in the photolysis of *n*-butyl, *n*-octyl and 4-phenyl-1-butyl azides, either in ether or in cyclohexane, the major product was that of hydrogen migration (**318**)[227]. Products due to hydrogen abstraction (**319**) (up to 15%), 1,2-alkyl migration (**320**) and solvent insertion (**321**) were obtained in minor amounts. In addition, in the case of 4-phenyl-1-butyl azide (**322**), a

$$RCH_2N_3 \xrightarrow[C_6H_{12}]{h\nu} \underset{\textbf{(318)}}{R-CH=NH} + \underset{\textbf{(319)}}{RCH_2NH_2} + \underset{\textbf{(320)}}{RN=CH_2} + \underset{\textbf{(321)}}{RCH_2NHC_6H_{11}} \qquad (143)$$

7–8% yield of 2-phenylpyrrolidine (**323**) was said to be formed, apparently by intramolecular C—H insertion of the corresponding nitrene[227]. More recently, however, it has been shown[228] that at low

$$H_2C(PhCH_2)-CH_2-CH_2N_3 \xrightarrow[C_6H_{12}]{h\nu} \text{2-phenylpyrrolidine} \quad (144)$$

(**322**) (**323**)

(5%) conversions, no intermolecular reactions (insertion, hydrogen abstraction) nor intramolecular insertions (**322** ↛ **323**) took place.

In agreement with the latter results, it was found that photolysis of tertiary alkyl azides in cyclohexane gave, in addition to small amounts of hydrocarbons (C—N fission, ca. 1%) products corresponding only to rearrangement[89,90,99]. No intramolecular cyclization, solvent insertion or hydrogen abstraction products were detected (< 0·2%).

$$R^1-C(R)_2-N_3 \xrightarrow{h\nu} (R^1)(R)C{=}N-R + (R)_2C{=}N-R^1 \quad (145)$$

Table 11 presents the phenyl to methyl migratory ratios derived from the decomposition of 2-phenyl-2-propyl azide and 1,1-diphenyl-

TABLE 11. Phenyl/methyl migratory ratios[a] from the decomposition of tertiary alkyl azides[99]

	2-Phenyl-2-propyl azide	1,1-Diphenylethyl azide
Thermal	4·05	2·36
Photochemical	0·96	2·18[b], 1[c]

[a] Corrected for statistical preference.
[b] Measured at high conversions. More recent value[c] obtained from measurements at low conversions.
[c] Revised value[229].

ethyl azide. Table 12 gives recent quantitative data on the migratory aptitudes of various groups, these being derived from thermolysis and photolysis of tertiary alkyl azides[88,90]. It is quite significant that in both thermal and photochemical decompositions, a methyl group migrates to an appreciable extent. These results may be compared with the observation that methyl groups do not migrate to a radical

TABLE 12. Migratory aptitudes[a] in the decomposition of tertiary alkyl azides[89, 90]:

	Ph/Me[b]	Ar/Me[c]	Ar/Ph[e]	$PhCH_2CH_2$/Me[f]
Thermal	1·9	1·9	1·1	—[g]
Photochemical	0·75	0·69[d]	0·44	0·89

[a] Corrected for statistical preference.
[b] From 2-phenyl-2-propyl azide[90].
[c] Ar = 2-biphenylyl, from 1-biphenyl-2-yl-1-methylethyl azide[89].
[d] Revised, previously reported as 0·43[89], and 0·34[78].
[e] Ar = 2-biphenylyl, from biphenyl-2-yldiphenylmethyl azide[89,90].
[f] From 2-azido-2-methyl-4-phenylbutane[90].
[g] No product of β-phenethyl migration observed, only methyl migration.

centre. For example, only phenyl migration occurred (57–63%) when a neophyl radical (**325**) was produced from the aldehyde (**324**) (equation 146)[230]. In formolysis of neophyl derivatives (**326**) (migra-

$$\underset{(\mathbf{324})}{Ph{-}C(Me)_2{-}CH_2CHO} \xrightarrow[130^\circ,\ -H^\cdot,\ -CO]{(t\text{-BuO})_2} \underset{(\mathbf{325})}{Ph{-}C(Me)_2{-}\dot{C}H_2} \longrightarrow PhCH_2\dot{C}Me_2 \xrightarrow{[H]} PhCH_2CH_2Me \qquad (146)$$

tion to a cationic centre), phenyl/methyl migratory ratios were close to 1000:1 or greater, depending on the leaving group X (equation 147)[231]. Nitrous acid deamination of 3-phenyl-2-butylamine (**327**)

$$\underset{(\mathbf{326})}{Ph{-}C(Me)_2{-}CH_2X} \longrightarrow Me{-}C(Me){-}CH_2{-}X^{\delta-}\ \text{(bridged phenonium, } \delta+\text{)} \longrightarrow \text{product} \qquad (147)$$

led to appreciable amounts of methyl and hydrogen migration in addition to phenyl migration, depending on which diastereoisomer was used[232]. For example, the (+)-*threo* isomer (**327**) gave 32% methyl (**328**), 24% phenyl (**329**) and 24% hydrogen (**330**) migration (equation 148). In contrast, 6% methyl, 68% phenyl and 20% hydrogen migration was observed with the *erythro* isomer. The results

$$\text{(327)} \xrightarrow[\text{HOAc}]{\text{HONO}} \underset{\textbf{(328)}}{Me_2CHCH(OAc)Ph} + \underset{\textbf{(329)}}{MePhCHCH(OAc)Me} + \underset{\textbf{(330)}}{MeCH_2C(OAc)PhMe} \qquad (148)$$

were in agreement with migration to a little-solvated, high energy, very short-lived open carbonium ion, in which the ground state conformations greatly influenced the product distribution (Hammond postulate)[233]. It has been found that 1,2-alkyl migrations are preferred in the oxygen diradicals (**332**) generated thermally from β-peroxylactones (**331**)[234–236]. It was felt that the rearrangement was

$$\textbf{(331)} \xrightarrow{\Delta} \textbf{(332)} \longrightarrow \text{transition states} \xrightarrow{-CO_2} PhCH_2COR \text{ or } RCH_2COPh \qquad (149)$$

synchronous with elimination of carbon dioxide, and that a 'push-pull' mechanism was operative, in which the β-scission and carbonyl formation were as important as the 'pulling' action of the leaving carbon dioxide[235]. Preferred conformations probably play an important role in determining migratory aptitudes here as well[236]. Finally, it has been found that the phenyl to methyl migratory ratio was about 10 to 1 in the *thermal* decomposition of 1-diazo-2-methyl-2-phenylpropane (**333**) (equation 150)[237].

$$\underset{\textbf{(333)}}{Ph{-}CMe_2{-}CH{=}N_2} \xrightarrow[-N_2]{\Delta} PhCH{=}CMe_2 + MeCH{=}CMePh + \text{other products} \qquad (150)$$

In the light of the above discussion, it seems clear that if an assisted elimination of nitrogen were occurring in the thermolysis of tertiary alkyl azide, the preference for aryl group migration would be much greater than is actually observed, by analogy with solvolyses[231]. Thus, in the thermolysis of alkyl azides, the migratory aptitudes are best rationalized in terms of a highly electrophilic singlet nitrene (**334**), which is then stabilized by migration of one of the groups on the α-carbon to the nitrogen.

$$R^1\text{—}C(R)(R^2)\text{—}N_3 \longrightarrow R^1\text{—}C(R)(R^2)\text{—}\ddot{\underset{\cdot\cdot}{N}} \longrightarrow \text{products} \qquad (151)$$

(**334**)

The situation as regards the photolysis of alkyl azides does not appear to be as straightforward. Evidence has been presented that triplet sensitized decomposition of *p*-X-phenyldiphenylmethyl azides led to migratory aptitudes close to unity[238], that were similar but not identical with those obtained in the direct photolyses[99] (Table 13). It was concluded, therefore, that direct photolysis involved triplet azide, and because of the lack of selectivity in the subsequent migrations, probably a triplet nitrene. There exists some doubt, however, as to whether or not the triplet azide was indeed formed in the sensitized photolyses (Table 13), since some of the sensitizers used (tri-

TABLE 13. Migratory aptitudes[a] Ar/Ph in the direct and triphenylene sensitized photolysis of $p\text{-}XC_6H_4(C_6H_5)_2CN_3$

X in $p\text{-}XC_6H_4$	Direct photolysis[b]	Sensitized photolysis[e]
NO_2	1·03[c]	1·07
Cl	1·26[d]	0·97
CH_3	—	0·89
OCH_3	1·16	1·11
$N(CH_3)_2$	—	1·08

[a] Corrected for statistical preference.
[b] Hexane solvent, room temperature[99].
[c] Hexane–ether (9:1, v/v).
[d] Average of two values from different amounts of conversions[99].
[e] Benzene solvent, triphenylene sensitizer[238].

phenylene, naphthalene and pyrene) and assumed to give rise to triplet energy transfer have recently been used as singlet sensitizers in the photolysis of aryl[239,240] and alkyl azides[241]. Sensitizers which undoubtedly give rise to triplet energy transfer (e.g. benzophenone, acetophenone, cyclopropyl phenyl ketone) do lead to decomposition of alkyl azides, but have not been used in the determination of migration aptitudes[238,242]. Thus, there is at present no information concerning migratory aptitudes derived from unambiguously triplet sensitized decompositions of alkyl azides.

In a later paper, Lewis and Saunders[229] observed that triplet quenchers (*cis*-piperylene, oxygen) failed to affect the course of the direct photolysis of alkyl azides, from which it was concluded that the photolysis proceeded via a singlet azide and singlet nitrene. This was further supported by the observation that hexyl azide acted as an efficient quencher of aromatic hydrocarbon fluorescence, and that this singlet sensitization of hexyl azide led to the decomposition of the azide with an efficiency similar to that of direct photolysis[241]. Thus, although triplet sensitization leads to decomposition of alkyl azides, it appears that direct photolysis proceeds by way of an excited singlet azide without intersystem crossing to the triplet.

Evidence has been presented to show that both triplet[242] and singlet[241] sensitization could result from a non-classical energy transfer, perhaps through a vertical excitation of a vibrationally excited (bent) azide ground state. Using the technique of sensitization, it was estimated that the lowest alkyl azide triplet energy lies at 77–78 kcal/mole[242], and the first excited singlet energy at 91–92 kcal/mole[241]. The latter is in agreement with the ultraviolet absorption spectrum of alkyl azides which is quite broad and has a maximum corresponding in energy to 99 kcal/mole[243].

It seems reasonable to assume that the initial absorption of a photon of light leads to an electronically excited singlet azide (**335**) which, *a priori*, may react in several ways. It could decompose in a concerted or stepwise process to give product; alternatively, it might undergo intersystem crossing (ISC) to the triplet species (**336**) (equation 152). This could then decompose to product, again in a concerted or a step-

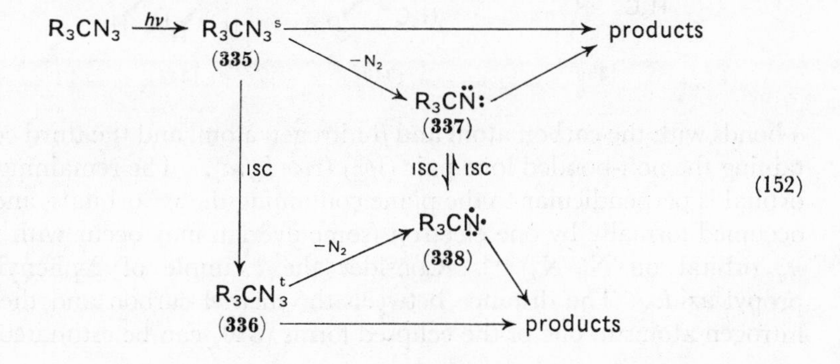

(152)

wise process. In the presence of a triplet sensitizer, triplet azide (**336**) would be produced directly. It is also possible that in the stepwise process intersystem crossing of the nitrenes (**337**) and (**338**) might occur, but the conversion **338** → **337** would be rather unlikely.

The photolytic migratory aptitudes (Tables 11 and 12 and Moriarity and Reardon's[228] unpublished results) are difficult to rationalize on the basis of a nitrene intermediate (**337** or **338**). The possibility was considered that, if a nitrene were formed, migration could occur at a diffusion-controlled rate i.e. $v \propto M^{-1/2}$, where v = velocity of migration and M = molecular weight of the migrating group. While this would explain the observed 2-biphenyl/methyl migratory ratio, no correlation was found between the observed and calculated migratory ratios in the other cases. Since the nitrene nitrogen is cylindrically symmetrical, whether it is in the singlet or triplet state, no steric preference for migration would be expected. Certainly there appears to be little evidence of intramolecular acceleration by an aryl group (*vide supra*), since methyl migration is preferred relative to phenyl and 2-biphenylyl migration in the photolyses studied so far.

Moriarity[228] has suggested that ground state conformation is important in determining migratory aptitudes in alkyl azide photolysis. We can extend this concept to develop a theory to explain the results obtained so far[90]. It is first necessary to consider the likely non-bonded interactions involved in the various possible conformations of the ground state of alkyl azides. The structure of methyl azide has been determined to be as in **339**[244]. The α-nitrogen atom is considered to be sp^2 hybridized, with two of the sp^2 hybrid orbitals forming

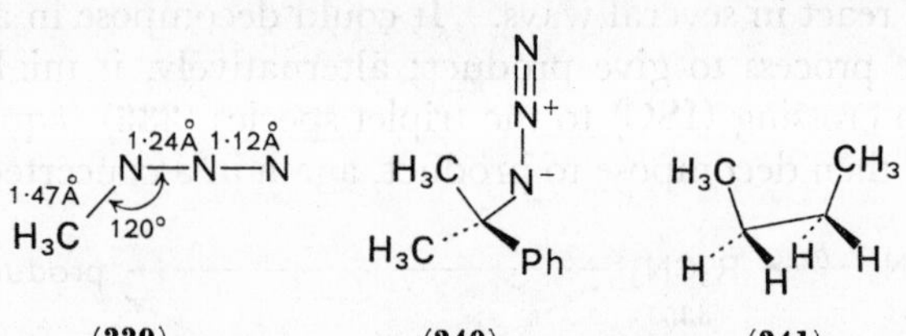

1·24Å 1·12Å
N—N—N
1·47Å
H₃C 120°

(**339**) (**340**) (**341**)

σ bonds with the carbon atom and β-nitrogen atom, and the third containing the non-bonded lone pair (sp_x^2) (*vide infra*). The remaining p_y orbital is perpendicular to the plane containing the sp^2 orbitals, and is occupied formally by one electron (some overlap may occur with the π_y orbital on N_β–N_α)[243]. Consider the example of 2-phenyl-2-propyl azide. The distance between the methyl carbon and the β-nitrogen atoms in one of the eclipsed forms (**340**) can be estimated to be about 2·7 Å (assuming normal bond lengths and angles). In comparison, the distance between the methyl carbon atoms in the eclipsed form of butane (**341**) is estimated to be 2·54 Å. Thus, although the C—N_α–N_β is greater than the C_2—C_3—C_4 angle in butane, the C—N_α bond is shorter than a C—C bond, so that the Me—N_β distance is still quite short, and appreciable non-bonded interactions might be expected in the ground state. It is further assumed that an N_2–Ph interaction would be greater than an N_2–Me repulsion, and that these in turn are greater than the corresponding interactions between these substituents and the N_α lone pair (sp_x^2). On that basis, the Newman projections **342–345** of some of the possible conformations of the ground state of 2-phenyl-2-propyl azide indicate that conformations **342** and **343** should be lower in energy than those in which the large phenyl group is closer to the N_2 group (**344** and **345**), with the skew (**343**) possibly the more stable of the two.

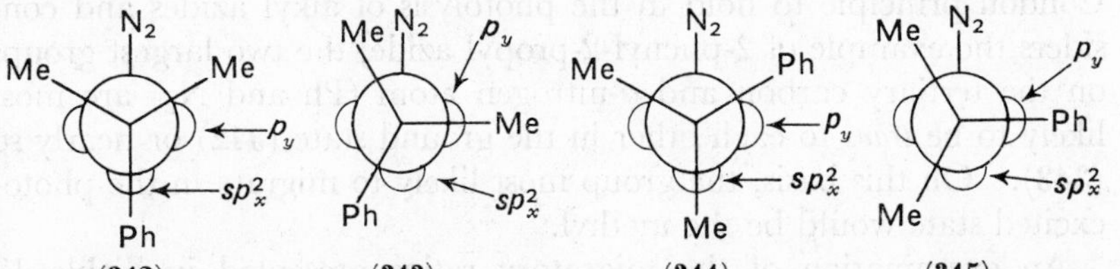

(**342**) (**343**) (**344**) (**345**)

In contrast to thermolysis, photolysis leads to an electronically excited state in which the α-nitrogen atom is probably electron-deficient. Clossen and Gray[243] have described the electronic transition normally involved in photolysis (287 nm) as a $\pi_y \rightarrow \pi_x^*$ transition [**346** → **347**] (equation 153). This would leave the p_y orbital on the

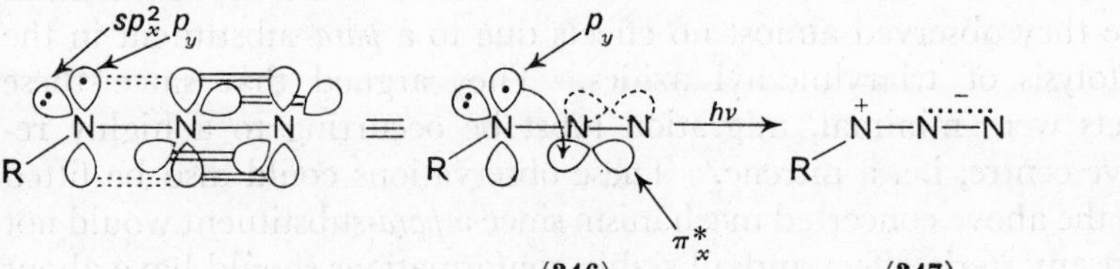

(153)

(**346**) (**347**)

α-nitrogen atom electron-deficient. Although the situation is not completely analogous, molecular orbital calculation on phenyl azide do show that in both the ππ* and nπ* excited states, the α-nitrogen atom is electron deficient relative to the ground state[245]. In the $\pi_y\pi_x^*$ state, a concerted rearrangement and elimination of nitrogen could occur leading to an excited-state imine (**348**). One consequence of this is that a concerted migration–elimination would not involve a backside (*trans*) attack (which would require migration to a filled sp^2 orbital), but rather the migrating group and departing nitrogen molecule would be in mutually perpendicular planes (equation 154). Thus, the relevant orbitals at the migration origin and the p_y orbital (migration terminus) would be coplanar. If one assumes the Franck–

(**348**) (154)

Condon principle to hold in the photolysis of alkyl azides and considers the example of 2-phenyl-2-propyl azide, the two largest groups on the tertiary carbon and α-nitrogen atom (Ph and N_2) are most likely to be *trans* to each other in the ground state (**342**) or nearly so (**343**). On this basis, the group most likely to migrate in the photo-excited state would be the methyl.

An examination of the migratory ratios presented in Table 12 indicates that steric effects are important in migratory aptitudes in photochemical decompositions. In every case, the above theory predicted correctly which group would migrate most readily. The observations that methyl migration is favoured over phenethyl and aryl group (phenyl and 2-biphenylyl) migration, and that a 2-biphenylyl group migrates less readily than phenyl are thus accounted for. Saunders and co-workers[99,229] favoured a nitrene mechanism since they observed almost no effects due to a *para*-substituent in the photolysis of triarylmethyl azides. They argued that since these effects were minimal, migration must be occurring to a highly reactive centre, i.e. a nitrene. These observations could also be fitted into the above concerted mechanism since a *para*-substituent would not exert any steric effect and all stable conformations should have about the same energies. Being cylindrically symmetrical, a nitrene should not lead to the observed selectivity (Table 12). Finally, the above theory accounts quite nicely for the observations that aromatic substitution, aliphatic C—H insertion, and hydrogen abstraction have not been unambiguously shown to take place in the photolysis of alkyl azides[89,90,228].

The concerted mechanism may also explain why attempts to observe triplet alkylnitrenes by direct irradiation of alkyl azides (e.g. n-PrN_3, 2-$OctN_3$ and $C_6H_{11}N_3$) at 77°K were unsuccessful, although weak signals were observed at 4°K[132,246]. Perhaps at 77°K a concerted decomposition occurs in which no nitrene is formed on expulsion of nitrogen, but at 4°K there may be insufficient molecular bending motion to permit migration. The singlet azide could then undergo intersystem crossing to the triplet azide and then the triplet nitrene. Triplet sensitized (benzophenone) photolysis of alkyl mono- and diazides (*vide infra*) did give the triplet nitrenes at 77°K, which were stable for days at this temperature[246].

Very recently, evidence has been presented to show that photolysis of a highly fluorinated alkyl azide, 2*H*-hexafluoropropyl azide, can lead to a nitrene. In cyclohexane, cyclohexene or methylcyclohexane the insertion products were formed.

In contrast to thermolysis, photolysis of tertiary alkyl azides leads only to very small amounts of hydrocarbon products (C—N fission)[89,90]. For example, irradiation of 2-phenyl-2-propyl azide (**341**) gave α-methylstyrene in 1·5% yield[90]. Lewis and Saunders[100] obtained evidence that photolysis of triphenylmethyl azide (**349**) led to the formation of some triphenylmethyl radicals (**350**). Using

$$\underset{(\mathbf{349})}{Ph_3C{-}N{=}\overset{+}{N}{=}\overset{-}{N}^*} \xrightarrow{h\nu} \underset{(\mathbf{350})}{Ph_3C^{\bullet}} + N_3^{\bullet} \longrightarrow \qquad (155)$$

$$Ph_3C{-}N{=}\overset{+}{N}{=}\overset{-}{N}^* + Ph_3CN{-}N^*{=}\overset{+}{N}{=}\overset{-}{N}$$

N* stands for ^{15}N

^{15}N-labelled **349** the e.s.r. spectrum of **350** was observed and, from the amount of ^{15}N-scrambling, it was estimated that C—N fission occurred to the extent of about 21%.

Photolysis of *gem*-alkyldiazides has been studied[246–248]. E.s.r. experiments at 77°K showed that photosensitized decomposition of **351** gave a triplet nitrene (**352**)[246]. Prolonged irradiation gave the carbene (**353**). The results of direct photolysis of **351** at room

$$\underset{(\mathbf{351})}{Ph_2C(N_3)_2} \xrightarrow{h\nu} \underset{(\mathbf{352})}{Ph_2C(\ddot{N}^{\bullet})N_3} \longrightarrow Ph_2C{=}N_2 \longrightarrow \underset{(\mathbf{353})}{Ph_2C\!:} \qquad (156)$$

temperature illustrate the danger of comparing photosensitized to direct photolyses on the one hand, and low temperature solid matrix studies to ambient temperature solution studies on the other. It was found that **351** gave three products: **354, 355** and **356** on direct photolysis in benzene (equation 157)[248]. No products derived from diphenylcarbene (**353**) were observed.

$$\mathbf{351} \xrightarrow{h\nu} \text{Ph(N}_3\text{)C=NPh} \xrightarrow{-N_2} \text{Ph(N)C=NPh} \longrightarrow (\text{PhN=C=NPh})_3 \; (\mathbf{356}) \qquad (157)$$

Ph(N₃)C=NPh → (**354**) (1,5-diphenyltetrazole) $\xrightarrow[-N_2]{h\nu}$ (**355**) (2-phenylbenzimidazole); Ph(N)C=NPh → (**355**)

B. Aryl Azides

Photolysis of phenyl and *o*-trifluoromethylphenyl azide in solid matrices led to the triplet nitrene as detected by e.s.r.[130]. The actual processes involved in aromatic azide photolyses have been the subject of much study[85,245,249–251]. The electronic spectra of the nitrenes were measured by photolysis of a number of aryl azides in organic matrices at 77°K[249]. These species were stable indefinitely at this temperature, no change being observed in the spectra for hours. The photolysis of diazides at 77°K, whether conjugated (e.g. *p*-diazidobenzene) or not [e.g. bis(*p*-azidophenyl)methane] proceeded in two distinct steps to the dinitrene. The second step was about two to three times as efficient as the first ($\phi_2/\phi_1 \sim 2$–3)[250].

$$\text{Diazide} \xrightarrow[\phi_1]{h\nu} \text{azidonitrene} \xrightarrow[\phi_2]{h\nu} \text{dinitrene} \qquad (158)$$

In order to determine whether the photolysis of aryl azides in solid matrices at 77°K and the solution phase photolysis at 25° involved the same processes, a series of eight substituted aryl azides were studied under both sets of conditions. The quantum yields ranged from 0·37 to 1·00 for the various azides, and it was found that without exception the quantum yields obtained from solid matrix and solution photolyses were identical within experimental error[245]. For example, 3-azidobiphenyl in hexane at 25°C gave a quantum yield of 0·37 ± 0·05 and of 0·36 ± 0·05 in methylcyclohexane isopentane at 77°K.

Aromatic azide photolytic decomposition is thought to be a special case of the more general process known as aromatic side chain photolysis[245]. The excited states are those of the parent hydrocarbon and are not due to an electronic transition in the azido group but rather are $\pi \rightarrow \pi^*$ in the aromatic nucleus. In addition there is a low intensity long wavelength band due to transition of a non-bonding electron localized on nitrogen to a π^* orbital. Reiser and Marley[245] felt that the best explanation of the photolysis was that an absorption in the main band in the aromatic azide populates a $\pi\pi^*$ excited singlet state in a vibrational level not sufficient to induce bond dissociation. The excited molecule would then undergo rapid internal conversion to lower excited states and finally to a vibrationally excited ground state in which transmission of vibrational momentum and energy from the aromatic skeleton to the azide side chain would lead to dissociation of the N—N bond to give the nitrene.

One of the problems that has received some attention concerns the nature of the electronic state of the nitrene produced on photolysis. Reiser and co-workers[245] approached this question by comparing the photolysis of 2-azidobiphenyl (**356**) in a solid matrix and in solution. It was found that photolysis gave the nitrene (**357**) as identified by its electronic absorption spectrum, and that the reaction could be followed by ultraviolet spectroscopy, monitoring azide disappearance, nitrene appearance and disappearance, and finally, the

N_3 (**356**) $\xrightarrow[77°K,\ -N_2]{h\nu}$ N (**357**) $\xrightarrow{h\nu}$ (**358**) N–H (159)

appearance of carbazole (**358**). It was necessary to irradiate **357** in order to obtain the spectrum of **358**. In ethanol at 25° the quantum yield for nitrene formation was 0·44 ± 0·02, using the quantum yield of carbazole as a monitor, as compared with a quantum yield of 0·43 ± 0·05 in a solid matrix at 77°K. Thus, it appeared that the same intermediate was involved at both temperatures. The question was still unanswered, however, as to whether the intermediate was a singlet or a triplet. E.s.r. evidence showed the ground state to be the triplet[130], but the fact still remained that the nitrene had to be irradiated to form **358**. The quantum yield for the second step was low (0·01–0·02). One possible interpretation was that a ground state triplet nitrene was being excited to a higher triplet state which, by internal conversion, could supply additional energy in a particular vibrational mode required for cyclization. This would complete with rapid redistribution of energy within the molecule, and hence cause a low quantum yield. It was pointed out, however, that an excited singlet nitrene could be populated via the excited triplet, and this could give rise to the cyclization. The work of Swenton and co-workers[239,240,252] throws much light on this subject. Direct photolysis of 2-azidobiphenyl (**356**) in benzene gives rise to carbazole (**358**) and the corresponding azo compound (**359**) in yields of 68–71% and 8–11%, respectively[239]. Triplet sensitizers (acetophenone, benzophenone)

$$\mathbf{356} \xrightarrow[C_6H_6]{h\nu} \mathbf{358} + \underset{(\mathbf{359})}{\text{ArN=NAr}} \qquad (160)$$

Ar = 2-biphenylyl

drastically reduced the yield of **358** (<2%) and increased the yield of **359** (ca. 40%). On the other hand, singlet sensitizers such as triphenylene and naphthalene gave about the same product distribution as direct irradiation. With pyrene as the sensitizer carbazole was formed in 95% yield and the azo compound in less than 1%. This marked decrease in yield of **359** and increase in **358** could be evidence that the production of triplet nitrene occurs by way of the triplet azide and not from the singlet nitrene by intersystem crossing. Ground state pyrene is acting as a quencher to azo compound formation (i.e. a triplet quencher), as well as a singlet sensitizer[240]. To explain the increased production of carbazole and decreased azo compound production one can postulate that the triplet nitrene (T_0) arises predominantly via pathway *a* (equation 161) so that this route could be blocked by a triplet quencher[240]. If the triplet nitrene were formed

$$\begin{array}{ccccc} \text{Azide } (S_0) & \xrightarrow{h\nu} & \text{Azide } (S_1) & \xrightarrow{a} & \text{Azide } (T_0) \\ & & b \big\downarrow -N_2 & & -N_2 \big\downarrow d \\ & & \text{Nitrene } (S_0) & \xrightarrow{c} & \text{Nitrene } (T_0) \end{array} \qquad (161)$$

by pathway *b*, *c*, a triplet quencher would not have the effect of recycling azide to the singlet nitrene, and thus increasing the yield of carbazole. The evidence thus indicates that photolysis of aryl azides leads to aryl nitrenes; the singlet species is responsible for aromatic substitution (cyclization) and the triplet gives rise to azo compounds.

By using flash photolysis, in which the rates of disappearance of nitrene and formation of product were followed by ultraviolet spectroscopy, the rates of three basic processes involved in aryl azide photolyses were measured[85]. It was found that flash photolysis of 1-azidonaphthalene with high intensity light (with which high concentrations of nitrenes could be realized) gave high yields of 1,1′-azonaphthalene (dimerization or reaction of nitrene with unreacted azide) at a diffusion-controlled rate. When 1-azidoanthracene was photolysed with a flash of reduced intensity such that about 5% nitrene appeared, it was found that 10% of the azide had disappeared by the end of the process. This must mean that the azide was being consumed by the nitrene to give the azo compound and nitrogen. It was calculated that this bimolecular process also occurs at a diffusion-controlled rate. Flash photolysis of 4,4′-diazidobiphenyl in ethanol gave an absolute bimolecular rate constant for hydrogen abstraction, $k_2 = 1{\cdot}3 \times 10^{-2}$ l mole^{-1} sec^{-1}, referred to the concentration of secondary C—H bonds in the reaction mixture. No evidence was presented, however, that the nitrene was indeed reacting with the secondary C—H and not the O—H[85].

Photolysis of 4-phenyl- and 4-methoxyphenyl azides gave good yields of azo compounds, whereas phenyl, 4-nitrophenyl and 4-chlorophenyl azide gave undefined products[253]. Photolysis of 4-methoxyphenyl azide in benzene gave only 18% yield of azo compound (4-azidobiphenyl gave 81%), but in solvents such as tetrahydrofuran and methylsulphide, the yields of the azo compound were 80–90%. It was felt that complexing of the nitrene with solvent lone pairs of electrons stabilized this species and enhanced the probability of azo compound formation[253].

The photolysis of substituted phenyl azides in the presence of nucleophiles has been studied. Irradiation of phenyl azide in diethylamine resulted in a 34% yield of 2-diethylamino-3*H*-azepine (**360**), whereas in liquid ammonia, the corresponding aminoazepine (25%) was formed, and in the presence of aniline in triethylamine the anilino-azepine (**154**, 2%) 'dibenzamil'[104], (see section III.B.2) was obtained. These products arose via intermediate **361**, which could also be trapped with hydrogen sulphide, albeit in low yields (5%), to give 1,2-dihydro-2-thienoketo-3*H*-azepine (**362**)[106]. Similar results were

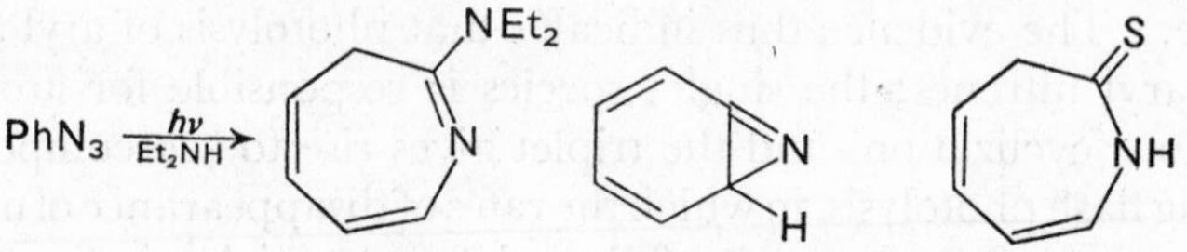

$$PhN_3 \xrightarrow[Et_2NH]{h\nu} \text{(360)} \quad \text{(361)} \quad \text{(362)} \qquad (162)$$

reported for *p*-methoxy- and *p*-chlorophenyl azides[254]. On the other hand, photolysis of *p*-cyanophenyl azide (**363**) in dimethylamine gave a 70% yield of 1,1-dimethyl-2-(4-cyanophenyl)hydrazine (**364**) and *p*-cyanoaniline (**365**) (5%)[254]. When a triplet sensitizer (xanthen-9-

$$\underset{(\mathbf{363})}{p\text{-}CNC_6H_4N_3} \xrightarrow[Me_2NH]{h\nu} \underset{(\mathbf{364})}{p\text{-}CNC_6H_4NHNMe_2} + \underset{(\mathbf{365})}{p\text{-}CNC_6H_4NH_2} \qquad (163)$$

one) was used the relative product yields were reversed. Thus, it may be inferred that the singlet nitrene gives **364** and the triplet **365**. It was felt that two factors might enhance an intermolecular route at the expense of the intramolecular (azepine) pathway; the cyano group could decrease the rate of singlet nitrene ring closure to the cyano derivative of **361**; also, because of the increase in electrophilic character of the nitrene, the rate of reaction of nitrene with amine could be increased.

Photolysis of 2,2′-diazidobiphenyl (**366**) in *n*-heptane led to traces of benzo[*c*]cinnoline (**368**) and 4-azidocarbazole (**369**) (50%)[255]. It was concluded that the C—H bond in the *o*′-position was a better nitrenophile than the azido group in the same position. This is interesting in light of the observation[85] that combination of a nitrene and an azido group occurs at diffusion controlled rates. A possible explanation is that the nitrene (**367**) is formed in the singlet state, whereas azo compound formation is due to the triplet nitrene. It would be of interest to determine whether a triplet sensitizer would increase the yields of **368**.

Ferrocenyl azide (**370**) has been photolysed in various solvents in the presence and absence of oxygen[148,256]. The results show that two

$$\xrightarrow[n\text{-}C_7H_{16}]{h\nu} \qquad (164)$$

N_3 N_3 (**366**) → [N N_3 (**367**)] → N=N (**368**) + N_3 N H (**369**)

processes are involved in the photolysis (and thermolysis, see section III.B.6) of this azide: the major pathway involves N—N cleavage to give the nitrene (**372**); the minor process gives rise to C—N cleavage

$$\underset{(\mathbf{370})}{FcN_3} \xrightarrow[-N_3\cdot]{h\nu \text{ or } \Delta} \underset{(\mathbf{371})}{Fc\cdot} \xrightarrow{SH} FcH + FcS$$

$$\underset{(\mathbf{370})}{FcN_3} \xrightarrow[-N_2]{h\nu \text{ or } \Delta} \underset{(\mathbf{372})}{FcN} \xrightarrow{SH} FcNH_2 + FcNHS \qquad (165)$$

$$\underset{(\mathbf{372})}{FcN} \longrightarrow FcN{=\!=}NFc$$

$$\underset{(\mathbf{372})}{FcN} \xrightarrow{O_2} FcNO_2$$

Fc = Ferrocenyl
SH = Solvent

to give ferrocenyl (**371**) and azide radicals (equation 165). Table 14 summarizes the results of the decomposition of **370** in benzene, cyclohexane and cyclohexene in the presence and absence of oxygen[256]. Of particular interest is the formation of nitroferrocene (ca. 21%) when **370** was photolysed in the presence of oxygen either in benzene or in cyclohexane. This product is, presumably, the result of the reaction of **372** (in the triplet state) with oxygen, but could conceivably arise also from triplet azide and oxygen followed by loss of nitrogen. Ferrocenyl amine does not give nitroferrocene under these conditions. This is probably the best preparation available for nitroferrocene[256].

Photolysis of 4-azidopyridine-1-oxide in acetone under nitrogen gave a 37% yield of 4,4′-azopyridine-1,1′-dioxide. When the photolysis was carried out in the presence of oxygen 4,4′-azoxypyridine-1,1′-dioxide was obtained in 27% yield.

TABLE 14. Decomposition of ferrocenyl azide in various solvents in the presence and absence of oxygen[256]

Solvent	Δ or $h\nu$	FcN =NFc	$FcNH_2$	$FcNO_2$	FcPh	FcH	FcNHS[a]	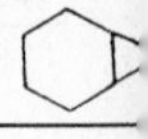
C_6H_6	Δ	17·8	16·9	—	1·9	13·6	—	—
	$h\nu$	16·4	16·1	—	2·8	34·9	—	—
C_6H_6/O_2	Δ	3·5	—	4·1	—	3·4	—	—
	$h\nu$	4·5	—	21·2	—	4·8	—	—
C_6H_{12}	Δ	20·5	—	—	—	7·3	—	—
	$h\nu$	42·2	—	—	—	9·0	—	—
C_6H_{12}/O_2	Δ	5·4	—	9·4	—	Trace	—	—
	$h\nu$	6·2	—	21·2	—	3·3	—	—
C_6H_{10}	Δ	4·0	35·2	—	—	1·4	2·4	4·8
	$h\nu$	6·2	8·2	—	—	4·4	2·9	4·4
C_6H_{10}/O_2	Δ	2·7	—	1·3	—	Trace	1·3	1·9
	$h\nu$	4·8	—	8·2	—	Trace	2·2	3·4

[a] $S=C_6H_9$, allylic insertion.

A study of the effect of temperature on direct and sensitized photolysis of *o*-propylphenyl azide (**373**) was carried out[257]. Direct photolysis of **373** in *i*-octane at room temperature gave 2-methyl-2,3-dihydroindole (**374**) in trace quantities along with several other products. When xanthen-9-one was used as a triplet sensitizer at room temperature, **374** was obtained in 36% yield. When **373** was photolysed directly at 99°, however, **374** was obtained in 49% yield

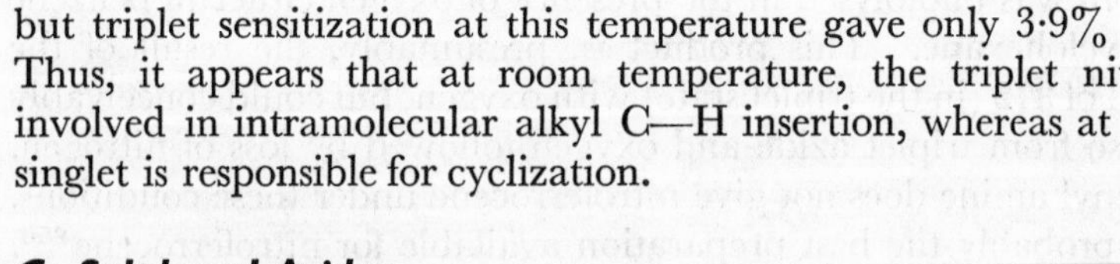

N_3 (**373**) $\xrightarrow[i\text{-octane}]{h\nu}$ (**374**) Me H N H + other products (166)

but triplet sensitization at this temperature gave only 3·9% of **374**. Thus, it appears that at room temperature, the triplet nitrene is involved in intramolecular alkyl C—H insertion, whereas at 99° the singlet is responsible for cyclization.

C. Sulphonyl Azides

Unlike the photolysis of alkyl and aryl azides, photolytic decomposition of sulphonyl azides in solvents such as cyclohexane, cyclohexene, benzyl alcohol, pyridine and thiophene gives insoluble polymeric materials[181,258,259]. For example, irradiation of methanesulphonyl azide in benzene gave a yellow amorphous material which did not melt below 290° and formed a gum with boiling ethanol[259]. When deposition of polymer on the sides of the quartz flask was prevented, a very small amount of nitrene was formed which reacted with the benzene.

Photolysis of sulphonyl azides in the presence of nucleophilic trapping agents such as dimethyl sulphoxide or dimethyl sulphide gave imine derivatives. For example, irradiation of *p*-toluenesulphonyl azide in dimethyl sulphide gave a 54% yield of *N*-(*p*-toluenesulphonyl) dimethylsulphimine (**375**)[181]. The use of dimethyl sulphoxide resulted in lower yields of the trapped nitrene (**377**) (13–32%), and

$$p\text{-MeC}_6\text{H}_4\text{SO}_2\text{N}_3 + \text{Me}_2\text{S} \xrightarrow[-\text{N}_2]{h\nu} \underset{(\mathbf{375})}{p\text{-MeC}_6\text{H}_4\text{SO}_2\text{N}{=}\text{SMe}_2} \qquad (167)$$

a concerted mechanism, perhaps involving an intermediate such as **376** cannot be ruled out. It is probable that a nitrene was involved in the formation of **375**.

$$\text{RSO}_2\text{N}_3 + \text{Me}_2\text{SO} \longrightarrow \underset{(\mathbf{376})}{\text{RSO}_2\text{N}\langle\text{N=N}\rangle\text{O--SMe}_2} \xrightarrow[-\text{N}_2]{h\nu} \underset{(\mathbf{377})}{\text{RSO}_2\text{N}{=}\overset{\text{O}}{\overset{\|}{\text{S}}}\text{Me}_2} \qquad (168)$$

Photolysis of *p*-toluenesulphonyl azide in methanol gave, as the main product, *N*-(*p*-toluenesulphonyl)-*O*-methylhydroxylamine (**378**) (44%)[181]. Irradiation of benzenesulphonyl azide in methanol gave

$$p\text{-MeC}_6\text{H}_4\text{SO}_2\text{N}_3 \xrightarrow[\text{MeOH},\ -\text{N}_2]{h\nu} \underset{(\mathbf{378})}{p\text{-MeC}_6\text{H}_4\text{SO}_2\text{NHOMe}} \qquad (169)$$

a product (**380**) (28%) which corresponded to a Curtius rearrangement of the starting azide, in addition to the hydroxylamine derivative (**379**) and benzenesulphonamide[260]. (Five other minor products

$$\text{PhSO}_2\text{N}_3 \xrightarrow[\text{MeOH},\ -\text{N}_2]{h\nu} \text{PhSO}_2\text{NH}_2 + \underset{(\mathbf{379})}{\text{PhSO}_2\text{NHOMe}} + \underset{(\mathbf{380})}{\text{PhNHSO}_2\text{OMe}} \qquad (170)$$

were obtained). It was felt that hydrogen-bonding with the solvent was important and a free nitrene might not be involved. An intermediate or transition state such as **381** was proposed. Although Horner and Christmann[181] did not report any rearrangement in the study of *p*-toluenesulphonyl azide, no search was made for this product[258].

$$PhSO_2N_3 \xrightarrow[MeOH]{h\nu} \text{(381)} \xrightarrow{-N_2} O{=}S({\cdots}HOMe){=}NPh \longrightarrow \mathbf{380} \quad (171)$$

Direct or sensitized photolysis of methanesulphonyl azide in isopropanol gave high yields of methanesulphonamide, acetone and nitrogen[261]. Direct photolysis showed a marked induction period

$$MeSO_2N_3 \xrightarrow[Me_2CHOH]{h\nu} MeSO_2NH_2 + Me_2CO + N_2 \quad (172)$$

which could be completely eliminated by the addition of a small amount of acetone to the reaction mixture. The average quantum yield over the initial 20% of the reaction ranged from 20 to 75 and the values calculated from the instantaneous rate at 20% completion ranged from 110 to 150. A radical chain mechanism which is accelerated by triplet sensitization was proposed. The marked induction period in direct photolysis was due to the initial lack of sensitizer (acetone). The benzophenone-sensitized reaction probably had two propagation sequences one of which involved benzophenone and hence was absent in the direct photolysis.

In contrast to the problems encountered on photolysis of alkyl and arylsulphonyl azides, it has been found that ferrocenylsulphonyl azide (**382**) is decomposed smoothly by 350 nm light in cyclohexane or benzene to give ferrocene, ferrocenylsulphonamide (**383**) and the

$$\mathbf{382} \xrightarrow[350\text{ mm}]{h\nu} FcH + FcSO_2NH_2\ (\mathbf{383}) + \mathbf{384} \quad (173)$$

bridged [2]-ferrocenophanethiazine-1,1-dioxide (**384**)[262]. The yield of **384** varied with the nature of the solvent, being 67% in benzene, 13·3% in cyclohexane and 0% in dimethyl sulphoxide or dimethyl sulphoxide–benzene[263]. The cyclization appeared to be a singlet reaction since the yield of bridged product (**384**) in benzene was essentially unaffected by oxygen or hydroquinones[263].

VI. TRANSITION METAL-CATALYSED DECOMPOSITIONS

A. *Alkyl Azides*

The decomposition of methyl azide in the presence of di-iron nonacarbonyl has been described briefly[264]. The principal product of this decomposition was the complex (**385**), in addition to a 20% yield of

$$MeN_3 + Fe_2(CO)_9 \longrightarrow (CO)_3Fe\cdots Fe(CO)_3 \text{ bridged by } MeN{-}C(=O){-}NMe \ (\mathbf{385}) + Me{-}N{=}N{-}N{=}N{-}Me \rightarrow Fe \longleftrightarrow Me{-}N{-}N{=}N{-}N{-}Me \rightarrow Fe \ (\mathbf{386}) \qquad (174)$$

a stable, volatile solid, which was identified as a metal-stabilized tetrazadiene (**386**). Numerous other minor products were obtained.

B. *Aryl Azides*

Phenyl azide and di-iron nonacarbonyl react rapidly in benzene at room temperature[264] (as compared with thermolysis of PhN_3 alone which occurs at temperatures of 140–170°[4]). The principal product was the orange phenyl nitrene-complex (**387**) which decomposed spontaneously in solution to give the urea-based complex (**388**). Also obtained in low yield was the orange complex (**389**). The yield of azobenzene was reported to be negligible. On the other hand, when the decomposition was carried out in benzene under reflux in the presence of $Fe_3(CO)_{12}$, a significant amount of azobenzene was found[264].

Campbell and Rees[265] investigated the possibility of metal-catalysed decomposition of aryl azides being a useful method for cyclizations. It was found that although decomposition takes place readily

$$PhN_3 + Fe_2(CO)_9 \xrightarrow{C_6H_6} (CO)_3Fe(PhN{-}NPh)Fe(CO)_3\ (\mathbf{387}) + (CO)_3Fe(PhNH)(PhNH)Fe(CO)_3\ (\mathbf{389}) \quad (175)$$

$$\mathbf{387} \longrightarrow (CO)_3Fe[PhN{-}C(O){-}NPh]Fe(CO)_3\ (\mathbf{388})$$

under very mild conditions (benzene solution at 20° in the dark), a variety of products were obtained. When 2-azidobiphenyl was decomposed in the presence of di-iron nonacarbonyl, carbazole (**389**) was formed in only 1% yield. Other products were *N*,*N′*-bis(2-biphenylyl)urea (**390**) (34%), 2-aminobiphenyl (**391**) (10%) and the dimeric arylnitrene bis(tricarbonyl) iron complex (**392**) (5%). It was concluded that although the decomposition of aryl azides is

$$\text{2-azidobiphenyl} \xrightarrow{Fe_2(CO)_9} \text{carbazole}\ (\mathbf{389}) + (ArNH)_2CO\ (\mathbf{390}) + ArNH_2\ (\mathbf{391}) + (CO)_3Fe(ArN{-}NAr)Fe(CO)_3\ (\mathbf{392}) \quad (176)$$

Ar = 2-biphenylyl

markedly catalysed by di-iron nonacarbonyl, the reactions are complex and do not offer an attractive alternative route to nitrene cyclization products.

The decomposition of 2-azido-4,6-dimethylpyrimidine (**186**) (in equilibrium with the tetrazolo compound **185**) in the presence of copper acetylacetonate in cyclohexane at 140° gave the hydrogen abstraction product (**187**) (46%) and C—H insertion product (**188**) (8·5%)[266]. The same products were obtained in essentially the same yields by thermolysis of **186** at 185°[140], (see section III.B.5). It was felt that the reaction proceeded via a copper-nitrene complex. The copper-catalysed decomposition of **186** and also 2-azidopyridine (in

(185) ⇌ (186) $\xrightarrow[\text{Cu(acac)}_2,\ -N_2]{C_6H_{12}}$ (187) + (188) (177)

equilibrium with the corresponding tetrazolo tautomer) was investigated in the presence of various other substrates[266]. For example **186** gave a mixture of the *trans*- (**393**) and *cis*-aziridine (**394**) in yields of 40% and 3%, respectively, with *trans*-stilbene at 120°. The same products were obtained at 160° without copper via the triazoline intermediate. When 2-pyridyl azide (**396**) was heated in the presence of

186 + PhCH=CHPh $\xrightarrow[\text{8 hr, Cu}]{120°}$ (393) + (394) (178)

copper powder with benzonitrile at 120°, 2-phenyl-*s*-triazolo[1,5-*a*]-pyridine (**398**) was obtained in 62% yield, the nitrenoid species (**397**) behaving as a 1,3-dipole.

(395) ⇌ (396) $\xrightarrow[\text{Cu}\ -N_2]{120°}$ [(397)] $\xrightarrow{PhC\equiv N}$ (398) (179)

C. Sulphonyl Azides

Decomposition of benzenesulphonyl azide in the presence of freshly reduced copper in boiling methanol gave a quantitative yield of nitrogen, together with benzenesulphonamide (ca. 80%) and minor amounts of methylene bis(benzenesulphonamide) (**400**) and 1,3,5-tris(benzenesulphonyl)hexahydro-*s*-triazine (**401**) [267]. The latter two

$$PhSO_2N_3 \xrightarrow[-N_2,\ MeOH]{Cu} Cu \Longleftarrow NSO_2Ph \rightleftharpoons Cu^{\circ} \ldots \cdot\ddot{N}SO_2Ph \quad (\mathbf{399})$$

$$\mathbf{399} \xrightarrow{MeOH} Cu^{\circ} + PhSO_2NH_2 + H_2CO \longrightarrow (PhSO_2NH)_2CH_2\ (\mathbf{400}) + \mathbf{401} \qquad (180)$$

apparently arose from condensation of benzenesulphonamide with formaldehyde formed by dehydrogenation of methanol. It was felt that the reaction proceeded by way of a copper-nitrene complex (**399**). When the reaction was carried out in the presence of a slight excess of dimethyl sulphoxide, a fourfold increase in the rate of nitrogen evolution was observed, and the only product obtained was *N*-benzenesulphonylsulphoximine (**404**) (97%). The rate acceleration was

$$\mathbf{402} \xrightarrow{DMSO} \mathbf{403} \xrightarrow{-N_2} Me_2S(=O)=NSO_2Ph\ (\mathbf{404}) \qquad (181)$$

thought to result from the interaction of dimethyl sulphoxide with the copper-azide complex (**402**) to give the oxathiatriazoline (**403**) and then **404** with loss of nitrogen.

The copper-catalysed decomposition of benzenesulphonyl azide in cyclohexene gave a variety of products (equation 182) [268]. In the presence of hydroquinone, cyclohexyl azide was not observed. It was assumed that cyclohexanone came from the hydrolysis of the imine (**405**). The copper-catalysed decomposition of 2-biphenylylsulphonyl

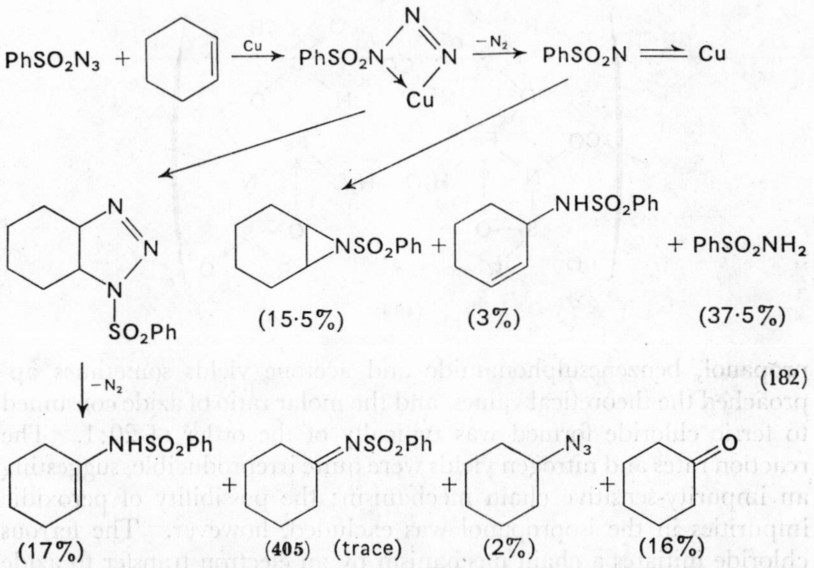

$$PhSO_2N_3 + \text{cyclohexene} \xrightarrow{Cu} PhSO_2N(N{=}N)Cu \xrightarrow{-N_2} PhSO_2N \rightleftharpoons Cu$$

(15·5%) (3%) (37·5%)

$+ PhSO_2NH_2$ $-N_2$ (182)

(17%) (405) (trace) (2%) (16%)

azide and methanesulphonyl azide in benzene and cyclohexane solution at 80° led to the corresponding sulphonamides (67–82%) as the only products isolated[269]. The reactions were slow at this temperature, requiring 3–10 days for about 50% completion.

Decomposition of methanesulphonyl azide in aromatic solvents (methyl benzoate or benzotrifluoride), in the presence of transition metal compounds (e.g. copper(II) acetylacetonate, manganese(II) acetylacetonate, di-cobalt octacarbonyl, tri-iron dodecacarbonyl, and iron pentacarbonyl) led to a marked decrease in the aromatic substitution product compared with thermolysis, and, with the iron carbonyls, to an increased yield of methanesulphonamide[155]. In addition, the aromatic substitution products shifted from mainly *ortho*-substitution with no additives to mainly *meta*-substitution in the presence of the additives mentioned above.

Sulphonyl azides are decomposed by $Fe_2(CO)_9$ at room temperature or by $Fe(CO)_5$ at 80° in non-polar non-protic solvents such as benzene or cyclohexane to give novel stable complexes having the probable structure (**406**) (or the cyclic dimer) in which all the carbonyl groups on iron have been displaced[156].
A related complex is obtained from methyl 2-azidophenyl sulphone[270].

The decomposition of various sulphonyl azides using aqueous ferrous chloride–hydrochloric acid mixtures as catalysts was studied by Reagan and Nickon[261]. When the reaction was carried out in iso-

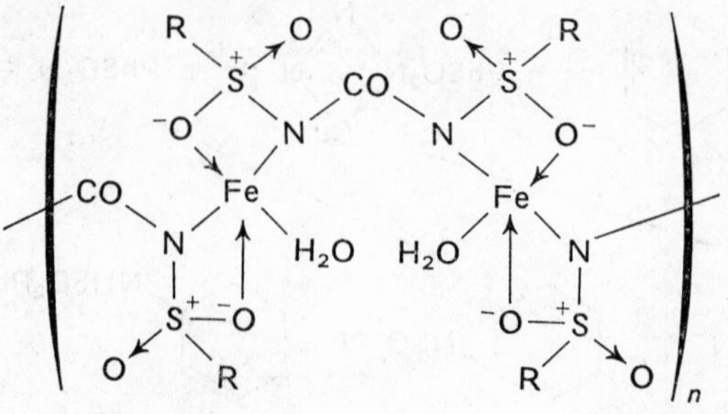

(**406**)

propanol, benzenesulphonamide and acetone yields sometimes approached the theoretical values, and the molar ratio of azide consumed to ferric chloride formed was typically of the order of 20:1. The reaction rates and nitrogen yields were quite irreproducible, suggesting an impurity-sensitive chain mechanism; the possibility of peroxidic impurities in the isopropanol was excluded, however. The ferrous chloride initiates a chain mechanism by an electron transfer to azide to form a radical anion intermediate. Loss of nitrogen and capture of a solvent proton would then produce the sulphonimido radical (**407**), a chain propagating species.

$$RSO_2N_3 + FeCl_2 + HCl \longrightarrow RSO_2\dot{\underset{\bullet\bullet}{N}}H + FeCl_3 + N_2 \qquad (183)$$

(**407**)

VII. REFERENCES

1. J. H. Boyer and F. C. Canter, *Chem. Rev.*, **54**, 1 (1954).
2. P. A. S. Smith in *Molecular Rearrangements*, Part 1 (Ed. P. de Mayo), Interscience Publishers, New York, 1963, pp. 462–479.
3. R. A. Abramovitch and B. A. Davis, *Chem. Rev.*, **64**, 149 (1964).
4. G. L'abbé, *Chem. Rev.*, **69**, 345 (1969).
5. T. Curtius and A. Darapsky, *J. Prakt. Chem.*, [2] **63**, 428 (1901).
6. T. Curtius and A. Darapsky, *Ber.*, **35**, 3229 (1902).
7. T. Curtius, *Ber.*, **45**, 1057 (1912).
8. K. W. Sherk, A. G. Houpt and A. W. Brown, *J. Am. Chem. Soc.*, **62**, 329 (1940).
9. J. H. Boyer, F. C. Canter, J. Hamer and R. K. Putney, *J. Am. Chem. Soc.*, **78**, 325 (1956).
10. W. Pritzkow and G. Mahler, *J. Prakt. Chem.*, [4] **8**, 314 (1959).
11. Y. Yukawa and K. Tanaka, *Mem. Inst. Sci. and Ind. Research, Osaka Univ.*, **14**, 199 (1957); *Chem. Abstr.*, **52**, 4513 (1958).

12. Y. Yukawa and K. Tanaka, *Mem. Inst. Sci. and Ind. Research, Osaka Univ.*, **14**, 205 (1957); *Chem. Abstr.*, **52**, 2795 (1958).
13. Y. Yukawa and K. Tanaka, *Nippon Kagaku Zasshi*, **78**, 1049 (1957); *Chem. Abstr.*, **54**, 5555 (1960).
14. J. H. Boyer and F. C. Canter, *J. Am. Chem. Soc.*, **77**, 3287 (1955).
15. C. L. Arcus, R. E. Marks and R. Vetterlein, *Chem. Ind.* (*London*), 1193 (1960).
16. M. M. Coombs, *J. Chem. Soc.*, 4200 (1958).
17. W. E. McEwen and N. B. Metha, *J. Am. Chem. Soc.*, **74**, 526 (1952).
18. R. F. Tietz and W. E. McEwen, *J. Am. Chem. Soc.*, **77**, 4007 (1955).
19. C. H. Gudmunsen and W. E. McEwen, *J. Am. Chem. Soc.*, **79**, 329 (1957).
20. D. E. Bublitz, J. Kleinberg and W. E. McEwen, *Chem. Ind.* (*London*), 936 (1960).
21. A. Berger, J. Kleinberg and W. E. McEwen, *Chem. Ind.* (*London*), 204 (1960).
22. A. Berger, W. E. McEwen and J. Kleinberg, *J. Am. Chem. Soc.*, **83**, 2274 (1961).
23. C. L. Arcus and R. J. Mesley, *J. Chem. Soc.*, 178 (1953).
24. C. L. Arcus and M. M. Coombs, *J. Chem. Soc.*, 4319 (1954).
25. C. L. Arcus and E. A. Lucken, *J. Chem. Soc.*, 1634 (1955).
26. C. L. Arcus, R. E. Marks and M. M. Coombs, *J. Chem. Soc.*, 4064 (1957).
27. C. L. Arcus and J. V. Evans, *J. Chem. Soc.*, 789 (1958).
28. M. M. Coombs, *J. Chem. Soc.*, 3454 (1958).
29. N. Koga and J.-P. Anselme, *Tetrahedron Letters*, 4773 (1969).
30. K. F. Schmidt, *Ber.*, **57**, 704 (1924).
31. H. Wolfe, *Org. React.*, **3**, 307 (1946).
32. C. L. Arcus, M. M. Coombs and J. V. Evans, *J. Chem. Soc.*, 1498 (1956).
33. C. L. Arcus and M. M. Coombs, *J. Chem. Soc.*, 3698 (1953).
34. C. L. Arcus and R. E. Marks, *J. Chem. Soc.*, 1627 (1956).
35. C. L. Arcus and G. Barrett, *J. Chem. Soc.*, 2098 (1960).
36. W. Pritzkow and A. Schuberth, *Chem. Ber.*, **93** 1725 (1960).
37. P. A. S. Smith, *J. Am. Chem. Soc.*, **70**, 320 (1948).
38. J. H. Boyer and J. Hamer, *J. Am. Chem. Soc.*, **77**, 951 (1955).
39. J. H. Boyer and L. R. Morgan, Jr., *J. Org. Chem.*, **24**, 561 (1959).
40. J. H. Boyer and L. R. Morgan, Jr., *J. Am. Chem. Soc.*, **80**, 2020 (1958).
41. J. H. Boyer and L. R. Morgan, Jr., *J. Am. Chem. Soc.*, **81**, 3369 (1959).
42. J. H. Boyer, W. E. Krueger and R. Modler, *J. Org. Chem.*, **34**, 1987 (1969).
43. W. Pritzkow and G. Pohl, *J. Prakt. Chem.*, [4] **20**, 132 (1963).
44. P. Griess, *Phil. Trans. Roy. Soc.* (*London*), **13**, 377 (1864).
45. P. Griess, *Ann. Chem.*, **137**, 39 (1866).
46. P. Griess, *Ber.*, **19**, 313 (1886).
47. P. Friedlaender and M. Zeitlin, *Ber.*, **27**, 192 (1894).
48. P. Friedlaender, *Ber.*, **28**, 1386 (1895).
49. K. Brass, E. Ferber and J. Stadler, *Ber.*, **57**, 121 (1924).
50. K. Brass and J. Stadler, *Ber.*, **57**, 128 (1924).
51. K. Brass and G. Nickel, *Ann. Chem.*, **441**, 217 (1925).
52. E. Bamberger, *Ann. Chem.*, **424**, 233 (1921).
53. C. K. Ingold, *Structure and Mechanism in Organic Chemistry*, 2nd ed., Cornell University Press, Ithaca, 1969, pp. 906–908.

54. P. A. S. Smith and B. B. Brown, *J. Am. Chem. Soc.*, **73**, 2438 (1951).
55. E. Bamberger and J. Brun, *Helv. Chim. Acta*, **7**, 112 (1924).
56. E. Bamberger and J. Brun, *Helv. Chim. Acta*, **6**, 942 (1923).
57. E. Bamberger, *Ann. Chem.*, **443**, 192 (1925).
58. E. Bamberger and J. Brun, *Helv. Chim. Acta*, **6**, 935 (1923).
59. H. E. Heller, E. D. Hughes and C. K. Ingold, *Nature*, **168**, 909 (1951).
60. T. Itai and S. Kamiya, *Chem. Pharm. Bull.* (*Tokyo*), **9**, 87 (1961); *Chem. Abstr.*, **55**, 27338c (1961).
61. D. Misiti, H. W. Moore and K. Folkers, *Tetrahedron Letters*, 1071 (1965).
62. H. W. Moore and H. R. Sheldon, *J. Org. Chem.*, **33**, 4019 (1968).
63. H. W. Moore and H. R. Sheldon, *Tetrahedron Letters*, 5431 (1968).
64. H. W. Moore, H. R. Sheldon and D. F. Shellhamer, *J. Org. Chem.*, **34**, 1999 (1969).
65. H. W. Moore, H. R. Sheldon and W. Weyler, Jr., *Tetrahedron Letters*, 1243 (1969).
66. H. W. Moore, H. R. Sheldon, D. W. Deters and R. J. Wikholm, *J. Am. Chem. Soc.*, **92**, 1675 (1970).
67. G. S. Sidhu, G. Thyagarajan and U. T. Bhalero, *Chem. Ind.* (*London*), 1301 (1966).
68. J. Goubeau, E. Allenstein and A. Schmidt, *Chem. Ber.*, **97**, 884 (1964).
69. R. Kreher and J. Jager, *Z. Naturforsch.*, **19b**, 657 (1964).
70. R. Kreher and J. Jager, *Angew. Chem. Int. Ed. Engl.*, **4**, 706 (1965).
71. R. Kreher and J. Jager, *Angew. Chem. Int. Ed. Engl.*, **4**, 952 (1965).
72. W. Borsche, *Ber.*, **75**, 1312 (1942).
73. W. Borsche and H. Hahn, *Chem. Ber.*, **82**, 260 (1949).
74. K. Hoegerle and P. E. Butler, *Chem. Ind.* (*London*), 933 (1964).
75. E. Halans, unpublished results, ref. 4, pp. 351–352.
76. L. Horner and A. Christmann, *Angew. Chem. Int. Ed. Engl.*, **2**, 599 (1963).
77. E. Lieber, J. S. Curtice and C. N. R. Rao, *Chem. Ind.*, (*London*), 586 (1966).
78. R. A. Abramovitch, *Chem. Soc. Special Publ.*, No. 24, Ch. 13, p. 323 (1970).
79. J. K. Senior, *J. Am. Chem. Soc.*, **38**, 2718 (1916).
80. H. C. Ramsperger, *J. Am. Chem. Soc.*, **51**, 2134 (1929).
81. J. A. Leermakers, *J. Am. Chem. Soc.*, **55**, 3098 (1933).
82. J. A. Leermakers, *J. Am. Chem. Soc.*, **55**, 2719 (1933).
83. F. O. Rice and C. J. Grelecki, *J. Phys. Chem.*, **61**, 830 (1957).
84. M. S. O'Dell, Jr. and B. deB. Darwent, *Can. J. Chem.*, **48**, 1140 (1970).
85. A. Reiser, F. W. Willets, G. C. Terry, V. Williams and R. Marley, *Trans. Faraday Soc.*, **64**, 3265 (1968).
86. G. Geiseler and W. König, *Z. Phys. Chem.*, **227**, 81 (1964).
87. P. Walker and W. A. Waters, *J. Chem. Soc.*, 1632 (1962).
88. W. Pritzkow and D. Timm, *J. Prakt. Chem.*, [4], **32**, 178 (1966).
89. R. A. Abramovitch and E. P. Kyba, *Chem. Comm.*, 265 (1969).
90. R. A. Abramovitch and E. P. Kyba, unpublished results.
91. B. Coffin and R. F. Robbins, *J. Chem. Soc.*, 1252 (1965).
92. H. Hart and E. A. Sedor, *J. Am. Chem. Soc.*, **89**, 2342 (1967).
93. J. H. Boyer and D. Straw, *J. Am. Chem. Soc.*, **74**, 4506 (1952).
94. J. H. Boyer and D. Straw, *J. Am. Chem. Soc.*, **75**, 1642 (1953).
95. J. H. Boyer and D. Straw, *J. Am. Chem. Soc.*, **75**, 2683 (1953).
96. N. Koga, G. Koga and J.-P. Anselme, *Can. J. Chem.*, **47**, 1143 (1969).

97. C. L. Currie and B. de B. Darwent, *Can. J. Chem.* **41**, 1552 (1963).
98. W. H. Saunders, Jr. and J. C. Ware, *J. Am. Chem. Soc.*, **80**, 3328 (1958).
99. W. H. Saunders, Jr. and E. A. Caress, *J. Am. Chem. Soc.*, **86**, 861 (1964).
100. F. D. Lewis and W. H. Saunders, Jr., *J. Am. Chem. Soc.*, **90**, 3828 (1968).
101. T. Curtius and K. Raschig, *J. Prakt. Chem.*, [2] **125**, 466 (1930).
102. T. Curtius and G. Ehrhart, *Ber.*, **55**, 1559 (1922).
103. A. L. Logothetis, *J. Am. Chem. Soc.*, **87**, 749 (1965).
104. L. Wolff, *Ann. Chem.*, **394**, 59 (1912).
105. R. Huisgen, D. Vossius and M. Appl, *Chem. Ber.*, **91**, 1 (1958).
106. W. von E. Doering and R. A. Odum, *Tetrahedron*, **22**, 81 (1966).
107. R. Huisgen and M. Appl, *Chem. Ber.*, **91**, 12 (1958).
108. M. Appl and R. Huisgen, *Chem. Ber.*, **92**, 2961 (1959).
109. J. I. G. Cadogan and M. J. Todd, *J. Chem. Soc.* (*C*), 2808 (1969).
110. P. A. S. Smith and J. H. Hall, *J. Am. Chem. Soc.*, **84**, 480 (1962).
111. K. Alder and G. Stein, *Ann. Chem.*, **485**, 211 (1931).
112. K. Alder and G. Stein, *Ann. Chem.*, **485**, 223 (1931).
113. R. Huisgen, L. Möbius, G. Müller, H. Stangl, G. Szeimies and J. M. Vernon, *Chem. Ber.*, **98**, 3992 (1965).
114. K. R. Henery-Logan and R. A. Clark, *Tetrahedron Letters*, 801 (1968).
115. T. F. Fagley, J. R. Sutter and R. L. Oglukian, *J. Am. Chem. Soc.*, **78**, 5567 (1956).
116. S. Patai and Y. Gotshal, *J. Chem. Soc.*, (*B*), 489 (1966).
117. L. K. Dyall and J. E. Kemp, *Australian J. Chem.*, **20**, 1625 (1967).
118. L. K. Dyall and J. E. Kemp, *J. Chem. Soc.* (*B*), 976 (1968).
119. E. A. Birkhimer, B. Norup and T. A. Bak, *Acta Chem. Scand.*, **14**, 1894 (1960).
120. E. Andersen, E. A. Birkhimer and T. A. Bak, *Acta Chem. Scand.*, **14**, 1899 (1960).
121. A. J. Boulton, A. C. G. Gray and A. R. Katritzky, *J. Chem. Soc.*, 5958 (1965).
122. P. A. S. Smith, B. B. Brown, R. K. Putney and R. F. Reinisch, *J. Am. Chem. Soc.*, **75**, 6335 (1953).
123. J. H. Hall and F. E. Behr, Abstracts, 2nd Intern. Congress of Heterocyclic Chemistry, Montpellier, July, 1969, paper B-1, p. 60.
124. R. A. Carboni and J. E. Castle, *J. Am. Chem. Soc.*, **84**, 2435 (1962).
125. R. A. Carboni, J. C. Kauer, J. E. Castle and H. E. Simmons, *J. Am. Chem. Soc.*, **89**, 2618 (1967).
126. R. A. Abramovitch and K. A. H. Adams, *Can. J. Chem.*, **39**, 2516 (1961).
127. J. H. Hall, J. G. Stephanie and D. K. Nordstrom, *J. Org. Chem.*, **33**, 2951 (1968).
128. R. H. Sundberg, L.-S. Lin and D. E. Blackburn, *J. Heterocycl. Chem.*, **6**, 441 (1969).
129. L. Krbechek and H. Takimoto, *J. Org. Chem.*, **29**, 1150 (1964).
130. G. Smolinsky, E. Wasserman and W. A. Yager, *J. Am. Chem. Soc.*, **84**, 3220 (1962).
131. G. Smolinsky, L. C. Snyder and E. Wasserman, *Rev. Mod. Phys.*, **35**, 576 (1963).
132. E. Wasserman, G. Smolinsky and W. A. Yager, *J. Am. Chem. Soc.*, **86**, 3166 (1964).

133. R. M. Moriarity, M. Rahman and R. J. King, *J. Am. Chem. Soc.*, **88**, 842 (1966).
134. D. S. Breslow, M. F. Sloan, N. R. Newburg and W. B. Renfrow, *J. Am. Chem. Soc.*, **91**, 2273 (1969).
135. D. S. Breslow and E. I. Edwards, *Tetrahedron Letters*, 2123 (1967).
136. D. S. Breslow, T. J. Prosser, A. F. Marcantonio and C. A. Genge, *J. Am. Chem. Soc.*, **89**, 2384 (1967).
137. W. Lwowski, *Angew. Chem. Int. Ed. Engl.*, **6**, 897 (1967).
138. A. Bertho, *Ber.*, **57**, 1138 (1924).
139. J. H. Hall, J. W. Hill and H. Tsai, *Tetrahedron Letters*, 2211 (1965).
140. R. Huisgen and K. v. Fraunberg, *Tetrahedron Letters*, 2595 (1969).
141. J. H. Hall, J. W. Hill and J. M. Fargher, *J. Am. Chem. Soc.*, **90**, 5313 (1968).
142. G. Smolinsky and B. I. Feuer, *J. Am. Chem. Soc.*, **86**, 3085 (1964).
143. G. Smolinsky and B. I. Feuer, *J. Org. Chem.*, **31**, 3882 (1966).
144. G. Smolinsky, *J. Am. Chem. Soc.*, **83**, 2489 (1961).
145. O. Meth-Cohn, R. K. Smalley and H. Suschitzky, *J. Chem. Soc.*, 1666 (1963).
146. G. Smolinsky, *Trans. N.Y. Acad. Sci.*, [2], **30**, 511 (1968).
147. R. K. Smalley and H. Suschitzky, *J. Chem. Soc.*, 5922 (1964).
148. R. A. Abramovitch, C. I. Azogu and R. G. Sutherland, in *Progress in Organometallic Chemistry* (ed. M. I. Bruce and F. G. A. Stone), Proc. 4th Int. Conference Organometallic Chem., Bristol, 1969, Abstract G. 4.
149. P. A. S. Smith and B. B. Brown, *J. Am. Chem. Soc.*, **73**, 2435 (1951).
150. G. Smolinsky, *J. Am. Chem. Soc.*, **82**, 4717 (1960).
151. L. Krbechek and H. Takimoto, *J. Org. Chem.*, **33**, 4286 (1968).
152. R. A. Abramovitch, J. Roy and V. Uma, *Can. J. Chem.*, **43**, 3407 (1965).
153. R. A. Abramovitch and V. Uma, *Chem. Comm.*, 797 (1968).
154. W. Lwowski and R. L. Johnson, *Tetrahedron Letters*, 891 (1967).
155. R. A. Abramovitch, G. N. Knaus and V. Uma, *J. Am. Chem. Soc.*, **91**, 7532 (1969).
156. R. A. Abramovitch and E. E. Knaus, unpublished results.
157. R. A. Abramovitch and E. F. V. Scriven, *Chem. Comm.*, 787 (1970).
158. W. D. Crow and C. Wentrup, *Tetrahedron Letters*, 4379 (1967).
159. W. D. Crow and C. Wentrup, *Chem. Comm.*, 1026 (1968).
160. W. D. Crow and C. Wentrup, *Chem. Comm.*, 1082 (1968).
161. E. Hedaya, M. E. Kent, D. W. McNeil, F. P. Lossing and T. McAllister, *Tetrahedron Letters*, 3415 (1968).
162. W. D. Crow and C. Wentrup, *Tetrahedron Letters*, 5569 (1968).
163. W. D. Crow and C. Wentrup, *Tetrahedron Letters*, 6149 (1968).
164. C. Wentrup, *Chem. Comm.*, 1386 (1969).
165. W. D. Crow and C. Wentrup, *Chem. Comm.*, 1387 (1969).
166. R. P. Bennett and W. P. Hardy, *J. Am. Chem. Soc.*, **90**, 3295 (1968).
167. G. Ribaldone, G. Capara and G. Borsotti, *Chim. Ind.* (*Milan*), **50**, 1200 (1968); *Chem. Abstr.*, **70**, 37326g (1969).
168. J. I. G. Cadogan and M. Cameron-Wood, *Proc. Chem. Soc.* (*London*), 361 (1962).
169. R. K. Smalley and H. Suschitzky, *J. Chem. Soc.*, 5571 (1963).
170. R. Garner, E. B. Mullock and H. Suschitzky, *J. Chem. Soc.* (*C*), 1980 (1966).

171. E. B. Mullock and H. Suschitzky, *J. Chem. Soc.* (*C*), 1937 (1968).
172. S. Bradbury, C. W. Rees and R. C. Storr, *Chem. Comm.*, 1428 (1969).
173. T. Shingaki, *Sci. Rep. Coll. Gen. Educ. Osaka Univ.*, **11**, 67, 81 (1963); *Chem. Abstr.*, **60**, 6733d (1964).
174. J. E. Leffler and H. H. Gibson, Jr., *J. Am. Chem. Soc.*, **90**, 4117 (1968).
175. L. Horner and G. Bauer, *Tetrahedron Letters*, 3573 (1966).
176. H. W. Moore, W. Weyler, Jr. and H. R. Sheldon, *Tetrahedron Letters*, 3947 (1969).
176a. H. W. Moore and W. Weyler, Jr., *J. Am. Chem. Soc.*, **92**, 4132 (1970).
177. J. D. Hobson and J. R. Malpass, *J. Chem. Soc.* (*C*), 1645 (1967).
178. J. D. Hobson and J. R. Malpass, *J. Chem. Soc.* (*C*), 1499 (1969).
179. R. A. Abramovitch and R. G. Sutherland, *Fortschr. chem. Forsch.*, **16**(1), 1(1970).
180. D. S. Breslow, *Sulfonyl Nitrenes* in *Nitrenes* (ed. W. Lwowski), John Wiley and Sons, New York, 1970, pp. 245–303.
181. L. Horner and A. Christmann, *Chem. Ber.*, **96**, 388 (1963).
182. J. E. Leffler and Y. Tsuno, *J. Org. Chem.*, **28**, 190 (1963).
183. J. E. Leffler and Y. Tsuno, *J. Org. Chem.*, **28**, 902 (1963).
184. G. P. Balabanov, Y. I. Dergunov and V. A. Gal'perin, *J. Org. Chem. U.S.S.R.*, **2**, 1797 (1966).
185. K. Takemato, R. Fujita and M. Imoto, *Makromol. Chem.*, **112**, 116 (1968).
186. G. P. Balabanov, Y. I. Dergunov and V. G. Golov, *Zh. Fiz. Khim.*, **40**, 2171 (1966); *Chem. Abstr.*, **65**, 19974 (1966).
187. M. F. Sloan, W. B. Renfrow and D. S. Breslow, *Tetrahedron Letters*, 2905 (1964).
188. R. A. Abramovitch and W. D. Holcomb, *Chem. Comm.*, 1298 (1969).
189. R. A. Abramovitch, C. I. Azogu and I. T. McMaster, *J. Am. Chem. Soc.*, **91**, 1219 (1969).
190. W. T. Reichle, *Inorg. Chem.*, **3**, 402 (1964).
191. D. S. Breslow, personal communication to R. A. Abramovitch.
192. J. F. Heathcock and M. T. Edmison, *J. Am. Chem. Soc.*, **82**, 3460 (1960).
193. T. Curtius and J. Rissom, *Angew. Chem.*, **26**, 134 (1913).
194. T. Curtius and F. W. Haas, *J. Prakt. Chem.* [2], **102**, 85 (1921).
195. T. Curtius and F. Schmidt, *Ber.*, **55**, 1571 (1922).
196. T. Curtius, *J. Prakt. Chem.*, [2], **125**, 303–424 (1930).
197. P. K. Datta, *J. Indian Chem. Soc.*, **24**, 109 (1947).
198. J. N. Ashley, G. L. Buchanan and A. P. T. Eason, *J. Chem. Soc.*, 60 (1947).
199. R. A. Abramovitch and T. Takaya, unpublished results.
200. T. Curtius and J. Rissom, *J. Prakt. Chem.*, [2], **125**, 311 (1930).
201. J. E. Franz and C. Osuch, *Tetrahedron Letters*, 837 (1963).
202. J. E. Franz, C. Osuch and M. W. Dietrich, *J. Org. Chem.*, **29**, 2922 (1964).
203. L. H. Zalkow and A. C. Oehlschlager, *J. Org. Chem.*, **28**, 3303 (1963).
204. R. Huisgen, R. Knorr, L. Möbius and G. Szeimies, *Chem. Ber.*, **98**, 4014 (1965).
205. G. Bianchetti, P. D. Croce and D. Pocar, *Tetrahedron Letters*, 2043 (1965).
206. R. Fusco, G. Bianchetti, D. Pocar and R. Ugo, *Chem. Ber.*, **96**, 802 (1963).
207. R. Huisgen, L. Möbius and G. Szeimies, *Chem. Ber.*, **98**, 1138 (1965).
208. R. F. Bleiholder and H. Schechter, *J. Am. Chem. Soc.*, **90**, 2131 (1968).
209. T. Curtius and W. Klavehn, *J. Prakt. Chem.*, [2], **112**, 65 (1926).

210. M. Regitz, *Angew. Chem. Int. Ed. Engl.*, **6**, 733 (1967).
211. W. Fischer and J.-P. Anselme, *J. Am. Chem. Soc.*, **89**, 5284 (1967).
212. J.-P. Anselme and W. Fischer, *Tetrahedron*, **25**, 855 (1968).
213. W. Fischer and J.-P. Anselme, *Tetrahedron Letters*, 877 (1968).
214. H. Staudinger and J. Meyer, *Helv. Chim. Acta*, **2**, 635 (1919).
215. H. Staudinger and E. Hauser, *Helv. Chim. Acta*, **4**, 861 (1921).
216. L. Horner and A. Gross, *Ann. Chem.*, **591**, 117 (1955).
217. E. Bergmann and A. Wolff, *Ber.*, **63**, 1176 (1930).
218. J. E. Leffler, U. Honsberg, Y. Tsuno and I. Forsblad, *J. Org. Chem.*, **26**, 4810 (1961).
219. J. E. Leffler and R. D. Temple, *J. Am. Chem. Soc.*, **89**, 5235 (1967).
220. J. Goerdeler and H. Ullmann, *Chem. Ber.*, **94**, 1067 (1961).
221. J. E. Franz and C. Osuch, *Tetrahedron Letters*, 841 (1963).
222. T. J. Maricich, *J. Am. Chem. Soc.*, **90**, 7179 (1968).
223. E. Koch, *Tetrahedron*, **23**, 1747 (1967).
224. D. H. R. Barton and L. R. Morgan, Jr., *Proc. Chem. Soc.*, 206 (1961).
225. D. H. R. Barton and L. R. Morgan, Jr., *J. Chem. Soc.*, 622 (1962).
226. D. H. R. Barton and A. N. Starratt, *J. Chem. Soc.*, 2444 (1965).
227. R. M. Moriarity and M. Rahman, *Tetrahedron*, **21**, 2877 (1965).
228. R. M. Moriarity and R. C. Reardon, personal communication to R. A. Abramovitch.
229. F. D. Lewis and W. H. Saunders, Jr., *J. Am. Chem. Soc.*, **90**, 7031 (1968).
230. F. H. Seubold, Jr., *J. Am. Chem. Soc.*, **75**, 2532 (1953).
231. W. H. Saunders, Jr. and R. H. Paine, *J. Am. Chem. Soc.*, **83**, 882 (1961).
232. D. J. Cram and J. E. McCarty, *J. Am. Chem. Soc.*, **79**, 2866 (1957).
233. G. S. Hammond, *J. Am. Chem. Soc.*, **77**, 334 (1955).
234. F. D. Greene, W. Adam and G. A. Knudsen, Jr., *J. Org. Chem.*, **31**, 2087 (1966).
235. W. Adam and Y. M. Cheng, *J. Am. Chem. Soc.*, **91**, 2109 (1969).
236. W. Adam, Y. M. Cheng, C. Wilkerson and W. A. Zaidi, *J. Am. Chem. Soc.*, **91**, 2111 (1969).
237. H. Phillip and J. Keating, *Tetrahedron Letters*, 523 (1961).
238. F. D. Lewis and W. H. Saunders, Jr., *J. Am. Chem. Soc.*, **89**, 645 (1967).
239. J. S. Swenton, T. J. Ikeler and B. H. Williams, *Chem. Comm.*, 1263 (1969).
240. J. S. Swenton, T. J. Ikeler and B. H. Williams, *J. Am. Chem. Soc.*, **92**, 3103 (1970).
241. F. D. Lewis and J. C. Dalton, *J. Am. Chem. Soc.*, **91**, 5260 (1969).
242. F. D. Lewis and W. H. Saunders, Jr., *J. Am. Chem. Soc.*, **90**, 7033 (1968).
243. W. D. Closson and H. B. Gray, *J. Am. Chem. Soc.*, **85**, 290 (1963).
244. R. L. Livingston and C. N. R. Rao, *J. Phys. Chem.*, **64**, 756 (1960).
245. A. Reiser and R. Marley, *Trans. Faraday Soc.*, **64**, 1806 (1968).
246. L. Barash, E. Wasserman and W. A. Yager, *J. Am. Chem. Soc.*, **89**, 3931 (1967).
247. R. M. Moriarity, J. M. Kliegman and C. Shovlin, *J. Am. Chem. Soc.*, **89**, 5958 (1967).
248. R. M. Moriarity and J. M. Kliegman, *J. Am. Chem. Soc.*, **89**, 5959 (1967).
249. A. Reiser, G. Bowes and R. J. Horne, *Trans. Faraday Soc.*, **62**, 3162 (1966).
250. A. Reiser, H. M. Wagner, R. Marley and G. Bowes, *Trans. Faraday Soc.*, **63**, 2403 (1967).

251. A. Reiser, H. Wagner and G. Bowes, *Tetrahedron Letters*, 2635 (1966).
252. J. S. Swenton, *Tetrahedron Letters*, 3421 (1968).
253. L. Horner, A. Christmann and A. Gross, *Chem. Ber.*, **96**, 399 (1963).
254. R. A. Odum and A. M. Aaronson, *J. Am. Chem. Soc.*, **91**, 5680 (1969).
255. J. H. Boyer and G. J. Mikol, *Chem. Comm.*, 734 (1969).
256. R. A. Abramovitch, C. I. Azogu and R. G. Sutherland, Abstracts of papers presented at Am. Chem. Soc., 160th National Meeting, Chicago, 1970; *Chem. Comm.* 134 (1971).
257. R. B. Trattner and R. A. Odum, Abstracts of papers presented at Am. Chem. Soc., 158th National Meeting, New York, Sept. 7, 1969, p. ORGN. 89.
258. W. Lwowski and E. Scheiffele, *J. Am. Chem. Soc.*, **87**, 4359 (1965).
259. R. A. Abramovitch and V. Uma, unpublished results.
260. W. Lwowski, R. DeMauriac, T. W. Mattingly, Jr. and E. Scheiffele *Tetrahedron Letters*, 3285 (1964).
261. M. T. Reagan and A. Nickon, *J. Am. Chem. Soc.*, **90**, 4096 (1968).
262. R. A. Abramovitch, C. I. Azogu and R. G. Sutherland, *Chem. Comm.*, 1439 (1969).
263. R. A. Abramovitch, C. I. Azogu and R. G. Sutherland, submitted for publication.
264. M. Dekker and G. R. Knox, *Chem. Comm.*, 1243 (1967).
265. C. D. Campbell and C. W. Rees, *Chem. Comm.*, 537 (1969).
266. K. v. Fraunberg and R. Huisgen, *Tetrahedron Letters*, 2599 (1969).
267. H. Kwart and A. A. Kahn, *J. Am. Chem. Soc.*, **89**, 1950 (1967).
268. H. Kwart and A. A. Kahn, *J. Am. Chem. Soc.*, **89**, 1951 (1967).
269. R. A. Abramovitch and T. Chellathurai, unpublished results.
270. R. A. Abramovitch, G. N. Knaus and R. W. Stowe, unpublished results.

CHAPTER 6

Azides as synthetic starting materials

TUVIA SHERADSKY

The Hebrew University, Jerusalem, Israel

I. INTRODUCTION 332
II. TRANSFORMATION OF AZIDO GROUPS INTO OTHER FUNCTIONAL GROUPS 333
A. Syntheses of Amines 333
1. By reduction of azides 333
a. Reduction methods 333
b. Utility 335
2. By acidolysis of azides 338
3. Via nitrene intermediates 339
a. Hydrogen abstraction 339
b. Intermolecular insertion 340
c. The Curtius rearrangement 341
B. Syntheses of Azomethines 342
1. By acidolysis of azides 342
2. By rearrangement of nitrenes 343
3. By decomposition of azide adducts 345
C. Syntheses of Nitriles 348
D. Syntheses of Isocyanates 349
E. Syntheses of Diazo Compounds 350
1. By diazo transfer 350
2. By cleavage of azide adducts 353
F. Syntheses of Azo Compounds 354
1. By diazo transfer 354
2. By dimerization of nitrenes 355
G. Syntheses of Iminophosphoranes and Iminosulphuranes . 356
H. Syntheses of Triazenes 357
I. Syntheses of Azides 358
III. AZIDES AS STARTING MATERIALS IN SYNTHESES OF HETEROCYCLES 358
A. General Considerations 358

B. Syntheses of Three-membered Rings 359
1. Aziridines 359
2. Azirines 361
3. Diaziridines 362
C. Syntheses of Five-membered Rings containing One Nitrogen 362
1. By intramolecular nitrene insertions and related reactions 362
2. By other methods 366
D. Syntheses of Five-membered Rings containing Two Nitrogens 367
1. Imidazoles 367
2. Pyrazoles 368
E. Syntheses of Five-membered Rings containing Nitrogen and Oxygen 369
1. Oxazoles 369
2. Isoxazoles 371
3. Oxadiazoles 371
F. Syntheses of Five-membered Rings containing Three Nitrogens 373
1. 1,2,3-Triazolines 373
a. By addition of azides to olefins 373
b. By other methods 376
2. 1,2,3-Triazoles 377
a. By addition of azides to olefins, followed by elimination 377
b. By addition of azides to triple bonds . . . 377
c. By nucleophilic attacks on azides, followed by cyclization 379
d. By decomposition of *o*-azidoazobenzenes . . . 381
e. By reductive cyclization of azides 382
G. Syntheses of Tetrazoles 382
1. By addition of azides to nitriles 382
2. By decomposition of geminal diazides 383
3. The azidoazomethine–tetrazole equilibrium . . . 383
H. Syntheses of Six-membered Rings 384
1. By intramolecular insertion of nitrenes 384
2. By ring expansions of five-membered rings . . . 385
3. By dimerization reactions 386
I. Syntheses of Seven-membered Rings 387
IV. References 389

I. INTRODUCTION

The azido group possesses all the qualities required to make it an excellent starting group for synthesis. It can be easily introduced into aliphatic, aromatic and heterocyclic nuclei, even in the presence of various other functional groups. The dipolar character and the relative instability of the azido group enable it to react in numerous fashions, depending on the structure of the molecule, the reagents and the conditions.

The most important reaction types of azides are the following:

(1) Attacks by electrophilic reagents at the α-nitrogen atom

$$RN_3 + E \longrightarrow R{-}\underset{\underset{E^-}{|}}{N}{-}\overset{+}{N}{\equiv}N$$

(2) Attacks by nucleophilic reagents at the azide terminus

$$RN_3 + B{:} \longrightarrow R{-}\bar{N}{-}N{=}N{-}B^+$$

(3) 1,3-Dipolar cycloadditions to polarophiles

$$\text{R}{-}\ddot{\bar{N}}{-}N{=}\overset{+}{N} \;+\; A{=}B \longrightarrow \text{R}{-}N\langle N{=}N{-}B{-}A\rangle \text{ (triazoline ring)}$$

(4) Decomposition reactions, resulting in the very highly reactive nitrenes

$$RN_3 \longrightarrow R{-}N + N_2$$

(5) Reductions

$$RN_3 \xrightarrow{[H]} R{-}N{=}N{-}NH_2 \longrightarrow RNH_2 + N_2$$

The above equations show the formation of primary intermediates which in most cases react further and can yield a very large variety of products. Azides were in fact utilized for the synthesis of most types of open-chain and heterocyclic nitrogen-containing organic molecules.

Although the chemistry of azides has been reviewed several times[1–4], a first attempt was made in this chapter to arrange azide reactions according to products rather than to reaction types or starting materials. This seemed preferable in order to classify the synthetic applications and to point out the various possibilities.

II. TRANSFORMATION OF AZIDO GROUPS INTO OTHER FUNCTIONAL GROUPS

A. Syntheses of Amines

1. By reduction of azides

a. Reduction methods. Azides can be reduced to amines by most reduction methods known. The method of choice generally depends on the other functional groups present in the molecule.

The most common method is catalytic hydrogenation. It is usually performed at low hydrogen pressure, with catalysts such as platinum oxide, palladium on carbon or Raney nickel. The reaction course cannot be followed volumetrically, since hydrogen uptake is balanced by nitrogen evolution, and the completion of the hydrogenation can only be estimated. In cases involving selective hydrogenation, however, this is sometimes possible, e.g. the hydrogenation of phenacyl azide (**1**) could be either stopped (without pressure drop) at α-aminoacetophenone (**2**) or continued (with pressure drop) to 2-amino-1-phenylethanol (**3**)[5]

$$\underset{(\mathbf{1})}{PhCOCH_2N_3} \xrightarrow{H_2/Pd} \underset{(\mathbf{2})}{PhCOCH_2NH_2} \xrightarrow{H_2/Pd} \underset{(\mathbf{3})}{PhCH(OH)CH_2NH_2}$$

As yields in the catalytic hydrogenation are usually almost quantitative and the work-up is very simple, this is the method most widely in use. However in certain cases a non-catalytic reduction is preferable. A very useful reagent is lithium aluminium hydride (equation 1).

$$4\,RN_3 + LiAlH_4 \longrightarrow (RNH)_4AlLi + 4\,N_2 \longrightarrow 4\,RNH_2 \tag{1}$$

The reagent reduces all types of azides, yields in monofunctional azides being 80–90%[6]. Azido ketones yield amino alcohols[6] while unsaturated azides retain the double bonds, thus reduction of 3-azido-4-hydroxy-1-butene (**4**) gave 3-amino-4-hydroxy-1-butene (**5**)[7].

$$CH_3CH_2CH(NH_2)CH_2OH \xleftarrow{H_2/Pd} \underset{(\mathbf{4})}{CH_2{=}CHCH(N_3)CH_2OH} \xrightarrow{LiAlH_4} \underset{(\mathbf{5})}{CH_2{=}CHCH(NH_2)CH_2OH}$$

Sodium borohydride has been found to be a much less satisfactory reagent. Monofunctional aliphatic and aromatic azides were not reduced, while azido ketones yielded azido alcohols[8]. Later work showed that phenyl azide can be reduced to aniline by $NaBH_4$ upon employment of harsher conditions[9]. In carbohydrates the reduction of azides by $NaBH_4$ proceeded without difficulty in excellent yields[10].

Additional reduction methods reported in the literature but seldom used include the use of sodium bisulphite[11], sodium sulphide[12], sodium arsenite[13], titanous chloride[14] tin-hydrochloric acid[15] and sodium–ethanol[16].

The reaction of aryl azides with two equivalents of sodium in dry ether[17] or liquid ammonia[18] yielded the disodio salt of the corresponding aniline **6**. This could be either decomposed by water to the

$$ArN_3 \xrightarrow{2Na} ArN(Na)_2 \; (\mathbf{6}) + N_2$$

$$\mathbf{6} \xrightarrow{H_2O} ArNH_2 \; (\mathbf{7}) \qquad \mathbf{6} \xrightarrow{2Rx} ArNR_2 \; (\mathbf{8})$$

$$(Ar = Ph,\ p\text{-}MeC_6H_4,\ o\text{-}MeC_6H_4,\ R = Me,\ C_2H_5)$$

aniline **7** or dialkylated with alkyl halides to **8**[17,18]. A series of azido sugars were reduced to the corresponding amino sugars by zinc dust in boiling dimethylformamide[19].

b. Utility. The introduction of amino groups into organic nuclei via azides is one of the most important and of most frequent use. The reaction sequences

$$Hal \longrightarrow N_3 \longrightarrow NH_2 \quad \text{and} \quad OH \longrightarrow OSO_2R \longrightarrow N_3 \longrightarrow NH_2$$

are now standard methods for converting halides and alcohols to amines. The azide route is in many cases preferable over direct amination with ammonia, as the higher nucleophility and lower

$$\text{(F)}\text{-}CH_2CH(Br)COOH \; (\mathbf{9}) \xrightarrow{NH_4OH} H_2N\text{-(F)-}CH{=}CHCOOH \; (\mathbf{10})$$

$$\mathbf{9} \xrightarrow{NaN_3} \text{(F)}\text{-}CH_2CH(N_3)COOH \; (\mathbf{11}) \xrightarrow{H_2/Pd} \text{(F)}\text{-}CH_2CH(NH_2)COOH \; (\mathbf{12})$$

basicity of the azide ion decreases the propensity for side reactions, such as β-eliminations and ring fissions.

In the synthesis of pentafluorophenylalanine (**12**), attempted amination of 2-bromo-3-(pentafluorophenyl)propionic acid (**9**) yielded only the cinnamic acid derivative **10**. The desired **12** was however obtained from the azido acid **11** by reduction[20]. Another example is the synthesis of 2-amino-4,4,4-trifluorobutyric acid (**16**) from **13**. This was achieved only via the azido ester **15**, as direct amination yielded the 3-aminoamide **14**[21]. Examples of avoidance of ring

$$\underset{\textbf{(13)}}{CF_3CH_2\underset{\displaystyle Br}{\underset{|}{C}}HCOOC_2H_5} \xrightarrow{NH_4OH} \underset{\textbf{(14)}}{CF_3\underset{\displaystyle NH_2}{\underset{|}{C}}HCH_2CONH_2}$$

$$\downarrow NaN_3$$

$$\underset{\textbf{(15)}}{CF_3CH_2\underset{\displaystyle N_3}{\underset{|}{C}}HCOOC_2H_5} \xrightarrow[2.\ OH^-]{1.\ H_2/Pd} \underset{\textbf{(16)}}{CF_3CH_2\underset{\displaystyle NH_2}{\underset{|}{C}}HCOOH}$$

openings are the syntheses of α-aminocaprolactam (**18**), prepared from the azide **17**[22] and of α-amino-γ-butyrolactone[23].

$$\text{(caprolactam)-Cl} \xrightarrow{NaN_3} \underset{\textbf{(17)}}{\text{(caprolactam)-}N_3} \xrightarrow{H_2/\text{Raney Ni}} \underset{\textbf{(18)}}{\text{(caprolactam)-}NH_2}$$

The use of azido acids for the introduction of aminoacyl groups eliminates the need of protecting the amino group during the acylation. The azidoacyl derivatives obtained are reduced directly to the aminoacyl ones. This method was used in the preparation of aminopenicillins (**19**) from azidoacyl chlorides and 6-aminopenicillanic acid[24]. Azido-

$$R\underset{\displaystyle N_3}{\underset{|}{C}}HCOCl + HAPA \longrightarrow R\underset{\displaystyle N_3}{\underset{|}{C}}HCOAPA \longrightarrow \underset{\textbf{(19)}}{R\underset{\displaystyle NH_2}{\underset{|}{C}}HCOAPA}$$

$$\left(APA = \begin{array}{l} -NH-CH-HC^{\diagup S \diagdown}C(CH_3)_2 \\ \qquad\ \ |\qquad\ \ |\qquad\qquad |\\ \qquad CO-\!\!-N-\!\!-\!\!-CHCOOH \end{array} \right)$$

acylation has been employed also in the preparations of aminotriglycerides[25] and of glycopeptides[26].

A very important advantage of the azide route to amines is its high stereospecificity. The introduction of azides by nucleophilic substitution[27] and by ring openings of epoxides[7] or aziridines[28] proceeds with inversion of configuration, while the reduction step proceeds with retention. This stereospecificity is well demonstrated in the synthesis

cis epoxide $\xrightarrow{NaN_3}$ *threo* azido alcohol $\xrightarrow{1.\ CH_3SO_2Cl;\ 2.\ NaN_3}$ *erythro* diazide $\xrightarrow{H_2/Pd}$ *erythro* diamine (**20**)

trans epoxide $\xrightarrow{NaN_3}$ *erythro* azido alcohol $\xrightarrow{1.\ CH_3SO_2Cl;\ 2.\ NaN_3}$ *threo* diazide $\xrightarrow{H_2/Pd}$ *threo* diamine (**21**)

of the stereoisomeric vicinal diamines **20** and **21**[29]. Sterically controlled introduction of the amino group is of special importance in certain fields of natural products chemistry, such as steroids, terpenes and carbohydrates. The azide route has been used for conversion of cholestanol to 3α-aminocholestane by the reaction sequence: equatorial alcohol → equatorial tosylate → axial azide → axial amine[30]. The same sequence has also been used for introduction of axial amino groups into terpenes[31]. In the carbohydrate field the method is in frequent use in the preparation of aminosugars. Thus 4-amino-4-deoxy-α-D-glucopyranoside (**22b**) was readily obtained from the corresponding 4-mesyl-α-D-galactopyranoside (**22a**)[32]. Other recent

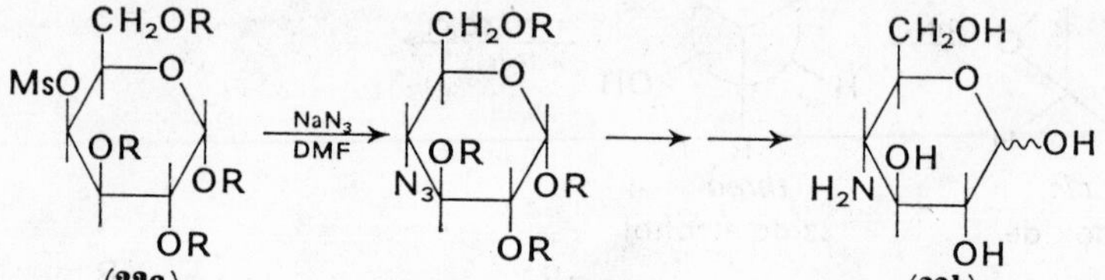

(**22a**) $\xrightarrow{NaN_3/DMF}$ → → (**22b**)

examples are the syntheses of 6-amino-6-deoxy-D-mannose[33] and of 3-amino-3-deoxy-D-ribose[34].

In nucleotides, azides have been used for the introduction of amino groups into either the purine moiety, as in the synthesis of 8-aminoadenosine[35], or into the sugar moiety, as in the synthesis of 5-amino-5-deoxy uridine[36].

2. By acidolysis of azides

Azides undergo decomposition upon treatment with strong concentrated acids, one of the main products being primary amines. The reaction of aryl azides with concentrated sulphuric acid yielded complex mixtures[37], probably because of nucleophilic attacks on the positively charged intermediates. Amines could be isolated in low yields only, and the reaction has no real preparative value.

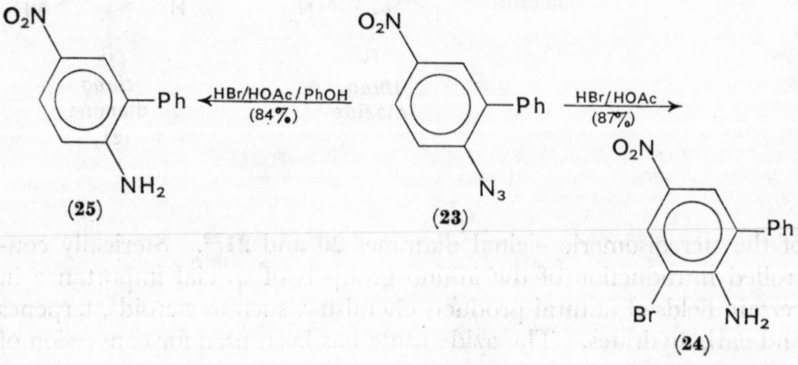

(**25**) $\xleftarrow[(84\%)]{HBr/HOAc/PhOH}$ (**23**) $\xrightarrow[(87\%)]{HBr/HOAc}$ (**24**)

Smooth conversion of azides to amines was achieved by using hydrogen bromide in acetic acid solution[38]. The reduction is, when possible, accompanied by bromination by the liberated bromine. The bromination can be prevented by addition of a bromine acceptor. 2-Azido-5-nitrobiphenyl (**23**) thus yielded 2-amino-3-bromo-5-nitrobiphenyl (**24**), but in the presence of phenol the product was 2-amino-5-nitrobiphenyl (**25**)[38]. The method has also been used with amino acids, acetone serving as a bromine acceptor[39].

The reaction of azides with aluminium chloride in benzene produces, beside other products and much tar, substituted anilines in low yields[40]. The reaction has been applied to α-azidoketones; thus azidoacetone (**26**) yielded 35% of α-anilinoacetone (**27**)[41].

$$\underset{(26)}{MeC(=O)CH_2N_3} \xrightarrow[C_6H_6]{AlCl_3} \underset{(27)}{MeC(=O)CH_2NHC_6H_5}$$

In the presence of sulphuric acid, azides react with aromatic aldehydes[42]. The first step is probably a condensation of the azide with the protonated aldehyde, to give the intermediate **28** which can either lose nitrogen to give the substituted benzamide **29** (path *a*, in the presence of an excess benzaldehyde), or rearrange with phenyl migration to the nitrogen to give, after decarbonylation, the substituted aniline **30** (path *b*). Yields in both cases were low (<30%)[42].

$$PhCHO + RN_3 \xrightarrow{H^+} \underset{(28)}{PhCH(OH)N(R)—N_2^+}$$

$$\text{path } b: \underset{(30)}{PhNHR} + CO + N_2 + H^+ \qquad \text{path } a: \underset{(29)}{PhC(=O)NHR} + N_2 + H^+$$

3. Via nitrene intermediates

a. Hydrogen abstraction. Nitrenes, generated from azides by pyrolysis or photolysis, yield stable products by several reactions, one of which is hydrogen abstraction from the environment to give primary amines

$$RN_3 \xrightarrow{\Delta \text{ or } h\nu} RN \xrightarrow{[H]} RNH_2$$

The amine-forming pathway is not the most favoured one. In the case of aliphatic azides, primary amines are very minor products[43,44]. In aromatic azides, where rearrangement is impossible, amine formation is more frequently encountered, sometimes in nearly quantitative yields[45]. The reaction course can be solvent dependent, thus pyrolysis of *o*-azidodiphenylmethane (**31**) in decalin, which is a good hydrogen donor, yielded the abstraction product, *o*-aminodiphenylmethane (**32**), while pyrolysis in 1,2,3-trichlorobenzene yielded exclusively azepino [2,1-*a*]-11*H*-indole (**310**)[46]. Some cases of intra-

$$\text{(31)}\ o\text{-}N_3C_6H_4CH_2C_6H_5 \xrightarrow[\text{decalin}]{\Delta} o\text{-}NH_2C_6H_4CH_2C_6H_5\ \text{(32)}$$

molecular hydrogen abstraction have also been reported. Pyrolysis of *o*-azidobenzyl alcohol (**33**) produced *o*-aminobenzaldehyde (**34**) in 60% yield[47].

$$\text{(33)}\ o\text{-}N_3C_6H_4CH_2OH \xrightarrow{\Delta} o\text{-}NH_2C_6H_4CHO\ \text{(34)}$$

b. Intermolecular insertion. The nitrene intermediates may undergo insertion into a molecule of the solvent, to give secondary amines. By

$$RN_3 \longrightarrow RN \xrightarrow{R'H} RNHR'$$

this process low yields of *N*-alkylanilines were obtained either by photolysis of alkyl azides in benzene[43] or pyrolysis of phenyl azide in aliphatic hydrocarbons[48].

Better yields have been obtained in the insertion reactions of carboalkoxy nitrenes. Pyrolysis of *n*-octadecylazidoformate (**35**) in cyclohexane gave 60% of (*N*-cyclohexyl)-*n*-octadecyl carbamate(**36**) along with 23% of the abstraction product **37**[49].

$$\underset{(35)}{Me(CH_2)_{17}OC(=O)N_3} \xrightarrow[130^\circ]{\text{cyclohexane}} \underset{(36)}{Me(CH_2)_{17}OC(=O)NH\text{-}C_6H_{11}} + \underset{(37)}{Me(CH_2)_{17}OC(=O)NH_2}$$

Similar results were obtained in the photolysis of ethyl azidoformate[50] and in the decomposition of sulphonyl azides, which yielded sulphonamides[51].

Heating of aryl azides in acetic anhydride resulted in insertion of the nitrene into the anhydride molecule. The phenylhydroxylamine derivative (**38**) formed readily rearranged to *N,O*-diacetyl-*o*-hydroxyaniline (**39**)[52].

$$PhN_3 \xrightarrow{Ac_2O} PhN(Ac)OC(=O)CH_3 \; (\mathbf{38}) \longrightarrow o\text{-}AcO\text{-}C_6H_4\text{-}NHAc \; (\mathbf{39})$$

c. The Curtius rearrangement. The behaviour of acyl azides upon heating differs from that of the other azide types described above. In this case migration of an alkyl or aryl group onto the nitrene nitrogen gives isocyanates, which are usually allowed to react further to yield primary amines, either directly or via urethans (equation 2). The

$$RCN_3(=O) \longrightarrow [RC(=O)N] \longrightarrow RNCO \xrightarrow{R'OH} RNHC(=O)OR \xrightarrow[\text{or } OH^-]{H^+} RNH_2 \quad (2)$$

$$RNCO \xrightarrow{H_2O/\,H^+ \text{ or } OH^-} RNH_2$$

reaction is one of the most useful methods for converting carboxylic acids to amines or amine derivatives containing one carbon less. A tabular compilation of Curtius reactions published up to 1945[53] lists hundreds of examples and shows the wide scope of the reaction. It has been carried out successfully on aliphatic, alicyclic, aromatic and

$$\text{Carbazole-N-C(=O)N}_3 \; (\mathbf{40}) \xrightarrow[\text{10 days}]{t\text{-BuOH, reflux}} \text{Carbazole-N-NHCOO } t\text{-Bu} \longrightarrow \text{Carbazole-N-NH}_2 \; (\mathbf{41})$$

heterocyclic acids, on saturated and unsaturated acids and on acids containing various functional groups.

Other types of carbonyl azides, such as azidoformates and carbamoyl azides, previously believed not to undergo the rearrangement, can also be induced to rearrange by photolysis in alcohols[54]. Diarylcarbamoyl azides have been found to rearrange even upon heating in *t*-butanol and this reaction was used for the synthesis of 1,1-diarylhydrazines, thus *N*-aminocarbazole (**41**) was prepared in 80% yield from **40**[55].

B. Syntheses of Azomethines

I. By acidolysis of azides

The action of strong acids on aliphatic azides causes decomposition with migration of either a hydrogen or an alkyl group onto the nitrogen, to give aldimines or ketimines respectively (equation 3). The

$$RR'CHN_3 \xrightarrow{H^+} RR'C{=}NH + RCH{=}NR' + R'CH{=}NR \qquad (3)$$

produced imines usually cannot be isolated and the reaction course can be determined only by detection of the amines and the carbonyl compounds formed by their hydrolysis. Acidolyses of benzyl azide[56] or ethyl azide[57] indeed yielded mixtures, but in higher azides, because of the increased difficulty of the larger alkyl groups to migrate, only hydrogen migration was detected, by isolation of the corresponding aldehyde. In this manner hexyl azide and dodecyl azide yielded *n*-caproaldehyde (73%) and lauraldehyde (84%) respectively[58]. The reaction can actually be regarded as a synthesis of carbonyl compounds and was suggested as the best method for converting ketones to α-diketones, as shown in the synthesis of 1,2-cyclohexanedione (**43**) from 2-azidocyclohexanone (**42**) (63% yield)[59].

2-azidocyclohexanone (**42**) $\xrightarrow{H_2SO_4}$ 2-oxocyclohexyl–$NH{-}\overset{+}{N}{\equiv}N$ (H) $\longrightarrow$ 2-oxocyclohexyl–NH^+ (H) $\longrightarrow$ 2-(NH_2^+=)cyclohexanone $\longrightarrow$ 1,2-cyclohexanedione (**43**)

It is interesting to note that acidolysis of β-azidoketones also yields α-diketones, probably via an aziridine intermediate. In this manner 1-azidobutane-3-one (**44**) yielded biacetyl (**45**)[60].

$$\underset{(\mathbf{44})}{MeCOCH_2CH_2N_3} \xrightarrow{H^+} Me{-}\underset{\underset{O}{\|}}{C}{-}\overset{H}{\underset{H}{C}}{-}\underset{NH{-}\overset{+}{N}{\equiv}N}{CH_2} \longrightarrow$$

$$MeC(=O){-}\overset{H}{C}{\underset{NH_2^+}{-}}CH_2 \longrightarrow MeC(=O){-}C(={}^+NH_2)CH_3 \longrightarrow \underset{(\mathbf{45})}{MeC(=O){-}C(=O)CH_3}$$

Acid treatment of benzhydryl azides and of 1,1-diarylethyl azides resulted in aryl migration[61], the migration aptitude being promoted by electron-releasing substituents and retarded by electron-attracting ones[61].

Treatment of methyl azide with antimony pentachloride yielded a crystalline adduct (**46**), which decomposed under the action of dry hydrogen chloride, to give formaldimine hexachloroantimonate (**47**)[62].

$$CH_3N_3 \xrightarrow{SbCl_5} \underset{(\mathbf{46})}{CH_3\underset{SbCl_5^-}{N}{-}N_2^+} \xrightarrow{HCl} \underset{(\mathbf{47})}{[CH_2{=}NH_2]^+\ SbCl_6^-}$$

Completely analogous to the protonation of azide is the attack by carbonium ions. This could be accomplished only by using very powerful carbonium ion donors. The reaction of ethyl azide and triethyloxonium fluoroborate thus yielded the imine **48**[63]. Other examples of this reaction have been reported[64].

$$C_2H_5N_3 \xrightarrow{(C_2H_5)_3O^+BF_4^-} C_2H_5{-}\underset{C_2H_5}{N}{-}N_2^+BF_4^- \longrightarrow \underset{(\mathbf{48})}{[C_2H_5NH{=}CHCH_3]^+\ BF_4^-}$$

2. By rearrangement of nitrenes

Aliphatic nitrenes having an α-hydrogen can form imines by 1,2-hydrogen shift (equation 4). This rearrangement has been found to

$$RCH_2N_3 \longrightarrow RCH_2N \longrightarrow RCH{=}NH \qquad (4)$$

be the main stabilization process both in the photolysis[44] and the gas phase pyrolysis[65] of alkyl azides. As in acidolysis, usually only the corresponding carbonyl compounds could be isolated; however, imines have been obtained in several cases. Heating of the azide **49** yielded the imine **50**[66] and photolysis of **51** yielded the imine **52**[67].

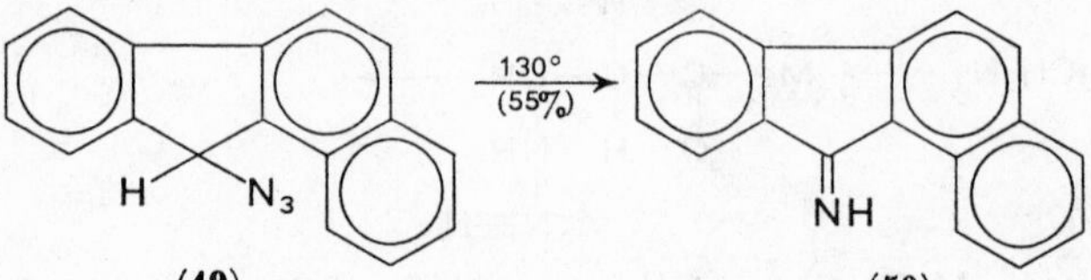

H N$_3$ —130° (55%)→ NH

(49) **(50)**

CH_2N_3 —$h\nu$ (35%)→ CH=NH

O, OAc, AcO, OCH_3, OAc

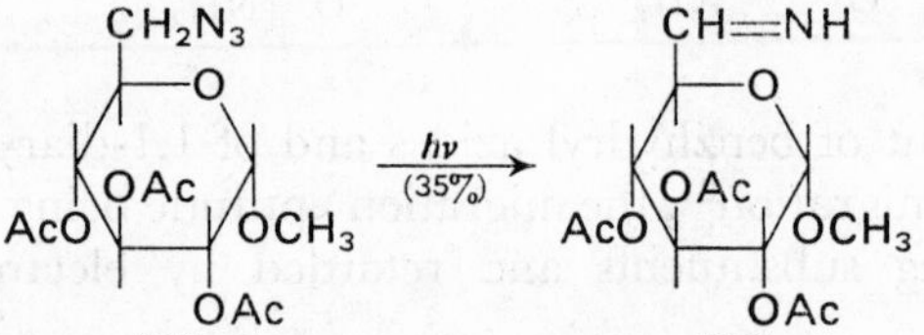

(51) **(52)**

When the azido group is linked to a tertiary carbon atom, an alkyl or aryl shift occurs, and the resulting *N*-substituted azomethines can often be isolated. Trityl azide (**53**) yielded, by action of heat[68] or light[69], benzophenone anil (**54**).

$$\underset{\textbf{(53)}}{Ph_3CN_3} \xrightarrow{\text{or } h\nu} \underset{\textbf{(54)}}{Ph_2C{=}NPh}$$

As the migrating aptitudes of the aryl groups are only slightly influenced by substituents in the thermal reaction and not at all in the photochemical one, trityl azides containing differently substituted aryl groups always yield mixtures of anils.

Some α-keto anils have been obtained in this manner by heating the suitable azidoketones, but yields were low. The imine **56** was prepared by heating of α-azido-α-phenylpropiophenone (**55**) in 19% yield[70].

$$\underset{\textbf{(55)}}{PhCO{-}\overset{\displaystyle Ph}{\overset{|}{C}}Me{-}N_3} \longrightarrow \underset{\textbf{(56)}}{PhCOCMe{=}NPh}$$

Vinyl nitrenes usually do not produce keteneimines. However, in the case of 1-azido-2,2-dicyanoethylenes (**57**) the major products were the dicyanoketeneimines **58**. Due to their instability only their addition products [i.e. with ethanol (**59**)] were isolated[71].

$$(NC)_2C{=}C(R)N_3 \ \textbf{(57)} \xrightarrow{\Delta} (NC)_2C{=}C{=}NR \ \textbf{(58)} \xrightarrow{C_2H_5OH} (NC)_2C{=}C(NHR)OC_2H_5 \ \textbf{(59)}$$

(R = H, Me, Ph)

3. By decomposition of azide adducts

Organic azides react with alkenes via 1,3-dipolar cycloaddition[72]. The resulting 1,2,3-triazolines are thermally unstable and eliminate nitrogen, to give azomethines (equation 5) beside other products such as aziridines.

$$>C{=}C(H)< + N_3R \longrightarrow \text{triazoline} \longrightarrow \text{(loss of } N_2^+\text{)} \longrightarrow >CH{-}C{=}NR \quad (5)$$

The course of the decomposition reaction is much dependent on the relative stereochemistry of the substituents on the C—C bond of the triazoline[73]. At intermediate temperatures (40–90°) the stereochemistry is conserved and usually only one product is detected. Cyclopentene, cycloheptene and *cis*-cyclooctene reacted with phenyl azide to give the corresponding *N*-phenylimino derivatives in excellent yields, while cyclohexene and *trans*-cyclooctene did not yield imines at all, aziridines being the sole products[73]. When higher temperatures are used, the stereochemical selectivity is apparently lost and a mixture of products is generally observed. In the reaction of norbornene

$$\text{norbornene} + C_6H_5N_3 \longrightarrow \textbf{(60)} \xrightarrow{160°} \text{norbornan-2-one } NC_6H_5 \ \textbf{(61)} + \text{aziridine } N{-}C_6H_5 + \text{other products}$$

with phenyl azide, decomposition of the adduct **60** yielded five products, with their ratio being dependent on the solvent used. The 2-phenylimine derivative **61** accounted for 46% of the products in dimethylformamide, and for only 18% in decalin[74].

Triazolines formed from azides attached to electron-withdrawing groups are much less stable and the decomposition products are obtained directly. *N*-Picryl imines were obtained in the reaction of picryl azide with a series of olefins: yields were 70% with cyclopentene, 58% with cycloheptene and only 20% with cyclohexene. No imine was formed with norbornene[75].

Allenes have been reported to undergo a carbon skeletal rearrangement on reacting with picryl azide. Thus tetramethylallene (**62**) yielded the imine **63**[76].

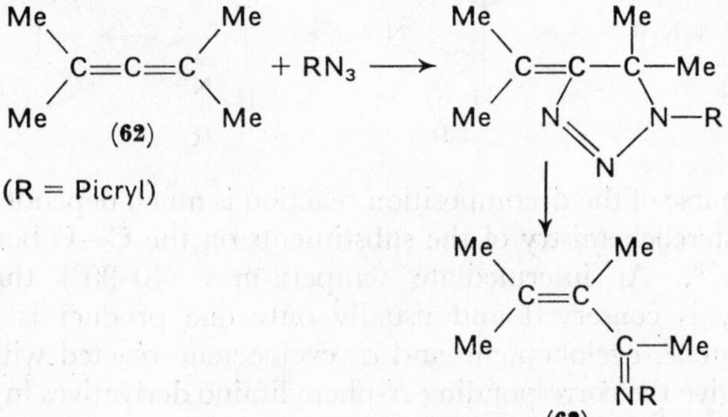

$$\mathrm{Me_2C{=}C{=}CMe_2}\ (\mathbf{62}) + \mathrm{RN_3} \longrightarrow \text{triazoline} \longrightarrow \mathrm{Me_2C{=}C(Me){-}C(Me){=}NR}\ (\mathbf{63})$$

(R = Picryl)

Cyclic enol ethers can be converted by this method to iminolactones. Compound **66** was obtained from 2,3-dihydro-2*H*-pyran (**64**) by heating the triazoline **65** in boiling toluene[77].

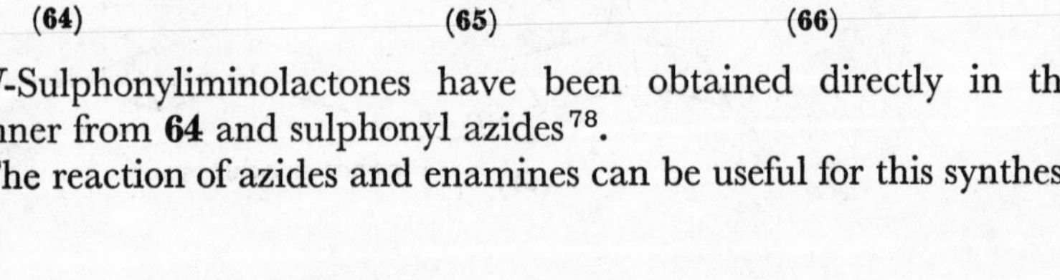

$$(\mathbf{64}) + p\text{-}\mathrm{BrC_6H_4N_3} \longrightarrow (\mathbf{65})\ (\mathrm{N{-}C_6H_4Br}\text{-}p) \longrightarrow (\mathbf{66})\ \mathrm{{=}N{-}C_6H_4Br}\text{-}p$$

N-Sulphonyliminolactones have been obtained directly in this manner from **64** and sulphonyl azides[78].

The reaction of azides and enamines can be useful for this synthesis of amidines. The amidine **68** was the final product in the reaction of 1-morpholino-1-butene (**67**) and *p*-nitrophenyl azide[79]. When tosyl

$$C_2H_5CH{=}CH{-}N(\text{morpholino}) \; (\mathbf{67}) + p\text{-}N_3C_6H_4NO_2 \longrightarrow \text{triazoline} \; (C_2H_5, \text{morpholino}, N\text{-}C_6H_4NO_2\text{-}p) \longrightarrow n\text{-}PrC(\text{morpholino}){=}N{-}C_6H_4{-}NO_2 \; (\mathbf{68})$$

azide is employed in the reaction with enamines, the triazoline may, in certain cases, decompose via a 1,3-elimination. The benzoyl enamine **69** thus yielded the enamine **70** and diazo-acetophenone[80]. A similar cleavage was also observed in the reaction of some enol ethers with tosyl azide[81].

$$PhCOCH{=}CHN(Me)Ph \; (\mathbf{69}) + TosN_3 \longrightarrow \text{triazoline} \; (PhOC, N(Me)Ph, N\text{-}Tos) \longrightarrow (Me)(Ph)NCH{=}NTos \; (\mathbf{70}) + C_6H_5COCHN_2$$

Azomethines have also been obtained in the reaction of azides with thioketones. Thiobenzephenone and phenyl azide thus yielded benzophenone anil (**72**), probably via the cyclic intermediate **71**[82].

$$Ph_2C{=}S + PhN_3 \longrightarrow \underset{\textbf{(71)}}{\text{Ph}_2\text{C—S—N=N—N(Ph) (ring)}} \longrightarrow \underset{\textbf{(72)}}{Ph_2C{=}NPh}$$

C. Syntheses of Nitriles

Terminal vinyl azides give, upon decomposition, vinylnitrenes which can stabilize by rearrangement to nitriles. The pyrolysis of β-styryl azide (**73**) afforded in this manner phenylacetonitrile (**74**) (74%)[83]. Nitriles were also obtained from α-azidoketenes (**75**) generated *in situ* by treatment of α-azidoacyl chloride with triethylamine[84]. The postulated mechanism is a cyclization to an azirine (**76**) which undergoes decarbonylation to the nitrile **77**. Yields were 40–85%[84].

$$\underset{\textbf{(73)}}{PhCH{=}CHN_3} \longrightarrow PhCH{=}CHN \longrightarrow \underset{\textbf{(74)}}{PhCH_2C{\equiv}N}$$

$$\underset{}{RCH(N_3)COCl} \xrightarrow{(Et)_3N} \underset{\textbf{(75)}}{RC(N_3){=}C{=}O} \longrightarrow \underset{\textbf{(76)}}{R{-}C(=N){-}C{=}O \text{ (azirine)}} \longrightarrow \underset{\textbf{(77)}}{RC{\equiv}N}$$

(R = Et, *n*-Bu)

Cyclic vicinal diazides undergo a thermal ring fission with formation of dinitriles. In this manner 1,2-diazidobenzenes yielded (in ca. 80% yield) *cis, cis*-1,4-dicyano-1,3-butadienes (**78**), and 1,2-diazidonaphthalene yielded *cis*-2-cyanocinnamonitrile (**79**)[85].

$$\text{R-C}_6\text{H}_3(N_3)_2 \xrightarrow{\Delta} \underset{\textbf{(78)}}{\text{R-CH=CH-CH=CH(CN)... (dicyanobutadiene)}}$$

$$\text{1,2-diazidonaphthalene} \xrightarrow{\Delta} \underset{\textbf{(79)}}{\text{2-(CN)C}_6\text{H}_4\text{-CH=CH-CN}}$$

(R = H, Me, OCH_3, Cl)

Other examples are the fissions of 2,3-diazido-1,4-naphthoquinone to phthaloyl cyanide (**80**)[86] and of 2,3-diazido-*N*-phenylmaleimide to

N,*N*-bis (cyanocarbonyl) aniline (**81**)[87]. Another type of ring fission

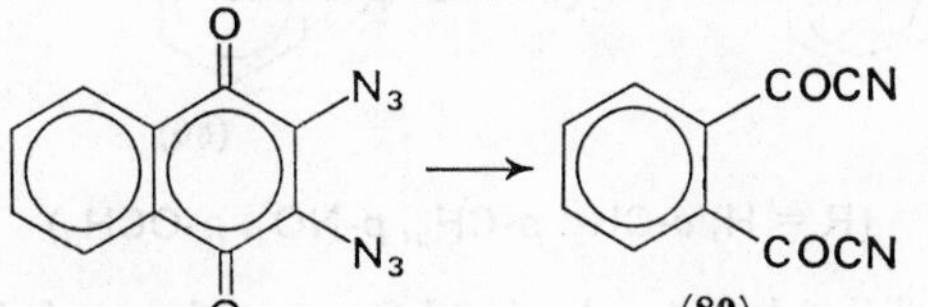

(**80**)

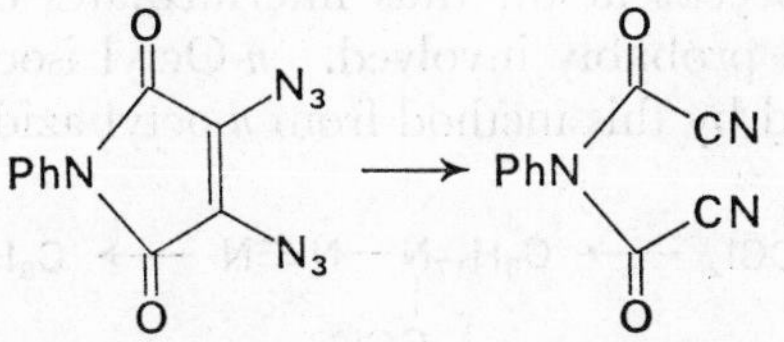

(**81**)

with nitrile formation has been observed on heating 2-azidotropone (**82**). The cyanoketene **83** formed cyclized to salicylonitrile (**84**) or could be trapped by reaction with nucleophiles, yielding their acyl derivatives **85**[88].

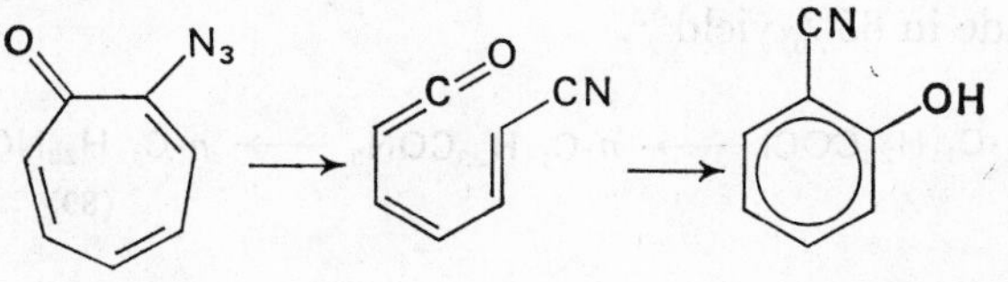

(**82**) (**83**) (**84**)

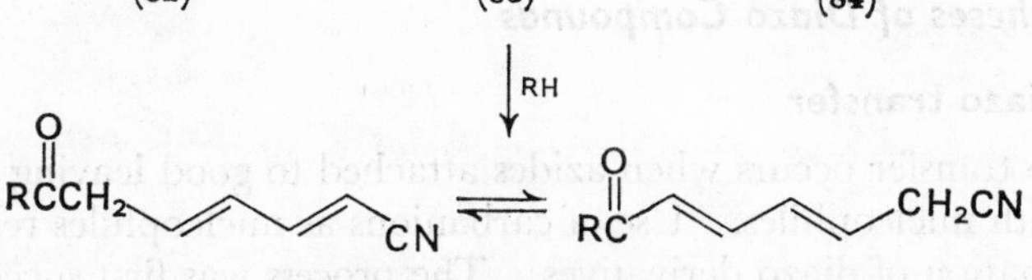

(**85**) (R = OMe, NHC_6H_5)

D. Syntheses of Isocyanates

Isocyanates can be prepared from azides by reaction with carbon monoxide. The reaction has been at first reported to proceed only with catalysis of rhodium or iridium carbonyl complexes[89]. Later work has however shown that aryl azides and carbon monoxide interact without catalysis at temperatures of 160–180° and pressures of 200–300 atm, yielding aryl isocyanates (**86**) in good yields. Ethyl azidoformate yielded under these conditions ethoxyisocyanate[90].

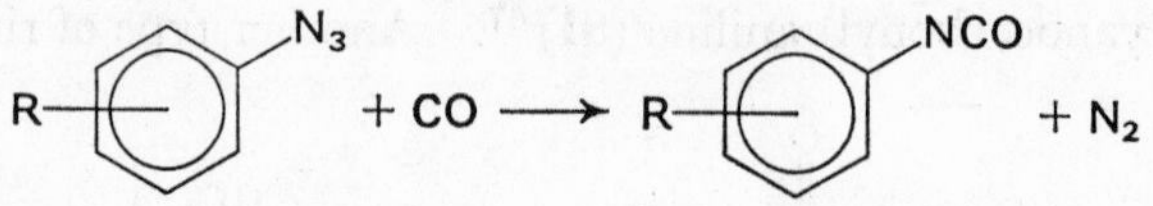

$$R\text{-}C_6H_4N_3 + CO \longrightarrow R\text{-}C_6H_4NCO + N_2$$

(86)

(R = H, *o*-CH_3, *p*-CH_3, *p*-NO_2, *p*-OCH_3)

Isocyanate dichlorides are obtained from azides and dichlorocarbene. The reaction proceeds at 0°, thus intermediates of type **87** rather than nitrenes are probably involved. *n*-Octyl isocyanate dichloride (**88**) was prepared by this method from *n*-octyl azide in 89% yield[91].

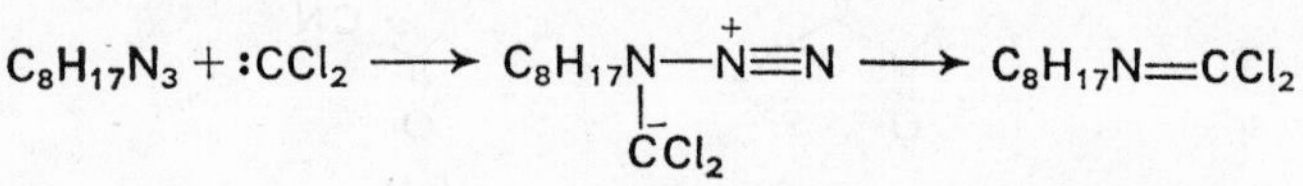

$$C_8H_{17}N_3 + :CCl_2 \longrightarrow C_8H_{17}N(\bar{C}Cl_2)\text{—}\overset{+}{N}{\equiv}N \longrightarrow C_8H_{17}N{=}CCl_2$$

(87) **(88)**

Another approach to the preparation of isocyanates is the Curtius rearrangement. The reaction can be stopped at the isocyanate stage when carried out in an inert solvent such as benzene. Undecyl isocyanate (**89**) has been prepared from the corresponding acyl chloride via the azide in 85% yield[92].

$$n\text{-}C_{11}H_{23}COCl \longrightarrow n\text{-}C_{11}H_{23}CON_3 \longrightarrow n\text{-}C_{11}H_{23}NCO$$

(89)

E. Syntheses of Diazo Compounds

I. By diazo transfer

Diazo transfer occurs when azides attached to good leaving groups react with nucleophiles. Use of carbanions as nucleophiles results in the formation of diazo derivatives. The process was first successfully utilized by Doering and DePuy[93], who reacted cyclopentadienyl lithium with tosyl azide and obtained diazocyclopentadiene (**90**) in

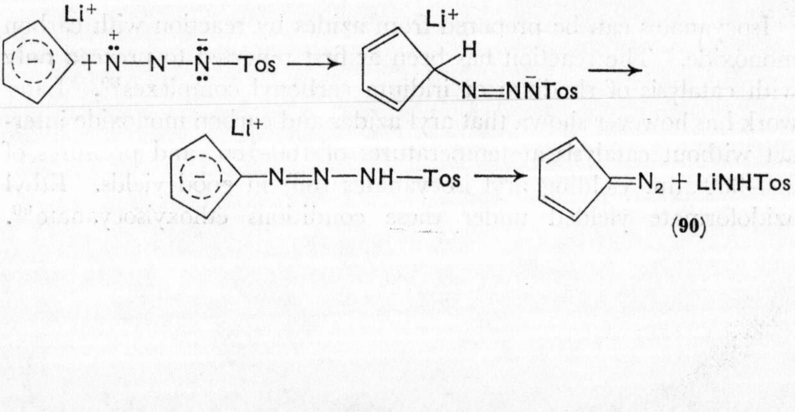

$$C_5H_5^- Li^+ + \overset{+}{N}{-}N{-}\bar{N}{-}Tos \longrightarrow C_5H_5(H)N{=}N\bar{N}Tos\ Li^+ \longrightarrow$$

$$C_5H_4^- Li^+\text{—}N{=}N{-}NH{-}Tos \longrightarrow C_5H_4{=}N_2 + LiNHTos$$

(90)

35% yield. The reaction has since been used for the preparation of various types of diazo compounds and has been recently reviewed[94].

The synthesis of diazocyclopentadiene and its substituted derivatives has been improved by using weaker bases such as diethyl amine[95,96] or piperidine[97]. The yield of **90** was increased to 85% and that of its triphenyl derivative to 95%. 2,5-Diphenylcyclopentadiene (**91**) yielded a mixture of **92** (58%) and **93** (30%), as a result of the corresponding mesomeric forms of the carbanion[97].

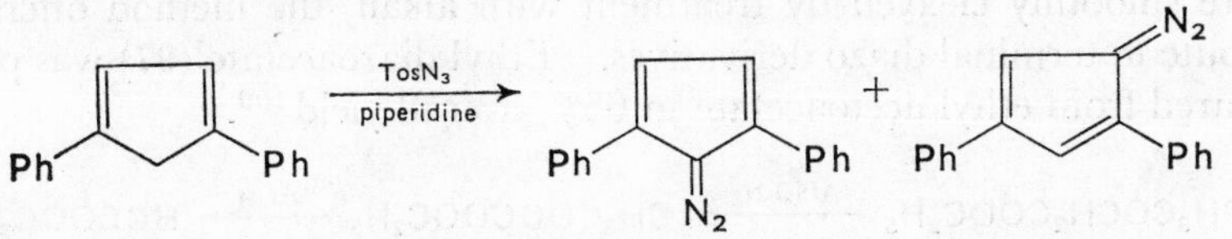

TosN$_3$ / piperidine; Ph; Ph; N$_2$; +

(**91**) (**92**) (**93**)

Diazoanthrone (**94**) was obtained in the same manner from anthrone in 94% yield[98].

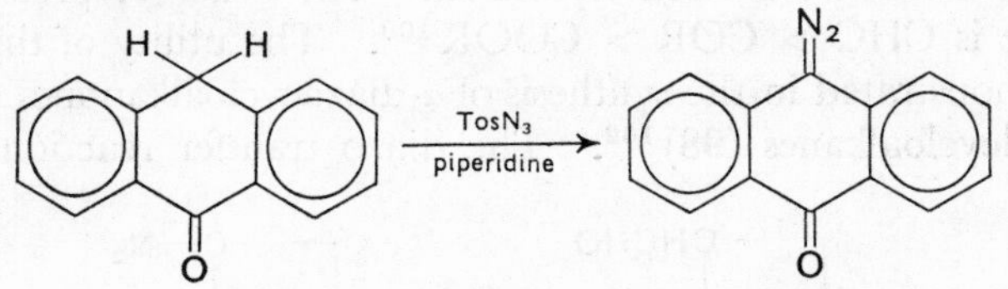

H H; O; TosN$_3$ / piperidine; N$_2$; O

(**94**)

Excellent yields of diazo derivatives can be obtained by the reaction of the active methylene group in 1,3-dicarbonyl compounds. In this

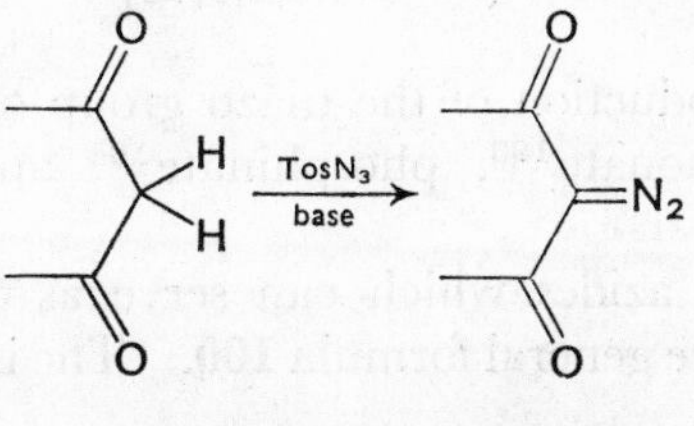

O; H; H; O; TosN$_3$ / base; =N$_2$

manner 1,3-diketones, β-keto esters and dialkyl malonates[99,100] were transformed into the corresponding diazo compounds, the most effective reagent being tosyl azide. Picryl azide gave substantially poorer results[100].

Methylene groups activated by only one carbonyl group usually do not react, but β-aryl ketones are reactive. Side products formed via a Wolf rearrangement are also obtained, thus phenylacetone yielded the expected **95** together with **96**[100,101]. As α-diazoketones

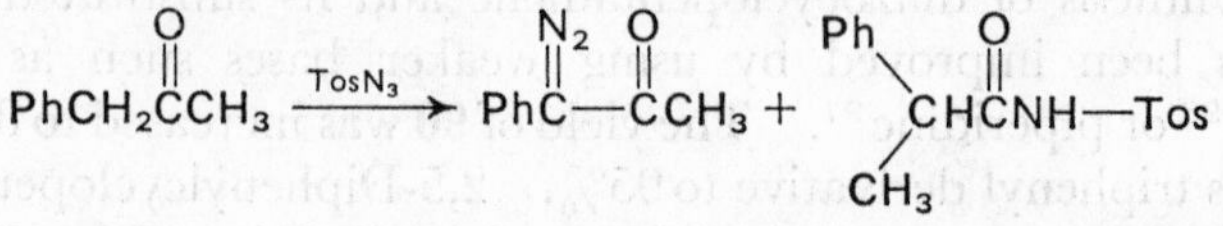

$$PhCH_2\overset{O}{\overset{\|}{C}}CH_3 \xrightarrow{TosN_3} PhC(=N_2)-\overset{O}{\overset{\|}{C}}CH_3 + Ph(CH_3)CH\overset{O}{\overset{\|}{C}}NH-Tos$$

(95) **(96)**

are smoothly cleaved by treatment with alkali, the method offers a route to terminal diazo derivatives. Ethyl diazoacetate (**97**) was prepared from ethyl acetoacetate in 65% overall yield[100].

$$CH_3COCH_2COOC_2H_5 \xrightarrow{ArSO_2N_3} CH_3COC(=N_2)COOC_2H_5 \xrightarrow{NaOH} HC(=N_2)COOC_2H_5$$

(97)

The cleavage of β-carbonyl groups occurs readily when the α-position contains only one hydrogen. The order of preference for cleavage is CHO > COR > COOR[100]. The utility of this process was demonstrated in the synthesis of α-diazocycloalkanones (**99**) from α-formylcycloalkanes (**98**)[102]. The diazo transfer reaction has also

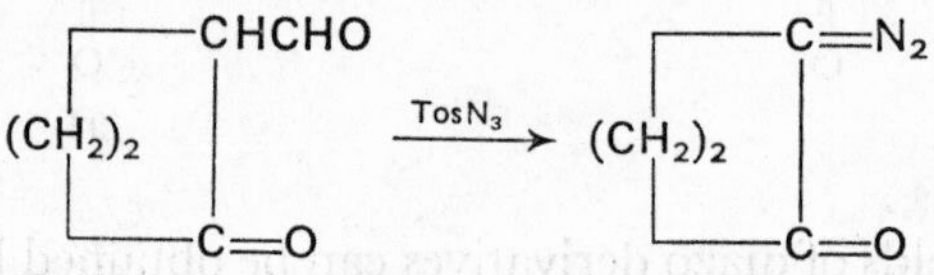

$$\underset{(98)}{\overbrace{(CH_2)_n\langle CHCHO, C=O\rangle}} \xrightarrow{TosN_3} \underset{(99)}{\overbrace{(CH_2)_n\langle C=N_2, C=O\rangle}}$$

(n = 3,4,5,6,9,10)

served for the introduction of the diazo group on methylene groups flanked by phosphonate[103], phosphinate[104] and sulphonate[105,106] groups.

Another type of azides which can serve as diazo donors is the azidinium salts of the general formula **100.** The important advantage

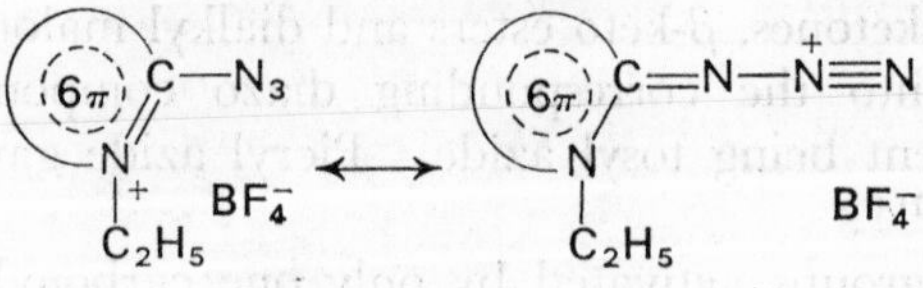

$$(6\pi)\,C{-}N_3{=}\overset{+}{N}(C_2H_5)\ BF_4^- \longleftrightarrow (6\pi)\,C{=}N{-}\overset{+}{N}{\equiv}N,\ N(C_2H_5)\ BF_4^-$$

(100)

of these reagents over sulphonyl azides is their ability to function in a non-basic medium, thus avoiding side reactions. An example is the reaction of α-cyanoacetophenone (**101**) with 2-azido-3-ethylbenzothiazolium fluoroborate (**102**), which yielded α-diazo-α-cyanoacetophenone (**103**) in 88% yield[107]. The method served for the first preparation α-diazo-α-nitro compounds[108].

$$\text{2-}N_3\text{-3-Et-benzothiazolium } BF_4^- \ (\mathbf{102}) + H_2C(CN)COPh \ (\mathbf{101}) \longrightarrow \text{2-}NH_2\text{-3-Et-benzothiazolium } BF_3^- + N_2{=}C(CN)COPh \ (\mathbf{103})$$

2. By cleavage of azide adducts

The cleavage of the triazolines formed by the addition of azides to oxo-enamines has already been mentioned in section II.B.3. This cleavage can be applied to the synthesis of diazo compounds unavailable otherwise. α-Diazobutyraldehyde (**105**) was obtained for the first time from α-ethyl-β-dimethylaminoacraldehyde (**104**) and tosyl or picryl azide[109].

$$(CH_3)_2NCH{=}C(C_2H_5){-}CHO \ (\mathbf{104}) + RN_3 \longrightarrow$$

$$(CH_3)_2N{-}HC{-}C(C_2H_5)(CHO) \text{ [triazoline ring: } R{-}N{-}N{=}N] \longrightarrow CH_3CH_2C({=}N_2)CHO \ (\mathbf{105}) + (CH_3)_2N{-}CH{=}NR$$

The triazolines formed from aryl azides and α,β-unsaturated esters or ketones can be opened by bases to α-aminodiazo compounds. Methyl acrylate and phenyl azide thus yielded ethyl 2-diazo-3-phenylaminopropionate (**106**)[110].

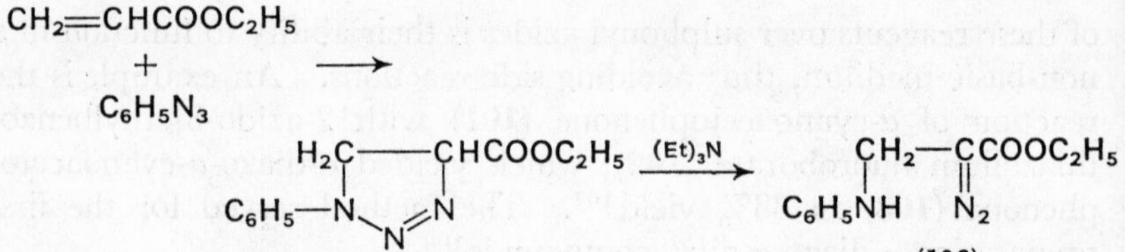

$$CH_2{=}CHCOOC_2H_5 + C_6H_5N_3 \longrightarrow \text{triazoline } (H_2C{-}CHCOOC_2H_5,\ C_6H_5{-}N{-}N{=}N) \xrightarrow{(Et)_3N} C_6H_5NH{-}CH_2{-}C(=N_2){-}COOC_2H_5$$

(106)

Imino diazo compounds are obtained directly by reaction of sulphonyl azides with ethoxyacetylene. With tosyl azide the product was **107**[111].

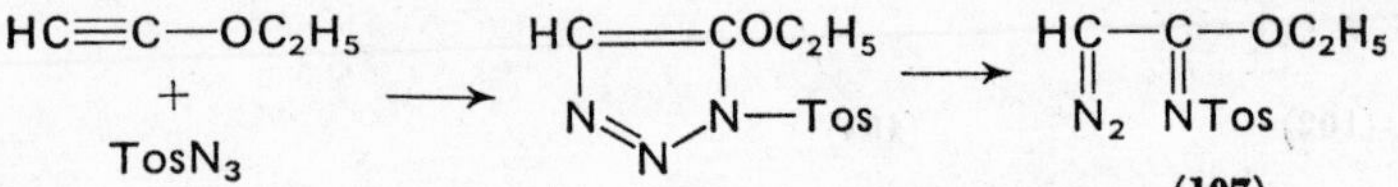

$$HC{\equiv}C{-}OC_2H_5 + TosN_3 \longrightarrow \text{triazole } (HC{=}COC_2H_5,\ N{=}N{-}N{-}Tos) \longrightarrow HC(=N_2){-}C(=NTos){-}OC_2H_5$$

(107)

F. Syntheses of Azo Compounds

I. By diazo transfer

Diazo compounds couple with active methylene groups under basic conditions. Using one mole of the azide with two moles of methylene

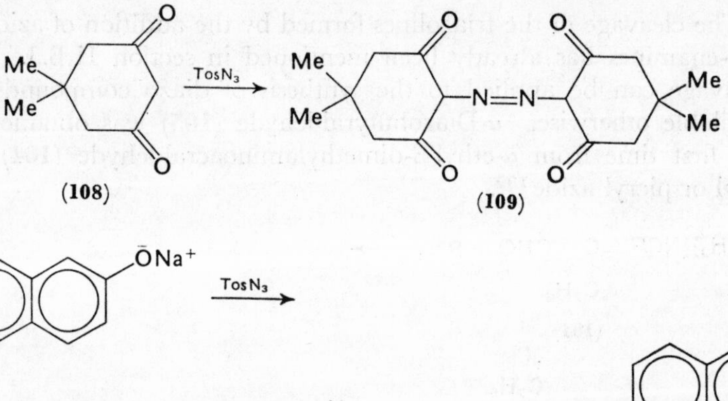

$$\text{(108)} \xrightarrow{TosN_3} \text{(109)}$$

(108) (109)

$$\text{2-naphthol sodium salt } (\bar{O}Na^+) \xrightarrow{TosN_3} \text{1-diazonio-2-naphthoxide } (N_2^+,\ O^-) \longrightarrow \text{(110)}$$

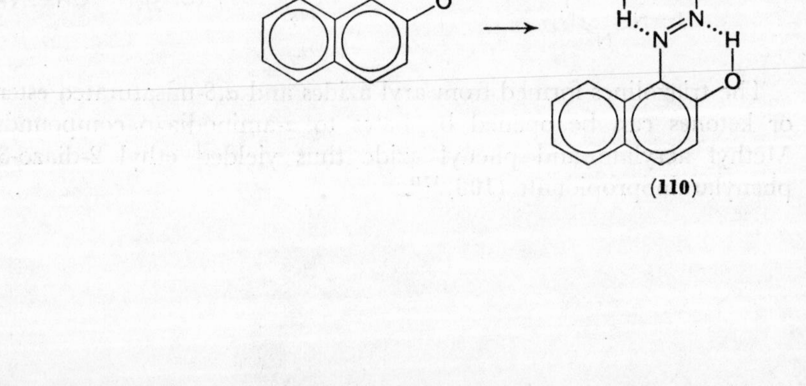

(110)

compound in the usual diazo transfer reaction furnishes the azo derivatives directly. Dimedone (**108**) afforded the azo compound **109** in 70% yield[112]. Azo derivatives are also obtained in the reaction of tosyl azide with phenoxide ions. Sodium β-naphthoxide yielded 2,2′-dihydroxy-1,1′-azonaphthalene (**110**)[113].

2. By dimerization of nitrenes

One of the stabilization modes of nitrenes is a coupling to form azo compounds. This dimerization has been observed both in the pyrolysis[47] and the photolysis[114] of aryl azides. The pyrolysis of *o*-trifluoromethylphenyl azide in the gas phase yielded 80% of bis (*o*-trifluoromethyl) azobenzene (**111**).

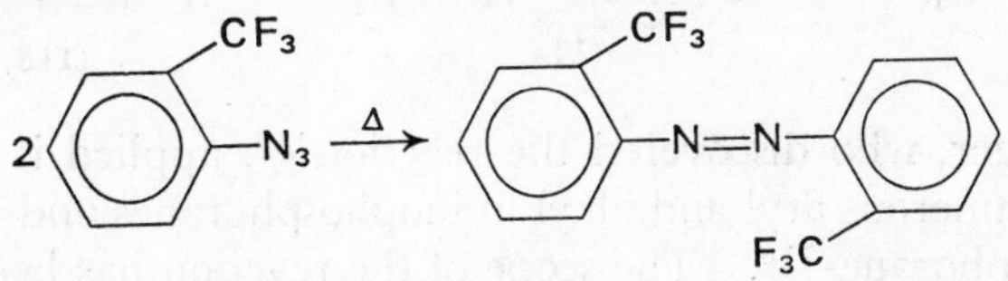

CF_3 N_3 Δ CF_3 —N=N— F_3C

(**111**)

Photolysis of *p*-methoxy and *p*-phenyl phenyl azides afforded the corresponding azo derivatives in 94% and 81% respectively. A mixture of these two azides yielded the three possible azo products in about equal amounts[114]. The tendency towards dimerization on photolysis is strengthened by addition of sensitizers. 2-Azidobiphenyl yielded in the presence of acetophenone mainly the azo derivative (**112**), with only minor amounts of carbazole[115]. Azidoformates can also form

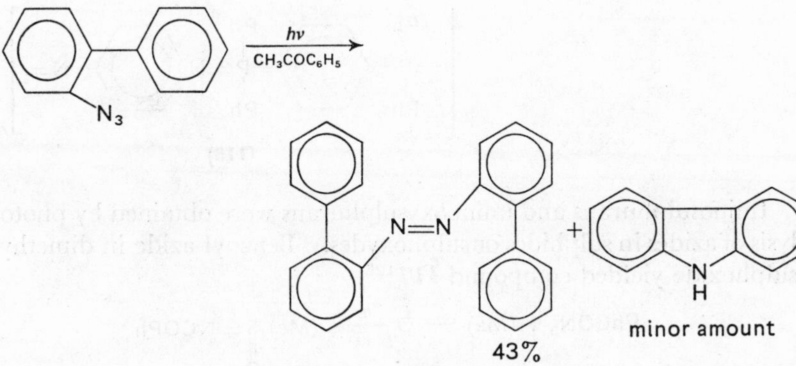

N_3 $h\nu$ $CH_3COC_6H_5$

N=N + N H

minor amount

43%

(**112**)

azo compounds, but in lower yields. The photolysis of neat ethyl azidoformate thus yielded diethyl azodiformate (**113**)[116,117].

$$2\ C_2H_5O\overset{O}{\overset{\|}{C}}N_3 \xrightarrow{h\nu} C_2H_5O\overset{O}{\overset{\|}{C}}N{=}N\overset{O}{\overset{\|}{C}}OC_2H_5$$

(**113**)

G. Syntheses of Iminophosphoranes and Iminosulphuranes

Iminophosphoranes (**115**) are produced in the reaction of azides with tertiary phosphines. The reaction proceeds via the phosphatriazenes **114**.

$$RN_3 + R'_3P \longrightarrow R{-}N{=}N{-}N{=}PR'_3 \longrightarrow R{-}N{=}PR'_3 + N_2$$

(**114**) (**115**)

Staudinger, who discovered the reaction[118], applied it to the synthesis of numerous aryl and alkyl iminophosphoranes and to benzoyliminophosphoranes[119]. The scope of the reaction has been widened to include sulphonyl azides[120] and carbamoyl azides[121].

A polymeric iminophosphorane (**116**) has been prepared by the reaction of 1,4-diazidobenzene with 1,4-bis(diphenylphosphino)benzene[122]. The iminophosphoranes are useful intermediates for the synthesis of a variety of nitrogen compounds from azides[123,124].

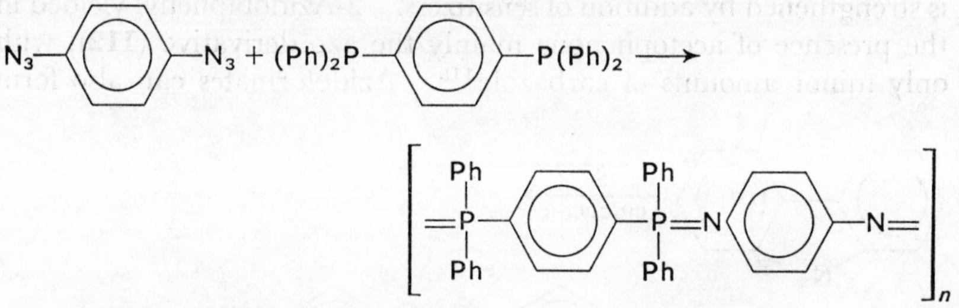

$$N_3{-}C_6H_4{-}N_3 + (Ph)_2P{-}C_6H_4{-}P(Ph)_2 \longrightarrow \left[{=}P(Ph)_2{-}C_6H_4{-}P(Ph)_2{=}N{-}C_6H_4{-}N{=} \right]_n$$

(**116**)

Iminosulphurans and iminooxysulphurans were obtained by photolysis of azides in sulphides or sulphoxydes. Benzoyl azide in dimethyl sulphoxide yielded compound **117**[125].

$$PhCON_3 + (Me)_2S{=}O \xrightarrow{h\nu} (Me)_2\underset{\underset{O}{\|}}{S}{=}NCOPh$$

(**117**)

H. Syntheses of Triazenes

The conversion of azides to mono substituted triazenes by reduction is a very difficult operation, owing to the instability of the products. Phenyltriazene (**118**) was prepared by the reduction of phenyl azide by stannous chloride in ether at $-20°$ [126].

$$C_6H_5N_3 \xrightarrow{SnCl_2} C_6H_5N{=}N{-}NH_2$$

(118)

1,3-Disubstituted triazenes are formed in the reaction of azides with nucleophiles, the attack always occurring at the azide terminus. With Grignard reagents, triazenes of the type **119** are obtained.

$$RN_3 + R'MgX \longrightarrow R{-}\underset{MgX}{\underset{|}{N}}{-}N{=}N{-}R' \xrightarrow{H_2O} R{-}NH{-}N{=}N{-}R'$$

(119)

The reaction was discovered by Dimroth [127], who synthesized a series of diaryl, dialkyl and aryl alkyl triazenes [128]. Vinyl triazenes were prepared in the same manner, thus phenyl azide and styrylmagnesium bromide afforded phenylstyryl triazene (**120**) in 54% yield [129].

$$C_6H_5N_3 + C_6H_5CH{=}CHMgBr \longrightarrow C_6H_5{-}NH{-}N{=}N{-}CH{=}CHC_6H_5$$

(120)

The reaction of azides with cyanide ions yields, after acidification, cyanotriazenes (**121**). Due to their instability these were isolated as sodium or silver salts [130].

$$C_6H_5N_3 + CN^- \longrightarrow C_6H_5{-}N{=}N{-}NHCN$$

(121)

The isolation of phosphatriazenes ($R{-}N{=}N{-}N{=}PR_3'$), the intermediates in the synthesis of iminophosphoranes, has been reported in several cases [131–133].

Triazenes such as **122** are obtained in the reaction of azidinium salts with sodium azide. The reaction proceeds via an intermediate

(3-ethyl-2-azidobenzothiazolium BF_4^-) $+ NaN_3 \longrightarrow$ (benzothiazolium–$N{=}N{-}N{=}$benzothiazoline triazene, N-C_2H_5, BF_4^-)

(122)

tetrazene, which combines with an additional molecule of the starting azidinium salt with elimination of $2N_2$[134].

I. Syntheses of Azides

Tosyl azide has been used as a starting material for the synthesis of alkyl and aryl azides by acting as a diazo donor to amine metal or Grignard compounds. Yields were 25–50%[100,135,136]. An example is the conversion of cyclohexylamine to cyclohexyl azide (**123**) (35%)[136].

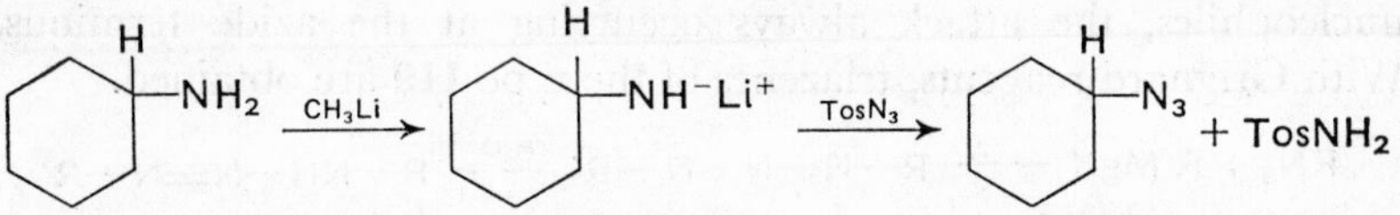

$$C_6H_{11}NH_2 \xrightarrow{CH_3Li} C_6H_{11}NH^-Li^+ \xrightarrow{TosN_3} C_6H_{11}N_3 + TosNH_2$$

(**123**)

Another approach utilized the reaction of tosyl azide with aryl Grignard reagents and fragmentation of the resulting tosyltriazene salts by aqueous sodium pyrophosphate. Mesityl azide (**124**) was thus obtained from mesityl bromide in 63% yield[137].

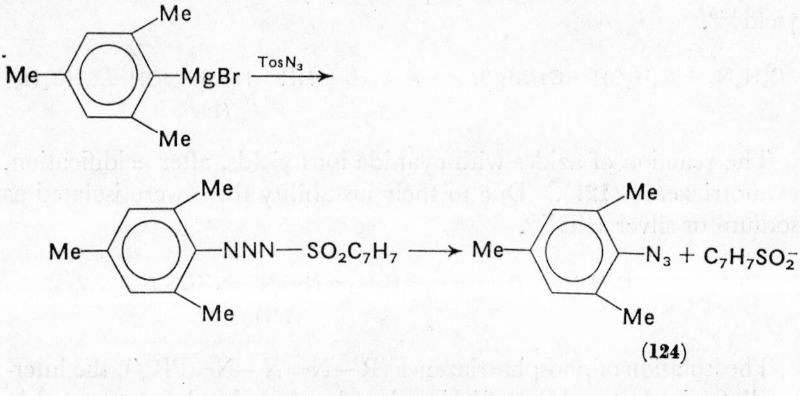

$$2,4,6\text{-}Me_3C_6H_2MgBr \xrightarrow{TosN_3}$$

$$2,4,6\text{-}Me_3C_6H_2\text{—NNN—}SO_2C_7H_7 \longrightarrow 2,4,6\text{-}Me_3C_6H_2\text{—}N_3 + C_7H_7SO_2^-$$

(**124**)

III. AZIDES AS STARTING MATERIALS IN SYNTHESIS OF HETEROCYCLES

A. General Considerations

The azido group can be utilized for the synthesis of heterocyclic compounds in several ways, the most important being the following:

(1) Syntheses of functional groups described above, which involve a bimolecular reaction, can take place in an intramolecular fashion when the azide group is placed in a suitable molecule.

The newly formed functional group thus becomes a part of a ring.

(2) Azide reactions can be carried out on azides containing suitable additional functional groups to yield products which can undergo cyclization to form heterocycles.

(3) Heterocycles may be formed directly by addition reaction of azides.

Applying these approaches azides have been widely used as starting materials in the synthesis of heterocycles of various types and sizes, some of which are unavailable by other routes.

B. Syntheses of Three-membered Rings

I. Aziridines

The photochemical decomposition of the triazolines formed by the addition of organic azides to double bonds is a very efficient method for the preparation of aziridines (equation 6).

$$\text{C=C} + RN_3 \longrightarrow \text{triazoline} \xrightarrow{h\nu} \text{aziridine (N—R)} \qquad (6)$$

$$CH_2{=}CH{-}CONH_2 + C_6H_5N_3 \xrightarrow{(36\%)} \text{triazoline} \xrightarrow{h\nu} \text{N—}C_6H_5 \text{ aziridine} \quad (\mathbf{125})$$

$$\text{methylenecyclohexane} + p\text{-}BrC_6H_4N_3 \xrightarrow{(24\%)} \text{spiro-triazoline} \xrightarrow{h\nu} \text{spiro-aziridine, N—}C_6H_4Br\text{-}p \quad (\mathbf{126})$$

$$\text{cyclohexadiene} + p\text{-}BrC_6H_4N_3 \xrightarrow{(71\%)} \text{triazoline} \xrightarrow{h\nu} \text{N—}C_6H_4Br\text{-}p \quad (\mathbf{127})$$

Aziridines are obtained almost exclusively and contamination with azomethines is very small if any[138]. The course of the photolysis is not affected by the nature of the substituents on the triazoline nucleus, thus the reaction has quite a wide scope. Recent examples are the syntheses of 2-carboxamido-1-phenylaziridine (**125**), 1-*p*-bromophenyl-1-azaspiro[2,5]-octane (**126**) and 7-*p*-bromophenyl-7-azabicyclo[4,1,0]hept-2-ene (**127**)[139]. Yields in the irradiation step were almost quantitative.

Although thermolysis of triazoline can also, in certain cases[73], lead to aziridines, mixtures with azomethines usually result, and the photochemical process is preferable. The triazolines formed from acyl[140] or sulphonyl azides[141] are unstable and the corresponding aziridines can be obtained directly. The reaction of norbornene and benzoyl azide at 40° yielded the aziridine (**128**) directly[140].

$$\text{norbornene} + C_6H_5CON_3 \xrightarrow{40\%} \text{N—}COC_6H_5$$

(**128**)

A related reaction is the addition of nitrenes to olefins. Irradiation of ethyl azidoformate in cyclohexene yielded 7-carbethoxy-7-azabicyclo[2,1,0] heptane (**129**) (50%)[50].

$$\text{cyclohexene} + N_3COOC_2H_5 \xrightarrow{h\nu} \text{NCOOC}_2H_5$$

(**129**)

As the two components do not react in the dark, a triazoline intermediate is not probable. Other *N*-carbethoxyaziridines were formed via carbethoxy nitrene when ethyl azidoformate was irradiated in dihydropyran[142] and in enolacetates[143]. The stereospecificity of the addition is high. The photolysis of *cis* and *trans* 2-butene at $-20°$ yielded mainly the *cis* (**130**) and *trans* (**131**) aziridines respectively[144]. This stereospecificity has been found to diminish upon dilution[145].

$$H_3C\text{—CH=CH—}CH_3 + N_3COOC_2H_5 \xrightarrow{h\nu}$$

CH₃ CH₃ / H— —H / N—$COOC_2H_5$ (87% + 13% of **131**)

(**130**)

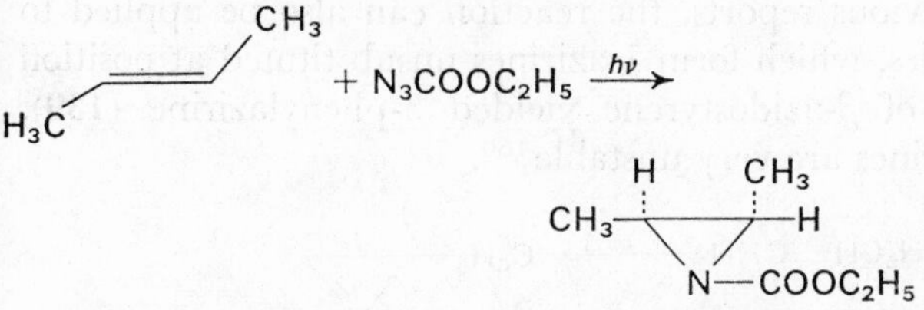

$$CH_3CH{=}CHCH_3 + N_3COOC_2H_5 \xrightarrow{h\nu}$$ (131) (92% + 8% of 130)

CH₃, H, CH₃, H, N—COOC₂H₅

The photolysis of 1-azido-2-phenylprop-2-ene (**132**) yielded 3-phenyl-1-azabicyclo[1,1,0]butane (**133**), contaminated with the imine **134**, probably via an intramolecular addition of the allylic nitrene[146].

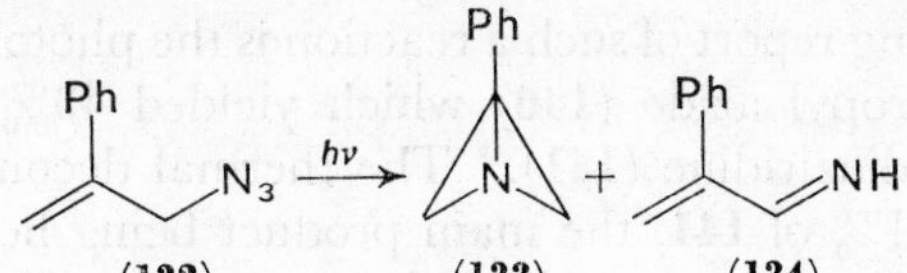

(132) $\xrightarrow{h\nu}$ (133) + (134)

2. Azirines

Thermolysis or photolysis of vinyl azides gives 1-azirines, usually in good yields. Heating α-azidostyrene in the gas phase produced 2-phenyl-1-azirine (**135**) in 80% yield[147]. The carbethoxy azirine

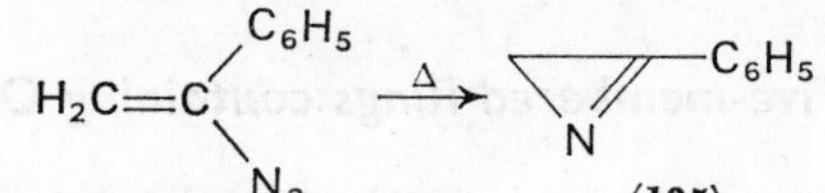

$$H_2C{=}C(C_6H_5)N_3 \xrightarrow{\Delta}$$ (135)

137 was obtained in 65% yield by photolysis of the azide **136**[148].

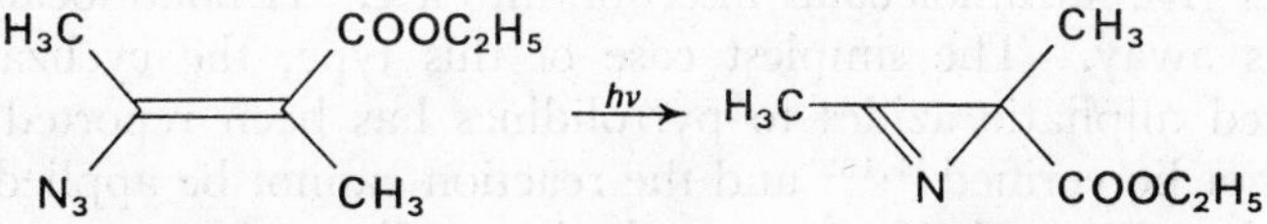

(136) $\xrightarrow{h\nu}$ (137)

The development of newer methods for synthesis of vinyl azides enabled the preparation of series of azirines, including fused ones such as **138**[149].

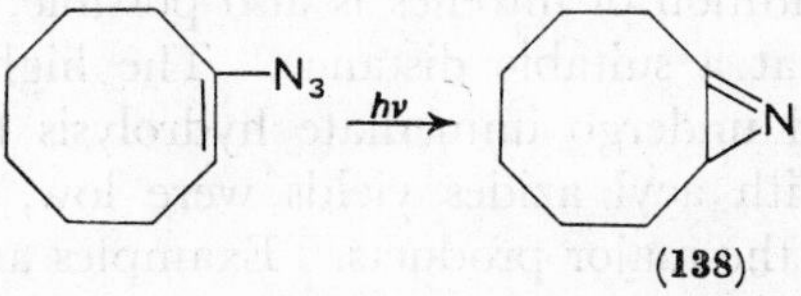

$\xrightarrow{h\nu}$ (138)

Contrary to previous reports, the reaction can also be applied to terminal vinyl azides, which form 1-azirines unsubstituted at position 2. Both isomers of β-azidostyrene yielded 3-phenylazirine (**139**), however these azirines are very unstable[150].

$$C_6H_5CH{=}CHN_3 \xrightarrow{h\nu \text{ or } \Delta} C_6H_5\text{—azirine}$$

cis or *trans* (**139**)

3. Diaziridines

Nitrene insertions into α-amino groups should lead to diaziridines. The only existing report of such a reaction is the photolysis of 2-amino-hexafluoroisopropyl azide (**140**) which yielded 43% of 3,3-bis (tri-fluoromethyl)-diaziridine (**141**). The thermal decomposition of **140** yielded only 11% of **141**, the main product being hexafluoroacetone hydrazone[151].

$$(CF_3)_2C(NH_2)(N_3) \xrightarrow{h\nu} (CF_3)_2C\langle NH{-}NH \rangle$$

(**140**) (**141**)

C. Syntheses of Five-membered Rings containing One Nitrogen

1. By intramolecular nitrene insertions and related reactions

Five-membered rings are formed from azides when the respective nitrenes give intramolecular insertion into a C—H bond located four carbons away. The simplest case of this type, the cyclization of saturated aliphatic azides to pyrrolidines has been reported[43], but could not be verified[44,152] and the reaction cannot be applied to the synthesis of completely saturated rings. Exceptions are cases in which a suitable steric arrangement enables easy accessibility of a methyl group. Photolysis of 1,1-dimethyl-*trans*-decalin-10-carbonyl azide (**142**) yielded 14% of the pyrrolidone **143** along with the δ-lactam **144** and rearrangement and hydrogen abstraction products[153].

An internal addition of nitrenes is also possible, when the double bond is located at a suitable distance. The highly strained fused aziridines formed undergo immediate hydrolysis to hydroxymethyl pyrrolidines. With acyl azides yields were low, as the respective isocyanates were the major products. Examples are the conversions

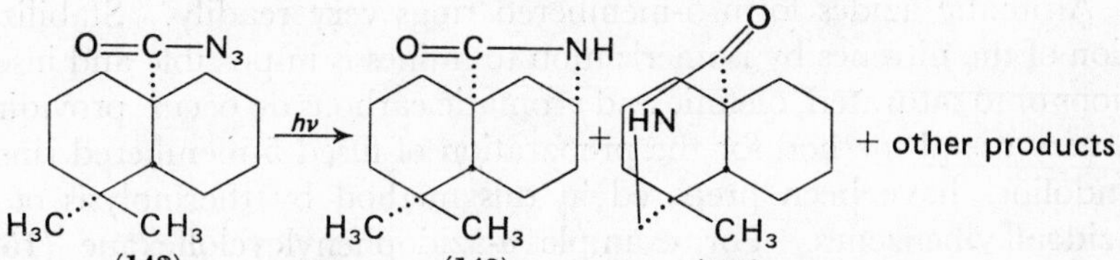

O=C—N$_3$ ... $\xrightarrow{h\nu}$ O=C—NH + HN ... O + other products

H$_3$C CH$_3$ (142) H$_3$C CH$_3$ (143) CH$_3$ (144)

of *o*-vinylbenzoyl azide (**145**) to 3-hydroxymethyl-1,2-dihydroisoindoline (**146**) (yield 10%) and of *endo*-norbornene-5-carbonylazide (**147**) to the pyrrolidone **148** (yield 20%)[154].

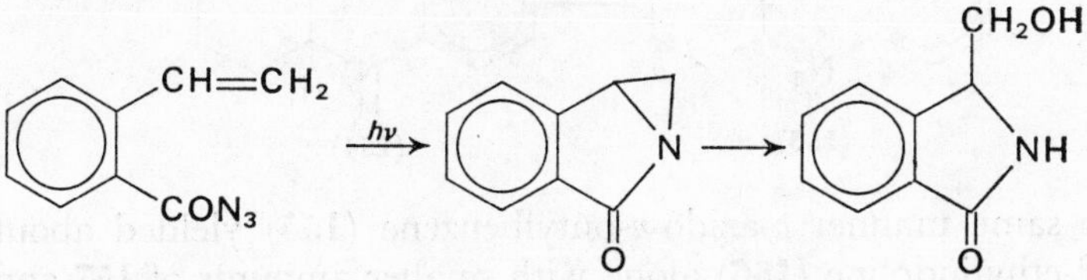

CH=CH$_2$ CON$_3$ (145) $\xrightarrow{h\nu}$ N O $\longrightarrow$ CH$_2$OH NH O (146)

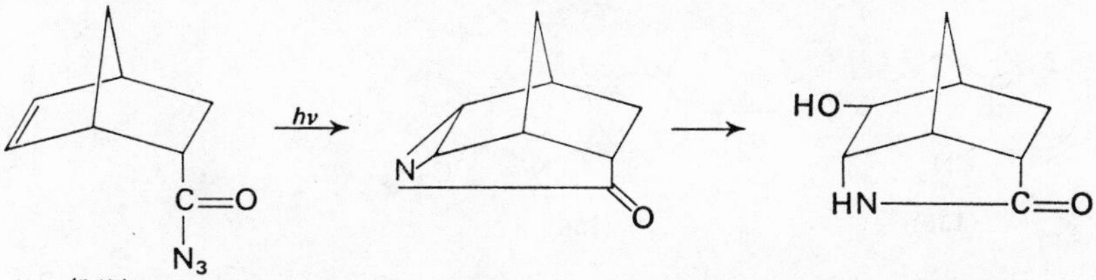

C=O N$_3$ (147) $\xrightarrow{h\nu}$ N O $\longrightarrow$ HO HN—C=O (148)

Five-membered rings have also been obtained from unsaturated alkyl azides. Heating of 5-azido-5-methyl-1-hexene (**149**) yielded a mixture of 2,5,5-trimethyl-1-pyrroline (**151**) and 2,2-dimethyl-1-azabicyclo[3,1,0]hexane (**152**) (total yield 70%, products ratio 2·2:1). This reaction however proceeds via the isolable triazoline **150** and no nitrene intermediates are involved[155].

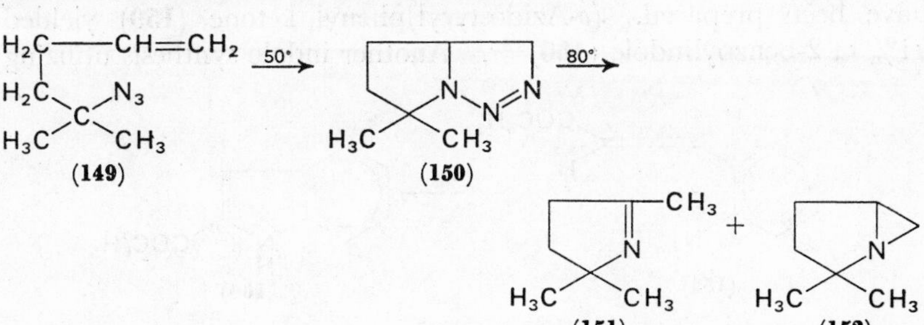

H$_2$C—CH=CH$_2$ H$_2$C N$_3$ C H$_3$C CH$_3$ (149) $\xrightarrow{50°}$ N N=N H$_3$C CH$_3$ (150) $\xrightarrow{80°}$

CH$_3$ N H$_3$C CH$_3$ (151) + N H$_3$C CH$_3$ (152)

Aromatic azides form 5-membered rings very readily. Stabilization of the nitrenes by isomerization to imines is impossible and insertions onto saturated, olefinic and aromatic carbons do occur, providing a very useful method for the preparation of fused 5-membered rings. Indolines have been prepared in this method by thermolysis of *o*-azidoalkylbenzenes. For example *o*-azidophenylcyclohexane (**153**) yielded 94% of hexahydrocarbazole (**154**) (about equal amounts of the *cis* and *trans* isomers)[156].

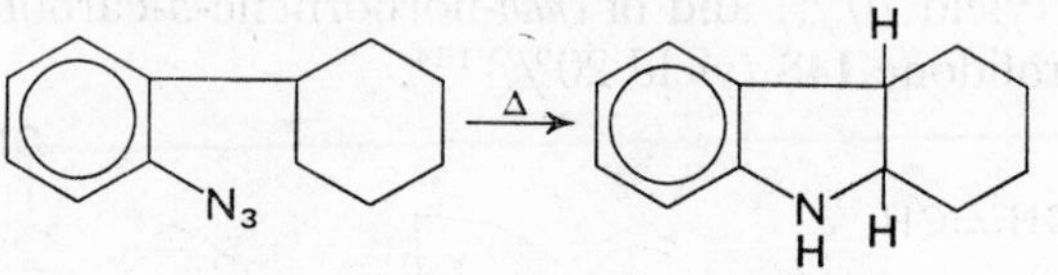

N_3 Δ H N H H

(**153**) (**154**)

In the same manner *o*-azido-*n*-butylbenzene (**155**) yielded about 40% of 2-ethylindoline (**156**) along with smaller amounts of **157** and **158**[157].

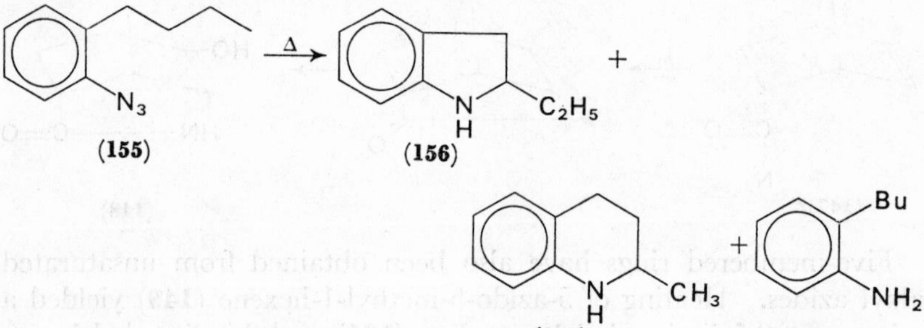

N_3 Δ N H C_2H_5 +

(**155**) (**156**)

N H CH_3 + Bu NH_2

(**157**) (**158**)

The thermolysis of *o*-azidostyrenes gives good yields of indoles. By this method 2-alkyl, 2-aryl and the relatively inaccessible 2-acylindoles have been prepared. (*o*-Azidostyryl)phenyl ketone (**159**) yielded 71% of 2-benzoylindole (**160**)[158]. Another indole synthesis utilizing

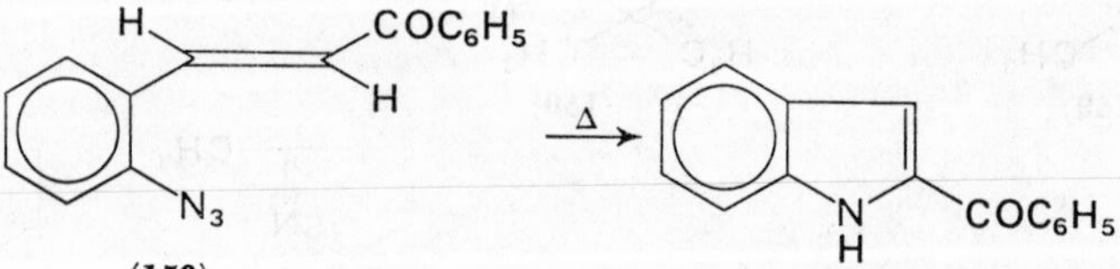

H COC_6H_5 H N_3 Δ N H COC_6H_5

(**159**) (**160**)

azides is the thermolysis of β-azidostyrenes: α- and β-methyl-β-azidostyrene (**161** and **162**) yielded 3- and 2-methylindole (**163** and **164**) respectively[87, 159].

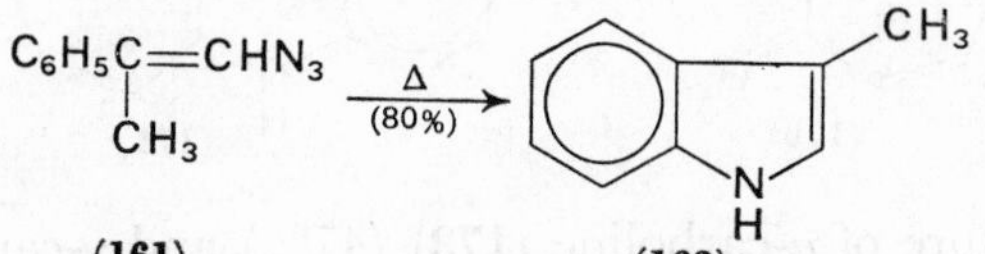

$C_6H_5C{=}CHN_3$ (with CH_3 on C) $\xrightarrow[(80\%)]{\Delta}$ 3-methylindole

(161) **(163)**

$C_6H_5CH{=}CN_3$ (with CH_3 on C) $\xrightarrow[(84\%)]{\Delta}$ 2-methylindole

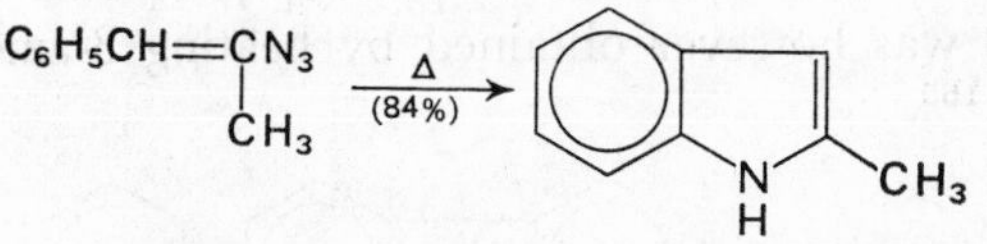

(162) **(164)**

The most versatile reaction of this type is the insertion of aromatic nitrenes into aromatic rings to form tricyclic systems. Pyrolysis or photolysis of *o*-azidobiphenyl (**165**) afforded a good yield of carbazole (**166**)[160]. The reaction has been utilized for the synthesis of a number

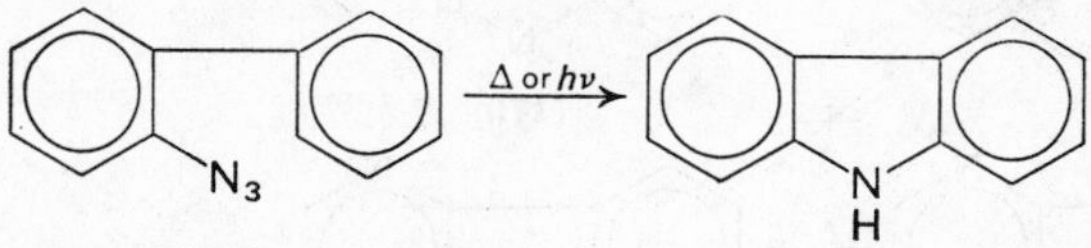

N_3 $\xrightarrow{\Delta \text{ or } h\nu}$ N H

(165) **(166)**

of substituted carbazoles, including fused ones[160,161]. 1,2-Benzocarbazole (**169**) could be prepared from either **167** or **168** (yield 94% in both routes)[161]. Non-benzenoid aromatic azides react as well.

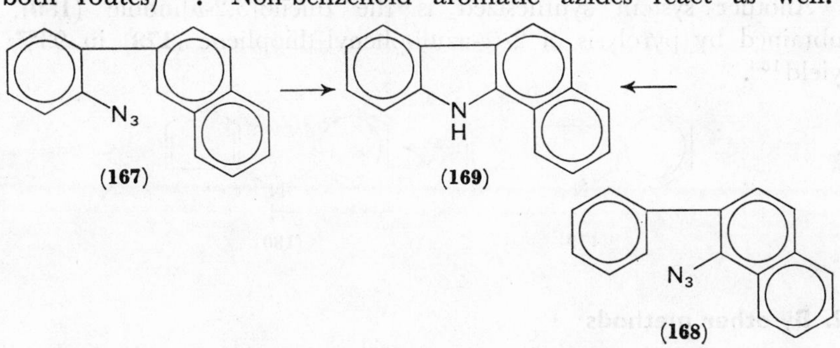

N_3 → N H ← N_3

(167) **(169)** **(168)**

Pyrolysis of 4-phenyl-5-azidotropolone (**170**) yielded 50% of indolo-[3,2-*d*]tropolone (**171**)[162]. The reaction has also been applied to the synthesis of carbolines. Pyrolysis of 3(*o*-azidophenyl)pyridine (**172**)

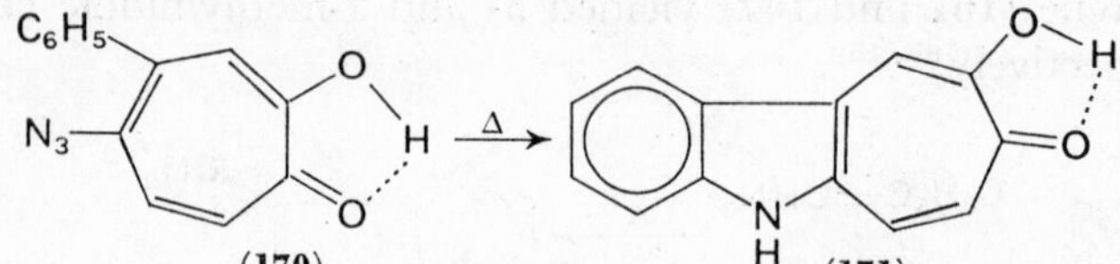

(170) **(171)**

yielded a mixture of α-carboline (**173**) (47%) and γ-carboline (**174**) (23%)[163]. Heating of 2-(*o*-azidophenyl)pyridine (**175**) did not afford δ-carboline (**178**) but cyclized to pyrido[1,2-*b*]indazole (**176**)[164]. Carboline **178** was however obtained by heating 3-azido-2-phenylpyridine (**177**)[165].

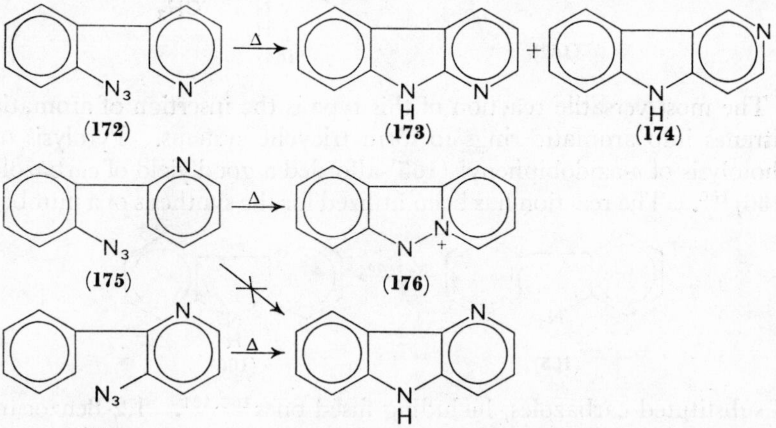

(172) **(173)** **(174)**

(175) **(176)**

(177) **(178)**

Another system synthesized is the thieno[3,2-*b*]indole (**180**), obtained by pyrolysis of 2-(*o*-azidophenyl)thiophene (**179**) in 93% yield[163].

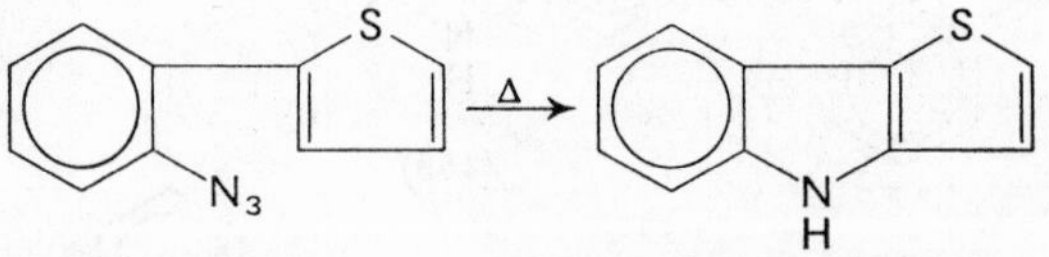

(179) **(180)**

2. By other methods

A special type of substituted pyrolidines has been obtained when aryl azides were reacted at 90° with ethylenes carrying two electron-withdrawing groups on one carbon. Phenyl azide and ethyl benzalcyanoacetate (**181**) yielded the pyrrolidine **183.** The reaction probably proceeded via the aziridine **182** which underwent 1,3-dipolar addition with another molecule of the ethylene. Yields were 20–80%[166].

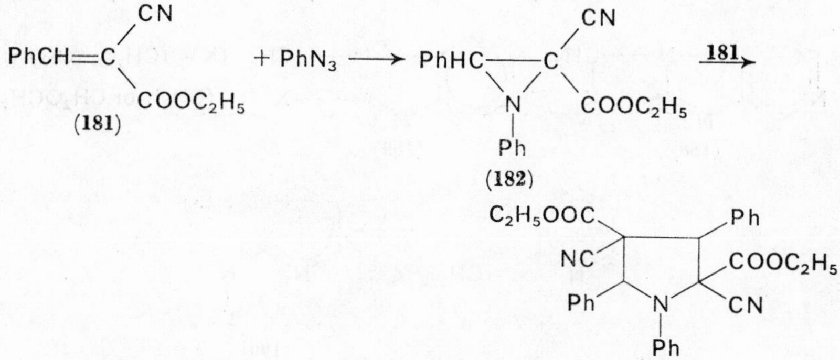

$$\text{PhCH}{=}\text{C(CN)COOC}_2\text{H}_5\ (\textbf{181}) + \text{PhN}_3 \longrightarrow \text{PhHC}\text{—}\text{C(CN)(COOC}_2\text{H}_5\text{)}\ (\text{aziridine, N–Ph})\ (\textbf{182}) \xrightarrow{\textbf{181}}$$

(183): pyrrolidine, N–Ph, with substituents C_2H_5OOC, NC, Ph, Ph, $COOC_2H_5$, CN

D. Syntheses of Five-membered Rings containing Two Nitrogens

I. Imidazoles

Imidazoles have been obtained by pyrolysis of phenacyl azides, probably by dimerization of the intermediate imines. 2-Benzoyl-4(5)phenylmidazole (**184**) was obtained from phenacyl azide in 64% yield[167]. A more general method, which produces fused imidazoles,

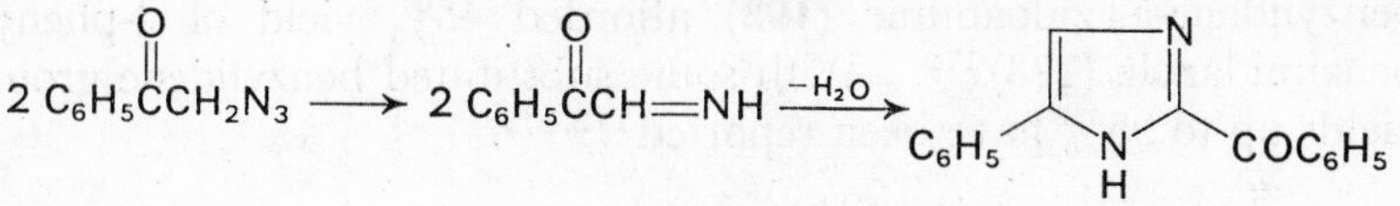

$$2\ C_6H_5\overset{O}{\overset{\|}{C}}CH_2N_3 \longrightarrow 2\ C_6H_5\overset{O}{\overset{\|}{C}}CH{=}NH \xrightarrow{-H_2O} \textbf{(184)}\ (\text{imidazole: } C_6H_5,\ NH,\ COC_6H_5)$$

is the insertion of aromatic nitrenes into nitrogen containing side chains. For example pyrolysis of *N*-(*o*-azidophenyl)piperidine (**185**) resulted in cyclization to the benzimidazoline (**186**). This eliminated hydrogen to yield piperidino[1,2-*a*]benzimidazole (**187**) (30%)[168].

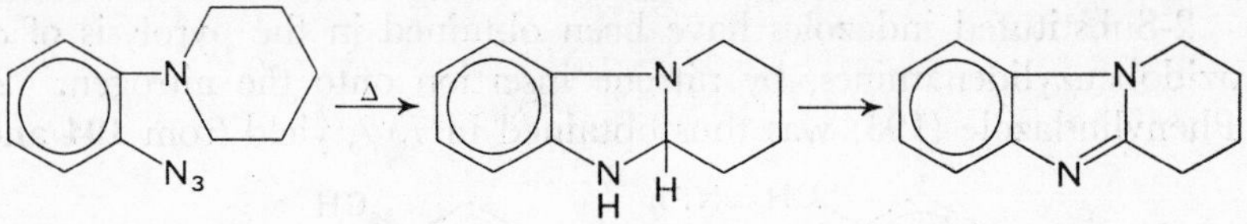

$$\textbf{(185)} \xrightarrow{\Delta} \textbf{(186)} \longrightarrow \textbf{(187)}$$

Similar reactions have been carried out with *N*-(*o*-azidophenyl)-morpholine[168] and 1-(*o*-azidophenyl)-4-acetylpiperazine[169]. The 2-

and 4-substituted 3-azidopyridines yielded in this manner the imidazolopyridines of types **188** and **189** respectively[170]. With azidodimethylaminopyridines the methylated imidazolopyridines **190** and **191** were produced in low yields[171]. Better yields are usually obtained

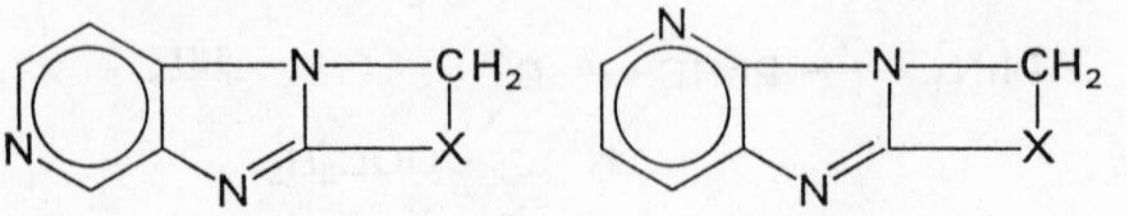

(X = $(CH_2)_2$, $(CH_2)_3$, $(CH_2)_4$ or CH_2OCH_2)

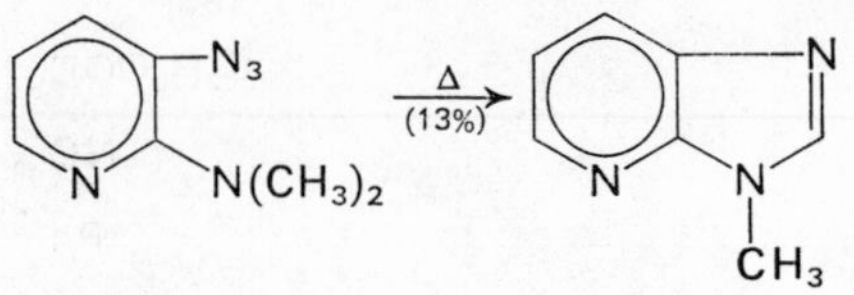

(**188**) (**189**)

(**190**)

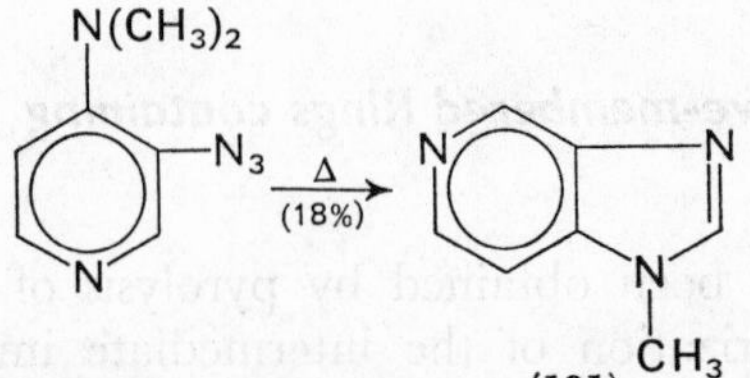

(**191**)

when the nitrene can attack at an unsaturated position. Pyrolysis of benzylidene-*o*-azidoaniline (**192**) afforded 45% yield of 2-phenylbenzimidazole (**193**)[172]. With some substituted benzylidene groups yields up to 95% have been reported[172,173].

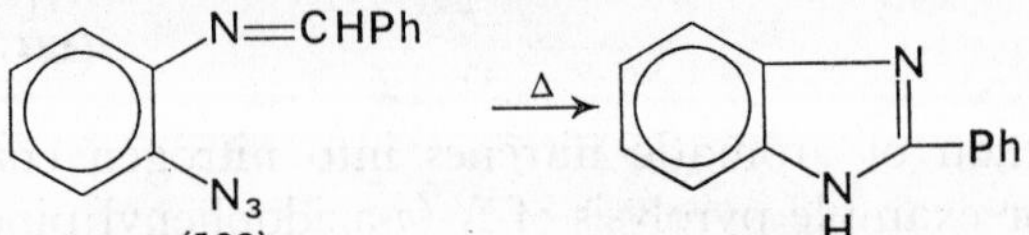

(**192**) (**193**)

2. Pyrazoles

2-Substituted indazoles have been obtained in the pyrolysis of *o*-azidobenzylidenamines, by nitrene insertion onto the nitrogen. 2-Phenylindazole (**195**) was thus obtained in 75% yield from **194** and

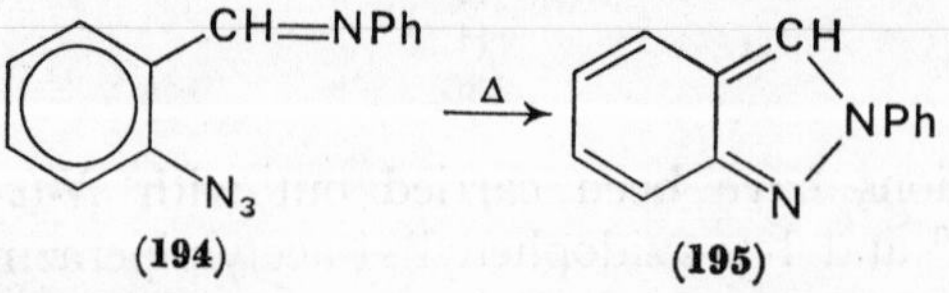

(**194**) (**195**)

2,2'-bi indazole (**197**) in 90% from *o*-azidobenzaldehyde azine (**196**)[174].

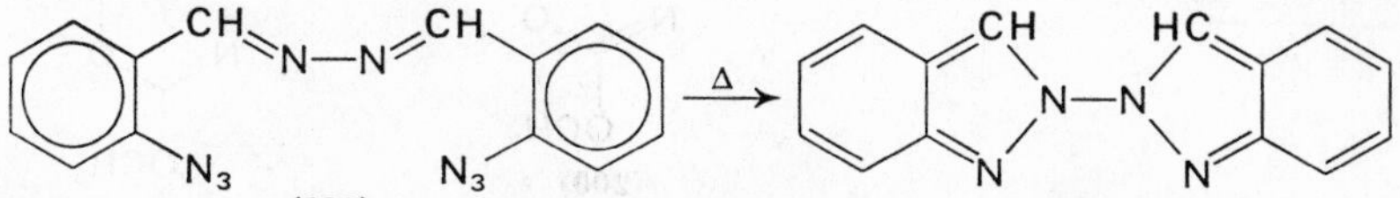

(**196**) (**197**)

E. Syntheses of Five-membered Rings containing Nitrogen and Oxygen

1. Oxazoles

Azidoformates, containing a C—H bond at the right distance from the azido group, cyclize upon thermolysis or photolysis to yield 2-oxazolidones. 5,5-Dimethyl-2-oxazolidone (**198**) was obtained in this manner in 60–75% yield from *t*-butyl azidoformate[175,176]. Heating

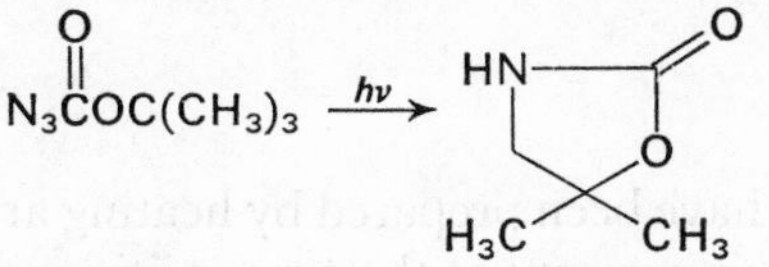

$$N_3COC(CH_3)_3 \xrightarrow{h\nu}$$ (**198**)

of ethyl, isopropyl and *t*-butyl azidoformates at 300° afforded the corresponding 2-oxazolidones in 45–75% yield[177].

The thermal reaction (130°) of ethyl azidoformate with acetylenes yields, by 1,3-dipolar addition of the nitrene, 2-ethoxyoxazoles. With diphenylacetylene, 4,5-diphenyl-2-ethoxyoxazole (**199**) was obtained in 45% yield. With methyl propiolate or dimethyl acetylenedicarboxylate, addition of the azide was faster than the nitrene formation, so triazoles were the major products. In these cases oxazoles were formed only in 3% yield[178]. The photochemical reaction of methyl

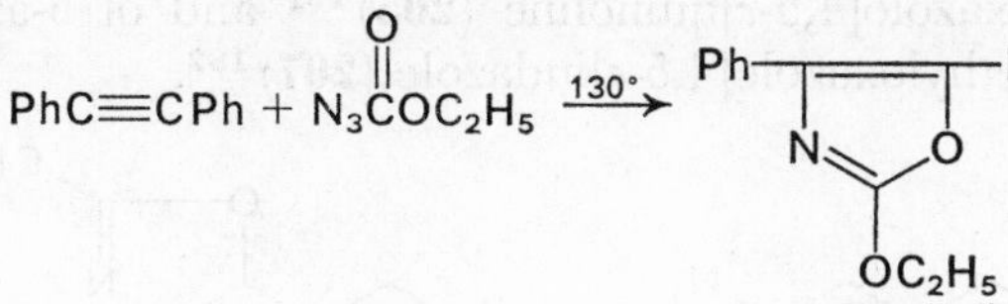

$$PhC{\equiv}CPh + N_3COC_2H_5 \xrightarrow{130^\circ}$$ (**199**)

azidoformate with 2-butyne yielded 12% of the oxazole **200** and 30% of a 2:1 adduct which rearranged to the oxazoline **201**[179].

$$CH_3C{\equiv}CCH_3 + N_3\overset{O}{\overset{\|}{C}}OCH_3 \xrightarrow{h\nu}$$ **(200)** + **(201)**

Oxazole derivatives can also be obtained from azides by reactions which involve acidolysis. β-Hydroxy azides react with aromatic aldehydes in sulphuric acid, thus 2-azidoethanol and benzaldehyde yielded (71%) 2-phenyl-Δ^2-oxazoline (**202**) [180].

$$C_6H_5CHO + N_3CH_2CH_2OH \xrightarrow[-H_2O]{H^+}$$ **(202)**

Fused oxazoles have been prepared by heating aryl azides containing electron-withdrawing groups at the *para* position with carboxylic acids in polyphosphoric acid. *p*-Nitrophenyl azide and acetic acid yielded (83%) 2-methyl-6-nitrobenzoxazole (**203**) [181].

$$+ CH_3COOH \xrightarrow{P.P.A.}$$ **(203)**

Heteroatoms of heterocyclic rings can also serve as the electron-withdrawing group, as in the conversions of 3-azidoquinoline (**204**) to 2-methyloxazolo[4,5-*c*]quinoline (**205**) [181] and of 5-azidoindazole (**206**) to 2-methyloxazolo[4,5-*e*]indazole (**207**) [182].

(204) $\xrightarrow[P.P.A.]{AcOH}$ **(205)**

(206) → AcOH, P.P.A. (62%) → (207)

2. Isoxazoles

Formation of an isoxazole system has been reported in the case of the pyrolysis of *o*-azidoketones. 2-Azidobenzophenone (**208**) gave a good yield of 3-phenylanthranil (**209**)[183].

(208) → Δ → (209)

3. Oxadiazoles

The 1,3,4-oxadiazole system is formed on cycloaddition of carboalkoxy nitrenes to nitriles. The photolysis of ethyl azidoformate in acetonitrile yielded 2-ethoxy-5-methyl-1,3,4-oxadiazole (**210**) (yields 52–60%)[175,184,185]. With benzonitriles yields were much lower[185].

$CH_3C{\equiv}N + N_3COC_2H_5 \xrightarrow{h\nu}$ (210)

o-Nitrophenyl azides cyclize upon heating to give good yields of benzofuroxans. The unsubstituted benzofuroxan **212** was obtained on heating a toluene solution of **211** on a water bath (yield 77–85%)[186].

(211) → (212)

It should be noted that pyrolysis of either of the substituted nitrophenylazides **213** and **214** yields the same single compound[187]. Thus a tautomerism must occur between the products **216** and **217**, probably via the dinitroso derivative **215**, to give the more stable isomer[188]. In the case where X = Cl or Br this has been shown, by an X-ray crystallographic study, to be the 5-substituted benzofuroxan **217**[189].

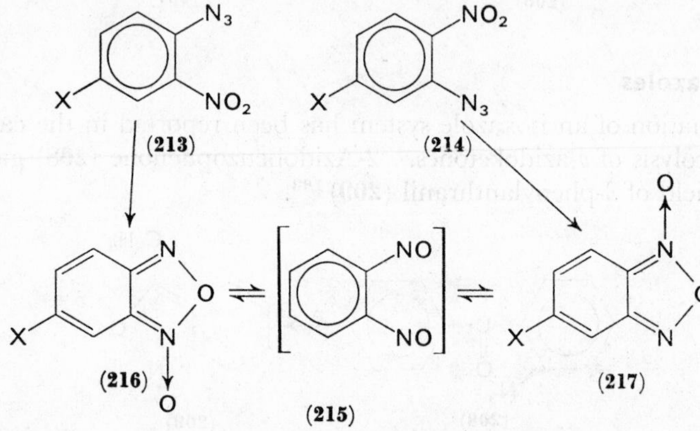

(**213**) (**214**)

(**216**) (**215**) (**217**)

The low reaction temperature of benzofuroxan formation from azides can be rationalized by assuming a concerted mechanism, rather than nitrene intermediates. Benzofuroxan formation is preferred over the products obtained usually from nitrenes. Heating of 2-azido-3-nitrobiphenyl (**218**) yielded exclusively the furoxan **219** and not the carbazole **220**[160].

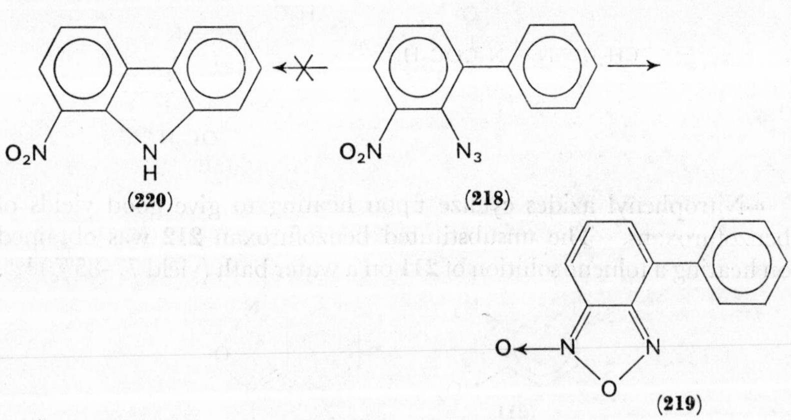

(**220**) (**218**)

(**219**)

A completely analogous reaction is the cyclization of *o*-nitroso-azides to benzofurazans. 4-Chlorobenzofurazan (**221**) was prepared in this manner in 93% yield[190].

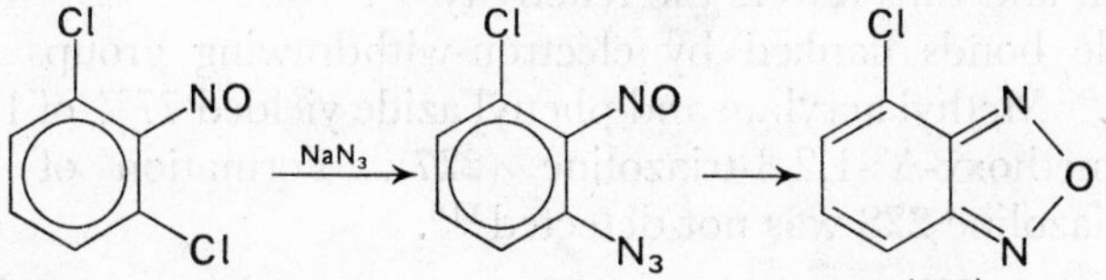

(**221**)

F. Syntheses of Five-membered Rings containing Three Nitrogens

I. 1,2,3-Triazolines

a. By addition of azides to olefins. The dipolar character of the azido group enables it to undergo 1,3-cycloaddition with olefins, forming Δ^2-1,2,3-triazolines (**222**)[72].

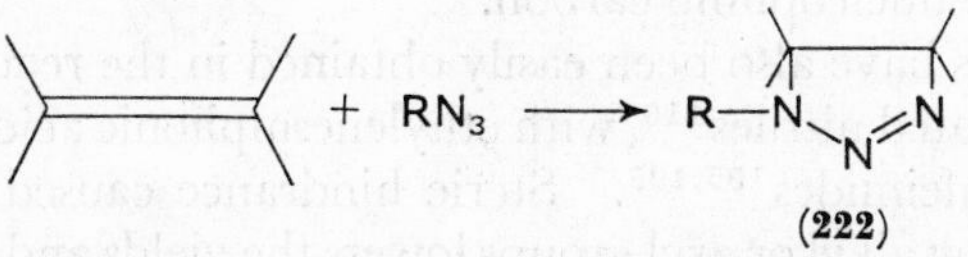

+ RN_3 ⟶ (**222**)

Double bonds which are a part of strained bicyclic systems are particularly reactive. Norbornene (**223**) for example, yielded quantitatively in a very fast reaction, the triazoline **224**, the addition occurring at the less hindered *exo* side[191]. The importance of the strain is

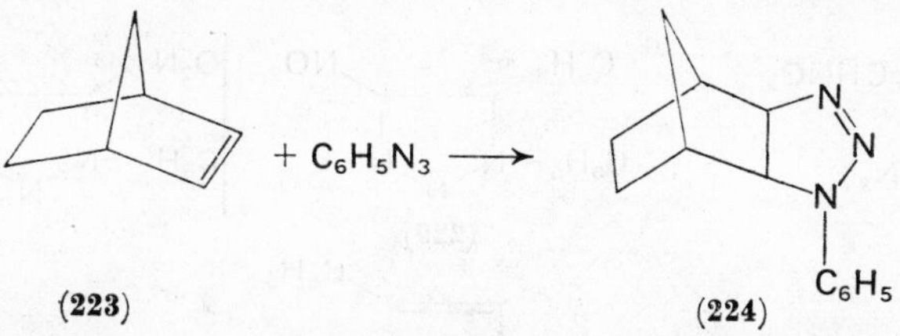

(**223**) + $C_6H_5N_3$ ⟶ (**224**)

evident in the reactivity of dicyclopentadiene (**225**). Even on using an excess of phenyl azide only the double bond of the norbornene moiety reacted to yield **226**[192].

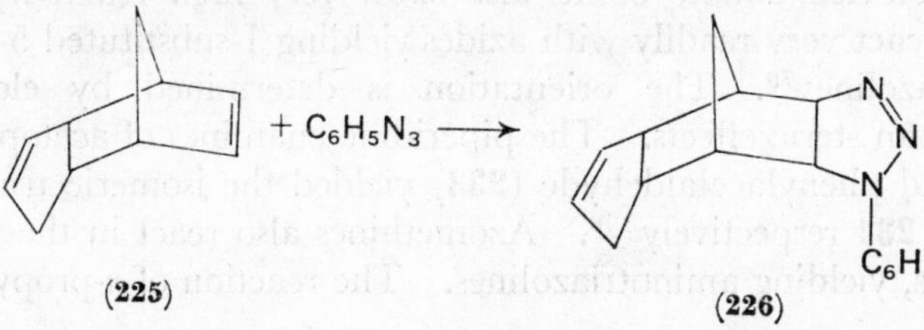

(**225**) + $C_6H_5N_3$ ⟶ (**226**)

Another type of strained olefins which add azides are the medium-sized *trans*-cycloolefins. In this series increase in the ring size relieves the strain and thus lowers the reactivity[193].

Double bonds flanked by electron-withdrawing groups are also reactive. Methyl acrylate and phenyl azide yielded 77% of 1-phenyl-4-carbomethoxy-Δ^2-1,2,3-triazoline (**227**). Formation of the isomeric triazoline **228** was not detected[110].

$CH_2{=}CHCOOCH_3$ + $C_6H_5N_3$ → (**227**) ((**228**))

The exclusive formation of **227** indicates that the orientation of the addition is controlled by electronic effects, the azide terminus attacking the more nucleophilic carbon.

Triazolines have also been easily obtained in the reactions of azides with unsaturated nitriles[110], with ethylenesulphonic acid derivatives[194] and with maleimides[195,196]. Steric hindrance caused by additional substitution by alkyl or aryl groups lowers the yields and also influences the orientation. The reaction of phenyl azide with β-nitrostyrene proceeded only at 130° and yielded only 20% of the expected 1,5-diphenyl-4-nitrotriazoline (**229**), together with 1,4-diphenyltriazole (**230**), resulting from addition in the opposite direction[197].

$C_6H_5CH{=}CHNO_2$ + $C_6H_5N_3$ → (**229**) + [ ] → (**230**)

Electron-rich double bonds also show very high reactivity. Enamines react very readily with azides yielding 1-substituted 5-amino-1,2,3-triazolines[79]. The orientation is determined by electronic rather than steric effects. The piperidine enamines of acetophenone (**231**) and phenylacetaldehyde (**233**) yielded the isomeric triazolines **232** and **234** respectively[198]. Azomethines also react in the form of enamines, yielding aminotriazolines. The reaction of *n*-propylidene-

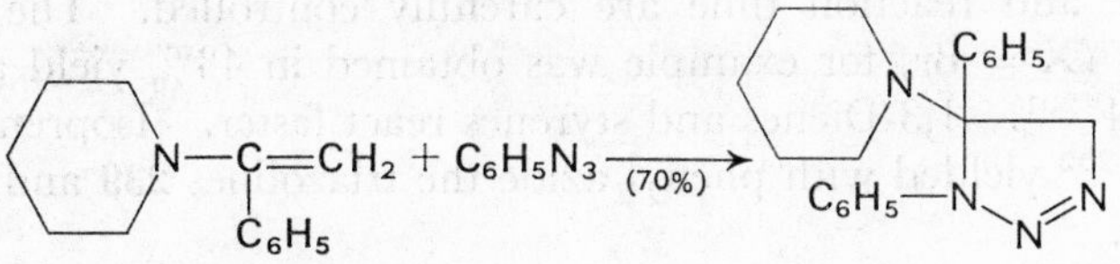

$$\text{C}_5\text{H}_{10}\text{N—C(C}_6\text{H}_5\text{)=CH}_2 + \text{C}_6\text{H}_5\text{N}_3 \xrightarrow{(70\%)} \text{232}$$

(231) (232)

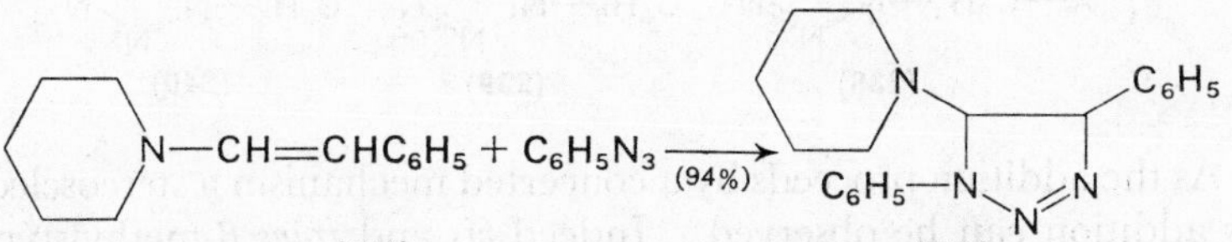

$$\text{C}_5\text{H}_{10}\text{N—CH=CHC}_6\text{H}_5 + \text{C}_6\text{H}_5\text{N}_3 \xrightarrow{(94\%)} \text{234}$$

(233) (234)

propylamine (**235**) with *p*-nitrophenyl azide produced compound **236**[199].

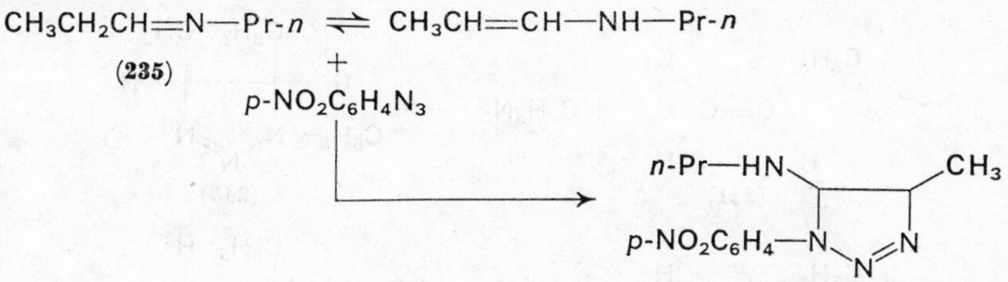

$$\text{CH}_3\text{CH}_2\text{CH=N—Pr-}n \rightleftharpoons \text{CH}_3\text{CH=CH—NH—Pr-}n$$

(235)

$$+\ p\text{-NO}_2\text{C}_6\text{H}_4\text{N}_3 \longrightarrow \text{236}$$

(236)

Vinyl ethers react in the same manner, the reaction being somewhat slower. α-Methoxystyrene and *p*-nitrophenyl azide yielded (93%) the triazoline **237**[200].

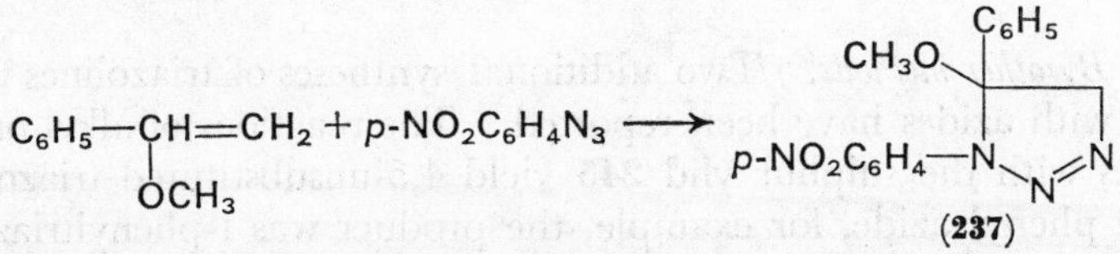

$$\text{C}_6\text{H}_5\text{—C(OCH}_3\text{)H=CH}_2 + p\text{-NO}_2\text{C}_6\text{H}_4\text{N}_3 \longrightarrow \text{237}$$

(237)

The reaction of simple unstrained and unactivated olefins with azides had been considered until recently to proceed too sluggishly to have any synthetic utility. The formation of the triazoline **238** (X = Cl) from the aryl azides and 1-hexene was complete (89%) at room temperature only after 5·5 months. At elevated temperatures (>80°) extensive decomposition was observed. However it was possible to obtain acceptable yields of triazolines provided that the temperature and reaction time are carefully controlled. The triazoline **238** (X = Br) for example was obtained in 43% yield after 3 days at 64°[201]. 1,3-Dienes and styrenes react faster. Isoprene[201] and styrene[202] yielded with phenyl azide the triazolines **239** and **240** respectively.

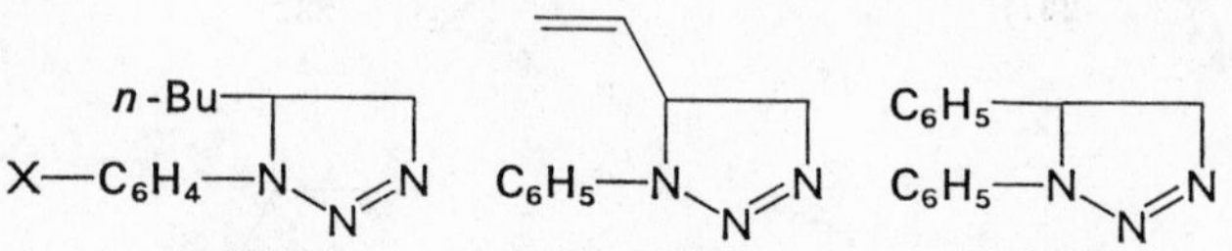

(238) (239) (240)

As the addition proceeds by a concerted mechanism a stereoselective *cis* addition can be observed. Indeed *cis* and *trans* β-methylstyrenes (**241** and **242**) yielded exclusively *cis* and *trans* 4-methyl-1,5-diphenyltriazoline **243** and **244** respectively[203]. The same stereoselectivity has also been observed in the addition reaction of vinyl ethers[204].

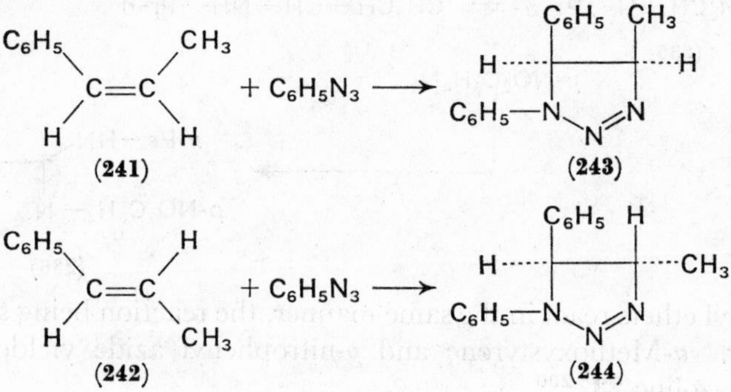

$C_6H_5CH{=}CHCH_3$ (241) + $C_6H_5N_3$ ⟶ (243)

$C_6H_5CH{=}CHCH_3$ (242) + $C_6H_5N_3$ ⟶ (244)

b. By other methods. Two additional syntheses of triazolines which start with azides have been reported. The reactions of alkyl or aryl azides with the sulphur ylid **245** yield 4,5-unsubstituted triazolines. With phenyl azide, for example, the product was 1-phenyltriazoline (**246**). The mechanism involves two consecutive methylene transfers followed by cyclization. Yields were 65–90%[205,206].

5-Hydroxytriazoles have been prepared by the reaction of aliphatic

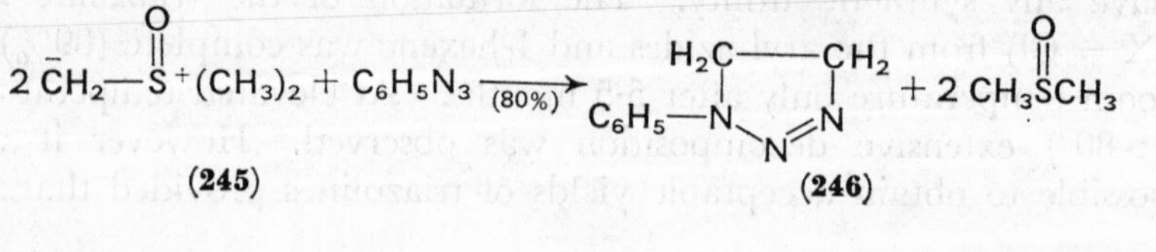

$$2\,\bar{C}H_2{-}\overset{O}{\overset{\|}{S}}{}^{+}(CH_3)_2 + C_6H_5N_3 \xrightarrow[(80\%)]{} (246) + 2\,CH_3\overset{O}{\overset{\|}{S}}CH_3$$

(245) (246)

ketones with alkyl or aryl azides in the presence of potassium *t*-butoxide. The reaction proceeds via a nucleophilic attack of the α-anion of the ketone on the azide terminus, followed by cyclization. Methyl ethyl ketone and phenyl azide thus yielded the triazoline **247**[207].

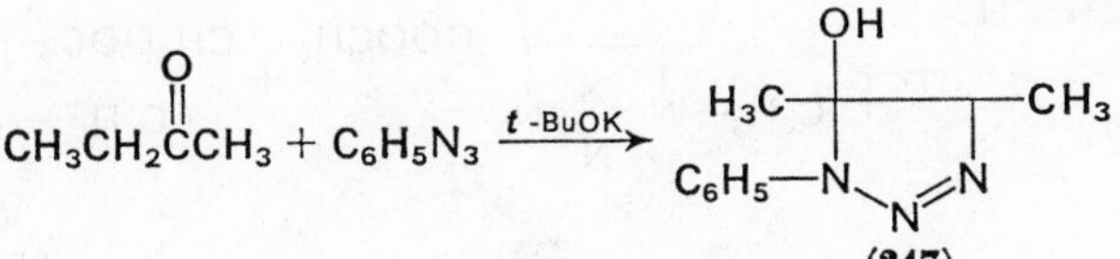

$$CH_3CH_2COCH_3 + C_6H_5N_3 \xrightarrow{t\text{-BuOK}} \textbf{(247)}$$

2. 1,2,3-Triazoles

a. By addition of azides to olefins, followed by elimination. Most hydroxy- and amino-triazoles, prepared by addition of azides to vinyl ethers or enamines, can be induced to eliminate a molecule of alcohol or amine and form the corresponding triazole[198,200]. 1,5-Diphenyl-1,2,3-triazole (**249**) was thus obtained from the acetophenone enamine **248** and phenyl azide[198].

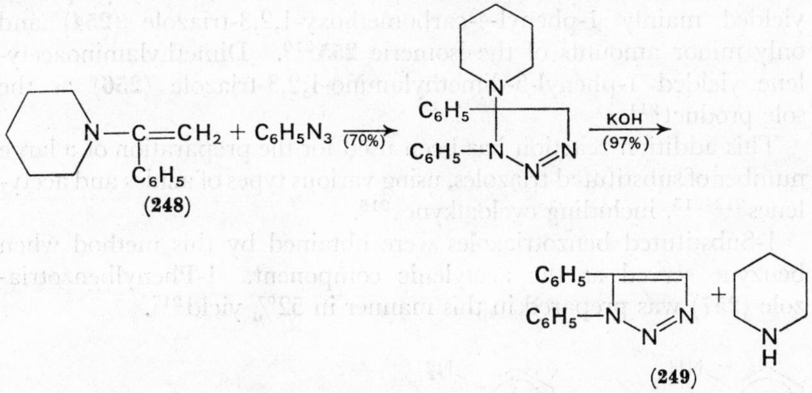

$$\textbf{(248)} + C_6H_5N_3 \xrightarrow{(70\%)} \xrightarrow[(97\%)]{KOH} \textbf{(249)} + \text{piperidine}$$

b. By addition of azides to triple bonds. The reaction of azides and acetylenes yields 1,2,3-triazoles directly. In the synthesis of the parent ring **251** benzyl azide was reacted with acetylene yielding 1-benzyl-1,2,3-triazole (**250**) (85%), which was converted to 1,2,3-triazole (**251**) by catalytic hydrogenolysis[208].

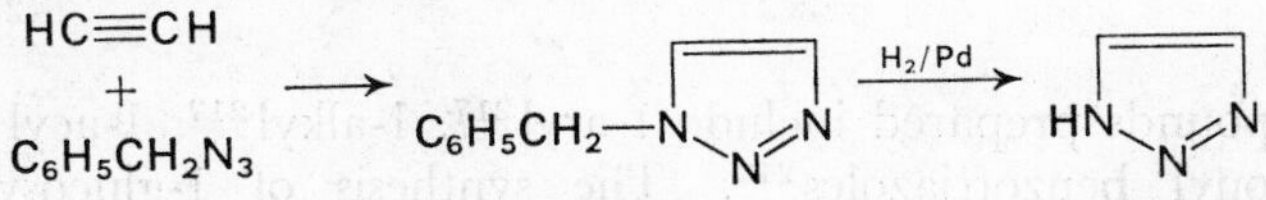

$$HC{\equiv}CH + C_6H_5CH_2N_3 \longrightarrow \textbf{(250)} \xrightarrow{H_2/Pd} \textbf{(251)}$$

$$C_6H_5C{\equiv}CH + C_6H_5N_3 \longrightarrow \textbf{(252)} + \textbf{(253)} \quad \text{(ratio 1·2:1)}$$

$$CH_3OOCC{\equiv}CH + C_6H_5N_3 \longrightarrow \textbf{(254)} + \textbf{(255)} \quad \text{(ratio 7:1)}$$

$$(CH_3)_2NC{\equiv}CH + C_6H_5N_3 \longrightarrow \textbf{(256)}$$

In the addition to unsymmetrical acetylenes the orientation is controlled by electronic effects. Thus, while in the reaction of phenyl azide with phenylacetylene the two isomeric diphenyltriazoles **252** and **253** were obtained in almost equal amounts[209], methyl propiolate yielded mainly 1-phenyl-4-carbomethoxy-1,2,3-triazole (**254**) and only minor amounts of the isomeric **255**[210]. Dimethylaminoacetylene yielded 1-phenyl-5-dimethylamino-1,2,3-triazole (**256**) as the sole product[211].

This addition reaction has been used for the preparation of a large number of substituted triazoles, using various types of azides and acetylenes[212–215], including cycloalkynes[216].

1-Substituted benzotriazoles were obtained by this method when benzyne served as the acetylenic component. 1-Phenylbenzotriazole (**257**) was prepared in this manner in 52% yield[217].

$$\text{2-}H_2NC_6H_4COOH \longrightarrow \text{2-}N_2^+C_6H_4COOH \longrightarrow C_6H_4 \xrightarrow{C_6H_5N_3} \textbf{(257)}$$

Compounds prepared include 1-aryl[217], 1-alkyl[217], 1-acyl[218] and 1-sulphonyl benzotriazoles[218]. The synthesis of 1-glucosylbenzotriazoles from glucosyl azides and benzyne has also been recently reported[219].

c. By nucleophilic attacks on azides, followed by cyclization. The base-catalysed reaction of alkyl and aryl azides with active methylenes provides a very useful triazole synthesis. The reaction proceeds via a nucleophilic attack of the carbanion on the azide terminus, followed by cyclization on the adjacent functional group. The reactions of benzyl azide with ethyl acetoacetate and with diethyl malonate yielded 1-benzyl-4-carbethoxy-5-methyl-1,2,3-triazole (**258**) and 1-benzyl-4-

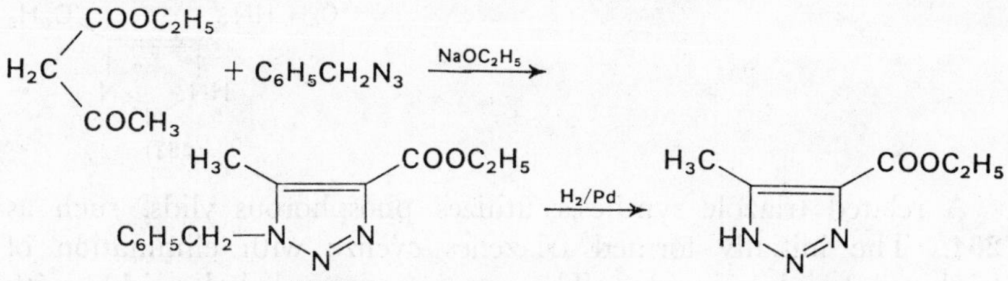

$$H_2C(COOC_2H_5)(COCH_3) + C_6H_5CH_2N_3 \xrightarrow{NaOC_2H_5}$$

(**258**) $\xrightarrow{H_2/Pd}$ (**259**)

$$H_2C(COOC_2H_5)_2 + C_6H_5CH_2N_3 \xrightarrow{NaOC_2H_5}$$

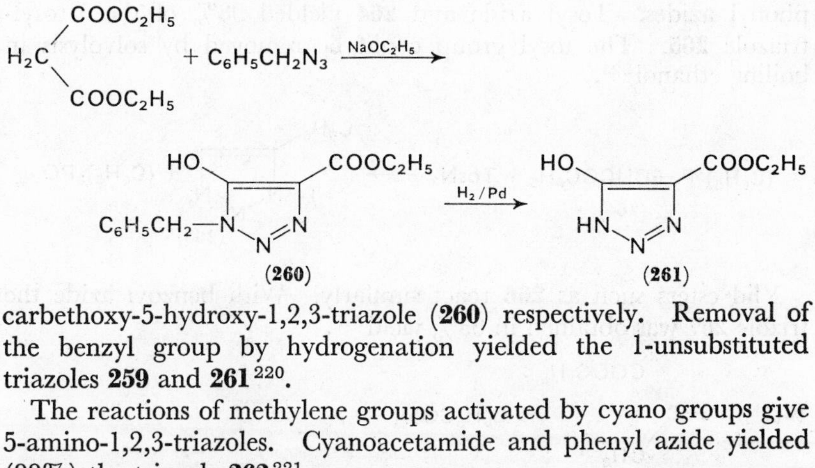

(**260**) $\xrightarrow{H_2/Pd}$ (**261**)

carbethoxy-5-hydroxy-1,2,3-triazole (**260**) respectively. Removal of the benzyl group by hydrogenation yielded the 1-unsubstituted triazoles **259** and **261**[220].

The reactions of methylene groups activated by cyano groups give 5-amino-1,2,3-triazoles. Cyanoacetamide and phenyl azide yielded (88%) the triazole **262**[221].

With phenylacetonitrile, 1-substituted-4-phenyl-5-amino-1,2,3-triazoles are produced in excellent yields. Compound **262a** has been

$$H_2NCOCH_2CN + C_6H_5N_3 \xrightarrow{NaOC_2H_5}$$ (**262**)

obtained in 99% yield from phenylacetonitrile and phenyl azide[222]. With *n*-hexyl azide a yield of 98% has been achieved[223]. The 5-aminotriazoles rearrange very readily on treatment with pyridine, with substituent migration from position 1 to the 5-amino group. Compound **262a** was thus transformed into **263** in 92% yield[222].

$$C_6H_5CH_2CN + C_6H_5N_3 \longrightarrow \textbf{(262a)} \xrightarrow{\text{pyridine}} \textbf{(263)}$$

(262a): 1-C_6H_5, 5-H_2N, 4-C_6H_5-1,2,3-triazole; (263): 5-C_6H_5HN, 4-C_6H_5-1,2,3-triazole (N-H)

A related triazole synthesis utilizes phosphorous ylids, such as **264**. The initially formed triazenes cyclize with elimination of triphenylphosphine oxide. The reaction proceeded sluggishly with phenyl azide, but good results have been obtained with acyl or sulphonyl azides. Tosyl azide and **264** yielded 98% of the 1-tosyltriazole **265**. The tosyl group could be removed by solvolysis in boiling ethanol[224].

$$\underset{\textbf{(264)}}{(C_6H_5)_3P{=}CHCOC_6H_5} + TosN_3 \longrightarrow \textbf{(265)} + (C_6H_5)_3PO$$

(265): 1-TosN, 5-C_6H_5-1,2,3-triazole

Ylid esters such as **266** react similarly. With benzoyl azide the trizole **267** was obtained in 63% yield[225].

$$\underset{\textbf{(266)}}{(C_6H_5)_3P{=}C(COOC_2H_5)(CH_3)} + C_6H_5CON_3 \longrightarrow \textbf{(267)} + (C_6H_5)_3PO$$

(267): 1-C_6H_5CON, 5-C_2H_5O, 4-CH_3-1,2,3-triazole

It should be noted that prolonged heating of aryl azides with sodium ethoxide alone also produces triazoles. The reaction requires two equivalents of the azide, one of which is reduced to aniline[226]. This reaction has been recently used for the synthesis of a number of 1-aryltriazoles[227]. Phenyl azide, for example, yielded (45%) 1-phenyl-1,2,3-triazole (**268**). Use of sodium *n*-propoxide yielded the 4-methyl derivatives[227].

$$2\,C_6H_5N_3 + NaOC_2H_5 \longrightarrow C_6H_5\text{-}N_3C_2H_2\ (\mathbf{268}) + C_6H_5NH_2 + NaOH$$

d. By decomposition of o-*azidoazobenzenes.* 2-Substituted benzotriazoles are obtained by the thermal decomposition of *o*-azidoazobenzenes. The conversion of **269** to 2-(4-dimethylaminophenyl)benzotriazole (**270**) was affected in quantitative yield by reflux in dioxan[228]. The low decomposition temperature indicates a concerted mechanism.

(269) → (270)

The reaction has been utilized for syntheses in the tetraazapenetalene series. Heating of *o,o'*-diazidoazobenzene (**271**) gave 1,3*a*,4,6*a*-tetraazapenetalene (**273**) in 93% yield. The intermediate benzotriazole **272** could be isolated when the reaction was carried out at 58°. The second step which probably involves a nitrene intermediate required a temperature of 170°[229].

(271) —58°→ (272) —170°→ (273)

e. By reductive cyclization of azides. This type of cyclization has been observed during the hydrogenation of α-azidodiphenylacetonitrile (**274**). A probable intermediate is the cyanotriazene **275** which cyclized to the aminotriazolenine **276** (59% yield). The normal hydrogenation product **277** was however also obtained (30%)[230]. The scope of the reaction has been extended to include also the reductive cyclization of α-azido esters to hydroxytriazolenines[231].

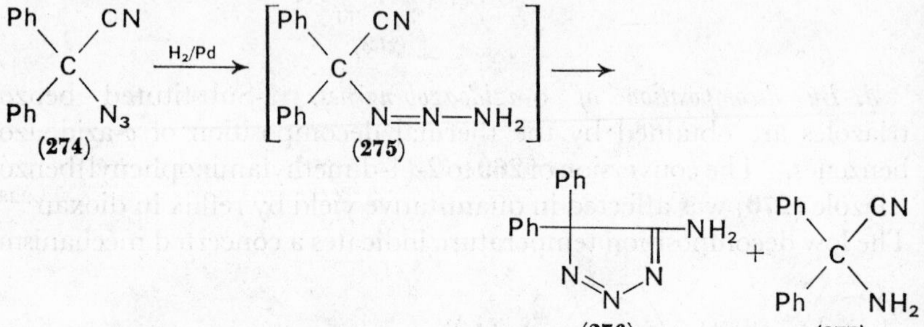

$$\mathrm{Ph_2C(CN)N_3}\ (\mathbf{274}) \xrightarrow{H_2/Pd} \left[\mathrm{Ph_2C(CN)N{=}N{-}NH_2}\right]\ (\mathbf{275}) \longrightarrow (\mathbf{276}) + \mathrm{Ph_2C(CN)NH_2}\ (\mathbf{277})$$

(**274**) (**275**) (**276**) (**277**)

G. Syntheses of Tetrazoles

I. By addition of azides to nitriles

The uncatalysed addition of organic azides to nitriles to form tetrazoles can be affected only if the nitriles are activated by electron-withdrawing groups. The reaction of octyl azide and trifluoroacetonitrile at 120–150° yielded (96%) the tetrazole **278**. Such good yields can be obtained only when the azide employed is stable at the high reaction temperature. With the less stable phenyl azide the yield was only 22%[232].

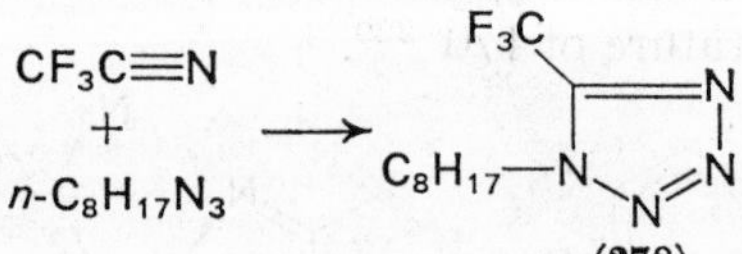

$$CF_3C{\equiv}N + n\text{-}C_8H_{17}N_3 \longrightarrow (\mathbf{278})$$

(**278**)

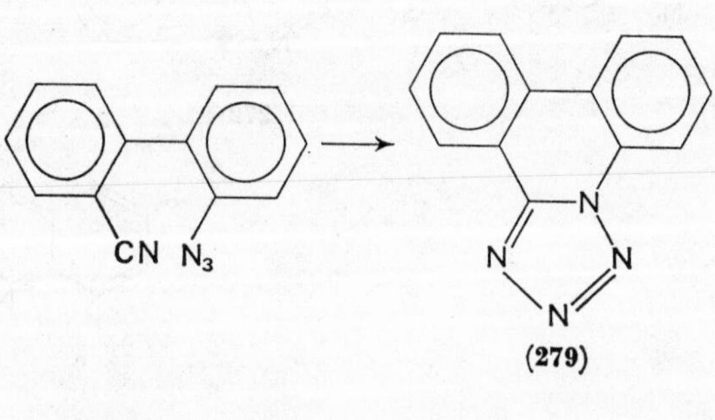

(**279**)

An intramolecular uncatalysed addition has been observed upon heating of 2-azido-2′-cyanobiphenyl which yielded the tetrazolophenanthridine **279**[161]. The cyclization of azidonitriles of the general formula **280** has been carried out with acid catalysis, and yielded the fused tetrazoles **281**[232].

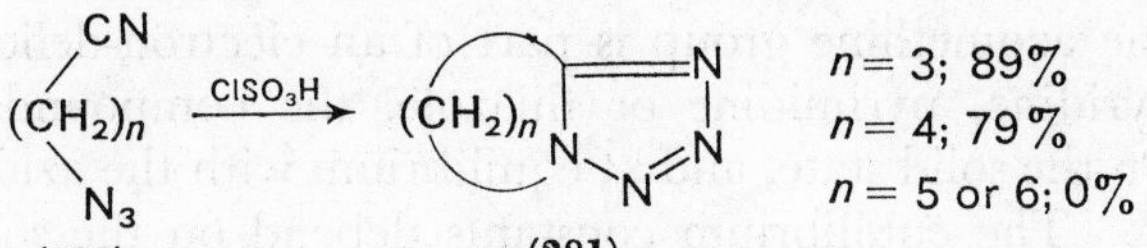

$$NC-(CH_2)_n-N_3 \xrightarrow{ClSO_3H} (CH_2)_n\text{-fused tetrazole}$$

(**280**) (**281**)

$n = 3$; 89%
$n = 4$; 79%
$n = 5$ or 6; 0%

2. By decomposition of geminal diazides

The thermolysis of benzophenone diazide (**282**) has been reported to produce 1,5-diphenyltetrazole (**283**) in 90% yield[233].

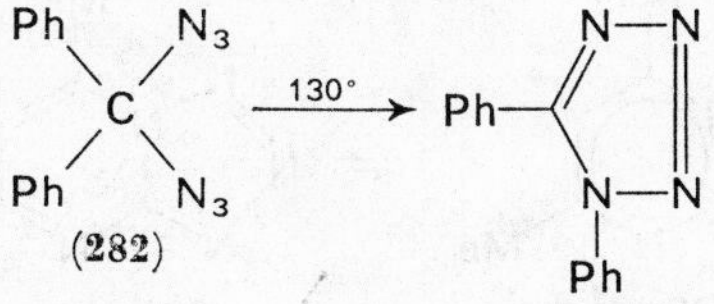

$$Ph_2C(N_3)_2 \xrightarrow{130^\circ} \text{1,5-diphenyltetrazole}$$

(**282**) (**283**)

Photolysis of **282** yielded only 14% of **283**[234]. With diethyldiazidomalonate the yield of 1,5-dicarbethoxytetrazole was 48%[235].

3. The azidoazomethine–tetrazole equilibrium

Azidoazomethines (**284**), usually cannot be isolated as they readily isomerize to tetrazoles. Thus the azides in this case have to be considered as intermediates, rather than starting materials. The isomerization has been applied to the synthesis of large series of 1,5-disubstituted tetrazoles (**285**; R, R′ = aryl or alkyl)[236,237].

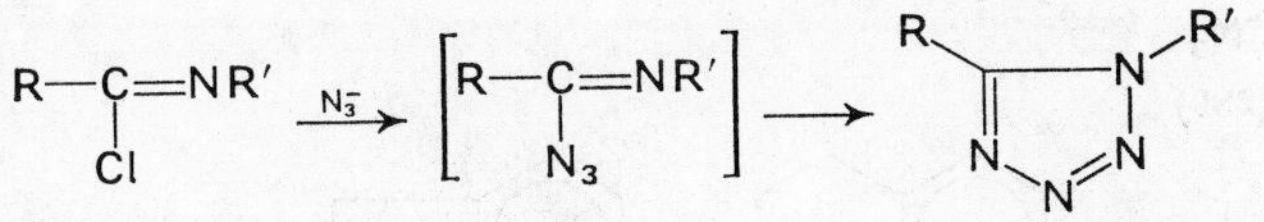

$$R-C(Cl)=NR' \xrightarrow{N_3^-} \left[R-C(N_3)=NR'\right] \longrightarrow \text{5-R-1-R'-tetrazole}$$

(**284**) (**285**)

In some special cases the azides could be isolated. One example is the stable nitroguanyl azide (**286**), which served for the synthesis of 5-nitraminotetrazole (**287**)[238].

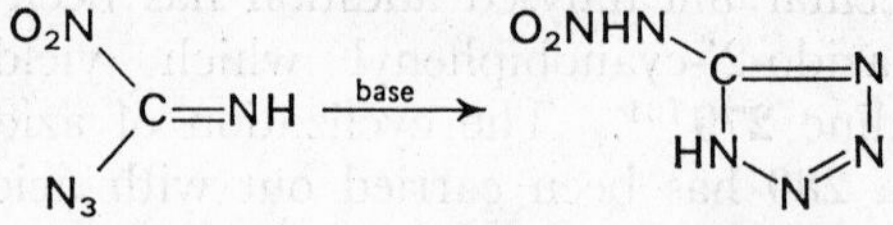

(286) (287)

When the azomethine group is part of an electron-deficient ring, such as pyridine, pyrimidine or thiazole, the compounds exist as tetrazoles in the solid state, and at equilibrium with the azido form in solution[239]. The equilibrium constants depend on the solvent, the nature of the substituents and the temperature[240]. 2-Azido-4,6-dimethylpyrimidine (**288a**) thus exists in equilibrium with tetrazolopyrimidine (**288b**). Its chemical behaviour is, however, in accord with the azide structure **288a**, including dipolar addition reactions[241] and nitrene reactions[242].

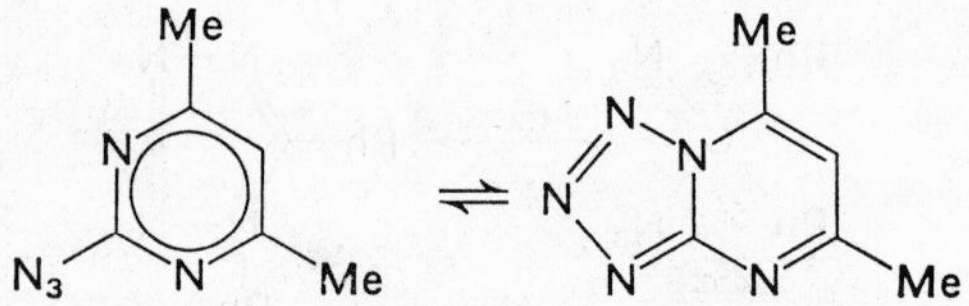

(288a) (288b)

H. *Syntheses of Six-membered Rings*

1. By intramolecular insertion of nitrenes

Cases in which decomposition of an azide results in the formation of a 6-membered ring are quite rare. In the thermolysis of (*o*-azidophenyl)butane-*n* (**289**), 2-methyl-1,2,3,4-tetra-hydroquinoline (**290**) is only a minor product (10–20%)[157].

Transformations of the type **291** → **292** were successful when

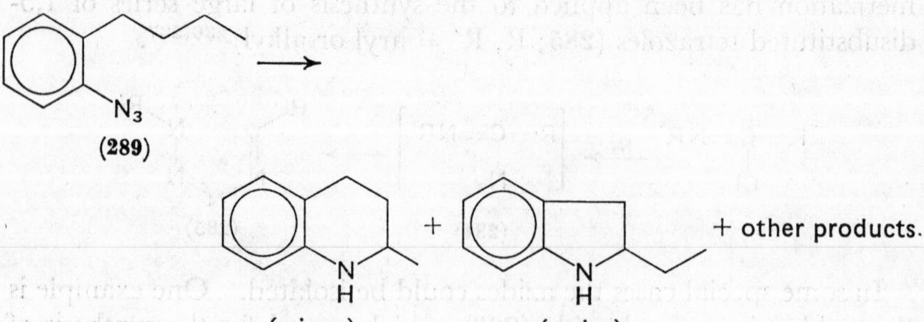

(289)

(minor) (290) (major) + other products.

applied to the synthesis of phenothiazine (X = S, 32%) and its dioxide (X = SO_2, 42%)[183] and failed when X was O, NH or CH_2[183,46].

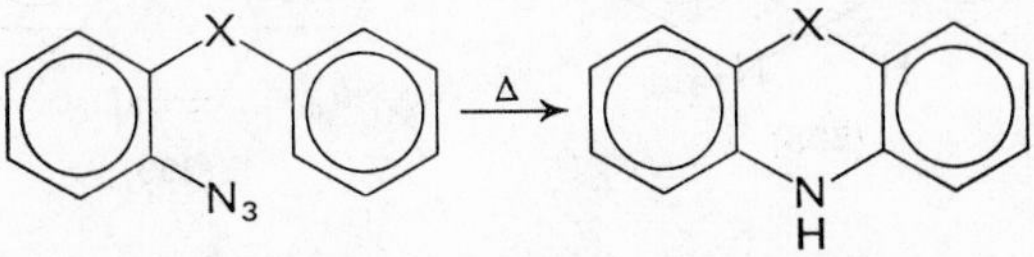

(291) (292)

Pyrolysis of the azidobiphenyl **293** also led to a 6-membered ring, yielding, along with other products, 48% of 2,4-dimethylphenanthridine (**294**)[243].

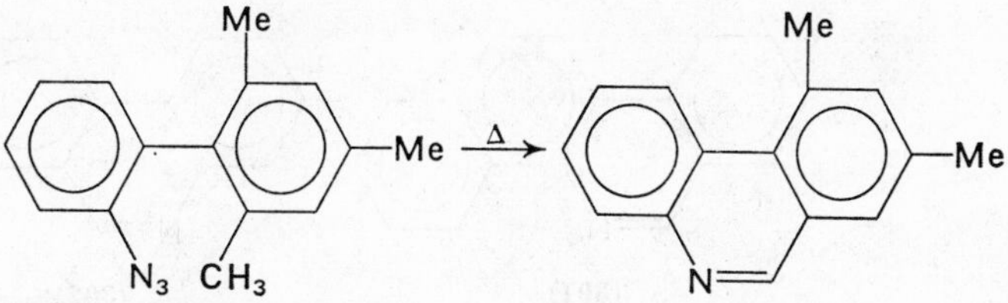

(293) (294)

A suitable spatial arrangement of a molecule, which allows an easy access of the azido group to a C—H bond (as in **295**) can promote the formation of a 6-membered ring. Photolysis of the perhydrophenanthrenene derivative **296** yielded (25%) the lactam **297**[244].

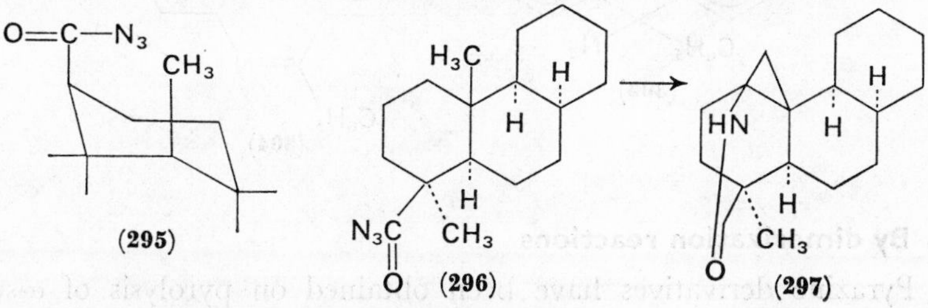

(295) (296) (297)

2. By ring expansions of five-membered rings

Five-membered ring azides can rearrange upon decomposition to form 6-membered ones. Acid decomposition of 9-azidofluorene (**298**) yielded in this manner phenanthridine (**299**)[245]. Unsymmetrical fluorenes produce mixtures, thus both **301** and **302** have been obtained from **300**[66].

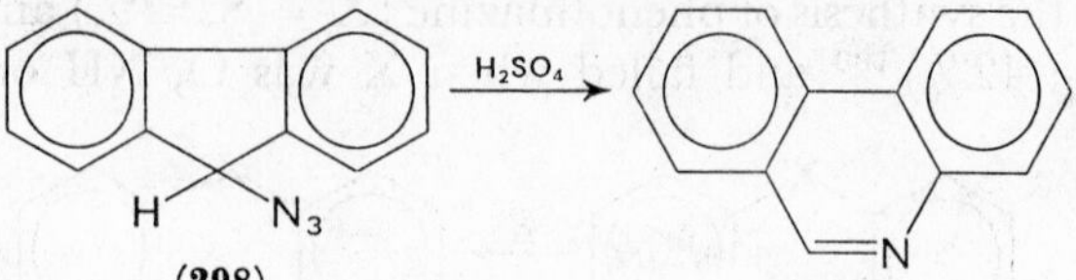

(298) (299)

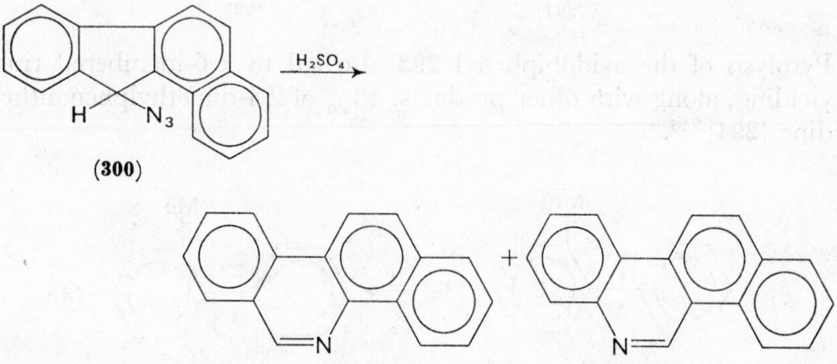

(300)

(301) (302)

9-Azido-9-substituted fluorenes yield the corresponding phenanthridines upon thermolysis. Thus 9-azido-9-phenylfluorene (**303**) yielded **304**[66].

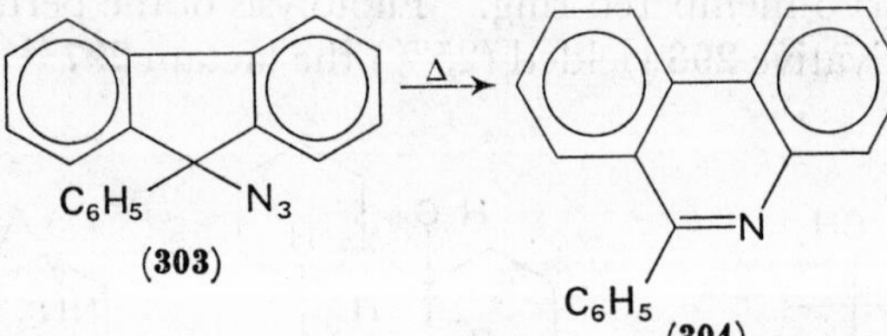

(303) (304)

3. By dimerization reactions

Pyrazine derivatives have been obtained on pyrolysis of α-substituted vinyl azides in ethanol. Thus 1-azido-2-phenylpropene (**305**) produced 2,5-dimethyl-2,5-diphenyldihydropyrazine (**306**) (48%) on

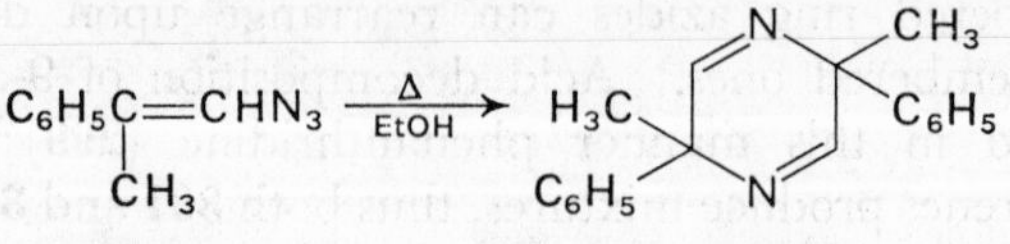

(305) (306)

pyrolysis in ethanol. No trace of **306** was detected when the pyrolysis was carried out in mesitylene[87].

The reaction of α-azidoketones with triphenylphosphine has recently been reported to give pyrazines in good yields. In this manner 2,5-diphenylpyrazine (**307**) was obtained in 75% yield from phenacyl azide. The reaction probably proceeds by dimerization followed by oxidation[246].

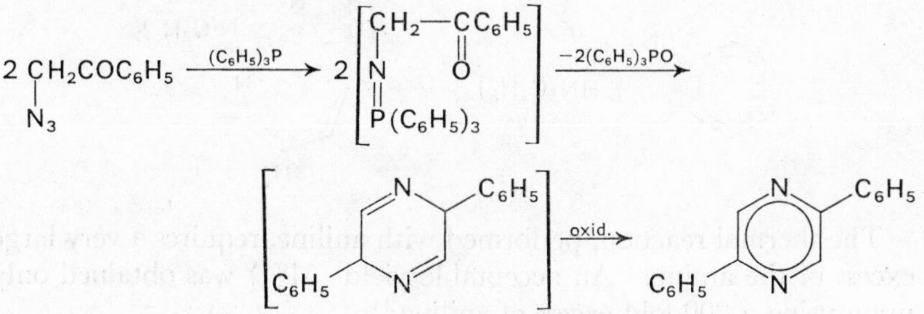

$$2\,\underset{\displaystyle N_3}{CH_2}COC_6H_5 \xrightarrow{(C_6H_5)_3P} 2\left[\begin{array}{l} CH_2{-}C(=O)C_6H_5 \\ | \\ N{=}P(C_6H_5)_3 \end{array}\right] \xrightarrow{-2(C_6H_5)_3PO}$$

$$\left[\text{2,5-diphenyldihydropyrazine}\right] \xrightarrow{\text{oxid.}} \text{2,5-diphenylpyrazine}$$

(307)

I. Syntheses of Seven-membered Rings

Carboalkoxy nitrenes add to benzene rings yielding, after valence-bond isomerization of intermediate aziridines, *N*-carboalkoxy azepines. *N*-Carbethoxy azepine (**308**) was thus synthesized by photo-

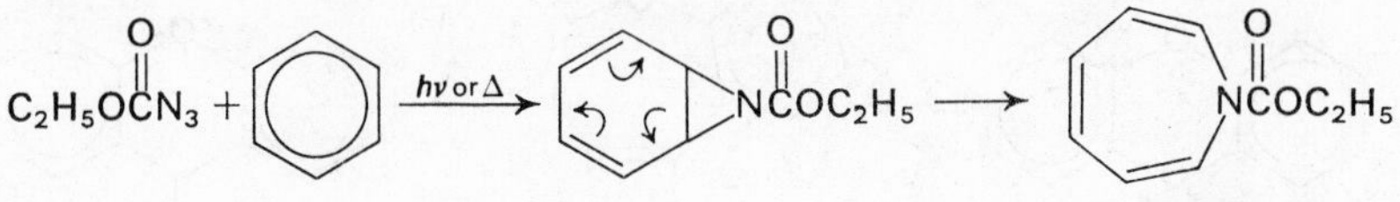

$$C_2H_5OC(=O)N_3 + C_6H_6 \xrightarrow{h\nu \text{ or } \Delta} \text{(aziridine)}\,NC(=O)OC_2H_5 \longrightarrow \text{(azepine)}\,NC(=O)OC_2H_5$$

(308)

lysis (75% yield)[247] or thermolysis (41% yield)[248] of ethyl azidoformate in benzene solution. Some substituted azepines have also been prepared by this method[249].

In one case a similar reaction of an aryl azide has been observed. The thermal decomposition of *o*-azidodiphenylmethane (**309**) yielded (66%) azepino[2,1-*a*]-*H*-indole (**310**)[46].

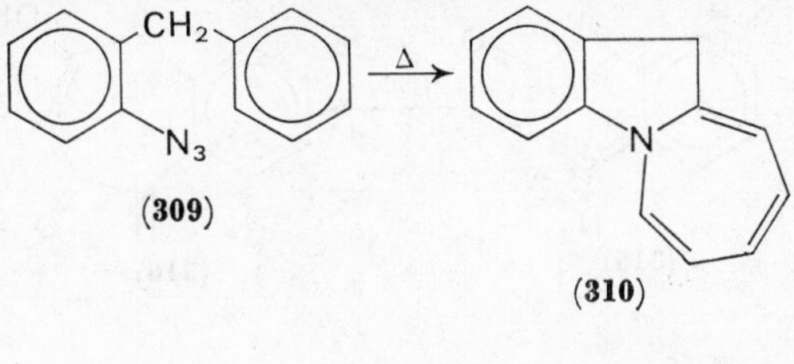

$$o\text{-}N_3C_6H_4CH_2C_6H_5 \xrightarrow{\Delta} \text{azepino[2,1-}a\text{]-}H\text{-indole}$$

(309) **(310)**

Another synthesis of azepines from azides involved the photolysis or thermolysis of aryl azides in the presence of nucleophiles. The photolysis of phenyl azide in diethylamine yielded (34%) 2-diethylamino-3*H*-azepine (**311**)[250]. In the same manner 2-substituted azepines were obtained from phenyl azide and liquid ammonia, aniline and hydrogen sulphide[250].

$$C_6H_5N_3 + HN(C_2H_5)_2 \xrightarrow{h\nu}$$ (**311**)

The thermal reaction, performed with aniline, requires a very large excess of the amine. An acceptable yield (54%) was obtained only upon using a 200-fold excess of aniline[251].

The ring expansion route has also been applied to the synthesis of 7-membered rings, however, imines were also formed in every case. Pyrolysis of 9-azido-9-phenylxanthene (**312**) yielded a mixture of the desired dibenzoxazepine (**313**) and the anil **314**. Similar results have been observed with the corresponding thioxanthene[252].

(**312**) $\xrightarrow{\Delta}$ (**313**) + (**314**)

A ring expansion has also been observed upon acid treatment of 4-azido-1,2-naphthoquinone (**315**). 4-Hydroxy-benzazepine-2,5-dione (**316**) was obtained in 82% yield[253].

(**315**) $\xrightarrow{H_2SO_4}$ (**316**)

IV. REFERENCES

1. C. Grundmann in *Methoden der Organischen Chemie* (Houben–Weyl), Vol. 10/3, 1965, pp. 777–836.
2. P. A. S. Smith, *Open Chain Nitrogen Compounds*, Vol. 2, A. Benjamin, New York and Amsterdam, 1966, pp 211–268.
3. G. L'abbe, *Chem. Rev.*, **69,** 345 (1969).
4. G. L'abbe, *Indust. Chim. Belge*, **34,** 519 (1969).
5. H. Bretschneider and H. H. Hormann, *Monatsh.*, **84,** 1021 (1953).
6. J. H. Boyer, *J. Am. Chem. Soc.*, **73,** 5865 (1951).
7. C. A. VanderWerf, R. Y. Heisler and W. E. McEwen, *J. Am. Chem. Soc.*, **76,** 1231 (1954).
8. J. H. Boyer and S. E. Ellzey, *J. Org. Chem.*, **23,** 127 (1958).
9. P. A. S. Smith, J. H. Hall and R. O. Kan, *J. Am. Chem. Soc.*, **84,** 485 (1962).
10. Y. Ali and A. L. Richardson, *Carbohydrate Res.*, **5,** 441 (1967).
11. R. Adams and D. C. Blomstrom, *J. Am. Chem. Soc.*, **75,** 3405 (1953).
12. M. O. Forster and K. A. N. Rao, *J. Chem. Soc.*, 1943 (1926).
13. I. Ugi, H. Perlinger and L. Behringer, *Chem. Ber.*, **91,** 2330 (1958).
14. H. Rathsbung, *Ber.*, **54,** 3183 (1921).
15. G. Barger and A. J. Ewins, *J. Chem. Soc.*, **97,** 2253 (1910).
16. T. Curtius, *J. Prakt. Chem.*, **52,** 210 (1895).
17. T. Kauffmann and S. M. Hage, *Angew. Chem. Int. Ed.*, **2,** 156 (1963).
18. W. Büchner and R. Dufaux, *Angew. Chem. Int. Ed.*, **5,** 586 (1966).
19. H. Ohrui and S. Emoto, *Carbohydrate Res.*, **10,** 221 (1969).
20. R. Filler, N. R. Ayyangar, W. Gustowski and H. H. Kang, *J. Org. Chem.*, **34,** 534 (1969).
21. H. M. Walborsky and M. E. Baum, *J. Org. Chem.*, **21,** 538 (1956); D. F. Loncrini and H. M. Walborsky, *J. Med. Chem.*, **7,** 369 (1964).
22. M. Brenner and H. R. Rickenbacher, *Helv. Chim. Acta.*, **41,** 181 (1958).
23. M. Frankel, Y. Knobler and T. Sheradsky, *J. Chem. Soc.*, 3642 (1959).
24. B. Erkstrom, A. Gomez-Revilla, R. Mollberg, H. Thelin and B. Sjoberg, *Acta Chim. Scan.*, **19,** 281 (1965).
25. W. F. Huber, *J. Am. Chem. Soc.*, **77,** 112 (1955).
26. A. Bertho and E. Strecker, *Ann. Chem.*, **607,** 194 (1957).
27. P. Brewster, F. Hiron, E. D. Hughes, C. K. Ingold and P. A. Rao, *Nature*, **166,** 178 (1950).
28. R. D. Guthrie and D. Murphy, *J. Chem. Soc.*, 3828 (1965).
29. G. Swift and D. Swern, *J. Org. Chem.*, **32,** 511 (1967).
30. A. K. Bose, J. F. Kistner and L. Farber, *J. Org. Chem.*, **27,** 2925 (1962).
31. A. K. Bose, S. Harrison and L. Farber, *J. Org. Chem.*, **28,** 1223 (1963).
32. E. J. Reist, R. R. Spencer, B. R. Baker and L. Goodman, *Chem. Ind.*, 1794 (1962).
33. D. Horton and A. E. Leutzow, *Carbohydrate Res.*, **7,** 101 (1968).
34. J. Defeye and A. M. Miquel, *Carbohydrate Res.*, **9,** 250 (1969).
35. R. E. Holmes and R. K. Robins, *J. Am. Chem. Soc.*, **87,** 1772 (1965).
36. J. P. Horwitz, A. J. Tomson, J. A. Urbanski and J. Chua, *J. Org. Chem.*, **27,** 3045 (1962).
37. E. Bamberger, *Ann. Chem.*, **443,** 192 (1925).
38. P. A. S. Smith and B. B. Brown, *J. Am. Chem. Soc.*, **73,** 2438 (1951).

39. T. Wieland and H. Urbach, *Ann. Chem.*, **613,** 84 (1958).
40 W. Borsche and H. Hahn, *Chem. Ber.*, **82,** 260 (1949).
41. R. Kreher and G. Jager, *Angew. Chem. Int. Ed.*, **4,** 952 (1965).
42. J. H. Boyer and L. R. Morgan, *J. Org. Chem.*, **24,** 561 (1959).
43. D. H. R. Barton and L. R. Morgan, *J. Chem. Soc.*, **622,** (1962).
44. R. M. Moriarty and M. Rahman, *Tetrahedron*, **21,** 2877 (1965).
45. P. A. S. Smith and J. H. Hall, *J. Am. Chem. Soc.*, **84,** 480 (1962).
46. L. Krbechek and H. Takimoto, *J. Org. Chem.*, **33,** 4286 (1968).
47. G. Smolinsky, *J. Org. Chem.*, **26,** 4108 (1961).
48. J. H. Hall, J. W. Hill and H. C. Tsai, *Tetrahedron Letters*, 2211 (1965).
49. T. J. Prosser, A. F. Marcantonio, C. A. Genge and D. S. Breslow, *Tetrahedron Letters*, 2483 (1964).
50. W. Lwowski and T. W. Mattingly, *J. Am. Chem. Soc.*, **87,** 1947 (1965).
51. M. F. Sloan, W. B. Renfrow and D. S. Breslow, *Tetrahedron Letters*, 2905 (1964).
52. R. K. Smalley and H. Suschitzky, *J. Chem. Soc.*, 5571 (1963).
53. P. A. S. Smith, *Org. Reactions*, **3,** 337 (1946).
54. W. Lwowski, R. DeMauriac, T. W. Mattingly and E. Scheiffele, *Tetrahedron Letters*, 3284 (1964).
55. N. Koga and J.-P. Anselme, *J. Org. Chem.*, **33,** 3963 (1968).
56. T. Curtius and G. Ehrhart, *Ber.*, **55,** 1559 (1922).
57. K. W. Sherk, A. G. Houpt and A. W. Browne, *J. Am. Chem. Soc.*, **62,** 329 (1940).
58. J. H. Boyer, F. C. Canter, J. Hamer and R. K. Putney, *J. Am. Chem. Soc.*, **78,** 325 (1956).
59. O. E. Edwards and K. K. Purushothaman, *Can. J. Chem.*, **42,** 712 (1964).
60. A. J. Davies, A. S. R. Donald and R. E. Marks, *J. Chem. Soc.* (*C*), 2109 (1967).
61. G. H. Gudmundsen and W. E. McEwen, *J. Am. Chem. Soc.*, **79,** 329 (1957).
62. J. Goubeau, E. Allenstein and A. Schmidt, *Chem. Ber.*, **97,** 884 (1964).
63. N. Wiberg and K. H. Schmid, *Angew. Chem. Int. Ed.*, **3,** 444 (1964).
64. W. Pritzkow and G. Pohl, *J. Prakt. Chem.* [*4*], **20,** 132 (1963).
65. W. Pritzkow and D. Timm, *J. Prakt. Chem.* [*4*], **32,** 178 (1966).
66. C. L. Arcus, R. E. Marks and M. M. Coombs, *J. Chem. Soc.*, 4064 (1957).
67. R. L. Whistler and A. K. M. Anisuzzaman, *J. Org. Chem.*, **34,** 3823 (1969).
68. W. H. Saunders and J. C. Ware, *J. Am. Chem. Soc.*, **80,** 3328 (1958).
69. W. H. Saunders and E. A. Caress, *J. Am. Chem. Soc.*, **86,** 861 (1964).
70. J. H. Boyer and D. Straw, *J. Am. Chem. Soc.*, **75,** 1642 (1953).
71. K. Friedrich, *Angew. Chem. Int. Ed.*, **6,** 959 (1967).
72. R. Huisgen, R. Grashey and J. Sauer in *The Chemistry of Alkenes*, (Ed. S. Patai), Interscience, London 1963, p. 835.
73. K. R. Henery-Logan and R. A. Clark, *Tetrahedron Letters*, 801 (1968).
74. R. S. McDaniel and A. C. Oehlschlager, *Tetrahedron*, **25,** 1381 (1969).
75. A. S. Bailey and J. J. Wedgwood, *J. Chem. Soc.* (*C*), 682 (1968).
76. R. F. Bleiholder and H. Shechter, *J. Am. Chem. Soc.*, **90,** 2131 (1968).
77. P. Scheiner, *J. Org. Chem.*, **32,** 2022 (1967).
78. D. L. Rector and R. E. Harmon, *J. Org. Chem.*, **31,** 2837 (1966).
79. R. Fusco, G. Bianchetti and D. Pocar, *Gazz. Chim. Ital.*, **91,** 849, 933 (1961).
80. R. Fusco, G. Bianchetti, D. Pocar and R. Ugo, *Chem. Ber.*, **96,** 802 (1963).

81. J. E. Franz, M. W. Dietrich, A. E. Henshall and C. Osuch, *J. Org. Chem.*, **31,** 2847 (1966).
82. A. Schönberg and W. Urban, *J. Chem. Soc.*, 530 (1935).
83. J. H. Boyer, W. E. Krueger and G. J. Mikol, *J. Am. Chem. Soc.*, **89,** 5504 (1967).
84. A. Hassner, R. J. Isbister, R. B. Greenwald, J. T. Klug and E. C. Taylor, *Tetrahedron*, **25,** 1637 (1969).
85. J. H. Hall and E. Patterson, *J. Am. Chem. Soc.*, **89,** 5856 (1967).
86. J. A. VanAllan, W. J. Priest, A. S. Marshall and G. A. Reynolds, *J. Org. Chem.*, **33,** 1100 (1968).
87. G. Smolinsky and C. A. Pryde, *J. Org. Chem.*, **33,** 2411 (1968).
88. J. D. Hobson and J. R. Malpass, *J. Chem. Soc.* (*C*), 1645 (1967).
89. J. P. Collman, M. Kubota, J. Y. Sun and F. Vastine, *J. Am. Chem. Soc.*, **89,** 169 (1967).
90. R. P. Bennett and W. B. Hardy, *J. Am. Chem. Soc.*, **90,** 3295 (1968).
91. J. E. Baldwin and J. E. Patrick, *Chem. Comm.*, 968 (1968).
92. C. F. H. Allen and A. Bell, *Organic Synthesis*, Coll. Vol. **3,** 846 (1955).
93. W. von E. Doering and C. H. DePuy, *J. Am. Chem. Soc.*, **75,** 5955 (1953).
94. M. Regitz, *Angew. Chem. Int. Ed.*, **6,** 733 (1967).
95. T. Weil and M. Cais, *J. Org. Chem.*, **28,** 2472 (1963).
96. D. Lloyd and F. I. Wasson, *J. Chem. Soc.* (*C*), 408 (1966).
97. M. Regitz and A. Liedhegener, *Tetrahedron*, **23,** 2701 (1967).
98. M. Regitz, *Chem. Ber.*, **97,** 2742 (1964).
99. M. Regitz and A. Liedhegener, *Chem. Ber.*, **99,** 3128 (1966).
100. J. B. Hendrickson and W. A. Wolf, *J. Org. Chem.*, **33,** 3610 (1968).
101. M. Regitz, *Chem. Ber.*, **98,** 1210 (1965).
102. M. Regitz, F. Menz and J. Rüter, *Tetrahedron Letters*, 739 (1967).
103. M. Regitz, A. Anschütz, and A. Liedhegener, *Chem. Ber.*, **101,** 3734 (1968).
104. M. Regitz and A. Anschütz, *Chem. Ber.*, **102,** 2216 (1969).
105. A. M. van Leusen, P. M. Smid and J. Strating, *Tetrahedron Letters*, 337 (1965).
106. M. Regitz, *Chem. Ber.*, **98,** 36 (1965).
107. H. Balli and V. Muller, *Angew. Chem. Int. Ed.*, **3,** 644 (1964).
108. H. Balli and R. Löw, *Tetrahedron Letters*, 5821 (1966).
109. J. Kucera and Z. Arnold, *Tetrahedron Letters*, 1109 (1966).
110. R. Huisgen, G. Szeimies and L. Mobius, *Chem. Ber.*, **99,** 475 (1966).
111. P. Grünanger, P. V. Finzi and C. Scotti, *Chem. Ber.*, **98,** 623 (1965).
112. M. Regitz and D. Stadler, *Ann. Chem.*, **687,** 214 (1965).
113. J. M. Tedder and B. Webster, *J. Chem. Soc.*, 4417 (1960).
114. L. Horner, A. Christmann and A. Gross, *Chem. Ber.*, **96,** 399 (1963).
115. J. S. Swenton, *Tetrahedron Letters*, 3421 (1968).
116. J. Hancock, *Tetrahedron Letters*, 1585 (1964).
117. W. Lwowski, T. W. Mattingly and T. J. Maricich, *Tetrahedron Letters*, 1591 (1964).
118. H. Staudinger and J. Meyer, *Helv. Chim. Acta*, **2,** 635 (1919).
119. H. Staudinger and E. Hauser, *Helv. Chim. Acta*, **4,** 861 (1921).
120. J. Goerdeler and H. Ullmann, *Chem. Ber.*, **94,** 1067 (1961).
121. R. Neidlein, *Angew. Chem. Int. Ed.*, **5,** 314 (1966).
122. D. L. Herring, *J. Org. Chem.*, **26,** 3998 (1961).

123. L. Horner and A. Gross, *Ann. Chem.*, **591,** 117 (1955).
124. A. Messmer, I. Pinter and F. Szego, *Angew. Chem. Int. Ed.*, **3,** 228 (1964).
125. L. Horner and A. Christmann, *Chem. Ber.*, **96,** 388 (1963).
126. O. Dimroth, *Ber.*, **40,** 2376 (1907).
127. O. Dimroth, *Ber.*, **36,** 909 (1903).
128. O. Dimroth, *Ber.*, **38,** 670 (1905); **39,** 3905 (1906).
129. G. S. Akimova, I. G. Kolkoltseva, V. N. Christokletov and A. A. Petrov, *Zh. Org. Khim.*, **4,** 954 (1968); English translation *J. Org. Chem. U.S.S.R.*, **4,** 927 (1968).
130. H. Bretschneider and H. Rager, *Monatsh.*, **81,** 981 (1950).
131. J. E. Leffler, U. Honsberg, Y. Tsuno and I. Forsblad, *J. Org. Chem.*, **26,** 4810 (1961).
132. J. E. Franz and C. Osuch, *Tetrahedron Letters*, 841 (1963).
133. H. Bock and W. Wiegrabe, *Angew. Chem. Int. Ed.*, **2,** 484 (1964).
134. H. Balli and F. Kersting, *Ann. Chem.*, **663,** 96 (1963).
135. W. Fischer and J.-P. Anselme, *J. Am. Chem. Soc.*, **89,** 5284 (1967).
136. J.-P. Anselme and W. Fischer, *Tetrahedron*, **25,** 855 (1969).
137. P. A. S. Smith, C. D. Rowe and L. B. Bruner, *J. Org. Chem.*, **34,** 3430 (1969).
138. P. Scheiner, *J. Org. Chem.*, **30,** 7 (1965).
139. P. Schiener, *Tetrahedron*, **24,** 2757 (1968).
140. R. Huisgen, L. Möbius, G. Müller, H. Stangl, G. Szeimies and J. M. Vernon, *Chem. Ber.*, **98,** 3992 (1965).
141. J. E. Franz, C. Osuch and M. W. Dietrich, *J. Org. Chem.*, **29,** 2922 (1964).
142. I. Brown and E. O. Edwards, *Can. J. Chem.*, **43,** 1266 (1965).
143. J. F. W. Keana, S. B. Keana and D. Beetham, *J. Org. Chem.*, **32,** 3057 (1967).
144. K. Hafner, W. Kaiser and R. Puttner, *Tetrahedron Letters*, 3953 (1964).
145. J. S. McConaghy and W. Lwowski, *J. Am. Chem. Soc.*, **89,** 4450 (1967).
146. A. G. Hortman and J. E. Martinelli, *Tetrahedron Letters*, 6205 (1968).
147. G. Smolinsky, *J. Org. Chem.*, **27,** 3557 (1962).
148. G. R. Harvey and K. W. Ratts, *J. Org. Chem.*, **31,** 3907 (1966).
149. A. Hassner and F. W. Fowler, *J. Am. Chem. Soc.*, **90,** 2869 (1968).
150. K. Isomura, M. Okada and H. Taniguchi, *Tetrahedron Letters*, 4073(1969).
151. W. J. Middleton and C. G. Krespan, *J. Org. Chem.*, **30,** 1398 (1965).
152. D. H. R. Barton and A. N. Starratt, *J. Chem. Soc.*, 2444 (1965).
153. W. L. Meyer and A. S. Levinson, *J. Org. Chem.*, **28,** 2859 (1963).
154. I. Brown, O. E. Edwards, J. M. McIntosh and D. Vocelle, *Can. J. Chem.*, **47,** 2751 (1969).
155. A. L. Logothetis, *J. Am. Chem. Soc.*, **87,** 749 (1965).
156. G. Smolinsky, *J. Am. Chem. Soc.*, **83,** 2489 (1961).
157. G. Smolinsky and B. I. Feuer, *J. Org. Chem.*, **29,** 3097 (1964).
158. R. J. Sundberg, L. S. Lin and D. E. Blackburn, *J. Heterocyclic Chem.*, **6,** 441 (1969).
159. K. Isomura, S. Kobayashi and H. Taniguchi, *Tetrahedron Letters*, 3499 (1968).
160. P. A. S. Smith and B. B. Brown, *J. Am. Chem. Soc.*, **73,** 2435 (1951).
161. P. A. S. Smith, J. M. Clegg and J. H. Hall, *J. Org. Chem.*, **23,** 524 (1958).
162. T. Nozoe, H. Horinu and T. Toda, *Tetrahedron Letters*, 5349 (1967).
163. P. A. S. Smith and J. H. Boyer, *J. Am. Chem. Soc.*, **73,** 2626 (1951).

164. R. A. Abramovitch and K. A. H. Adams, *Can. J. Chem.*, **39,** 2516 (1961).
165. R. A. Abramovitch, K. A. H. Adams and A. D. Notation, *Can. J. Chem.*, **38,** 2152 (1960).
166. F. Texier and R. Carrie, *Tetrahedron Letters*, 823 (1969).
167. J. H. Boyer and D. Straw, *J. Am. Chem. Soc.*, **74,** 4506 (1952).
168. K. H. Saunders, *J. Chem. Soc.*, 3275 (1955).
169. J. Schmutz and F. Kunzle, *Helv. Chim. Acta*, **39,** 1144 (1956).
170. O. Meth-Cohn, R. K. Smalley and H. Suschitzky, *J. Chem. Soc.*, 1666 (1963).
171. R. K. Smalley, *J. Chem. Soc. (C)*, 80 (1966).
172. L. Krbechek and H. Takimoto, *J. Org. Chem.*, **29,** 3630 (1964).
173. J. H. Hall and D. R. Kamm, *J. Org. Chem.*, **30,** 2092 (1965).
174. L. Krbechek and H. Takimoto, *J. Org. Chem.*, **29,** 1150 (1964).
175. R. Puttner and K. Hafner, *Tetrahedron Letters*, 3119 (1964).
176. R. Kreher and G. H. Bockhorn, *Angew. Chem. Int. Ed.*, **3,** 589 (1964).
177. R. Kreher and D. Kuhling, *Angew. Chem. Int. Ed.*, **4,** 69 (1965).
178. R. Huisgen and H. Blaschke, *Chem. Ber.*, **98,** 2985 (1965).
179. J. Meinwald and D. H. Aue, *J. Am. Chem. Soc.*, **88,** 2849 (1966).
180. J. H. Boyer and J. Hamer, *J. Am. Chem. Soc.*, **77,** 951 (1955).
181. R. Garner, E. B. Mullock and H. Suschitzky, *J. Chem. Soc. (C)*, 1980 (1966).
182. E. B. Mullock and H. Suschitzky, *J. Chem. Soc. (C)*. 1937 (1968).
183. P. A. S. Smith, B. B. Brown, R. K. Putney, and R. F. Reinisch, *J. Am. Chem. Soc.*, **75,** 6335 (1953).
184. W. Lwowski, A. Hartenstein, C. deVita and R. L. Smick, *Tetrahedron Letters*, 2497 (1964).
185. R. Huisgen and H. Blaschke, *Ann. Chem.*, **686**, 145 (1965).
186. P. A. S. Smith and J. H. Boyer, *Org. Synthesis* Coll. Vol. **IV**, 75 (1963).
187. M. O. Forster and M. F. Barker, *J. Chem. Soc.*, **103,** 1918 (1913).
188. D. L. Hammick, W. E. Edwardes and E. R. Steiner, *J. Chem. Soc.*, 3308 (1931).
189. D. Britton and W. E. Noland, *J. Org. Chem.*, **27,** 3218 (1962).
190. A. J. Boulton, P. B. Ghosh and A. R. Katritzky, *Tetrahedron Letters*, 2887 (1966).
191. K. Alder and G. Stein, *Ann. Chem.*, **515,** 185 (1935).
192. K. Alder and G. Stein, *Ann. Chem.*, **485,** 223 (1931).
193. K. Ziegler, H. Sauer, L. Bruns, H. Froitzheim-Kühlhorn and J. Schneider, *Ann. Chem.*, **589,** 122 (1954).
194. C. S. Rondestvedt and P. K. Chang, *J. Am. Chem. Soc.*, 77, 6532 (1955).
195. S. J. Davies and C. S. Rondestvedt, *Chem. Ind.*, 845 (1956).
196. A. Mustafa, S. M. A. D. Zayed and S. Khattab, *J. Am. Chem. Soc.*, **78,** 145 (1956).
197. G. Rembarz, B. Kirchhoff and G. Dongowski, *J. Prakt. Chem.* (*4*), **33,** 199 (1966).
198. M. E. Munk and Y. K. Kim, *J. Am. Chem. Soc.*, **86,** 2213 (1964).
199. G. Bianchetti, P. D. Croce and D. Pocar, *Tetrahedron Letters*, 2043 (1965).
200. R. Huisgen, L. Mobius and G. Szeimies, *Chem. Ber.*, **98,** 1138 (1965).
201. P. Scheiner, *Tetrahedron*, **24,** 349 (1968).
202. G. D. Buckley, *J. Chem. Soc.*, 1850 (1954).
203. P. Scheiner, *J. Am. Chem. Soc.*, **88,** 4759 (1966).

204. R. Huisgen and G. Szeimies, *Chem. Ber.*, **98,** 1153 (1965).
205. G. Guardino, A. Umani-Ronchi, P. Bravo and M. Acampora, *Tetrahedron Letters*, 107 (1967).
206. G. Guardino, C. Ticozzi, A. Umani-Ronchi and P. Bravo, *Gazz. Chim. Ital.*, **97,** 1411 (1967).
207. C. E. Olsen and C. Pedersen, *Tetrahedron Letters*, 3805 (1968).
208. H. Gold, *Ann. Chem.*, **688**, 205 (1965).
209. W. Kirmse and L. Horner, *Ann. Chem.*, **614,** 1 (1958).
210. R. Huisgen, R. Knorr, L. Mobius and G. Szeimies, *Chem. Ber.*, **98,** 4014 (1965).
211. R. Fuks, R. Buijle and H. G. Viehe, *Angew Chem. Int. Ed.*, **5,** 585 (1966).
212. J. C. Sheehan and C. A. Robinson, *J. Am. Chem. Soc.*, **73,** 1207 (1951).
213. F. Moulin, *Helv. Chim. Acta*, **35,** 167 (1952).
214. R. H. Wiley, N. R. Smith, D. M. Johnson and J. Moffat, *J. Am. Chem. Soc.*, **77,** 3412 (1955).
215. J. A. Durden, H. A. Stansbury and W. H. Catlette, *J. Chem. Eng. Data*, **9,** 228 (1964).
216. G. Wittig and A. Krebs, *Chem. Ber.*, **94,** 3260 (1961).
217. G. A. Reynolds, *J. Org. Chem.*, **29,** 3733 (1964).
218. W. Ried and M. Schon, *Chem. Ber.*, **98,** 3142 (1965).
219. G. Garcia-Munoz, J. Iglesias, M. Lora-Tamayo and R. Mandronero, *J. Heterocyclic Chem.*, **5,** 699 (1968).
220. J. R. E. Hoover and A. R. Day, *J. Am. Chem. Soc.*, **78,** 5832 (1956).
221. A. Dornow and J. Helberg, *Chem. Ber.*, **93,** 2001 (1960).
222. E. Lieber, T. S. Chao and C. N. R. Rao, *J. Org. Chem.*, **22,** 654 (1957).
223. E. Lieber, C. N. R. Rao and T. V. Rajkumar, *J. Org. Chem.*, **24,** 134 (1959).
224. G. R. Harvey, *J. Org. Chem.*, **31,** 1587 (1966).
225. G. L'abbe, P. Ykman and G. Smets, *Tetrahedron*, **25,** 5421 (1969).
226. A. Bertho, *Ber.*, **58,** 859 (1925).
227. H. El Khadem, H. A. R. Mansour and M. H. Meshreki, *J. Chem. Soc. (C)*, 1329 (1968).
228. J. H. Hall, *J. Org. Chem.*, **33,** 2954 (1968).
229. R. A. Carboni, J. C. Kauer, J. E. Castle and H. E. Simmons, *J. Am. Chem. Soc.*, **89,** 2618 (1967).
230. K. Hohenlohe-Oehringen, *Monatsh.*, **89,** 557 (1958).
231. K. Hohenlohe-Oehringen, *Monatsh.*, **89,** 562 (1958).
232. W. R. Carpenter, *J. Org. Chem.*, **27,** 2085 (1962).
233. S. Gotzky, *Ber.*, **64,** 1555 (1931).
234. R. M. Moriarty and J. M. Kliegman, *J. Am. Chem. Soc.*, **89,** 5959 (1967).
235. R. M. Moriarty, J. M. Kliegman and C. Shovlin, *J. Am. Chem. Soc.*, **89,** 5958 (1967).
236. J. Von Braun and W. Rudolf, *Chem. Ber.*, **74,** 264 (1941).
237. R. M. Herbst, C. W. Roberts, H. T. F. Givens and E. K. Harvill, *J. Org. Chem.*, **17,** 262 (1952).
238. E. Lieber, E. Sherman, R. A. Henry and J. Cohen, *J. Am. Chem., Soc.*, **73,** 2327 (1951).
239. J. H. Boyer and E. J. Miller, *J. Am. Chem. Soc.*, **81,** 467 (1959).
240. J. H. Boyer and H. W. Hyde, *J. Org. Chem.*, **25,** 458 (1960).

241. R. Huisgen, K. Fraunberg and H. J. Strum, *Tetrahedron Letters*, 2589 (1969).
242. R. Huisgen and K. Fraunberg, *Tetrahedron Letters*, 2595 (1969).
243. G. Smolinsky, *J. Am. Chem. Soc.*, **82,** 4717 (1960).
244. J. W. ApSimon and O. E. Edwards, *Can. J. Chem.*, **40,** 896 (1962).
245. C. L. Arcus and R. J. Mesley, *J. Chem. Soc.*, 178 (1953).
246. E. Zbiral and J. Stroh, *Ann. Chem.*, **727,** 231 (1969).
247. K. Hafner and C. König, *Angew Chem. Int. Ed.*, **2,** 796 (1963).
248. R. J. Cotter and W. F. Beach, *J. Org. Chem.*, **29,** 751 (1964).
249. K. Hafner, D. Zinser and K. L. Moritz, *Tetrahedron Letters*, 1733 (1964).
250. W. von E. Doering and R. A. Odum, *Tetrahedron*, **22,** 81 (1966).
251. R. Huisgen, D. Vossius and M. Appl, *Chem. Ber.*, **91,** 1 (1958).
252. R. H. B. Galt, J. D. Loudon and A. D. B. Sloan, *J. Chem. Soc.*, 1588 (1958).
253. H. W. Moore, H. R. Sheldon and W. Weyler, *Tetrahedron Letters*, 1243 (1969).

CHAPTER 7

Rearrangements involving azido groups

D. V. BANTHORPE

University College, London

I. INTRODUCTION 398

II. THE CURTIUS REARRANGEMENT OF ACYL AZIDES 398
A. Thermal Reactions 398
B. Acid-Catalysed Reactions 402
C. Photochemical Reactions 402
D. Unusual Substrates 404

III. THE SCHMIDT REACTION 405
A. Reactions of Carbonyl Compounds 405
B. Secondary Reactions 412
C. Reactions of Other Compounds 414

IV. ACID-CATALYSED REARRANGEMENTS OF ALKYL AND ARYL AZIDES 416

V. THERMAL REARRANGEMENTS OF ALKYL AND ARYL AZIDES . 421

VI. PHOTOCHEMICAL REARRANGEMENTS OF ALKYL AND ARYL AZIDES 426

VII. MISCELLANEOUS REARRANGEMENTS 428
A. Reactions forming Azides 428
B. Acyclic 1,2,3-Triazenes 430
C. Cyclic 1,2,3-Triazenes 431

VIII. REFERENCES 434

I. INTRODUCTION

Rearrangements involving azides can be grouped into several classes showing some mutual overlap. In the following sections the status of these classes at January, 1970, is summarized mainly from the viewpoint of their reaction mechanisms. Neither historical development nor complete coverage is attempted: the former is adequately described in several articles and the latter is precluded by the quantity and diversity of recent work. Details of reaction conditions, beyond bare descriptions as photolyses, thermolyses, etc., are given only when of mechanistic or other significance since conditions often vary from substrate to substrate for a particular type of reaction so as to make generalizations or inferences from individual examples of little value.

The subject matter of the present article has been discussed with varying degrees of completeness in reviews on the general reactions of organic azides[1–7] and on nitrenes[8–11].

II. THE CURTIUS REARRANGEMENT OF ACYL AZIDES

A. Thermal Reactions

In 1890 Curtius discovered that acyl azides readily lost nitrogen on heating in an inert solvent and formed isocyanates, some requiring refluxing for many hours while others reacted near room temperature. In the more frequently used alcoholic solvents urethanes were produced by addition of alcohol to the initially formed products. Yields were usually good and often greater than 80%. Since the starting materials can be easily synthesized in high yield by treatment of an acid chloride with azide ion and as both isocyanates and urethanes can be efficiently hydrolysed by mineral acids to amines, the reaction forms an important route for the conversion of a carboxyl group into an amino group (reaction 1) and is especially convenient for decarboxylation of tracer-containing material of biosynthetic origin as part of degradation schemes.

$$\mathrm{RCOOH} \longrightarrow \mathrm{ROCl} \longrightarrow \mathrm{RCON_3} \xrightarrow{\Delta} \left\{\begin{matrix}\mathrm{RNCO} \\ \mathrm{RNHCOOR}\end{matrix}\right. \xrightarrow{\mathrm{H^+}} \mathrm{RNH_2} \quad (1)$$

A less used route to the substrate is the treatment of an acyl hydrazine $RCONHNH_2$ (formed from an ester and hydrazine) with nitrous acid. Detailed compilations and reviews[4,12,13] list the many hundreds of examples of the rearrangement recorded up to 1961. The procedure is extremely successful for aliphatic, alicyclic, aromatic and heterocyclic azides of various sizes and complexity: all rearrange uneventfully except for one case when a cyclopropyl ring is opened[14] and even here rearrangement took place normally and ring-scission occurred at a later step. In some compounds migration of groups from nitrogen and oxygen to nitrogen[43] and from sulphur to nitrogen[44] were found.

The rearrangement has a mechanism similar to those of the Hofmann rearrangement of amides, the Lossen rearrangement of acylhydroxamic esters, the Schmidt rearrangement of carbonyl compounds and the Wolff rearrangement of diazoketones. Evidence concerning the mechanism of one can often be applied to the others, and the whole family has been reviewed briefly[15]. Sometimes the distinction is made that the conversion of an acyl azide into an isocyanate or urethane is the Curtius *rearrangement* whereas the overall sequence is the Curtius *reaction*, but usually the former name is used for both processes.

There appears no doubt that the migrating group never becomes kinetically free in this and the related rearrangements. Optically active amines of the same specific rotation were isolated from the rearrangements of the azide or amide derived from (+)-2-methyl-3-phenylpropionic acid and correlations of the rotations of starting material and products in this and in other examples led to the conclusion that the configuration of the migrating group was preserved[16] with typically greater than 99% retention of optical purity[17–19]. In another example[20], optical activity was retained in a biphenyl where the asymmetry was due to hindered rotation (reaction 2) and separation of the migrating group (if any) was not sufficient to allow free rotation about the central bond. Recently, geometrical isomerism has also been found[21] to be retained in the rearrangement

2-Me-6-NO_2-2'-CON_3-biphenyl $\xrightarrow{\Delta}$ 2-Me-6-NO_2-2'-NH_2-biphenyl (2)

$$cis\text{-PhCH{=}CHCON}_3 \xrightarrow{\Delta} cis\text{-PhCH{=}CHNH}_2 \qquad (3)$$

(reaction 3). Before the fact of intramolecularity had been conclusively established the possibility was considered that the migrating group was a carbonium ion or a free radical. The former species was eliminated for the Hofmann rearrangement by observations that groups very susceptible to Wagner–Meerwein rearrangements, e.g. neopentyl and 2,2,2-triphenylethyl, migrated to the nitrogen with their structure unaltered[22] and the latter was ruled out when the Curtius rearrangement was found to be unaffected by the addition of sources of triphenylmethyl radicals to the reaction medium[23,24] and also when acrylonitrile added to the medium was not polymerized[25]. The Curtius rearrangement has nitrogen both in the leaving group and at the migration terminus but studies with ^{15}N-tracer have shown that far spanning migrations do not occur[26] (cf. reaction 4) and a 1,2-shift appears universally likely.

$$PhCONHNH_2 + H^{15}NO_2 \longrightarrow PhCO{-}\bar{N}{-}\overset{+}{N}{\equiv}{}^{15}N \xrightarrow[(ii)\,H^+]{(i)\,\Delta} ArNH_2 + N{\equiv}{}^{15}N \quad (4)$$

The driving force for the family of rearrangements is undoubtedly the tendency of an electron-deficient nitrogen to acquire electrons from a β-linked carbon which is thus induced to migrate, but there has been considerable controversy as to whether rearrangement occurs in one stage of concerted migration and nitrogen loss or in two stages with the intermediate formation of a nitrene. These alternatives for the Curtius rearrangement are shown (reaction 5). One possible method of deciding between them is to observe the effect of variation of R on the rates: the concerted route should be dependent on

$$R{-}C({=}O){-}\bar{N}{-}\overset{+}{N}{\equiv}N \equiv R{-}C({=}O){-}\bar{N}{-}\overset{+}{N}{\equiv}N \longrightarrow R{-}N{=}C{=}O$$

$$R{-}C({=}O){-}\bar{N}{-}\overset{+}{N}{\equiv}N \longrightarrow R{-}C({=}O){-}\ddot{N}\!: \longrightarrow R{-}N{=}C{=}O \quad (5)$$

the migratory aptitude of R whereas a route involving formation of nitrene in the almost certain rate-determining step would be unaffected by this aptitude. In consequence, several investigations have been made of the kinetics of rearrangement of *o*, *m* and *p*-substituted benzoyl azides.

The reaction of benzoyl azide itself was first-order, was facilitated in polar solvents and showed an unusually large ($\Delta H^\ddagger = 25$–32 kcal mole^{-1}) dependence on temperature[27–8]. Electron-releasing *p*-substituents decreased the rate compared with the unsubstituted parent in a way unexpected for either of the routes of reaction (5) and at variance with the corresponding situation for the Hofmann and Lossen rearrangements. Presumably acyl azides are uniquely stabilized by conjugation between the aromatic and triazene groups which more than counteracts even the migratory ability impressed on the former by its *p*-substituents[4,29,30]. *o*-Substituted groups of any polarity increased the rate by some 50–100-fold above that of the unsubstituted parent[31] and this, which is similar to findings for the Beckmann rearrangement[15], has been reasonably attributed to steric inhibition of conjugation between the triazene group and the aromatic ring destabilizing the molecule and in addition promoting its orientation for effective overlap with the empty orbital of the nitrogen to which it is migrating. Electron-releasing *m*-groups increased the rate of rearrangement[29,30] but the effects were small and no correlation with the Hammett equation could be demonstrated. These last results have been considered consistent with either the nitrene[8] or the concerted[4] mechanism.

The former route was also supported by measurements of the volumes of activation of the Hofmann rearrangement of benzamide carried out at high pressures[32] but the consensus of opinion for the Curtius and related rearrangements definitely favours the concerted mechanism[7,11]. There are several reasons for this decision. The disproof of the long advocated intermediacy of monovalent nitrogen derivatives in the Beckmann rearrangement cast doubts on the existence of analogous species in the rearrangement under consideration and all attempts to trap nitrenes in the Curtius rearrangement have failed[33–4]: e.g. no hydroxamic acid could be detected by sensitive tests when the reaction was carried out in aqueous solvents[35]. More decisively, large ^{14}C-isotope effects were found for the Curtius, Hofmann, and Lossen rearrangements of compounds labelled in the migrating groups[11] and this is compatible only with the concerted mechanism. Also little variation in yields was found when the Curtius rearrangement was carried out in a range of solvents[36] which would have almost certainly trapped intermediate nitrenes to differing extents. However, relatively few reactions have been studied to date and the possibility remains of a spectrum of mechanisms with the concerted and two-step routes as extremes: the latter may yet be characterized for suitably elaborated structures and chosen conditions.

An interesting cyclization promoted by heat or ultraviolet irradiation has been found to accompany rearrangement of a few olefinic acyl azides[47,48] (cf. reaction 6).

$$\mathrm{PhCH{=}CHCON_3} \xrightarrow[\mathrm{H_2O}]{\Delta \text{ or } h\nu} \text{PhHC—CH}_2\text{ / NO—N—N(H)—CO (ring)} \tag{6}$$

B. Acid-Catalysed Reactions

After consideration of the mechanism of the Schmidt reaction of carboxylic acids (cf. section III.A) the prediction was made and verified that the Curtius rearrangement would be subject to acid catalysis[37]. The charge and bond distribution in the substrate can almost certainly be best represented by **1** and protonation (reaction 7) was believed to form **2** in which the positive charge on the nitrogen of the

$$\underset{(1)}{\mathrm{RCO{-}\overset{-}{N}{-}\overset{+}{N}{\equiv}N}} \longrightarrow \underset{(2)}{\mathrm{RCO{-}NH{-}\overset{+}{N}{\equiv}N}} \tag{7}$$

potential leaving group favoured the tendency of this group to split off. Acid catalysis by sulphuric acid in acetic acid as solvent was demonstrated for the rearrangement of benzoyl azides and the small but significant effect of electron-donating *m*-substituents which could be correlated with the Hammett equation ($\rho - 1{\cdot}1$) was consistent with a concerted mechanism[38]. Protonation of the oxygen of the acyl group rather than the innermost nitrogen has been suggested by analogy with the situation in the Schmidt reaction[39], but acid-catalysed hydration to $RC(OH)_2N_3$ is not a necessary preliminary to rearrangement as anhydrous acids are efficient catalysts[4]: in addition, an isocyanate has been isolated in one instance for the related Schmidt process[40] and this could not reasonably be derived from the hydrated species.

Lewis acids may also act as catalysts with an efficiency roughly parallel to that found in Friedel–Crafts and related reactions[41,42]. These rearrangements are first-order in acid and a substrate-acid adduct is believed to be rapidly formed which rearranges in a slow step with concomitant release of catalyst. Adducts of benzoyl azide with boron trichloride and trifluoride have been isolated at low temperatures[42].

C. Photochemical Reactions

Rearrangements induced by ultraviolet radiation (cf. reactions 8 and 9) are well known[25,45] and neighbouring groups can be impli-

$$PhCON_3 \xrightarrow{h\nu} PhNCO \quad (8)$$

$$PhCH_2CON_3 \xrightarrow[EtOH]{h\nu} PhCH_2NHCOOEt \quad (9)$$

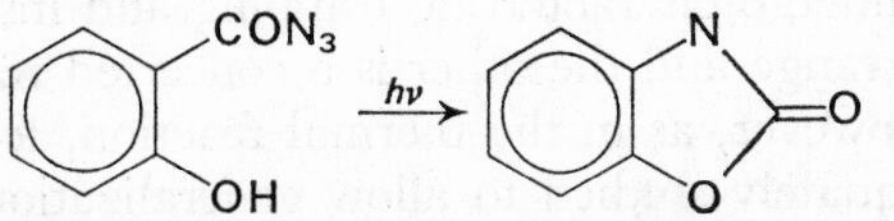

(10)

cated[45] (reaction 10). Triplet nitrenes have been proposed as intermediates which lead to various products of interaction with alcoholic solvents[8,46]. This is reasonable since irradiation could well provide sufficient energy for N—N bond fission without the need for neighbouring-group participation leading to a concerted reaction. Reactions of synthetic utility have been derived from these supposed nitrenes[49–51]: thus the photolysis shown in reaction (11) gave the lactam whereas thermolysis of the same starting material gave the isocyanate in a conventional Curtius rearrangement. In another example the

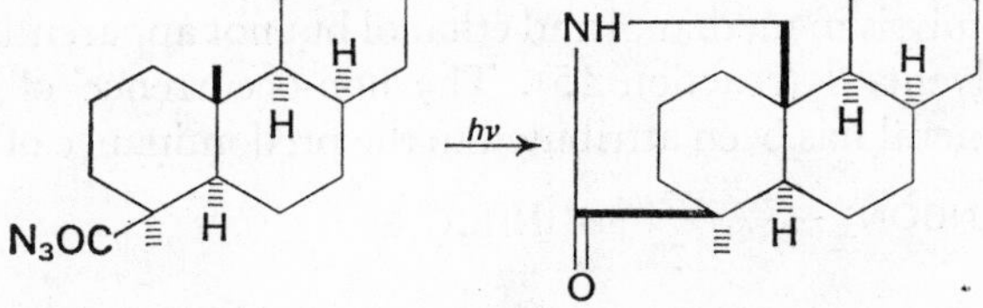

(11)

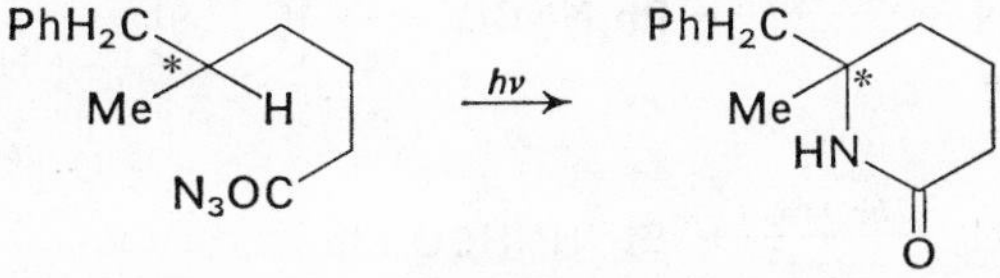

(12)

acyl azide t-BuCoN$_3$ gave both products of rearrangement and of insertion into the solvent on photolysis whereas thermolysis gave only the former[52]. The intramolecular insertion (reaction 12) led to a product with complete retention of configuration at the starred carbon atom[54] in a reaction with a postulated singlet nitrene as intermediate. Nitrenes that were presumed to be intermediates of photochemical Curtius rearrangements have been trapped with dimethyl sulphoxide[46,53] and other scavengers[34].

Nevertheless, although nitrenes undoubtedly can exist under conditions where rearrangement can occur, the observation of ^{14}C-isotope effects on rates and products for photochemical as well as for thermal rearrangements[11] and the small effect of solvent in altering the yield of rearrangement product obtained from photolysis of pivaloyl azide[55] strongly suggest that two independent reactions are operating under these conditions of photolysis[11] at least. One reaction leads to nitrenes that undergo characteristic trapping and insertion reactions but do not rearrange and the other is a concerted rearrangement to isocyanate. However, as in the thermal reaction, too few examples have been adequately studied to allow generalizations of any wide-spread validity to be made.

D. Unusual Substrates

Various types of so-called 'rigid' azides, e.g. carbamoyl azides R_2NCON_3, sulphonazides RSO_2N_3, and diazides of carbonic and sulphinic acids were once thought not to undergo the Curtius re-arrangement[56] and rationalizations were proposed on structural and electronic grounds[57]. Certain *N,N*-diaryl or *N,N*-alkyl-aryl carba-moyl azides have, nevertheless, been found to rearrange on heating in appropriate solvents[5,58,59] (reactions 13, 14) and other types do so under photolysis in methanol and ethanol but not apparently in hydro-carbon solvents[60] (reaction 15). The non-occurrence of rearrange-ment in general has been attributed to the predominance of competing

$$Ph_2NCON_3 \xrightarrow[EtOH]{\Delta} PhNHNHCO_2Et \qquad (13)$$

$$Ph_2NCON_3 \xrightarrow{\Delta,\ toluene} Ph_2NNCO \longrightarrow \text{(benzo-fused ring with NPh, NH, C=O)} \qquad (14)$$

$$EtNHCON_3 \xrightarrow{h\nu,\ MeOH} EtNHNHCO_2Me \qquad (15)$$

solvolysis[61], but the former is usually only of very minor extent even when carbamoyl azides are heated in an inert solvent until general decomposition sets in.

Heating benzenesulphonyl azide at 100° does not cause rearrange-ment although products of reaction of the derived nitrene with the solvent can be isolated[62]: if rearrangement in general involved the nitrene route some would surely be expected to intrude here. In contrast, photolysis in methanol leads to phenylsulphamic esters in a reaction analogous to the Curtius rearrangement[60,63]. Other sul-phonyl azides probably decompose to nitrenes on thermolysis as products typically derived from such intermediates, i.e. of hydrogen abstraction and insertion, are found[11,64,65].

Azidoformate esters, N_3CO_2R, also can show the Curtius rearrangement. A low yield (ca. 10%) of a trimer of methoxyisocyanate was isolated after irradiation of methylazidoformate in aprotic media[66], but generally both photolysis and thermolysis lead to similar products of secondary reactions that are characteristic of the formation of nitrenes[66] with the rearrangement product either not present or in very minor yield. In aromatic solvents insertions can lead[67] to

$$\text{(3): azepine-}NCO_2R \qquad OC(N_3)-O-CR_2CH_3 \xrightarrow{h\nu \text{ or } \Delta} OC(HN)-O-CR_2-CH_2 \quad (16)$$

interesting azepines (**3**), and in inert solvents cyclizations (reaction 16) can occur[68,69]. The necessity of azidoformates and sulphonyl azides to form nitrenes in the step of nitrogen loss provides the only reasonable explanation for their great difference in reactivity compared with acyl azides. The former sluggishly undergo transformations at high temperatures, typically over 120°, whereas the latter can usually rearrange cleanly at much lower temperatures by utilizing a concerted mechanism[7].

III. THE SCHMIDT REACTION

This name is given to a rather loosely defined family of reactions of hydrazoic acid with various types of organic compounds in the presence of a strong acid, usually concentrated sulphuric acid, that are usually carried out in an inert solvent at about 40°. In all types, the crucial step is a rearrangement and the reaction has close similarities to, and indeed some overlap with, the Curtius rearrangement. The latter differs in that acyl azides form a distinct stage either as starting materials or as isolable intermediates and acid catalysis is not obligatory although it can occur.

A. Reaction of Carbonyl Compounds

The most important group of Schmidt reactions are those carried out with carbonyl compounds. Aromatic and aliphatic acids and ketones form amines and amides respectively (reactions 17 and 18), but aldehydes and a few ketones give mainly cyanides together sometimes with very poor yields of formyl derivatives of amines, RNHCHO.

$$\text{RCOOH} \xrightarrow{\text{N}_3\text{H}} [\text{RNHCOOH}] \xrightarrow{\text{H}^+} \text{RNH}_2 + \text{CO}_2 \qquad (17)$$

$$\text{RCOR}^1 \xrightarrow{\text{N}_3\text{H}} \text{RNHCOR}^1 \longrightarrow \text{RNH}_2 + \text{R}^1\text{COOH} \qquad (18)$$

The reaction is of widespread application and has been summarized and compended[4,13,70] up to 1961 and is extensively used in synthesis[71], but the scope is somewhat less than that of the Curtius rearrangement, owing both to the forcing conditions isomerizing sensitive substrates and products and to the multi-step nature of the reaction that gives scope for the formation of side products. For some classes of substrate milder conditions can be used: concentrated hydrochloric acid will affect rearrangement of dialkyl ketones and molten trichloroacetic acid will suffice for alkyl aryl ketones[72]. Another limitation is that although nitrogen-containing heterocyclic substrates can undergo reaction, most are protonated on the ring and the protonation on the side chain oxygen or nitrogen that is necessary to induce rearrangement is thereby hindered. Despite these restrictions, few anomalous reactions are known[73] and phenyl bonded to phosphorus, sulphur and other metalloids can migrate to nitrogen[74].

The reaction was discovered in 1923 in interesting circumstances. Products from the decomposition of hydrazoic acid in various solvents were investigated and the decomposition catalysed by sulphuric acid in benzene was found to lead to aniline Schmidt considered an imine radical NH to be the species responsible and on attempting to trap this with benzophenone he obtained benzamide. The reaction was generalized and rapidly exploited, mainly by its discoverer. The currently accepted mechanism was proposed in outline shortly afterwards[75] and is shown in its modern form in scheme (19): for carboxylic

$$\text{RCOR}^1 \xrightarrow{\text{H}^+} \underset{(4)}{\text{R—}\overset{\overset{+}{\text{OH}}}{\overset{\|}{\text{C}}}\text{—R}^1} \xrightarrow[(a)]{\text{N}_3\text{H}} \underset{(5)}{\text{R—}\underset{\text{NH}\overset{+}{\text{N}}_2}{\overset{\text{OH}}{\text{C}}}\text{—R}^1} \xrightarrow{(b)} \underset{(6)}{\text{R—}\underset{\text{N—N}_2^+}{\overset{\|}{\text{C}}}\text{—R}^1} \xrightarrow{(c)} \underset{(7)}{\overset{+}{\text{C}}\text{—R}^1 = \text{N—R}} \longrightarrow \underset{(8)}{\overset{+}{\text{H}}\text{O=C(—R}^1\text{)—NH—R}} \longrightarrow \text{O=C(—R}^1\text{)—NH—R} \qquad (19)$$

acids ($R^1 = OH$) the similarity of **6** to the intermediate in the acid-catalysed Curtius rearrangement is evident. This scheme is mainly derived from secondary evidence and analogy with related reactions.

The broad outline was apparent when the intramolecularity was shown to be as for the Curtius rearrangement by the conversion of optically active acids into amines without racemization[17–19] and sequences such as that in reaction (20) showed that despite the presence of the strong acid no skeletal rearrangement occurred in the

$$PhCH_2{}^{14}CH_2COOH \longrightarrow PhCH_2{}^{14}CH_2NH_2 \longrightarrow PhCOOH + {}^{14}CO_2 \quad (20)$$

migrating group[76]. The only intermediate that has been unambiguously identified is the isocyanate formed in the reaction of phenanthrene-4-carboxylic acid[40]: the general unaccessibility of isocyanates is not surprising as they would rapidly decompose in the strongly acid conditions usually employed and in the above mentioned case atypically mild conditions of conducting the reaction in trifluoroacetic acid had to be used. Direct conversion of **5** into **8** is ruled out by the inability of alkyl and aryl azides to replace hydrazoic acid: the short-cut would be equally possible with all these reagents whereas the sequence **5** to **6**, **7** and **8** can only be followed if hydrazoic acid is employed.

It is convenient to consider separately how the observed kinetics and products are consistent with scheme (19) rather than with a route involving an iminium ion, $RR^1C{=}N^+$, that could in theory be derived from **6**. Typical reactions of ketones follow a rate equation that is first-order each in hydrazoic acid, in substrate, and in Hammett's h_0 function[4,77] and they are also catalysed by Lewis acids, but it is difficult to characterize the rate-determining step or to provide unambiguous evidence for a particular scheme from kinetics alone. It has been argued that the effect of substitution on the relative rates and on the Eyring parameters $\Delta H^{\ddagger}$ and ΔS^{*} indicate that some step other than the rearrangement stage (*c*), scheme (19), is rate-determining[4], but the rate-determining step for release of nitrogen could be either steps (*a*), (*b*) or (*c*), and in addition the dehydration (*b*) could be unimolecular or could involve solvent or more of the catalysing acid. Consequently, a complex rate expression could result in the general case and the overall rate could be relatively insensitive to the migratory aptitudes and polarity of electron-releasing substituents.

Nevertheless, rearrangements of *m*- and *p*-substituted benzoic acids were sensitive to polar influences[78,79] as shown by their correlation in a Hammett relationship with ρ-1·76 and since the decomposition of aryl azides was insensitive to such substitution (cf. section IV) it is tempting to consider both the sensitivity to polar groups and the higher rates in the former set of compounds to be attributable to migration concerted with splitting-off of nitrogen. *o*-Substituted benzoic acids reacted faster than the unsubstituted parent regardless of the polarity of the substituent presumably for the same reasons as outlined for the Curtius rearrangement (section IIA). Di-*o*-substituted benzoic acids reacted even more readily — often undergoing rearrangement some 20° lower than their parents — and consequently selective reaction at one group of a dicarboxylic acid may be possible despite steric hindrance to reaction at the chosen site[37] (reaction 21). This

$$HO_2C-C_6H_2(CH_3)_2-CO_2H \xrightarrow[H^+]{N_3H} HO_2C-C_6H_2(CH_3)_2-NH_2 \qquad (21)$$

enhanced reactivity is reasonably attributed to the ease of formation of an acylium ion, RCO^+, at the hindered centre[80] which led to rapid formation of RCO^+N_3H. It could not be determined whether the other benzoic acids (*m*, *p* and unsubstituted) react through a very small concentration of such an ion or through the more plentiful but less reactive conjugate acid of scheme (19).

There appears to be considerable scope for kinetic studies using fast reaction techniques capable of detecting the various transient intermediates in these sets of reactions and measuring the rates of their interconversions.

The Schmidt-type rearrangement of ketones puts two possible migrating groups R and R^1 (reaction 19), into competition. Such migration in the iminium ion $RR^1C{=}N^+$ would lead to essentially equal quantities of products of migration of each group as little discrimination would be expected for movement to such a reactive centre. Actually, for the crucial cases (see later) of competition between groups of different sizes, different migration aptitudes were found. In reaction (22) the product distribution A:B was 9:95 for R = Me; 15:85 for R = Et; and 50:50 for R = *i*-Pr[83]. This is inconsistent

$$C_6H_5COR \longrightarrow \underset{(A)}{C_6H_5CONHR} + \underset{(B)}{C_6H_5NHCOR} \qquad (22)$$

with the intermediacy of iminium ions but the Ph:Alkyl migration aptitude is considerably smaller than that found in the Beckmann and pinacol rearrangements[15]. In contrast, mixtures of isomeric benzanilides from various mono-*p*-substituted benzophenones were found in nearly equal proportions regardless of the steric or polar nature of the substituent[81–83], these results being very similar to those obtained for the Beckmann rearrangement of equilibrated oximes. All these data can be readily rationalized if it is assumed that the rearrangement step (*c*) in reaction (19) involves a concerted migration and nitrogen loss, when an *anti*-orientation of the migrating and leaving groups is necessary as in the Beckmann and other 1,2-rearrangements, and that equilibration of the isomers **9** and **10** is more rapid than the rearrangement. The last assumption seems reasonable as by analogy with

$$\underset{\textbf{(9)}}{\begin{array}{c}R-C-R^1\\ \| \\ N-N_2^+\end{array}} \qquad \underset{\textbf{(10)}}{\begin{array}{c}R-C-R^1\\ \| \\ {}^+N_2-N\end{array}}$$

oximation[84] the only irreversible step in the reaction sequence is probably the rearrangement. The products of rearrangement are in consequence controlled by the relative populations of **9** and **10** rather than by intrinsic migration aptitudes and the more favoured of the latter pair is that where the larger group R and the bonded nitrogen molecule occupy an *anti*-configuration. This results in the smaller migrating group generally having the least tendency to shift. This analysis accommodates the bulk of the available data but direct evidence for the importance of intrinsic migratory aptitudes in the presumed concerted rearrangement of aliphatic ketones is the kinetic and product effects observed on reaction of 1-^{14}C-acetone[85]. Here the methyl group containing the heavy isotope migrated more slowly and less readily than its isotopically normal partner and steric effects on populations of intermediates can be ruled out.

Probably the safest summary of the situation is that both steric effects and migratory aptitudes can govern the direction of migration depending on the structure and reaction conditions, but that the migration is definitely concerted. Since the order of intrinsic ability to migrate, i.e. Me < Et < *i*-Pr < *t*-Bu, as found in the pinacol rearrangement, is the same as the order of increasing bulk, the empirical rule applies that, regardless of the cause, the group with the largest bulk in the neighbourhood of the reaction centre will preferentially migrate.

Similar arguments apply to many alicyclic ketones[86,87]: the substituted carbon of the ring of 2-substituted cyclohexanones and cyclopentanones showed the major ability to migrate to nitrogen[88] (reaction

23), but norcamphor and cyclopentanonorcamphor gave lactams resulting from migration of the unsubstituted side[89], although the significance of this result is limited by the low material balances that were achieved. Cyclic aryl-alkyl ketones behaved similarly: when the α-methylene group was unsubstituted: reaction led to lactams with the NH group attached to the aromatic residue[90,91], but when the α-methylene was substituted this orientation could be reversed[92].

Several types of Schmidt reaction merit special mention: (a) An interesting consequence of the factors governing migratory aptitudes is that acetanilides are the major (typically >95%) products of rearrangement of acetophenones. This leads to a practical alternative to nitration and reduction as a preparative route to aromatic amines[93,94], cf. reaction (24). (b) Product analyses showed that

$$ArCOCH_3 \longrightarrow ArNHCOCH_3 \longrightarrow ArNH_2 \quad (24)$$

predominant migration of the unsubstituted group of mono-*o*-substituted benzophenones, *o*-XC_6H_4COPh, occurred when X was methyl ethyl, isopropyl or halogen, and similarly when *o*-substitution was located in a naphthyl ring such as that of 1-benzoylnaphthalene[82,95]. But when X was methoxy nearly equal proportions of the two rearrangement products were found and when X was nitro or carboxy there was predominant migration of the substituted ring. A correlation of products with neither size nor polarity of substituents was possible and the results were especially puzzling as *o*-substituted acetophenones rearranged as expected with predominant aryl migration. The pathway for the *o*-carboxy compound has been rationalized[96] on the grounds of reaction of the lactol form of the substrate and an oxazirine has been isolated from the reaction mixture (reaction 25). A similar explanation has been proposed for the nitro-compound[82] and may apply to other apparent anomalies[97] in the

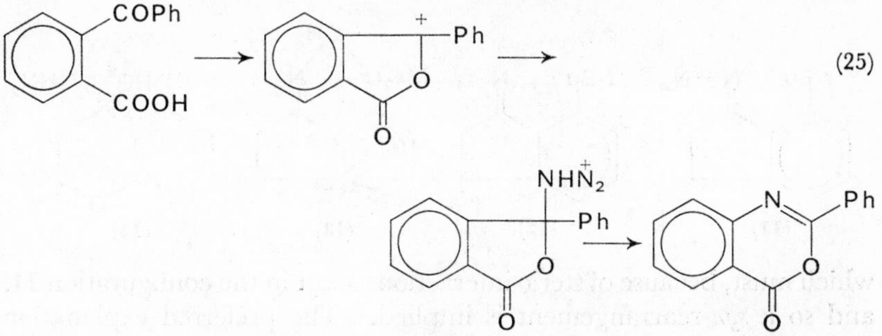

(25)

literature. The remaining results can be convincingly explained[4,82] if it is appreciated that only one of the aromatic rings can be in conjugation with the triazene system while the other must be forced out of the plane containing this ring and the carbon-linked triazene group: application of simple electronic theory allows prediction of which ring is not conjugated with the triazene system and the relative migrating tendencies of the two aromatic rings. The latter tendencies are governed by the relative populations of ions with configurations **9** and **10**. (c) In another study, the anomalous situation was found that although methyl groups typically migrate to a lesser extent than other alkyl groups, unsaturation reverses the effect: e.g. benzalacetone reacted as in (26). The explanation for this is obscure, but it may be

$$PhCH{=}CHCOCH_3 \longrightarrow PhCH{=}CHCONHCH_3 \quad (26)$$

a general occurrence that the styryl group has little tendency to migrate. Cyclopropyl groups which have chemical similarities to vinyl groups also show a reduced migratory aptitude in the reactions of some ketones and a low aptitude for cyclopropyl as compared with *n*-butyl has been demonstrated in rearrangements of *t*-carbinylazides[98]. (d) A route apparently involving the iminium cation $R_2C{=}N^+$ was discovered in certain Beckmann and Schmidt reactions[99,100]. For the latter, reaction (27) was observed to give predominantly

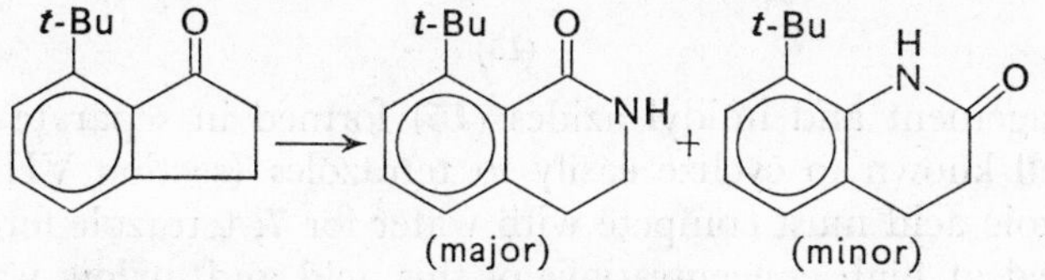

(major) (minor) (27)

products of alkyl, rather than of aryl, migration. If the usual route is followed, the reaction must pass through the imidodiazonium ion

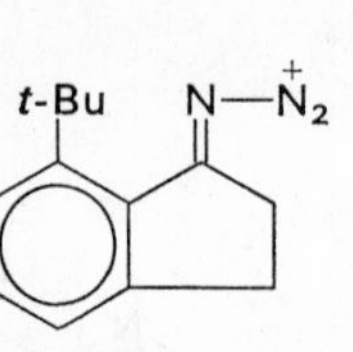

(11) 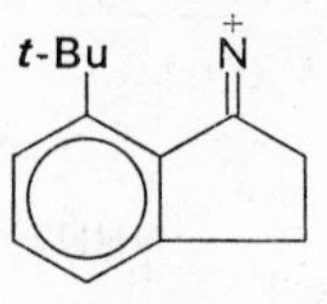 (12) 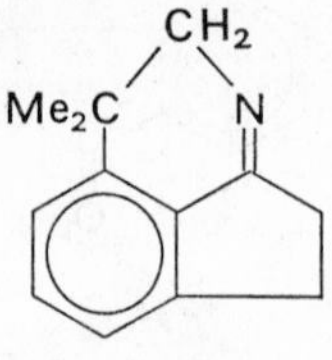 (13) 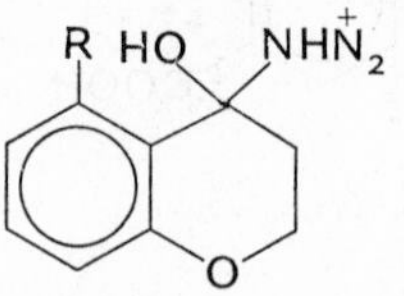 (14)

which must, because of steric interactions, exist in the configuration **11**, and so a *syn*-rearrangement is implied. The preferred explanation was that the ion **12** was an intermediate, and this view was favoured by the isolation of **13** as a side product. When the *t*-butyl group in the substrate was replaced by hydrogen the predominant product was that of aryl migration, as expected on the conventional theory. Rearrangements of chromanones and other alicyclic aryl ketones gave similar results and an azidohydrin, **14**, was considered to be implicated as a direct precursor of rearranged product in these and other strained systems[101,102]. The predominance of either alkyl or aryl migration in several benzocycloalkenones was strongly influenced by substitution in the aromatic ring and by the acidity of the medium, whereas these factors had little effect on reactions of flexible ketones such as acetophenone when aryl migration predominated[103]. The reasons for these phenomena have certainly not been adequately explained.

B. *Secondary Reactions*

Ancillary evidence for scheme (19) is provided by the nature of products formed in addition to the main rearrangement. Excess of hydrazoic acid can intercept the iminocarbonium ion **7** to form a tetrazole in either a two-step addition, reaction (28), or a concerted 1,3-addition[4]. It is well established that amides, once formed, cannot react further to give such products under the conditions of the

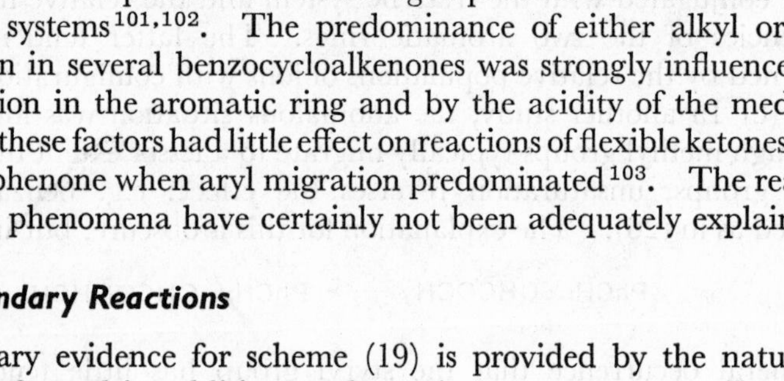

$$R^1{-}\overset{+}{C}{=}NR \;(\mathbf{7}) \longrightarrow R^1{-}C(=NR){-}N{=}\overset{+}{N}_2H \;(\mathbf{15}) \longrightarrow \text{tetrazole} \qquad (28)$$

rearrangement and imidyl azides (**15**) formed in separate reactions are well known to cyclize easily to tetrazoles (section VII). Since hydrazoic acid must compete with water for **7**, tetrazole formation is favoured at high concentrations of this acid and at low water (and hence high catalysing acid) concentrations. These conditions can be made the basis of useful synthetic methods: thus **16** (reaction 29) was formed in good yield in typical Schmidt conditions under catalysis by concentrated sulphuric acid, whereas **17** resulted if sodium azide was slowly added to a solution of the starting material in trichloroacetic acid[90,91]. If aqueous acids such as concentrated hydrochloric acid can be used as catalysts, tetrazole formation is easily avoided[72]: conversely this route can be enhanced by use of aprotic Lewis acids[104,105].

(29)

(16) **(17)**

Good evidence[106] for the existence of **15** is the isolation of substituted ureas (**19**), that are presumably derived from carbodiimides (**18**) reaction (30), from some rearrangement mixtures, and the last

$$R^1-C(NHN_2^+)=N-R \longrightarrow R^1\overset{+}{N}H=C=NR \xrightarrow{H_2O} R^1NHCONHR \qquad (30)$$

(15) **(18)** **(19)**

named can also lead to aminotetrazoles (**20**) by further reaction with hydrazoic acid[107,108], reaction (31). Iminocarbonium ions should

$$R^1\overset{+}{N}H=C=NR \xrightarrow{N_3H} \left[R^1NH-C(NHN_2^+)=NPh\right] \longrightarrow R^1NH-C\text{—}N\text{—}R \ (\text{tetrazole ring}) \qquad (31)$$

(18) **(20)**

theoretically be capable of attacking a suitably activated reactant or additive but no example of such an intermolecular process appears to have been recorded. Intramolecular attack, as in reaction (32*a*) is possible[95,109] although an alternative route to products that perhaps fits in with the fine details in a more satisfactory manner is reaction (32*b*). Closely related cyclizations occurred in the rearrangement of acetylacetone and benzoylacetone to form oxazoles[110], reaction (33).

Side reactions at the azidohydrin stage **5** have not been unequivocally identified but the subsequently formed iminodiazonium ion **6** can fragment to nitriles and species derived from a carbonium ion[111,112], reaction (34), and such a mode of reaction accounts for most of the

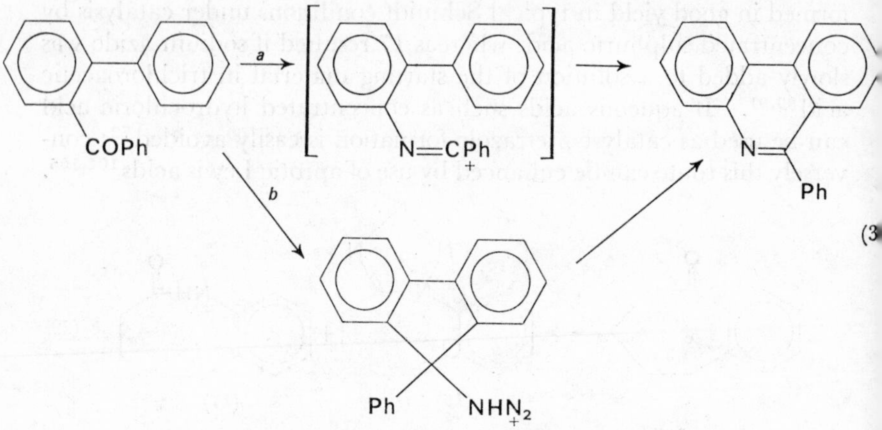

COPh —a→ [N═CPh (+)] → N═ Ph

b ↓ Ph NHN$_2^+$ ↗

(3

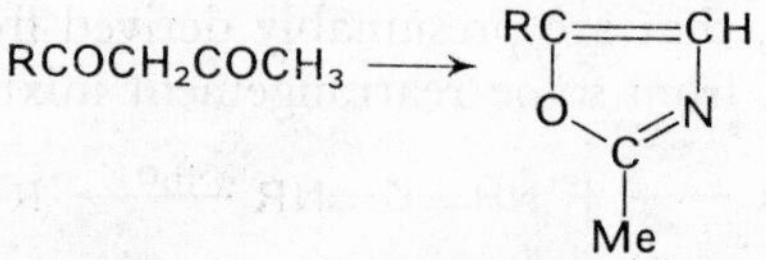

$$RCOCH_2COCH_3 \longrightarrow \begin{array}{c} RC{=\!=}CH \\ O\quad N \\ C \\ | \\ Me \end{array} \tag{33}$$

Schmidt reaction of aldehydes and of a few ketones[82,113]. Reactions secondary to the main process may also follow from skeletal rearrangement of the initially protonated carbonyl compound[111], **4** in scheme

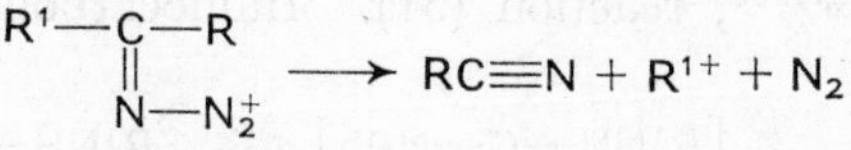

$$\underset{\textbf{(6)}}{\begin{array}{c} R^1{-}C{-}R \\ \| \\ N{-}N_2^+ \end{array}} \longrightarrow RC{\equiv}N + R^{1+} + N_2 \tag{34}$$

(19), and these protonated species may self-condense or attack products of rearrangement[114] very often to form intractible tar-like mixtures. Such undesirable processes can often be eliminated by using the weakest acid that is capable of catalysing the rearrangement.

C. Reactions of Other Compounds

Rearrangements are known to follow treatment of other types of compounds with hydrazoic acid in the presence of strong mineral acids[7]. Most can be regarded (as are the additions to iminocarbonium ions discussed in the last section) as either concerted 1,3-additions to unsaturated bonds or as 1,2-additions followed by cyclization. Triazene or triazole (the names appear to be used synonymously) derivatives that can exist in tautomeric forms which are interconvertible by prototrophy are produced.

Addition to cyanides such as HCN, $(CN)_2$, BrCN, NH_2CN, RNC, NaONC, EtO_2CCN and Et_2NCN, reaction (35), were discovered in the period 1912 to 1918 and have been briefly reviewed[1]. These reactions are synthetically useful and would repay more detailed study using modern techniques. Alkyl and aryl cyanides react[115] to give

$$XC{\equiv}N \xrightarrow{N_3H,\ H^+} XC(=NH)N_3 \longrightarrow \text{X–C=N–N=N–NH (5-X-tetrazole ring)} \quad \text{(and tautomers)} \tag{35}$$

$$RC{\equiv}N \xrightarrow{N_3H,\ H^+} NH{=}C{=}NR \xrightarrow{N_3H} \text{5-}NH_2\text{-1-R-tetrazole} \tag{36}$$

5-aminotetrazoles presumably as shown in reaction (36). Analogous additions of hydrazoic acid to olefins with the double bond conjugated with a nitro, carbonyl, aryl or azomethine group lead to substituted azides, the decomposition of which falls within the province of section IV. Such addition to unconjugated olefins is less common[7]. Addition to acetylenes generally leads to triazoles rather than vinyl azides and the latter were claimed to be ruled out as intermediates by the demonstration that when separately prepared they did not cyclize. However, the conditions of these experiments did not quite simulate those of the direct addition as some hydrochloric acid was present in the former that was missing from the latter[1,3,5]. Another type of Schmidt reaction that will be more fully discussed in section VII.C is reaction (37), and an entirely different sub-class is represented[116] by reaction (38).

$$Ph{-}C(=NH)OEt \xrightarrow{N_3H} Ph{-}C(=NH){-}N_3 \longrightarrow \text{5-Ph-tetrazole (Ph–C=N–N=N–NH)} \tag{37}$$

$$Cl(CH_2)_3CN \xrightarrow[H^+]{NaN_3} H_2C\langle CH_2{-}C{=}N{-}N{=}N{-}N{-}CH_2 \rangle \text{ (fused tetrazole)} \tag{38}$$

Many attempts have been made to carry out the Schmidt reaction with alkyl or aryl azides in place of hydrazoic acid in the presence or absence of catalysing acids and either in the cold or on heating. Most have been fruitless[3,7] although a few apparently valid claims are recorded. One is that treatment of acetophenone with methyl azide led to a trace of acetanilide[72] and since the reagent was strictly free of hydrazoic acid and was not hydrolysed or otherwise split under the reaction conditions some demethylation must have occurred at an unknown intermediate stage of the procedure. Addition of phenyl azide to olefins[117] is considered to be a concerted 1,3-dipolar addition to give triazole rather than a step-wise 1,3-addition and a similar route has been proposed[118] for addition of this reagent to acetylenes in acetic acid, although a bipolar complex of indefinite structure was claimed to be detectable at low temperatures.

IV. ACID-CATALYSED REARRANGEMENTS OF ALKYL AND ARYL AZIDES

Although the rearrangement of alkyl and aryl azides on treatment with acids, reaction (39), was discovered by Curtius[119], it is sometimes called a Schmidt reaction, although Schmidt did not work with these

$$R_2CHN_3 \xrightarrow{H^+} RCH{=}NR \tag{39}$$

substrates as such and his main discoveries came some 20 years after Curtius' investigations. The confusion is due to the fact that Schmidt carried out reactions, under conditions typical of those used for carbonyl compounds, with olefins and alcohols as starting materials and the sequences (40) and (41) resulted. Careful treatment of the substrates with hydrazoic acid and catalyst sometimes gave the azides in

$$R_3COH \xrightarrow[H^+]{N_3H} R_3CN_3 \xrightarrow{H^+} R_2C{=}NR \tag{40}$$

$$R_2C{=}CR_2 \xrightarrow{N_3H} R_2CH{-}CR_2N_3 \xrightarrow{H^+} R_2CHCR{=}NR \tag{41}$$

good yield with the virtual exclusion of rearrangement products and subsequent addition of stronger acid or an increase in temperature then brought about the rearrangement. The properties of isomeric products from such direct or interrupted Schmidt reactions were found to be the same as those from the decomposition under similar conditions of separately prepared azides, and so the various adaptions on the same mechanistic theme can be discussed together[120,121]. In this application of the Schmidt reaction the acid catalyst performs the dual function of generating the carbonium ion necessary to form the azide and to protonate the latter to initiate rearrangement. Protonation of the azide in either the direct treatment of azides with acid or in the appropriate step of the Schmidt reaction must, as in the Curtius process, form the species $RNHN_2^+$ in order to lead to the products by reasonable mechanisms.

Although the outlines of these reactions are well understood, the details of the mechanisms have not been clarified. It is tacitly and reasonably assumed that all are intramolecular and that the migrating group retains its configuration, but the usual controversy as to concerted or nitrene-forming reactions applies. A dependence of rate on the h_0 acidity function has been found for several reactions[4,122] and a detailed study of the kinetics of rearrangement of *m*- and *p*-substituted benzhydryl and 1,1-diaryl azides gave rate data that obeyed a Hammett relationship and was deemed to support a nitrene-forming route[121]. However the same data have been accommodated on the contrary premise of a concerted mechanism[4]. The difficulty of interpretation of such results is that substituents not only affect the migration step but also the position of equilibrium protonation of the starting material, and so it is not easy to pin point the cause of variations in rate for different substrates.

Migratory aptitudes have also been studied in attempts to elucidate the mechanism. A marked influence of substitution on the direction of ring expansion from 9-fluorenols and 9-fluorenylazides (reaction 42) was found[123] even when the substituent was too far removed from the 9-position to have any conceivable steric effect. 2- and 3-nitro and 2-amino substituents (the last presumably largely existing in the protonated form in the reaction conditions) led to 94–100% (normalized

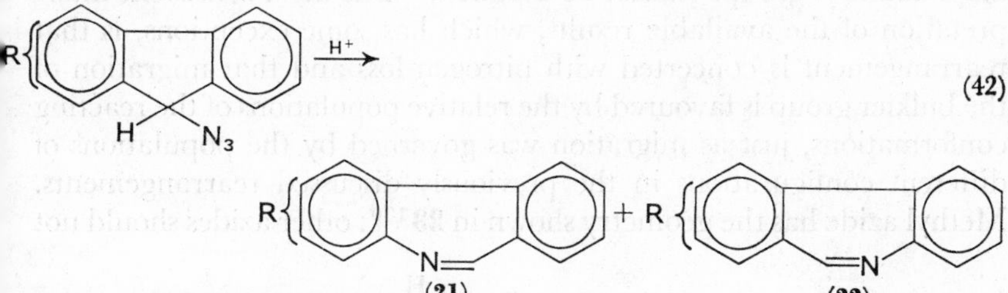

R, H, N_3 —H^+→ (42)

(21) + **(22)**

product ratios) of **22** whereas the 2-methyl and 2-methoxy compounds gave comparable amounts of **21** and **22** and the 3-methyl compound gave about 100% of **21**. Thus electron-releasing substituents promoted migration of that ring to which they were attached and electron-withdrawing substituents showed the reverse effect. Analogous findings were reported for the Schmidt reactions of 1,1-diarylethylenes[124–127] and benzhydrols[79]. Studies on diarylcarbinyl azides[121] showed that aryl groups migrated exclusively and no product corresponding to methyl migration could be detected (reaction 43); furthermore the order of ease of migration of the aryl groups was in the order of electron-release of their substituents; *viz*, *p*-OMe > *p*-Me > *p*-Ph > H > *p*-Cl > *p*-NO_2. On the other hand both phenyl and

$$MeO-C_6H_4-C(CH_3)(N_3)-C_6H_5 \xrightarrow{H^+} MeO-C_6H_4-N{=}C(CH_3)-C_6H_5 \quad (43)$$

alkyl migration occurred in the reaction of 1-alkyl-1-phenylethyl alcohols under Schmidt conditions[128,129] (reaction 44), with the

$$Ph-C(OH)(CH_3)-R \longrightarrow Ph-C(CH_3){=}NR + R-C(CH_3){=}NPh \quad (44)$$

migratory aptitudes: *i*-Pr, cyclohexyl > Ph ≫ Me, Et; in other examples either aryl and hydrogen migration could compete on equal terms[4] (reaction 45), or phenyl migration could occur exclusively,

$$PhCH_2N_3 \xrightarrow{H^+} PhCH{=}NH\ (50\%) + PhN{=}CH_2\ (50\%) \quad (45)$$

$$Ph_2CHN_3 \xrightarrow{H^+} Ph_2C{=}NH\ (0\%) + PhCH{=}NPh\ (100\%) \quad (46)$$

(reaction 46). Thus a universally valid migration order for hydrogen, alkyl and aryl groups cannot be assigned. The most consistent interpretation of the available results, which has some exceptions, is that rearrangement is concerted with nitrogen loss and that migration of the bulkier group is favoured by the relative populations of the reacting conformations, just as migration was governed by the populations of different configurations in the previously discussed rearrangements. Methyl azide has the geometry shown in **23**[130]: other azides should not

$$R-\overset{-}{N}-\overset{+}{N}{\equiv}N \ (120°) \qquad R-CR_2-N(H)-\overset{+}{N}{\equiv}N$$

(23) **(24)**

differ greatly and protonation on the negative nitrogen would undoubtedly result in a more acute CNN angle that favoured the orbital correlations necessary to affect a concerted rearrangement with 'inversion' at the innermost nitrogen, cf. (**24**). Formation of an iminium ion and of the derived nitrene would be considerably more difficult than that of the isoelectronic carbonium ion and so there would be a greater tendency for the former to be concerted: this is consistent with the appreciable sensitivity of rate and migration aptitudes to ring-substitution in benzyhydryl azides but the insensitivity of migration aptitudes to substitution in the deamination of benzyhydrylamines[131].

Triarylmethyl azides, Ar_3CN_3, are generally rather inert to acids perhaps owing to the electron-withdrawal by each aryl group lowering the intrinsic mobility of the others[4]. There is some evidence that such azides decompose in sulphuric acid to form carbonium ions rather than to yield rearranged products[132]: e.g. whereas the 9-phenylxanthyl azide (**25**) rearranges as expected on heating (reaction 47) on treatment with acid it gives a solution with the properties of the 9-phenylxanthyl cation[4]. Although the sulphur analogue behaves similarly, its dioxide rearranges normally with acids (reaction 48) the sulphoxy group apparently destabilizing the xanthyl cation.

O Ph N_3 (**25**) $\xrightarrow{\Delta}$ O NPh (47)

SO_2 Ph OH $\xrightarrow{N_3H,\ H^+}$ SO_2 NPh (48)

The acid-promoted rearrangement of purely aliphatic azides has not been adequately studied. Ethyl azide, directly prepared or generated in a Schmidt reaction, on treatment with hot fuming sulphuric acid gave methylamine, formaldehyde, acetaldehyde and ammonia that resulted from fission of the products formed by both methyl and hydrogen migration[133] and *n*-butyl azide behaved similarly[134,135]. However *n*-hexyl azide and homologues only gave products corresponding to hydrogen migration, and increasing the length of the alkyl chain apparently selectively retards alkyl migration: the reason for this is not clear. Few examples of reactions of secondary and tertiary alkyl azides have been studied that would permit comparison of the migration tendencies of secondary and tertiary alkyl groups. The kinetics of rearrangement of several azidoheptanes have been measured[136] and although the results were not rigorously evaluated an h_0 dependence seems probable. The migratory aptitudes were: *n*-Pen > Me and *n*-Bu > Et. Analogous Schmidt reactions on the isomeric heptanols[110] produced complicated products due to extensive rearrangements at the carbonium ion stages before formation of azides. This illustrates that the generally assumed, and observed, concordance of products between these reactions and acid-treatment of separately prepared azides depends on comparing structures that are resistant to rearrangement in the former conditions.

1-Alkylcycloalkyl azides readily underwent ring expansion some 4 to 10 times faster than alkyl shift to give heterocyclic products (reaction 49), that of the 5-membered rings being most rapid[129,134,137]. Parallel results were found for Schmidt reactions on the corresponding types of alcohols[135] and olefins[138,139] and interesting azepine derivatives were formed from suitable starting materials[140] (reaction 50).

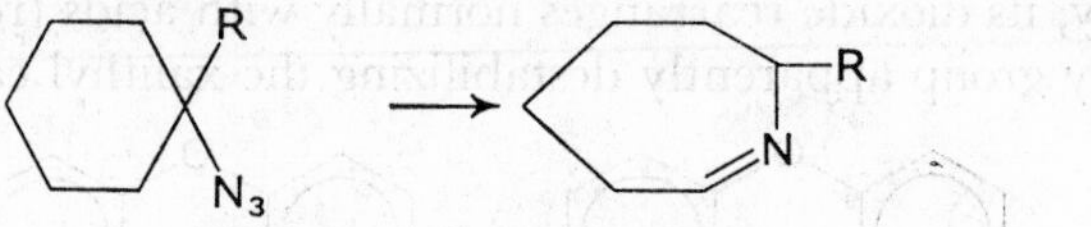

(49)

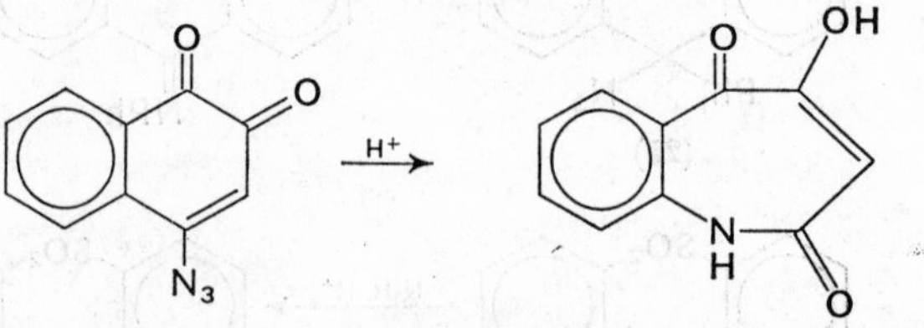

(50)

In all these examples the general rule applies that the bulkiest group migrates predominantly and concerted routes are presumed to occur.

A few random observations on acid-catalysed rearrangements are worth recording. Lewis acids catalyse the decomposition of aryl and

$$\mathrm{RNH{-}C(NHR^1){-}N_3 \xrightarrow{H^+} RNHC{=}NNHR^1} \quad (51)$$

(26)

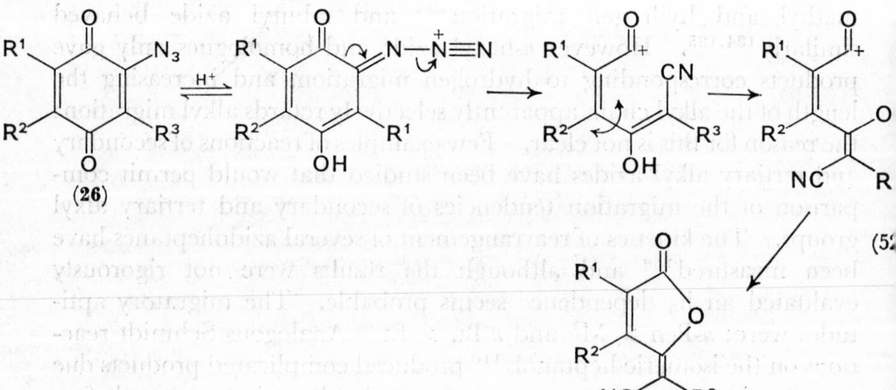

(52)

alkyl azides forming a complex that decomposes on warming to give some products of rearrangement[141,142]. Guanyl azides allow nitrogen to nitrogen migration[143] (reaction 51). Ferrocenyl phenyl carbinyl azides show no migration of the iron-containing group[144], presumably because protonation of a lone pair of the iron atom greatly reduces its migrating ability. The quinonoid azide **26** shows an exotic rearrangement, the mechanism of which is believed[145] to follow the route shown. The ring substituents can be alkyl, aryl, hydroxy, amino groups etc., or hydrogens, and the products are formed in 60–95% yield.

V. THERMAL REARRANGEMENTS OF ALKYL AND ARYL AZIDES

This class of reactions has been quite extensively studied and the earlier work has been well reviewed[3]. Although both hydrogen and phenyl migration occurred on treatment of benzyl azide with acid (reaction 45), heating in the absence of catalyst led to decomposition but not rearrangement[146]. Paradoxically, the triarylmethyl azides that are virtually inert to acid can undergo thermal rearrangement (reaction 53) although the process is not smooth and yields of anils are much lower than those of nitrogen[147]. A monovalent nitrogen intermediate was suggested for this reaction at an early date[148], no doubt stimulated by the then prevailing speculations as to the mechanism of the Beckmann rearrangement. A few alkyl azides have been found[149] to rearrange on gas-phase pyrolysis (reaction 54).

$$Ar_3CN_3 \xrightarrow{\Delta} Ar_2C{=}NAr \tag{53}$$

$$t\text{-}BuN_3 \xrightarrow{\Delta} Me_2C{=}NMe \tag{54}$$

A detailed study[150] of the rearrangement of *p*-substituted triarylmethyl azides to benzophenone anils showed that the rate was sensitive to substitution and varied over a range from 2–10 fold greater than that of the unsubstituted parent for substituents with polarities ranging from that of nitro to dimethylamino groups: the migratory aptitudes were in the same order but varied from 0·2–7·0 along the series relative to the parent at 1·0. As all substituents increased the rate, no correlation with a Hammett equation could be achieved and simple inductive control was rejected. The variations of rates with solvent were small, but still the rates were significantly faster in polar media. Analysis of the parameters $\Delta H^{\ddagger}$ and ΔS^{*} for the series led to the conclusion that the rearrangement was concerted[4], but the alternative view that a singlet nitrene was involved that did show a small discrimination, despite its great electrophilicity, towards the available β-linked groups has been held to be also consistent with the data[8]. The issue is undecided, although the considerable difference towards the sensitivities of rates to substituents in the acid-catalysed and thermally induced reactions suggests that different mechanisms may be operative, and as the former is probably concerted the latter can be inferred to involve nitrenes.

Thermally-induced ring expansion is found in certain reactions[152,153] (reaction 55) and the relative ease and cleanness of these compared with the reactions of triarylmethyl azides can probably be

Ar N3 —Δ→ Ar =N (55)

O, Ph N3 (27) —Δ→ O, NPh (major) + O, Ph =N (minor) (56)

traced to the relief of strain in the cyclopentyl ring and more importantly to the formation of an extended conjugated system. When the 9-aryl group is replaced by hydrogen ring expansion did not occur and fluorenone imines resulted[153,154]. Expansion of a 6-membered ring is not as ready as that of a 5-membered ring, although it can occur (reaction 56). The thiaxanthyl azides analogous to **27** give only anils[114].

Suitably chosen structures can exhibit intramolecular cyclization. Decomposition of *o*-nitroaryl azides gave furoxans in excellent yields at temperatures some 70–100° below those necessary to induce decomposition of unsubstituted aryl azides[155,156] (reaction 57) and an extensive kinetic study[157] of the arrangement of **28**, with widely differing substituents X in different solvents, led to the deduction of a concerted mechanism in which the nitro group lent considerable anchimeric assistance to loss of nitrogen. Similar conclusions follow for reactions with the participation of *o*-acetyl, *o*-benzoyl and other groups[161] and formation of a nitrene followed by cyclization is definitely excluded. Other α-azido compounds can also cyclize in a similar fashion[158–160] (reactions 58, 59, 60) to provide good synthetic routes to the heterocyclic products[8].

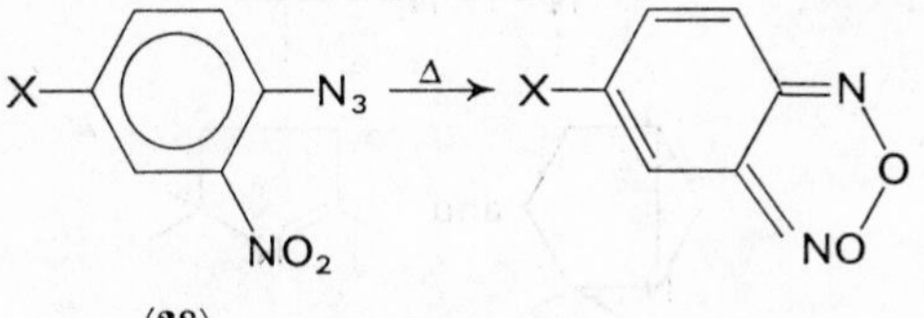

$$X\text{-}C_6H_3(NO_2)\text{-}N_3 \xrightarrow{\Delta} X\text{-}C_6H_3(=N\text{-}O\text{-}NO) \quad (57)$$

(28)

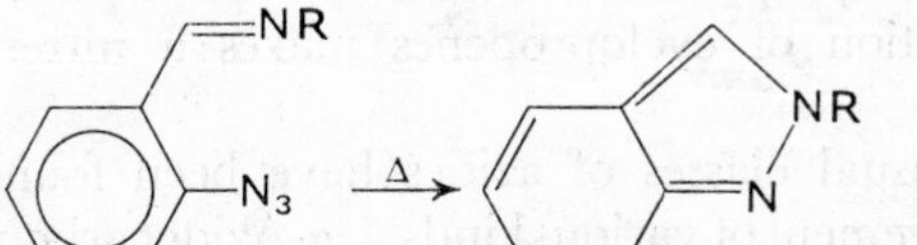

$$o\text{-}(RN{=}CH)C_6H_4N_3 \xrightarrow{\Delta} \text{2-R-indazole} \quad (58)$$

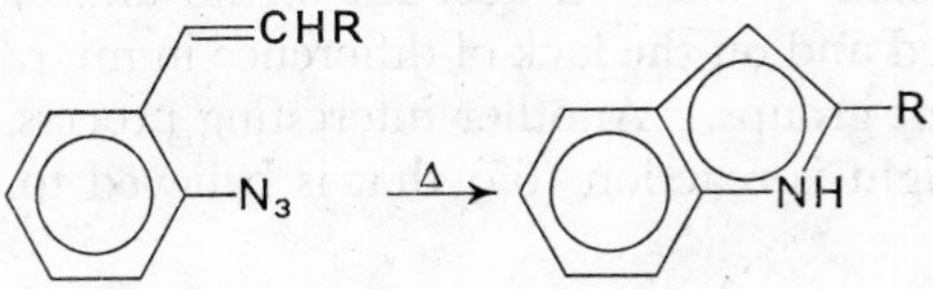

$$o\text{-}(RHC{=}CH)C_6H_4N_3 \xrightarrow{\Delta} \text{2-R-indole} \quad (59)$$

$$o\text{-}(C_6H_5)C_6H_4N_3 \xrightarrow{\Delta} \text{carbazole} \quad (60)$$

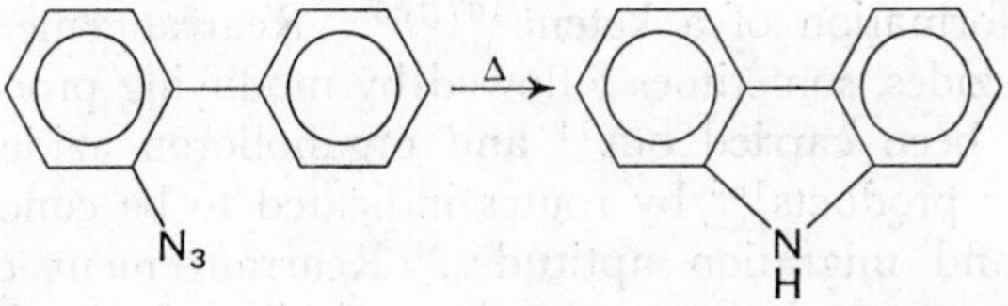

Another type of intramolecular interaction that could be classed as neighbouring group participation occurs in the thermolysis of vinyl azides[162,163,165] (reactions 61 and 62) and other olefinic azides such as **31** (reaction 63). The reaction of the last is believed to proceed by the route shown[164] and intermediates have been isolated. 'Normal'-type rearrangement products such as **30** only account for some 5% of the total reaction of **29**, with R=Ph. Although the mechanisms of for-

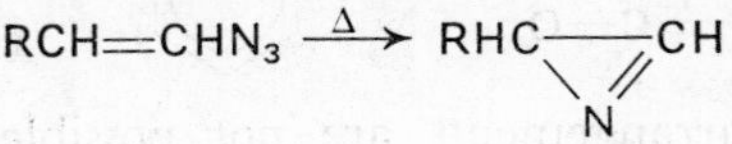

$$RCH{=}CHN_3 \xrightarrow{\Delta} RHC\overset{N}{—}CH \quad (61)$$

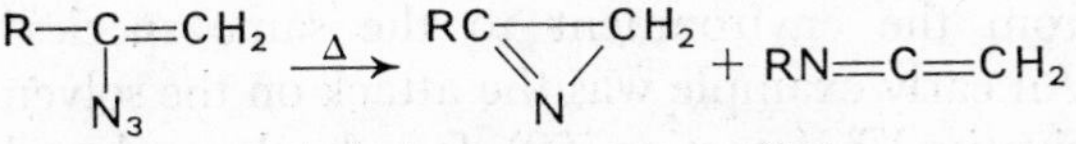

$$R\text{—}C(N_3){=}CH_2 \xrightarrow{\Delta} RC\overset{N}{—}CH_2 + RN{=}C{=}CH_2 \quad (62)$$

(29) **(30)**

$$CH_2{=}CH(CH_2)_2CMe_2N_3 \xrightarrow{\Delta} \text{[cyclic triazoline]} \longrightarrow \overset{+}{N_2}CH_2\text{–[pyrrolidine anion]} \longrightarrow \text{[bicyclic amine] and [pyrroline]} \quad (63)$$

(**31**)

mation of azacyclopropenes have not been established, the similarity with the formation of cyclopropenes makes a nitrene route very likely[8].

Several less usual classes of azides have been found to undergo thermal rearrangement of various kinds. α-Azidocarbonyl compounds rearrange as expected at ca. 200° (reaction 64) and a nitrene mechanism was suggested[166] both on account of the unusually high temperature required and on the lack of difference in migration aptitudes between different groups. Another interesting process, promoted by either heat or light is reaction (65) that is believed to involve inter-

$$RCOCR^1R^2N_3 \xrightarrow{\Delta} RCOCR^1{=}NR^2 + RCOCR^2{=}NR^1 \quad (64)$$

mediate formation of a ketene[167,168]. Rearrangements of organometallic azides, sometimes followed by modifying processes (reaction 66) have been carried out[74] and organoboron azides give similar oligomeric products[169] by routes indicated to be concerted by both kinetics and migration aptitudes. Rearrangement of the carbon skeleton is reported to occur to the exclusion of migration of fluorine in a fluorocarbon azide[170] (reaction 67) although cyclization involving migration of fluorine to nitrogen can occur when perfluoropropenyl azide is heated[171] (reaction 68).

$$\text{[2-azidotropone]} \xrightarrow[\text{or } \Delta]{h\nu} \text{[CN-substituted ketene, C=O]} \longrightarrow \text{[o-cyanobenzaldehyde, CN, CHO]} \quad (65)$$

When conventional rearrangements are not possible numerous reactions attributable to insertion of nitrenes or of abstraction of hydrogen from the environment to the same species have been reported. An early example was the attack on the solvent to give an azepine derivative[172] (reaction 69) found when phenyl azide was

$$Ph_3SiN_3 \xrightarrow[\text{or } \Delta]{h\nu} Ph_2Si{=}NPh \longrightarrow \begin{matrix} Ph_2Si{-}NPh \\ | \quad\quad | \\ PhN{-}SiPh_2 \end{matrix} \quad (66)$$

$$CF_3CF_2CHFCF_2N_3 \xrightarrow{\Delta} CF_2{=}NCHFCF_2CF_3 \quad (67)$$

$$CF_3CF{=}CFN_3 \xrightarrow{\Delta} \begin{matrix} F_2C{-}CF \\ | \quad\quad \| \\ FN{-}CF \end{matrix} \quad (68)$$

$$PhN_3 \longrightarrow PhN \xrightarrow{PhNH_2} \text{2-anilino-3H-azepine (NHPh, N, H, H)} \quad (69)$$

heated in aniline. The effect of substituents on the rate was consistent with the formation of a nitrene in the rate-determining step[173]. Heating phenyl or other aryl azides in an inert solvent usually results in an amorphous mass, apparently polymeric, together with amines formed by hydrogen-abstraction: the nitrenes that are undoubtedly generated[174] may be efficiently trapped by carrying out the reaction under carbon monoxide when isocyanates are formed[175]. Gas-phase decompositions of either phenyl azide or benztriazole led to a variety of products of which that of reaction (70) is of particular interest[176].

$$\text{benzotriazole (N, N, N–H) or } PhN_3 \xrightarrow{\Delta} PhN \longrightarrow \text{bicyclic azirine (N, H)} \longrightarrow \text{cyanocyclopentadiene (CN)} \quad (70)$$

Similar products were obtained from other aryl azides but the reactions, which may be heterogeneous, are complicated by unexpected migrations, fragmentations and ring expansions and a full analysis will require extensive isotopic studies[177].

Nitrenes can undoubtedly occur under the conditions leading to rearrangement products but whether they are obligatory for formation of these products is controversial[7,9,10]. If nitrenes do occur their assignment as singlets or triplets is uncertain and the multiplicity achieved probably depends on the nature of the reactant and the reaction conditions.

VI. PHOTOCHEMICAL REARRANGEMENTS OF ALKYL AND ARYL AZIDES

As a consequence of the recent upsurge in interest in organic photochemistry a considerable literature has accumulated on this topic. The rearrangements follow a few well defined pathways and it is generally considered that nitrenes, usually in the triplet state, are intermediates[8,178,182].

A typical rearrangement of an alkyl azide is reaction (71) to form

$$\text{cyclo-}C_5H_9{-}N_3 \xrightarrow{h\nu} \text{cyclo-}C_5H_8{=}NH \; (55\%) + \text{piperidine (NH)} \; (45\%) \qquad (71)$$

products in quite different proportions[179] from those of the acid-catalysed reaction[134]. Another example is reaction (72) where a supposed nitrene intermediate inserts into a CH bond[179]. Nitrenes

$$(CH_3)_3CN_3 \xrightarrow{h\nu} (CH_3)_2C{<}(CH_2)(NH){>} \qquad (72)$$

formed in other photolyses were considered to parallel carbenes in isomerizing to imines, in abstracting hydrogen from the solvent and in undergoing insertion reactions with the environment. Claims to effect cyclizations, generalized in reaction (73), were made[179] and were stated to lead to efficient synthesis of proline, connessine and other pyrrolidines, but such reactions could either not be repeated at all by later workers[180] or gave the cyclic products in minute and

$$RCH_2{\cdots}N_3 \xrightarrow{h\nu} RCH_2{\cdots}\dot{N}{:} \longrightarrow R\dot{C}H{\cdots}H\dot{N} \longrightarrow R{-}\text{pyrrolidine (NH)} \qquad (73)$$

extremely irreproducible yields[181] and the main products were found to be imines in all cases.

Photolysis of the aryl–alkyl azide $CH_3CPh_2N_3$ showed that the migratory aptitudes of the methyl and phenyl groups were almost identical[183] and this and the formation of triphenylmethyl amine from irradiation of triphenylmethyl azide in the presence of efficient hydrogen donors were taken to confirm the existence of discrete nitrene intermediates. Although the occurrence of a triplet-sensitized decomposition from alkyl azides[184] and triphenylmethyl azides[185] (the latter showing very similar migratory aptitudes for phenyl groups with *p*-nitro, *p*-methyl and other *p*-groups) shows that a triplet azide and presumably a triplet nitrene can also be involved, attempts to detect triplet azide in the direct, unsensitized photolysis by quenching experiments all failed. This failure and the high quantum yields observed strongly suggested that part or all of the direct path occurred via a singlet azide and a singlet nitrene. An earlier investigation compared photolyses of 1,1-diphenylethyl and 2-phenyl-2-propyl azides in hexane[151] with thermolyses of the same compounds[150] and revealed the small but significant differences in migrating ability of aryl and methyl groups (Ph:Me aptitudes about 2:1) in the latter reactions that did not occur in the former. These data and comparable studies on triarylmethyl azides were rationalized by the assumption that the photolysis involved a triplet nitrene that, as a free radical, did not show appreciable preference for attack on one substituted phenyl group rather than another, whereas the thermal reaction led to a singlet nitrene that, although a highly reactive electrophile, showed a slight but noticeable discrimination.

Phenyl azide can yield benzotriazole on irradiation[186], but more generally products of self-insertion or of attack on the solvent are found to lead to azepines and related compounds[187–189]. Photolysis of *o*-azidobiphenyls formed carbazoles by intramolecular cyclization similar to that found in thermolysis[190] and an analogous intramolecular cyclization involving, e.g. a nitro group could occur (reaction 74) in direct photolysis but not apparently in photosensitized reactions when only azobiphenyl was formed in appreciable yield[191]. The

N_3 NO_2 $\xrightarrow{h\nu}$ N NO O (74)

direct and sensitized reactions are presumed to lead to singlet and triplet nitrenes respectively and such a difference in multiplicity can account for differences in products of photolysis in other reactions[192]. Irradiation of vinyl azides gave major products similar to those formed on heating[186], cf. reaction (75), together with some different

N_3 $\xrightarrow{h\nu}$ N + $PhN{=}C{=}CH_2$ (75)

(32)

minor products[193] and detailed investigations of alkyl and aryl-substituted compounds of this class are well documented[165,194,195]. Sometimes a product of rearrangement **32** is formed in small yield. Scheme (76) was proposed for the photolysis of **33** in benzene[196,197]: two stereoisomeric nitrenes were postulated to exist on the basis of the observation that irradiation of **34** that had been separately prepared gave only **35** and none of the trimer **36**.

$$Ph_2C(N_3)_2 \ (\mathbf{33}) \xrightarrow{h\nu} Ph_2C(N_3)\text{—}\ddot{N}\text{:} \longrightarrow (Ph)(N_3)C{=}N\text{—}Ph \ \text{(two stereoisomers)} \tag{76}$$

$$(Ph)(N_3)C{=}NPh \longrightarrow Ph\text{—}C{=}N\text{—}N{=}N\text{—}NPh \ (\mathbf{34}) \longrightarrow (Ph)(\ddot{N}\text{:})C{=}NPh \longrightarrow \text{2-phenylbenzimidazole} \ (\mathbf{35})$$

$$(Ph)(N_3)C{=}NPh \longrightarrow (Ph)(\ddot{N}\text{:})C{=}NPh \longrightarrow (PhN{=}C{=}NPh)_3 \ (\mathbf{36})$$

VII. MISCELLANEOUS REARRANGEMENTS

A. Reactions forming Azides

Most rearrangements so far described have involved azides as starting materials but several interesting reactions involve rearrangements en route to azides. A good route to arylazides is treatment of an arylhydrazine with nitrous acid in the presence of a mineral acid. If ^{15}N-labelled nitrous acid was used, some 7% Ph—N—^{15}N—N (omitting the charges and bond orders for clarity of presentation) was detected by suitable degradations[198] in addition to the expected Ph—N—N—^{15}N, and a similar unusual location of tracer was found for the analogous formation of diazoacetate esters from glycine esters[199]. Presumably some reaction as outlined in scheme (77) must have occurred, but as intermediate **37** would have led to equal

$$PhNHNH_2 \xrightarrow{^{15}NO^+} Ph\text{—}N(^{15}NO)\text{—}NH_2 \longrightarrow Ph\text{—}N\text{—}N \ (\text{ring with } N_{15}) \ (\mathbf{37}) \longrightarrow Ph\text{—}N\text{—}^{15}N\text{—}N + Ph\text{—}N\text{—}N\text{—}^{15}N \tag{77}$$

amounts of the isotope-position isomers the main component of the reaction must have been reaction (78). 2,4-Dinitrophenylhydrazine

$$PhNHNH_2 \xrightarrow{^{15}NO^+} PhNHNH^{15}NO \longrightarrow Ph\text{—}N\text{—}N\text{—}^{15}N \quad (78)$$

and benzoic acid hydrazide gave only that azido compound on treatment with ^{15}N-nitrous acid in which the terminal nitrogen was tagged[200] and the neighbouring electrophilic groups must have directed attack onto the β-nitrogen to give a nitroso derivative analogous to that in reaction (78) which spontaneously decomposed into the azido-compound with a labelled γ-nitrogen.

A simpler and more certain route to aryl azides is treatment of a diazonium salt with hydrazoic acid. This was long supposed to be an S_N1 process but tracer studies have revealed a much more complicated sequence of reactions involving rearrangements[201,202]. If the nitrogens involved are numbered 1–4 the observed results can be summarized as in scheme (79); again charges and bond orders have been omitted. No direct S_N1 route could be detected. Introduction of substituents into the aryl group showed that the major route was

$$Ar\text{—}N^1\text{—}N^2 + N^3\text{—}N^4\text{—}N^3 \longrightarrow \begin{cases} Ar\text{—}N^3\text{—}N^4\text{—}N^3 + N^1\text{—}N^2 (0\%) \\ Ar\text{—}N^1\text{—}N^3\text{—}N^4 + N^2\text{—}N^3 \text{ (ca.15\%)} \\ Ar\text{—}N^1\text{—}N^2\text{—}N^3 + N^3\text{—}N^4 \text{ (ca.85\%)} \end{cases} \quad (79)$$

favoured by increasing the electrophilic nature of the diazonium ion and kinetic and product studies[203–206] revealed the details of the mechanism. Nitrogen was found to be evolved in two steps: on treatment of benzenediazonium chloride with lithium azide in methanol at −40°, some 76% of the theoretical quantity was liberated in a first-order process with a half-life of 5 minutes whereas the rest had to be driven off at 0° in a reaction that was also first-order with a half-life of about 14 minutes. These and the tracer results led to the

$$Ar\text{—}N\text{—}\overset{*}{N} + N\text{—}N\text{—}N \xrightarrow{f} Ar\text{—}N\text{—}\overset{*}{N}\text{—}N\text{—}N\text{—}N \xrightarrow{-40^\circ} Ar\text{—}N\text{—}\overset{*}{N}\text{—}N + N\text{—}N$$

$$\downarrow \quad (80)$$

$$Ar\text{—}N\langle\text{cyclic }\overset{*}{N}\text{—}N,\ N\text{—}N\rangle \xrightarrow{0^\circ} Ar\text{—}N\text{—}\overset{*}{N}\text{—}N + N\text{—}N$$

$$\xrightarrow{0^\circ} Ar\text{—}N\text{—}N\text{—}N + \overset{*}{N}\text{—}N$$

postulation of scheme (80), and subsequently intermediates with the pentazole skeleton were isolated under mild conditions.

Aliphatic azides undergo a little known allylic-type rearrangement (reaction 81). This has been studied in a variety of solvents for R = methyl and hydrogen and the percentages of the isomers at equilibrium have been measured[207]. In contrast to the analogous isomerizations of allylic chlorides the rates of rearrangement are insensitive both to alkyl substitution and to solvent and no accompanying

$$R_2C(N_3)CH{=}CH_2 \rightleftharpoons \left[\text{(38)}\right] \rightleftharpoons R_2C{=}CHCH_2N_3 \quad (81)$$

solvolysis was detected. The reaction must thus be intramolecular with a transition state of negligible polarity and recent studies[208] have confirmed the earlier conclusion[207] that a cyclic transition state (**38**), akin to that encountered in 1,3-dipolar addition, is utilized: such a mechanism is also in keeping with the large negative entropies of activation. Alkyl azides in their ground states possess a linear array of the three nitrogens but simple Hückel molecular orbital calculations show that the energy differences between linear and bent configurations are not large[209] and extended Hückel calculations suggest that the latter configurations may be even more stable in the first excited state[184]: consequently transition states of the type **38** should be readily attainable.

B. Acyclic 1,2,3-Triazenes

Straight chain triazenes may be considered as substituted azides and so will be mentioned here very briefly. The best known rearrangements involving such compounds are the acid-promoted conversions of diazoaminoarenes into *p*-aminoazoarenes, which are universally considered to be intermolecular (reaction 82). This reaction has been recently reviewed[15] and only two points need adding. Firstly, that the intermolecular route is supported by ^{15}N-tracer studies[210–212] that reveal, for instance, that step (*a*) comes to equilibrium before the recoupling (*b*) occurs[213]; and secondly, that analysis

$$PhNHN{=}NPh \overset{H^+}{\rightleftharpoons} Ph\overset{+}{N}H_2N{=}NPh \overset{(a)}{\rightleftharpoons} PhNH_2 + \overset{+}{N}_2Ph \xrightarrow{(b)} p\text{-}NH_2C_6H_4N{=}NPh \quad (82)$$

of products of certain reactions yielding *o*-aminoazo compounds (which are formed when the *p*-positions are blocked) have led to the conclusion that such processes are intramolecular[214]. The arguments for the last case seem very weak[215].

Acyclic triazenes can take part in prototropic equilibria (reaction 83) and many attempts have been made to measure the proportions

$$RN{=}NNHR^1 \rightleftharpoons RNHN{=}NR^1 \quad (83)$$

of the isomers by trapping by chemical means, e.g. by acid reduction, acid hydrolysis, halogenation, complexing, alkylation, etc. It is now realized that such methods give meaningful information only if the trapping reaction is much faster than the rate of interconversion of the tautomers and almost all of these early attempts are valueless[216]. The only reliable means of measurement are physical methods that do not perturb the equilibrium. On theoretical grounds it can be predicted that negative groups in the 1-aryl ring of a 1,3-diaryltriazene should favour that tautomer in which the hydrogen atom is at position 3. Presumably the solid triazene is one or other of the tautomers or a solid solution of the two, but there appears to be no information concerning this.

Interesting rearrangements involving acyclic triazenes are reactions (84) and (85)[217–218].

$$PhCH_2NHNMeNO \xrightarrow{\Delta} PhCH_2NNONHMe \quad (84)$$

$$Me\text{-}C_6H_4\text{-}N{=}NNMe_2 \xrightarrow{PhLi} Me\text{-}C_6H_4\text{-}N{=}N\text{-}N(\bar{C}H_2)Me \longrightarrow Me\text{-}C_6H_3(CH_2\bar{N}Me)\text{-}N{=}NH \longrightarrow Me\text{-}C_6H_4\text{-}CH_2NHMe \quad (85)$$

C. Cyclic 1,2,3-Triazenes

Many attempts have been made to isolate the individual tautomers of cyclic 1,2,3-triazenes (or triazoles) but although a number of claims have been made all have been rejected and no authenticated case is known[219].

Skeletal rearrangement of these compounds have been well reviewed[2]. One characteristic rearrangement of *C*-aminotriazoles is ring opening and intercharge of amino groups and the reaction is often reversible (reaction 86): another well studied reaction[220] often

$$\text{RNH–(1-}R^1\text{-1,2,3-triazol-5-yl)} \underset{}{\overset{\Delta}{\rightleftharpoons}} R^1\text{NH–(1-R-1,2,3-triazol-5-yl)} \qquad (86)$$

used in the synthesis of these compounds is reaction (87) and again

$$\begin{matrix} CHN_2 \\ | \\ OCNH_2 \end{matrix} \xrightarrow[OH^-]{\Delta} \begin{matrix} CHN_2 \\ | \\ HOC{=}NH \end{matrix} \longrightarrow \text{HO–(1}H\text{-1,2,3-triazole)} \qquad (87)$$

this can be reversed under appropriate conditions[221,222,224]. One type of ring opening of triazoles gave a feasible preparative route to aromatic azides from *o*-aminobenzaldoximes[223]: the sequence (reaction 88) involved careful hydrolysis of the cyclic species to allow

$$o\text{-}C_6H_4(CR{=}NOH)(NH_2) \xrightarrow{\Delta} \text{(benzotriazine N-oxide, CR, NO, N=N)} \xrightarrow{OH^-} o\text{-}C_6H_4(COR)(N_3) \qquad (88)$$

isolation of a hydroxytriazene which, on standing, lost water to form azide.

Tetrazole formation in the Schmidt reaction has been discussed in section III.B but considerable work has centred on the interconversions of straight chain azides and tetrazoles outside this special context. Probably the best known reactions are those centred on imidyl azides[1], $RNN_3{=}NR^1$. Only a few compounds of this structure have been isolated[225,227] as most rearrange under the conditions of their preparation to 1,5-tetrazoles (reaction 89) and a good route to

$$\begin{matrix} RC{=}NR^1 \\ | \\ NHNH_2 \end{matrix} \xrightarrow{NO^+} \begin{matrix} RC{=}NR^1 \\ | \\ N_3 \end{matrix} \xrightarrow{\Delta} \text{(1-}R^1\text{-5-R-tetrazole)} \qquad (89)$$

[R = Me, Ph, OH etc.]

the latter is treatment of $RN{=}CClR^1$ with azide ion[226]. At higher temperatures, sometimes in the presence of acid, tetrazoles may equilibrate with the open chain structure and rearrangement sometimes followed by irreversible decomposition may occur[1,143] (reactions

$$\text{Ph-tetrazole-NPh} \xrightarrow[\Delta]{H^+} [\text{Ph—C(=NPh)—}N_3] \longrightarrow 2\ PhNH_2 + N_2 + CO_2 \quad (90)$$

$$\text{PhNH-tetrazole-NH} \xrightarrow[\Delta]{H^+} [\text{PhNH—C(=NH)—}N_3] \longrightarrow PhNH_2 + NH_2NH_2 \quad (91)$$

$$\text{Ph-tetrazole-NPh} \xrightarrow{\Delta} [\text{Ph—C(=NPh)—}N_3] \longrightarrow PhN{=}C{=}NPh + \text{Ph-benzimidazole} \quad (92)$$

90, 91); nevertheless under milder conditions carbodiimides and cyclic products may be formed[228] (reaction 92). Tetrazole formation can also occur in aromatic systems[1,226] (reaction 93).

$$\text{2-chloropyridine} \longrightarrow \text{2-}NHNH_2\text{-pyridine} \xrightarrow{NO^+} \text{tetrazolo[1,5-a]pyridine} \quad (93)$$

The tendency of compounds $RCN_3{=}X$ to cyclize depends on the nature of X. When it is oxygen only open chain compounds are detected[5], and similarly for hydrazidic acids, $RCN_3{=}NNHAr$, attempts to synthesize the cyclic form were failures[230]. The latter type of compounds rearranged on treatment with acid to form semicarbazides with retention of configuration. Guanyl azides can be isolated in open chain forms that cyclize on heating[147] (reaction 94).

$$NH_2\text{—C(=NH)—}N_3 \xrightarrow{\Delta} \text{5-aminotetrazole} \quad (94)$$

Attempts to synthesize thioacylazides $RCN_3{=}S$ by treatment of thiosemicarbazides with nitrous acid give the cyclic isomers[231–233] but these can be opened by thermolysis or photolysis to give rearrangement products: thus heating **39**, (reaction 95), gives a nitrile without

$$\text{Ph}\text{—}\underset{(\mathbf{39})}{\text{thiatriazole}} \longrightarrow \left[\text{Ph—C(=S)—N}_3\right] \longrightarrow \left[\text{Ph—C(=S)—N}\right] \quad (95)$$

$$\left[\text{Ph—C(=S)—N}_3\right] \xrightarrow{h\nu} \text{PhNCS}; \quad \left[\text{Ph—C(=S)—N}\right] \xrightarrow{\Delta} \text{PhCN + S}$$

rearrangement but irradiation with ultraviolet light gives a low (ca. 10%) yield of isothiocyanate[225,234]. The perennial question as to whether concerted or nitrene-forming mechanisms occur in this family of reactions is open[228], but the thermally and photochemically induced rearrangements of tetrazoles (reaction 92), give widely differing proportions of acyclic and cyclic products and may follow the two respective routes. Nitrenes have actually been trapped under the photolysis conditions[225,229]. On the other hand, nitrenes of different multiplicity may be formed in the two cases.

An odd reaction is the well known, once industrially used, preparation of 5-aminotetrazole from thiohydantoin by treatment with sodium azide and lead oxide in an atmosphere of carbon dioxide[235] (reaction

$$\text{OC—NH—CS—NH—H}_2\text{C (thiohydantoin)} \xrightarrow[\text{C}_2]{\text{PbO, N}_3\text{H}} \text{NH}_2\text{—C(=N—N=N—NH) (5-aminotetrazole)} \quad (96)$$

$$\text{Ph}_2\text{CCl}_2 \xrightarrow{\text{N}_3^-} \text{Ph}_2\text{C(N}_3)_2 \xrightarrow{\Delta} \left[\text{Ph}_2\text{C—C(=N)—N}_3\right] \longrightarrow \text{Ph, Ph—N (triazole/tetrazole ring)} \quad (97)$$

96). Another curio[236,238] is the rearrangement followed by cyclization of the azide derived from the dichloride of benzophenone (reaction 97). An example of ring opening of a triazole leading to an amidine is reaction (98)[237].

$$\text{Et—(triazoline, HN, NPh)—N(morpholine)} \xrightarrow{\Delta} n\text{-Pr—C(=NPh)—N(morpholine)} \quad (98)$$

VIII. REFERENCES

1. F. R. Benson, *Chem. Rev.*, **41**, 1 (1947).
2. F. R. Benson and W. L. Savell, *Chem. Rev.*, **46**, 1 (1950).
3. J. H. Boyer and F. C. Canter, *Chem. Rev.*, **54**, 1 (1954).

4. P. A. S. Smith in *Molecular Rearrangements* (Ed. P. de Mayo), Interscience, New York, 1963, p. 457.
5. E. Lieber, R. L. Minnis and C. N. R. Rao, *Chem. Rev.*, **65**, 377 (1965).
6. E. Lieber, J. S. Curtice and C. N. R. Rao, *Chem. Ind.* (*London*), 586 (1966).
7. G. L'Abbe, *Chem. Rev.*, **69**, 345 (1969).
8. R. A. Abramovitch and B. A. Davis, *Chem. Rev.*, **64**, 149 (1964).
9. J. H. Boyer in *Mechanisms of Molecular Migrations*, Vol. 2 (Ed. B. S. Thyagarajan), Interscience, New York, 1969, p. 267.
10. W. L. Lwowski, *Nitrenes*, Wiley, New York, 1968.
11. T. L. Gilchrist and C. W. Rees, *Carbenes, Nitrenes and Arynes*, Nelson, London, 1969.
12. P. A. S. Smith, *Org. Reactions*, **3**, 337 (1946).
13. F. Moller in *Methoden der Organischen Chemie*, Vol. 11 (2), 4th edn. (Ed. Houben-Weyl), Thieme, Stuttgart, 1958, p. 876.
14. W. von E. Doering and M. J. Goldstein, *Tetrahedron*, **5**, 53 (1959).
15. D. V. Banthorpe in *The Chemistry of the Amino Group* (Ed. S. Patai), Interscience, New York, 1968, p. 585.
16. J. Kenyon, H. Phillips and V. P. Pittman, *J. Chem. Soc.*, 1072 (1935).
17. J. Kenyon and D. P. Young, *J. Chem. Soc.*, 263 (1941).
18. A. Campbell and J. Kenyon, *J. Chem. Soc.*, 25 (1946).
19. A. W. Schrecker, *J. Org. Chem.*, **22**, 33 (1957).
20. F. Bell, *J. Chem. Soc.*, 835 (1934).
21. M. B. Hocking, *Can. J. Chem.*, **46**, 2275 (1968).
22. F. C. Whitmore and A. H. Homeyer, *J. Am. Chem. Soc.*, **54**, 3435 (1932).
23. G. Powell, *J. Am. Chem. Soc.*, **51**, 2436 (1929).
24. E. S. Wallis, *J. Am. Chem. Soc.*, **51**, 2982 (1929).
25. L. Horner, E. Spietschka and A. Gross, *Ann. Chem.*, **573**, 17 (1951).
26. A. A. Bothner-By and L. Friedman, *J. Am. Chem. Soc.*, **73**, 5391 (1951).
27. M. S. Newman, S. H. Lee and A. B. Garrett, *J. Am. Chem. Soc.*, **69**, 113 (1947).
28. C. W. Porter and L. Young, *J. Am. Chem. Soc.*, **60**, 1497 (1938).
29. Y. Yukawa, W. S. Durrell and R. D. Dresdner, *J. Am. Chem. Soc.*, **82**, 4553 (1960).
30. Y. Yukawa and Y. Tsuno, *J. Am. Chem. Soc.*, **79**, 5530 (1957).
31. Y. Yukawa and Y. Tsuno, *J. Am. Chem. Soc.*, **80**, 6346 (1958).
32. J. R. Brower, *J. Am. Chem. Soc.*, **83**, 4370 (1961).
33. R. Huisgen, *Chem. Weekblad*, **8**, 59 (1963).
34. R. K. Smalley and T. E. Bingham, *J. Chem. Soc. C.*, 2481 (1969).
35. C. R. Hauser and S. W. Kantor, *J. Am. Chem. Soc.*, **72**, 4284 (1950).
36. S. Linke, G. T. Tisue and W. Lwowski, *J. Am. Chem. Soc.*, **89**, 6308 (1967).
37. M. S. Newman and H. L. Gildenhorn, *J. Am. Chem. Soc.*, **70**, 317 (1948).
38. Y. Yukawa and Y. Tsuno, *J. Am. Chem. Soc.*, **81**, 2007 (1959).
39. J. Hine, *Physical Organic Chemistry*, 2nd edn., McGraw-Hill, New York, 1962, p. 336.
40. K. G. Rutherford and M. S. Newman, *J. Am. Chem. Soc.*, **79**, 213 (1957).
41. R. A. Coleman, M. S. Newman and A. B. Garrett, *J. Am. Chem. Soc.*, **76**, 4534 (1954).
42. E. Fahr and Z. Neumann, *Angew. Chem. Intern. Ed. Engl.*, **4**, 595 (1965).

43. W. Lwowski, R. DeMauric, T. W. Mattingly and E. Scheiffele, *Tetrahedron Letters*, 3285 (1964).
44. W. M. Prince and C. M. Orlando, *Chem. Comm.*, 818 (1967).
45. A. Schonberg, *Preparative Organic Photochemistry*, 2nd edn. (English), Springer-Verlag, Berlin, 1968, p. 327.
46. L. Horner, G. Bauer and J. Dorges, *Chem. Ber.*, **98**, 2631 (1965).
47. J. v. Braun, *Ber.*, **67**, 218 (1934).
48. T. Curtius and O. Hofmann, *J. Prakt. Chem.*, **96**, 202 (1918).
49. J. ApSimon and O. E. Edwards, *Can. J. Chem.*, **40**, 896 (1962).
50. R. F. C. Brown, *Australian J. Chem.*, **17**, 47 (1964).
51. S. Huneck, *Chem. Ber.*, **98**, 2305 (1965).
52. W. Lwowski and G. T. Tisue, *J. Am. Chem. Soc.*, **87**, 4022 (1965).
53. L. Horner and A. Christmann, *Chem. Ber.*, **96**, 388 (1963).
54. S. I. Yamada and S. Terashima, *Chem. Comm.*, 511 (1969).
55. W. Lwowski, *Angew. Chem. Intern. Edn. Engl.*, **6**, 897 (1967).
56. A. Bertho, *J. Prakt. Chem.*, **228**, 89 (1929).
57. F. L. Scott, A. Koczarski and J. Reilly, *Nature*, **170**, 922 (1952).
58. T. Curtius and F. Schmidt, *Ber.*, **55**, 1571 (1922).
59. R. Stolle and M. Merkle, *Ann. Chem.*, **227**, 275 (1928).
60. W. Lwowski and E. Scheiffele, *J. Am. Chem. Soc.*, **87**, 4354 (1965).
61. F. L. Scott, *Chem. Ind.* (*London*), 959 (1954).
62. D. C. Dermer and M. T. Edmison, *J. Am. Chem. Soc.*, **77**, 70 (1955).
63. W. Lwowski and E. Scheiffele, *J. Am. Chem. Soc.*, **87**, 4359 (1965).
64. R. A. Abramovitch and V. Uma, *Chem. Comm.*, 797 (1968).
65. R. A. Abramovitch, J. Roy and V. Uma, *Can. J. Chem.*, **43**, 3407 (1965).
66. R. Puttner, W. Kaiser and K. Hafner, *Tetrahedron Letters*, 4315 (1968).
67. L. E. Chapman and R. F. Robbins, *Chem. Ind.* (*London*), 1266 (1966).
68. R. Kreher and D. Kuhling, *Angew. Chem.*, **77**, 42 (1965).
69. R. Puttner and K. Hafner, *Tetrahedron Letters*, 3119 (1964).
70. H. Wolff, *Org. Reactions*, **3**, 307 (1946).
71. P. Uyeo, *Pure Appl. Chem.*, **7**, 269 (1963).
72. P. A. S. Smith, *J. Am. Chem. Soc.*, **70**, 320 (1948).
73. A. J. Davies and R. E. Marks, *J. Chem. Soc. C.*, 2703 (1968).
74. W. T. Reichle, *Inorg. Chem.*, **3**, 402 (1964).
75. E. Oliveri-Mandala, *Gazzetta*, **55** (1), 271 (1925).
76. C. C. Lee, G. P. Slater and J. W. T. Spinks, *Can. J. Chem.*, **35**, 276 (1957).
77. T. A. Bak, *Acta Chem. Scand.*, **8**, 1733 (1954).
78. L. H. Briggs and J. W. Lyttleton, *J. Chem. Soc.*, 421 (1943).
79. R. F. Tietz and W. E. McEwen, *J. Am. Chem. Soc.*, **77**, 4007 (1955).
80. A. T. Blomquist and R. D. Spencer, *J. Am. Chem. Soc.*, **70**, 30 (1948).
81. P. A. S. Smith and B. Ashby, *J. Am. Chem. Soc.*, **72**, 2503 (1950).
82. P. A. S. Smith and E. P. Antoniades, *Tetrahedron*, **9**, 210 (1960).
83. P. A. S. Smith and J. P. Horowitz, *J. Am. Chem. Soc.*, **72**, 3718 (1950).
84. W. P. Jencks, *J. Am. Chem. Soc.*, **81**, 475 (1959).
85. G. A. Ropp, W. A. Bonner, M. T. Clark and V. F. Raaen, *J. Am. Chem. Soc.*, **76**, 1710 (1954).
86. R. Fusco and S. Rossi, *Gazzetta*, **81**, 511 (1951).
87. R. T. Conley, *Chem. Ind.* (*London*), 438 (1958).
88. H. Schechter and J. C. Kirk, *J. Am. Chem. Soc.*, **73**, 3087 (1951).

89. R. C. Elderfield and E. T. Losin, *J. Org. Chem.*, **26**, 1703 (1961).
90. R. Huisgen, *Ann. Chem.*, **574**, 171 (1951).
91. P. A. S. Smith and W. L. Berry, *J. Org. Chem.*, **26**, 27 (1961).
92. K. H. Hoffmann, H. J. Schmid and A. Hunger, *U.S. Patent* 2785159; *Chem. Abstr.*, **51**, 14828a (1957).
93. L. R. Dice and P. A. S. Smith, *J. Org. Chem.*, **14**, 179 (1949).
94. N. Campbell, W. K. Leadill and J. F. K. Wiltshire, *J. Chem. Soc.*, 1404 (1951).
95. P. A. S. Smith, *J. Am. Chem. Soc.*, **76**, 431 (1954).
96. C. L. Arcus and M. M. Coombs, *J. Chem., Soc.*, 3698 (1953).
97. S. Palazzo, *Ann. Chim.* (Roma), **49**, 835 (1959); *Chem. Abstr.*, **54**, 24509i (1960).
98. S. C. Bunce and J. B. Clarke, *J. Am. Chem. Soc.*, **76**, 2244 (1954).
99. P. T. Lansbury and N. R. Mancuso, *J. Am. Chem. Soc.*, **88**, 1205 (1966).
100. P. T. Lansbury and N. R. Mancuso, *Tetrahedron Letters*, 2445 (1965).
101. G. Dimaio and V. Permutti, *Tetrahedron*, **22**, 2059 (1966).
102. U. T. Bhalerao and G. Thyagarajan, *Can. J. Chem.*, **46**, 3367 (1968).
103. M. Tomita, S. Minami and S. Uyeo, *J. Chem. Soc. C.*, 183 (1969).
104. E. K. Harvill, R. M. Herbst, E. C. Schreiner and C. W. Roberts, *J. Org. Chem.*, **15**, 662 (1950).
105. N. B. Chapman, H. McCombie and B. C. Saunders, *J. Chem. Soc.*, 929 (1945).
106. P. A. S. Smith, *J. Am. Chem. Soc.*, **76**, 436 (1954).
107. K. F. Schmidt, *Friedländers Fortsh. der Teerfarb.*, **15**, 333 (1930): *Chem. Abstr.*, **25**, 2603 (1931).
108. K. F. Schmidt and P. Zutarein, *U.S. Patent* 1889323; *Chem. Abstr.*, **27**, 1361 (1933).
109. L. Birkofer and I. Storch, *Chem. Ber.*, **86**, 749 (1953).
110. W. Pritzkow and A. Schuberth, *Chem. Ber.*, **93**, 1725 (1960).
111. H. D. Zook and S. C. Paviak, *J. Am. Chem. Soc.*, **77**, 2501 (1955).
112. G. M. Badger and J. H. Seidler, *J. Chem. Soc.*, 2329 (1954).
113. W. E. McEwen, W. E. Conrad and C. A. Van der Werf, *J. Am. Chem. Soc.*, **74**, 1168 (1952).
114. R. H. B. Galt, J. D. Loudon and A. B. Sloan, *J. Chem. Soc.*, 1588 (1958).
115. J. v. Braun and W. Keller, *Ber.*, **65**, 1677 (1932).
116. Knoll A.G., Chemische Fabriken, *DRP* 521870; *Chem. Abstr.*, **25**, 3364 (1931).
117. F. D. Chattaway and G. D. Parkes, *J. Chem. Soc.*, 113 (1926).
118. G. S. Akimova, V. N. Chistokletov and A. A. Petrov, *Z. Org. Khim.*, **4**, 389 (1968).
119. T. Curtius and S. Darapsky, *J. Prakt. Chem.*, **62**, 428 (1901).
120. C. L. Arcus and M. M. Coombs, *J. Chem. Soc.*, 4319 (1954).
121. G. H. Gudmundsen and W. E. McEwen, *J. Am. Chem. Soc.*, **79**, 329 (1957).
122. C. L. Arcus and J. V. Evans, *J. Chem. Soc.*, 789 (1958).
123. C. L. Arcus and E. A. Lucken, *J. Chem. Soc.*, 1634 (1955).
124. W. E. McEwen and N. Mehta, *J. Am. Chem. Soc.*, **74**, 526 (1952).
125. W. E. McEwen, M. Gillilard and B. I. Sparr, *J. Am. Chem. Soc.*, **72**, 3212 (1950).
126. L. P. Kuhn and J. DiDomenico, *J. Am. Chem. Soc.*, **72**, 5777 (1950).

127. D. R. Nielsen and W. E. McEwen, *J. Am. Chem. Soc.*, **76**, 4042 (1954).
128. Y. Yukawa and K. Tanaka, *Mem. Inst. Sc. Ind. Res., Osaka Univ.*, **14**, 205 (1957); *Chem. Abstr.*, **52**, 4513 (1958).
129. Y. Yukawa and K. Tanaka, *Nippon Kagaku Zasshi*, **78**, 1049 (1957); *Chem. Abstr.*, **54**, 555 (1960).
130. R. L. Livingston and C. N. R. Rao, *J. Phys. Chem.*, **64**, 756 (1960).
131. J. G. Burr and L. S. Ciereszko, *J. Am. Chem. Soc.*, **74**, 5426 (1952).
132. M. M. Coombs, *J. Chem. Soc.*, 4200 (1958).
133. K. W. Shark, A. G. Houpt and A. W. Browne, *J. Am. Chem. Soc.*, **63**, 329 (1940).
134. J. H. Boyer, F. C. Canter, J. Hamer and R. K. Putney, *J. Am. Chem. Soc.*, **78**, 325 (1956).
135. J. H. Boyer and F. C. Canter, *J. Am. Chem. Soc.*, **77**, 3287 (1955).
136. W. Pritzkow and G. Mahler, *J. Prakt. Chem.*, **8**, 314 (1959).
137. K. Dietzsch, *J. Prakt. Chem.*, **27**, 34 (1965).
138. C. L. Arcus, R. E. Marks and R. Vetterlein, *Chem. Ind.* (*London*), 1193 (1960).
139. K. Biemann, G. Büchi and B. H. Walker, *J. Am. Chem. Soc.*, **79**, 5558 (1957).
140. H. W. Moore, R. H. Shelden and W. Weyler, *Tetrahedron Letters*, 1243, (1969).
141. J. Goubeau, E. Allenstein and A. Schmidt, *Chem. Ber.*, **97**, 884 (1964).
142. R. Kreher and G. Jäger, *Angew. Chem. Intern. Ed. Engl.*, **4**, 952 (1965).
143. E. Lieber, R. A. Henry and W. E. Finnegan, *J. Am. Chem. Soc.*, **75**, 2023 (1953).
144. A. Berger, W. E. McEwen and J. Kleinberg, *J. Am. Chem. Soc.*, **83**, 2274 (1961).
145. H. W. Moore and H. R. Shelden, *Tetrahedron Letters*, 5431 (1968).
146. T. Curtius and G. Ehrhart, *Ber.*, **55**, 1559 (1922).
147. A. Hantzsch and A. Vogt, *Ann. Chem.*, **314**, 261 (1901).
148. J. K. Senior, *J. Am. Chem. Soc.*, **38**, 2718 (1916).
149. W. Pritzkow and D. Timm, *J. Prakt. Chem.*, **32**, 178 (1966).
150. W. H. Saunders and W. C. Ware, *J. Am. Chem. Soc.*, **80**, 3328 (1958).
151. W. H. Saunders and E. A. Caress, *J. Am. Chem. Soc.*, **86**, 861 (1964).
152. L. A. Pink and H. E. Hilbert, *J. Am. Chem. Soc.*, **59**, 8 (1937).
153. C. L. Arcus, R. E. Marks and M. M. Coombs, *J. Chem. Soc.*, 4064 (1957).
154. C. L. Arcus and R. J. Mesley, *J. Chem. Soc.*, 178 (1953).
155. E. A. Birkhimer, B. Norup and T. A. Bak, *Acta Chem. Scand.*, **14**, 1594 (1960).
156. T. F. Fagley, J. R. Sutter and R. L. Oglukian, *J. Am. Chem. Soc.*, **78**, 5567 (1956).
157. S. Patai and Y. Gotshal, *J. Chem. Soc., B.*, 489 (1966).
158. L. Krbechek and H. Takimoto, *J. Org. Chem.*, **29**, 3630 (1964).
159. R. J. Sundberg, S. Lin and D. E. Blackburn, *J. Heterocyclic Chem.*, **6**, 441 (1969).
160. P. A. S. Smith and J. H. Hall, *J. Am. Chem. Soc.*, **84**, 480 (1962).
161. L. K. Dyall and J. E. Kemp, *J. Chem. Soc., B.*, 976 (1968).
162. G. Smolinsky and C. A. Pryde, *J. Org. Chem.*, **33**, 2411 (1968).
163. J. S. Meek and J. S. Fowler, *J. Org. Chem.*, **33**, 3418 (1968).

164. A. Logothetis, *J. Am. Chem. Soc.*, **87**, 749 (1965).
165. G. Smolinsky, *J. Org. Chem.*, **27**, 3557 (1962).
166. J. H. Boyer and D. Straw, *J. Am. Chem. Soc.*, **75**, 2683 (1953).
167. J. A. van Allan, W. J. Priest, A. S. Marshall and E. A. Reynolds, *J. Org. Chem.*, **33**, 1100 (1968).
168. J. D. Hobson and J. R. Malpass, *J. Chem. Soc. C.*, 1645 (1967).
169. P. Paetzad, *Angew. Chem. Intern. Edn. Engl.*, **6**, 572 (1967).
170. I. L. Knunyants, E. G. Bykhorskaya and V. N. Frosin, *Dokl. Akad. Nauk. SSSR*, **132**, 357 (1960).
171. I. L. Knunyants and E. G. Bykhorskaya, *Dokl. Akad. Nauk. SSSR*, **131**, 1338 (1960).
172. R. Huisgen, D. Vossius and M. Appl, *Chem. Ber.*, **91**, 12 (1958).
173. M. Appl and R. Huisgen, *Chem. Ber.*, **92**, 2961 (1959).
174. P. Walker and W. A. Waters, *J. Chem. Soc.*, 1632 (1962).
175. G. Ribaldone, G. Caprara and G. Borsotti, *Chem. Ind (Milano)*, **50**, 1200 (1968); *Chem. Abstr.*, **70**, 37326 (1969).
176. W. D. Crow and C. Wentrup, *Tetrahedron Letters*, 4379 (1967).
177. W. D. Crow and C. Wentrup, *Tetrahedron Letters*, 5569 (1968).
178. G. Smolinsky, *Trans. N.Y. Acad. Sci.*, **30**, 511 (1968).
179. D. H. R. Barton and L. R. Morgan, *J. Chem. Soc.*, 622 (1962).
180. D. H. R. Barton and A. N. Starratt, *J. Chem. Soc.*, 2444 (1965).
181. R. M. Moriarty and M. Rahman, *Tetrahedron*, **21**, 2877 (1965).
182. L. Horner and A. Christmann, *Angew. Chem.*, **75**, 707 (1963).
183. F. O. Lewis and W. H. Saunders, *J. Am. Chem. Soc.*, **90**, 7031 (1968).
184. F. O. Lewis and W. H. Saunders, *J. Am. Chem. Soc.*, **90**, 7033 (1968).
185. F. O. Lewis and W. H. Saunders, *J. Am. Chem. Soc.*, **89**, 645 (1967).
186. L. Horner, A. Christmann and A. Gross, *Chem. Ber.*, **96**, 399 (1963).
187. W. von E. Doering and R. A. Odum, *Tetrahedron*, **22**, 81 (1966).
188. L. Krbechek and H. Takimoto, *J. Org. Chem.*, **33**, 4286 (1968).
189. L. A. Paquette and R. J. Halvska, *Chem. Comm.*, 1371 (1968).
190. B. Coffin and R. F. Robbins, *J. Chem. Soc.*, 1252 (1965).
191. J. S. Swenton, *Tetrahedron Letters*, 3421 (1968).
192. I. Brown and O. E. Edwards, *Can. J. Chem.*, **45**, 2599 (1967).
193. F. P. Woerner, H. Remlinger and D. R. Arnold, *Angew. Chem. Intern. Edn. Engl.*, **7**, 130 (1968).
194. A. Hassner and F. W. Fowles, *J. Am. Chem. Soc.*, **90**, 2869 (1968).
195. G. R. Harvey and K. W. Ratts, *J. Org. Chem.*, **31**, 3907 (1966).
196. R. M. Moriarty and J. M. Kliegmann, *J. Am. Chem. Soc.*, **89**, 5959 (1967).
197. R. M. Moriarty, J. M. Kliegmann and C. Shovlin, *J. Am. Chem. Soc.*, **89**, 5958 (1967).
198. K. Clusius and H. R. Weisser, *Helv. Chim. Acta*, **35**, 1548 (1952).
199. K. Clusius and U. Lüthi, *Helv. Chim. Acta*, **40**, 445 (1957).
200. K. Clusius and K. Schwarzenbach, *Helv. Chim. Acta*, **41**, 1413 (1958).
201. K. Clusius and H. Hurzeler, *Helv. Chim. Acta*, **37**, 798 (1954).
202. K. Clusius and M. Vecchi, *Helv. Chim. Acta*, **39**, 1469 (1956).
203. R. Huisgen and I. Ugi, *Chem. Ber.*, **90**, 2914 (1957).
204. I. Ugi, R. Huisgen, K. Clusius and M. Vecchi, *Angew. Chem.*, **68**, 753 (1956).
205. I. Ugi and R. Huisgen, *Chem. Ber.*, **91**, 531 (1958).

206. I. Ugi, H. Perlinger and L. Behringer, *Chem. Ber.*, **91**, 2324 (1958).
207. A. Gagneux, S. Winstein and W. G. Young, *J. Am. Chem. Soc.*, **82**, 5956 (1960).
208. C. A. Van derWerf and V. L. Heasley, *J. Org. Chem.*, **31**, 3534 (1966).
209. J. D. Roberts, *Chem. Ber.*, **94**, 273 (1961).
210. K. Clusius and H. R. Weisser, *Helv. Chim. Acta*, **35**, 1524 (1952).
211. S. Weckherlin and W. Lüttke, *Ann. Chem.*, **700**, 59 (1966).
212. M. M. Shemyakin, V. I. Maimind and E. Gomes, *J. Gen. Chem. USSR*, **27**, 1906 (1957).
213. K. Clusius and H. R. Weisser, *Helv Chim. Acta*, **35**, 400 (1952).
214. V. M. Berezovskii and L. S. Tulchinskaya, *J. Gen. Chem. USSR*, **31**, 3371 (1961).
215. H. J. Shine, *Aromatic Rearrangements*, Elsevier, 1967, p. 220.
216. T. W. Campbell and B. F. Day, *Chem. Rev.*, **48**, 299 (1951).
217. A. Angeli, *Atti R. Accad. Sci. Torino*, **5** [vi], 732 (1927).
218. F. Klages and W. Mesch, *Chem. Ber.*, **88**, 388 (1955).
219. A. R. Katritzky and J. M. Lagowski, *Adv. Heterocyclic Chem.*, **2**, 27 (1963).
220. M. Regitz and H. Schwall, *Ann. Chem.*, **728**, 99 (1969).
221. B. R. Brown and D. L. Hammick, *J. Chem. Soc.*, 1384 (1947).
222. R. Huisgen, G. Szeimies and L. Moebius, *Chem. Ber.*, **99**, 475 (1966).
223. J. Miesenheimer, O. Senn and P. Zimmerman, *Ber.*, **60**, 1736 (1927).
224. P. Grünanger, P. Vitafinzi and C. Scotti, *Chem. Ber.*, **98**, 623 (1965).
225. R. Huisgen, *Angew. Chem.*, **72**, 359 (1960).
226. J. von Braun and J. Rudolf, *Ber.*, **74**, 264 (1941).
227. W. P. Norris and R. A. Henry, *J. Org. Chem.*, **29**, 650 (1964).
228. J. Vaughan and P. A. S. Smith, *J. Org. Chem.*, **23**, 1909 (1958).
229. W. Kirmse, *Angew. Chem.*, **71**, 537 (1959).
230. A. F. Hegerty, J. B. Aylward and F. L. Scott, *J. Chem. Soc. C.*, 2587 (1967).
231. E. Lieber, E. Offendahl and C. N. R. Rao, *J. Org. Chem.*, **28**, 164 (1963).
232. D. Martin and W. Mücke, *Chem. Ber.*, **98**, 2059 (1965).
233. E. Lieber, C. N. R. Rao, C. N. Pillai, J. Ramachandran and R. D. Hites, *Can. J. Chem.*, **36**, 801 (1958).
234. W. Kirmse, *Chem. Ber.*, **93**, 2353 (1960).
235. R. Stolle and O. Roser, *J. Prakt. Chem.*, **136**, 314 (1933).
236. G. Gotzky, *Ber.*, **64**, 1555 (1931).
237. R. Fusco, G. Bianchetti and D. Pocar, *Gazzetta*, **91**, 933 (1961).
238. P. Kereszty and A. P. L. Wolff, *U.S. Patent*, 2020937; *Chem. Abstr.* **30**, 575a (1936).

CHAPTER 8

Photochemistry of the azido group

A. REISER and H. M. WAGNER

Research Laboratory, Kodak Limited, Harrow, Middlesex, England

I. INTRODUCTION 442

II. SPECTRA AND EXCITED STATES 444

III. QUANTUM YIELD AND MECHANISM OF THE PRIMARY PHOTOLYTIC STEP 447
A. Direct Photolysis 447
B. Sensitized Photolysis 455

IV. PROPERTIES AND GENERAL REACTIONS OF NITRENES . . . 456
A. Electronic Structure and Spectra 456
B. General Reactions of Nitrenes 461
1. Recombination 461
2. Electrophilic attack on bonding pairs 461
a. Insertion into C—H bonds 461
b. Insertion into O—H bonds 462
c. Insertion into N—H bonds 462
d. Hydrogen abstraction 463
3. Addition to multiple bonds 464
a. Addition to C=C double bonds 464
b. 1,3-Cycloaddition 465
4. Electrophilic attack on non-bonding pairs . . . 466

V. PHOTOREACTIONS OF ORGANIC AZIDES 468
A. Alkyl Azides 468
B. Vinyl Azides 475
C. Acyl Azides 479
1. Azidoformates 479
2. Acetyl azide 486
3. Pivaloyl azide 487
4. Benzoyl azide 489
D. Aryl Azides 491

VI. REFERENCES 498

I. INTRODUCTION

A salient feature of the photochemistry of organic azides is the facile elimination of molecular nitrogen from the azido group on irradiation.

$$RN_3 + h\nu \longrightarrow RN + N_2 \tag{1}$$

The molecular fragment RN which is the primary product of this process is formally a derivative of monovalent nitrogen and has been variously termed imidogen, azene, azylene, imene and electron deficient nitrogen; in recent years the term nitrene seems to have gained general acceptance (see however Abramovitch[1] and Horner and Christmann[2]). Although various other mechanisms which do not involve nitrenes can be made to account for the final products in many azide reactions, the intermediacy of a reactive species RN suggested itself to chemists very early on. Thus, a carbonylnitrene R—CO—N was postulated by Thiemann[3] (1891) in his interpretation of the Lossen rearrangement, and Curtius[4] (1922) referred to a 'short-lived intermediate R—CO—N . . .' to account for the reactions of some carbonyl azides. Bertho[5], in 1924, made phenyl nitrene responsible for the production of aniline and bis-xylenyl in the decomposition of phenyl azide.

At the time, the real existence of monovalent nitrogen may have been in doubt, but support for the nitrene hypothesis was soon to be forthcoming. Beckman and Dickenson[6] identified H_2, N_2 and NH_3 as the main products of the photodecomposition of hydrazoic acid and noted the absence of any signs of a chain reaction in the process. Irradiating gaseous HN_3 with a mercury line at 199 nm they found that independently of pressure the quantum yield of nitrogen evolution was $3{\cdot}0 \pm 0{\cdot}5$[7]. This led Beckman to propose the following reaction

$$\begin{aligned} HN_3 + h\nu &\longrightarrow HN + N_2 \\ HN + HN_3 &\longrightarrow H_2N_2 + N_2 \\ H_2N_2 + HN_3 &\longrightarrow NH_3 + 2\,N_2\ (80\%) \\ HN + HN_3 &\longrightarrow H_2 + 2\,N_2\ (20\%) \end{aligned} \tag{2}$$

scheme, in which the primary formation of the imino radical (hydrogen nitrene) provides a simple rationalization of the final products.

In the same year Gleu[8] observed that irradiation of an aqueous solution of HN_3 (at 254 nm) produced hydroxylamine, nitrogen and small amounts of ammonia and of hydrazine, again pointing to the imino radical as an intermediate.

$$HN + H_2O \longrightarrow H_2NOH \tag{3}$$

The Beckman mechanism has since been confirmed by Thrush[9] who was able, by flash spectroscopy, to observe directly not only the imino radical, but also the radicals NH_2 and N_3, attributed to the secondary reactions

$$\begin{aligned} HN + HN_3 &\longrightarrow NH_2 + N_2 \\ NH_2 + HN_3 &\longrightarrow NH_3 + N_3 \end{aligned} \tag{4}$$

While the imino radical itself has been observed in a wide range of conditions[10–12] and its spectroscopy is by now fully documented (see for example the references in Okabe[13]), the existence of its organic analogues, the nitrenes, and their role in azide chemistry is by no means a foregone conclusion. Although organic chemists have postulated nitrenes as intermediates for some time, the final products of most azide reactions are equally well accounted for by other mechanisms, such as the initial formation of a triazoline-adduct between azide and substrate, or quite generally by a reaction where the expulsion of molecular nitrogen is concerted with the formation of a new bond. There have been numerous attempts to decide this fundamental point by chemical means. Since these have been concerned with thermal azide reactions the reader is referred to some of the original papers[14–21].

Physical proof for the existence of organic nitrenes came as late as 1962 when Smolinsky, Wasserman and Yager[22] reported the e.s.r. signals of stable triplet species obtained by irradiating solid solutions of phenyl azide and related molecules at 77°K. The e.s.r. spectra were characteristic of two strongly interacting unpaired spins localized essentially on a single atom, and were assigned to the triplet ground state of the aromatic nitrenes. Similar spectra of alkyl nitrenes were obtained two years later in the photolysis of matrix isolated alkyl azides at 4°K[23]. Direct observation of a nitrene at room temperature was first reported in the flash photolysis of 1-azidoanthracene[24,25]. On the strength of these observations and of a considerable amount of chemical evidence it seems now well established that in the majority of cases the formation of a nitrene is the first step in azide photolysis.

We propose to discuss the mechanism of the primary photolytic step with reference to the excited states of the azides, to survey briefly the properties and reactions of organic nitrenes and finally to review the photoreactions of alkyl, vinyl, acyl and aryl azides.

II. SPECTRA AND EXCITED STATES

The spectra of organic azides are of interest here because they identify the excited states which arise in the act of light absorption and are potentially involved in the photolytic reaction.

The solution spectrum of HN_3 has two absorption bands in the accessible u.v.: a weak band at 260 nm ($\varepsilon \sim 40$) and a somewhat stronger band at about 200 nm ($\varepsilon \sim 500$)[26]. The band at 260 nm is also found in solid HN_3[27], the band at 200 nm appears in the spectrum of gaseous HN_3 as a diffuse absorption with a shallow maximum at 190 nm[13]. Closson and Gray[26] have interpreted this spectrum in terms of molecular orbital theory. The atomic *p*-orbitals of HN_3 are shown diagrammatically in Figure 1. The first nitrogen, bonded to

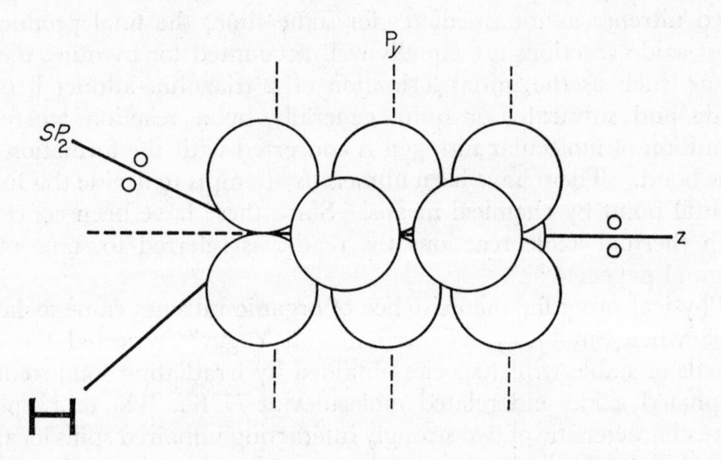

FIGURE 1. Atomic *p*-orbitals of HN_3.

hydrogen is sp_2-hybridized, the second and third are *sp*-hybridized; non-bonding pairs of electrons are indicated in the diagram by dots. The five *p*-orbitals of the nitrogens give rise to five delocalized molecular orbitals π_y, π_y^n, π_y^*, π_x, π_x^*, which in the ground state of HN_3 are occupied in order of increasing energy as follows:

$$(\sigma_{23})^2(2s)^2(\sigma_{12})^2(\pi_y)^2(\pi_x)^2(sp_2)^2(\pi_y^n)^2$$

Closson and Gray attribute the absorption band at 260 nm to the transition of an electron from the highest occupied π_y^n-orbital to the lowest unoccupied π_x^*-orbital. This transition leads to the excited electronic configuration $\ldots(sp)^2(\pi_y^n)^1(\pi_x^*)^1$. The band at 200 nm is assigned to the transition $sp_2 \rightarrow \pi_y^*$ and produces the excited configuration $\ldots(\pi_x)^2(sp_2)^1(\pi_y^n)^2(\pi_y^*)^1$. Both transitions are symmetry forbidden, hence the low extinction, both correlate with the $^1\Sigma_g^+ \rightarrow {}^1\Delta_u$ transition of the N_3^- ion[28]. Since the $^1\Delta_u$-state of N_3^- is strongly bent[29–31] it can be inferred that the first and second excited states of HN_3 will have a similar geometry. Intuitively one can see that the occupation of the antibonding π_x^*-orbital will force the HN_2—N bond out of the trigonal plane of the first nitrogen, so as to reduce repulsive overlap of the orbitals. This change in equilibrium geometry between the ground state and the excited states plays, we believe, an important role in the photodissociation of the molecule (see section III).

The spectra of the alkyl azides have the same structure as that of HN_3[26,32–36]. A weak band at 287 nm ($\varepsilon \sim 25$) is assigned to the $\pi_y^n \rightarrow \pi_x^*$ transition, a stronger band at 216 nm ($\varepsilon \sim 500$) to the transition $sp_2 \rightarrow \pi_y^*$. The position and extinction of these bands are independent of the structure of the alkyl group and are also remarkably insensitive to solvent changes[26] (Table 1). In the acyl azides, conjugation with the carbonyl group causes a small blue shift of the low energy band and a slight enhancement of its extinction. For the

TABLE 1. The low energy absorption band in several alkyl and acyl azides RN_3

R	Solvent	λ_m(nm)	ε	Reference
Ethyl	Ethanol	286	26	32
n-Butyl	iso-Octane	286	24	26
t-Butyl	iso-Octane	288	—	26
s-Heptyl	Heptane	289	23	41
Cyclohexyl	iso-Octane	287	25	26
2-Phenethyl	Heptane	289	25	41
2-Chlorethyl	iso-Octane	283	34	26
2-Hydroxyethyl	iso-Octane	283	—	26
2-Acetoxyethyl	iso-Octane	284	25	26
Acetyl	Ethanol	288	25	32
Ethoxycarbonyl	iso-Octane	280	25	26
Methoxycarbonyl	Ethanol	260 (s)	40	32
H	iso-Octane	264	—	26
	Water	260	43	26

spectra of some azido substituted thio-ethers and amines see Lieber and Rao[33].

While the absorption spectrum of the alkyl azides is that of the iso-

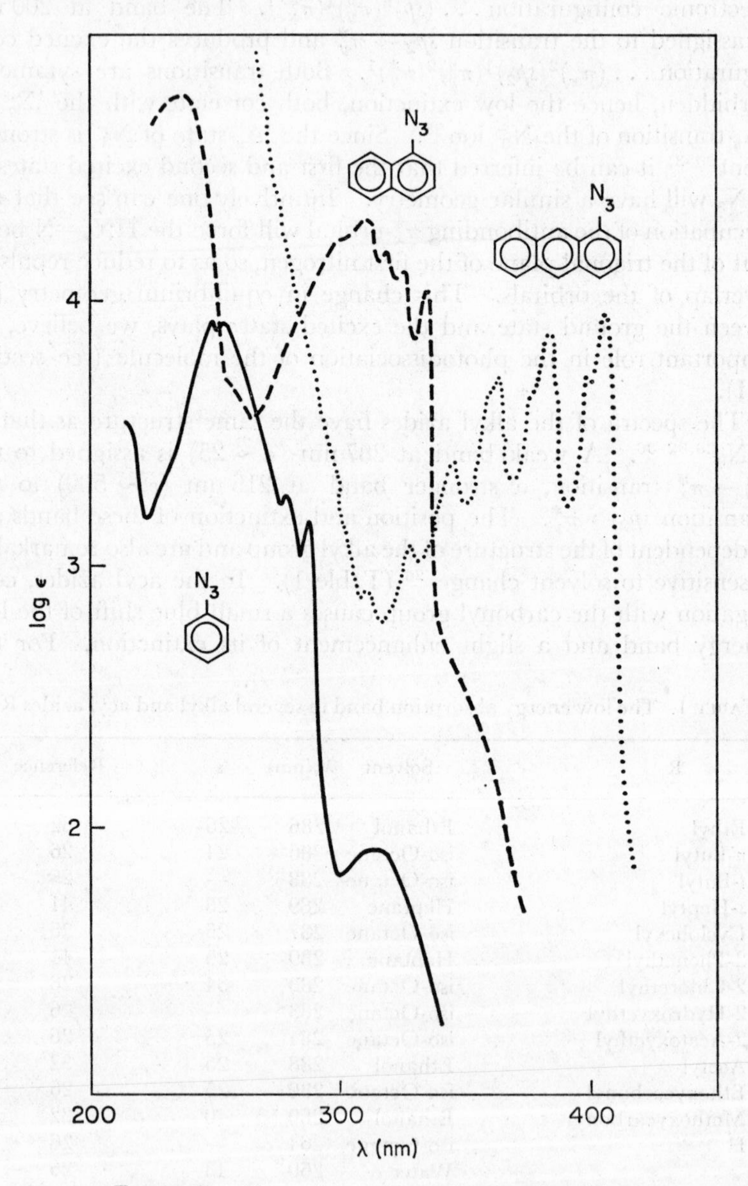

FIGURE 2. Absorption spectra of aromatic azides.

lated azido group, the spectra of aromatic azides are essentially those of the parent hydrocarbon, only a weak additional band due to the azido group appears as a shoulder on the long wavelength side of the hydrocarbon spectrum[37] (Figure 2). The coupling of the azido group with the aromatic system corresponds to a charge flow from nitrogen towards the aromatic ring (the azido group is electron donating with a Hammett constant $\sigma_p^+ = -0{\cdot}54$[38]). This lowers the energy of the non-bonding π_y^n-orbital below the level of the sp_2-orbital; consequently the azido band is assigned to the transition of lowest energy, namely $sp_2 \rightarrow \pi_y^*$, where the π_y^*-orbital extends now over the whole of the aromatic system.

The azido group affects the spectrum of the hydrocarbon itself in two ways: it causes a red shift of the ^1L_a-band and a smaller red shift of the other bands (the size of the aromatic system is increased); it also reduces the symmetry of the molecule, thus enhancing the extinction of the symmetry forbidden transition ^1L_b at the expense of the associated ^1B-transition (intensity borrowing[39]). From the size of the effect it can be inferred that the inductive interaction of the azido group with the ring is comparable with, if somewhat weaker than, that of the amino group[37]. For the spectra of aromatic diazides see reference 40.

III. QUANTUM YIELD AND MECHANISM OF THE PRIMARY PHOTOLYTIC STEP

A. Direct Photolysis

The mechanism of nitrogen elimination from the azido group has been studied in some detail for *hydrazoic acid*. The overall quantum yield of the gas phase photodecomposition of HN_3 implies a quantum yield of unity for the primary photolytic step[6]. As a consequence it has been generally assumed that the excited states of HN_3 are spontaneously dissociative[42]. Indeed, the energy absorbed on excitation is more than sufficient to break the HN—N_2 bond. The dissociation of ground state HN_3 into an imino radical and nitrogen, both in their ground states[13] requires an enthalpy change of only 9 kcal mole^{-1}.

$$HN_3 \longrightarrow NH + N_2(^1\Sigma_g^+), \quad \Delta H = -9 \text{ kcal mole}^{-1} \tag{5}$$

This reaction, however, infringes the spin conservation rule. The nearest spin allowed process is:

$$HN_3 \longrightarrow NH\ (a\ ^1\Delta) + N_2\ (^1\Sigma_g^+) \quad (6)$$

which requires 46 kcal mole^{-1} (see also reference 43), still only a fraction of the 110 kcal mole^{-1} of the absorbed photon.

The singlet state NH (a $^1\Delta$) has never been actually intercepted in the decomposition of HN_3, but Welge[44] was able to observe by flash spectroscopy in vacuum u.v. the related process

$$HN_3^* \longrightarrow NH(c\ ^1\Pi,\ v = 0,1) + N_2\ (^1\Sigma_g^+) \quad (7)$$

Okabe[13] has confirmed the identity of the fragments in scheme (7) by analysing the fluorescence emitted on irradiation of HN_3. He assigns the emission to the process

$$NH\ (c\ ^1\Pi) \longrightarrow NH\ (a\ ^1\Delta) \quad (8)$$

with possibly a small contribution from

$$NH\ (A\ ^3\Pi_i) \longrightarrow NH\ (X\ ^3\Sigma^-) \quad (9)$$

The imino radical and nitrogen appear in equation (7) as singlets, the dissociating species must therefore be a singlet excited molecule of hydrazoic acid. While this has been established only for the vacuum u.v. photolysis of hydrazoic acid, it is assumed that photolysis in the lowest absorption band (260 nm) of HN_3 proceeds also through an excited singlet state which decomposes by predissociation into a singlet excited imino radical and ground state nitrogen,

$$HN_3^*\ (^1A'') \longrightarrow NH\ (a\ ^1\Delta) + N_2\ (^1\Sigma_g^+) \quad (10)$$

The sequence of processes leading finally to decomposition may then be described as follows: absorption of a photon promotes the molecule to the singlet excited state $HN_3^*(^1A'')$ which in the first instance appears in the linear configuration of the ground state (Frank–Condon principle). Since the equilibrium configuration of $HN_3^*(^1A'')$ is angular (see the preceding section) the act of light absorption leads of necessity to some vibrational excitation as well. The vibrational energy thus released ($\sim$5 kcal mole^{-1}) is not in itself sufficient to break the HN—N_2 bond, but it promotes the interaction ('mixing') of state $^1A''$ with other energetically accessible states. At least one of these is repulsive in the critical bond coordinate and brings about the dissociation of the molecule.

In analogy with HN_3 the decomposition of the *alkyl azides* is believed to go through a singlet state. In the gas phase photolysis of methyl azide the primary step is the formation of methyl nitrene[45]

$$CH_3N_3 + h\nu \longrightarrow CH_3N + N_2 \qquad (11)$$

which then reacts with methyl azide.

$$\begin{aligned} CH_3N + CH_3N_3 &\longrightarrow CH_3NHCH_2N + N_2 \\ \text{and } CH_3N + CH_3N_3 &\longrightarrow CH_3N{=}NCH_3 + N_2 \end{aligned} \qquad (12)$$

The secondary products react further with nitrene and with azide to produce a polymer of overall composition $[CH_3N]_x$ and nitrogen.

$$\begin{aligned} CH_3NHCH_2N + CH_3N_3 &\longrightarrow [CH_3N]_3 + N_2 \\ CH_3N + [CH_3N]_3 &\longrightarrow [CH_3N]_4 \text{ etc.} \end{aligned} \qquad (12a)$$

From the data of Currie and Darwent[45] the quantum yield of the primary photolytic step is estimated to be between 0·7 and 1·0. The overall yield of nitrogen evolution varies with the wavelength of irradiation from 1·6 at 313 nm to 2·4 at 254 nm. Free radical scavengers reduce the overall quantum yield considerably, but an inert gas (CO_2) has no effect. From that it is concluded that the wavelength dependence of the quantum yield reflects the higher energy of the absorbed photon and not the greater reactivity of a 'hot' nitrene.

In solution, the photolysis yield of alkyl azides is independent of the wavelength of irradiation and insensitive to solvent changes[46] (Table 2).

The photolytic decomposition of *carbonyl azides* produces generally, but not exclusively, isocyanates[47–49]. Lwowski has investigated this photorearrangement in pivaloyl azide[50,51]. He finds an isocyanate yield of 40% in a variety of conditions quite independently of the

TABLE 2. Quantum yield (ϕ) of decomposition of alkyl azides RN_3

R	Solvent	Wavelength (nm)	ϕ	Reference
CH_3	Gas	254–313	2·4–1·6 (1·0–0·7)	45
C_2H_5	Methanol	354	0·88	46
	Methanol	313	0·88	46
	Ethanol	313	0·89	46
n-C_3H_7	Methanol	313	0·83	46
	Heptane	313	0·79	46
n-C_4H_9	Hexane	313	0·79	46
n-C_6H_{13}	Methanol	254	0·86	46
	Methanol	313	0·71	46
	Ether	313	0·71	46
	Heptane	313	0·69	46
cyclo-C_6H_{11}	Heptane	313	0·68	46
$(C_6H_5)_3C$	Methanol	313	0·80	46

solvent. In contrast, the yield of the other products (amines and aziridines) increases from 1% in neopentane and 13% in cyclopentane to 47% in cyclohexene. Evidently the isocyanate cannot have been formed via the same intermediate (nitrene) as the amines. Lwowski concludes that the photo-Curtius rearrangement, like its thermal counterpart, is probably a concerted process not involving a discrete nitrene intermediate.

Of particular mechanistic interest is the photochemistry of ethyl azidoformate which does not normally undergo the Curtius rearrangement (the migratory aptitude of the ethoxy group is low), and where the nitrene can be efficiently trapped, e.g. by acetylenes[52] and nitriles[53,54]. Lwowski[55] was able to establish the spin state of carbethoxynitrene (ethoxycarbonyl nitrene*) by an elegant method adapted from the work of Skell on carbenes[56–58]. It is based on the assumption, fully justified by the results, that a singlet species deficient in two electrons will add stereospecifically to a double bond. A triplet reactant can accept only one electron at a time and will close the bond with the second electron only after one of the spins has

* Ethoxycarbonyl nitrene has been termed until recently carbethoxynitrene. We are retaining the older nomenclature in this article, to conform with the extensive work of Lwowski.

inverted. The waiting period between the first and the second step is long enough to allow equilibration of the conformers of the transition state. Consequently, addition of a triplet nitrene is not stereospecific.

The reaction sequence following the production of a singlet nitrene is described by the following scheme:

$$\begin{array}{ccccccc} \text{azide} & \xrightarrow{k_1} & \text{singlet nitrene} & \xrightarrow{k_2} & \text{triplet nitrene} & \xrightarrow{k_5} & \text{by-products} \\ & & \downarrow k_3 & & \downarrow k_4 & & \\ & & \text{stereospecific addition product} & & \text{non-stereospecific addition product} & & \end{array}$$

Provided the singlet nitrene is sufficiently reactive towards the olefin the yield of the stereospecific adduct will be a direct measure of the concentration of singlet nitrene in the steady state, and hence an indication of the distribution of spin states in the primary products of azide decomposition.

Lwowski used *cis*- and *trans*-4-methylpent-2-ene as the olefin[59,60] and produced carbethoxynitrene in three independent ways[61–63]: by thermal decomposition of ethyl azidoformate, by the alkali induced decomposition of *N*-(*p*-nitrobenzenesulphonoxy)-urethane (an α-elimination reaction) and by the direct photolysis of ethyl azidoformate. All three routes produced the same aziridine in good yield. In thermolysis and in α-elimination the addition of the nitrene to the double bond is almost completely stereospecific, if conducted in the neat olefin; *trans*-aziridine is formed from *trans*-pentene, *cis*-aziridine from the *cis* isomer. This is the expected result, since only a singlet nitrene can be generated in a thermal reaction from the singlet ground state of the azide. If the olefin is diluted with an inert solvent (dichloromethane), an increasing amount of non-stereospecific addition is observed as more singlet nitrenes have had time to decay to the triplet ground state.

In contrast to these results, the photolytic decomposition of ethyl azidoformate[64] produces a mixture of the stereospecific and the non-stereospecific addition product corresponding to 70% of the singlet and 30% of the triplet species. Interception experiments in cyclohexane point equally to a triplet component of 30% in photogenerated carbethoxynitrene[65].

Lwowski's work throws considerable light on the mechanism of the primary photolytic step. Since the thermal experiments have shown that singlet carbethoxynitrene is quantitatively intercepted in the neat olefin, the triplet nitrene observed in the photolytic reaction must originate in an excited triplet state of ethyl azidoformate. Intersystem crossing from the singlet excited azide to the triplet state competes successfully with dissociation of the singlet excited azide. The rate of decomposition is therefore comparable with the rate of intersystem crossing in a carbonyl compound (greater than 10^{10} litre $mole^{-1} sec^{-1}$). Both the singlet and the triplet excited states of the azido group must be capable of dissociation, a conclusion which is borne out by sensitization experiments.

Cyanogen azide is an abundant source of the spectroscopically interesting symmetric cyanonitrene N=C=N[66]. Both the thermal[67] and the photolytic[68] decomposition of N_3CN lead to singlet excited nitrenes, as evidenced by stereospecific interception and by spectroscopy. The singlet nitrene decays subsequently to the triplet ground state of the species[69].

In *aromatic azides*[70] the mechanism of the primary step is slightly different from that of the isolated azido group. Here the upper excited states are those of the parent hydrocarbon and, on irradiation, energy is absorbed in the aromatic system as a whole. Decomposition is therefore preceded by a transfer of excitation from the hydrocarbon to the azido group. How this comes about will be described on the example of 1-azidonaphthalene (see Figure 3 where the potential energy curves of the RN—N_2 bond are shown for the singlet states of this molecule).

On absorption of a quantum of say 100 kcal $mole^{-1}$, 1-azidonaphthalene is promoted to the naphthalene state ^1L_a. It will transmit its excess vibrational energy to the medium until the lowest vibrational level of the lowest naphthalene state ^1L_b is reached. At this stage the molecule will enter the excited $n\pi^*$-azide state (excitation transfer) and will either decompose by the mechanism described for HN_3 or cross over to the triplet state and disintegrate here in a similar way. There are indications that both alternatives apply in aromatic azides[71] and that therefore the rate of intersystem crossing is comparable with the rate of dissociation in the singlet excited state. High concentrations of triplet nitrenes have indeed been observed in the flash photolysis of aromatic azides[72]. This, however, is not conclusive evidence, and the spin state of photogenerated aromatic nitrenes is still in question.

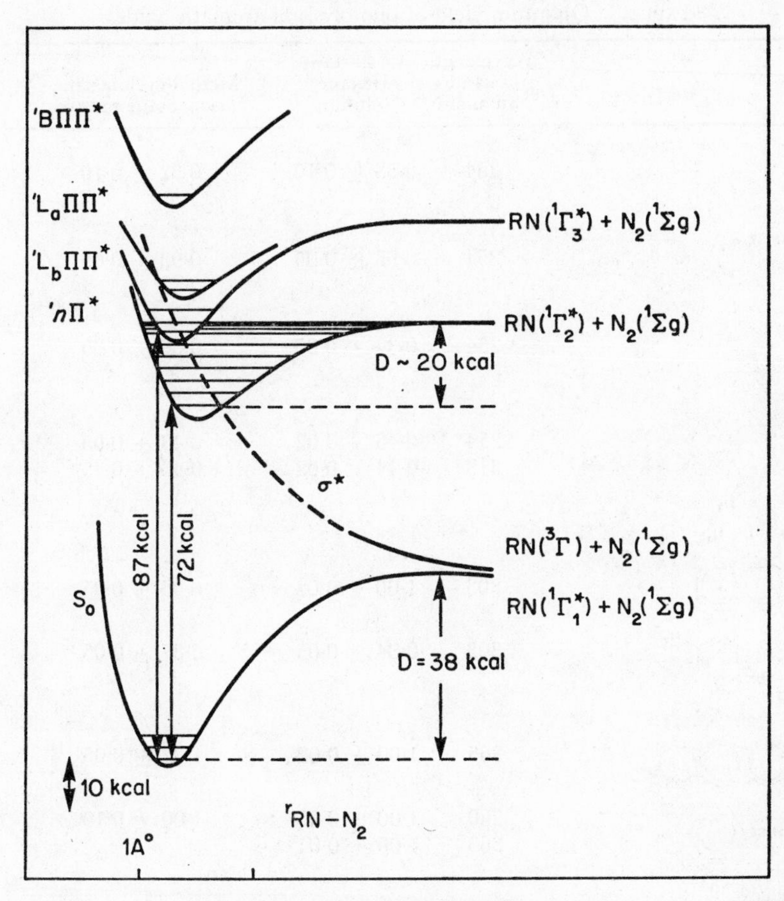

FIGURE 3. Potential energy diagram of the RN—N_2 bond in an aromatic azide.

Because of the transfer of excitation energy from the hydrocarbon to the azido group at some stage of the reaction, the structure of the aromatic system has some effect on the quantum yield of photolysis (see Table 3). Thus, the photolytic quantum yield is higher for the azides of polynuclear aromatic hydrocarbons than for phenyl azide and its derivatives, although the energy of the absorbed photon is larger in the latter compounds. In 4,4'-diazidobiphenyl and its vinylogues the quantum yield of photolysis decreases with increasing chain length, in parallel with the energy of the absorbed quantum.

In molecules with two or more azido groups, such as the diazides in

TABLE 3. Quantum yield of photolysis of aromatic azides

	Wavelength of irradiation nm	25°C Hexane solution	77°K Methyl cyclohexane–isopentane matrix	
$-N_3$	254	0·53 ± 0·10	0·52 ± 0·10	
$-N_3$	254	0·96 ± 0·05	0·98 ± 0·05	
N_3	254	0·37 ± 0·05	0·36 ± 0·05	
N_3	254	0·46 ± 0·02	0·43 ± 0·05	
	313	0·44 ± 0·02	0·43 ± 0·45	
N_3	303	1·00 ± 0·02	0·95 ± 0·05	
N_3, N	303	0·84 ± 0·05	0·87 ± 0·05	
N_3	365	1·00 ± 0·02	1·02 ± 0·05	
CH_3CO, $-N_3$	280	1·00 ± 0·05	1·00 ± 0·10	
	365	1·00 ± 0·01		
			ϕ_1	ϕ_2
N_3-, $-N_3$	254		0·45 ± 0·10	1·00 ± 0·10
	303		0·45 ± 0·10	1·00 ± 0·10
N_3-, $-N_3$	303		0·40 ± 0·05	1·00 ± 0·05
N_3-, $-CH{=}CH-$, $-N_3$	365		0·37 ± 0·05	1·00 ± 0·05
N_3-, $-N{=}N-$, $-N_3$	365		0·30 ± 0·10	0·80 ± 0·10
N_3-, $-(CH{=}CH)_3-$, $-N_3$	405		0·20 ± 0·10	0·60 ± 0·10
N_3-, $-C(=O)-$, $-N_3$	303	1·00 ± 0·10		
	365	1·00 ± 0·10		

ϕ_1 is the quantum yield of decomposition of the diazide (first azido group); ϕ_2 is the decomposition yield of the azidonitrene (second azido group).

Table 3, these groups decompose independently of each other. This applies even to *p*-phenylene diazide where a coupling of the dissociative processes may possibly have been anticipated[40].

B. Sensitized Photolysis

The sensitization of the photolysis of azides by triplet sensitizers is of potential importance in some photographic applications of organic azides. Also, it opens a route to the study of triplet nitrenes uncontaminated by the singlet species.

ApSimon and Edwards[48] have attempted the sensitization of azidodihydropimaric acid, Moriarty and Rahman[73] have sensitized *n*-butyl azide with benzophenone. The first conclusive experiments are those by Lwowski and Mattingly[61] who sensitized the photolysis of ethyl azidoformate with acetophenone. They found that in cyclohexene the reaction produced almost exclusively ethoxyurethane. Direct photolysis of ethyl azidoformate in the same solvent led mainly to addition (50% aziridine) and insertion (12% ethoxycyclohexenyl urethane) and produced no urethane at all if oxygen was excluded from the solution. These results demonstrate clearly the different chemical behaviour of singlet and triplet nitrenes: the singlet, a strong electrophile, adds to double bonds and undergoes insertion into C—H bonds, the triplet, a biradical, will mainly abstract hydrogen.

Swenton[71] found a somewhat analogous behaviour in the photodecomposition of 2-azidobiphenyl. Direct photolysis resulted in ring closure to carbazole (71%), acetophenone sensitized photolysis produced 43% of the azo compound and less than 8% carbazole. The presence of piperylene, a triplet quencher, during photolysis reduced the formation of azo compound to 4% and enhanced the carbazole yield to 89%.

Lewis and Saunders[74] have studied the sensitized photorearrangement of ring-substituted triphenylmethyl azides,

$$Ar_3C{-}N_3 + h\nu \longrightarrow Ar_2C{=}NAr + N_2 \qquad (13)$$

The reaction was noticeably sensitized even by sensitizers with low triplet energies (for example by pyrene, which has a triplet energy of 48 kcal mole^{-1}). In the sensitized reaction, as well as in direct photolysis, the migratory aptitude was unaffected by substituents on the phenyl rings, in contrast to the thermal rearrangement, where the migratory aptitude of aryl groups depends markedly on the electronic character of ring substituents. Lewis and Saunders concluded from this that the thermal rearrangement is a concerted process whereas the photoreaction proceeds through a nitrene intermediate.

A striking, but isolated, observation was reported by Horner[75]. He found that direct photolysis of benzoyl azide in ethanol or isopropanol led to the Curtius rearrangement (44%), to insertion (31%) and to some hydrogen abstraction (23%). However, in the presence of small concentrations of benzophenone the photoreaction produced exclusively benzamide, with a quantum yield far in excess of unity. The high yield indicated a chain reaction, benzophenone playing more the part of a photocatalyst than that of a conventional sensitizer.

Photosensitization is of interest also as a means of establishing the triplet energies of azides. These are not accessible by phosphorescence measurements because azides do not luminesce[70]. With this aim in mind Lewis and Saunders[76] have investigated the sensitized photolysis of several aliphatic azides, phenyl azide and ethyl azidoformate. From the concentration dependence of the quantum yield they were able to determine the rate of energy transfer between sensitizer and azide. With acetophenone and sensitizers of similar triplet energy the transfer process was reasonably efficient, but did not reach a diffusion controlled rate even at the highest azide concentrations. The authors interpreted this as a case of non-classical energy transfer, similar to the mechanism proposed by Hammond for energy transfer from sensitizers to the triplet state of stilbene[77].

IV. PROPERTIES AND GENERAL REACTIONS OF NITRENES

A. Electronic Structure and Spectra

The nitrogen in the nitrenes is bonded to a single carbon centre and is consequently *sp*-hybridized. One of the five valence electrons takes part in the σ-bond to carbon, one pair occupies the non-bonding *sp*-orbital and the two remaining electrons are in the unhybridized orbitals p_x and p_y (Figure 4). The two p-orbitals are equivalent in energy and will carry one electron each, both electrons having the same spin: the ground state of the nitrenes is expected to be a triplet. Smolinsky and co-workers[22,23] have observed e.s.r. signals of triplets on irradiating solid solutions of azides at 77°K. They assigned these to the ground state of the photogenerated nitrenes. By analysing the e.s.r. spectra in terms of the zero-field splitting parameters D and E it

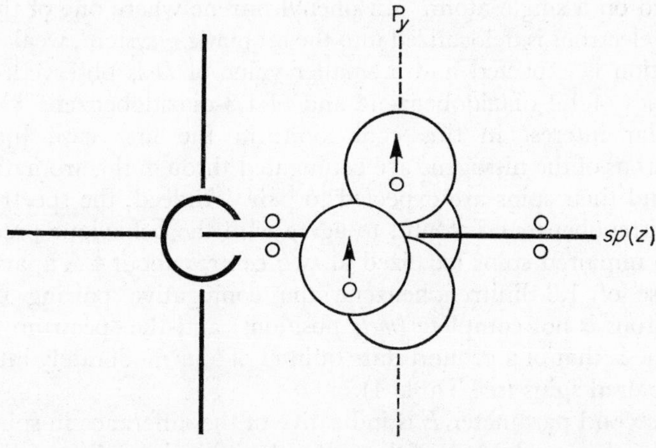

FIGURE 4. Electronic configuration of triplet nitrene.

was possible to obtain detailed information on the electronic structure of the nitrenes[22,23,78,79].

The parameter D is a measure of the interaction between the two unpaired spins of the triplet. The exceptionally high value of D in the spectra of the nitrenes (Table 4) indicates that the two spins are

TABLE 4. Zero-field splitting parameters in the triplet ground states of some nitrenes

$$\mathscr{H} = g3\vec{H}.\vec{S} + DS_z^2 + E(S_x^2 - S_y^2)$$

Nitrene	Experimental condition	D (cm⁻¹)	E (cm⁻¹)	Reference
HN	U.V. spectrum of HNCO	1·86		142
n-Propyl	Matrix, 4°K	1·607	0·0034	23
2-Octyl	Matrix, 4°K	1·616	0·0019	23
Cyclohexyl	Matrix, 4°K	1·599	0·002	23
Cyclopentyl	Matrix, 4°K	1·575	0·002	23
t-Butyl	Matrix, 4°K	1·625	0·002	23
p-Ethoxycarbonylbenzyl	Matrix, 4°K	1·659	0·002	23
Phenyl	Matrix, 77°K	1·33	0·30	22
Phenyl	Glassy state, 110°K	0·999	0	143
p-Phenylene dinitrene	Matrix, 77°K	0·0675	0	144
m-Phenylene dinitrene	Matrix, 4°K	0·156	0·029	79

localized on a single atom. In phenyl nitrene where one of the two radical electrons is delocalized into the aromatic π-system, weaker spin interaction is expected and a smaller value of D is observed. The dinitrenes of 1,4-diazidobenzene and of 1,3-diazidobenzene[79] are of particular interest in this connexion: in the first case the two p_y-electrons of the nitrogens are conjugated through the aromatic network and their spins are expected to pair. Indeed, the spectrum of 1,4-dinitrenobenzene is found to agree with the behaviour predicted for two unpaired spins localized at two centres about 4 Å apart. In the case of 1,3-dinitrenobenzene the conjugative pairing of the p_y-electrons is not complete (*meta* positions) and the spectrum of the dinitrene is that of a ground state quintet of four moderately interacting unpaired spins (see Table 4).

The second parameter E is indicative of the difference in spin density along the x and y axes of the molecule. The low value of E in the alkyl nitrenes is a consequence of the cylindrical symmetry of the p-orbitals about the C—N bond. In phenyl nitrene the p_x-orbital is locked in the plane of the phenyl ring and the equivalence of the two p-orbitals is removed. The spin distribution is no longer symmetrical about the z-axis and the value of E is consequently quite high.

Electronic absorption spectra have been obtained for several aromatic nitrenes[37,80] and dinitrenes[40]. They confirm in every respect the structure assigned to the triplet ground state of the nitrenes. For example, phenyl nitrene has seven π-electrons in its aromatic shell. Its lower (triplet) excited states originate in transitions of an electron from the highest occupied to the lowest unoccupied (or half-occupied) MO-levels (Figure 5). In the Hückel approximation the two excited states, ψ_1, ψ_2 are nearly degenerate and will be split by configuration interaction into an upper state ψ_- and a lower state ψ_+, giving rise to a fairly strong band $\psi_0 \rightarrow \psi_-$ at 314 nm and a weaker band $\psi_0 \rightarrow \psi_+$ at 402 nm. Such a band pair is typical of an open aromatic shell and is observed quite generally in all aromatic π-radicals. The spectrum of phenyl nitrene is in fact very similar to that of the isoelectronic benzyl radical[81].

Aromatic dinitrenes in which the nitrogens are conjugated through the central π-system have closed shell spectra, indicating the spin-pairing of the p_y-electrons. The spectra of 1,4-diazidobenzene and of its photoproducts[40] are here of some interest because they illustrate the alternation of closed and open aromatic shells (even and odd numbers of π-electrons) in the successive stages of the photolytic process (Figure 6). The spectrum of the diazide (curve I_a) is that of a sub-

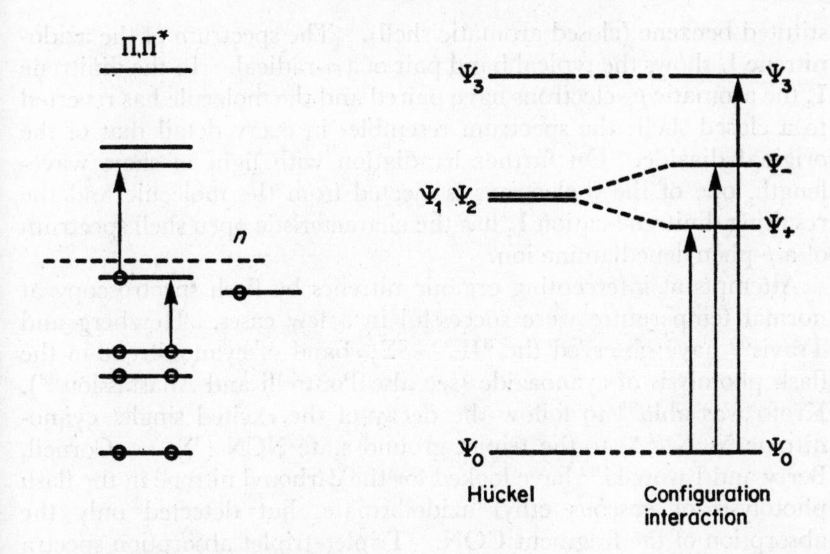

FIGURE 5. (a) Hückel MO-levels of phenyl nitrene; (b) Energy levels of excited states of phenyl nitrene.

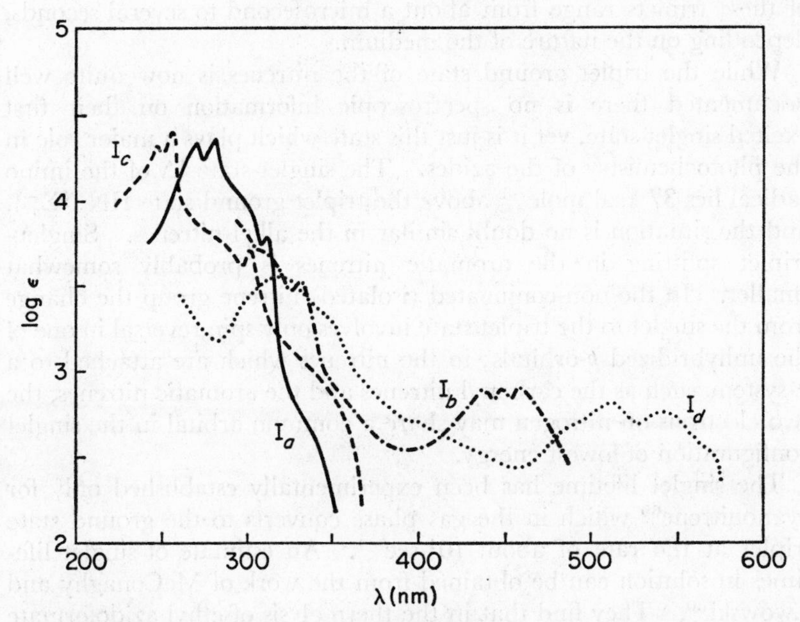

FIGURE 6. Absorption spectra of 1,4-diazidobenzene (I) and its photoproducts; matrix isolated at 77°K. I_a, diazide; I_b, azido-nitrene; I_c, dinitrene; I_d, dinitrene cation.

stituted benzene (closed aromatic shell). The spectrum of the azido-nitrene I_b shows the typical band pair of a π-radical. In the dinitrene I_c the aromatic p_y-electrons have paired and the molecule has reverted to a closed shell; the spectrum resembles in every detail that of the original diazide. On further irradiation with light of short wavelength, one of the π-electrons is ejected from the molecule and the resulting dinitrene-cation I_d has the characteristic open shell spectrum of a *p*-phenylenediamine ion.

Attempts at intercepting organic nitrenes by flash spectroscopy at normal temperature were successful in a few cases. Herzberg and Travis[66] have observed the $^3\Pi_u \rightarrow {}^3\Sigma_g^-$ band of cyanonitrene in the flash photolysis of cyanoazide (see also Pontrelli and Anastassiou[68]). Kroto was able[69] to follow the decay of the excited singlet cyanonitrene NCN ($^1\Delta$) to the triplet ground state NCN ($^3\Sigma_g^-$). Cornell, Berry and Lwowski[82] have looked for the carbonyl nitrene in the flash photolysis of gaseous ethyl azidoformate, but detected only the absorption of the fragment CON. Triplet–triplet absorption spectra of the ground states of several aromatic nitrenes were observed in the flash photolysis of the corresponding azides[72]. First-order lifetimes of these triplets range from about a microsecond to several seconds, depending on the nature of the medium.

While the triplet ground state of the nitrenes is now quite well documented there is no spectroscopic information on their first excited singlet state, yet it is just this state which plays a major role in the photochemistry of the azides. The singlet state $^1\Delta$ of the imino radical lies 37 kcal mole^{-1} above the triplet ground state HN ($^3\Sigma_g^-$), and the situation is no doubt similar in the alkyl nitrenes. Singlet–triplet splitting in the aromatic nitrenes is probably somewhat smaller. In the non-conjugated (isolated) nitrene group the change from the singlet to the triplet state involves only spin reversal in one of the unhybridized *p*-orbitals; in the nitrenes which are attached to a π-system, such as the carbonyl nitrenes and the aromatic nitrenes, the two electrons on nitrogen may share a common orbital in the singlet configuration of lowest energy.

The singlet lifetime has been experimentally established only for cyanonitrene[69] which in the gas phase converts to the ground state triplet at the rate of about 10^4 sec^{-1}. An estimate of singlet lifetimes in solution can be obtained from the work of McConaghy and Lwowski[60]. They find that in the thermolysis of ethyl azidoformate in the presence of *cis*-4-methylpent-2-ene the rate of spin reversal of the nitrene is about 10 times slower than the rate of addition to the double bond. Assuming that addition is an order of magnitude slower than a diffusion controlled process one obtains for the nitrene a spin conversion rate of about 10^7 sec^{-1}, corresponding to a half-life of 10^{-7} sec. It is therefore not surprising that singlet nitrenes have so far escaped detection in conventional flash spectroscopy.

B. *General Reactions of Nitrenes*

1. Recombination

The recombination of two nitrenes to form an azo compound is spin

$$2\ RN \longrightarrow R{-}N{=}N{-}R \tag{14}$$

allowed for both the singlet and the triplet species. The reaction will be rarely observed in the continuous photolysis of azides because of the low concentration of free nitrene at any time. However, in flash photolysis where a high local nitrene concentration is produced instantaneously, recombination is the preferred reaction path, provided the solvent is sufficiently inert[72]. Nitrene recombination is a facile process requiring little activation energy, and it is therefore the only reaction to occur at very low temperatures.

2. Electrophilic attack on bonding pairs

a. Insertion into C—H *bonds.* Insertion into C—H bonds is exclusively a reaction of singlet nitrenes. Both thermally and photolytically generated nitrenes undergo insertion, the yield of secondary

$$R{-}\underline{\overline{N}} + H{-}\overset{|}{\underset{|}{C}}{-} \longrightarrow R{-}\underset{H}{\underset{|}{N}}{-}\overset{|}{\underset{|}{C}}{-} \tag{15}$$

amine depending on the substrate and on the rate of competing processes. Barton[83] has claimed intramolecular nitrene insertion in the photocyclization of aliphatic azides, but his results are in some doubt[73]. It appears now that in the alkyl nitrenes the insertion reaction is overtaken by the faster hydrogen migration process which results in imines. When hydrogen migration and other rearrangements are inhibited, insertion becomes important. For example, carbethoxynitrene inserts into cyclohexane with a yield of 50%[61]. *t*-Butylcarbonyl nitrene (pivaloyl nitrene) inserts into cyclohexane with a yield of 20%, into cyclopentane with a yield of 13%, into neopentane (primary C—H bonds only) with a yield of 0·7%. The insertion yields of pivaloyl nitrene into primary, secondary and tertiary C—H bonds are in the ratios 1:9:160. Photolysis of pivaloyl azide in cyclohexene yields 12% of the insertion products[84,85].

Intramolecular insertion in substituted aromatic nitrenes has been used as a means of ring closure to pyrrolidines[18,86].

Aromatic nitrenes are as selective in their insertion reactions as carbonyl nitrenes. In phenyl nitrene the insertion yields into primary, secondary and tertiary C—H bonds are in the approximate ratios 1:10:100[87]. Aromatic nitrenes insert into aromatic C—H bonds about as efficiently as into secondary C—H bonds of aliphatic hydrocarbons: photolysis and thermolysis of 2-azidobiphenyl produces

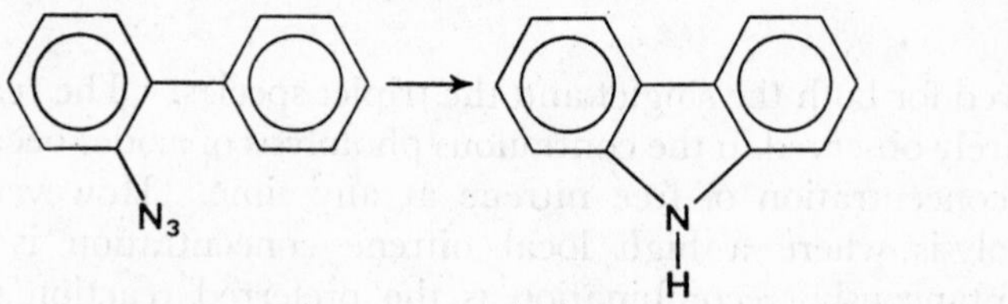

N_3 → N—H (16)

carbazole in 78% yield[88,89]. The analogous reaction with a cyclohexane produces ring closure in 86% yield[17].

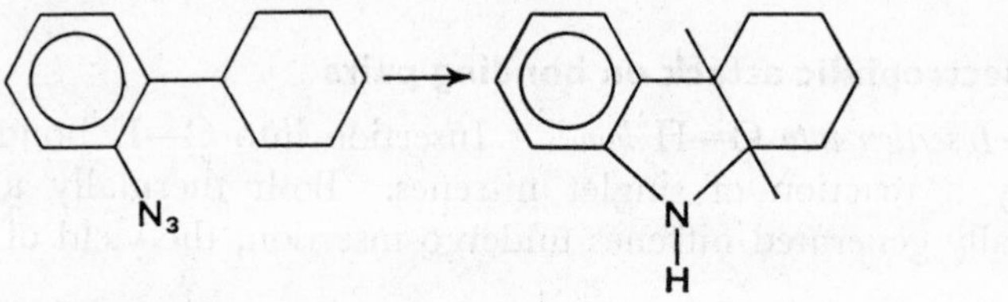

N_3 → N—H (17)

b. Insertion into O—H *bonds.* Insertion into O—H bonds with the formation of hydroxylamine derivatives is observed in carbonyl nitrenes. The photolytic and the thermal decomposition of ethyl azidoformate in *t*-butanol leads to the insertion product in 55%

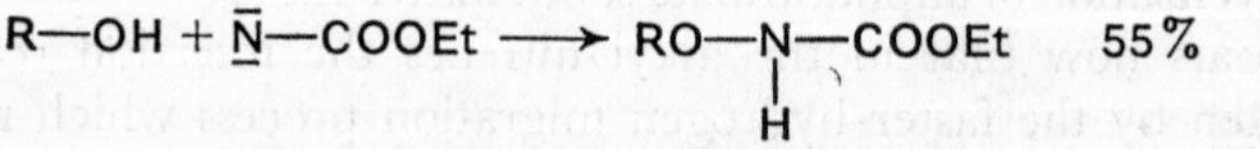

$$\text{R—OH} + \underline{\bar{\text{N}}}\text{—COOEt} \longrightarrow \text{RO—N(H)—COOEt} \quad 55\% \qquad (18)$$

yield[90]. Photogenerated benzoyl nitrene inserts into alcohols and appears also to insert into the O—H bond of acetic acid.

c. Insertion into N—H *bonds.* The photolysis and the thermolysis of azides in the presence of amines produce small yields of hydrazines which can formally be accounted for by insertion of the nitrene into

$$\text{Ph—CO—N}_3 \xrightarrow[\text{PhNH}_2]{h\nu} \text{Ph—CO—NH—NH—Ph} \quad 14\% \qquad (19)$$

the N—H bond of the amine[75]. It is, however, probable that these reactions proceed via a primary attack by the nitrene on the non-bonding pair of the amino group followed by rearrangement[91].

$$\text{Ph—NH}_2 + \overline{\text{N}}\text{—CO—Ph} \longrightarrow \text{Ph—N}^{+}\text{H}_2\text{—}\overline{\text{N}}^{-}\text{—CO—Ph} \longrightarrow \text{Ph—}\overline{\text{N}}\text{H—}\underline{\text{N}}\text{H—CO—Ph} \qquad (20)$$

d. Hydrogen abstraction. Hydrogen abstraction is possibly the most general reaction of triplet nitrenes. Two separate abstraction steps are required to saturate the electron deficiency of the triplet. In the first step hydrogen is abstracted from the substrate leaving a carbon

$$\text{—}\overset{|}{\underset{|}{\text{C}}}\text{—}\dot{\text{N}}{\uparrow\uparrow} + \text{H—}\overset{|}{\underset{|}{\text{C}}}\text{—} \longrightarrow \text{—}\overset{|}{\underset{|}{\text{C}}}\text{—}\overset{\uparrow}{\dot{\text{N}}}\text{—H} + {\uparrow}\dot{\text{C}}\text{—} \qquad (21)$$

radical behind and turning the nitrene into an amino-radical. The two radicals have at this stage correlated (unpaired) spins and cannot couple unless one of their spins is reversed. The time required for spin reversal is usually sufficient to allow the two radicals to diffuse away from each other. The amino-radical will then abstract a

$$\text{—}\overset{|}{\underset{|}{\text{C}}}\text{—}\dot{\text{N}}\text{H} + \text{H—}\overset{|}{\underset{|}{\text{C}}}\text{—} \longrightarrow \text{—}\overset{|}{\underset{|}{\text{C}}}\text{—NH}_2 + \dot{\text{C}}\text{—} \qquad (22)$$

second hydrogen and form a primary amine. There is a certain probability of radical recombination leading to secondary amines (pseudo-insertion) hydrazines and hydrocarbon dimers (equation 23).

In suitable solvents even alkyl nitrenes produce small amounts of primary amines[73]. Carbethoxynitrene abstracts hydrogen from hydrocarbons[63] (12%) and more efficiently from alcohols[90] (90%). Aromatic nitrenes are in general less reactive than carbonyl nitrenes, they abstract hydrogen, provided the attack on unreacted azido groups

$$-\overset{|}{\underset{|}{C}}-\dot{N}H + \cdot\overset{|}{\underset{|}{C}}- \longrightarrow -\overset{|}{\underset{|}{C}}-\overset{H}{\overset{|}{N}}-\overset{|}{\underset{|}{C}}-$$

$$-\overset{|}{\underset{|}{C}}-\dot{N}H + H\dot{N}-\overset{|}{\underset{|}{C}}- \longrightarrow -\overset{|}{\underset{|}{C}}-\underset{H}{\underset{|}{N}}-\overset{H}{\overset{|}{N}}-\overset{|}{\underset{|}{C}}- \qquad (23)$$

$$-\overset{|}{\underset{|}{C}}\cdot + \cdot\overset{|}{\underset{|}{C}}- \longrightarrow -\overset{|}{\underset{|}{C}}-\overset{|}{\underset{|}{C}}-$$

can be controlled[72]. Also recombination of amino-radicals and carbon radicals (pseudo-insertion) occurs in sterically favourable conditions, for example in the formation of carbazole from 2-azido-biphenyl where the triplet nitrene has been shown to be an intermediate[25]. In the photolysis of azides in polymer matrices up to 90% of secondary amines are formed, the reaction yield depending on the rigidity of the matrix[72].

Proof for the triplet character of the hydrogen abstracting species is obtained in sensitization experiments. Thus, direct photolysis of ethyl azidoformate in cyclohexene results in the formation of only 3% of the hydrogen abstraction product (urethane) and a trace of *bis*-cyclohexene[61]. In the sensitized photoreaction, where triplet nitrene is produced directly, the urethane yield increases to 74%, the yield of *bis*-cyclohexene to 63%.

In a similar way, direct photolysis of benzoyl azide in alcohols produces about 25% of the urethane[75]. On sensitization with benzophenone, urethane is the sole product of the reaction.

3. Addition to multiple bonds

a. Addition to C=C *double bonds.* Addition to double bonds is a reaction of both singlet and triplet nitrenes. However, the formation of stable 3-membered aziridine rings has been positively established only for some carbonyl nitrenes. The addition of singlet nitrene

$$-\overset{|}{\underset{|}{C}}-N + \begin{matrix} \overset{|}{C}-R \\ \| \\ \underset{|}{C}-R \end{matrix} \longrightarrow -\overset{|}{\underset{|}{C}}-N\!\!<\!\begin{matrix} \overset{|}{C}-R \\ | \\ \underset{|}{C}-R \end{matrix} \qquad (24)$$

proceeds by a one-step mechanism and the aziridine retains the steric configuration of the olefinic substrate. Triplet nitrenes add to double bonds without retention of configuration. The reaction has been used by Lwowski[60,64] and by Hafner[92] to identify the spin state of carbethoxynitrene[55].

While carbethoxyaziridines are usually quite stable (they withstand the injection temperature of gas chromatography) other aziridines react further and are therefore not detected in the final products. For example, the aziridine presumably formed from pivaloyl nitrene and cyclohexene rearranges to a Schiff's base and to an amine[51]:

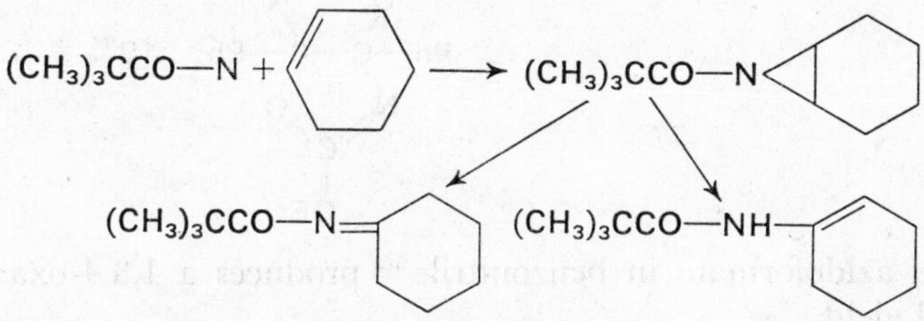

$(CH_3)_3CCO—N$ + cyclohexene ⟶ $(CH_3)_3CCO—N$< (7-azabicyclo[4.1.0]heptane) ⟶ $(CH_3)_3CCO—N$= (cyclohexylidene) and $(CH_3)_3CCO—NH$— (cyclohexenyl) (25)

Carbethoxynitrene undergoes 1,2-addition to the double bonds of dienes[94], but no 1,4-cycloaddition product has been detected.

1,2-Addition to aromatic systems is probably the primary step in a number of more complex nitrene reactions such as the formation of azepines from carbethoxynitrene and benzene[62,93] and a number of

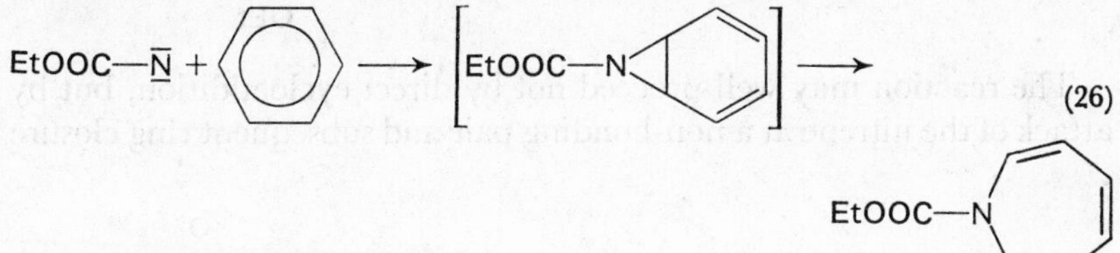

EtOOC—N̲ + benzene ⟶ [EtOOC—N< (azanorcaradiene)] ⟶ EtOOC—N (azepine) (26)

interesting rearrangements induced by carbethoxynitrene in some 5-membered heterocycles[94]. Azepines are also formed on irradiation of phenyl azide in the presence of strong nucleophiles[95].

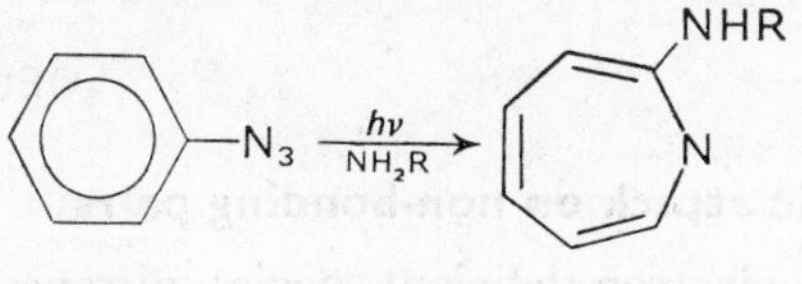

C_6H_5—N_3 $\xrightarrow[NH_2R]{h\nu}$ 2-NHR-azepine (27)

b. 1,3-Cycloaddition. The cycloaddition of carbonyl nitrenes to triple bonds has been observed by Huisgen in the reaction of carbethoxynitrene with tolan[52] and with phenylacetylene[47]. Photolysis

$$\text{EtO—C(=O)—N} + \text{Ph—C}\equiv\text{C—Ph} \rightarrow \text{(2-ethoxy-4,5-diphenyloxazole, 33\%)} + \text{(bis-oxadiazoline, 19\%)} \quad (28)$$

of ethyl azidoformate in benzonitrile[53] produces a 1,3,4-oxadiazole in good yield.

$$\text{EtO—C(=O)—N} + \text{R—C}\equiv\text{N} \rightarrow \text{R—C}=\text{N—N}=\text{C(OEt)—O (ring)} \quad (29)$$

The reaction may well proceed not by direct cycloaddition, but by attack of the nitrene at a non-bonding pair and subsequent ring closure

$$\text{R—C}\equiv\text{N|} + \overline{\text{N}}\text{—COOEt} \longrightarrow \text{R—C}\equiv\overset{+}{\text{N}}\text{—}\underline{\text{N}}=\text{C(O}^-\text{)(OEt)} \rightarrow \text{R—C}=\text{N|—N|}=\text{C(OEt)—O (ring)} \quad (30)$$

4. Electrophilic attack on non-bonding pairs

As a strongly electron deficient species, nitrenes will react with non-bonding electron pairs. Thus, carbethoxynitrene does attack the lone pair of amino groups[91] with the formation of hydrazine derivatives and polyamines. In the reaction of carbethoxynitrene with

$$\text{C}_6\text{H}_5\text{N(R)H} + \bar{\text{N}}\text{—COOEt} \longrightarrow \text{C}_6\text{H}_5\overset{+}{\text{N}}(\text{R})(\text{H})\text{—}\bar{\underline{\text{N}}}\text{—COOEt} \longrightarrow \text{R—N(C}_6\text{H}_4\text{)—N(H)—COOEt} \quad (31)$$

$$\text{C}_6\text{H}_5\text{N(R)CH}_3 + \text{N—COOEt} \longrightarrow \text{C}_6\text{H}_5\text{N(R)(CH}_3\text{)—N—COOEt} \longrightarrow \text{C}_6\text{H}_5\text{N(R)—CH}_2\text{—NH—COOEt} \quad (32)$$

pyridine, the intermediate zwitterion-betaine has been actually isolated[91].

$$\text{C}_5\text{H}_5\text{N} + \bar{\text{N}}\text{—COOEt} \longrightarrow \text{C}_5\text{H}_5\overset{+}{\text{N}}\text{—}\overset{-}{\bar{\text{N}}}\text{—COOEt} \quad 67\% \quad (33)$$

In a similar way, benzoyl nitrene is trapped by strong nucleophiles with lone pairs, such as dimethyl sulphoxide[96], but here the reaction

$$\text{C}_6\text{H}_5\text{CO—}\bar{\text{N}} + (\text{CH}_3)_2\text{S=O} \longrightarrow \text{C}_6\text{H}_5\text{CO—N=S(CH}_3)_2\text{=O} \quad 28\% \quad (34)$$

of the nitrene has to compete with the Curtius rearrangement of the azide precursor[75].

One of the most ubiquitous reactions is the formation of azo compounds by the attack of nitrene on the azido group. This process accounts for the high overall quantum yield of nitrogen evolution in the gas phase photolysis of hydrazoic acid[6] and of methyl azide[45]. It has also been demonstrated in the gas phase pyrolysis of phenyl azide and of *ortho*-trifluoromethyl azide[18] where azobenzenes are the major reaction products. Azo compounds are formed in near quantitative yield in the solution photolysis of *p*-methoxyphenyl azide and

4-azidobiphenyl[97] (see equations 82, 84). The nitrene mechanism of these reactions is supported by experiments where mixtures of *p*-methoxyphenyl azide and 4-azidobiphenyl were photolysed and it was found that the 'mixed' azo compound 4-methoxy-4′-phenylazobenzene and the symmetrical 4,4′-dimethoxyazobenzene and 4,4′-diphenylazobenzene were formed in about equal proportions[97].

The reaction of a nitrene with the lone pair of a nitro group in 2-azido-3-nitrobiphenyl leads to the formation of furoxanes (equation 80), reaction with a sterically accessible carbonyl group in 2-azidobenzophenone produces oxazole derivatives[98] (equation 81).

V. PHOTOREACTIONS OF ORGANIC AZIDES

A. Alkyl Azides

Methyl azide (**1**) and deuterated methyl azide (CD_3N_3) were photolysed at 4°K and at 50°K and the products analysed by infrared spectrometry[99,100]. Methylene imine (**3**) and the corresponding deuterium compound were observed, presumably formed by a 1,2-hydrogen shift in the nitrene intermediate (**2**). The vapour phase

$$\underset{(\mathbf{1})}{CH_3N_3} \longrightarrow \underset{(\mathbf{2})}{[CH_3\bar{N}]} \longrightarrow \underset{(\mathbf{3})}{CH_2{=}NH} \qquad (35)$$

photolysis of methyl azide[45] produced mainly nitrogen, together with some hydrogen (5–10%) and small quantities of methane, ethane, ethylene and a polymer analysing for $(CH_3N)_x$. The effect of carbon dioxide, azomethane and ethylene on the quantum yield of nitrogen evolution indicated a radical mechanism involving methyl nitrene. Reaction scheme (12) was proposed to account for these observations.

The photoreactions of higher alkyl azides were studied by Barton and co-workers[83,101] in connexion with the synthesis of the alkaloid conessine[102]. They reported that irradiation of *n*-butyl azide (**4**) in ethanol or ether produced pyrrolidine (**5**) in yields of 22% and 14% respectively, together with the imine (**6**). *n*-Heptyl azide was reported

$$\underset{(\mathbf{4})}{CH_3(CH_2)_3N_3} \xrightarrow{h\nu} [CH_3(CH_2)_3N] \longrightarrow \underset{(\mathbf{5})}{\text{pyrrolidine (N–H)}} + \underset{(\mathbf{6})}{CH_3(CH_2)_2CH{=}NH} \qquad (36)$$

to form *n*-propylpyrrolidine (15%) and *n*-octyl azide to form *n*-butylpyrrolidine in yields of 5–35%, depending on the solvent.

The possibility of ring closure was expected to depend on molecular geometry. Thus, photolysis of *n*-propyl azide (**7**) in cyclohexane produced only the imine (**8**) in 59% yield, isolated as the 2,4-dinitrophenylhydrazone (**9**). Irradiation of phenylethyl azide (**10**) resulted

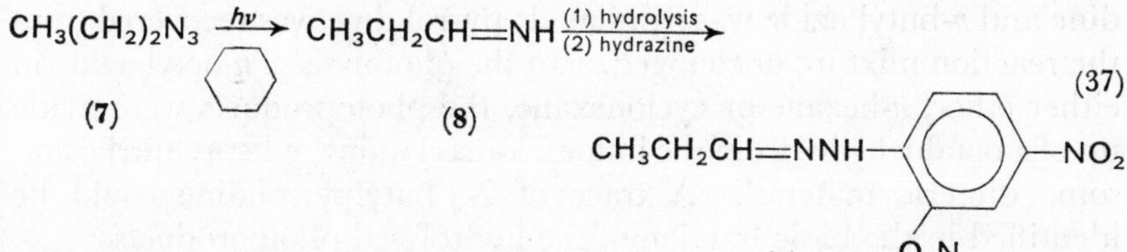

$$CH_3(CH_2)_2N_3 \xrightarrow[\text{cyclohexane}]{h\nu} CH_3CH_2CH{=}NH \xrightarrow[\text{(2) hydrazine}]{\text{(1) hydrolysis}} CH_3CH_2CH{=}NNH{-}C_6H_3(NO_2)_2 \quad (37)$$

(**7**) (**8**) (**9**)

in the imine (**11**), identified by reduction to the amine (**12**) but 3-

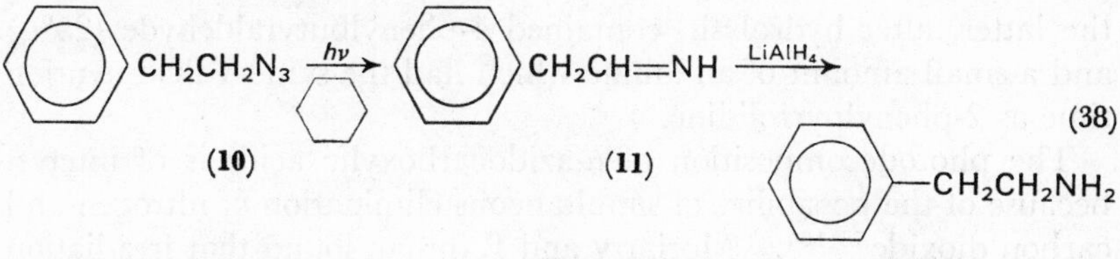

$$C_6H_5CH_2CH_2N_3 \xrightarrow[\text{cyclohexane}]{h\nu} C_6H_5CH_2CH{=}NH \xrightarrow{LiAlH_4} C_6H_5CH_2CH_2NH_2 \quad (38)$$

(**10**) (**11**) (**12**)

phenylpropyl azide (**13**) was reported to cyclize under similar conditions to tetrahydroquinoline (**14**) in 21% yield. Barton concluded

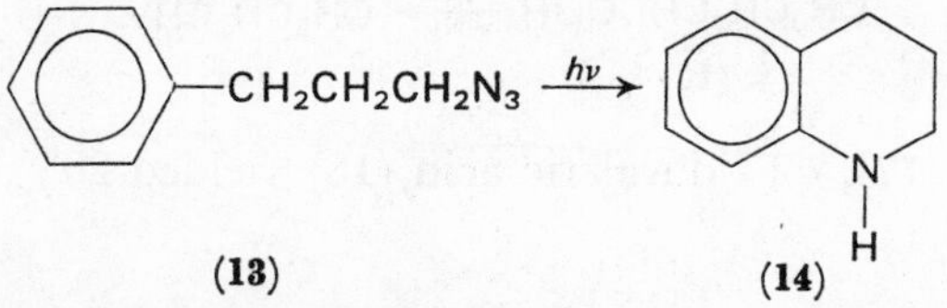

$$C_6H_5CH_2CH_2CH_2N_3 \xrightarrow{h\nu} \text{tetrahydroquinoline (N–H)} \quad (39)$$

(**13**) (**14**)

from these results that a necessary requirement for ring closure was the possibility of formation of a 6-membered cyclic transition state (**15**).

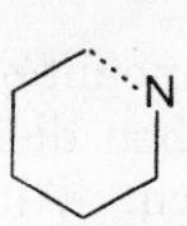

(**15**)

In view of the considerable importance of a photochemical ring closure reaction in preparative heterocyclic chemistry attempts were made by others[23,86] and by Barton himself[103] to repeat the synthesis of pyrrolidines by this route. The results were disappointing although a variety of conditions and light sources were tried. Moriarty[73] finally established, in a careful re-examination, that in the photolysis of alkyl azides, cyclization does not occur to any appreciable extent. Irradiation of *n*-butyl azide (**4**) in ether produced 70% of the imine, some amine (5%) and polymer, but no pyrrolidine. To check that any pyrrolidine formed had not reacted with the imine, a mixture of pyrrolidine and *n*-butyl azide was irradiated; pyrrolidine was recovered from the reaction mixture unchanged. In the photolysis of *n*-octyl azide in either ether, *n*-hexane or cyclohexane, the photoproducts were made up of *n*-octaldehyde (from the imine), *n*-octylamine, *n*-heptylnitrile and some dimeric material. A trace of 2-*n*-butylpyrrolidine could be identified in the basic fraction of the hydrolysed photoproducts.

To study the effect of the presence of an aromatic ring, Moriarty photolysed 4-phenylbutyl azide. The products were separated into a basic and a neutral fraction. No pyrrolidine was found in the former, the latter, after hydrolysis, contained 4-phenylbutyraldehyde (25%) and a small amount of an amine which had the same VPC retention time as 2-phenylpyrrolidine.

The photodecomposition of α-azidocarboxylic acids is of interest because of the possibility of simultaneous elimination of nitrogen and carbon dioxide[75,104]. Moriarty and Rahman found that irradiation of α-azidobutyric acid (**16**) in methanol produced 25% of propionaldehyde imine (**17**), α-azidovaleric acid (**18**) yielded 20% of butyraldehyde imine (**19**) and a small amount of α-aminovaleric acid (**20**). In these reactions the loss of carbon dioxide may have been concerted with the elimination of nitrogen. Alternatively, a nitrene may have been formed first, with subsequent loss of carbon dioxide. To decide this point Moriarty and Rahman photolysed α-deutero-α-azidobutyric acid. The reaction products were isolated as 2,4-dinitrophenylhydrazones. 58% of the recovered propionaldehyde-2,4-dinitro-

$$\underset{(\mathbf{16})}{CH_3CH_2\overset{\displaystyle N_3}{\overset{|}{C}}HCOOH} \xrightarrow[-N_2,\ -CO_2]{h\nu} \underset{(\mathbf{17})}{CH_3CH_2CH{=}NH} \qquad (40)$$

$$\underset{(\mathbf{18})}{CH_3CH_2CH_2\overset{\displaystyle N_3}{\overset{|}{C}}HCOOH} \xrightarrow[-N_2,\ -CO_2]{h\nu} \underset{(\mathbf{19})}{CH_3(CH_2)_2CH{=}NH} + \underset{(\mathbf{20})}{CH_3(CH_2)_2\overset{\displaystyle NH_2}{\overset{|}{C}}HCOOH} \qquad (41)$$

phenylhydrazone contained deuterium on the aldehydic carbon, thus implying concerted decarboxylation to the extent of about 50%.

The possibility of alkyl or aryl group migration in competition with the loss of carbon dioxide was investigated in the photolysis of α-azido-iso-butyric acid (**21**) and azidodiphenylacetic acid (**23**). In the first case acetone (**22**) was the only product, and no trace of acetaldehyde could be found; methyl-migration had not taken place. Evidence of phenyl-migration was obtained in the photolysis of **23**. The reaction products were identified as a mixture of the 2,4-dinitrophenylhydrazones of benzophenone and benzaldehyde obtained from the imines **24** and **25**. The latter was thought to have been formed by phenyl-migration according to the following reaction scheme

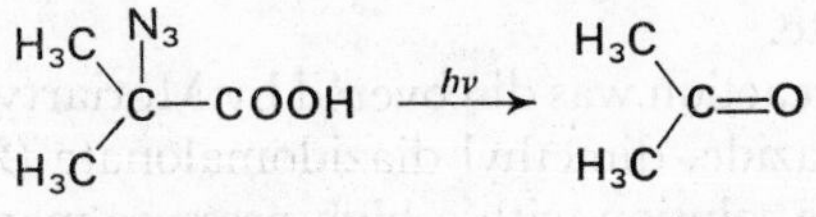

$$(CH_3)_2C(N_3)COOH \xrightarrow{h\nu} (CH_3)_2C{=}O \qquad (42)$$

(**21**) → (**22**)

$$(C_6H_5)_2C(N_3)COOH \xrightarrow{h\nu} \left[(C_6H_5)_2C(\dot{N})COOH\right] \longrightarrow (C_6H_5)_2C{=}NH \qquad (43)$$

(**23**) → (**24**)

$$\left[(C_6H_5)_2C(\dot{N})COOH\right] \longrightarrow \left[C_6H_5N{=}C(C_6H_5)COOH\right] \xrightarrow{-CO_2} C_6H_5N{=}CHC_6H_5$$

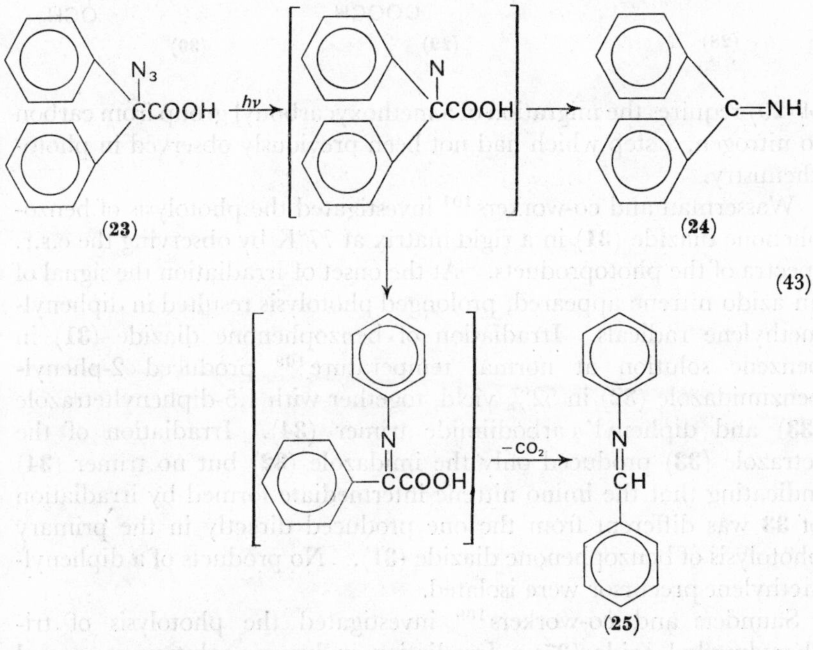

(**25**)

Photolysis of the ester of α-azidobutyric acid (**26**) did not result in decarboxylation. Instead, ethyl-α-iminobutyrate (**27**) was obtained.

$$\underset{(\mathbf{26})}{CH_3CH_2\overset{\displaystyle N_3}{\overset{|}{C}}HCOOC_2H_5} \longrightarrow \underset{(\mathbf{27})}{CH_3CH_2\overset{\displaystyle NH}{\overset{\|}{C}}COOC_2H_5} \qquad (44)$$

Ethyl α-azidovalerate gave, under similar conditions, a 15% yield of ethyl α-oxovalerate.

An interesting reaction was discovered by Moriarty in the photolysis of the geminal diazide, dimethyl diazidomalonate (**28**)[105,106]. Irradiation in benzene solution with a high-pressure mercury lamp led to the evolution of one mole nitrogen and produced 1,5-dimethoxycarbonyltetrazole (**29**) in 48% yield. Longer irradiation of **28** or separate photolysis of **29** resulted in a 32% yield of **30**. The formation

$$\underset{(\mathbf{28})}{(CH_3OOC)_2C(N_3)_2} \xrightarrow[-N_2]{h\nu} \underset{(\mathbf{29})}{CH_3OOC\text{-}C\overset{N-N}{\underset{N-N(COOCH_3)}{\;}}} \longrightarrow \underset{(\mathbf{30})}{CH_3OOC\text{-}C\overset{N-O}{\underset{N=C(OCH_3)}{\;}}} \qquad (45)$$

of (**29**) requires the migration of a methoxycarbonyl group from carbon to nitrogen, a step which had not been previously observed in photochemistry.

Wasserman and co-workers[107] investigated the photolysis of benzophenone diazide (**31**) in a rigid matrix at 77°K by observing the e.s.r. spectra of the photoproducts. At the onset of irradiation the signal of an azido nitrene appeared, prolonged photolysis resulted in diphenylmethylene radicals. Irradiation of benzophenone diazide (**31**) in benzene solution at normal temperature[108] produced 2-phenylbenzimidazole (**32**) in 52% yield, together with 1,5-diphenyltetrazole (**33**) and diphenyl carbodiimide trimer (**34**). Irradiation of the tetrazole (**33**) produced only the imidazole (**32**) but no trimer (**34**) indicating that the imino nitrene intermediate formed by irradiation of **33** was different from the one produced directly in the primary photolysis of benzophenone diazide (**31**). No products of a diphenylmethylene precursor were isolated.

Saunders and co-workers[109] investigated the photolysis of triphenylmethyl azide (**35**). Irradiation in hexane solution at normal

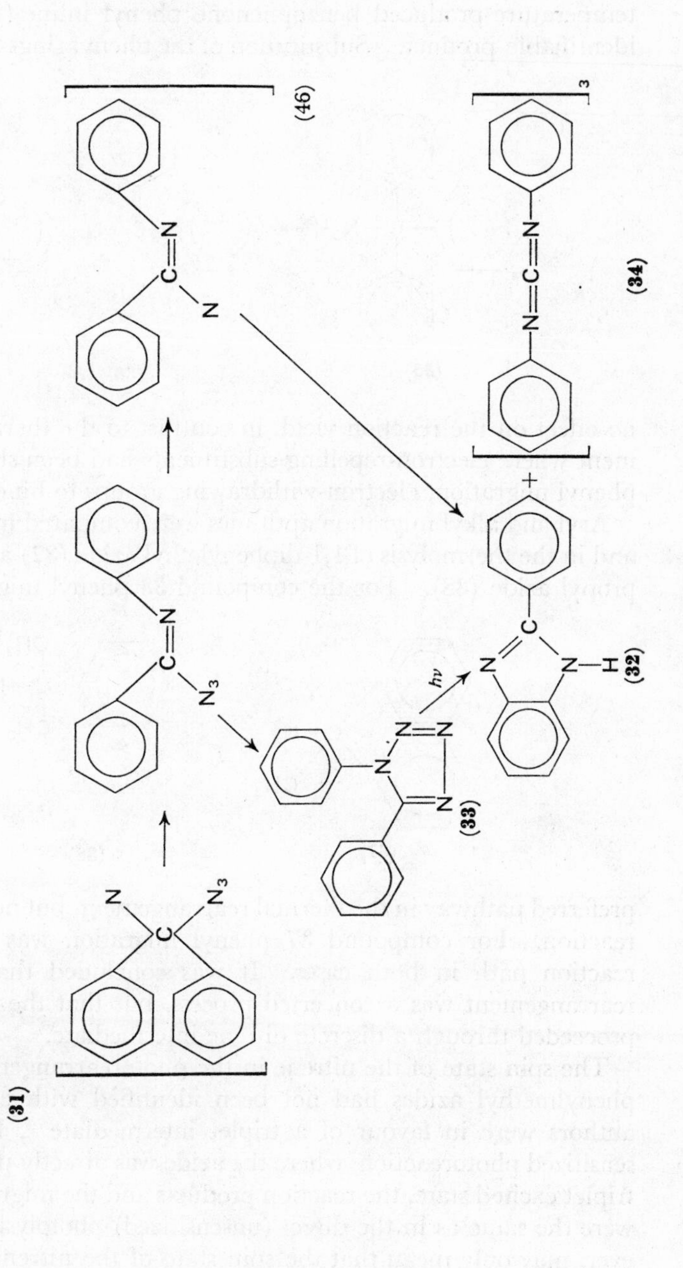
(31)
hν
(46)
(33)
hν
(32)
(34)

temperature produced benzophenone phenyl imine (**36**) as the only identifiable product. Substitution of the phenyl rings had practically

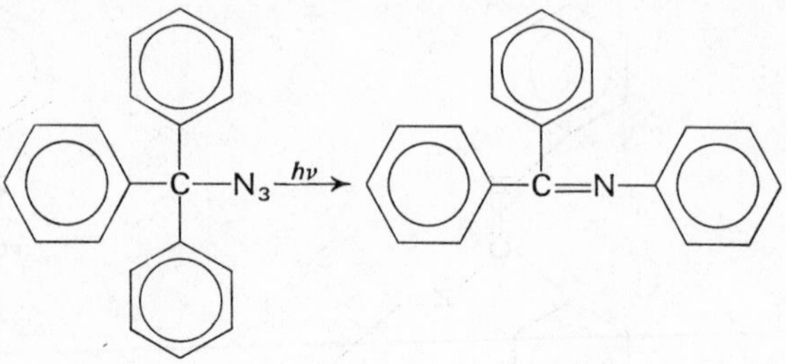

(47)

(**35**) (**36**)

no effect on the reaction yield, in contrast to the thermal rearrangement where electron-repelling substituents had been shown to favour phenyl migration, electron-withdrawing groups to hinder it.

Aryl and alkyl migration aptitudes were compared in the photolysis and in the thermolysis of 1,1-diphenylethyl azide (**37**) and 2-phenyl-2-propyl azide (**38**). For the compound **38** phenyl migration was the

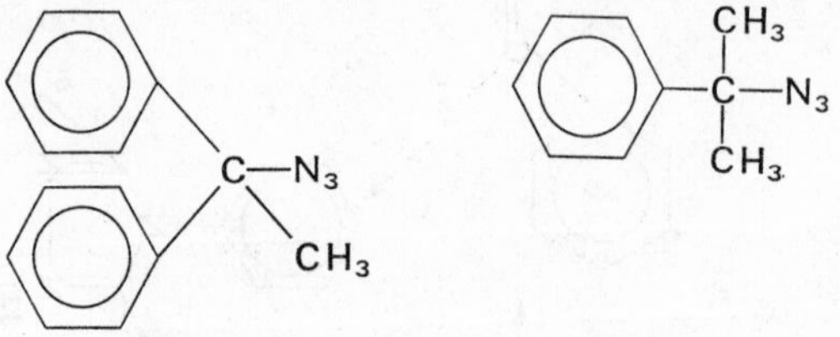

(**37**) (**38**)

preferred pathway in the thermal rearrangement, but not in the photo-reaction. For compound **37** phenyl migration was the dominant reaction path in both cases. It was concluded that the thermal rearrangement was a concerted process, but that the photo-reaction proceeded through a discrete nitrene intermediate.

The spin state of the nitrene in the photorearrangement of the triphenylmethyl azides had not been identified with certainty. The authors were in favour of a triplet intermediate[74], because in the sensitized photoreaction, where the azide was directly promoted to the triplet excited state, the reaction products and the migration aptitudes were the same as in the direct (unsensitized) photolysis. This, however, may only mean that the spin state of the nitrene has no effect on the migratory aptitudes of the substituents.

B. Vinyl Azides

The photodecomposition of vinyl azides is of interest as a potential synthetic route to the highly strained azirine ring system. Smolinsky[110] was the first to prepare compounds of this class by the pyrolysis of vinyl azides (**39**) and Hassner and Fowler[111,112] have shown that the corresponding photo-reaction also produces 1-azirines (**40**) in excellent yield. Three possible mechanisms were proposed by these authors, one involving a nitrene intermediate **41**.

N (41) N + (41a) N_3 (39) N N N N=N N N (40) (48)

The same general method was applied to the synthesis of aminoketones directly from vinyl azides (**39**). In the presence of 0·5% sodium methoxide in methanol the reaction produced the amino-

N_3 R R¹ (39) $\xrightarrow[CH_3OH]{h\nu}$ CH_3O OCH_3 R¹ R H NH_2 (42) ⟶ O R—C R¹ $NH_3^+Cl^-$ (43) (49)

N R¹ R (45) $\xrightarrow{CH_3OH}$ H R N R¹ CH_3O (44)

dimethylketal (**42**). This was not isolated, but was hydrolysed to the aminoketone hydrochloride (**43**). The aziridine (**44**) was prepared from the azirine (**45**) by treatment with methanol. Further reaction of the aziridine with methanol led again to the ketal **42**.

The reaction was applied successfully to cyclic vinyl azides to produce fused azirine rings. 1-Azido-cyclooctene (**46**) was converted on irradiation in 93% yield into 9-azabicyclo-[6.1.0]-non-1-(9)-ene (**47**). The azirine could again be hydrolysed to the aminoketone (**48**)

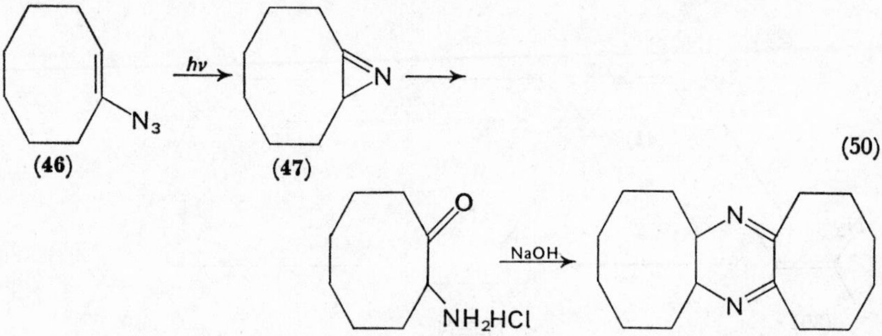

N_3 (**46**) $\xrightarrow{h\nu}$ N (**47**) $\longrightarrow$ (50)

O, NH_2HCl (**48**) $\xrightarrow{NaOH}$ N, N (**49**)

which on treatment with base gave the dihydropyrazine (**49**). Another pyrazine, namely 2,5-di-*t*-butylpyrazine (**50**) was obtained in good yield on irradiation of 1-azido-3,3-dimethyl-1-butene (**51**) in methanol[113]. Photolysis of the azide in an inert solvent resulted only in polymeric materials.

4-Azido-1,2-dihydronaphthalene (**52**) was irradiated in an attempt

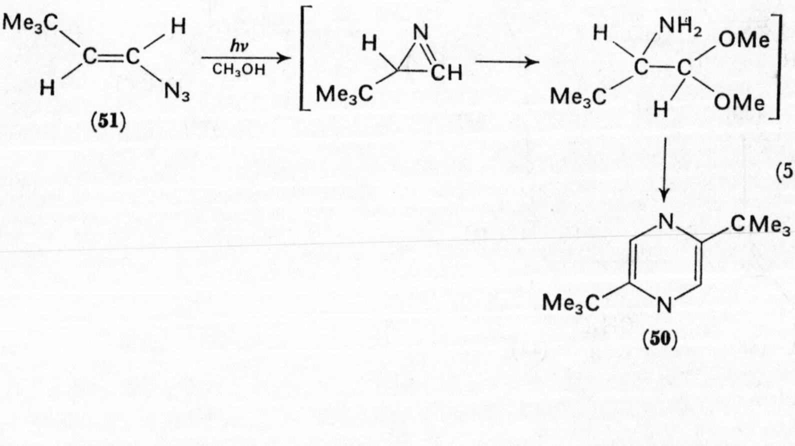

Me_3C, H, C=C, H, N_3 (**51**) $\xrightarrow[CH_3OH]{h\nu}$ [H, N, CH, Me_3C $\longrightarrow$ H, NH_2, OMe, C—C, Me_3C, H, OMe] (51)

$\downarrow$

N, CMe_3, Me_3C, N (**50**)

to prepare azirines fused to smaller ring systems, but the reaction produced only unidentifiable polymers. However, the presence of an azirine during photolysis could be demonstrated by conducting the reaction in the presence of sodium methoxide in methanol. In these conditions the ketal **53** was formed and could be hydrolysed to the aminoketone **54** in good yield.

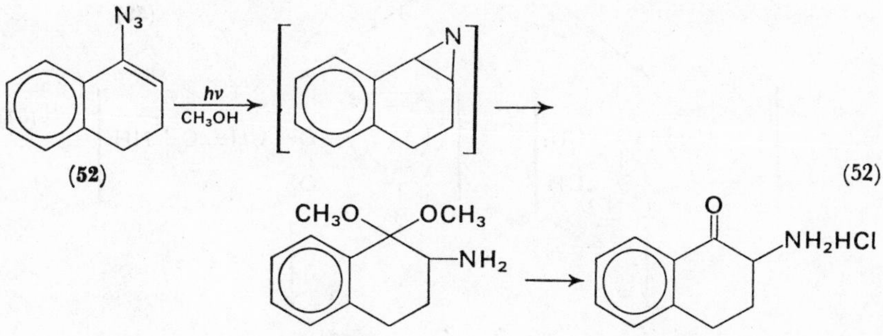

N_3 hν CH_3OH N

(52) (52)

CH_3O OCH_3 NH_2 O NH_2HCl

(53) **(54)**

An interesting new heterocyclic system was prepared by photolysis of α-azidostyrene (**55**) in benzene solution[114]. The main product of the reaction was 2-phenylazirine (**56**), but 4-phenyl-3-phenylimino-1-azabicyclo-[2.1.0]-pentane (**57**) was also isolated in 10% yield. The

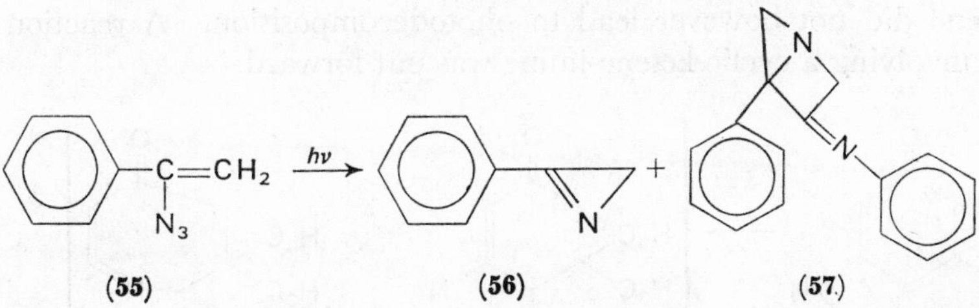

$C{=}CH_2$ N_3 hν N + N N (53)

(55) **(56)** **(57)**

structure of the new compound was established by n.m.r. analysis. It was not produced in the pyrolysis of α-azidostyrene (**55**), but it was formed on irradiation of 2-phenylazirine (**56**). This seems to point to a photocycloaddition mechanism for reaction (53).

The photolysis of β-azidovinylphenyl ketone (**58**)[115] in benzene did not give the expected 3-benzoylazirine (**59**) but only a 20% yield of benzoylacetonitrile (**60**). A mechanism for this reaction was suggested as follows:

$$C_6H_5\text{–}\underset{\text{O}}{\overset{\|}{C}}\text{–}CH{=}CHN_3 \rightarrow \left[C_6H_5\text{–}\underset{\text{O}}{\overset{\|}{C}}\text{–}CH{=}CHN\right] \not\rightarrow C_6H_5\text{–}\underset{\text{O}}{\overset{\|}{C}}\text{–}(\text{azirine})$$

(58) (59)

or

$$\left[C_6H_5\text{–}C(=O)\text{–}CH{=}CH\text{–}N\right] \rightarrow \text{5-phenylisoxazole} \xrightarrow{CH_3ONa} C_6H_5\text{–}\underset{\text{O}}{\overset{\|}{C}}\text{–}CH_2\text{–}CN \quad (54)$$

$$\left[C_6H_5\text{–}\underset{\text{O}}{\overset{\|}{C}}\text{–}CH{=}C{=}NH\right] \rightarrow C_6H_5\text{–}\underset{\text{O}}{\overset{\|}{C}}\text{–}CH_2\text{–}CN$$

(60)

Photolysis of the cyclic vinyl ketone 5,5-dimethyl-3-azido-2-cyclohexene-1-one (**61**) in aqueous tetrahydrofuran resulted in ring enlargement to the azepine (**62**). Irradiation of the same azide in benzene did not however lead to photodecomposition. A reaction path involving a cyclic ketene-imine was put forward.

$$\text{(61)} \xrightarrow{h\nu} [\text{nitrene} \rightarrow \text{cyclic ketene-imine}] \rightarrow \text{(62)} \quad (55)$$

Irradiation of allyl azide produced the imine of acrolein[83], but photolysis of 1-azido-2-phenylprop-2-ene (**63**) in cyclohexane solution resulted in a small yield of the azacyclobutane (**64**) and 2-phenylpropenalimine (**65**)[116]. This is the first observation of an 1-azabicyclo [1.1.0]-butane ring in the products of photodecomposition of an allylic azide.

$$\text{C}_6\text{H}_5\text{C}(=\text{CH}_2)\text{CH}_2\text{N}_3 \ (\mathbf{63}) \xrightarrow{h\nu} \mathbf{64} + \text{CH}_2{=}\text{C}(\text{C}_6\text{H}_5){-}\text{CH}{=}\text{NH} \ (\mathbf{65}) \quad (56)$$

C. Acyl Azides

The outstanding thermal reaction of the acyl azides is their rearrangement to isocyanates. This Curtius rearrangement occurs also in the photodecomposition of the acyl azides, but the yield of isocyanate is lower than in the corresponding thermal process, and other products appear which can be accounted for by a nitrene precursor. A considerable amount of work has been directed towards an elucidation of the mechanism of the photo-Curtius rearrangement and the bulk of the evidence seems to favour the view that it[120] is a concerted process not involving a discrete nitrene intermediate. Nitrenes are nevertheless formed in the photolysis of acyl azides[120] and the products of this alternative route will now be discussed.

1. Azidoformates

The photochemistry of the azidoformates has received particular attention in recent years because these compounds belong to the group of 'rigid' azides (Curtius[117,118]) which do not normally undergo the Curtius rearrangement to isocyanates. The best known example is ethyl azidoformate which has been thoroughly studied by Lwowski[55], Huisgen[128], Hafner[91,123] and others[129].

Irradiation of ethyl azidoformate (**66**) at 2537 Å in cyclohexene results in five identifiable products: 7-ethoxycarbonyl-7-azabicyclo [4.1.0]-heptane (**67**) (50%), *N*-2-cyclohexenylurethane (**68**) (9%), *N*-3-cyclohexenylurethane (**69**) (3%), urethane (**70**) (3%) and bi(cyclohex-2-enyl) (**71**) (1–7%). Photolysis of ethyl azidoformate

$C_2H_5OOCN_3$ (66) $\xrightarrow{h\nu}$ (cyclohexene) → $NCOOC_2H_5$ (67), $NHCOOC_2H_5$ (68), $NHCOOC_2H_5$ (69), $H_2NCOOC_2H_5$ (70), (71) (57)

$C_2H_5OOCN_3$ (66) $\xrightarrow{h\nu}$ (cyclohexane) → $NHCOOC_2H_5$ (72), $H_2NCOOC_2H_5$ (70), $(C_2H_5OOCNH)_2$ (73)

in cyclohexane gave mainly *N*-cyclohexylurethane (**72**), urethane (**70**) and trace amounts of *N,N'*-diethoxycarbonylhydrazine (**73**).

All these products are compatible with the formation of a nitrene, yet on their own they do not constitute conclusive proof for the real existence of a discrete intermediate. It was only when Lwowski was able to show that the nitrene produced by α-elimination from *N*-*p*-nitrobenzenesulphonoxyurethane[55,122] (**74**) led to the same products as the photolysis of ethyl azidoformate that the evidence for a photolytic nitrene mechanism became entirely convincing, the more

$$C_2H_5OOCNHOSO_2-C_6H_4-NO_2 \xrightarrow{(C_2H_5)_3N} [C_2H_5OOCN] \quad (58)$$

(74) → with cyclohexene (67); with cyclohexane (72)

so, as also the product distribution was nearly identical in both cases.

The photolysis of ethyl azidoformate (**66**) in aromatic solvents leads to the formation of azepines[96,123] (**75**). Toluene, xylene, mesitylene

$$C_2H_5OOCN_3 \ (\mathbf{66}) \xrightarrow[\text{benzene}]{h\nu} \left[C_6H_6{>}N{-}COOC_2H_5 \right] \longrightarrow \text{azepine-}N{-}COOC_2H_5 \ (\mathbf{75}) \tag{59}$$

and durene behave in this way and there is no evidence of nitrene attack on the side-chain of the hydrocarbon. With chlorobenzene a mixture of about equal quantities of the 2,3- and 4-chloro-4-ethoxycarbonylazepines is obtained; the substituent has apparently little influence on the locus of nitrene attack on the ring. From the effect of dilution of the aromatic solvent with cyclohexane on the azepine yield Lwowski has concluded that azepine formation is a reaction of the excited singlet nitrene only[93].

In the photolysis of neat ethyl azidoformate (**66**) diethyl azodiformate (**76**) is produced in the early stages of the process, to be later replaced by the triethyl nitrilotriformate (**77**) which is produced in 58% yield[124]. A detailed investigation showed that the nitrilo compound (**77**) is apparently not formed by the action of a nitrene, and the reaction between an azide and an excited azodiformate was suggested as a possible pathway. Hancock[125] has proposed an alternative mechanism which involves triazacyclopropene (see, however, Huisgen[128]).

$$N_3COOC_2H_5 \ (\mathbf{66}) \xrightarrow{h\nu} [NCOOC_2H_5] \longrightarrow C_2H_5OOCN{=}NCOOC_2H_5 \ (\mathbf{76}) \longrightarrow C_2H_5OOCN(COOC_2H_5)_2 \ (\mathbf{77}) \tag{60}$$

Irradiation of ethyl azidoformate in the presence of nitriles, such as acetonitrile, isobutyronitrile and 3-ethoxypropionitrile produces oxadiazoles (**78**) in good yields. In acrylonitrile 1-ethoxycarbonyl-

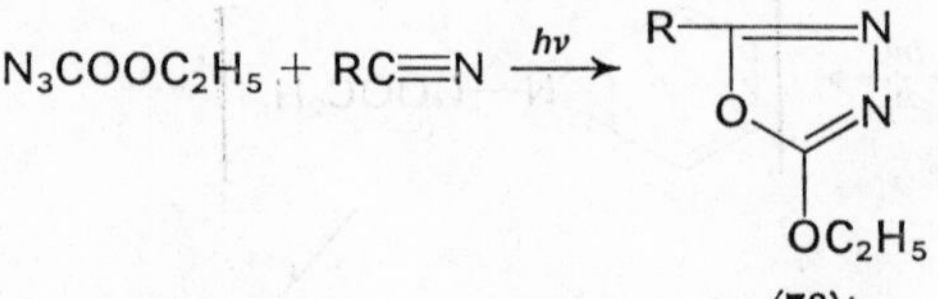

$$N_3COOC_2H_5 + RC{\equiv}N \xrightarrow{h\nu} \text{(78)} \qquad (61)$$

R, N, N, O, OC_2H_5

(**78**)

2-cyanoaziridine (**79**) is the main product. It is not clear, whether

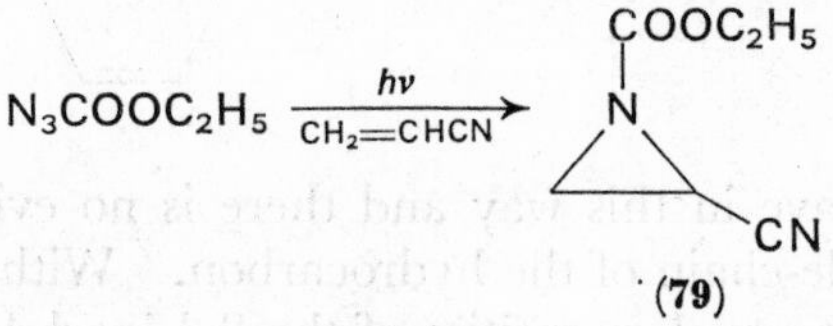

$$N_3COOC_2H_5 \xrightarrow[CH_2{=}CHCN]{h\nu} \text{(79)} \qquad (62)$$

$COOC_2H_5$, N, CN

(**79**)

the oxadiazole ring is formed in these reactions through true cycloaddition or via a nitrilimine[53]. For other photochemical syntheses of oxadiazoles see Puttner and Hafner[54] and Huisgen[128].

Photolysis of ethyl azidoformate (**66**) in alcohols leads to urethanes as the main products[129]. Thus *N-t*-butoxyurethane (**80**) is produced

$$\underset{\textbf{(66)}}{N_3COOC_2H_5} + (CH_3)_3COH \longrightarrow \underset{\textbf{(80)}}{C_2H_5OOCNHOC(CH_3)_3} \qquad (63)$$

in *t*-butyl alcohol. Photolysis of *t*-butylazidoformate (**81**) in *t*-butyl alcohol does not only produce the expected carbamate (**82**), but also the oxazolidone (**83**) by intramolecular insertion of the nitrene:

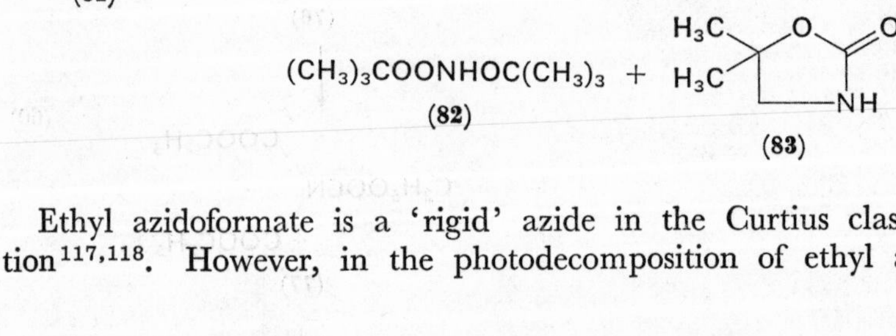

$$\underset{\textbf{(81)}}{(CH_3)_3COOCN_3} + (CH_3)_3COH \xrightarrow{h\nu} \qquad (64)$$

$$\underset{\textbf{(82)}}{(CH_3)_3COONHOC(CH_3)_3} + \text{(83)}$$

H_3C, H_3C, O, O, NH

(**83**)

Ethyl azidoformate is a 'rigid' azide in the Curtius classification[117,118]. However, in the photodecomposition of ethyl azidoformate in methanol at low temperature (−10°C) the rearrangement product methyl *N*-ethoxyaminoformate (**84**) is obtained in 13% yield, together with the insertion product ethyl *N*-methoxyaminoformate (**85**)[50,130]. A similar behaviour is observed in the photolysis of methyl

$$\underset{(\mathbf{66})}{N_3COOC_2H_5} + CH_3OH \longrightarrow \underset{(\mathbf{85})}{C_2H_5OOCNHOCH_3} + \underset{(\mathbf{84})}{C_2H_5ONHCOOCH_3} \quad (65)$$

azidoformate. Other rigid azides which undergo the Curtius rearrangement on irradiation are phenylcarbamoyl azide (**86**) which yielded *N'*-methoxycarbonyl-*N*-phenylhydrazine (**87**), and ethyl-

$$\underset{(\mathbf{86})}{C_6H_5\text{—}NHCON_3} \xrightarrow[CH_3OH]{h\nu} \underset{(\mathbf{87})}{C_6H_5\text{—}NHNHCOOCH_3} \quad (66)$$

carbamoyl azide (**88**), which gave *N*-ethyl-*N'*-methoxycarbonylhydrazine (**89**).

$$\underset{(\mathbf{88})}{C_2H_5NHCON_3} \xrightarrow[CH_3OH]{h\nu} \underset{(\mathbf{89})}{C_2H_5NHNHCOOCH_3} \quad (67)$$

Most of our present knowledge of the effect of the spin state of the nitrene on the final products of azide photolysis is derived from Lwowski's elegant experiments on azidoformates[122,127]. He was able to show conclusively that the addition of singlet nitrenes to suitable double bonds proceeds with retention of molecular configuration whereas the analogous reaction of the triplet nitrene is non-stereospecific. Thus, photolysis of ethyl azidoformate in neat *cis*-4-methyl-pent-2-ene at −30°C produces 87% *cis*-*N*-ethoxycarbonyl-2-iso-propyl-3-methylaziridine. Addition to the *trans*-pentene results in 92% of the corresponding *trans* isomer. On dilution of the olefin with an inert solvent (dichloromethane) an increasing proportion of non-stereospecific addition is observed as more singlet nitrenes have time to revert to the triplet ground state. From the detailed results Lwowski concluded that in photolysis about 30% of the nitrenes are directly generated in the triplet state, 70% in the singlet state. (See also Skell[131].)

The different behaviour of singlet and triplet nitrenes is also well demonstrated in the system ethyl azidoformate-cyclohexene[65]. Direct photolysis (which produces predominantly singlet nitrenes) leads to about 50% addition, 12% insertion and to only a trace of the hydrogen abstraction product urethane, if oxygen is excluded from the solution.

$$N_3COOC_2H_5 + \text{cyclohexene} \rightarrow \text{(7-azabicyclo[4.1.0]heptane)}NCOOC_2H_5 \ (50\%) + \text{cyclohexenyl}{-}NHCOOC_2H_5 \ (12\%) + H_2NCOOC_2H_5 \ (\text{trace}) \quad (68)$$

However, on sensitization with acetophenone, when only triplet nitrenes are produced, the main photoproduct is urethane (74%) and the corresponding bi(cyclohex-2-enyl) (63%). Insertion is evidently

$$N_3COOC_2H_5 + \text{cyclohexene} \xrightarrow[PhCOMe]{h\nu} H_2NCOOC_2H_5 + \text{bi(cyclohex-2-enyl)} \quad (68a)$$

a reaction of the singlet niti ene whereas the triplet will predominantly abstract hydrogen. This conclusion is further confirmed by the photolytic behaviour of optically active azidoformates[133]. Thus, the optically active azidoformate (**90**) produced on irradiation the optically active 2-oxazolidinone (**91**) by intramolecular insertion with retention of configuration. Similarly (−)-5-methyl-6-phenylhexanoyl azide[134] (**92**) was transformed on photolysis into (−)-6-benzyl-6-methyl-2-piperidone (**93**), which is evidence for a single step insertion reaction.

$$R{-}\overset{*}{C}H(CH_3){-}CH_2OOCN_3 \ (\mathbf{90}) \xrightarrow{h\nu} [R{-}\overset{*}{C}H(CH_3){-}CH_2OOCN] \rightarrow \text{4-R-4-methyl-2-oxazolidinone} \ (\mathbf{91}) \quad (69)$$

$(CH_2)_3CON_3$ $H_3C—C—H$ CH_2 (92) $\xrightarrow{h\nu}$ [$(CH_2)_3CON$ $H_3C—C—H$ CH_2] ⟶ H_2C H_3C HN O (93) (70)

Another reaction of singlet nitrenes is the formation of azepines in aromatic solvents[93]. The spin state of the nitrene in these reactions has again been demonstrated by dilution experiments. Photolysis of ethyl azidoformate in benzene–cyclohexane mixtures resulted in azepines (from the benzene) and cyclohexylurethane (from the cyclohexane). On dilution with an inert solvent (dichloromethane) the yields of azepine and urethane decreased, but their ratio remained constant. In the system benzene–cyclohexene the products of photolysis were azepine and the expected aziridine. On dilution with dichloromethane the yield of azepine declined sharply relative to the yield of aziridine and their ratio approached zero at high dilution. It is concluded that only singlet nitrenes form azepines and insert into C—H bonds, but that both the singlet and the triplet will add to double bonds although with not quite the same efficiency. It appears that the singlet nitrene does not discriminate between structurally different double bonds, while the triplet nitrene prefers double bonds which favour a more stable diradical intermediate. Ethyl azidoformate (**66**) will react photolytically with butadiene[92] to give the 1-ethoxycarbonyl-2-vinylaziridine (**94**) and no 1,4 addition product is formed. In isoprene (**95**)[132] two aziridines (**96** and **97**) are formed in nearly equal proportion. However, on dilution with dichloromethane the ratio of the yields of **96** and **97** increases from 1·17 to 2·13, indicating that the triplet nitrene favours reaction with the double bond bearing the methyl group. Structural changes have only a small effect on the reaction between nitrenes and double bonds. Photolysis of ethyl azidoformate in a mixture of cyclohexene and isoprene in 2:1 proportion results in a mixture of products containing about 1·6 times as much of the isoprene adduct as the cyclohexene adduct, no matter how much the system is diluted with an inert solvent: isoprene is therefore only about twice as reactive towards carbethoxynitrene as cyclohexene.

$$N_3COOC_2H_5 + CH_2{=}CH{-}CH{=}CH_2 \longrightarrow CH_2{=}CH{-}HC\underset{N-COOC_2H_5}{\diagdown\!\diagup}CH_2 \quad \textbf{(94)} \qquad (71)$$

$$\underset{\textbf{(66)}}{N_3COOC_2H_5} + \underset{\textbf{(95)}}{CH_2{=}C(CH_3){-}CH{=}CH_2} \longrightarrow \textbf{(96)} + \textbf{(97)} \qquad (72)$$

(96): N-COOC$_2$H$_5$ aziridine bearing CH$_3$ and vinyl groups; (97): N-COOC$_2$H$_5$ aziridine bearing isopropenyl (CH$_3$) group.

2. Acetyl azide

Irradiation of acetyl azide (**98**) in benzonitrile produced a small yield of the 1,3-cycloaddition product, 2-methyl-5-phenyl-1,3,4-oxadiazole (**99**)[47]. Photolysis of **98** in the presence of phenylacetylene resulted in a small yield of the oxazole **101**. While the tetrazole

$$\underset{\textbf{(98)}}{CH_3CON_3} \xrightarrow{h\nu} \left[CH_3{-}\underset{\substack{\|\\O}}{C}{-}N \longleftrightarrow CH_3{-}\underset{\substack{|\\O^-}}{C}{=}N^+\right] \xrightarrow{C_6H_5CN} \textbf{(99)} \qquad (73)$$

$$\underset{\textbf{(102)}}{\text{1-acetyl-5-phenyltetrazole } (CH_3CON{-}N{=}N{-}N{=}C{-}C_6H_5)} \xrightarrow{h\nu} \textbf{(99)}$$

$$CH_3CON_3 + C_6H_5{-}C{\equiv}CH \xrightarrow{h\nu} \text{(101)} \quad (74)$$

(98) (100) (101) (103)

CH_3 N O CH_3CON N N

102 could be photolysed to the oxadiazole **99** the corresponding triazole **103** could not be photochemically converted to the oxazole **101** and there is now little doubt that a nitrene intermediate is involved in both reaction (73) and (74).

3. Pivaloyl azide

The thermal decomposition of pivaloyl azide (**104**) leads quantitatively to the isocyanate **105** (Curtius rearrangement); however the photoreaction produces also a number of addition and insertion products analogous to those obtained in the photolysis of ethyl azidoformate[51]. In cyclohexane solution *N*-cyclohexylpivalamide (**106**) is formed (20%), together with pivalamide (**111**) and a trace of *N,N'*-dipivaloylhydrazine (**112**). In cyclopentane the corresponding *N*-cyclopentylpivalamide, and in neopentane small amounts of *N*-neopentylpivalamide (**107**) have been identified, together with traces of 2-methylprop-1-ene (**107a**). Irradiation of pivaloyl azide in 2-methylbutane results in four insertion products; from their relative amounts the reactivity of the nitrene towards primary, secondary and tertiary C—H bonds can be inferred. It is found that in common with other nitrenes[62,122,87] pivaloyl nitrene prefers insertion into tertiary C—H bonds, to that into secondary and primary bonds. The experimental data fit in well with bond strength calculations.

Photolysis of pivaloyl azide in cyclohexene produces 41% isocyanate, 45% *N*-pivaloylaziridine (**108**), 1·5% of the allylic amide (**109**), a trace of bi(cyclohex-2-enyl) (**110**) and some pivalamide (**111**).

Lwowski's work on pivaloyl azide[85] has some interesting implications for the mechanism of the photo-Curtius rearrangement: photolysis of pivaloyl azide in a variety of solvents results invariably in about 40% isocyanate, irrespective of the medium in which the reaction is conducted. Even in mixed solvents the isocyanate yield is hardly affected

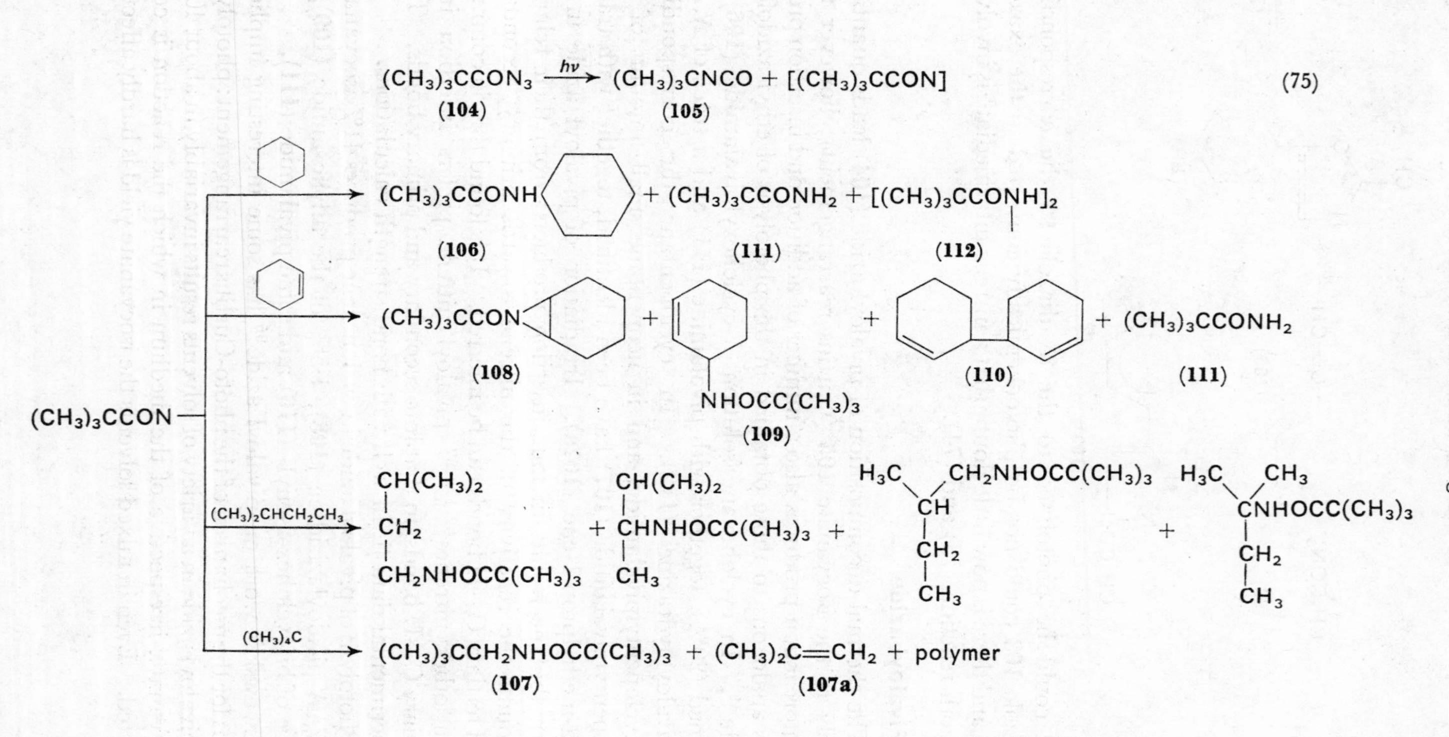

(CH3)3CCON3 —hν→ (CH3)3CNCO + [(CH3)3CCON] (75)
(104) (105)
(CH3)3CCON
(CH3)3CCONH + (CH3)3CCONH2 + [(CH3)3CCONH]2
(106) (111) (112)
(CH3)3CCON + + + (CH3)3CCONH2
(108) (110) (111)
NHOCC(CH3)3
(109)
(CH3)2CHCH2CH3
CH(CH3)2 CH2 CH2NHOCC(CH3)3 + CH(CH3)2 CHNHOCC(CH3)3 CH3 + H3C CH CH2 CH3 CH2NHOCC(CH3)3 + H3C CH3 CNHOCC(CH3)3 CH2 CH3
(CH3)4C
(CH3)3CCH2NHOCC(CH3)3 + (CH3)2C=CH2 + polymer
(107) (107a)

by changes in the solvent ratio. In contrast yield of the 'nitrene products' changes drastically with changes in solvent composition. Thus, in cyclopentane–neopentane mixtures the yield of the insertion product *N*-cyclopentylpivalamide decreases from 13% to 0·8% in going from 100 to 10 mole% cyclopentane. In cyclohexene–cyclopentane mixtures the yield of the double bond adduct [the aziridine (**108**)] remains fairly constant over a range of cyclohexene–cyclopentane ratios, but the yield of *N*-cyclopentyl pivalamide decreases rapidly with the cyclopentane content of the solvent. It is concluded that the isocyanate and the amides (or aziridines) do not have a common precursor, rearrangement and nitrene formation must be competing routes for the disappearance of the photo-excited azide. Nitrogen elimination in the photo-Curtius reaction is therefore probably concerted with the regrouping to isocyanate.

4. Benzoyl azide

Photolysis of benzoyl azide (**113**) in the presence of primary and secondary alcohols[54,75] leads not only to the urethanes (**114**), but also to *N*-alkoxyamide (**115**) and some benzamide (**116**) together with the

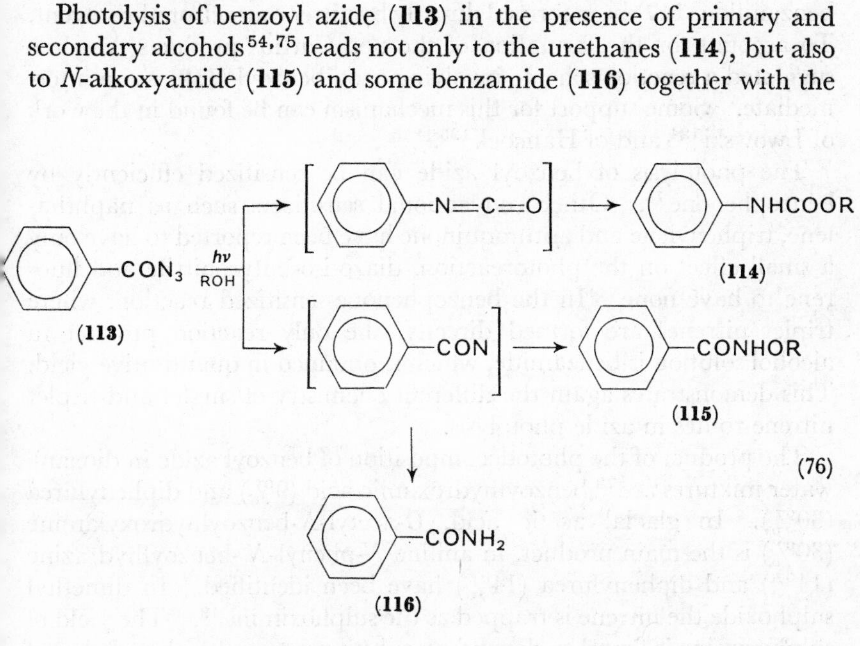

$$C_6H_5CON_3 \xrightarrow[ROH]{h\nu} [C_6H_5N{=}C{=}O] \longrightarrow C_6H_5NHCOOR$$

(**113**) (**114**)

$$C_6H_5CON_3 \xrightarrow[ROH]{h\nu} [C_6H_5CON] \longrightarrow C_6H_5CONHOR$$

(**115**)

$$[C_6H_5CON] \longrightarrow C_6H_5CONH_2 \qquad (76)$$

(**116**)

oxidation products of the alcohols (aldehydes, ketones). The urethanes are clearly formed via the isocyanate (Curtius rearrangement), the other products were thought to arise by a nitrene mechanism[75]. In ethanol, isopropanol and benzyl alcohol 90% of the reaction products were accounted for in this way; in methanol only 70% of the products can be recovered. The yield of the hydroxylamine (**115**) is independent of the nature of the alcohol, but the yields of urethane and of benzamide vary considerably. In alcohols with a low redox potential, the yield of benzamide and of the corresponding oxidation product of the alcohol is in general high. Tertiary alcohols do not lead to benzamide (**116**) but only to urethanes (**114**) and hydroxylamines (**115**).

While these experiments are suggestive of a nitrene mechanism for the non-Curtius products, they do not constitute entirely conclusive evidence. Anselme[126] re-examined some of Horner's earlier work[119] and found that photolysis of benzoyl azide in benzene followed by the addition of aniline to the photolysate produced 30% of benzanilide (**117**) and 56% of *N,N'*-diphenylurea (**118**). The same products were isolated when benzonitrile was the solvent, but no cyanobenzanilide was found, which should have been one of the products had benzanilide (**117**) been formed directly by nitrene attack on the solvent. To account for the formation of the non-Curtius products Anselme suggested a reaction scheme involving azodibenzoyl (**119**) as an intermediate. Some support for this mechanism can be found in the work of Lwowski[124] and of Hancock[125].

The photolysis of benzoyl azide can be sensitized efficiently by benzophenone[75]. Other conventional sensitizers such as naphthalene, triphenylene and anthraquinone have been reported to have only a small effect on the photoreaction, diazo-iso-butyronitrile and fluorene to have none. In the benzophenone-sensitized reaction, where triplet nitrenes are formed directly, the only reaction product in alcohol solution is benzamide, which is obtained in quantitative yield. This demonstrates again the different chemistry of singlet and triplet nitrene routes in azide photolysis.

The product of the photodecomposition of benzoyl azide in dioxan–water mixtures are[73] benzoylhydroxamic acid (9%) and diphenylurea (30%). In glacial acetic acid, *O*-acetyl-*N*-benzoylhydroxylamine (30%) is the main product, in aniline *N*-phenyl-*N'*-benzoylhydrazine (14%) and diphenylurea (14%) have been identified. In dimethyl sulphoxide the nitrene is trapped as the sulphoximine[49]. The yield of sulphoximine is found to depend on substituents on the phenyl ring of benzoyl azide, the unsubstituted benzoyl azide giving 20% imine, *p*-methoxybenzoyl azide 30%, *p*-nitrobenzoylazide 3%, and α-naphthoyl azide less than 1%. The electron-donating or withdrawing power of the substituent clearly plays a role in the reaction.

$$\text{PhCON}_3 \ (\mathbf{113}) \xrightarrow[\text{PhNH}_2]{h\nu} \text{PhCONHPh} \ (\mathbf{117}) + \text{PhNHCONHPh} \ (\mathbf{118}) \qquad (77)$$

(78)

PhCON=NOCOPh (**119**) → [$PhCON_3$ or PhCON] → PhC(=O)–N=N⁺–N⁻–C(=O)Ph with PhC=O → PhC(=O)–N=N⁺ ring with N–C(=O)Ph and ⁻O–C(Ph) → PhC(=O)–N=N–N(COPh)(COPh) → [H_2O] → PhCONHOC–Ph

(**119**) → [$PhNH_2$] → PhCONH–Ph (**117**); PhC(=O)–N=N–N(COPh)₂ → [$PhNH_2$] → (**117**)

D. Aryl Azides

Smith and Brown[88,89] were the first to study the photodecomposition of an aromatic azide and were able to synthesize a number of carbazole (**119**) derivatives by the photolysis of 2-azidobiphenyls (**118**).

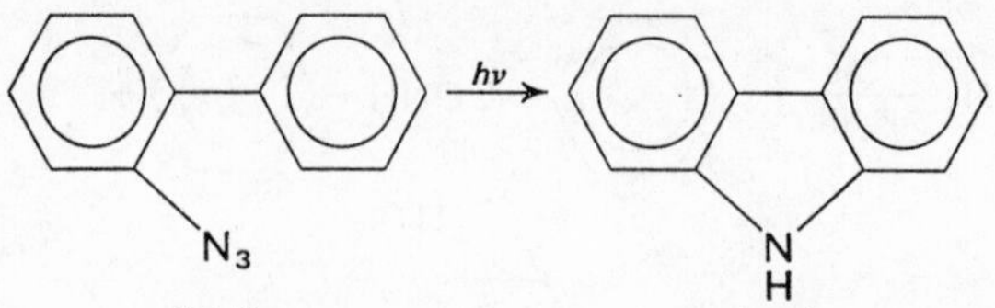

(79)

(118) (119)

Carried out at high dilution in tetralin the reaction led to the satisfactory preparation of carbazole itself (80%) and of several substituted carbazoles (e.g. 2-azido-3,5-dibromobiphenyl produced 1,3-dibromocarbazole in 57% yield). However, if the 2-azidobiphenyl was substituted in a neighbouring position to the azide by a group capable of undergoing addition to a nitrene, the azide reacted preferentially with this group. For example, 2-azido-3-nitrobiphenyl (**120**) cyclizes to 4-phenylbenzfuroxan (**121**) and 2-azidobenzophenone (**122**)

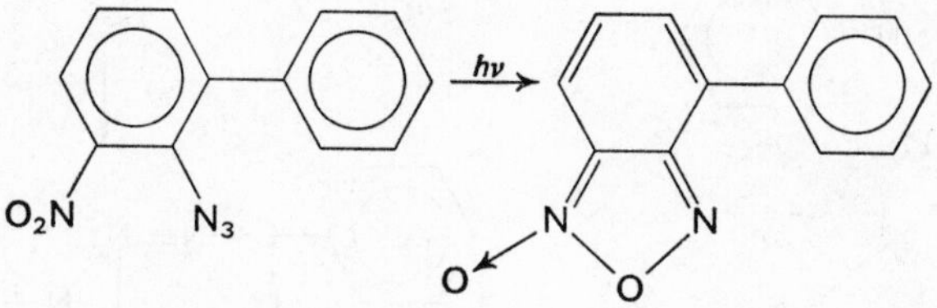

(80)

(120) (121)

undergoes ring closure to the 3-phenylbenzisooxazole (**123**)

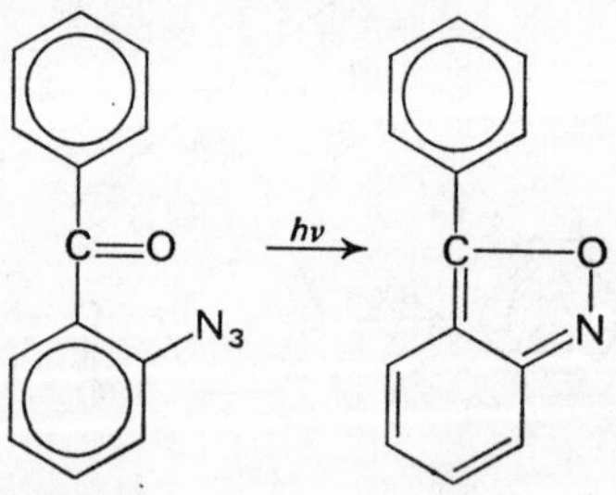

(81)

(122) (123)

Horner has undertaken a systematic study of aryl azide photolysis[97]. He found that substituted phenyl azides produce on irradiation in solution high yields of the corresponding azo compounds. *p*-Methoxyphenyl azide (**124**) gave 18% of the azo compound (**125**)

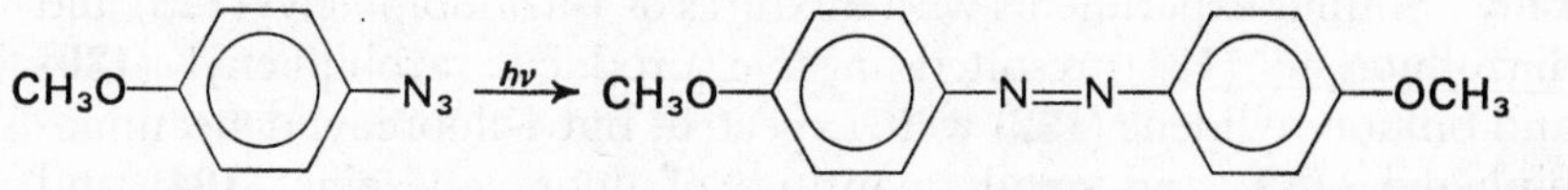

$$CH_3O-C_6H_4-N_3 \xrightarrow{h\nu} CH_3O-C_6H_4-N{=}N-C_6H_4-OCH_3 \qquad (82)$$

(124) (125)

on photolysis in benzene, 82% in tetrahydrofuran and in acetonitrile, and 91% in dimethylsulphide. 4-Azidobiphenyl gave 81% of the azo derivative in benzene, 72% in ethylacetoacetate. No azo derivative was isolated in the photolysis of phenyl azide itself or of *p*-chlorophenyl azide. Small yields of azo compound were later obtained in the photolysis of phenyl azide by others[95].

Irradiation of phenyl azide in acetic acid produced small yields of *p*-acetoxyacetanilide (**126**) and *o*-hydroxyacetanilide (**127**)[95,97].

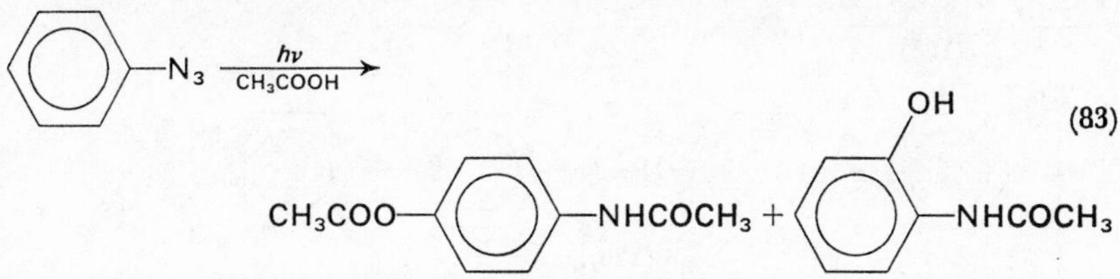

$$C_6H_5N_3 \xrightarrow[CH_3COOH]{h\nu} CH_3COO-C_6H_4-NHCOCH_3 + o\text{-}HO-C_6H_4-NHCOCH_3 \qquad (83)$$

(126) (127)

The photolysis of mixtures of aromatic azides in benzene produced mixed azo derivatives and this was taken as evidence for a nitrene mechanism in the formation of azo compounds[97]. 4-Azidobiphenyl (**128**) and *p*-methoxyphenyl azide (**124**) yielded the mixed azo compound **129** as well as the expected symmetrical derivatives **125** and

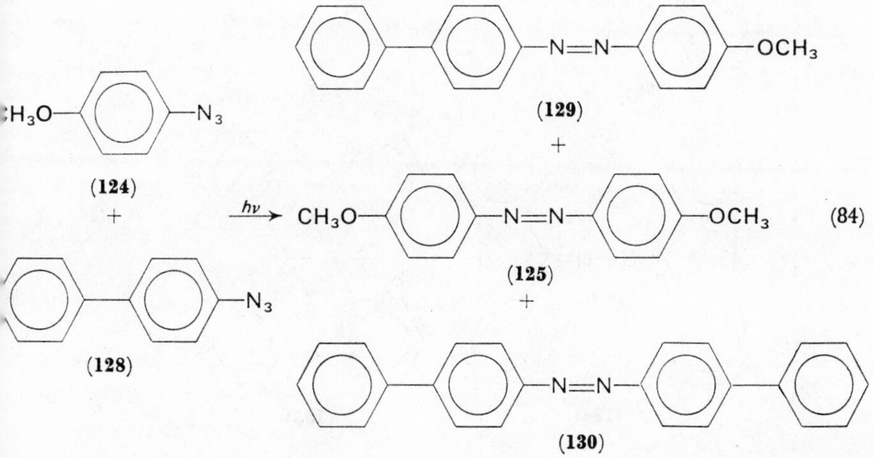

$$CH_3O-C_6H_4-N_3\ (124) + C_6H_5-C_6H_4-N_3\ (128) \xrightarrow{h\nu} C_6H_5-C_6H_4-N{=}N-C_6H_4-OCH_3\ (129) + CH_3O-C_6H_4-N{=}N-C_6H_4-OCH_3\ (125) + C_6H_5-C_6H_4-N{=}N-C_6H_4-C_6H_5\ (130) \qquad (84)$$

130. Similar experiments with mixtures of 4-azidobiphenyl (**128**) and diazofluorene (**131**) resulted in five products: azobiphenyl (**130**) and bifluorenylidene (**132**) were expected, but 4-fluorenylideneamino-biphenyl (**133**) and small quantities of fluorenoneazine (**134**) and bifluorenyl (**135**) were also obtained. The presence of substantial amounts of **133** in the photoproducts seems to emphasize the close similarity in the behaviour of the nitrene of **128** and the carbene of **131** which are assumed to be the intermediates in this reaction.

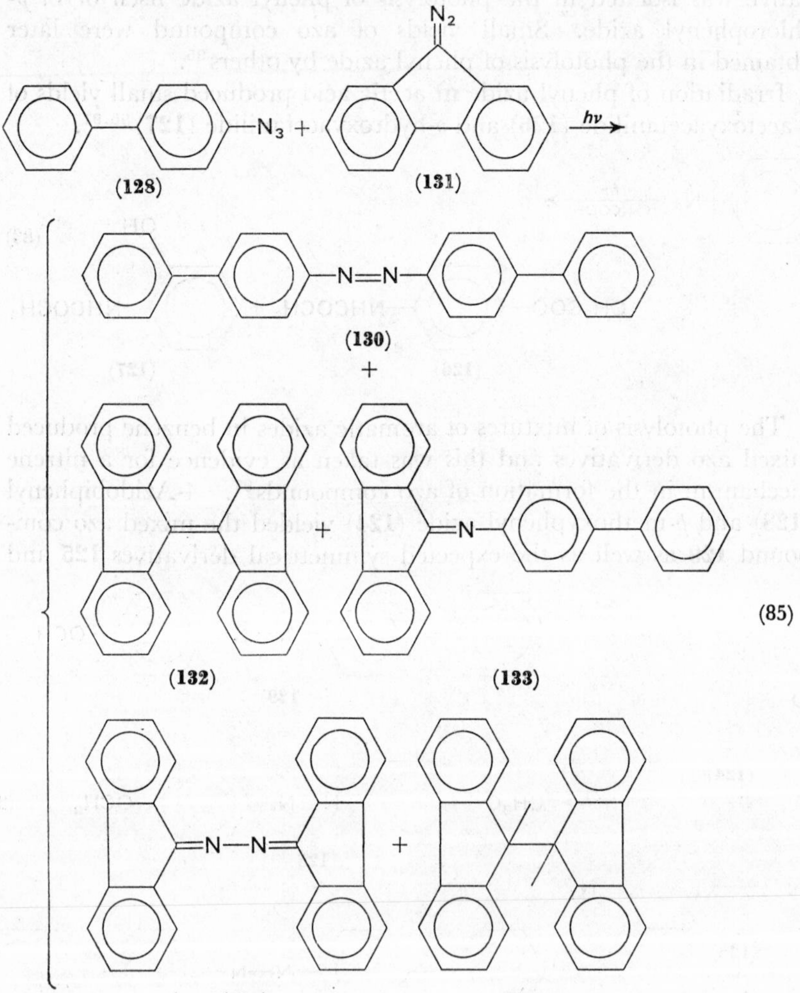

N_3 + N_2 $\xrightarrow{h\nu}$

(**128**) (**131**)

N=N

(**130**)

+

+ =N

(**132**) (**133**)

(85)

=N—N= +

(**134**) (**135**)

A by-product of the study of aryl azide photolysis was the solution of a problem of long standing in heterocyclic chemistry. Wolff[135], in 1910, had pyrolysed phenyl azide in the presence of aniline and obtained a product which he formulated as 'dibenzamil' (**136**).

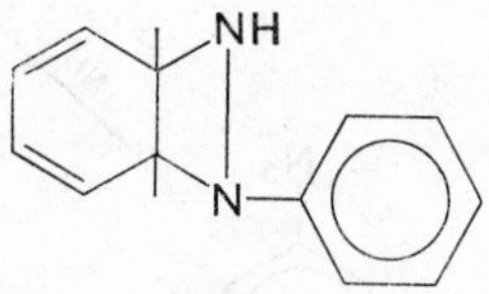

(**136**)

Huisgen and co-workers[15,136,137,138] were able to show that Wolff's dibenzamil was really a 7-membered ring amidine, either the 2-anilino-3H-azepine (**137**) or 2-anilino-7H-azepine (**138**).

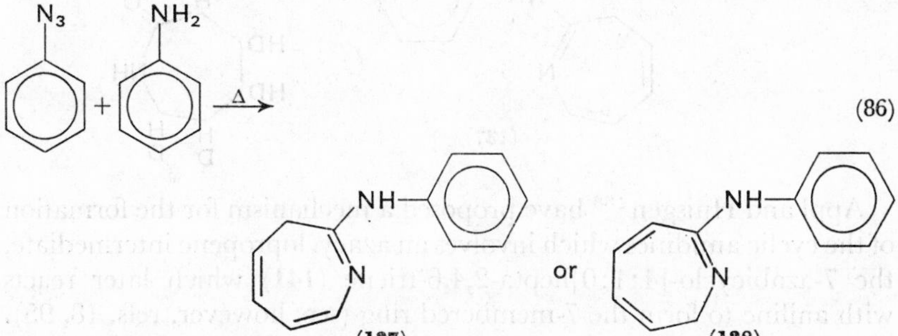

(86)

(**137**) or (**138**)

Doering and Odum[95] have examined the photoreaction of phenyl azide in the presence of bases and other nucleophiles. On irradiation of phenyl azide in either aniline, diethylamine or liquid ammonia, 2-anilino-3H-azepine (**137**), 2-diethylamino-3H-azepine (**139**) and 2-amino-3H-azepine (**140**) were isolated respectively. Treatment of phenyl azide with hydrogen sulphide gave a very small yield of 2-thio-3H-azepine. The structure of the cyclic amidines was established chemically by hydrogenation and subsequent hydrolysis to the ε-aminocaproic acid and the corresponding base. The u.v. spectra were similar to that of acetamidine and also the absence of —NH— absorption in the infrared together with a strong absorption at 1600 cm^{-1}* are in accord with the proposed structure. Most important, the n.m.r. spectrum of the product proved unequivocally the presence of a 3H-azepine.

* *N,N,N'*-trimethyl-benzamidine absorbs at 1621 cm^{-1}.

$N(C_2H_5)_2$ N (139) $NH(C_2H_5)_2$ NH_2 N (140) NH_3 H_2O N_3 O NH S H_2S NH $PhNH_2$ NH N (137) D_2 Pt H_2 O HD HD NH H D H D (87)

Appl and Huisgen[138] have proposed a mechanism for the formation of the cyclic amidines which involves an azacyclopropene intermediate, the 7-azabicyclo-[4:1:0]hepta-2,4,6-triene (**141**) which later reacts with aniline to form the 7-membered ring (see, however, refs. 18, 95).

N_3 → N → N (141) $PhNH_2$ NH N (137) (88)

The photochemistry of 2-azidotropone (**142**) a non-benzenoid aromatic azide, was investigated by Hobson and Malpass[139]. The primary reaction product is a ketene (**143**) which in the absence of protic solvents rearranges to 2-cyanophenol (**144**). In methanol the ester (**145**) is obtained. Photolysis of the tropone in aniline results in the corresponding anilide (**146**); photolysis in ether produces 2-cyanophenol (**144**) and the ester (**148**). It was not possible to decide

(142) —hν→ HC=C=O / CHCN (143) → OH, CN (144)

(143) —ROH or RNH_2→ $ROCCH_2$–CH=CH–CH=CH–CN, R = OCH_3, (**145**); R = NH–C_6H_5 (**146**) → $HOOCCH_2(CH{=}CH)_2COOH$ (**147**)

(143) —$C_2H_5OC_2H_5$→ $OOCCH_2(CH{=}CH)_2CN$ on 2-cyanophenyl (**148**) (89)

whether the reaction proceeded through a nitrene, or by a concerted process where nitrogen elimination is synchronous with the cleavage of the 1,2-bond of the tropone.

(**142**) → [O^- … $N{-}\overset{+}{N}{\equiv}N$] → (**143**) (90)

The photolysis of 2,3-diazidonaphthoquinone (**149**) in benzene produces phthaloylcyanide (**150**) [140].
In aqueous alcohol or in dioxan a red dye is formed which acts as an internal filter and effectively stops the photoreaction. In ether photolysis produced only polymers, possibly via the radical (**151**) [141].

$$\text{(149)} \xrightarrow{h\nu} \text{(150)} \quad (91)$$

(149) (150)

(151)

VI. REFERENCES

1. R. A. Abramovitch and B. A. Davies, *Chem. Revs.*, **64**, 149 (1964).
2. L. Horner and A. Christmann, *Angew. Chem. intern. Edit.*, **2**, 599 (163).
3. F. Thiemann, *Ber.*, **24**, 4162 (1891).
4. T. Curtius and F. Schmidt, *Ber.*, **55**, 1571 (1922).
5. A. Bertho, *Ber.*, **57**, 1138 (1924).
6. A. O. Beckman and R. G. Dickenson, *J. Am. Chem. Soc.*, **50**, 1870 (1928).
7. A. O. Beckman and R. G. Dickenson, *J. Am. Chem. Soc.*, **52**, 124 (1930).
8. K. Gleu, *Ber.*, **61**, 702 (1928).
9. B. A. Thrush, *Proc. Roy. Soc.*, **A235**, 143 (1956).
10. M. E. Jacox and D. E. Milligan, *J. Am. Chem. Soc.*, **85**, 278 (1963).
11. D. E. Milligan and M. E. Jacox, *J. Chem. Phys.*, **41**, 2838 (1964).
12. K. Rosengren and G. C. Pimentel, *J. Chem. Phys.*, **43**, 507 (1965).
13. H. Okabe, *J. Chem. Phys.*, **49**, 2726 (1968).
14. R. Huisgen, *Angew. Chem.*, **67**, 756 (1955).
15. R. Huisgen and M. Appl, *Chem. Ber.*, **91**, 1, 12 (1958).
16. G. Smolinsky, *J. Am. Chem. Soc.*, **82**, 4717 (1960).
17. G. Smolinsky, *J. Am. Chem. Soc.*, **83**, 2489 (1961).
18. G. Smolinsky, *J. Org. Chem.*, **26**, 4108 (1961).
19. J. F. Heacock and M. T. Edmison, *J. Am. Chem. Soc.*, **82**, 3460 (1960).
20. P. A. S. Smith and J. H. Hall, *J. Am. Chem. Soc.*, **84**, 480 (1962).
21. P. Walker and J. A. Waters, *J. Chem. Soc.*, 1632 (1962).
22. G. Smolinsky, E. Wasserman and W. A. Yager, *J. Am. Chem. Soc.*, **84**, 3220 (1962).
23. E. Wasserman, G. Smolinsky and W. A. Yager, *J. Am. Chem. Soc.*, **86**, 3166 (1964).
24. A. Reiser, G. C. Terry and F. W. Willets, *Nature*, **211**, 410 (1966).
25. A. Reiser, H. M. Wagner and G. Bowes, *Tetrahedron Letters*, 2635 (1966).

26. W. D. Closson and H. B. Gray, *J. Am. Chem. Soc.*, **85**, 290 (1963).
27. H. A. Papazian and A. P. Margozzi, *J. Chem. Phys.*, **44**, 843 (1966).
28. I. Burak and A. Treinin, *J. Chem. Phys.*, **39**, 189 (1963).
29. A. D. Walsh, *J. Chem. Soc.*, 2260 (1953).
30. R. S. Mullikan, *Can. J. Chem.*, **36**, 10 (1958).
31. S. D. Peyerimhoff and R. J. Buenker, *J. Chem. Phys.*, **47**, 1953 (1966).
32. Ju. N. Sheinker, *Dokl. Akad. Nauk SSSR*, **77**, 1043 (1951).
33. E. Lieber, C. N. Rao, T. S. Chao and W. H. Wahl, *J. Scient. Ind. Res.*, **16B**, 95 (1957).
34. E. Lieber and A. E. Thomas, *Appl. Spectrosc.*, **15**, 144 (1961).
35. P. Grammaticacis, *Compt. Rend.*, **144**, 1517 (1957).
36. G. A. Reynolds, J. A. Allen and J. F. Tinker, *J. Org. Chem.*, **24**, 1205 (1959).
37. A. Reiser, G. Bowes and R. J. Horne, *Trans. Faraday Soc.*, **62**, 3162 (1966).
38. P. A. S. Smith, J. H. Hall and R. O. Kan, *J. Am. Chem. Soc.*, **84**, 485 (1962).
39. J. N. Murrell, *The Theory of the Electronic Spectra of Organic Molecules*, Methuen, London, 1963.
40. A. Reiser, H. M. Wagner, R. Marley and G. Bowes, *Trans. Faraday Soc.*, **63**, 2403 (1967).
41. P. A. Levene and A. Rothen, *J. Chem. Phys.*, **5**, 985 (1937).
42. I. Burak and A. Treinin, *J. Am. Chem. Soc.*, **87**, 4031 (1965).
43. B. L. Evans, A. D. Yoffe and P. Gray, *Chem. Revs.*, **59**, 515 (1959).
44. K. H. Welge, *J. Chem. Phys.*, **45**, 4373 (1966).
45. C. L. Currie and B. de B. Darwent, *Can. J. Chem.*, **41**, 1552 (1963).
46. F. D. Lewis and W. H. Saunders, *J. Am. Chem. Soc.*, **90**, 7031 (1968).
47. R. Huisgen and J. P. Anselme, *Chem. Ber.*, **98**, 2998 (1965).
48. J. W. ApSimon and O. E. Edwards, *Can. J. Chem.*, **40**, 896 (1962).
49. L. Horner and A. Christmann, *Chem. Ber.*, **96**, 388 (1963).
50. W. Lwowski, R. DeMauriac, T. W. Mattingly and E. Scheiffele, *Tetrahedron Letters*, 3285 (1964).
51. W. Lwowski and G. T. Tisue, *J. Am. Chem. Soc.*, **87**, 4022 (1965).
52. R. Huisgen and H. Blaschke, *Tetrahedron Letters*, 1409 (1964).
53. W. Lwowski, A. Hartenstein, C. de Vita and R. L. Smick, *Tetrahedron Letters*, 2497 (1964).
54. R. Puttner and K. Hafner, *Tetrahedron Letters*, 3119 (1964).
55. W. Lwowski, *Angew. Chem. intern. Edit.*, **6**, 897 (1967).
56. P. S. Skell and R. C. Woodworth, *J. Am. Chem. Soc.*, **78**, 4494 (1956).
57. R. C. Woodworth and P. S. Skell, *J. Am. Chem. Soc.*, **81**, 3383 (1959).
58. P. P. Gaspar and G. S. Hammond, in W. Kirmse, *Carbene Chemistry*, Academic Press, New York, 1964, pp. 259 ff.
59. W. Lwowski and J. S. McConaghy, *J. Am. Chem. Soc.*, **87**, 5490 (1965).
60. J. S. McConaghy and W. Lwowski, *J. Am. Chem. Soc.*, **89**, 2357 (1967).
61. W. Lwowski and T. W. Mattingly Jr., *J. Am. Chem. Soc.*, **87**, 1947 (1965).
62. W. Lwowski, T. J. Maricich and E. W. Mattingly Jr., *J. Am. Chem. Soc.*, **85**, 1200 (1963).
63. W. Lwowski and T. J. Maricich, *J. Am. Chem. Soc.*, **87**, 3630 (1965).
64. J. S. McConaghy and W. Lwowski, *J. Am. Chem. Soc.*, **89**, 4450 (1967).
65. W. Lwowski and F. P. Woerner, *J. Am. Chem. Soc.*, **87**, 5491 (1965).
66. G. Herzberg and D. N. Travis, *Can. J. Phys.*, **42**, 1658 (1964).

67. A. G. Anastassiou, *J. Am. Chem. Soc.*, **89**, 3184 (1967).
68. G. J. Pontrelli and A. G. Anastassiou, *J. Chem. Phys.*, **42**, 3735 (1965).
69. H. W. Kroto, *J. Chem. Phys.*, **44**, 831 (1966).
70. A. Reiser and R. Marley, *Trans. Faraday Soc.*, **64**, 1806 (1968).
71. J. S. Swenton, *Tetrahedron Letters*, 3421 (1969).
72. A. Reiser, F. W. Willets, G. C. Terry, V. Williams and R. Marley, *Trans. Faraday Soc.*, **64**, 3265 (1968).
73. R. M. Moriarty and M. Rahman, *Tetrahedron*, **21**, 2877 (1965).
74. F. D. Lewis and W. H. Saunders Jr., *J. Am. Chem. Soc.*, **89**, 645 (1967).
75. L. Horner, G. Bauer and J. Doerges, *Chem. Ber.*, **89**, 2631 (1965).
76. F. D. Lewis and W. H. Saunders Jr., *J. Am. Chem. Soc.*, **90**, 7033 (1968).
77. G. S. Hammond, J. Saltiel, A. A. Lamola, N. J. Turro, J. S. Bradshaw, D. O. Cowan, V. Vogt and C. Dalton, *J. Am. Chem. Soc.*, **86**, 3197 (1964).
78. A. M. Trozzolo, R. W. Murray, G. Smolinsky, W. A. Yager and E. Wasserman, *J. Am. Chem. Soc.*, **85**, 2526 (1963).
79. E. Wasserman, R. W. Murray, W. A. Yager, A. M. Trozzolo and G. Smolinsky, *J. Am. Chem. Soc.*, **89**, 5076 (1967).
80. A. Reiser and V. Fraser, *Nature*, **208**, 682 (1965).
81. G. Porter and E. Strachan, *Spectrochim. Acta*, **13**, 299 (1958).
82. D. W. Cornell, R. S. Berry and W. Lwowski, *J. Am. Chem. Soc.*, **87**, 3626 (1965).
83. D. H. R. Barton and L. R. Morgan Jr., *J. Chem. Soc.*, 622 (1962).
84. G. T. Tisue, S. Linke and W. Lwowski, *J. Am. Chem. Soc.*, **89**, 6303 (1967).
85. S. Linke, G. T. Tisue and W. Lwowski, *J. Am. Chem. Soc.*, **89**, 6308 (1967).
86. G. Smolinsky and B. I. Feuer, *J. Am. Chem. Soc.*, **86**, 3085 (1964).
87. J. H. Hall, J. W. Hill and Hu-Chu Tsai, *Tetrahedron Letters*, 2211 (1965), and J. H. Hall, J. W. Hill and J. M. Fargher, *J. Am. Chem. Soc.*, **90**, 5513 (1968).
88. P. A. S. Smith and B. B. Brown, *J. Am. Chem. Soc.*, **73**, 2435, 2438 (1951).
89. P. A. S. Smith, J. M. Clegg and J. H. Hall, *J. Org. Chem.*, **32**, 524 (1958).
90. R. Puttner and K. Hafner, *Tetrahedron Letters*, 3119 (1964).
91. K. Hafner, D. Zinser and K. L. Moritz, *Tetrahedron Letters*, 1733 (1964).
92. K. Hafner, W. Kaiser and R. Puttner, *Tetrahedron Letters*, 3953 (1964).
93. W. Lwowski and R. L. Johnson, *Tetrahedron Letters*, 891 (1967).
94. K. Hafner and W. Kaiser, *Tetrahedron Letters*, 2185 (1964).
95. W. von E. Doering and R. A. Odum, *Tetrahedron*, **22**, 81 (1966).
96. L. Horner and A. Christman, *Chem. Ber.*, **96**, 388 (1963).
97. L. Horner, A. Christman and A. Gross, *Chem. Ber.*, **96**, 399 (1963).
98. P. A. S. Smith, B. B. Brown, R. K. Putney and R. F. Reinisch, *J. Am. Chem. Soc.*, **73**, 6335 (1953).
99. D. E. Milligan, *J. Chem. Phys.*, **35**, 1491 (1961).
100. D. E. Milligan and M. E. Jacox, *J. Chem. Phys.*, **39**, 712 (1963).
101. D. H. R. Barton and L. R. Morgan Jr., *Proc. Chem. Soc.*, **206** (1961).
102. D. H. R. Barton, *South African Ind. Chemist.*, **15**, 299 (1961).
103. D. H. R. Barton and A. N. Starratt, *J. Chem. Soc.*, 2444 (1965).
104. R. M. Moriarty and M. Rahman, *J. Am. Chem. Soc.*, **87**, 2519 (1965).
105. R. M. Moriarty, J. M. Kliegman and C. Shovlin, *J. Am. Chem. Soc.*, **89**, 5958 (1967).

106. R. M. Moriarty, J. M. Kliegman and C. Shovlin, *J. Am. Chem. Soc.*, **90**, 5948 (1968).
107. L. Barash, E. Wasserman and W. A. Yager, *J. Am. Chem. Soc.*, **89**, 3932 (1967).
108. R. M. Moriarty and J. M. Kliegman, *J. Am. Chem. Soc.*, **89**, 5959 (1967).
109. W. H. Saunders Jr. and E. A. Caress, *J. Am. Chem. Soc.*, **86**, 861 (1964).
110. G. Smolinsky, *J. Org. Chem.*, **27**, 3557 (1962).
111. A. Hassner and F. W. Fowler, *Tetrahedron Letters*, 1545 (1967).
112. A. Hassner and F. W. Fowler, *J. Am. Chem. Soc.*, **90**, 2869 (1968).
113. F. W. Fowler, A. Hassner and L. A. Levy, *J. Am. Chem. Soc.*, **89**, 2077 (1967).
114. F. P. Woerner, H. Reimlinger and D. R. Arnold, *Angew. Chem. intern. Edit.*, **7**, 130 (1968).
115. S. Sato, *Bull. Chem. Soc. Japan*, **41**, 2524 (1968).
116. A. G. Hortmann and J. E. Martinelli, *Tetrahedron Letters*, 6205 (1968).
117. T. Curtius, *Z. Angew. Chem.*, **27**, 111, 213 (1914).
118. A. Bertho, *J. Prakt. Chem.*, **120**, 89 (1928).
119. L. Horner, E. Spietschka and A. Gross, *Ann.* **573**, 17 (1951).
120. R. Huisgen, *Angew. Chem.*, **72**, 359 (1960).
121. J. W. ApSimon and O. E. Edwards, *Proc. Chem. Soc.*, 461 (1961).
122. W. Lwowski and T. J. Maricich, *J. Am. Chem. Soc.*, **86**, 3164 (1964).
123. K. Hafner and C. Koenig, *Angew. Chem. intern. Edit.*, **2**, 96 (1963).
124. W. Lwowski, T. W. Mattingly Jr. and T. J. Maricich, *Tetrahedron Letters*, 1591 (1964).
125. J. Hancock, *Tetrahedron Letters*, 1585 (1964).
126. J.-P. Anselme, *Chem. Ind.*, 1794 (1966).
127. R. S. Berry, D. Cornell and W. Lwowski, *J. Am. Chem. Soc.*, **85**, 1199 (1963).
128. R. Huisgen and H. Blaschke, *Ann.*, **686**, 145 (1965).
129. R. Kreher and G. H. Brockhorn, *Angew. Chem. intern. Edit.*, **3**, 589 (1964).
130. R. Puttner, W. Kaiser and K. Hafner, *Tetrahedron Letters*, 4315 (1968).
131. R. M. Ether, H. S. Skovronek and P. S. Skell, *J. Am. Chem. Soc.*, **81**, 1008 (1959).
132. A. Mishra, S. N. Rice and W. Lwowski, *J. Org. Chem.*, **33**, 481 (1968).
133. S. Terashima and S. Yamada, *Chem. Pharm. Bull.* (*Japan*), **16**, 1953 (1968).
134. S. Yamada and S. Terashima, *Chem. Com.*, 511 (1969).
135. L. Wolff, *Ann.*, **394**, 59 (1912).
136. R. Huisgen, *Congress Handbook: XIII IUPAC Congress*, Zurich, 1955, p. 109.
137. R. Huisgen, D. Vosius and M. Appl, *Angew. Chem.*, **67**, 756 (1955).
138. M. Appl and R. Huisgen, *Chem. Ber.*, **92**, 2961 (1959).
139. J. D. Hobson and J. R. Malpass, *J. Chem. Soc.*, 1645 (1967).
140. J. A. van Allen, W. J. Priest, A. S. Marshall and G. A. Reynolds, *J. Org. Chem.*, **33**, 1100 (1968).
141. V. F. Raaen, *J. Org. Chem.*, **31**, 3310 (1966).
142. R. N. Dixon, *Can. J. Phys.*, **37**, 1171 (1959).
143. R. M. Moriarty, M. Rahman and G. J. King, *J. Am. Chem. Soc.*, **88**, 842 (1966).
144. A. M. Trozzolo, R. W. Murray, G. Smolinsky, W. A. Yager and E. Wasserman, *J. Am. Chem. Soc.*, **85**, 2526 (1963).

CHAPTER 9

Acyl azides

WALTER LWOWSKI

New Mexico State University, Las Cruces, New Mexico, U.S.A.

I. ACYL AZIDES—TYPES AND INDIVIDUAL REACTIONS . . . 504
A. Alkanoyl, Alkenoyl and Aroyl Azides 504
1. Preparation and properties 504
B. Azides of the Type C—C(=X)—N_3 507
1. Thiocarbonyl azides C—C(=S)—N_3 507
2. Imidoyl azides, RC(=NX)—N_3 510
3. Imidoyl azides containing an open-chain function C=NX 510
a. Azidoximes 512
b. Diaryltetrazoles (imidoyl azides C—C(=NH)—N_3) 513
c. Guanyl azides 513
d. Cyanoguanyl azides 516
e. Azidoformamidinium salts 516
f. Hydrazidic azides 517
g. Azidobenzalazines 517
4. Imidoyl azides in which the C=N function is part of a ring 518
C. Azides of the Type $XCON_3$ 520
1. Carbamoyl azides $RR'NCON_3$ 520
a. Preparation of carbamoyl azides 520
b. Reactions of carbamoyl azides 521
c. Dissociation of carbamoyl azides 521
d. Curtius rearrangement of carbamoyl azides . . . 522
e. Reactions possibly involving a carbamoylnitrene . 524
2. Azidoformates $ROCON_3$ 527
3. Mercaptocarbonyl azides $RSCON_3$ 528
II. REACTIONS COMMON TO ALL CLASSES OF ACYL AZIDES . . 529
A. 1,3-Dipolar Additions of Acyl Azides 529
B. Additions to the Terminal Nitrogen of Acyl Azides . . 531
1. Phosphinimines and phosphazides 532
2. Nucleophilic attack on the terminal nitrogen by ylides . 533
3. Triazene formation 534
C. Nucleophilic Displacement of the Azido Group . . . 534
D. Reactions of Acyl Azides with Acids 535

E. Radical-induced Decomposition of Acyl Azides 535
F. Nitrene Formation from Acyl Azides 536
1. Dissociation of acylnitrenes 538
2. Insertion into C—H bonds 538
3. Insertion into O—H bonds 540
4. Insertion into N—H bonds 540
5. Addition to C=C double bonds 541
6. Reactions with aromatic and heteroaromatic systems . 542
7. Additions to alkynes. 543
8. Additions to nitriles 543
9. Addition to isonitriles 544
10. Addition to sulphoxides 544
III. References 544

This chapter deals with acyl azides, R—C(=X)—N_3, where R may be a residue connected to the C=X function by a carbon, or a nitrogen or oxygen atom, and where X may be oxygen, sulphur or nitrogen. The different classes created by varying R and X are described first, together with the reactions particular to any one class. This is followed by a treatment of those reactions which are common to all or many of these classes.

I. ACYL AZIDES—TYPES AND INDIVIDUAL REACTIONS

A. *Alkanoyl, Alkenoyl and Aroyl Azides*

1. Preparation and properties

The preparation of azides in general is treated in Chapter 2 of this book, so all that is needed here are some remarks specifically pertaining to acyl azides. The most general and most often the easiest method is the nucleophilic displacement of a leaving group, most often chloride ion, from acyl derivatives (see Chapter 2, section III.A)[1–6]. The method is critically discussed in References 3, 4 and 6. Aqueous–organic solvent mixtures are often used, such as water and acetone[7–10], dioxan[7,10,11], acetic acid[5,8,15,16], methanol[10], ethanol[13,14], or dimethylformamide[17]. Tertiary amines or amine-*N*-oxides may be employed as catalysts[12]. 'Dry' methods[6,18] use suspensions of sodium azide in a variety of solvents, usually those that boil high enough to permit thermolysis of the acyl azide *in situ*, to make isocyanates by the Curtius rearrangement. Sodium azide, the most commonly employed reagent, is but little soluble in organic solvents. At room temperature, 43·6 g of NaN_3 are dissolved by 100 g of water, 17·95 g by 100 g of 40% aqueous ethanol, and 0·81 g by 100 g of 95·5% ethanol[19]. For the 'dry', heterogeneous process, the sodium azide should be activated, such as by trituration with hydrazine and precipitation by acetone[6,20,21]. Lithium azide[22] can be used in dimethylformamide[21]. Ammonium azide has also been employed[23] and, when made from sodium azide and ammonium chloride *in situ*[24], it seems to offer advantages. A number of ionic azides, soluble in less polar solvents, are available: dimethyldioctadecylammonium azide[25] (made from the bromide by ion exchange), teramethyl-, tetraethyl-, tetra-*n*-propyl- and tetra-*n*-butylammonium azides[26] (made by neutralizing the ammonium hydroxides with HN_3 and extraction into acetone or acetonitrile). Some of the latter four azides are quite stable thermally, their decomposition temperatures are: $Me_4N^+N_3^-$: 255°; $Et_4N^+N_3^-$: 250°; $Pr_4N^+N_3^-$: 216°; $Bu_4N^+N_3^-$: 80°[26]. Dicyclohexylammonium azide[27] is somewhat soluble in tetrahydrofuran. Tetramethylguanidinium azide[28] has been used, in boiling chloroform, to make alkyl azides.

Where acid chlorides are difficult to obtain, Weinstock's method[29] of using mixed carboxylic–carbonic acid anhydrides can be helpful. The acid to be converted into its azide is treated with ethyl chloroformate and base, the ethyl carbonate moiety is then displaced by azide ion. The method has been used in the penicillin field[30]. It is not general, however: when the alkanoyl part of the mixed anhydride is sterically hindered, the azide ion displaces it instead of the carbonate moiety, and ethyl azidoformate is produced[31]. Cyclic anhydrides can be opened by azide ion, to give the salts of omega-azidocarbonyl acids, such as $Na^+\ {}^-OOCCH_2CH_2CON_3$[32].

The second most popular method for making acyl azides is treating hydrazides with nitrites (see Chapter 2, section V.D.). The discovery of hydrazoic azid stems from Curtius'[33] treating benzoyl hydrazide with sodium nitrite in acetic acid, and then hydrolysing the azide. This 'diazotization' method is widely used, but is not as general as the acyl halide–azide ion procedure[11,31,34]. Usually, a cold solution (aqueous or aqueous–organic) of the hydrazide is treated with sodium nitrite[6,35,36], and the azide is often extracted immediately into a layer of ether. Alcohols may be used when anhydrous conditions are desired, or when the hydrazide is insoluble in water[37–42]. Besides sodium nitrite amyl nitrite[39–42], nitrogen trioxide[43] and nitrosyl chloride[44] have been used. Aqueous or anhydrous acid may be the solvent[45–47]. Where acid-sensitive molecules are involved, mineral acid may be added to a solution of hydrazide and nitrite[48,49], but the higher pH thus maintained favours displacement of azide ion from $RCON_3$ by hydrazide anion, to give the undesired RCONHNHCOR[6]. Diazonium salts have also been used to convert hydrazides to azides[50,51].

Azides of the type $RCON_3$ are colourless liquids or solids. *Alkanoyl and alkenoyl azides* are often prepared, but rarely isolated because they undergo Curtius rearrangement already at or below room temperature. Aroyl azides are much more stable thermally.

In carbonyl azides, the azido group gives rise to i.r. absorption bands between 2137 and 2155 cm^{-1} (asym. stretch), 1210–1261 cm^{-1} (sym. stretch) and the carbonyl band is found from 1684–1730 cm^{-1} [52]. The azide band is often split, presumably due to Fermi resonance[52], in which case there are also strong bands at 983–1025 and 857–920 cm^{-1} [52]. Sheinker has measured the positions and intensities of a variety of azides[53,54].

The u.v. spectra of carbonyl azides show a band around 210–220 nm extending far towards long wavelength, so that photolysis is sometimes possible with light near 300 nm. The long wavelength band around 285 nm, found in alkyl azides, is not readily visible in carbonyl azides[55,56]. The electronic spectra are discussed in Chapter 8, section II.

Acetyl azide has been prepared only in solution[57]. *Propionyl azide*[58] is a colourless liquid of pungent odour, the inhalation of which causes severe headaches. *Higher fatty acid azides* have been obtained by distillation at 0° and 0·5 mm Hg and their infrared spectra measured neat (Table 1). n-*Hexanoyl azide*[63] shows i.r. bands at 2130 and 1720 cm^{-1} in cyclohexane solution. A number of more complex alkanoyl azides have also been prepared in solution[64,65,66] while the azide of tetra-acetyl-D-arabonic acid was isolated crystalline (m.p. 105–6° dec.)[67]. An example of a more exotic acyl azide is 2,3-diphenyl-cycloprop-2-ene-1-carbonyl azide[68]. The diazide of oxalic acid was isolated crystalline (from CCl_4) but exploded when touched with a spatula[69]. Succinamoyl azide[70], a crystalline substance with a decomposition point of 70–80°, undergoes slow Curtius rearrangement already at room temperature, just as the azides of monocarboxylic acids do.

Carbonylazide, N_3CON_3[71], detonated on touching when crystalline[72,73], but may be handled in small quantities in solution[74].

Diazoacetyl azide $N_2{=}CHCON_3$[75,84] m.p. 7–8° can be distilled in vacuo, but is very explosive.

Aromatic and heteroaromatic carbonyl azides are often low-melting solids which decompose, with Curtius rearrangement, at or somewhat above

TABLE 1. Infrared spectra of fatty acid azides

Azide	References	ν_{N_3} (cm^{-1})	$\nu_{C=O}$ (cm^{-1})
n-$C_3H_7CON_3$	60	2136	1717
n-$C_4H_9CON_3$	31	2138	1718
n-$C_5H_{11}CON_3$	31	2137	1718
n-$C_7H_{15}CON_3$	31	2136	1720
t-$C_4H_9CON_3$	59, 61	2135	1709
$Me_2CHCH_2CON_3$	62	2140	1720
$Me_3CCH_2CON_3$	31	2135	1715
$MeCH_2CMe_2CON_3$	31	2142	1716
$Me_3CCMe_2CON_3$	31	2130	1700
Cyclohexylcarbonyl azide	52	2146	1724

their melting points. Benzazide[33] melts at 32°, α-thiophenylcarbonyl azide at 37·2°[76], the nitrobenzazides[7] melt with decomposition at 39° (*ortho*), 68° (*meta*) and 71° (*para*). 2,4-Dinitrobenzazide decomposes at 68°, while *p*-methoxybenzazide melts at 69° and begins to decompose at 80°[7], and *meta*-methoxybenzazide melts at 22·5° and begins to decompose at 61°[7]. Many other such azides have been prepared, for example by Otsuji[77].

B. *Azides of the Type* C—C(=X)—N₃

1. Thiocarbonyl azides C—C(=S)—N₃

Treatment of thionhydrazides, RC(=S)—NH—NH_2 with nitrous acid should lead to thiocarbonyl azides, RC(=S)—N_3, and it seems likely that these azides are indeed formed as short-lived intermediates. The isolated products are 1,2,3,4-thiatriazoles. When thiosemicarbazides are nitrosated, 5-amino-1,2,3,4-thiatriazoles or/and 5-mercaptotetrazoles are obtained. Freund[78–81] formulated his products as such in 1895, while Oliveri-Mandala[82,83] considered them thiocarbamoyl azides on the basis of some chemical evidence. Lieber resolved this controversy by recording the infrared spectra of the compounds and demonstrating the absence of an azide band[84–86]. He did the same for the 5-alkylmercapto-1,2,3,4-thiatriazoles[87], which had earlier been reported as azido-dithiocarbonates[88], and for the products from azide ion and thiophosgene[89] and carbon disulphide[90], all of which possess thiatriazole structures. Lieber[85] also demonstrated the rearrangement of 5-anilino-1,2,3,4-thiatriazole to 1-phenyl-5-mercaptotetrazole, and Sheppard[91] described the conversion of 1-aryl-5-mercaptotetrazoles to 5-arylamino-1,2,3,4-thiatriazoles by heat, and the reversal of this rearrangement by treatment of the thiatriazole by base. These interconversions can be understood as

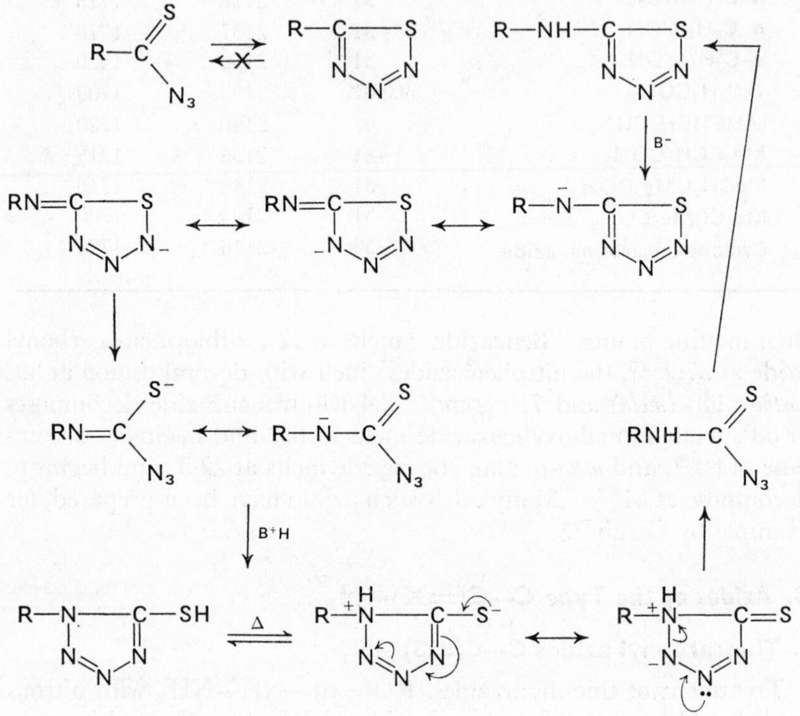

involving the thiocarbamoyl azide as a transient intermediate, with the anion cyclizing to the mercaptotetrazole anion, because of the greater acidity of the SH group compared to the RNH— group, while the neutral species cyclizes to the thiatriazole.

When Sheppard's thiatriazoles were heated in boiling benzene for 12 hours, nitrogen and sulphur were lost and arylcyanamides formed. In this manner, *o*-nitrophenylcyanamide can be made in high yield[91].

Smith and Kenny[92] attempted the preparation of thioacyl azides C—CS—N_3 (as opposed to the type C—NH—CS—N_3 dealt with above). They obtained only thiatriazoles again, as shown by the absence of an azide band in the i.r. spectra of their products. Thermolysis of their thiatriazoles gave no indication of an intermediate formation of thiocarbonylnitrenes, C—CS—N. Instead, sulphur, nitrogen and nitriles are formed. Jensen[93] attempted the synthesis of thioacyl azides by an interesting method: displacement of $^{-}SCH_2COO^{-}$ from $RCSSCH_2COO^{-}$ by azide ion. He obtained the corresponding thiatriazole.

Aryloxy- and alkyloxy-1,2,3,4-thiatriazoles have been prepared from azide ion and aryl chlorothionoformates[94–97] or alkyl chlorothionoformates[98,99], respectively. Alkyloxythiatriazoles are also made from alkyl thionocarbazates and nitrous acid[100–102]. The subject has been reviewed by Martin[103]. The 5-aryloxy- or 5-alkyloxy-1,2,3,4-thiatriazoles decompose thermolytically in a manner analogous to the decomposition of the other triazoles mentioned above; they give nitrogen, sulphur and cyanates. This is the most used route to alkyl and aryl cyanates, R—OCN. The mechanism of this decomposition could be concerted, or it could involve a thioacyl azide:

$$\text{R—X–(thiatriazole)} \longrightarrow \text{R—X—C}\equiv\text{N} + \text{S} + \text{N}_2 \quad \text{or}$$

$$\text{R—X–(thiatriazole)} \longrightarrow \text{R—X—C(=S)—}\ddot{\text{N}}\text{—N}\equiv\text{N:} \longleftrightarrow \longleftrightarrow$$

$$\downarrow -\text{N}_2$$

$$\text{R—X—C}\equiv\text{N} + \text{S} \longleftarrow \left[\text{R—X—C(=S)—}\ddot{\text{N}} \text{ or } \text{R—X—C}\!<\!\!\begin{smallmatrix}\text{N}\\ \text{S}\end{smallmatrix} \text{ or } \text{R—X—C}\!=\!\overset{+}{\text{N}}\text{(S}^{-}\text{)} \right]$$

The only data on hand to distinguish between the two mechanisms are the kinetic studies of Jensen[104] on a variety of 5-alkoxy- and on 5-phenoxy-1,2,3,4-thiatriazoles. The energies of activation and the activation entropies are nearly the same for all compounds studied: $\Delta E^{\ddagger} = 24{\cdot}6$ kcal mole^{-1} and $\Delta S^{\ddagger} = 4{\cdot}4$ e.u. Jensen interprets this as indicating a stepwise mechanism, in which thioacyl azide is formed in the rate-determining step, which is the breaking of the S—N bond in the thiatriazole ring. The thioacyl azide does not, however, build up to a concentration where its i.r. absorption can be observed in decomposing thiatriazoles[104].

Thioacyl azides or thiatriazoles or both might well be intermediates in reactions of fluorinated thiocarboxylic acids with sodium azide[105]. The photolysis of thiatriazoles leads to some rearrangement, forming isothiocyanates[106], perhaps by way of the thioacyl azide.

2. Imidoyl azides, RC(=NX)—N_3

Imidoyl azides, RC(=NX)—N_3, display ring-chain tautomerism, as do the thioacyl azides, RCS—N_3, discussed in the preceding section. Again, the cyclic valence tautomer is most often the more stable form, but in contrast to thioacyl azides, many imidoyl azides are known and are indefinitely stable at room temperature.

$$\text{R—C}\!\underset{\text{N=N—N}}{\overset{}{=}}\!\text{N—X (tetrazole ring)} \rightleftharpoons \text{R—C(=N—X)—N}_3$$

Three groups of these azides are prominent: imidoyl azides in which the function C=NX is not part of a ring, those in which it is, and a special class of the former group, hydrazidic azides, containing an open-chain function C=NNRR′.

3. Imidoyl azides containing an open-chain function C=NX

Eloy[107] has pointed out that many compounds described in the older literature as tetrazoles are really imidoyl azides, and vice versa. A few such instances had been clearly recognized before[108], but in general the structures were sorted out only after infrared spectroscopy became available, and the presence or absence of an azide band was used as the decisive criterion.

Where the tetrazoles or imidoyl azides are prepared by one of the usual methods for introducing an azide group (see Chapter 2 and above), it is reasonable to assume that the imidoyl azide is an intermediate in the tetrazole formation[24]. Indeed, there are numerous cases (such as those described by Stolle[108,109] (**1**) or by Woodward and Olofson[110] (**2**) in which the azide can be isolated and isomerized to the tetrazole by heating, indicating that the tetrazole is more stable than the azide (see also below, under 5-aminotetrazoles). In other cases, such as **3**[108], the azide shows no tendency to isomerize to the tetrazole.

N—N PhC CPh N_3 H **(1)** NCH$_3$ PhCOCH$_2$C N_3 **(2)** N—N PhC CPh N_3 N_3 **(3)**

Behringer and Fischer[111] obtained a compound (**4**) analogous to **3**, in which one of the azido groups is replaced by chlorine. **4** shows no azide band in its i.r. spectrum and is the corresponding tetrazole. From it the authors[111] could obtain, via the hydrazide and nitrosation, the bis-tetrazole (**5**) (the cyclic isomer of Stolle's **3**). It is a stable compound, m.p. 120–121° (dec.), as is **3**, m.p. 136–8° (dec.). Attempts to equilibrate **3** and **5** failed[111]: the bis-azide (**3**) was not changed by heating to 70°, while **5** was converted to the transient azidotetrazole (**6**), which then rearranged to the carbodiimide (**7**) which was trapped by ethanol to give **8**[111]. The bis-azide (**3**) undergoes an analogous conversion, rearrangement to the bis-carbodiimide followed by addition of ROH, when heated in methanol to 100° for 16 hours[108].

Ph C=N—N—CPh Cl N N N **(4)** PhC—N—N—CPh N N N N N N **(5)** $\xrightarrow{\Delta}$

PhC=N—N—CPh N_3 N N N **(6)** $\xrightarrow{-N_2}$ PhN=C=N—N—CPh N N N **(7)** $\xrightarrow{\text{EtOH}}$

PhN=C—NH—N—CPh EtO N N N **(8)**

The isomers **3** and **5** are clearly separated by a high activation barrier. The ease of cyclization of other imidoyl azides, and the calculations by Roberts[112], make it unlikely that this is due to the need of bending the azide group in the cyclization step. Perhaps it is the resonance interaction between the phenyl groups with the azide groups (in **3**) and with the tetrazole rings (in **5**) that is disrupted during the conversions from **3** to **5** and from **5** to **3**, causing a high activation barrier.

On the whole, it seems that the tetrazole resonance makes the cyclic structure energetically preferred, except for tetrazoles bearing a strongly electron-withdrawing group in the 1-position (see below). Further investigations will, no doubt, shed more light on the relative energies of the various valence tautomers possible, aiding our understanding of the tetrazole–imidoyl azide equilibrium. For example, Scott[113] found tetrazolyl-nitrosamines that may owe their stability to a

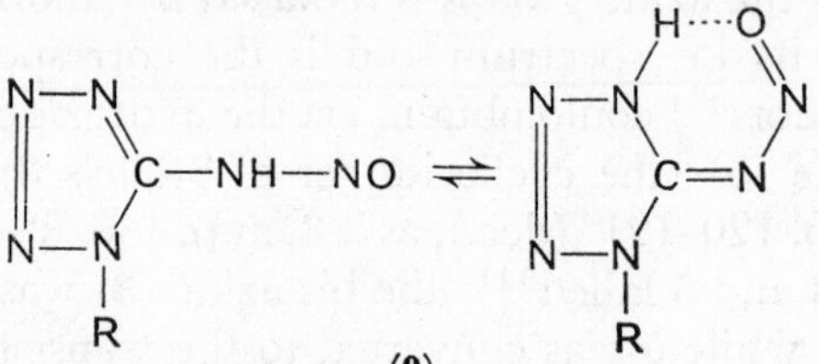

N—N, C—NH—NO, N—N, R ⇌ H····O, N—N, N, C=N, N—N, R

(9)

tautomerism which substantially alters the tetrazole resonance. A tautomer similar to **9** was also encountered by Scott[114] in the synthesis of [1,2,4]-triazolo-[4,3-*d*]-tetrazoles.

a. Azidoximes. RC(=NOH)N_3 have been isolated by several authors and vary widely in stability. Compounds obtained by Forster[115] (**10**) and Wieland[116] and thought to be tetrazoles have been shown by Eloy[107] to be oximino azides, as has **11**. Eloy[107] prepared a number

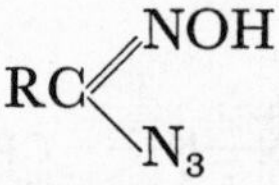

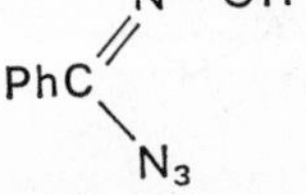

PhC(=N—OH)N_3

(10)

HO—N=C(N_3)—C(N_3)=N—OH

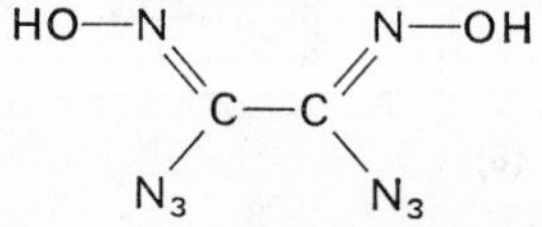

(11)

of oximino azides of the type **12** and found those with R=OH, OCH_3 or OCOPh to exist as azides of surprising stability. While they could be made to detonate, they could be recrystallized from

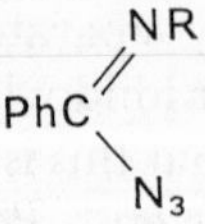

PhC(=NR)N_3

(12)

boiling ethanol and some could be heated to 200° before exploding. Chang and Matuszko[117] prepared seven pyridyl azidoximes and two terephthaldiazidoximes, all solids melting between 114° and 174°, none of them cyclizing to the corresponding tetrazoles. Several of the azides, however, decomposed on standing. Grundmann[118] prepared mesityl azidoxime by adding azide ion to mesitonitrile-*N*-oxide. The very unstable azide decomposed at 80° to give nitrogen, mesitonitrile and hyponitrous acid.

b. Diaryltetrazoles. These exist in the cyclic form, rather than as imidoyl azides. However, Smith[119,120] has demonstrated the existence of a ring-chain tautomerism in their thermolysis. Huisgen[121] demonstrated the same equilibrium by intercepting the azido isomer of 1,5-diphenyltetrazole with copper, which catalyses the azide decomposition (to form a copper–nitrene complex).

$$\text{1,5-diphenyltetrazole (PhC—NPh, N=N—N ring)} \longleftarrow \text{Ph—C(=NPh)N}_3 \xrightarrow{\text{Cu}} \text{PhC(=NPh)N(Cu)} \xrightarrow{\text{PhC}\equiv\text{N}} \text{PhC—NPh, N—C(Ph)=N ring}$$

c. Guanyl azides. Perhaps the most studied class of tetrazole–imidoyl azide isomerism is that involving *5-aminotetrazoles* and *guanyl azides* (*azidoformamidines*). Huisgen summarized the knowledge of this system in 1960[122]. The free guanyl azide cyclizes spontaneously to 5-aminotetrazole[123,124] but the salts $[(NH_2)_2C—N_3]^+X^-$ have the guanyl azide structure. Lieber[125] measured the activation energy for

$$\text{H}_2\text{N—C(=NH)N}_3 \rightleftharpoons \text{5-aminotetrazole (H}_2\text{N—C, NH, N=N—N ring)}$$

the cyclization of the free guanyl azide as 17·8 kcal mole^{-1}, and that of the corresponding *N*-nitroguanyl azide as 21·2 kcal mole^{-1}. The difference in energy (by heat of combustion) between 5-nitrotetrazole and the nitroguanyl azide is 11 kcal mole^{-1} in favour of the tetrazole[126]. Similarly, diazotization of $H_2NNHC({=}NH){-}CN$ gives 5-cyanotetrazole rather than the azide[127]. Amines displace azide ion from the salts of guanyl azides[128], although ring closure to the 5-aminotetrazoles is usually favoured.

5-Alkylamino- or 5-arylaminotetrazoles can be equilibrated with 1-alkyl or 1-aryl-5-aminotetrazoles respectively. Lieber[125,129] has shown that this isomerization goes through the corresponding guanyl azides, rather than through bicyclic intermediates discussed earlier

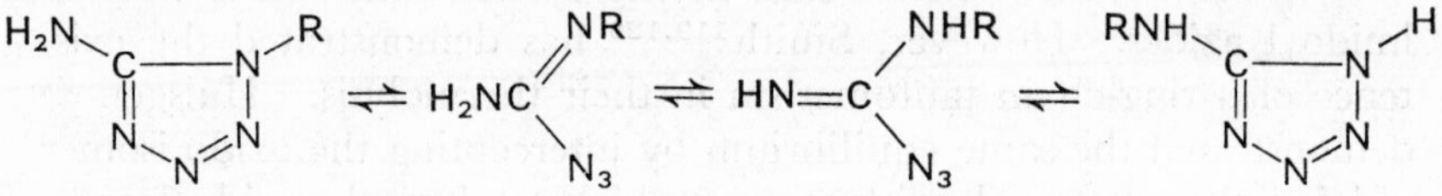

$$\text{1-R-5-}H_2N\text{-tetrazole} \rightleftharpoons H_2NC({=}NR)N_3 \rightleftharpoons HN{=}C(NHR)N_3 \rightleftharpoons \text{5-RNH-1H-tetrazole}$$

by Garbrecht[130]:

$$\text{1-R-5-}H_2N\text{-tetrazole} \rightleftharpoons \text{bicyclic NH-bridged intermediate} \rightleftharpoons \text{5-RNH-1H-tetrazole}$$

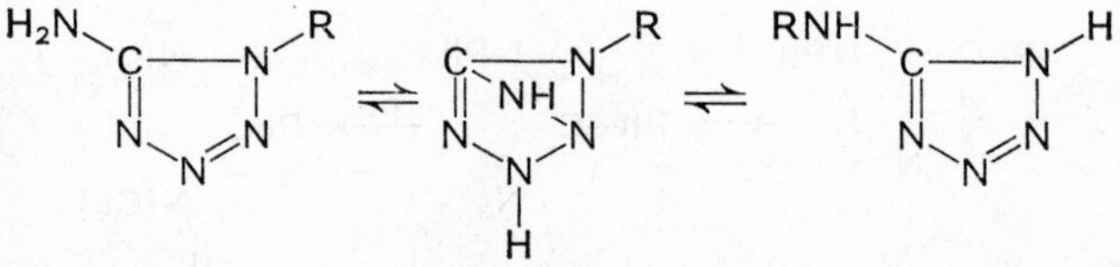

The guanyl azides are intermediates of higher energy than the tetrazoles in these isomerizations. Consequently, substituents that stabilize the azides also lower the energy of activation for the ring-opening of the tetrazoles, an expectation confirmed by Lieber's[125] kinetic data.

Nagy[131] and Jensen[132], working independently, solved an interesting riddle[133,134]: treatment of 5-aminotetrazole with arenesulphonyl chlorides gives, as the apparent primary products, arenesulphonylguanyl azides rather than arenesulphonamido-tetrazoles, and these guanyl azides isomerize, when treated with base, to 5-arenesulphonamidotetrazoles. Apparently, acylation on the ring, rather than on the 5-amino group, occurs first. The 1-arenesulphonyltetrazole then opens to the guanyl azide, which is separated from the 5-arenesulphonamidotetrazole by a high activation barrier and can be isolated. Treatment with base converts the azide to its conjugate base, which cyclizes readily. In accord with this mechanism, Jensen[132] found 1-methyl-5-aminotetrazole to be unaffected by tosyl chloride under his reaction conditions, while 5-dimethylaminotetrazole reacted (with the ultimate formation of 1,1-dimethylhydrazine, as expected from ring-opening of the 1-sulphonyltetrazole, followed by Curtius rearrangement and hydrolysis).

The acetylation of derivatives of 5-aminotetrazoles proceeds in a similar manner: Herbst[130,135,136] found that acetic anhydride in the cold converts 5-alkylaminotetrazoles into isolable 1-acetyl-5-alkylaminotetrazoles, which thermally rearrange to 5-acetylamido-1-aryltetrazoles. On the basis of these results, Nagy[131] was able to reinterpret observations of Veldstra and Wiardi[137], which involve an interesting conversion of an imidoyl azide to azide ion and cyanamide. The arenesulphonylguanyl azides of Nagy and Jensen have an

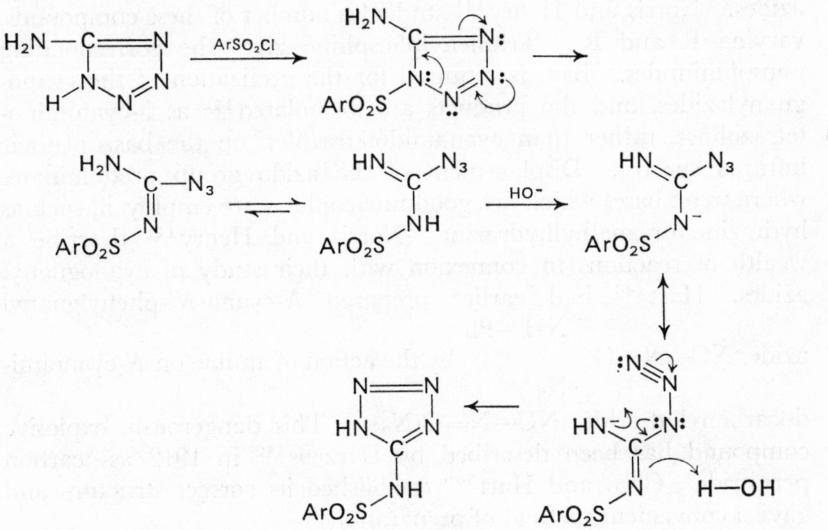

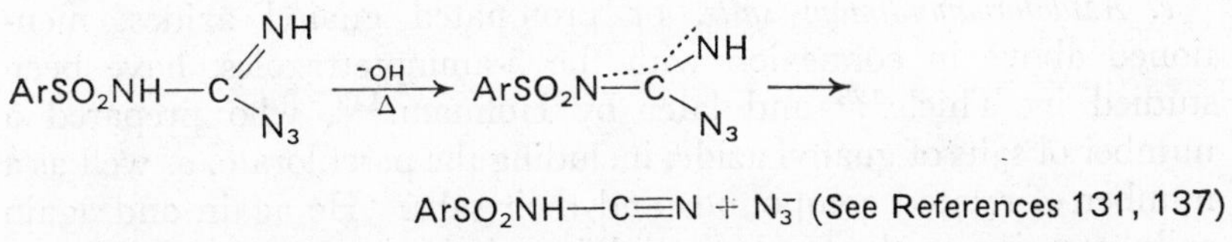

$$ArSO_2NH—C(=NH)—N_3 \xrightarrow[\Delta]{^-OH} ArSO_2N\overset{\dots}{=}C(NH)—N_3 \longrightarrow$$

$$ArSO_2NH—C{\equiv}N + N_3^- \text{ (See References 131, 137)}$$

unsubstituted imido group (C=NH). Neidlein[138] has prepared five analogous, but *N*-substituted, azides $ArSO_2NH(C{=}NR')—N_3$ by addition of hydrazoic acid to 1-sulphonyl-3-alkyl-carbodiimides $ArSO_2N{=}C{=}N{=}R'$. His compounds cyclized to 1-alkyl-5-arenesulphonamidotetrazoles upon heating in toluene. Neidlein also prepared[139] *C*-alkylmercapto-*N*-sulphonyl-imidoyl azide,

$$RSO_2N{=}C(N_3)—S—CH_3$$

(R = phenyl, *p*-tolyl, methyl). Because of the danger inherent in their distillation, he purified these compounds by column chromatography.

d. Cyanoguanyl azides. $RR'NC(=N{-}CN)N_3$ were prepared by Hart[140] in a manner analogous to that for preparing sulphonylguanyl azides: reactions of 5-aminotetrazoles with cyanogen bromide, to form the transient 1-cyano-5-aminotetrazoles, which open to the cyanoguanyl azides. Norris and Henry[141] studied a number of these compounds, varying R and R′. Triphenylphosphine gives the corresponding phosphinimides. Base is required for the cyclization of the cyanoguanylazides, and the products are formulated[141] as 5-cyanimino-tetrazolines, rather than cyanamidotetrazoles, on the basis of their infrared spectra. Displacement of the azido group predominates where weak bases which are good nucleophiles are employed, such as hydrazine or methylhydrazine. Norris and Henry[141] describe a wealth of reactions in connexion with their study of cyanoguanyl azides. Hart[140] had earlier prepared *N*-cyano-*N*′-phenylguanyl azide, $NC{-}N{=}C(NH{-}Ph)N_3$ by the action of aniline on *N*-cyanoimidocarbonyl diazide, $NC{-}N{=}C(N_3)_2$. This dangerously explosive compound has been described by Darzens[142] in 1912 as 'carbon pernitride', CN_4, and Hart[140] established its correct structure and gave a convenient method of preparation.

e. Azidoformamidinium salts, or protonated guanyl azides, mentioned above in connexion with the 5-aminotetrazoles, have been studied by Thiele[123] and later by Hofmann[143], who prepared a number of salts of guanyl azide, including the perchlorate, as well as a number of related compounds and their salts. He again and again calls attention to the 'truly terrible' explosive properties of his compounds. Schmidt[144] prepared guanyl azide salts, $[N_3C(NH_2)_2]^+X^-$ by treating chloroformamidinium salts with trimethylsilyl azide and made an extensive study of their i.r. spectra. In contrast to Hofmann's perchlorate, azidoguanidylium chloride, hexachloroantimonate, and azido-pentachloroantimonate are not particularly sensitive to shock or heat. The antisymmetric azide stretching frequency of the hexachloroantimonate $[(NH_2)_2CN_3]^+SbCl_6^-$ is at 2165 cm^{-1}, while that of $[H_2NC(N_3)_2]^+SbCl_6^-$ (prepared[144] from chloroformamidinium-azide-hexachlorantimonate and the dimer of tetrachloroantimony (V) azide) shows two strong bands at 2195 and 2170 cm^{-1}. Schmidt discusses at some length the resonance stabilization of his compounds. He also prepared[145] azidoimidinium salts of the more general type $Y—C(=^{+}NH_2)—N_3$, with Y=H—, Cl—, ClH_2C—, Cl_2HC—, Cl_3C— and RO—, and studied and analysed their i.r. spectra. Again, these compounds proved to be relatively insensitive towards shock or heat.

f. Hydrazidic azides, $RC(=NNHAr)—N_3$, have been obtained by Scott[146,147] from hydrazidic bromides and sodium azide. Treatment with nitrite of the hydrazidines corresponding to the bromides led only to the cyclized products[147,148]. Ten hydrazidic azides[147] could not be converted to the tetrazoles under a variety of conditions. They undergo acid-induced rearrangement through transient *N*-arylamino-carbodiimides to give semicarbazides:

$$RC(=NNHAr)N_3 \ (\mathbf{13}) \xrightarrow[-N_2]{H^+} [RN=C=N—NHAr] \xrightarrow{F_3CCOOH} RNH—C(OCOCF_3)=N—NHAr \longrightarrow RNHCONHN(Ar)COCF_3$$

$$[RN=C=N—NHAr] \xrightarrow{H_2O} RNHCONHNHAr$$

g. Azidobenzalazines. The *azido-benzalazine system* and its relation to the corresponding tetrazoles[108,109,111] was discussed above. The photochemistry of α-azido-benzalazine (**1**) has been studied by Lwowski and Grassmann[149]. Irradiating **1** with light of maximum intensity at 350 nm, in a pyrex vessel, leads to loss of nitrogen from the ^{15}N-labelled azide group and fragmentation to benzonitrile (82%) and phenyldiazomethane, which was decomposed with benzoic acid to give nitrogen completely free of excess ^{15}N label:

$$PhC(N=N=^{15}N)=N—N=CHPh \ (\mathbf{1}) \xrightarrow[h\nu]{350\,nm} PhCN \ (82\%) + {}^{15}N_2 + N=N=CHPh$$

4. Imidoyl azides in which the C=N function is part of a ring

The preceding section dealt with imidoyl azides in which the imido function is not part of a heteroaromatic (or otherwise highly conjugated) system. In these cases, the tetrazole turned out to be the thermodynamically favoured isomer. If the imido function is made part of a ring the azide form is often much more stable; in many other cases, equilibria between azide and tetrazole forms are established. Lattice energies can make one form exclusive in the crystalline state of compounds, which in solution display an azide-tetrazole equilibrium[150].

Stolle[151] already recognized that in 2,4-diazido quinoxaline only one azido group cyclizes to give the tetrazolo azido quinoxaline, while the other azido group (in the 4-position) remains. This result has been confirmed by a number of authors[152]. Boyer[150] studied equilibria **14** ⇌ **15**. He found the compounds with X = CH_2 and NH to be tetrazoles (**15**) both in the solid state and in solution, while that with X = S is a tetrazole in the solid and contains much of the azide form in solution. The latter is also true for **16**.

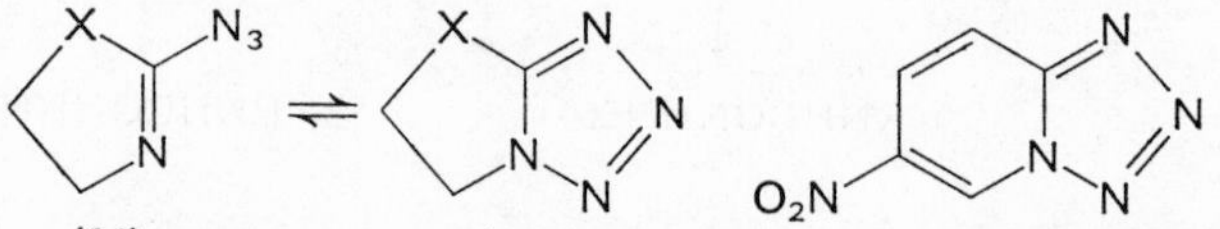

(14) **(15)** **(16)**

When the imido-function is made part of a heteroaromatic ring, one might expect the preference for the azide form to be even stronger. However, Reynolds[152,153] found systems of this kind, in which the preferred isomer depends on the nature of the heteroaromatic ring.

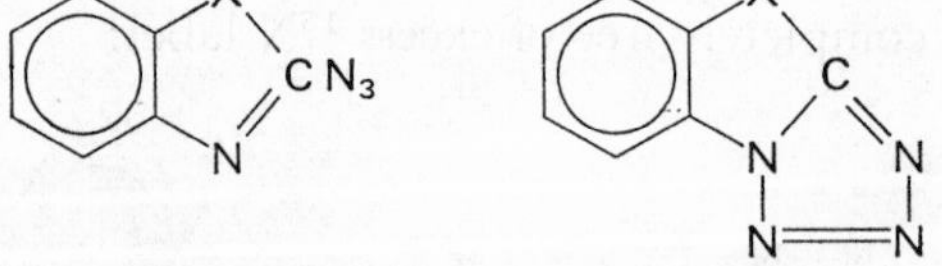

X = NH; O; SO_2 X = S; Se

They re-assigned, on the basis of spectral data, structure **17**[154] and confirmed **18**[155].

Fusco[156] examined spectroscopically tetrazolo[1,5,*a*]-pyridine, 5,7-dimethyltetrazolo[1,5,*a*]pyrimidine, 5,6-diphenyltetrazolo[1,5,*b*]-1,2,4-triazine, and tetrazolo[5,1,*b*]benzothiazole, and found all of

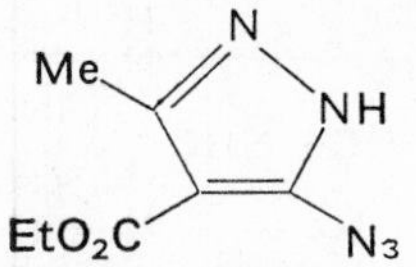

(17) 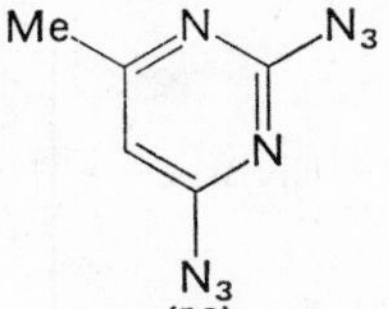 (18)

them to be tetrazoles in the solid state. However, the presence of the tautomeric azide in solution was demonstrated by addition of the azido group to an enamine[156].

Other tetrazole–azido–heteroaromatic equilibria have been studied where the heterocyclics to which the azide group is attached (or to which it is condensed as a tetrazole ring) are: 1,2,3-triazoles, pyrimidines[121,157–163] quinoxalines[164,165], naphthothiazoles[166], 1,3,5-triazones[167], purines[168–170].

Perhaps the most exotic imidoyl azide, containing the imido function as part of a heteroaromatic system, is 5-azidotetrazole, HCN_7, the potassium and sodium salts of which are obtained when 1,3-diaminoguanidine (imidocarbazide) is treated with alkali nitrites in acetic acid[123,143,171]. The alkali salts, Na^+ and $K^{+-}CN_7$, are extremely explosive, the free azide[171] less so. It shows the i.r. absorption of the azide group at 2151 cm^{-1}.

Some stable imidoyl azides, with the imido function a part of a heteroaromatic ring, are formed not by simple opening of a tetrazole but by more complicated processes. Scott[172,173] described the conversions of 5-*N*-tetrazolylbenzhydrazidic halides to 5-azido-3-aryl-

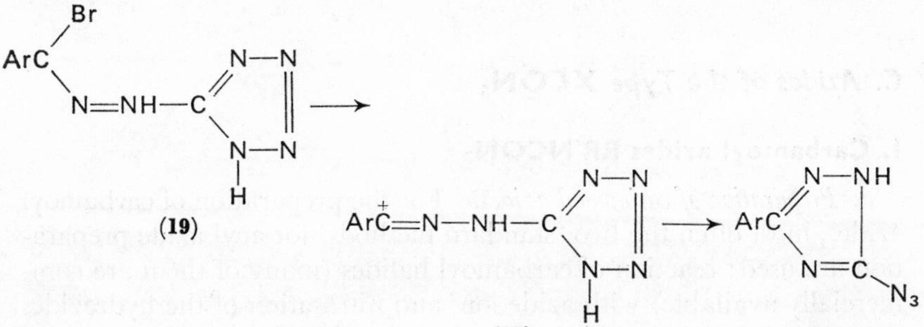

(19) (20) (21)

1,2,4-triazoles. The ρ value of $-1{\cdot}8$ of a Hammett plot using σ^+, of the rates of bromide formation suggests the intervention of the conjugate acid of a nitrilimine **20**. Henry[174], in his studies of cyanoguanyl azide chemistry (see above; ref. 141) treated bis-5-tetrazolyl-

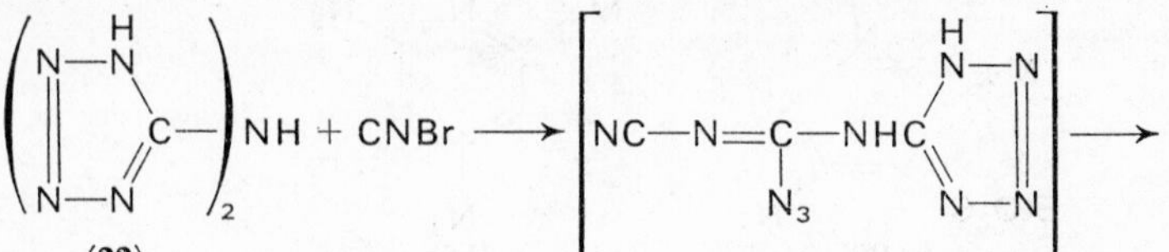

$$\left(\begin{array}{c}\text{H}\\ \text{N—N}\\ \| \quad \backslash \\ \quad \text{C—}\\ \| \quad / \\ \text{N—N}\end{array}\right)_2\text{NH} + \text{CNBr} \longrightarrow \left[\text{NC—N=C(N}_3\text{)—NHC}\begin{array}{c}\text{H}\\ \text{N—N}\\ \|\\ \text{N—N}\end{array}\right] \longrightarrow$$

(22)

$$\text{(iminotetrazolotriazine)} \rightleftharpoons \text{(aminotetrazolotriazine)} \rightleftharpoons \text{2-amino-4,6-diazido-1,3,5-triazine}$$

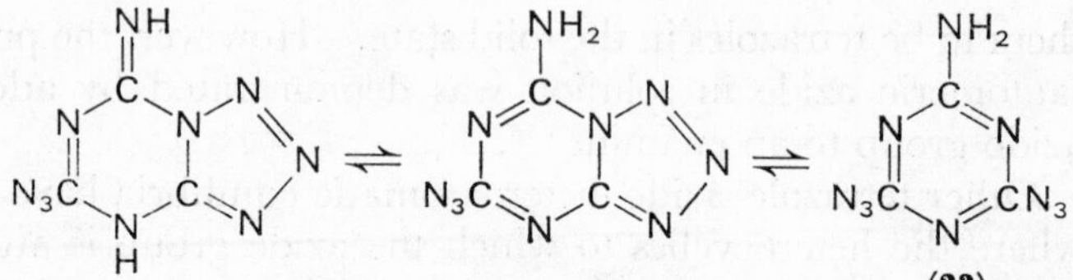

(23)

amine with cyanogen bromide to obtain 2-amino-4,6-diazido-1,3,5-triazine. Azides of methyltetrazolopyridazines can interconvert by closing the azide function to a tetrazole ring while opening the tetrazole ring to an azide group, an isomerism noticeable by the apparent shift of the alkyl substitutent[175]:

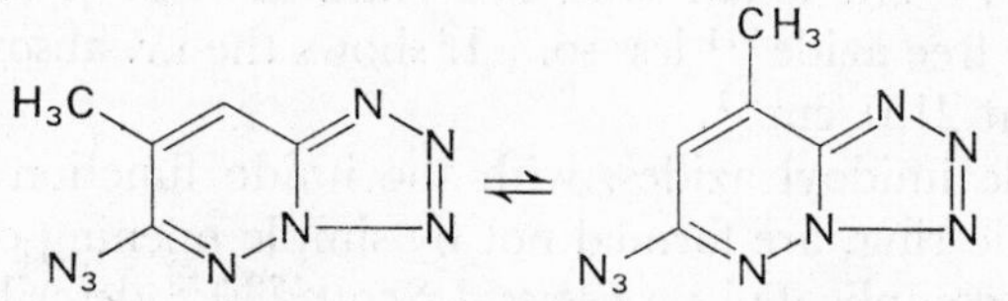

$$\mathbf{(24)} \rightleftharpoons \mathbf{(25)}$$

(24) **(25)**

C. *Azides of the Type* XCON$_3$

I. Carbamoyl azides RR′NCON$_3$

a. Preparation of carbamoyl azides. For the preparation of carbamoyl azides, most often the two 'standard methods' for acyl azide preparation are used: reaction of carbamoyl halides (many of them are commercially available) with azide ion, and nitrosation of the hydrazides corresponding to the azides. The latter method is especially valuable for making azides of the type RHN—CO—N$_3$. In Chapter 2, sections III.B and V.A.2. deal with these methods. Monoalkylcarbamoyl azides, RHN—CO—N$_3$ can also be made by adding hydrazoic acid to isocyanates, RN=C=O[82,176]. Oliveri-Mandala has prepared a substantial number of such azides this way[176,177]. He also studied the reaction of ketenes with hydrazoic acid[178], to give carbamoyl azides via alkanoyl azides and their Curtius rearrangement:

$$H_2C{=}C{=}O \xrightarrow{HN_3} H_3C{-}CO{-}N_3 \xrightarrow{\Delta} H_3C{-}NCO \xrightarrow{HN_3} H_3C{-}NH{-}CO{-}N_3$$

Carbamoyl azide, $H_2N{-}CO{-}N_3$, has been known since 1894[178–180], made by nitrosation of semicarbazide. It is also formed in the partial decomposition of carbonyl azide, $CO(N_3)_2$[74,181], and by partial Curtius rearrangement of oxalyl diazide[69]. Curtius[182] warns that carbamoyl azide, while 'relatively harmless' can 'explode with unparalleled violence' under certain conditions, such as contact with copper powder. Carbamoyl azide forms a highly explosive silver salt[124].

Many other carbamoyl azides are known, with mono- or disubstituted amino groups of all kinds, including more exotic ones, such as *allophanyl azide*[183], *guanidylcarbonyl azide*[183], *hydrazodicarbonyl azide*, $N_3CONHNHCON_3$[72,184,185]. Neidlein[186,187] prepared a number of N-*acyl* and N-*sulphonyl-carbamoyl azides* by adding hydrazoic acid to the corresponding isocyanates. Acylcarbamoyl azides upon photolysis cyclize to 5-alkyl-1,2-4-oxidiazolin-3-ones[188]. Lieber[189], in a review

$$RC(=O){-}NH{-}CO{-}N_3 \xrightarrow[-N_2]{h\nu} \text{(ring: } RC{=}N{-}CO{-}NH{-}O{-}\text{)}$$

that covers the literature into 1964, has given physical data and references for a large number of carbamoyl azides. The infrared spectra of carbamoyl azides show the azide group at 2200 to 2141 (asym. stretching) and 1225 to 1212 (sym. stretching) cm^{-1}, the carbonyl group at 1725 to 1625 cm^{-1} [52,189].

b. Reactions of carbamoyl azides. Nucleophilic displacement of azide ion takes place in carbamoyl azides[190], just as in other acyl azides. However, the amide carbonyl in carbamoyl azides is less electrophilic than the carbonyl in other acyl azides, so that this reaction is less likely to become a bothersome side reaction. Cold aqueous alkali will effect the displacement[176,178]. With amines, displacement may take place on warming[191,192].

c. Dissociation of carbamoyl azides. A factor of general importance in predicting or interpreting reactions of monosubstituted carbamoyl azides is their dissociation into isocyanates and hydrazoic acid, the

$$RNHCON_3 \rightleftharpoons RNCO + HN_3$$

reverse of the Oliveri-Mandala procedure for making them. The parent compound in particular, H_2NCON_3, dissociates easily[193–195], on which basis many of its reactions have been explained. For example, on heating H_2NCON_3 in a sealed tube to 120°, the trimer of cyanic acid, cyanuric acid, is obtained, along with hydrazoic acid, ammonium azide and urazole[194,195].

d. Curtius rearrangement of carbamoyl azides. The Curtius rearrangement of carbamoyl azides, $RR'NCON_3 \rightarrow N_2 + RR'NNCO$, was for a long time regarded as an exception—carbamoyl azides were supposed to be, by and large, 'rigid azides'[190,193] and to lose nitrogen without migration of the amino function, to give a nitrene: $RR'NCON_3 \rightarrow N_2 + RR'NCON$. Stolle recognized in 1924 that diarylcarbamoyl azides rearrange readily upon heating[194] and investigated a number of cases[195,196]. The special feature that leads to the formation of a readily isolated product, in high yield, is the electrophilic attack on one of the aryl groups by the amino-isocyanate function, no doubt aided by the second aryl group, which can delocalize by resonance the unshared electron pair on the (diaryl substituted) nitrogen, rendering it less negative and thus reducing electrostatic destabilization in a transition state in which the isocyanate nitrogen becomes partially negatively charged. Consequently *N*-aryl-*N*-alkyl carbamoyl azides gave only poor yields of benzpyrazolinones (indazolones)[195,196]. Stolle was unable to get more than traces of identifiable products from the thermolysis of dialkylcarbamoyl azides[195,196] (however, see below). Scott[191,192] intercepted the diphenylaminoisocyanate by running the thermolysis in ethanol, obtaining a 90% yield of *N*,*N*-diphenyl-*N'*-ethoxycarbonylhydrazine. The reaction has been used to prepare *N*,*N*-diphenylhydrazine[197]. Mono- and dialkylcarbamoyl azides rearrange smoothly to the corresponding aminoisocyanates when decomposed photolytically in protic solvent[198]. In methanol, $PhNHCON_3$ gave a 65% yield of $PhNHNHCOOCH_3$; $C_2H_5NHCON_3$ gave a 57% yield of $C_2H_5NHNHCOOCH_3$, and *N*,*N*-diethylcarbamoylazide gave *N*,*N*-diethyl-*N'*-methoxycarbonylhydrazine in 63% yield. The photo-induced Curtius rearrangement of dialkylcarbamoyl azides (and perhaps also monoalkyl and monoaryl carbamoylazides) also takes place in aprotic media. Diethylaminoisocyanate formed from diethylcarbamoyl azide (easily prepared from the chloride and sodium azide[199]) does not attack solvents such as cyclohexane, cyclohexene, benzene or acetonitrile[199]. Instead, a dimer is formed, analogous to a dimethylaminoisocyanate dimer obtained by Wadsworth and Emmons[200], who made dimethylaminoisocyanate *in situ* by a phosphorus-organic route. Yields of about 40% of 1:1 adducts with other isocyanates can be obtained by photolysing diethylcarbamoyl azide in methyl or ethyl isocyanate as the solvents[199,201]. Carbodiimides also form 1:1 adducts[202]. Chemical and spectroscopic evidence led to the recognition[200,203] of the isocyanate adducts (including the dimers) as cyclic amine-imides. Thermolysis of diethylcarbamoyl azide also leads to Curtius rearrangement, but the temperature required (180°) leads to further reaction of the isocyanate dimer: the loss of ethylene, also observed with the isolated dimer and with the isocyanate adducts at about 160°[203]. Elimination of ethylene, as well as migration of the ethyl group from the ammonium to the imide nitrogen, gives urazoles (1,2,4-triazolidin-3,5-diones). Wadsworth and Emmons observed such a migration with their *N,N*-dimethyl compound[200]. In the case of the *N,N*-diethyl compound, elimination of ethylene is preferred upon heating alone, but acid catalysis promotes rearrangement, to give 1,2-diethylurazoles[201]: presumably because of protonation of the imide nitrogen and consequent loss of the inductive stabilization of the dialkylammonium function.

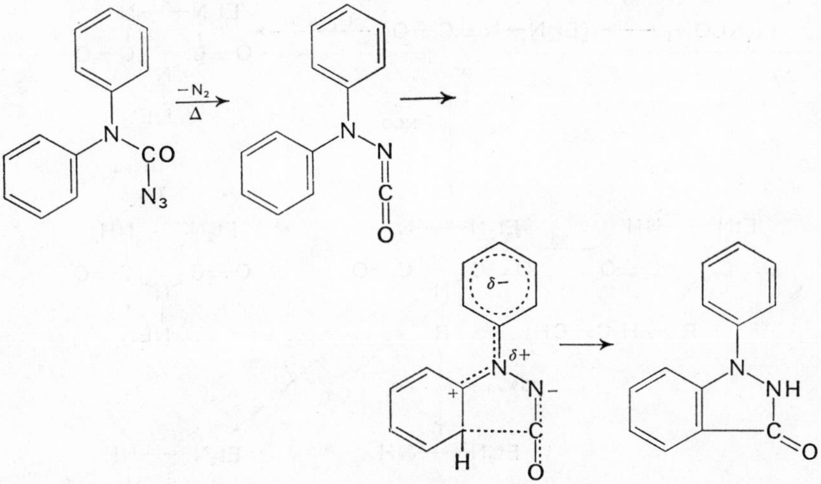

$Et_2NCON_3 \longrightarrow [Et_2N{-}N{=}C{=}O] \xrightarrow{Et_2NNCO}$ (urazole ylide: $Et_2\overset{+}{N}$—$\overset{-}{N}$, O=C, C=O, N–NEt_2) $\xrightarrow{H_2/Pd}$ (Et_2N, NH_2, O=C, C=O, N, NEt_2)

$[Et_2N{-}N{=}C{=}O] \xrightarrow{R{-}NCO}$ ($Et_2\overset{+}{N}$—$\overset{-}{N}$, O=C, C=O, N–R)

$\xrightarrow{130°}$ (EtN—NH, O=C, C=O, N–R) + $H_2C{=}CH_2$

$\xrightarrow{15\%\ aq.\ HCl,\ -CO_2}$ (Et_2N—NH, C=O, HN, Et)

$\xrightarrow{^-OMe/MeOH}$ (Et_2N—NH, MeOOC, C=O, N, Et)

($Et_2\overset{+}{N}$—$\overset{-}{N}$, O=C, C=O, N–R) $\xrightarrow{H^+,\ X^-}$ ($Et_2\overset{+}{N}$—NH X^-, O=C, C=O, N–R) $\longrightarrow$ (EtN—NEt, O=C, C=O, N–R)

isolated with R = Et, X = Cl

Traces of urazoles have been observed in the thermolysis of carbamoyl azide, H_2NCON_3 [193,194,195], and phenylcarbamoyl azide, $PhNHCON_3$ [204]. The mechanism of this urazole formation is not known beyond conjecture.

$2\ Et_2NNCO \longrightarrow$ (H_2C–$\overset{H_2}{C}$–H, EtN—N, O=C, C=O, N–NEt_2) $\xrightarrow{160°}$ $H_2C{=}CH_2$ + (EtN—NH, O=C, C=O, N–NEt_2)

e. Reactions possibly involving a carbamoylnitrene, RR′N—CO—N. The basis for classifying carbamoyl azides as 'rigid' (nonrearrangeing) [190,193] was a number of observations by Curtius and his school, in which the moiety RR′NCON was incorporated in adducts with a variety of substrates. Thermolysis of H_2NCON_3 in aromatic solvents gives, together with many side-products, arylureas, ArNHCO-NH_2[193,195,205]. This is most easily interpreted as a C—H insertion, or perhaps—in analogy with the reactions of sulphonylnitrenes with aromatic hydrocarbons[206–208]—as an addition, rearrangement reaction through an aza-norcaradiene or *N*-amidoazepine[209,210]. The

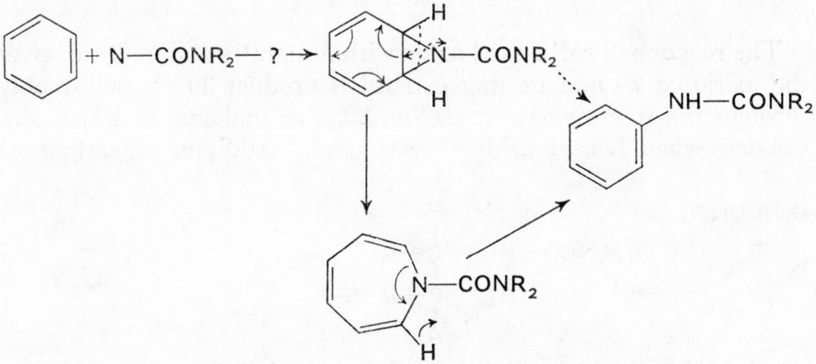

mechanistic evidence for this reaction, *in the case of carbamoyl azides*, does not rule out the possibility of a radical chain process in which a carbamoyl radical, RR′NCONH attacks the aromatic ring. The familiar intermediate in radical aromatic substitution so produced could then attack a carbamoyl azide molecule, generating a new carbamoyl radical and nitrogen.

Another argument for the formation of a carbamoylnitrene in the thermolysis of H_2NCON_3 is the apparent insertion into the methylene group of diethyl malonate[204], to give the mono- and the di-insertion products **26** and **27**. The analogous reaction fails, however, with

$H_2NCONHCH(COOEt)_2$ (**26**) $(H_2NCONH)_2C(COOEt)_2$ (**27**)

monophenylcarbamoyl azide[204]. In the insertion of ethoxycarbonylnitrene, the CH_2 group of diethyl malonate is particularly unreactive (relative reactivity per C—H 0·034 compared to cyclohexane = 1·00)[211], casting some doubt on the interpretation as a nitrene reaction. A radical chain process, involving a fairly nucleophilic urea or carbamoylazido radical, might be operative. Another possible mechanism is suggested by Fleury's[212] observation of triazene formation from acidic methylene compounds and toluenesulphonyl azide. The triazene (isolated in Fleury's case) could lose nitrogen to give the observed, formal insertion product. The reaction certainly deserves further study.

$$\mathrm{TosN_3 + {}^{-}CH(CN)_2 \longrightarrow TosNH{-}N{=}N{-}CH(CN)_2}$$

The reaction of carbamoyl azide with fumarates and maleates gives the aziridine **28** and its ring expansion product **30**[213], presumably through the intermediate triazoline **29**, the analogue of which was isolated when benzyl azide[214] was used. Addition of carbamoyl

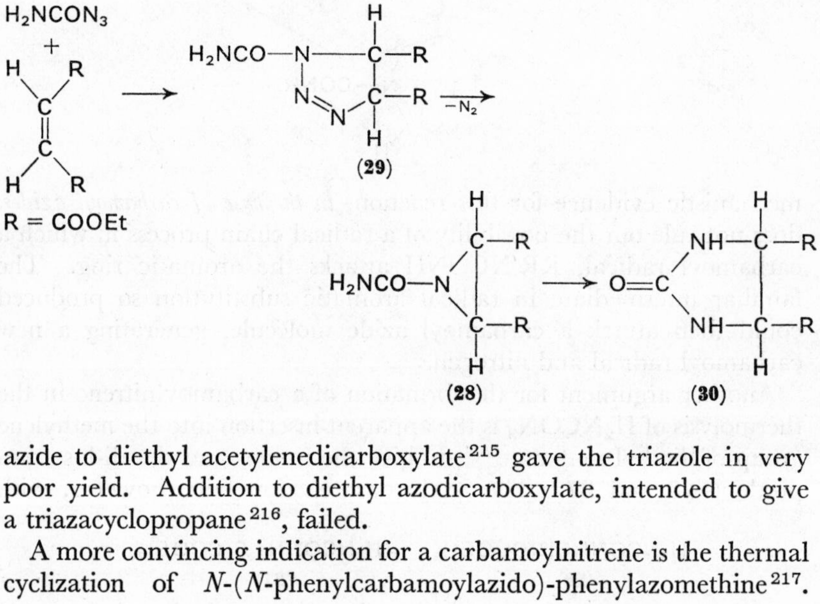

H_2NCON_3 + $HC(R){=}C(R)H$ ⟶ **(29)** $\xrightarrow{-N_2}$ **(28)** ⟶ **(30)**

R = COOEt

(29) **(28)** **(30)**

azide to diethyl acetylenedicarboxylate[215] gave the triazole in very poor yield. Addition to diethyl azodicarboxylate, intended to give a triazacyclopropane[216], failed.

A more convincing indication for a carbamoylnitrene is the thermal cyclization of *N*-(*N*-phenylcarbamoylazido)-phenylazomethine[217].

$$\mathrm{Ph{-}N(CON_3){-}N{=}CHPh \xrightarrow{\Delta} Ph{-}N{-}N{=}C(Ph){-}N{=}C(OH)}$$

(31) **(32)**

However, an intermediate bicylic [3.2.0] intermediate seems to be a possible alternative.

Another reaction that could involve a carbamoylnitrene was reported by Bauer[218]:

$$X-C_6H_4-SO_2^- Na^+ + N_3CONHR \longrightarrow X-C_6H_4-SO_2NHCONHR$$

(33) (34)

$X = CH_3; Cl.$ $R = n\text{-}C_4H_9;\ n\text{-}C_3H_7$

Kreher[219] obtained significant yields of *N*-alkoxyureas from photolyses of carbamoyl azide in a variety of alcohols, e.g.:

$$H_2NCON_3 + (CH_3)_3COH \xrightarrow{h\nu} H_2NCONHOC(CH_3)_3 \quad 45\%$$

He interprets this as a nitrene reaction, certainly an attractive explanation, as the insertion of alkoxycarbonylnitrenes into O—H bonds is well known[198,220,221]. A radical chain process seems to, at least, compete: while the yield of alkoxyurea was highest with *t*-butanol, it dropped to 8% in isopropanol, in which an 80% yield of the hydrogen abstraction product, urea, was obtained.

2. Azidoformates $ROCON_3$

The synthesis of azidoformates is strictly analogous to that of alkanoyl azides (see Chapter 2, sections III.B and V.D.). Azidoformates, $ROCON_3$, are usually more stable thermally than azides of the type C—CON_3, but will decompose at temperatures lower than the carbamoyl azides, $RR'NCON_3$. Methyl azidoformate[23,222] boils at 102° (but sometimes with explosion). Ethyl azidoformate[1], bp. 114° at 769 mm Hg also sometimes explodes when heated to that temperature and is better distilled at 2 mm Hg at 25° or, most conveniently, at aspirator pressure from a 40° bath into a receiver cooled to at least −15°. These azidoformates are shock sensitive. Tests done at the FMC Corporation Chemical Research and Development Center, Princeton, New Jersey, showed that ethyl azidoformate is ignited, on a drop weight tester, at 5 kilogram centimetres. This is approximately the shock sensitivity range of nitroglycerine[224]. Tertiary-butyl azidoformate, a useful reagent introduced by Carpino[225], boils at 73–76°/70 mm. However, Polzhofer[226] reports 'frequent explosions during distillation' and recommends distillation at 36·5–37·5°/10 mm Hg. He finds, as did others[227] with other volatile azidoformates, that the vapours cause severe headaches, nausea, shortness of breath and irritation of the mucous membranes. The vapours of ethyl azidoformate may cause fainting[228]. Quite a few other azidoformates have been reported, such as *t*-amyl azidoformate[229], the azidoformate of 3β-acetoxyandrost-5-en-17β-ol[230], benzhydryl azidoformate[231], *p*-methoxybenzyl azidoformate[232], tetramethylene bis-azidoformate[233], octadecyl azidoformate[234].

Sheinker[53,55] has studied the u.v. and i.r. spectra of azidoformates. The 285 nm band, seen in alkyl azides, was not found: perhaps it disappears into the short wavelength band due to a blue shift. The u.v. spectrum of ethyl azidoformate extends to nearly 300 nm (see Table 2) but its photolysis is most conveniently carried out with the mercury resonance line near 254 nm, for which efficient lamps are available[235]. Ethyl azidoformate shows i.r. absorption bands at 2185 and 2137 cm^{-1} (N_3), 1759 and 1730 (C=O) and 1242 (C—O) cm^{-1} [235]. The u.v. spectrum[235] is given in Table 2. The i.r. and

TABLE 2. U.v. spectrum of ethyl azidoformate in various solvents

Wavelength nm	Hexane ε	Methanol ε	Cyclohexene ε
250	88·1	85·1	—
260	66·9	63·1	—
270	40·6	36·9	41·9
280	17·4	15·3	19·0
290	6·9	6·0	8·1
300	2·7	2·1	3·6
310	1·2		
320	[0·1][a]		
330	[0·04][a]		
340	[0·02][a]		

[a] Obtained using extremely concentrated solutions.

u.v. spectra of methyl azidoformate are similar to that of the ethyl ester[236].

Azidoformates are used mainly for two purposes: the generation of alkoxycarbonylnitrenes (see section II.F) and the introduction of alkoxycarbonyl protecting groups in peptide synthesis (see section II.C).

3. Mercaptocarbonyl azides RSCON$_3$

To the author's knowledge, no mercaptoacyl azide has yet been isolated. However, Prince and Orlando[237] obtained decomposition products of $PhSCON_3$ when they treated PhSCOCl with sodium azide in acetone at 0°. Vigorous nitrogen evolution occurred, and Curtius rearrangement of the azide took place.

$$\underset{(\mathbf{35})}{PhSCO{-}N_3} \xrightarrow{-N_2} [PhS{-}N{=}C{=}O] \longrightarrow \underset{(\mathbf{36})}{PhSNHCONH_2} \xrightarrow{H_3O^+} PhSSPh$$

$$\underset{(\mathbf{36})}{PhSNHCONH_2} \xrightarrow[KMnO_4]{57\%} PhSO_2NHCONH_2$$

II. REACTIONS COMMON TO ALL CLASSES OF ACYL AZIDES

A. 1, 3-Dipolar Additions of Acyl Azides

The most common addition reaction of azides in general is the 1,3-dipolar cycloaddition to double and triple bonds. These reactions have been reviewed in this series of books[238] as well as elsewhere, primarily by Huisgen[239–241] and lately by L'abbe[242], and their mechanism has been discussed[240,243,244]. The 1,3-dipolar cycloaddition of azides is generally understood as a concerted process[240,244] in which the terminal nitrogen of the azido group binds to the atom that is more negative in the olefin, if the olefin happens to be electronically unsymmetric. As a consequence of its concerted nature, it is a stereospecific *cis* addition.

Examples involving acyl azides are not numerous. Confirmation of the rule that the terminal nitrogen of the azide group attaches itself to the more electron-rich carbon of C═C double bonds is found in Fusco's work[245]. Fusco[245] also discovered the rearrangement to

$$\xrightarrow{N_3COOR} (\mathbf{37})$$

$R = C_2H_5;\ CH_2Ph$

amidines of the 4-aminotriazolines so formed (see below). The orientation rule holds true for azidoformates as well as for imidoyl azides[156].

The triazolines formed from strained double bonds and acyl azides are often quite unstable. Huisgen and Müller[239,246] found benzoyl azide to add to norbornene, forming a triazoline that loses nitrogen at 40°, to give a *N*-benzoylaziridine, which in turn rearranges to the corresponding oxazoline. The rate of benzoyl azide addition at 25° in chloroform is $2{\cdot}1 \times 10^{-6}$ l mole^{-1} sec^{-1}[247]. Oehlschlager has studied[248] the mechanism of the rearrangement of the initial adduct, the triazoline, formed from norbornene and methyl azidoformate. Japanese authors studied the addition of ethyl azidoformate to norbornene, norbornadiene and benzonorbornadiene[249]. Isodrin adds *t*-butyl azidoformate to give an isolable triazoline[250], which decomposes to the aziridine upon chromatography on neutral alumina. Aldimines and ketimines seem to react in their tautomeric enamine form with benzoyl azide[251] to give the expected amidines, as do enamines themselves[245,251].

$$\mathrm{Me_2CH{-}CH{=}N{-}R} \rightleftharpoons \underset{(\mathbf{38})}{\mathrm{Me_2C{=}CH{-}NHR}} \xrightarrow{\mathrm{PhCON_3}} \text{[triazoline: } \mathrm{Me_2C{-}C(H){-}NHR,\ N{=}N{-}N{-}COPh]} \xrightarrow{-\mathrm{H^+}}$$

$$\mathrm{Me_2C{=}C(NHR)(NHCOPh)} \rightleftharpoons \underset{(\mathbf{39})}{\mathrm{Me_2CH{-}C({=}NR)(NHCOPh)}}$$

$R = CH_3$ or C_2H_5

The addition of ethyl azidoformate to dihydropyran, studied by Edwards[252], gives the triazoline **40**, which is hydrolysed to 2-hydroxy-2-ethoxycarbonamidopyran.

$$\text{dihydropyran} + \mathrm{N_3{-}CO_2Et} \longrightarrow \underset{(\mathbf{40})}{\text{triazoline } (\mathrm{N{-}CO_2Et})} \xrightarrow[-\mathrm{N_2}]{\mathrm{H_2O}} \text{2-(NH{-}CO}_2\text{Et)-2-OH-tetrahydropyran}$$

Ethyl azidoformate and 1-trimethylsilyl-1,4-dihydropyridine react at room temperature[253] to give the cyclic amidine **41**. Since the reaction takes place 60° below the decomposition temperature of ethyl azidoformate, it is unlikely that carbethoxynitrene intervenes, formation of the triazoline followed by a Fusco rearrangement being the more reasonable interpretation.

H_2 ... N_3—COOEt → ... $-N_2$, MeOH → ...

Me_3Si Me_3Si CO_2Et H CO_2Et

(**41**)

Triple bonds add azides to give 1,2,3-triazoles (from carbon–carbon triple bonds) and tetrazoles (from nitriles or isonitriles). The reactions are of very wide scope, but there are few examples involving acyl azides. The formation of triazoles from acetylenes and azides in general has been reviewed[242,254] and the scope of these reactions has been expanded by more recent work, a few examples of which are found in the references[255–258]. The triazole formation from acetylenes and carbonyl azides (carbamoyl azides and azidoformates) was first observed by Curtius[215]. Ried and Schön have added a number of aroyl azides to *benzyne*, obtaining benzotriazoles in yields of about 60%[259].

Additions to *nitriles*[260] presumably go via an intermediate imidoyl azide (see section I.B), which in most cases cyclizes to the tetrazole. For azides in general, the reaction has recently been reviewed[242,261]. There seem to be no examples of an addition of an acyl azide to a nitrile (to give a 2-acyltetrazole). Only nitriles with strongly electron-withdrawing substitutents add azides other than HN_3[262,263]. The rate constant of thermolysis of ethyl azidoformate is of the same order of magnitude in benzonitrile as it is in a series of other solvents, making initial reaction of the azide with the nitrile very unlikely[264]. Only in extreme cases might one expect the addition reaction of an acyl azide to a nitrile to compete successfully with other processes, such as formation of an acylnitrene.

B. *Additions to the Terminal Nitrogen of Acyl Azides*

Azides in general may form adducts formally derived from nucleophilic *attack on the terminal nitrogen.* Most prominent are the phosphine adducts, R—N=N—N=PR$'_3$, and the triazenes formed by carbanion attack, R—N=N—NH—R′.

I. Phosphinimines and phosphazides

Phosphinimines, (*Iminophosphoranes*) R_3P=N—R′, are usually the products isolated from the interaction of azides with phosphines: the so-called Staudinger reaction[265–267]. The reactions take place below the decomposition temperatures of the azides, so that a nitrene intermediate is unlikely. The intermediate adducts, containing three nitrogens, have indeed been isolated in an ever growing number of instances. Maltz, in 1959, isolated the benzoyl azide–phosphorus *tris*(di-*n*-propylamide) complex (**42**)[267], and some thirty other phosphazides (not derived from *acyl* azides) are known[268–270]. VanAllan and Reynolds[271] have described a number of phosphazides derived from imidoyl azides (with the imidoyl function a part of a heterocyclic ring), such as **43**. Leffler has studied the rates of the

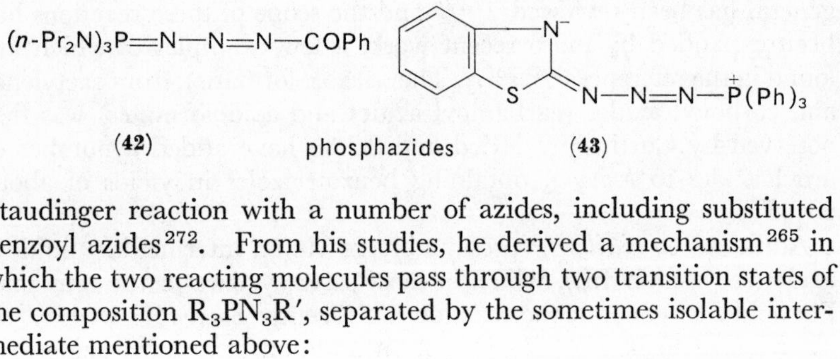

$(n\text{-}Pr_2N)_3P$=N—N=N—COPh (**42**)

phosphazides

(**43**) [benzothiazoline ring with N^- and S]=N—N=N—$\overset{+}{P}(Ph)_3$

Staudinger reaction with a number of azides, including substituted benzoyl azides[272]. From his studies, he derived a mechanism[265] in which the two reacting molecules pass through two transition states of the composition R_3PN_3R', separated by the sometimes isolable intermediate mentioned above:

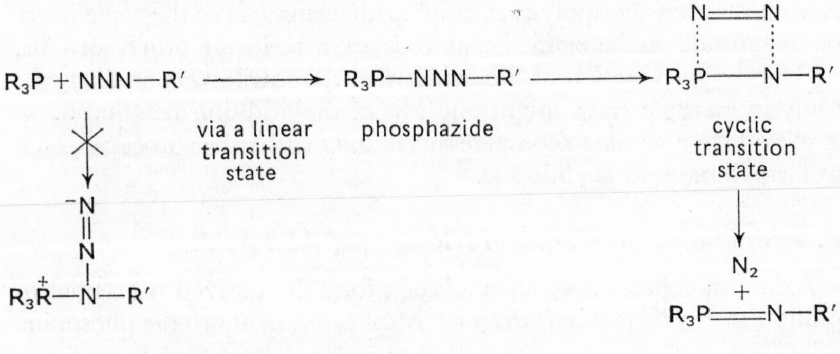

R_3P + NNN—R′ ⟶ (via a linear transition state) R_3P—NNN—R′ (phosphazide) ⟶ R_3P---N(—R′)---N=N (cyclic transition state) ⟶ N_2 + R_3P=N—R′

R_3P + NNN—R′ ⟶✗ $R_3\overset{+}{P}$—N(—R′)—N=N$^-$

2. Nucleophilic attack on the terminal azide nitrogen by ylides

Harvey[273] discovered the reactions of acylmethylene-triphenylphosphoranes with a variety of organic azides, including 3,4-dichloro-

$$Ph_3\overset{+}{P}-\overset{-}{C}H-\overset{O}{\overset{\|}{C}}-R + R'-N_3 \longrightarrow \left[\text{triazoline intermediate} \rightleftharpoons \text{betaine} \right] \longrightarrow Ph_3PO + \text{1-R'-5-R-1,2,3-triazole}$$

benzoyl azide. L'abbe[274] used the reaction to prepare 1-alkoxycarbonyl-1,2,3-triazoles (R' = EtOOC). Like Harvey[273], he observed a dependence of the course of the reaction on the nature of the acyl group in the phosphorane; weakly electrophilic acyl groups tend to give α-diazo esters rather than triazoles:

$$Ph_3\overset{+}{P}-\overset{-}{C}R'-\overset{O}{\overset{\|}{C}}-R + N_3-COOEt \longrightarrow [\text{intermediate}] \longrightarrow Ph_3PO + \text{R'C=C(R)-N(COOEt)-N=N (triazole)}$$

(a) R = CH_3; R' = H. (b) R = p-$O_2NC_6H_4$. (c) R = OEt; R' = CH_3.

But

$$Ph_3\overset{+}{P}-\overset{-}{C}HCOOEt + N_3-COOEt \longrightarrow [\text{Ph}_3\text{P}-\text{CHCOOEt}, \text{EtOOC}-\text{N}-\text{N}=\text{N}] \longrightarrow EtOOCN{=}PPh_3\ (75\%) + N_2{=}CHCOOEt\ (35\%)$$

A peculiar reaction of the terminal nitrogen in a carbamoyl azide is the conversion of phenylcarbamoyl azide to 1-phenyltriazolin-5-one under the influence of aluminium azide[275]:

$$PhNHCON_3 \xrightarrow[\text{THF, reflux}]{Al(N_3)_3} Ph-N-C{=}O,\ N{=}N-NH\ (\text{ring})$$

3. Triazene formation

Grignard reagents react with carbonyl azides not only by nucleophilic displacement of the azido group, but also by addition to the terminal nitrogen to form *triazenes*[222]:

$$H_3COCON_3 + PhMgBr \longrightarrow H_3COCONH{-}N{=}NPh$$

and

$$H_2NCON_3 + PhMgBr \longrightarrow H_2NCONH{-}N{=}NPh$$

and

$$\begin{matrix} HNCON_3 \\ | \\ HNCON_3 \end{matrix} + 2\,PhMgBr \longrightarrow PhN{=}N{-}NHCONHNHCONH{-}N{=}NPh$$

The reaction of an iridium complex with furanoyl azide[223] seems to involve coordination of the terminal nitrogen with the iridium, followed by dissociation to a stable iridium–nitrogen complex and furanoyl isocyanate (which incorporates a carbonyl that had been part of the iridium complex).

C. *Nucleophilic Displacement of the Azido Group*

The azido group is electron releasing[276,277], and decreases the electrophilicity of a carbonyl group to which it is attached. Thus, acyl azides are less electrophilic than acyl halides, the other substituent at the carbonyl group being equal. Still, azide ion is readily displaced from carbonyl azides, a reaction occasionally used where it is easier to make the azide than a halide. Examples are found in the heterocyclic[278] and natural products[279] field. The most important application is the introduction of alkoxycarbonyl groups in peptide synthesis. One method, mostly used for the activation of *C*-terminal amino acids in peptides, converts hydrazides of *N*-carbobenzoxypeptides to the (electrophilically reactive) azides[280] without racemization. For the protection of amino groups in peptide synthesis, *t*-butoxycarbonyl azide[225] is a useful and convenient reagent[280]. The corresponding acid chloride is too unstable for convenient use, while the azide reacts smoothly with the amino group of both free amino acids and their esters[281].

Carbanions can displace azide ion from carbonyl azides[282]. For displacements at guanyl azides see section I.B and at carbamoyl azides see section I.C.

D. Reactions of Acyl Azides with Acids

The Curtius rearrangement of carbonyl azides, $RCON_3 \rightarrow N_2 + R{-}N{=}C{=}O$, is catalysed by a variety of Lewis and proton acids[283,284], including $GaCl_3$, $AlBr_3$, $AlCl_3$, $FeCl_3$, $SbCl_5$, $TiCl_4$. At $-60°$, BF_3 adducts of acyl azides can be isolated[285,286]. These adducts show a normal azide absorption in the infrared spectrum, but the carbonyl absorption frequency is lowered drastically, to 1645–1667 cm^{-1}. At $-20°$, rearrangement to the isocyanates occurred[285]. Similar results were obtained using BCl_3[285]. Kreher[287] made similar observations. Using azidoformates, he observed loss of CO_2 and the formation of alkyl azides. The alkyl azides can rearrange to azo-

$$(CH_3)_3C{-}O{-}C({=}O{\cdots}AlCl_3){-}N_3 \xrightarrow{-CO_2} (CH_3)_3C{-}N_3{\cdot}AlCl_3 \xrightarrow{-N_2} (CH_3)_2C{=}NCH_3$$

methines, or, in the presence of benzene, take part in Lewis acid induced Friedel–Crafts reactions[288,289].

A reaction of phenylcarbamoyl azide with aluminium azide[275] has been mentioned above (section II.B).

Transition metal cyanides, such as $[Co(CN)_5]^{3-}$ have been observed to react with acyl azides with coordination of the metal to the α- and γ-nitrogens of the azide group, and ultimate formation of a new complex, containing two cobalt atoms bound to nitrogen[290]:

$$RCON_3 + 2\,[Co(CN)_5]^{3-} \longrightarrow [RCON(Co(CN)_5)_2]^{6-}$$

E. Radical-induced Decomposition of Acyl Azides

Acyl azides can be decomposed by reaction with free radicals. In methanol, ethoxycarbonyl azide requires one equivalent of radical initiator (diethyl peroxydicarbonate) to decompose one mole of the azide[235], but in alcohols that are better hydrogen donors, chain reactions occur: Horner[291] studied the dehydrogenation of isopropanol by aroyl, carbamoyl and alkoxycarbonyl azides:

$$RCON_3 + (CH_3)_2CHOH \longrightarrow RCONH_2 + N_2 + (CH_3)_2C{=}O$$

The reaction can be initiated by a variety of radical initiators, as well as by benzophenone-sensitized photolysis. In the latter case quantum yields up to 500 were observed[291]. This is consistent with a chain mechanism of the type:

$$X^{\bullet} + (CH_3)_2CHOH \longrightarrow (CH_3)_2C^{\bullet}OH$$
$$(CH_3)_2C^{\bullet}OH + RCON_3 \longrightarrow (CH_3)_2C{=}O + N_2 + RCONH^{\bullet}$$
$$RCONH^{\bullet} + (CH_3)_2CHOH \longrightarrow RCONH_2 + (CH_3)_2C^{\bullet}OH$$

A similar process is known for sulphonyl azides[292]. Leffler has discussed the reactions of other classes of azides with free radicals[293]. The chain mechanism observed in alcohols depends on the availability of two hydrogen atoms in the hydrogen donor. One can, however, write chain mechanisms with donors that make available only one hydrogen atom per donor molecule. All that is required is to introduce an additional step in which the dehydrogenated donor attacks another donor molecule, which then donates the second hydrogen atom to make $RCONH_2$ and a dehydro-dimer of the donor. While such processes have not been reported, one should not disregard their possible occurrence. Olefins seem to be likely candidates for such a reaction. Tri-*n*-butyltin hydride reduces benzoyl azide, in a radical induced reaction, to give a high yield of benzamide[294].

F. Nitrene Formation from Acyl Azides

Reactions of azides (of all types) that involve nitrogen evolution have often been assumed to proceed by a two-step 'nitrene mechanism', in which nitrogen is lost in the first step: $R{-}N_3 \rightarrow N_2 + R{-}N$. The nitrene R—N then leads to the observed reaction products. For azides in general, nitrene reactions are treated in Chapter 5, but the formation and the reactions of carbonylnitrenes, CO—N, will be dealt with here.

Carbonylnitrene intermediates have been postulated since 1891[295], but a problem recurs for every reaction which can be formulated as a nitrene reaction: one can always write a reasonable 'azide mechanism' that leads to the same products. One such alternative is a two-step process in which the azide reacts with the substrate in a first step, to give an adduct or an intermediate radical or ion pair, which loses nitrogen in a second step. Another alternative is a concerted process, in which loss of nitrogen and product formation are simultaneous. An example of the former azide mechanism is the aziridine formation from azides and olefins by way of an unstable triazoline. The Curtius rearrangement exemplifies the concerted azide reactions[296].

Other decompositions of carbonyl azides have been shown to be nitrene reactions. Here, the rate of azide disappearance (and nitrogen evolution) in the thermolysis of the azide is independent (for a given azide and temperature) of the nature and the concentration of the substrate[297–301]. This shows that the substrate is not involved in the azide decomposition, but must react with the nitrene in a second step. Alkoxycarbonylnitrenes have also been made by an independent route, by α-elimination from ions such as ROCO—$\bar{N}$—OSO_2Ar, which decompose to give ROCO—N and $^{-}O_3SAr$. The products observed by using this method were the same, quantitatively and qualitatively, as those obtained by azide decomposition in the same substrates[302,303].

Nitrenes can exist in the singlet and in the triplet states. The triplet state of alkoxycarbonylnitrenes has been shown to be the ground state by e.s.r. measurements at liquid helium temperature[304]. However, thermolysis of alkyl azidoformates, and the decomposition of the α-elimination precursor anions mentioned above, gives all of the alkoxycarbonylnitrene in the singlet state[305,306], while photolysis of ethyl azidoformate (at 38°) gave a 70:30 mixture of singlet and triplet nitrenes[307]. In the case of alkoxycarbonylnitrenes, the intersystem crossing to the triplet ground state is slow enough to allow intermolecular reactions of the singlet ROCO—N, and a large number of such singlet reactions are known[305–309]. Triplet alkoxycarbonylnitrenes also undergo intermolecular reactions. Often, the relative rates (intersystem crossing vs. singlet reaction rates) permit both singlet and triplet alkoxycarbonylnitrenes to operate in one and the same reaction mixture[305–307,310–312]. For the addition of singlet and triplet ethoxycarbonylnitrene to *cis*- and *trans*-4-methylpentene-2[305–307], and isoprene[313], the results agree quantitatively with the following general scheme:

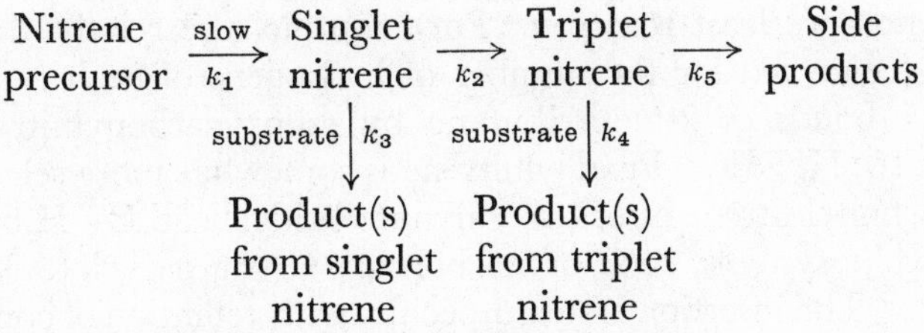

$$\text{Nitrene precursor} \xrightarrow[k_1]{\text{slow}} \text{Singlet nitrene} \xrightarrow[k_2]{} \text{Triplet nitrene} \xrightarrow[k_5]{} \text{Side products}$$

Singlet nitrene —substrate, k_3→ Product(s) from singlet nitrene

Triplet nitrene —substrate, k_4→ Product(s) from triplet nitrene

SCHEME 1. Singlet and triplet nitrene reactions

The 'sideproducts' (rate constant k_5) in this scheme include dissociation of the triplet nitrene[314,315] as well as possible return to singlet state.

Alkanoyl- and aroylnitrenes, C—CO—N, have so far been obtained only by photolysis of the corresponding azides[61,296,316–321] or nitrile oxides[322].

Typical reactions of carbonylnitrenes will be sketched in the following paragraphs.

I. Dissociation of acylnitrenes

Dissociation of acylnitrenes occurs, presumably from the triplet state, when carbonylnitrenes are produced in very unreactive media, or in a vacuum. Ethoxycarbonylnitrene, produced by flash photolysis in vacuo, dissociates to ˙NCO and EtO˙, the former being identified by its detailed and intense u.v. absorption spectrum. The ˙NCO radical is not produced, however, when cyclohexene vapour is present. In this case, the normal double bond adduct, 7-ethoxycarbonyl-7-azabicyclo[4.1.0]heptane is formed, as identified by its i.r. spectrum[314,315]. Pivaloylnitrene, *t*-BuCO—N, when produced in unreactive solvents, such as dichloromethane or neopentane, apparently dissociates in a similar manner. A polymer suggested as being the product from *t*-Bu˙ and ˙NCO is formed, and isobutene can be obtained in 22% yield (based on nitrene formed) when a stream of nitrogen is passed through the reaction mixture[61].

2. Insertion into C—H bonds

Insertion into C—H bonds[61,309,316,323,324] by carbonylnitrenes leads to alkylamides: RCO—N + H—C(—)(—)— → RCO—NH—C(—)(—)—

Tertiary C—H bonds are more reactive than secondary ones, primary C—H bonds are least reactive. For example, the relative rates for insertion (corrected for the number of hydrogens of each type) into the C—H bonds of 2-methylbutane by ethoxycarbonylnitrene are about 30:10:1[297,324]. Pivaloylnitrene is somewhat more selective[61]. Ethoxycarbonylnitrene inserts readily into bridgehead C—H bonds in small bicyclic systems, such as norbornane[325] or tricyclo[3.3.0.0^{2,6}] octane[326]. The insertion is stereospecific, with retention of configuration both in intramolecular nitrene cyclizations[327–331] and in the intermolecular insertion of ethoxycarbonylnitrene into the tertiary C—H bond of 3-methylhexane[309]. The deuterium isotope effect for the insertion of ethoxycarbonylnitrene into the C—H bonds of cyclohexane is $k_H/k_D = 1{\cdot}5$[303,332]. This value agrees well with the assumption of a triangular transition state for the singlet insertion[63,309]. It should be noted, however, that these data apply only to reactions of carbonylnitrenes with unactivated C—H bonds. The reaction of, for example, phenylnitrene with C—H bonds has a different mechanism, as indicated[332–334] by its incomplete stereospecificity and a kinetic isotope effect $k_H/k_D = 4{\cdot}1$ at 160°: almost the theoretical maximum at that temperature.

It appears that carbonylnitrenes insert at a detectable rate only in their singlet states, while phenylnitrene and cyanonitrene[335] can insert in both their singlet and triplet states. Also, activated C—H bonds (such as aromatic C—H bonds or those next to ether linkages) may react with triplet carbonylnitrenes[310–312].

Intermolecular C—H insertion, with cyclohexane as the solvent, gives yields of 20–78% with various carbonylnitrenes[230,235,297,301]. The yield of intramolecular insertion (cyclization) depends strongly on steric factors. In open-chain systems, the yields of γ-lactams rise dramatically when one or two sets of geminal methyl groups are built into the systems, to coil up the chain and bring the C—H bond to be attacked into the immediate vicinity of the nitrene function, as shown in Table 3. This is probably due to the short lifetime of the singlet nitrene. Unless a suitable C—H bond is encountered by the nitrene shortly after its creation, it decays to its triplet state, no longer capable of efficient insertion.

TABLE 3. Yields of γ-lactams from butyroyl azides[31]

Azide	Yield of γ-lactam %
Butyroyl azide	3·5
3,3-Dimethylbutyroyl azide	20·1
2,2,3,3-Tetramethylbutyroyl azide	56·2

In rigid systems, the stereochemistry may favour insertion in one or two particular C—H bonds, a fact utilized in several syntheses of natural products[316,317,322,337–341]. Edwards[316,317] reported a classical example:

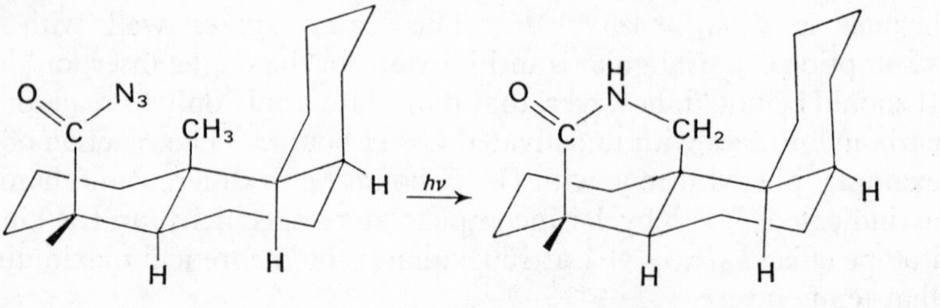

O N$_3$ C CH$_3$ H $h\nu$ → O H N C CH$_2$ H

H H H H

25% yield
+ a trace of a γ-lactam
+ isocyanate

3. Insertion into O—H bonds

Insertion into O—H bonds by alkoxycarbonylnitrenes gives *N*-alkoxycarbamates together with dehydrogenation products (aldehydes or ketones) and the adducts of the alcohol to alkoxyisocyanates, RO—NCO, produced by Curtius rearrangement of the alkoxycarbonyl azide[319,342–345]. Very high yields of dehydrogenation products have been reported by one author[343], perhaps because the radical-chain decomposition of the azide (see above) predominated. With *t*-butanol and ethoxycarbonyl azide, the yield of O—H insertion product, $C_2H_5OCONHOC(CH_3)_3$, is 60%[344]. The photolysis of benzazide in the presence of water[319] gave benzhydroxamic acid in 9% yield; a result significant in view of Hauser's[346] unsuccessful attempts to trap carbonylnitrenes with water in his attempts to show the intervention of nitrenes in the Curtius, Hofmann and Lossen rearrangements.

4. Insertion into N—H bonds

Insertion into N—H bonds by alkoxycarbonylnitrenes is a facile reaction, but is hampered by nucleophilic displacement of azide ion when alkoxycarbonyl azides are used as the nitrene precursor. Hafner obtained a 52% yield of phenylhydrazoformate, PhNHNH-$COOC_2H_5$, by thermolysis of ethyl azidoformate in aniline[347]. Aroyl azides have also been used: Horner[319] obtained a 14% yield of *N*-phenyl-*N'*-benzoylhydrazine by photolysing benzoyl azide in aniline. With more nucleophilic amines, it is better to use the α-elimination route to alkoxycarbonylnitrenes[302,303]. Using this method, yields of

51% of *t*-BuNHNHCOOEt, and 49% of *n*-BuNHNHCOOEt are easily obtained[348,349]. The reaction also works well with secondary amines.

5. Addition to C=C double bonds

Addition to C=C double bonds by alkoxycarbonylnitrenes is stereospecific (*cis*) for the singlet nitrenes, and non-stereospecific for the triplet species[305–307]. With 1,3-dienes, such as butadiene[350], isoprene[313], cyclopentadiene and cyclohexadiene[313], only 1,2-addition is observed. However, these primary products, vinylaziridines, can be thermally rearranged to pyrrolines (the apparent 1,4-addition products)[313,351].

RO_2C—N (*cis*) ← N—CO_2R (singlet) — *cis* — N—CO_2R (triplet) → RO_2C—N (*cis*) + RO_2CN (*trans*)

isoprene + EtO_2C—N → aziridine (N—CO_2Et) + aziridine (N—CO_2Et) —Δ→ pyrroline (N—CO_2Et)

Aziridine yields in intermolecular additions are good: addition of ethoxycarbonylnitrene to cyclohexene gave the aziridine in 56% yield at room temperature, and in 75% yield at Dry Ice temperature (azide photolysis)[235]. At a 0·2 mole% concentration of cyclohexene (in dichloromethane) a 35% yield of aziridine was obtained[352]. Other olefins give similar yields. The good yields at low olefin concentrations are possible because both the singlet and the triplet nitrenes will add readily, and the intersystem crossing of the nitrene does not interfere with the reaction (except for its stereospecificity, see above[305–307]). With isoprene[313], the singlet ethoxycarbonylnitrene does not measurably discriminate between the two double bonds, while the triplet species prefers the more substituted double bond by a factor of two.

The competing azide reaction, addition to give a triazoline, interferes little or not at all when carbonyl azides and unactivated olefins are used. However, it becomes important with enamines and enol derivatives (see above, section II.A).

Intramolecular addition of carbonylnitrenes to double bonds has been used to make a number of highly strained bi- and tricyclic systems[353]. Thus, 4-cycloheptenecarbonyl azide, upon irradiation, gave the tricyclic acylaziridine **44**, which is very readily hydrolysed to the hydroxylactam **45**. Another example of Edwards' intramole-

(44) (45)

cular aziridine formations starts with 5-endonorbornenylcarbonyl azide, to give the aziridine **46** and from it **47**:

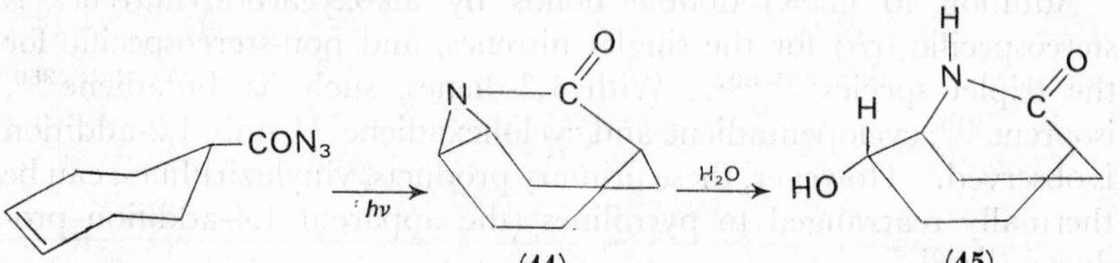

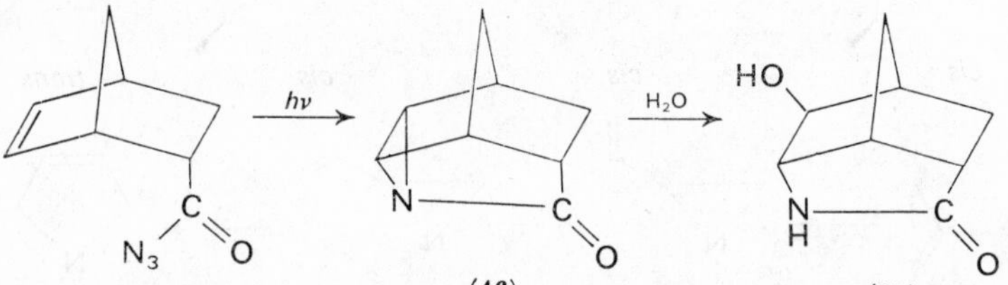

(46) (47)

6. Reactions with aromatic and heteroaromatic systems

Ethoxycarbonylnitrene, when generated in benzene or its derivatives, expands the ring to give *N*-alkoxycarbonylazepines[347,354–356].

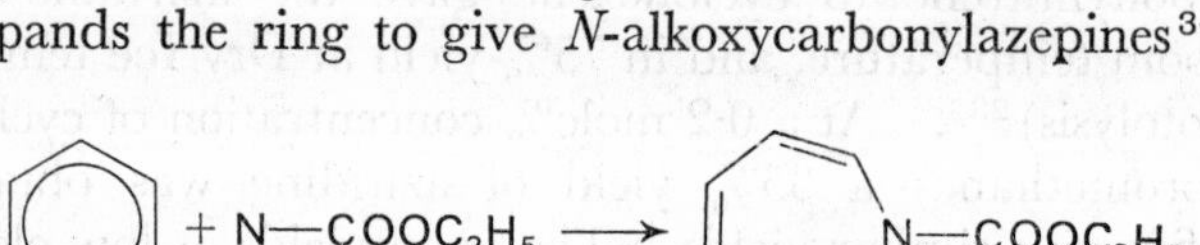

$$\text{C}_6\text{H}_6 + \text{N—COOC}_2\text{H}_5 \longrightarrow \text{(azepine)N—COOC}_2\text{H}_5$$

Analogously, carbonyl azide, N_3CON_3, gives *N*-azidocarbonylazepine[74] when thermolysed in benzene. Substituted benzenes give mixtures of azepines[347], which can be quite complex and intractable, because of prototropic and valence isomerizations, as observed in the reaction with phenol[357]. Condensed aromatics, such as naphthalene, anthracene and phenanthrene, give the apparent C—H insertion products instead of isolable azepines, at least in part due to isomerization of intermediary azepines[310,311]. The formation of the azepines is a reaction of the singlet carbonylnitrenes only[308,310,311,358].

Some 5-membered heteroaromatics react with ethoxycarbonylnitrene to form derivatives of *N*-ethoxycarbonylpyrroles[359]. For example, 2,5-dimethylthiophene gave, in 18% yield, 2,5-dimethyl-*N*-ethoxycarbonylpyrrole. This is most easily explained by postulating a 1,4-bridged intermediate:

Me, S, Me —($N—CO_2Et$)→ Me, S, Me, $N—CO_2Et$ —(−S)→ Me, N, Me, CO_2Et

7. Additions to alkynes

Alkynes add carbonylnitrenes to give 1,3-oxazoles[301,360], which may add another molecule of the nitrene to give the di-adduct **48**:

$$Ph—C{\equiv}C—Ph \xrightarrow[130°]{N_3CO_2Et} N_2 + \text{(Ph, Ph, N, O, OEt oxazole)} \xrightarrow{N—CO_2Et}$$

[OEt, O, N, Ph, Ph, N, O, OEt] ⟶ $Ph—C(=N—CO_2Et)—C(=N—CO_2Et)—Ph$

(**48**)

Meinwald[361] observed, besides oxazole formation, an adduct from two molecules of 2-butyne and one molecule of methoxycarbonylnitrene. Acetylnitrene also gave an oxazole when generated in the presence of phenylacetylene[362].

8. Additions to nitriles

Nitriles add carbonylnitrenes to form 1,3,4-oxadiazoles[264,343,362,363]. Alkoxycarbonylnitrenes (made either by azide photolysis, azide thermolysis or by the α-elimination route) give the 2-alkoxy-1,3,4-

oxadiazoles described earlier by Bacchetti[364], perhaps by way of an intermediate nitrilimine (through which Bacchetti had obtained the oxadiazoles). In accord with such an ionic mechanism (but also with

$$R—C{\equiv}N + N—COOR' \longrightarrow R—\overset{+}{C}{=}N—N{=}C(—O^-)—OR' \longrightarrow \text{2-R-5-OR'-1,3,4-oxadiazole}$$

a concerted cycloaddition) only the singlet nitrene gives oxadiazole, while triplet ethoxycarbonylnitrene reacts with acetonitrile to form another, as yet unidentified, product[365]. Good yields are obtained with many nitriles[363], but acrylonitrile reacts mostly at the C=C double bond, to give a 73% yield of cyanoaziridine and 14% oxadiazole[363].

9. Addition to isonitriles

Isonitriles add ethoxycarbonylnitrene to yield *N*-alkoxycarbonylcarbodiimides[366]:

$$t\text{-Bu}—NC + N—COOEt \longrightarrow t\text{-Bu}—N{=}C{=}N—COOEt$$

10. Addition to sulphoxides

Sulphoxides add carbonylnitrenes to give sulphoximines[298,318,319,367]:

$$ROOC—N + OSR'_2 \longrightarrow ROOC—N{=}\overset{O}{\overset{\|}{S}}R'_2$$

For a more comprehensive treatment of nitrene chemistry in general, the reader is referred to Chapter 5 of this book, and to reviews that have appeared recently[242,368,369].

III. REFERENCES

1. M. O. Forster and H. E. Fierz, *J. Chem. Soc.*, **93**, 72, 81 (1908).
2. G. Schroeter, *Ber.*, **42**, 3356 (1909).
3. M. O. Forster, *J. Chem. Soc.*, **95**, 433 (1909).
4. C. Naegli and G. Stefanovitch, *Helv. Chim. Acta*, **11**, 609 (1928).

5. H. Lindemann and W. Schultheiss, *Ann. Chem.*, **451**, 241 (1927).
6. P. A. S. Smith, in *Organic Reactions,* Vol. 3 (Ed. R. Adams), John Wiley, New York, 1946, p. 373ff.
7. C. Naegli, A. Tyabji and L. Conrad, *Helv. Chim. Acta,* **21**, 1127 (1938).
8. H. Lindemann and A. Pabst, *Ann. Chem.*, **462**, 29, 41 (1928).
9. G. Powell, *J. Am. Chem. Soc.*, **51**, 2436 (1929).
10. H. Ruschig, *Med. and Chem.*, **4**, 327 (1942).
11. A. Hofmann, H. Ott, R. Griot, P. A. Stadler and A. J. Frey, *Helv. Chim. Acta,* **46**, 2306 (1963).
12. R. K. Smalley and H. Suschitzky, *J. Chem. Soc.*, 755 (1964).
13. R. Stolle, *Ber.*, **57**, 1063 (1925).
14. R. Stolle, H. Nieland and M. Merkle, *J. Prakt. Chem.*, **116**, 192 (1927).
15. H. Lindemann and W. Wessel, *Ber.*, **58**, 1221 (1925).
16. C. Naegli and P. Lendorf, *Helv. Chim. Acta,* **15**, 49 (1932).
17. D. E. Horning and J. M. Muchowski, *Can. J. Chem.*, **45**, 1247 (1967).
18. B. Komppa and S. Beckmann, *Ann. Chem.*, **512**, 172 (1934).
19. E. Lieber, C. N. R. Rao, H. E. Dingle and J. Teetsov, *J. Chem. Eng. Data,* **11**, 105 (1966).
20. J. Nelles, *Ber.*, **65**, 1345 (1932).
21. M. Brown and R. E. Benson, *J. Org. Chem.*, **31**, 3849 (1966).
22. R. Huisgen and I. Ugi, *Chem. Ber.*, **90**, 2914 (1957).
23. T. Curtius and K. Heidenreich, *J. Prakt. Chem.*, **52**, 454 (1895).
24. W. G. Finnegan, R. A. Henry and R. Lofquist, *J. Am. Chem. Soc.*, **80**, 3908 (1958).
25. C. G. Swain and M. M. Kreevoy, *J. Am. Chem. Soc.*, **77**, 1122 (1955).
26. V. Gutmann, G. Hampel and O. Leitmann, *Monatsh. Chem.*, **95**, 1034 (1964).
27. F. Weygand and M. Reiher, *Chem. Ber.*, **88**, 26 (1955); F. Weygand and R. Geiger, *Chem. Ber.*, **90**, 634 (1957).
28. A. J. Papa, *J. Org. Chem.*, **31**, 1426 (1966).
29. J. Weinstock, *J. Org. Chem.*, **26**, 3511 (1961).
30. Y. G. Perron, L.-B. Crast, J. M. Esseny, R. R. Frazer, J. G. Godfrey, C. T. Holdrege, W. F. Minor, M. E. Neubert, R. A. Partyka and L. C. Chemey, *J. Med. Chem.*, **7**, 483 (1964).
31. W. Lwowski and S. Linke, unpublished results.
32. S. Marburg and P. A. Grieco, *Tetrahedron Letters,* 1305 (1966).
33. T. Curtius, *Ber.*, **23**, 3023 (1890).
34. W. O. Kermack and W. Muir, *J. Chem. Soc.*, 3089 (1931).
35. H. Meyer and E. R. v. Beck, *Monatsh. Chem.*, **36**, 731 (1915).
36. T. Curtius and O. Hofmann, *J. Prakt, Chem.*, **96**, 202 (1918).
37. cf. ref. 6, p. 369ff.
38. H. Jensen and L. Howland, *J. Am. Chem. Soc.*, **48**, 1988 (1926).
39. A. Windaus and H. Opitz, *Ber.* **44**, 1721 (1911).
40. A. Windaus and W. Vogt, *Ber.*, **40**, 3691 (1907).
41. R. Pschorr, H. Einbeck and O. Spangenberg, *Ber.*, **40**, 1998 (1907).
42. T. M. Sharp, *J. Chem. Soc.*, 1234 (1936).
43. O. Diels and F. Löflund, *Ber.*, **47**, 2351 (1914).
44. R. A. Clement, *J. Org. Chem.*, **27**, 1904 (1962).
45. L. C. Cheney and J. R. Piening, *J. Am. Chem. Soc.*, **66**, 1040 (1944).

46. W. Steinkopf, H. F. Schmitt and H. Fiedler, *Ann. Chem.*, **527**, 237 (1937).
47. W. S. Hingardner and T. B. Johnson, *J. Am. Chem. Soc.*, **52**, 3724 (1930).
48. R. Robinson and W. M. Todd, *J. Chem. Soc.*, 1743 (1939).
49. T. Curtius and E. Portner, *J. Prakt. Chem.*, **58**, 190 (1898).
50. T. Curtius, G. Struve and R. Radenhausen, *J. Prakt. Chem.*, **52**, 227 (1895).
51. T. Curtius and J. Jansen, *J. Prakt. Chem.*, **95**, 327 (1917).
52. E. Lieber, C. N. R. Rao, A. E. Thomas, E. Oftendahl, R. Minnis and C. V. N. Nambury, *Spectrochim. Acta*, **19**, 1135 (1963).
53. Yu. N. Sheinker and Ya. K. Syrkin, *Izvest. Akad. Nauk. SSSR, Ser. Fiz.*, **14**, 478 (1950); *Chem. Abstr.*, **45**, 3246g (1951).
54. Yu. N. Sheinker, *Dokl. Akad. Nauk. SSSR*, **77**, 1043 (1951); *Chem. Abstr.*, **45**, 6927b (1951).
55. Yu. N. Sheinker, L. B. Senyavina and V. N. Zheltova, *Dokl. Akad. Nauk. SSSR*, **160**, 215 (1965); *Chem. Abstr.*, **63**, 152h (1965).
56. W. D. Closson and H. B. Gray, *J. Am. Chem. Soc.*, **85**, 290 (1963).
57. R. Huisgen and J.-P. Anselme, *Chem. Ber.*, **98**, 2998 (1965).
58. T. Curtius and H. Hille, *J. Prakt. Chem.*, **64**, 208 (1901).
59. G. T. Tisue, Thesis, Yale University, 1966.
60. Y. Tsuno, *Mem. Inst. Sci. Res., Osaka U.*, **15**, 183 (1958), *Chem. Zbl.*, 15341 (1961): *Chem. Abstr.*, **53**, 1116b (1959).
61. G. T. Tisue, S. Linke and W. Lwowski, *J. Am. Chem. Soc.*, **89**, 6303 (1967).
62. T. Curtius and H. Hille, *J. Prakt. Chem.*, **64**, 401 (1901).
63. I. Brown and O. E. Edwards, *Can. J. Chem.*, **45**, 2599 (1967).
64. J. W. ApSimon and O. E. Edwards, *Can. J. Chem.*, **40**, 896 (1962).
65. W. L. Meyer and A. S. Levinson, *Proc. Chem. Soc.*, **15** (1963); *J. Org. Chem.*, **28**, 2589 (1963).
66. R. F. C. Brown, *Austral. J. Chem.*, **17**, 47 (1964).
67. R. Bognar, I. Farkas and I. F. Szabó, *Ann. Chem.*, **680**, 118 (1964).
68. N. C. Castellucci, M. Kato, H. Zenda and S. Masamune, *Chem. Comm.*, **3**, 473 (1967).
69. H. Roesky and O. Glemser, *Chem. Ber.*, **97**, 1710 (1964).
70. R. A. Clement, *J. Org. Chem.*, **27**, 1904 (1962).
71. T. Curtius and K. Heidenreich, *J. Prakt. Chem.*, **52**, 172, 454 (1895).
72. W. Kesting, *Ber.*, **57**, 1321 (1924).
73. T. Curtius and A. Bertho, *Ber.*, **59**, 565 (1926).
74. L. E. Chapman and R. F. Robbins, *Chem. Ind.*, 1266 (1966).
75. H. Neunhoeffer, G. Cuny and W. K. Franke, *Ann. Chem.*, **713**, 96 (1968).
76. T. Curtius and H. Thyssen, *J. Prakt. Chem.*, **65**, 1 (1902).
77. Y. Otsuji, Y. Koda, M. Kubo, M. Furukawa and E. Imoto, *Nippon Kagaku Zasshi*, **80**, 1307 (1959); *Chem. Abstr.*, **55**, 6477c (1961).
78. M. Freund and H. Hempel, *Ber.*, **28**, 74 (1895).
79. M. Freund and A. Schander, *Ber.*, **29**, 2500 (1896).
80. M. Freund and H. P. Schwarz, *Ber.*, **29**, 2506 (1896).
81. M. Freund and T. Paradies, *Ber.*, **34**, 3110 (1901).
82. E. Oliveri-Mandala and F. Noto, *Gazz. Chim. Ital.*, **43**, 304 (1913).
83. E. Oliveri-Mandala, *Gazz. Chim. Ital.*, **44**, 670 (1914).
84. E. Lieber, E. Oftedahl, C. N. Pillai and R. D. Hites, *J. Org. Chem.*, **22**, 441 (1957).

85. E. Lieber, C. Pillai and R. D. Hites, *Can. J. Chem.*, **35**, 832 (1957).
86. E. Lieber, C. N. R. Rao, C. N. Pillai, J. Ramachandran and R. D. Hites, *Can. J. Chem.*, **36**, 801 (1958).
87. E. Lieber, *J. Org. Chem.*, **22**, 1750 (1957).
88. G. B. L. Smith, F. Wilcoxon and A. W. Browne, *J. Am. Chem. Soc.*, **45**, 2604 (1923).
89. E. Lieber, C. B. Lawyer and J. P. Trivedi, *J. Org. Chem.*, **26**, 1644 (1961).
90. E. Lieber, E. Oftedahl and C. N. R. Rao, *J. Org. Chem.*, **28**, 194 (1963).
91. J. C. Kauer and W. A. Sheppard, *J. Org. Chem.*, **32**, 3580 (1967).
92. P. A. S. Smith and D. H. Kenny, *J. Org. Chem.*, **26**, 3580 (1961).
93. K. A. Jensen and C. Pedersen, *Acta Chim. Scand.*, **15**, 1104 (1961).
94. D. Martin, *Angew. Chem.*, **76**, 303 (1964); *Angew. Chem. Intl. Ed.*, **3**, 311 (1964).
95. D. Martin, *Chem. Ber.*, **97**, 2689 (1964).
96. M. Hedayatullah and L. Denivelle, *Compt. Rend.*, **260**, 2839 (1965).
97. P. Reich and D. Martin, *Chem. Ber.*, **98**, 2063 (1965).
98. D. Martin and W. Mucke, *Chem. Ber.*, **98**, 2059 (1965).
99. D. Martin, *Tetrahedron Letters*, 2829 (1964).
100. K. A. Jensen and A. Holm, *Acta Chim. Scand.*, **18**, 826 (1964).
101. K. A. Jensen, M. Due and A. Holm, *Acta Chim. Scand.*, **19**, 438 (1965).
102. K. A. Jensen, A. Holm and C. Wentrup, *Acta Chim. Scand.*, **20**, 2107 (1966).
103. D. Martin, *Z. Chemie*, **7**, 123 (1967).
104. K. A. Jensen, S. Burmester and T. A. Bak, *Acta Chim. Scand.*, **21**, 2792 (1967).
105. N. N. Yarovensko, L. I. Motornyi and L. I. Kirenskaya, *J. Gen. Chem. SSSR*, **27**, 2301 (1957): *Chem. Abstr.*, **52**, 8052b (1958).
106. W. Kirmse, *Angew. Chem.*, **71**, 537 (1959).
107. F. Eloy, *J. Org. Chem.*, **26**, 953 (1961).
108. R. Stolle and A. Netz, *Ber.*, **55**, 1297 (1922).
109. R. Stolle and F. Helwerth, *Ber.*, **47**, 1132 (1914).
110. R. B. Woodward and R. A. Olofson, *J. Am. Chem. Soc.*, **83**, 1007 (1961).
111. H. Behringer and H. J. Fischer, *Chem. Ber.*, **95**, 2546 (1962).
112. J. D. Roberts, *Chem. Ber.*, **94**, 273 (1961).
113. J. C. Tobin, R. N. Butler and F. L. Scott, *Chem. Comm.*, 112 (1970).
114. F. L. Scott, R. N. Butler and D. A. Cronin, *Angew. Chem.*, **77**, 963 (1965).
115. M. O. Forster, *J. Chem. Soc.*, **95**, 184 (1909).
116. H. Wieland, *Ber.*, **42**, 4199 (1909); cf. H. Heneka and P. Kurz in *Houben-Weyl. Methoden der Organischen Chemie*, Vol. 8, part III (Ed. E. Müller), Thieme Verlag, Stuttgart (1952), p. 696.
117. M. S. Chang and A. J. Matuszko, *J. Org. Chem.*, **28**, 2269 (1963).
118. C. Grundmann and H.-D. Frommeld, *J. Org. Chem.*, **31**, 157 (1966).
119. P. A. S. Smith, *J. Am. Chem. Soc.*, **76**, 436 (1954).
120. P. A. S. Smith and E. Leon, *J. Am. Chem. Soc.*, **80**, 4647 (1958).
121. K. v. Fraunberg and R. Huisgen, *Tetrahedron Letters*, 2599 (1969).
122. R. Huisgen, *Angew. Chem.*, **72**, 359 (1960).
123. J. Thiele, *Ann. Chem.*, **270**, 1, 54 (1892); **303**, 57 (1898).
124. A. Hantzsch and A. Vagt, *Ann. Chem.*, **314**, 339 (1901).

125. R. A. Henry, W. G. Finnegan and E. Lieber, *J. Am. Chem. Soc.*, **77**, 2264 (1955).
126. W. S. McEwan and M. M. Rigg, *J. Am. Chem. Soc.*, **73**, 4725 (1951).
127. K. Matsuda and L. T. Morin, *J. Org. Chem.*, **26**, 3783 (161).
128. F. L. Scott, F. C. Britten and J. Reilly, *J. Org. Chem.*, **21**, 1519 (1956).
129. R. A. Henry, W. Finnegan and E. Lieber, *J. Am. Chem. Soc.*, **76**, 88 (1954).
130. W. L. Garbrecht and R. M. Herbst, *J. Org. Chem.*, **18**, 1269 (1953).
131. H. K. Nagy, A. J. Tomson, and J. P. Horwitz, *J. Am. Chem. Soc.*, **82**, 1609 (1960).
132. K. A. Jensen and C. Pedersen, *Acta Chem. Scand.*, **15**, 991 (1961).
133. G. Tappi, *Rec. Trav. Chim.*, **62**, 207 (1943).
134. K. A. Jensen and O. R. Hansen, *Rec. Trav. Chim.*, **62**, 658 (1943).
135. R. M. Herbst and W. L. Garbrecht, *J. Org. Chem.*, **18**, 1283 (1953).
136. R. M. Herbst and J. E. Klingbeil, *J. Org. Chem.*, **23**, 1912 (1958).
137. H. Veldstra and P. W. Wiardi, *Rec. Trav. Chim.*, **62**, 660 (1943).
138. R. Neidlein and E. Heukelbach, *Ang. Chem.*, **78**, 548 (1966).
139. R. Neidlein and W. Haussmann, *Tetrahedron Letters*, 5401 (1966).
140. C. V. Hart, *J. Am. Chem. Soc.*, **50**, 1922 (1928).
141. W. P. Norris and R. A. Henry, *J. Org. Chem.*, **29**, 650 (1964).
142. G. Darzens, *Compt. Rend.*, **154**, 1232 (1912).
143. K. A. Hofmann, H. Hock and R. Roth, *Ber.*, **43**, 1087 (1910).
144. A. Schmidt, *Chem. Ber.*, **100**, 3725 (1967).
145. A. Schmidt, *Chem. Ber.*, **100**, 3319 (1967).
146. A. F. Hegarthy, J. B. Aylward and F. L. Scott, *Tetrahedron Letters*, 1259 (1967).
147. A. F. Hegarthy, J. B. Aylward and F. L. Scott, *J. Chem. Soc.* (*C*), 2587 (1967).
148. J. M. Burgess and M. S. Gibson, *Tetrahedron*, **18**, 1001 (1962).
149. W. Lwowski and D. Grassmann, Abstracts, Southwestern Regional Meeting, American Chemical Soc., Austin, Texas, December, 1968; Abstract no. 183, p. 91A.
150. J. H. Boyer and E. J. Miller, *J. Am. Chem. Soc.*, **81**, 4671 (1959).
151. R. Stolle and H. Storch, *J. Prakt. Chem.*, **135**, 128 (1932).
152. G. A. Reynolds, J. A. VanAllan and J. F. Tinker, *J. Org. Chem.*, **24**, 1205 (1959).
153. J. A. VanAllan, G. A. Reynolds and D. P. Maier, *J. Org. Chem.*, **34**, 1691 (1969).
154. H. Beyer, G. Wolter and H. Lemke, *Chem. Ber.*, **89**, 2554 (1956).
155. C. Benson, L. Hartzel and E. Otten, *J. Am. Chem. Soc.*, **76**, 1859 (1954).
156. R. Fusco, S. Rossi and S. Maiorana, *Tetrahedron Letters*, 1965 (1965).
157. C. Temple, Jr. and J. A. Montgomery, *J. Am. Chem. Soc.*, **86**, 2946 (1964).
158. C. Temple, Jr. and J. A. Montgomery, *J. Org. Chem.*, **30**, 826 (1965).
159. C. Temple, Jr., R. L. McKee and J. A. Montgomery, *J. Org. Chem.*, **30**, 829 (1965).
160. C. Temple, Jr., W. C. Coburn, Jr., M. C. Thorpe and J. A. Montgomery, *J. Org. Chem.*, **30**, 2395 (1965).
161. I. Ya. Postovskii and N. B. Smirnova, *Dokl. Akad. Nauk*, **166**, 1136 (1966); *Chem. Abstr.*, **64**, 17589g (1966).

162. R. Huisgen, K. v. Fraunberg and H. J. Sturm, *Tetrahedron Letters*, 2589 (1969).
163. R. Huisgen and K. v. Fraunberg, *Tetrahedron Letters*, 2595 (1969).
164. I. Ya. Postovskii and I. N. Goncharova, *Zh. Obshch. Khim.*, **33**, 2334 (1963); *J. Gen. Chem. SSSR*, **33**, 2274 (1963).
165. I. N. Goncharova and I. Ya. Postovskii, *Zh. Obshch. Khim.*, **33**, 2475 (1963); *J. Gen. Chem. SSSR*, **33**, 2413 (1963).
166. I. J. Postovskii, G. N. Tiurekova and L. F. Lipatova, *Dokl. Akad. Nauk*, **179**, 111 (1968): *Chem. Abstr.*, **69**, 67292q (1968).
167. CIBA, Ltd. Neth. Pat. App., 6,413,689; *Chem. Abstr.*, **64**, 741g (1966).
168. T. Itai and G. Ito, *Pharm. Chem. Bull. Japan*, **10**, 1141 (1962).
169. C. Temple, Jr., M. C. Thorpe, W. C. Coburn, Jr. and J. A. Montgomery, *J. Org. Chem.*, **31**, 935 (1966).
170. C. Temple, Jr., C. L. Kussner and J. A. Montgomery, *J. Org. Chem.*, **31**, 2110 (1966).
171. E. Lieber and D. R. Levering, *J. Am. Chem. Soc.*, **73**, 1313 (1951).
172. F. L. Scott and M. N. Holland, *Proc. Chem. Soc.*, **106**, (1962).
173. F. L. Scott and D. A. Cronin, *Tetrahedron Letters*, 715 (1963).
174. R. A. Henry, *J. Org. Chem.*, **31**, 1973 (1966).
175. B. Stanovic and M. Tisler, *Tetrahedron*, **23**, 3313 (1969).
176. E. Oliveri-Mandala and F. Noto, *Gazz. Chim. Ital.*, **43**, 514 (1913).
177. E. Oliveri-Mandala, *Gazz. Chim. Ital.*, **44**, 662 (1914).
178. J. Thiele and O. Stange, *Ann. Chem.*, **283**, 1 (1894).
179. J. Thiele and O. Stange, *Ber.*, **27**, 31 (1894).
180. T. Curtius and K. Heidenreich, *Ber.*, **27**, 55 (1894).
181. T. Curtius and K. Heidenreich, *J. Prakt. Chem.*, **52**, 454 (1895).
182. T. Curtius and F. Schmidt, *J. Prakt. Chem.*, **105**, 177 (1923).
183. J. Thiele and E. Uhlfelder, *Ann. Chem.*, **303**, 93 (1898).
184. R. Stolle, *Ber.*, **43**, 2468 (1910).
185. R. Stolle and K. Krauch, *Ber.*, **47**, 724 (1914).
186. R. Neidlein, *Angew. Chem.*, **78**, 333 (1966); *Angew. Chem. Intl. Ed.*, **5**, 314 (1966).
187. R. Neidlein, *Arch. Pharmazie*, **299**, 1003 (1966); *Chem. Abstr.*, **66**, 95135x (1967).
188. R. Neidlein, *Angew. Chem.*, **80**, 496 (1968).
189. E. Lieber and R. L. Minnis, *Chem. Rev.*, **65**, 377 (1965).
190. T. Curtius and A. Burkhardt, *J. Prakt. Chem.*, **58**, 205 (1898).
191. F. L. Scott, A. Koczarski and J. Reilly, *Nature*, **170**, 922 (1952).
192. F. L. Scott and M. T. Scott, *J. Am. Chem. Soc.*, **79**, 6077 (1957).
193. T. Curtius, *Angew. Chem.*, **27**/III, 213 (1914).
194. R. Stolle, *Ber.*, **57**, 1063 (1924).
195. R. Stolle, H. Nieland and M. Merkle, *J. Prakt. Chem.*, **116**, 192 (1927).
196. R. Stolle, H. Nieland and M. Merkle, *J. Prakt. Chem.*, **117**, 185 (1927).
197. N. Koga and J.-P. Anselme, *J. Org. Chem.*, **33**, 3963 (1968).
198. W. Lwowski, R. DeMauriac, T. W. Mattingly, Jr. and E. Scheiffele, *Tetrahedron Letters*, 3285 (1964).
199. R. DeMauriac, Thesis, Yale University, 1967.
200. W. S. Wadsworth and W. D. Emmons, *J. Org. Chem.*, **32**, 1279 (1967).
201. W. Lwowski and L. Luenow, unpublished results.

202. W. Lwowski and R. A. Murray, unpublished results.
203. W. Lwowski and R. DeMauriac, Abstracts, 155th Natl. Meeting, American Chemical Soc., San Francisco, April 1968; Abstract No. P-056.
204. T. Curtius, *Ber.*, **56** (1923); T. Curtius and H. Meier, *J. Prakt. Chem.*, **125**, 458 (1930).
205. A. Bertho, *J. Prakt. Chem.*, **120**, 29 (1929).
206. T. Curtius, *Angew. Chem.*, **28**/III, 5 (1915).
207. T. Curtius and F. Schmidt, *Ber.*, **55**, 1571 (1922).
208. T. Curtius and F. W. Haas, *J. Prakt. Chem.*, **102**, 85 (1921).
209. R. A. Abramovitch and V. Uma, *Chem. Comm.*, 797 (1968).
210. R. A. Abramovitch, G. N. Knauss and V. Uma, *J. Am. Chem. Soc.*, **91**, 7532 (1969).
211. W. Lwowski and K. Koyama, unpublished results.
212. J. P. Fleury, D. v. Assche and A. Bader, *Tetrahedron Letters*, 1399 (1965).
213. T. Curtius and W. Dörr, *J. Prakt. Chem.*, **125**, 425 (1930).
214. T. Curtius and K. Raschig, *J. Prakt. Chem.*, **125**, 466 (1930).
215. T. Curtius and W. Klavehn, *J. Prakt. Chem.*, **125**, 498 (1930).
216. T. Curtius and W. Sieber, *J. Prakt. Chem.*, **125**, 444 (1930).
217. R. Stolle and M. Merkle, *J. Prakt. Chem.*, **119**, 275 (1928).
218. V. J. Bauer, W. J. Fanshawe and R. S. Safir, *J. Org. Chem.* **31**, 3440 (1966).
219. R. Kreher and G. H. Berger, *Tetrahedron Letters*, 369 (1965).
220. R. Kreher and G. H. Bockhorn, *Angew. Chem.*, **76**, 681 (1964).
221. R. Puttner and K. Hafner, *Tetrahedron Letters*, 3119 (1964).
222. A. Bertho, *J. Prakt. Chem.*, **116**, 101 (1927).
223. J. P. Collman, M. Kubota, J. Y. Sun and F. Vastine, *J. Am. Chem. Soc.*, **89**, 169 (1967).
224. G. P. Volpp, R. Butler and H. M. Castrantas, personal communication; with permission of FMC Corporation.
225. L. A. Carpino, C. A. Giza and B. A. Carpino, *J. Am. Chem. Soc.*, **81**, 955 (1959).
226. K. P. Polzhofer, *Chimia*, **23**, 298 (1969).
227. T. W. Mattingly, Jr. Thesis, Yale University, 1964, p. 98.
228. Anonymous personal communication by the victim.
229. I. Honda, Y. Shimonishi and S. Sakakibara, *Bull. Chem. Soc. Japan*, **40**, 2415 (1957).
230. J. Hora, *Coll. Czech. Chem. Comm.*, **29**, 1079 (1964).
231. R. G. Hiskey and J. B. Adams, Jr., *J. Am. Chem. Soc.*, **87**, 3969 (1965).
232. F. Weygand and K. Hunger, *Chem. Ber.*, **59**, 1 (1962).
233. G. B. Feild and J. R. Lewis, *US Pat.* 3,211,677; *Chem. Abstr.*, **64**, 890d (1966).
234. T. J. Prosser, A. F. Marcantonio, C. A. Genge and D. S. Breslow, *Tetrahedron Letters*, 2483 (1964).
235. W. Lwowski and T. W. Mattingly, Jr., *J. Am. Chem. Soc.*, **87**, 1947 (1965).
236. S. M. A. Hai, Thesis, New Mexico State University, 1969.
237. M. Prince and C. M. Orlando, *Chem. Comm.*, 818 (1967).
238. R. Huisgen, R. Grashey and J. Sauer, in *The Chemistry of Alkenes* (Ed. S. Patai), Interscience, New York, 1964, p. 739, esp. pp. 835 ff.

239. R. Huisgen, *Angew. Chem.*, **75**, 604 (1963).
240. R. Huisgen, *Angew. Chem.*, **75**, 742 (1963).
241. R. Huisgen, *Angew. Chem.*, **80**, 329 (1968).
242. G. L'abbe, *Chem. Revs*, **69**, 345 (1969).
243. R. A. Firestone, *J. Org. Chem.*, **33**, 2285 (1968).
244. R. Huisgen, *J. Org. Chem.*, **33**, 2291 (1968).
245. R. Fusco, G. Bianchetti and D. Pocar, *Gazz. Chim. Ital.*, **91**, 933 (1961).
246. G. Müller, Dissertation, Univ. of Munich, 1962; as quoted in ref. 229.
247. A. S. Bailey and J. E. White, *Chem. and Ind.*, 1628 (1965).
248. A. C. Oehlschlager, P. Tillman and L. H. Zalkow, *Chem. Comm.*, 596 (1965).
249. K. Tori, K. Kitahanoki, Y. Takano, H. Tanida and T. Tsuji, *Tetrahedron Letters*, 869 (1965).
250. R. J. Stedman, A. C. Swift and J. R. E. Hoover, *Tetrahedron Letters*, 2525 (1965).
251. R. D. Burpitt and V. W. Goodlett, *J. Org. Chem.*, **30**, 4308 (1965).
252. I. Brown and O. E. Edwards, *Can. J. Chem.*, **43**, 1266 (1965).
253. E. J. Moriconi and R. E. Misner, *J. Org. Chem.*, **34**, 3672 (1969).
254. J. H. Boyer, in *Heterocyclic Compounds*, Vol. 7 (Ed. R. C. Elderfield), J. Wiley, New York, 1961, pp. 384 ff.
255. R. Huisgen, P. Knorr, L. Möbius and G. Szeimies, *Chem. Ber.*, **98**, 4014 (1965).
256. G. S. Akimova, V. N. Christokletov and A. A. Petrov, *Zh. Org. Khim.*, **3**, 968 (1967): *Chem. Abstr.*, **68**, 87243g (1968).
257. G. S. Akimova, V. N. Christokletov and A. A. Petrov, *Zh. Org. Khim.*, **4**, 389 (1968): *Chem. Abstr.*, **68**, 105100q (1968).
258. K. D. Berlin, R. Ranganathan and H. Haberlein, *J. Heterocycl. Chem.*, **5**, 813 (1968).
259. W. Ried and M. Schön, *Chem. Ber.*, **98**, 3142 (1965).
260. J. v. Braun and W. Rudolph, *Ber.*, **74**, 264 (1941).
261. F. R. Benson in *Heterocyclic Compounds*, Vol. 8 (Ed. R. C. Elderfield), J. Wiley, New York, 1967, pp. 11 ff.
262. R. Huisgen and L. Möbius, unpubl. results, see footnote 21 in ref. 254.
263. W. Carpenter, *J. Org. Chem.*, **27**, 2085 (1962).
264. R. Huisgen and H. Blaschke, *Ann. Chem.*, **686**, 145 (1945).
265. J. E. Leffler and R. D. Temple, *J. Am. Chem. Soc.*, **89**, 5235 (1967).
266. H. Staudinger and J. Meyer, *Helv. Chim. Acta*, **2**, 635 (1919); H. Staudinger and E. Hauser, *Helv. Chim. Acta*, **4**, 861 (1921).
267. H. Maltz, *German Patent* 1,104,958 (1959); *Chem. Abstr.*, **56**, 10041 (1962).
268. L. Horner and H. Oediger, *Ann. Chem.*, **627**, 142 (1959).
269. H. Bock and W. Wiegräbe, *Angew. Chem.*, **75**, 789 (1963).
270. J. E. Leffler, U. Honsberg, Y. Tsuno and I. Fosblad, *J. Org. Chem.*, **26**, 4810 (1961).
271. J. A. VanAllan and G. A. Reynolds, *J. Heterocyclic Chem.*, **5**, 471 (1968).
272. J. E. Leffler and Y. Tsuno, *J. Org. Chem.*, **28**, 902 (1963).
273. G. R. Harvey, *J. Org. Chem.*, **31**, 1587 (1966).
274. G. L'abbe and H. Bestmann, *Tetrahedron Letters*, 63 (1969).
275. E. Lieber and C. N. V. Nambury, *Chem. Ind.*, 883 (1959).

276. P. A. S. Smith, J. H. Hall and R. O. Kan, *J. Am. Chem. Soc.*, **84**, 485 (1962).
277. J. Ladik and A. Messmer, *Acta Chim. Hungar*, **34**, 7 (1962).
278. G. Palzzo and G. Strani, *Gazz. Chim. Ital.*, **90**, 1290 (1960).
279. L. Bernardi and O. Goffredo, *Gazz. Chim. Ital.*, **94**, 947 (1964).
280. R. A. Boissonnas, in *Advances in Organic Chemistry*, Vol. 3 (Ed. R. A. Raphael, E. C. Taylor and H. Wynberg) Interscience, New York, 1963, pp. 159 ff.
281. R. Schwyzer, P. Sieber and H. Kappeler, *Helv. Chim. Acta*, **42**, 2622 (1959).
282. R. Mertz and J. P. Fleury, *Compt. Rend.*, **262**, 571 (1966).
283. R. A. Coleman, M. S. Newman and A. B. Garrett, *J. Am. Chem. Soc.*, **76**, 4534 (1954).
284. M. S. Newman and H. L. Gildenhorn, *J. Am. Chem. Soc.*, **70**, 317 (1948).
285. E. Fahr and L. Neumann, *Angew. Chem.*, **77**, 591 (1965).
286. E. Fahr and N. Lutz, *Ann. Chem.*, **715**, 15 (1968).
287. R. Kreher and G. Jäger, *Angew. Chem.*, **77**, 730 (1965); *Angew. Chem. Intl. Ed.*, **4**, 706 (1965).
288. R. Kreher and G. Jäger, *Z. Naturforsch.*, **20b**, 276 (1965).
289. R. Kreher and G. Jäger, *Z. Naturforsch.*, **20b**, 1131 (1965).
290. W. C. Kaska, C. Sutton and E. Serros, *Chem. Comm.*, 100 (1970).
291. L. Horner and G. Bauer, *Tetrahedron Letters*, 5373 (1966).
292. M. Reagan and A. Nickon, *J. Am. Chem. Soc.*, **90**, 4096 (1968).
293. J. E. Leffler and H. H. Gibson, Jr., *J. Am. Chem. Soc.*, **90**, 4117 (1968).
294. M. Frankel, D. Wagner, D. Gertner and A. Zilkha, *J. Organomet. Chem.*, **7**, 518 (1967).
295. F. Tiemann, *Ber.*, **24**, 4126 (1891).
296. S. Linke, G. T. Tisue and W. Lwowski, *J. Am. Chem. Soc.*, **89**, 6308 (1967).
297. D. S. Breslow, T. J. Prosser, A. F. Marcantonio and C. A. Genge, *J. Am. Chem. Soc.*, **89**, 2384 (1967).
298. T. J. Prosser, A. F. Marcantonio and D. S. Breslow, *Tetrahedron Letters*, 2479 (1964).
299. T. J. Prosser, A. F. Marcantonio, C. A. Genge and D. S. Breslow, *Tetrahedron Letters*, 2483 (1964).
300. D. S. Breslow and E. I. Edwards, *Tetrahedron Letters*, 2123 (1967).
301. R. Huisgen and H. Blaschke, *Chem. Ber.*, **98**, 2985 (1965).
302. W. Lwowski, T. J. Maricich and T. W. Mattingly, *J. Am. Chem. Soc.*, **85**, 1200 (1963).
303. W. Lwowski and T. J. Maricich, *J. Am. Chem. Soc.*, **87**, 3630 (1965).
304. E. Wasserman, private communication.
305. W. Lwowski and J. S. McConaghy, *J. Am. Chem. Soc.*, **87**, 5490 (1965).
306. J. S. McConaghy, Jr. and W. Lwowski, *J. Am. Chem. Soc.*, **89**, 2357 (1967).
307. J. S. McConaghy, Jr. and W. Lwowski, *J. Am. Chem. Soc.*, **89**, 4450 (1967).
308. W. Lwowski and R. L. Johnson, *Tetrahedron Letters*, 891 (1967).
309. J. M. Simson and W. Lwowski, *J. Am. Chem. Soc.*, **91**, 5107 (1969).
310. A. L. J. Beckwith and J. W. Redmont, *Chem. Comm.*, 165 (1967); *J. Am. Chem. Soc.*, **90**, 1351 (1968).
311. A. L. J. Beckwith and J. W. Redmont, *Australian J. Chem.*, **19**, 1859 (1966).

312. H. Nozaki, S. Fujita, H. Takaya and R. Noyori, *Tetrahedron*, **23**, 45 (1967).
313. A. Mishra, S. N. Rice and W. Lwowski, *J. Org. Chem.*, **33**, 481 (1968).
314. R. S. Berry, D. W. Cornell and W. Lwowski, *J. Am. Chem. Soc.*, **85**, 1199 (1963).
315. D. W. Cornell, R. S. Berry and W. Lwowski, *J. Am. Chem. Soc.*, **87**, 3626 (1965).
316. J. ApSimon and O. E. Edwards, *Proc. Chem. Soc.*, 461 (1961).
317. J. ApSimon and O. E. Edwards, *Can. J. Chem.*, **40**, 896 (1962).
318. L. Horner and A. Christmann, *Chem. Ber.*, **96**, 388 (1963).
319. L. Horner, G. Bauer and J. Dörges, *Chem. Ber.*, **98**, 2631 (1965).
320. W. Lwowski and G. T. Tisue, *J. Am. Chem. Soc.*, **87**, 4022 (1965).
321. N. C. Castellucci, M. Kato, H. Zenda and S. Masamune, *Chem. Comm.*, 473 (1967).
322. G. Just and W. Zehetner, *Tetrahedron Letters*, 3389 (1967).
323. W. Lwowski and T. W. Mattingly, Jr., *Tetrahedron Letters*, 277 (1962).
324. W. Lwowski and T. J. Maricich, *J. Am. Chem. Soc.*, **86**, 3164 (1964).
325. D. S. Breslow, E. I. Edwards, R. Leone and P. v. R. Schleyer, *J. Am. Chem. Soc.*, **90**, 7097 (1968).
326. J. Meinwald and D. H. Aue, *Tetrahedron Letters*, 2317 (1967).
327. G. Smolinsky and B. I. Feuer, *J. Am. Chem. Soc.*, **86**, 3085 (1964).
328. G. Smolinsky, *Trans. N.Y. Acad. Sci., ser. II*, **30**, 511 (1968).
329. S.-I. Yamada, S. Terashima and K. Achiwa, *Chem. Pharm. Bull. Japan*, **13**, 751 (1965).
330. S. Terashima and S.-I. Yamada, *Chem. Pharm. Bull. Japan*, **16**, 1953 (1968).
331. S.-I. Yamada and S. Terashima, *Chem. Comm.*, 511 (1969).
332. J. H. Hall, 13th Annual Report, Petroleum Research Fund, for 1968, p. 145 (1969).
333. J. H. Hall, J. W. Hill and H.-C. Tsai, *Tetrahedron Letters*, 2211 (1965).
334. J. H. Hall, J. W. Hill and J. M. Fargher, *J. Am. Chem. Soc.*, **90**, 5313 (1968).
335. A. G. Anastassiou, *J. Am. Chem. Soc.*, **89**, 3184 (1967).
336. D. S. Breslow and E. I. Edwards, *Tetrahedron Letters*, 2123 (1967).
337. W. Antkowiak, O. E. Edwards, R. Howe and J. W. ApSimon, *Can. J. Chem.*, **45**, 2599 (1967).
338. S. Masamune, *J. Am. Chem. Soc.*, **86**, 288 (1964).
339. W. L. Meyer and A. S. Levinson, *J. Org. Chem.*, **28**, 2859 (1963).
340. R. F. C. Brown, *Australian J. Chem.*, **17**, 47 (1965).
341. S. Huneck, *Chem. Ber.*, **98**, 2305 (1965).
342. W. Lwowski, R. DeMauriac, T. W. Mattingly, Jr. and E. Scheiffele, *Tetrahedron Letters*, 3285 (1964).
343. R. Puttner and K. Hafner, *Tetrahedron Letters*, 3119 (1964).
344. R. Kreher and H. Bockhorn, *Angew. Chem.*, **76**, 681 (1964).
345. R. Puttner, W. Kaiser and K. Hafner, *Tetrahedron Letters*, 4315 (1968).
346. C. R. Hauser and S. W. Kantor, *J. Am. Chem. Soc.*, **72**, 4284 (1950).
347. K. Hafner, D. Zinser and K. L. Moritz, *Tetrahedron Letters*, 1733 (1964).
348. W. Lwowski and L. Selman, unpublished results; L. Selman, Thesis, Yale Univ., New Haven, Conn., 1966.

349. W. Lwowski and L. Selman, *Abstracts*, 150th Natl. Meeting, ACS, Atlantic City, Sept., 1965; Abstract S 25.
350. K. Hafner, W. Kaiser and R. Puttner, *Tetrahedron Letters*, 3953 (1964).
351. R. S. Atkinson and C. W. Rees, *Chem. Comm.*, 1232 (1967).
352. W. Lwowski and F. P. Woerner, *J. Am. Chem. Soc.*, **87**, 5491 (1965).
353. I. Brown, O. E. Edwards, J. M. McIntosh and D. Vocelle, *Can. J. Chem.*, **47**, 2751 (1969).
354. K. Hafner and C. König, *Angew. Chem.*, **75**, 89 (1963); *Angew. Chem. Intl. Ed.*, **2**, 96 (1963).
355. R. J. Cotter and W. F. Beach, *J. Org. Chem.*, **29**, 751 (1964).
356. W. Lwowski, T. J. Maricich and T. W. Mattingly, Jr., *J. Am. Chem. Soc.*, **85**, 1200 (1963).
357. W. Lwowski and U. Schumacher, unpublished results.
358. J. E. Baldwin and R. A. Smith, *J. Am. Chem. Soc.*, **89**, 1886 (1967); *J. Org. Chem.*, **32**, 3511 (1967).
359. K. Hafner and W. Kaiser, *Tetrahedron Letters*, 2185 (1964).
360. R. Huisgen and H. Blaschke, *Tetrahedron Letters*, 1409 (1964).
361. J. Meinwald and D. H. Aue, *J. Am. Chem. Soc.*, **88**, 2849 (1966).
362. R. Huisgen and J.-P. Anselme, *Chem. Ber.*, **98**, 2998 (1965).
363. W. Lwowski, A. Hartenstein, C. DeVita and R. L. Smick, *Tetrahedron Letters*, 2497 (1964).
364. T. Bacchetti, *Gazz. Chim. Ital.*, **91**, 866 (1961).
365. W. Lwowski and P.-L. Kao, unpublished results.
366. W. Lwowski, M. Grassmann and T. Shingaki, unpublished results.
367. P. Robson and P. H. R. Speakman, *J. Chem. Soc.* (*B*), 463 (1968).
368. W. Lwowski, *Angew. Chem.*, **79**, 922 (1967); *Angew. Chem. Intl. Ed.*, **6**, 897 (1967).
369. *Nitrenes* (Ed. W. Lwowski), Wiley–Interscience, New York, 1970.

CHAPTER **10**

The chemistry of vinyl azides

GERALD SMOLINSKY AND CORALIE A. PRYDE

Bell Telephone Laboratories, Murray Hill, New Jersey, U.S.A.

I. INTRODUCTION 555

II. PREPARATIVE METHODS 556
A. Addition of XN_3 to Alkynes 556
B. Elimination of HX from Suitably Constituted Azides . 557
C. Addition–Elimination Reactions 561
D. Use of Phosphoroörganic Compounds 562
E. Preparation of Diazides 563

III. REACTIONS OF VINYL AZIDES 563
A. Thermally and Photochemically Induced Expulsion . 564
1. Reactions of vinyl azides stable at room temperature . 564
2. Reactions of vinyl azides unstable at room temperature 571
3. Reactions assumed to involve vinyl azides as intermediates 573
B. Acid-Induced Loss of Nitrogen 576
C. Reactions not involving Loss of Molecular Nitrogen 576

IV. MECHANISM 577

V. REARRANGEMENTS OF AZIDOQUINONES 581

VI. REFERENCES 584

I. INTRODUCTION

Vinyl azides have been known since 1910 when the simplest representative, azidoethylene, was prepared by Forster and Newman[1]. During the next half century no new vinyl azides were isolated. The past decade, however, has witnessed a growing interest in these compounds and preparative methods have been developed which make available a wide variety of derivatives. Among the reasons for this renewed interest are an active interest in the mechanism of organic azide decompositions and the discovery that the decomposition of vinyl azides offers a simple route to azirines.

Vinyl azides are characteristically yellow compounds most readily identified by the presence of strong N_3 and moderate to weak C=C stretching frequencies near 4·7 μm and 6·1 μm respectively, in their infrared spectra.

Although Forster and Newman commented on the somewhat surprising stability of their 'vinylazoimide'[1], later authors[2] reported detonations of this material. Aryl or alkyl substituents appear to have a stabilizing effect since vinyl azides so substituted are generally stable at room temperatures and undergo decomposition smoothly at elevated temperatures. Vinyl azides with electron-withdrawing substituents are frequently unstable at room temperature and some, including the β-azidovinyl ketones[3], have been found to explode violently.

In this chapter we will present a review of the literature through 1969 on the preparation and reactions of vinyl azides, then discuss briefly the conclusions which may be drawn concerning the mechanism of vinyl azide decomposition. A short section will also be included on the chemistry of azidoquinones.

II. PREPARATIVE METHODS

A. *Addition of* XN_3 *to Alkynes*

The most obvious route to vinyl azide synthesis lies in the addition of hydrazoic acid to an alkyne. Unfortunately, this reaction has been found to be successful with only one group of compounds, the esters of acetylene dicarboxylic acid (equation 1)[4]. Acetylene mono-

$$RO_2CC{\equiv}CCO_2R + HN_3 \longrightarrow RO_2CC(N_3){=}CHCO_2R \qquad (1)$$

carboxylic acids and their derivatives, as well as alkyl and aryl substituted acetylenes, are unreactive.

A more successful approach is found in the addition of iodine or bromine azide to alkynes[5,6]. Iodine azide has been found to add regiospecifically* to 1-phenylpropyne (**1**) to give a mixture of the *cis* and *trans* isomers of 2-azido-1-iodo-1-phenylpropene (**2**)[5]. Similarly,

$$\underset{(\mathbf{1})}{PhC{\equiv}CMe} \xrightarrow{IN_3} \underset{(\mathbf{2})}{PhCI{=}C(N_3)Me}$$

phenyl(1-hydroxycyclopentyl)ethyne (**3**)[5] and phenylethynyl bromide (**5**)[6,8] react with iodine azide to give the corresponding vinyl iodo-azides **4** and **6**. Bromine azide, on the other hand, adds to **1** to give a

$$\underset{(\mathbf{3})}{PhC{\equiv}C\text{–}C_5H_8(OH)} \xrightarrow{IN_3} \underset{(\mathbf{4})}{PhCI{=}C(N_3)\text{–}C_5H_8(OH)}$$

$$\underset{(\mathbf{5})}{PhC{\equiv}CBr} \xrightarrow{IN_3} \underset{(\mathbf{6})}{PhC(N_3){=}CBrI}$$

mixture of the regioisomers **7** and **8**[5]. Presumably, **7** results from an

$$\underset{(\mathbf{7})}{PhCBr{=}C(N_3)Me} \qquad \underset{(\mathbf{8})}{PhC(N_3){=}CBrMe}$$

ionic addition of bromine azide analogous to the addition of iodine azide, whereas **8** is formed by a free radical reaction.

B. *Elimination of HX from Suitably Constituted Azides*

By far the most general method yet found for the preparation of vinyl azides is the elimination of HX from compounds of the type **9**. Indeed, Forster and Newman's preparation[1] of azidoethylene was

$$\underset{(\mathbf{9})}{\text{—}\overset{X}{\underset{|}{\overset{|}{C}}}\text{—}\overset{N_3}{\underset{H}{\overset{|}{\underset{|}{C}}}}\text{—}} \xrightarrow{-HX} \;\rangle C{=}C\langle^{N_3}$$

* The term *regiospecificity* was introduced by Hassner[7] in reference to the directional or orientational preference of bond formation. A reaction is said to be *regiospecific* if only one of two or more possible structural isomers is formed.

accomplished by treating 1-azido-2-iodoethane with potassium hydroxide in aqueous alcohol (equation 2). The azido group is generally

$$ICH_2CH_2N_3 \xrightarrow[\text{aq. alc.}]{\text{KOH}} CH_2{=}CHN_3 \qquad (2)$$

stable toward alkalis and base-induced elimination has been used successfully with numerous other β-azidoalkyl halides and sulphonates[2,9–13].

The major problem in this sequence of reactions is the preparation of the necessary precursors. Fortunately, the addition of iodine azide to olefins has provided a convenient route to many of these compounds. In the majority of the olefins studied, this addition has been found to be regiospecific and, assuming iodine to be the cation, in accordance with Markownikov's rule. Thus, addition of iodine azide to a terminal olefin followed by dehydroiodination leads to the formation of the internal vinyl azide (equation 3). Moreover, the base-

$$RCH{=}CH_2 \xrightarrow{IN_3} RCH(N_3)CH_2I \xrightarrow{\text{Base}} RC(N_3){=}CH_2 \qquad (3)$$

induced elimination of hydrogen iodide is in many cases both regio- and stereospecific. For example, the iodoazides derived from *cis*-2-butene (**10**) and *trans*-2-butene (**11**) give, respectively, the *trans*- and *cis*-2-azido-2-butenes (**12**) and (**13**)[11].

$$\underset{(\mathbf{10})}{(Me)(H)C{=}C(Me)(H)} \xrightarrow{IN_3} \underset{threo}{MeCH(N_3)CHIMe} \xrightarrow{-HI} \underset{(\mathbf{12})}{(Me)(H)C{=}C(N_3)(Me)}$$

$$\underset{(\mathbf{11})}{(Me)(H)C{=}C(H)(Me)} \xrightarrow{IN_3} \underset{erythro}{MeCH(N_3)CHIMe} \xrightarrow{-HI} \underset{(\mathbf{13})}{(Me)(H)C{=}C(Me)(N_3)}$$

Hassner and co-workers explained the regiospecificity of these eliminations by assuming a directive effect of the azido group, while the stereospecificity, they argued, results from a strong preference for *trans* elimination of the elements of hydrogen iodide[11]. In some cases these two effects may be incompatible. With iodoazides derived from cyclic olefins of intermediate size, for instance, the tendency toward *trans* elimination has been found to outweigh the directive effect of the azido group. Thus elimination of hydrogen iodide from *trans*-1-azido-2-iodocyclopentane (**14a**) and the analogous cyclohexane derivative **14b** gives allyl azides **15** rather than the vinyl compounds.

(**14**) → (**15**) (a) $n = 1$ (b) $n = 2$

However, the products of the reaction of medium sized cyclic olefins, such as cycloöctene, with iodine azide did decompose to give the vinyl azides[11].

The regiospecificity of the addition reactions has been accounted for by assuming the intermediacy of a cyclic iodonium ion **16**[11]. This would be expected to undergo ring-opening in such a way as to provide maximum stability for the incipient carbonium ion (i.e. to

$$RHC\overset{\overset{I}{+}}{—}CH_2 \longrightarrow R\overset{+}{C}HCH_2I$$
(**16**)

form the secondary rather than the primary ion). Thus the presence of a substituent such as phenyl, which can effectively stabilize a carbonium ion, results in completely regiospecific addition to give the Markownikov product, while strongly destabilizing substituents such as carbonyl result in regiospecificity in the opposite direction.

Anti-Markownikov addition may also result from steric hindrance[11,12]. Addition of iodine azide to the moderately hindered compound 3-methyl-1-butene (**17**) followed by dehydroiodination

$$i\text{-PrCH}{=}CH_2 \xrightarrow{IN_3} \xrightarrow{-HI} i\text{-Pr}\overset{N_3}{\overset{|}{C}}H{=}CH_2 + i\text{-PrCH}{=}CHN_3$$
(**17**) (**18**) (**19**)

gave a mixture of the regioisomers **18** and **19** with the internal azide **18** predominating[11]. Reaction of the very hindered olefin 3,3-dimethyl-1-butene (**20**) gave only the terminal azide **21**[11].

$$t\text{-BuCH}{=}\text{CH}_2 \xrightarrow{\text{IN}_3} t\text{-BuCHICH}_2\text{N}_3 \xrightarrow{\text{HI}} t\text{-BuCH}{=}\text{CHN}_3$$

(**20**) (**21**)

It is apparent, then, that precursors for terminal vinyl azides are obtained following iodine azide addition to olefins only when rather unusual steric or electronic factors are operating. A number of other reaction sequences have been investigated in the search for a general approach to the terminal compounds. The opening of a terminal epoxide by azide ion leads to an azidohydrin which can be readily dehydrated[13] (equation 4). Unfortunately, this reaction sequence

$$\text{RR}^1\text{C}\overset{\text{O}}{\frown}\text{CH}_2 \xrightarrow{\text{N}_3^-} \text{RR}^1\text{C(OH)CH}_2\text{N}_3 \xrightarrow{-\text{H}_2\text{O}} \text{RR}^1\text{C}{=}\text{CHN}_3 \qquad (4)$$

seems to be practical only when the substituents R and R^1 are bulky enough to ensure that the azide ion will attack only the terminal carbon. Thus, while 1,1-diphenylethylene oxide, on treatment with sodium azide and subsequent dehydration, gave a 50% overall yield of the desired vinyl azide[13], 1-methyl-1-phenylethylene oxide gave only a 25% yield[13], and an attempt to prepare β-azidostyrene (**22**) from styrene oxide was completely unsuccessful[14].

The preparation of **22** was accomplished[15] by a sequence involving the sodium borohydride reduction of phenacyl azide, followed by the conversion of the resulting azidohydrin to the azidochloride, which was then subjected to base-catalysed elimination (equation 5).

$$\text{PhCOCH}_2\text{N}_3 \longrightarrow \text{PhCH(OH)CH}_2\text{N}_3 \longrightarrow \text{PhCHClCH}_2\text{N}_3 \longrightarrow \text{PhCH}{=}\text{CHN}_3 \qquad (5)$$

(**22**)

A better method for the preparation of the terminal isomers has resulted from the observations that bromine azide, unlike iodine azide, may be induced to add to olefins by a free-radical mechanism[16]. Thus, when styrene was subjected to the addition of bromine azide in pentane a quantitative yield of the precursor **23** for the terminal azide was obtained, while reaction in a more polar solvent gave a very high yield of the opposite regioisomer **24**.

$$\underset{(\mathbf{23})}{PhCHBrCH_2N_3} \qquad \underset{(\mathbf{24})}{PhCH(N_3)CH_2Br}$$

C. Addition–Elimination Reactions

A useful, although somewhat limited, method of preparing vinyl azides involves the addition of azide ion to vinyl halides with subsequent elimination of halide ion. This reaction is only possible when the halogen is located β to a group which can effectively stabilize the intermediate carbanion. A number of alkyl and aryl substituted β-chlorovinyl ketones (**25**) have been found to react in this way[3,17]. In many cases, however, the resulting β-azidovinyl ketones (**26**) cannot

$$\underset{(\mathbf{25})}{ROCR^1{=}CR^2Cl} \xrightarrow{N_3^-} \left[\overset{O}{\overset{\|}{RC}}{\cdots}CR^1CR^2ClN_3 \right]^- \xrightarrow{-Cl^-} \underset{(\mathbf{26})}{RCOCR^1{=}CR^2N_3}$$

be separated from the starting material. Use of the tertiary ammonium salts (**27**) instead of the chlorides leads to more easily separable reaction mixtures[3].

$$\underset{(\mathbf{27})}{[RCOCR^1{=}CR^2N^+Me_3]\ Cl^-}$$

Analogous addition–elimination reactions have been carried out on (9-fluorenylidene)alkyl halides[13] and tosylates[18] (equation 6); 1,2-di-*p*-toluenesulphonylethylene[19] (equation 7) and 2,2-dicyano-vinyl chlorides[20] (equation 8).

$$\text{(9-fluorenylidene)}{=}C(R)X \xrightarrow{N_3^-} \text{(9-fluorenylidene)}{=}C(R)N_3 \tag{6}$$

$$TsCH{=}CHTs \xrightarrow{N_3^-} TsCH{=}CHN_3 \tag{7}$$

$$(NC)_2C{=}CClR \xrightarrow{N_3^-} (NC)_2C{=}C(N_3)R \tag{8}$$

Perfluorinated olefins have been found to give both vinyl azides and the saturated fluoroazides on reaction with triethylammonium azide[21–23]; apparently the intermediate carbanion tends to abstract a proton in this case. Reaction of hexafluoropropene (**28**) with sodium azide in DMF leads stereospecifically to the *trans* azide **29**.

$$\underset{(\mathbf{28})}{CF_3CF{=}CF_2} \xrightarrow{NaN_3} \underset{(\mathbf{29})}{(CF_3)(F)C{=}C(F)(N_3)}$$

In a reaction quite similar to the aforementioned addition–elimination reactions, conjugated vinyl azides have been obtained from certain terminally unsubstituted allenic esters (equation 9)[24].

$$CH_2{=}C{=}CRCO_2Et \xrightarrow{N_3^-} \left[CH_2{\cdots}C(N_3){\cdots}CR{\cdots}C(O)OEt \right]^- \longrightarrow MeC(N_3){=}CRCO_2Et \qquad (9)$$

D. Use of Phosphoroörganic Compounds

The cyclic β-azidovinyl carbonyl compound **30** was prepared[25] by

$$\text{Ph}_3\text{P}{=}\text{(3-ylidene-}\gamma\text{-butyrolactone)} \xrightarrow{AcCl} \text{3-(Ph}_3\overset{+}{\text{P}}\text{)-3-Ac-}\gamma\text{-butyrolactone} \xrightarrow{NaN_3} \text{[Ph}_3\text{P–O–C(Me)(N}_3\text{) spiro intermediate]} \longrightarrow \underset{(\mathbf{30})}{\text{3-(MeC(N}_3\text{)=)-}\gamma\text{-butyrolactone}} \qquad (10)$$

acylation of the corresponding α-acyl-α-alkyl-methylenetriphenylphosphorane followed by reaction with sodium azide (equation 10).

$MeC(N_3){=}C(CO_2Et)Me$ (*cis* and *trans*)

(31)

The β-azidoacrylester **31** was obtained[25] by a corresponding reaction sequence and also on direct treatment of the phosphorane with acetyl azide. Unfortunately, a number of other phosphoranes treated with either acetyl azide or acetyl chloride and sodium azide cyclized to 4,5-*N*-acetyl-1,2,3-triazoles[25].

E. Preparation of Diazides

Diazides may be prepared most easily by double addition–elimination reactions of azide ion with suitable geminal or vicinal dihaloölefins. For example, treatment of the dichloride **32** with sodium azide

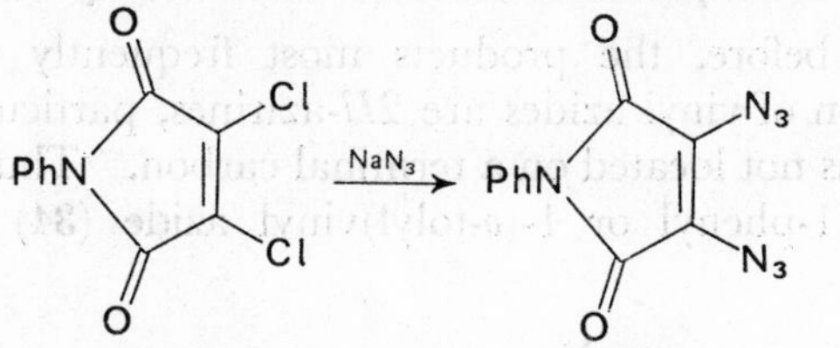

PhN(CO)C(Cl)=C(Cl)(CO) $\xrightarrow{NaN_3}$ PhN(CO)C(N_3)=C(N_3)(CO)

(32) **(33)**

gives 2,3-diazido-*N*-phenylmaleimide (**33**)[26]. A number of diazidoquinones have been prepared in this manner (section V).

III. REACTIONS OF VINYL AZIDES

The overwhelming majority of vinyl azide reactions involve the expulsion of molecular nitrogen and subsequent reorganization of the remainder of the molecule. Such decompositions may be initiated thermally or photochemically. Azirines are frequently isolated from the reaction mixtures, but a great variety of other products, including nitriles, dihydropyrazines, indoles and isoxazoles have also been obtained.

The decomposition method employed often determines the type of products isolated from a given azide. Such an observation can be explained as due to secondary reactions of unstable primary products, and does not necessarily indicate that more than one pathway is involved in the initial decomposition. For example, photo-induced, in contrast to thermally induced decompositions of terminal vinyl azides, frequently allow the isolation of azirines. In addition, the products obtained photochemically are generally free of the imino-ketenes or nitriles which are often present in the thermally generated product mixture. Unstable reaction intermediates will be mentioned in this section when there is reasonable physical or chemical evidence for their existence. A general discussion of mechanisms will be presented in section IV.

A limited number of reactions of vinyl azides are known in which molecular nitrogen is not extruded. These reactions, which will be discussed in Part C of this section, include nucleophilic displacement of the azido function and isomerization of the vinyl azide to a triazole.

A. Thermally and Photochemically Induced Expulsion

I. Reactions of vinyl azides stable at room temperature

As noted before, the products most frequently obtained from decomposition of vinyl azides are 2*H*-azirines, particularly when the azido group is not located on a terminal carbon. Thus vapour phase pyrolysis of 1-phenyl or 1-(*o*-tolyl)vinyl azide (**34**) gave azirines,

$$\underset{(\mathbf{34})}{\mathrm{Ar(N_3)C{=}CH_2}} \longrightarrow \underset{(\mathbf{35})}{\mathrm{ArC{=}N{-}CH_2}\ (\text{azirine})} + \underset{(\mathbf{36})}{\mathrm{ArN{=}C{=}CH_2}} \qquad \underset{(\mathbf{37})}{\text{Ph, NPh bicyclic}}$$

35, in about 80% yield[9]. A lower yield was obtained from 2-azido-1-hexene (**34**, R = *n*-butyl)[9]. Small quantities of iminoketenes **36** also were formed in these reactions as indicated by the infrared spectra of the crude pyrolysates[9]. Irradiation of very dilute solutions of α-azidostyrene (**34**, Ar = Ph) in benzene led to a better than 85% yield of the expected 2-phenylazirine (**35**, Ar = Ph) as well as 10% of 4-phenyl-3-phenylimino-1-azabicyclo [2.1.0] pentane (**37**)[27]. The latter compound is assumed to result from the photoinduced addition of *N*-phenyliminoketene (**36**, Ar = Ph) to the azirine.

Gas phase thermolysis of methyl azidofumarate (**38**) gave moderate-

$$\underset{(\mathbf{38})}{\mathrm{MeO_2C(H)C{=}C(N_3)CO_2Me}} \longrightarrow \underset{(\mathbf{39})}{\mathrm{N{=}CCO_2Me\ /\ CHCO_2Me}}$$

ly good yields of azirine **39**[13]. Photolysis of the ethyl azidocrotonates **40** (R = H, Me) in benzene solution gave nearly 80% yields of the

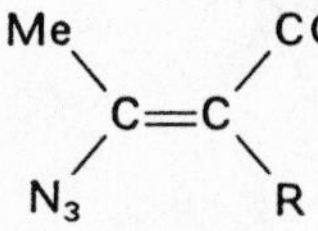

Me(N$_3$)C=C(CO$_2$Et)R **(40)**   MeC(=N)–C(CO$_2$Et)(R) **(41)**   MeN=C=C(CO$_2$Et)R **(42)**

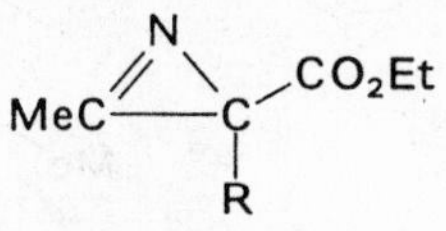

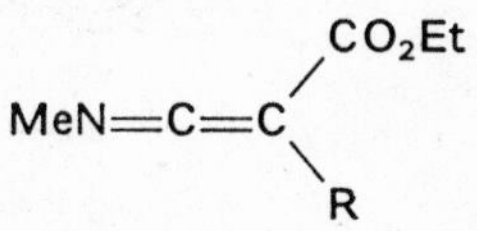

azirines **41** (R = H, Me) along with about 20% of the iminoketenes **42**[24].

On ultraviolet irradiation, the internal vinyl azides **43** gave good

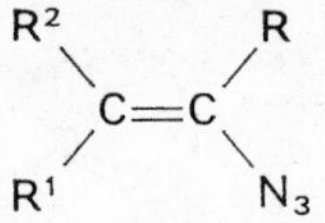

R^{2}R^{1}C=C(R)N$_3$ **(43)**   R^{1}R^{2}C–CR (=N) **(44)**

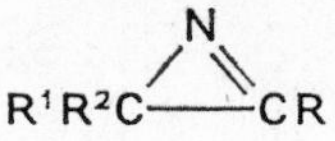

	a	*b*	*c*	*d*	*e*	*f*	*g*	*h*
R	Ph	Ph	Ph	Ph	PhCH$_2$	PhCH$_2$CH$_2$	*n*-Bu	Et
R^1	H	H	H	Ph	H	H	H	H
R^2	H	Me	Ph	H	H	H	H	Et

yields of azirines **44**[28]. Photolyses of the cyclic vinyl azides **45** and **46** yielded the bicyclic azirines **47** and **48**[28a].

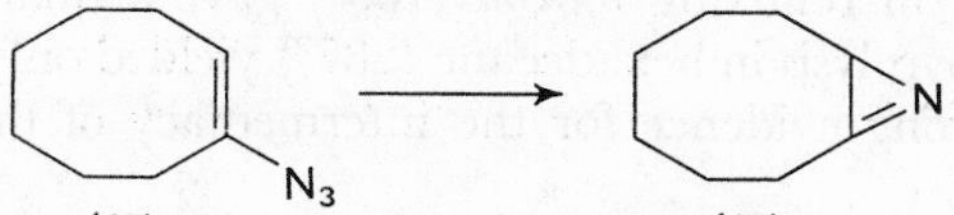

(45) → **(47)**

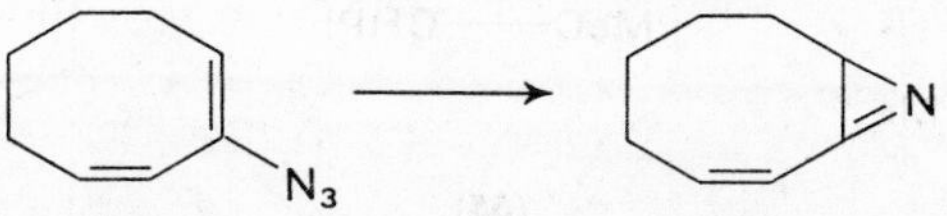

(46) → **(48)**

Decomposition of 9-(1-azidoethylidene)fluorene (**49**) in refluxing benzene led to high yields of the spiroazirine **50**[13]. In contrast, pyrolysis of the terminal vinyl azide, 9-(azidomethylene)fluorene (**51**) in benzene led to only one identifiable product, 9-(*N*,*N*-fluorenylidene-aminomethylene)fluorene (**52**), which was isolated in 25% yield[13].

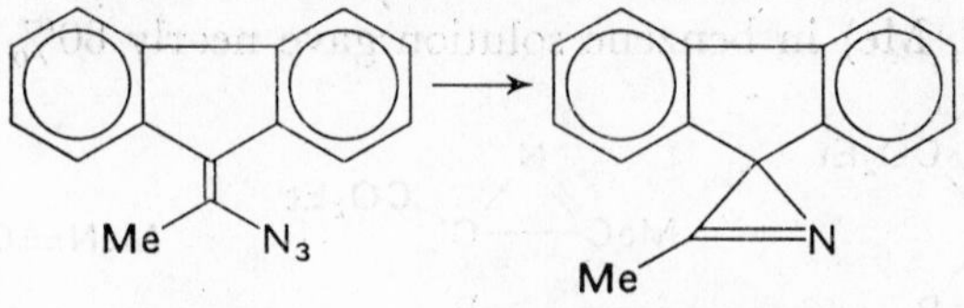

(49) (50)

However, photolytic decomposition[18] of **51** at −15° followed by a low-temperature work-up gave the spiroazirine **53**, which on warming in the presence of atmospheric oxygen reacted to give fluorenone and **52**.

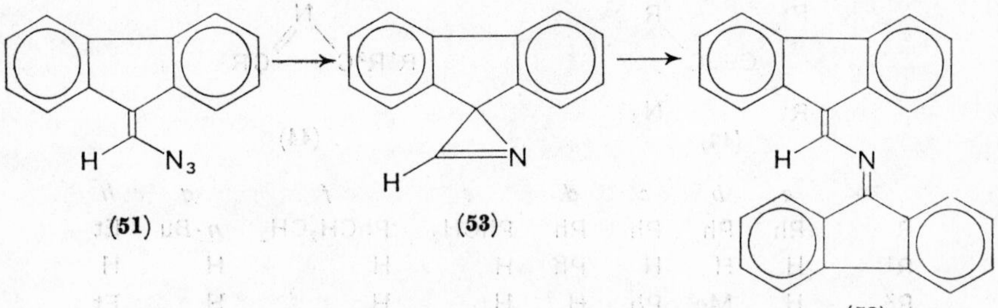

(51) (53) (52)

In some cases the azirines formed from internal azides may also be unstable to the pyrolysis conditions and undergo further rearrangement. Thus, while decomposition of *cis*- or *trans*-β-azido-β-methylstyrene (**54**) in refluxing ligroin (100°) gave a quantitative yield of azirine **55**, pyrolysis in hexadecane (287°) yielded only 2-methylindole (**56**)[28]. Strong evidence for the intermediacy of the azirine in the

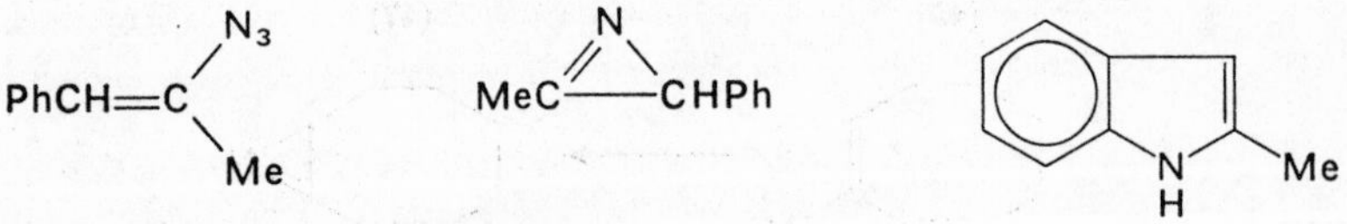

(54) (55) (56)

formation of **56** was obtained on conversion of **55** to the indole **56** by refluxing in hexadecane[29].

Pyrolysis[17] of the cyclic β-ketovinyl azide **57** at 50° gave the isoxazole **58** rather than the bicyclic azirine **59** which would be expected to possess a high degree of strain energy.

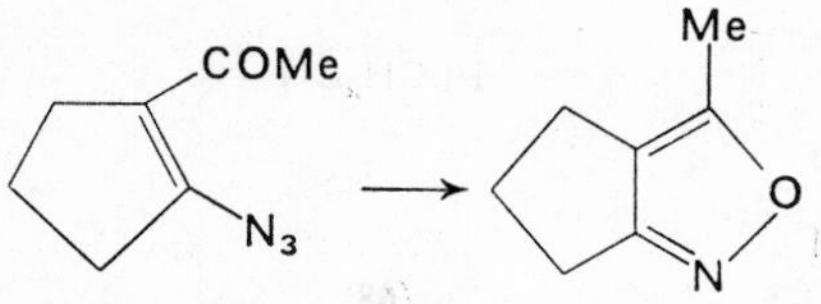

COMe N3 → Me O N COMe N

(57) (58) 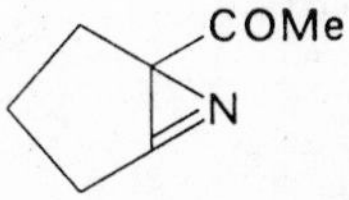 (59)

Thermolysis of terminal vinyl azides has generally led to the isolation of products other than azirines. A number of β-azidostyrene derivatives have, on pyrolysis, yielded indoles. Thus β-azido-α-methylstyrene, (**60**) when boiled in mesitylene, produced an 80% yield of 3-methylindole (**61**) and 9% of α-phenylpropionitrile (**62**)[13]. Decomposition of the same compound in ethanol, however, gave only a trace of the indole; the major product, isolated in about

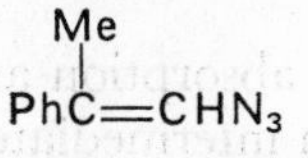

Me $PhC{=}CHN_3$ Me N H PhCH(CN)Me N Me Ph Ph Me N

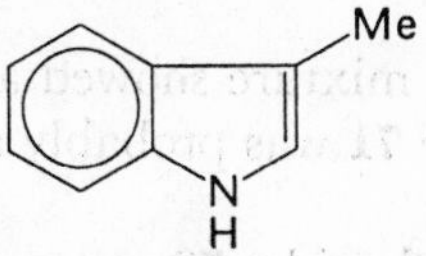

(60) (61) (62) (63)

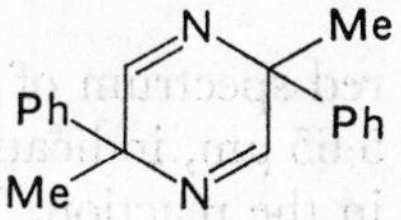

50% yield, was 2,5-dimethyl-2,5-diphenyldihydropyrazine (**63**). Similarly, 2-azido-1,1-diphenylethylene (**64**) gave 82% of 3-phenylindole (**65**) on decomposition in refluxing toluene, but only 53% of **65** and 20% of 2,2,5,5-tetraphenyldihydropyrazine (**66**) when the decomposition was carried out in ethanol[13].

$Ph_2C{=}CHN_3$ Ph N H N Ph Ph Ph Ph N

(64) (65) 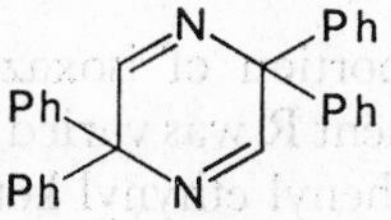 (66)

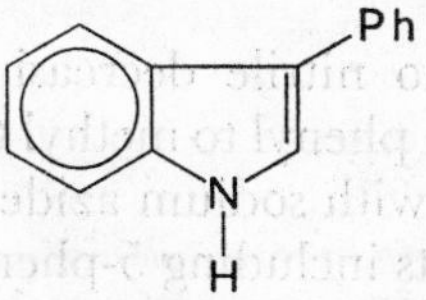

Pyrolysis of β-azidostyrene (**22**) in boiling hexadecane gave a 43% yield of indole (**67**) and an equal amount of phenylacetonitrile (**68**)[29]. When **22** was pyrolysed neat in the injection port of a gas chromatograph, the only product identified was the nitrile, isolated in 74% yield[15]. Only nitrile **70** was found when β-azidovinyl *p*-tolyl sulphone (**69**) was boiled in methanol[30]. Ultraviolet irradiation of **69**

(67) (68) $PhCH_2CN$

$TsCH{=}CHN_3$ (69) $TsCH_2CN$ (70)

in ethanol gave 2,3-di-*p*-toluenesulphonylaziridine (**72**)[30]. The infra-

TsHC—CH (azirine, **71**) $\xrightarrow{TsH}$ TsHC—CHTs (aziridine, NH, **72**)

red spectrum of the crude reaction mixture showed an absorption at 5·65 μm, indicating that the azirine **71** was probably an intermediate in the reaction.

Pyrolysis of the terminal ketovinyl azides **73** at temperatures below 110° led to the formation of isoxazoles **74** and nitriles **75** with the

$PhCOC(R){=}CHN_3$ (**73**) ⟶ isoxazole (**74**) + $PhCOCH(CN)R$ (**75**)

proportion of isoxazole to nitrile decreasing markedly as the substituent R was varied from phenyl to methyl to hydrogen[17]. Reaction of phenyl ethynyl ketone with sodium azide in an aqueous solution[31] gives a number of products including 5-phenylisoxazole (**74**, R = H). Since **74** was also obtained from **73**, it seems reasonable to suggest that **73** is an intermediate in this reaction. Azide **73** can be formed from the phenyl ethynyl ketone by addition of azide ion to the terminus of the triple bond resulting in a resonance-stabilized carbanion **76** which abstracts a proton from water.

Photo-induced decomposition of the β-azidovinyl ketones **73**[7,32] leads, as does the thermally induced reaction, to the isolation of isoxazoles **74** and nitriles **75**. It seems likely, however, that azirines may

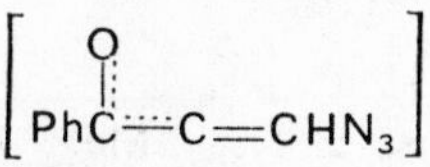

$$\left[PhC(=O)\cdots C{=}CHN_3 \right]$$
(76)

initially be formed in this reaction since other workers have shown[33a] that irradiation of 2-phenyl-3-benzoyl-1-azirine (78) with light of wavelength longer than 300 nm results in its conversion to 3,5-diphenyl isoxazole (77), a reaction which was reversed on irradiation

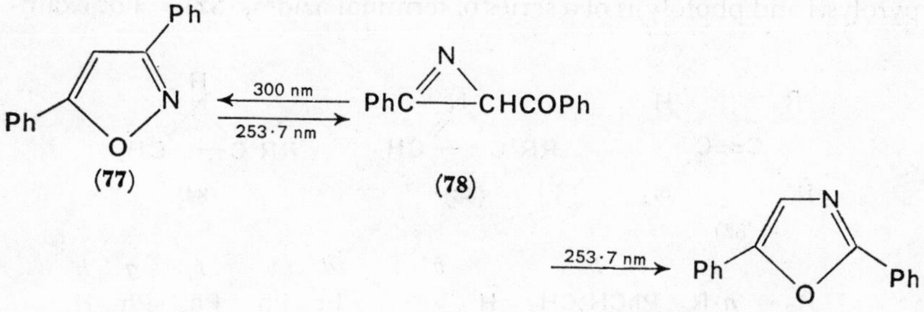

(77) $\xrightleftharpoons[253{\cdot}7\text{ nm}]{300\text{ nm}}$ PhC═N—CHCOPh (78) $\xrightarrow{253{\cdot}7\text{ nm}}$ (79)

with 253·7 nm light. Further irradiation of the azirine **78** at 253·7 nm converted it to the isomeric isoxazole **79**. Other aryl-substituted 3-benzoyl-1-azirines have also been found to undergo this type of transformation[33b]. In addition, it has been found that heating 3-phenyl-5-alkoxyisoxazoles to 200° converts them to 2-phenyl-3-carboalkoxy-1-azirines[34] (equation 11).

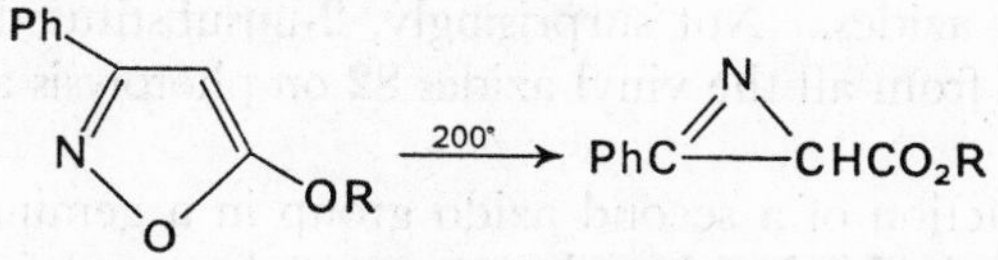

$$\text{3-Ph-5-OR-isoxazole} \xrightarrow{200^\circ} PhC{=}N{-}CHCO_2R \quad (11)$$

Irradiation of the cyclic azidovinyl ketone **80** in aqueous tetrahydrofuran was found to give the 7-membered keto lactam **81**[32]. This reaction appears to be analogous to the acid-induced decomposition of azido-*o*-quinones (see section V).

The recently reported results of Isomura and co-workers[35] suggest that azirines are generally formed on pyrolysis of terminal azides, but that 2-unsubstituted azirines are too unstable to survive most pyrolysis

(80) → (81)

conditions. They base their conclusion on the results of a study of the pyrolysis and photolysis of a series of terminal azides, (**82**). For example,

(82) RR¹C=CH–N_3; (83) RR¹C–CH azirine; (84) RR¹C–CH_2 aziridine (NH)

	a	*b*	*c*	*d*	*e*	*f*	*g*	*h*
R	*n*-R	$PhCH_2CH_2$	H	Et	Ph	Ph	Ph	H
R¹	H	H	$PhCH_2CH_2$	Et	Et	Ph	H	Ph

ple, they found that azirine **83***f* as well as the previously reported[13] indole **65** could be isolated when vinyl azide **82***f* was heated in refluxing benzene for one hour in the absence of oxygen. The intermediacy of azirines was further indicated by the observation that aziridines **84***b*, **84***c* and **84***e* were isolated after lithium aluminium hydride reduction of the crude pyrolysates obtained from the corresponding vinyl azides. Not surprisingly, 2-unsubstituted azirines **83** were obtained from all the vinyl azides **82** on photolysis at 365 nm at −50°.

The introduction of a second azido group in a geminal or vicinal position on the double bond might be expected to result in a relatively unstable compound. Such a prediction appears to be borne out by the fact that as yet no simple alkyl- or aryl-substituted geminal or vicinal diazide has been isolated. However, such compounds have been postulated on occasion as intermediates, as will be discussed in section III.A.3.

Decomposition of the relatively stable 2,3-diazido-*N*-phenylmaleimide (**33**) in refluxing benzene led to high yields of *N*,*N*-bis(cyanocarbonyl)aniline (**85**)[13]. This result is in keeping with the observation[36] that thermolysis of *ortho*-diazidobenzenes and 1,2-

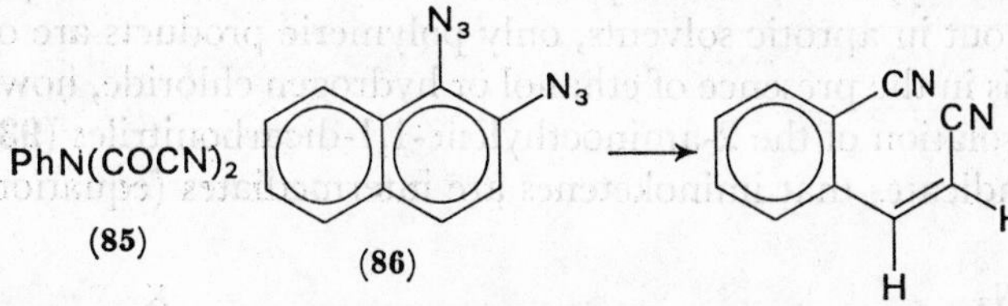

$PhN(COCN)_2$
(85)
(86)

diazidonapthalene (**86**) gives ring-opening to dinitriles. Similarly, diazide **87** formed phthaloyl cyanide (**88**) on decomposition[37].

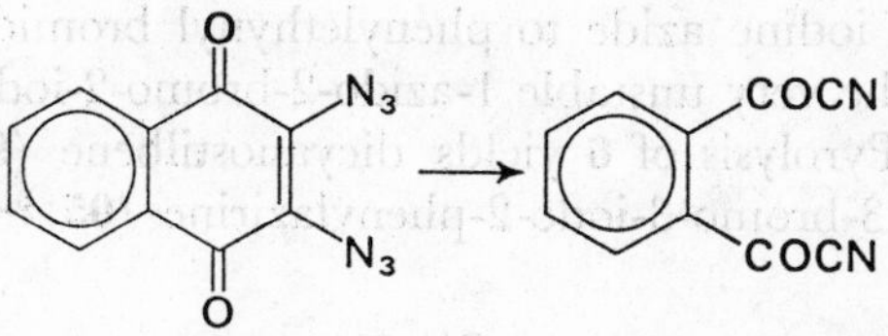

(87) (88)

2. Reactions of vinyl azides unstable at room temperature

The alkyl- and aryl-substituted vinyl azides discussed above are all reasonably stable at room temperature. However, the introduction of strongly electron-withdrawing substituents on the double bond appears to facilitate thermal decomposition of these compounds. A number of azides thus substituted have been found to decompose at room temperature or below.

trans-Perfluoropropenyl azide (**29**), for example, loses nitrogen smoothly at room temperature[22,23]. The major product is azirine (**89**) but variable amounts of the isomeric azirine **90** have also been reported. The latter compound probably results from isomerization of the initially produced **89** by hydrogen fluoride formed by the

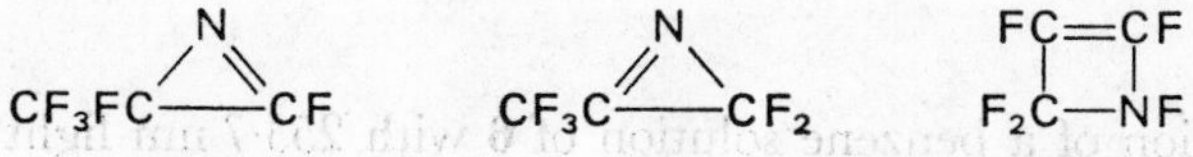

(89) (90) (91)

presence of trace amounts of water in the reaction mixture. A report[21] that **29** forms perfluoroazet-2-ene (**91**) could not be substantiated[22,23].

The geminal dicyanovinyl azides (**92**) also decompose, although rather slowly, at room temperature[20]. When the decomposition is carried out in aprotic solvents, only polymeric products are obtained. Pyrolysis in the presence of ethanol or hydrogen chloride, however, led to the isolation of the 2-aminoethylene-1,1-dicarbonitriles (**93**). This result indicates that iminoketenes are intermediates (equation 12).

$$\underset{(\mathbf{92})}{\overset{N_3}{\overset{|}{RC}}=C(CN)_2} \longrightarrow RN=C=C(CN)_2 \xrightarrow{HX} \underset{(\mathbf{93})}{RNH\overset{X}{\overset{|}{C}}=C(CN)_2} \qquad (12)$$

Addition of iodine azide to phenylethynyl bromide (**5**) results in formation of the very unstable 1-azido-2-bromo-2-iodo-1-phenylethylene (**6**)[6,8]. Pyrolysis of **6** yields dicyanostilbene (**94**) rather than the expected 3-bromo-3-iodo-2-phenylazirine (**95**)[6,8]. Reaction of

$$\underset{(\mathbf{94})}{Ph\overset{CN}{\overset{|}{C}}=\overset{CN}{\overset{|}{C}}Ph}$$

6 with aniline, however, gave *N,N'*-diphenylbenzamidine (**96**), which was postulated to have arisen by way of the azirine **95** (equation 13)[6].

$$\underset{(\mathbf{6})}{PhC(N_3)=CIBr} \longrightarrow \underset{(\mathbf{95})}{PhC\overset{N}{\diagup\diagdown}CIBr} \xrightarrow{PhNH_2} Ph\overset{N}{C}(\text{:}NHPh)\text{—}C\text{—}X \longrightarrow \underset{NPh}{\overset{\|}{PhC}}\text{—}CN \xrightarrow{PhNH_2} \underset{(\mathbf{96})}{\underset{NPh}{\overset{\|}{PhC}}\text{—}NHPh} \qquad (13)$$

Irradiation of a benzene solution of **6** with 253·7 nm light produced 9,10-dicyanophenanthrene (**97**) in about 20% yield[8]. When the photolysis was carried out in methanol, small amounts of methyl benzoate were also found. It has been suggested[8] that the dihaloazirine **95** is formed, then undergoes reaction with the alcohol to give first benzoyl cyanide and then the observed ester (equation 14).

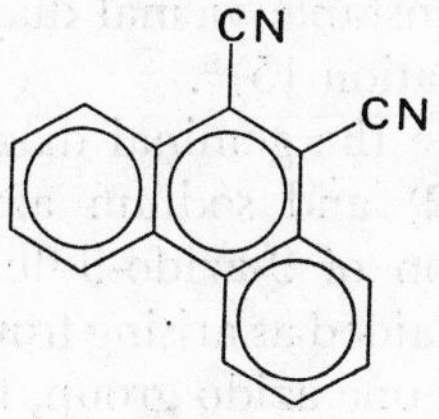

(97)

$$PhC\!\!\underset{}{\overset{N}{\triangle}}\!\!CIBr \xrightarrow{MeOH} PhCOCN \xrightarrow{MeOH} PhCO_2Me \quad (14)$$

(95)

Vinyl azides with two bulky substituents *cis* to each other on the double bond may be thermally quite unstable. Thus 1-azido-*cis*-stilbene (**98**, R = Ph) and methyl *β*-azido-*cis*-cinnamate (**98**, R = CO_2Me) were found, even at temperatures below 0°, to lose nitrogen and form 2,3-diphenylazirine (**99**, R = Ph) and 2-carbomethoxy-3-

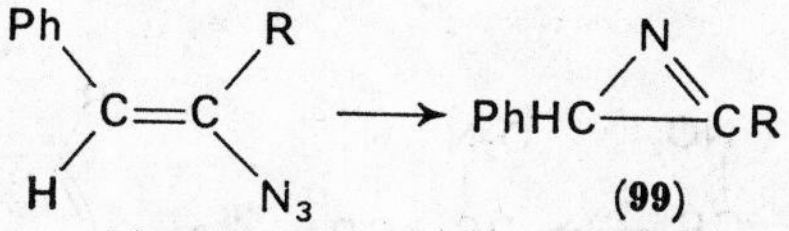

$$Ph(H)C{=}C(R)N_3 \longrightarrow PhHC\!\!\overset{N}{\triangle}\!\!CR$$

(98) (99)

phenylazirine (**99**, R = CO_2Me), respectively[11,12]. The *trans* isomer of 1-azidostilbene could be isolated and purified at room temperature[12].

3. Reactions assumed to involve vinyl azides as intermediates

Treatment of 1-phenylpropyne (**1**) with iodine azide yields a mixture of the *cis* and *trans* isomers of the vinyl iodoazide **2** along with about a 25% yield of benzonitrile (**100**)[5]. The nitrile is also formed when **2** is treated with iodine azide at 25°. These observations were rationalized by assuming a displacement of iodine by a second azide

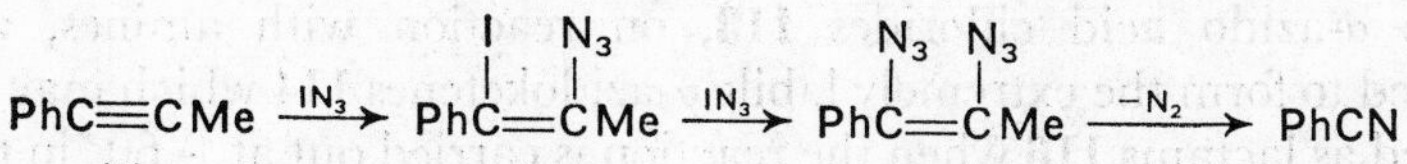

$$\underset{(1)}{PhC{\equiv}CMe} \xrightarrow{IN_3} \underset{(2)}{PhC(I){=}C(N_3)Me} \xrightarrow{IN_3} \underset{(101)}{PhC(N_3){=}C(N_3)Me} \xrightarrow{-N_2} \underset{(100)}{PhCN} \quad (15)$$

moiety resulting in an unstable vicinal diazide **101**, which is decomposed to the nitrile (equation 15)[5].

An attempt to prepare the geminal diazide **103** from 9-(dichloromethylene)fluorene (**102**) and sodium azide at room temperature resulted in the formation of 9-azido-9-fluorenecarbonitrile (**104**)[13]. This product can be explained as arising from the diazide **103** by loss of molecular nitrogen from one azido group, followed by reorganization of the remaining atoms.

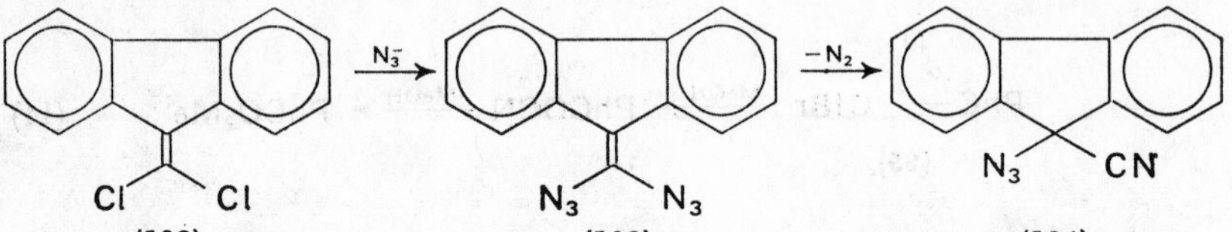

$$\textbf{(102)}\ \xrightarrow{N_3^-}\ \textbf{(103)}\ \xrightarrow{-N_2}\ \textbf{(104)}$$

The isolation of the furoxan **107** following reaction of 1,2-dinitroölefins **105** and sodium azide was explained[38] by assuming addition of azide ion to the double bond, followed by elimination of nitrite ion to form 1-azido-2-nitroölefins **106**. These vinyl azides could undergo loss of nitrogen concerted with cyclization to give **107**.

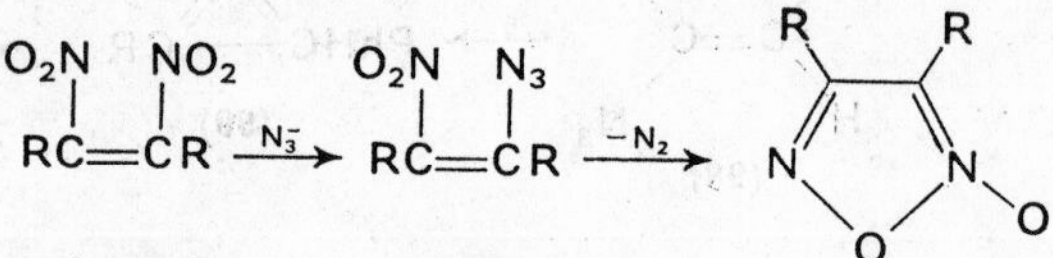

$$\underset{\textbf{(105)}}{RC(NO_2){=}C(NO_2)R}\ \xrightarrow{N_3^-}\ \underset{\textbf{(106)}}{RC(NO_2){=}C(N_3)R}\ \xrightarrow{-N_2}\ \textbf{(107)}$$

A vinyl azide intermediate has been invoked to account for the formation of 3,5-dibromo-2-hydroxybenzonitrile (**109**) in the pyrolysis of **108**[39]. The elimination of the elements of hydrazoic acid from **108** would give a terminal vinyl azide which might reasonably be expected to decompose and rearrange to the cyanide via an iminoketene (equation 16).

The formation of phenylcyanoketene (**112**) on reaction of 3-halo-4-phenylcyclobutenedione (**110**) with sodium azide most likely proceeds by way of the vinyl azide **111** which loses nitrogen and carbon monoxide forming **112**[40].

The α-azido acid chlorides **113**, on reaction with amines, are assumed to form the extremely labile α-azidoketenes **114** which may be trapped as lactams **116** when the reaction is carried out at −60° in the presence of Schiff bases[41]. Only the nitriles **115** were isolated fol-

(108) → (−HN₃) → (−N₂) → (109) (16)

(110) → (111) → (112)

lowing reactions of **113** with amines at temperatures above −30°. It has been suggested that azirinones **117** are the most plausible inter-

$$\text{RCH(N}_3\text{)COCl}\ (\mathbf{113}) \xrightarrow{Et_3N} \text{RC(N}_3\text{)=C=O}\ (\mathbf{114}) \longrightarrow \text{RCN} + N_2 + \text{CO}\ (\mathbf{115})$$

$$(\mathbf{114}) \xrightarrow{\text{PhCH=NPh}} (\mathbf{116})$$

mediates in the decomposition of the azidoketenes, but that triazolones **118** could be involved[41a].

(117) (118)

B. Acid-Induced Loss of Nitrogen

As do other azides, vinyl azides undergo loss of molecular nitrogen and rearrangement in the presence of strong mineral acids. Such acid-catalysed decompositions have not been extensively studied. One example, however, is the addition of hydrochloric acid to a glacial acetic acid solution of β-azidovinyl phenyl ketone (**73**, R = H), which led to the isolation of a high yield of cyanomethyl phenyl ketone (**75**, R = H) and a small amount of 5-phenylisoxazole (**74**, R = H) [3]. Thermolysis of this ketone gave very similar results [17,21]. (See section III.A.1.)

C. Reactions not involving Loss of Molecular Nitrogen

The products discussed in the preceding sections, despite their great variety, have one feature in common, i.e. in each case their formation involved loss of molecular nitrogen from the starting azide. However, a limited number of reactions of vinyl azides are known which do not involve the expulsion of nitrogen. Decomposition of azidoethylene at temperatures below 70° in the presence of free radical initiators, such as benzoyl peroxide or azobis*iso*butyronitrile, gives small amounts of a white polymer. Above 70° the polymer discolours and the elements of hydrazoic acid are evolved [2].

Azides are generally stable in the presence of bases and nucleophilic displacement of the azido group is uncommon, but displacements have been observed on a few vinyl azides. Reaction of β-azidovinyl phenyl ketone **73** (R = H) with piperidine gave phenyl-2-piperidino-vinyl ketone (**119**) and reaction of the same azide with sodium hydroxide in aqueous methanol gave the dimethyl acetal **120** [3].

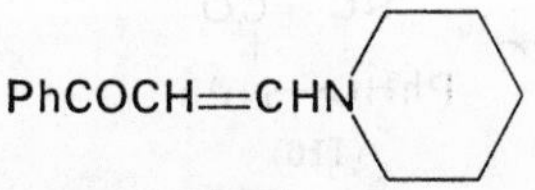

PhCOCH=CHN(piperidine ring) **(119)** $PhCOCH_2CH(OMe)_2$ **(120)** $PhNHCR{=}C(CN)_2$ **(121)**

Similarly, the 2-azidoethylene-1,1-dicarbonitriles (**92**) were found to react with aniline to give the corresponding anilino derivatives **121** [20].

Reaction of phenyl ethynyl ketone with sodium azide in an aqueous medium, it will be recalled, gave the isoxazole **74** (R = H). When this reaction was carried out in dimethylformamide, quite different results were found; in this case the sodium salt of 4-benzoyl-1,2,3-triazole was isolated in good yield [3]. The formation of this product could result from the cyclization of the carbanion **76** which would be formed by addition of azide ion to the ethyne (equation 17). Dime-

$$PhCOC{\equiv}CH \xrightarrow{N_3^-} PhC(O^-){=}C{=}CHN_3 \; (\mathbf{76}) \longrightarrow \text{4-(PhCO)-1,2,3-triazolide anion} \quad (17)$$

thyl sulphoxide solutions of β-azidovinyl *p*-tolyl sulphone containing bases such as azide, *p*-toluenesulphinate or *t*-butoxide ions produce 4 or 5-*p*-toluenesulphonyltriazole (**122**) at room temperature[19] (equation 18).

$$TsCH{=}CHN_3 \xrightarrow{base} Ts\bar{C}{=}CHN_3 \longrightarrow \text{4-Ts-1,2,3-triazolide anion} \longrightarrow \text{4-Ts-1H-1,2,3-triazole} \; (\mathbf{122}) \quad (18)$$

IV. MECHANISM

In most cases it is easy to rationalize the products obtained from vinyl azide reactions which do not involve the loss of molecular nitrogen. For example, the substitution of the azido group by a nucleophilic reagent in the 2-azidovinyl ketone **73** (R = H) and the β-dicyanovinyl

$$PhCOCH{=}CHN_3 \xrightarrow{X^-} PhC(O^-){=}CHCH(N_3)X \xrightarrow{-N_3^-} PhCOCH{=}CHX \quad (19)$$

$$RC(N_3){=}C(CN)_2 \; (\mathbf{92}) \xrightarrow{X^-} RC(N_3)X{-}\bar{C}(CN)_2 \xrightarrow{-N_3^-} RC(X){=}C(CN)_2 \quad (20)$$

compounds **92** are obviously addition–elimination reactions in which the addition step is favoured by resonance stabilization of the initially formed carbanion (equations 19 and 20). Cyclization of azidovinyl carbanions **123** to anions of triazoles **124** is quite obviously analogous to the well documented[19] isomerizations of the isoelectronic imidoyl azides and thioacyl azides (**125**, X = NH, S) to tetrazoles and thiatriazoles (**125**, X = NH, S), respectively.

(123) → **(124)** **(125)** → **(126)**

An explanation for the production of nitriles from terminal vinyl azides is found in a mechanism (equation 21) analogous to that for the Curtius rearrangement of acid azides to isocyanates and is consistent

$$>C{=}C(N{-}N_2)H \longrightarrow >C{=}C{=}NH \longrightarrow >CHCN \qquad (21)$$

with the production of *N*-substituted iminoketenes from the decomposition of internal vinyl azides.

Undoubtedly, acid-induced decomposition of vinyl azides involves the loss of molecular nitrogen concerted with new bond formation in a protonated azide species. For example, the acid catalysed decomposition of β-azidovinyl phenyl ketones **73** (R = H)[3] proceeds via two competitive pathways giving either nitriles (equation 22) or isoxazoles (equation 23).

$$PhCOCH{=}C(H){-}\overset{+}{N}H{-}N_2 \longrightarrow PhCOCH{=}C{=}\overset{+}{N}H_2 \longrightarrow PhCOCH_2CN + H^+ \qquad (22)$$

$$PhC(=O)CH{=}CH{-}\overset{+}{N}H{-}N_2 \longrightarrow PhC(=\overset{+}{O})CH{=}CH{-}NH \longrightarrow \text{5-phenylisoxazole} + H^+ \qquad (23)$$

Until recently it was difficult to envision a single, unified mechanistic scheme that would satisfactorily explain the formation of the many different products isolated from vinyl azide decomposition reactions.

Most products can be rationalized as arising either directly from the vinyl azide by concerted loss of molecular nitrogen and formation of the new bond, or through the intermediacy of a vinyl nitrene. (Arguments for and against vinyl nitrene formation will be discussed later in this section.) However, neither the concerted nor the vinyl nitrene mechanism explains why azirines have been so rarely isolated from the decomposition of terminal vinyl azides. Nor do they resolve the differences often found between thermal and photochemical decompositions of a given compound. These observations may be explained by assuming that iminoketenes and/or azirines are the primary products of vinyl azide decompositions, but that the azirines are often unstable toward the reaction conditions. Ring-opening of the azirine followed by reorganization could lead, where possible, to isoxazoles, indoles and perhaps dimerization to dihydropyrazines. However, formation of the latter compounds occurs most readily in alcoholic media[13,28b], thus their formation may actually involve prior addition[9,28b] of the alcohol across the C=N bond, ultimately opening the ring to an equivalent of an α-aminoketone or aldehyde, which are known precursors for dihydropyrazines[13].

It has been suggested[18] that 9-(*N*,*N*-fluorenylideneaminomethylene) fluorene (**52**) is formed by decomposition of the labile spiroazirine **53** to 9-fluorenylidene (**127**) which, in the absence of oxygen, adds to a second molecule of **53** to give **52** (equation 24). In the presence of

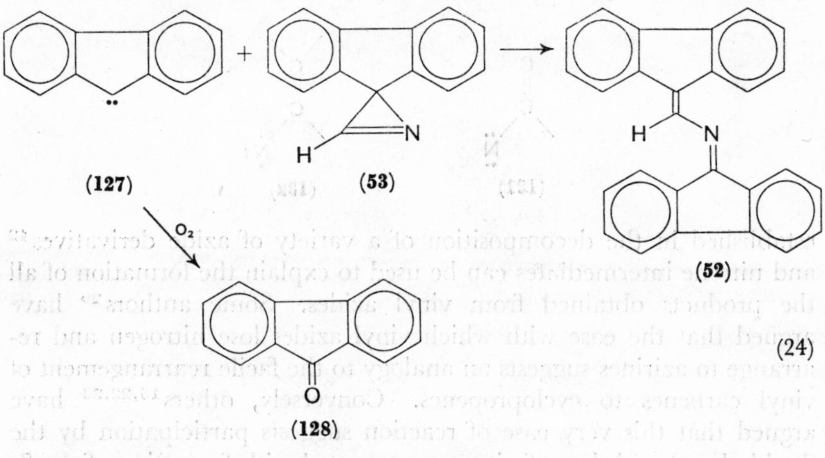

(127) + (53) → (52)

(127) —O_2→ (128)

(24)

oxygen **127** reacts to form fluorenone (**128**)[18].

The isolation[13] of 9-azido-9-fluorenecarbonitrile (**104**) from the reaction of 9-(dichloromethylene)fluorene (**102**) with sodium azide can best be rationalized by assuming formation of a labile geminal diazide **103**, which loses molecular nitrogen and cyclizes to azidoazirine **129**. Like all other iminoazides, **129** would be expected to be

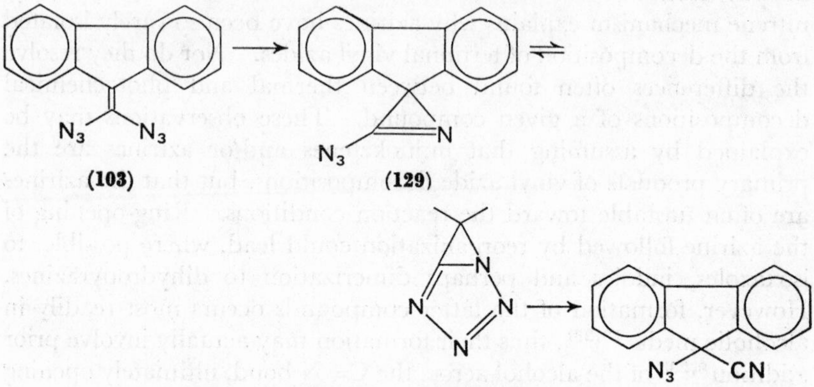

(**103**) (**129**)

(**130**) (**104**)

in equilibrium with its isomeric tetrazole **130**, which opens to the unstrained azidonitrile **104**.

In considering a mechanistic rationale for vinyl azide chemistry one is strongly tempted to invoke a vinyl nitrene intermediate in either its singlet, **131** or triplet, **132**, state. The existence of nitrenes has been

(**131**) (**132**)

established in the decomposition of a variety of azide derivatives[42] and nitrene intermediates can be used to explain the formation of all the products obtained from vinyl azides. Some authors[29] have argued that the ease with which vinyl azides lose nitrogen and rearrange to azirines suggests an analogy to the facile rearrangement of vinyl carbenes to cyclopropenes. Conversely, others[15,22,23] have argued that this very ease of reaction suggests participation by the double bond with loss of nitrogen concerted with formation of the 3-membered ring. No spectral evidence was found for the formation of vinyl nitrenes during the decomposition of vinyl azides[43]; furthermore, attempts to trap these intermediates have not been successful[23]. Therefore, we are inclined to believe that vinyl nitrenes are not formed directly from the decomposition of vinyl azides. They may very likely arise, however, from ring-opening of azirines.

V. REARRANGEMENTS OF AZIDOQUINONES

Although acid- and thermally induced rearrangements of azidoquinones have been known for several decades, only in the last few years have efforts been made to elucidate the mechanism of these reactions. Unlike most other vinyl azides discussed in this chapter, azidoquinone chemistry generally seems not to involve azirines. Instead the products result from expulsion of molecular nitrogen usually followed by rupture of a carbon–carbon bond.

Azidoquinones are best prepared from the corresponding haloquinone by reaction with sodium azide[44]. Certain aminoazidoquinones, however, have been prepared by adding two molar equivalents of hydrazoic acid to a quinone, thus giving a diazidohydroquinone which readily loses nitrogen and disproportionates into the product[45,46].

Moore and Sheldon[47] have treated a number of azido-*p*-quinones (**133**) with either trichloroacetic acid or sulphuric acid and observed the evolution of nitrogen and ring-contraction in good yield to γ-cyanoalkylidene-$\Delta^{\alpha\beta}$-butenolides (**134**). The reaction is highly stereoselective giving mainly isomer **134**. The proposed mechanism[45,47]

(133) → (134)

	R	*R*¹	*R*²	% *yield*
a	Me	H	*t*-Bu	87
b	Me	N_3	Me	83
c	Ph	OH	Ph	95
d	H	(—CH=CH—CH=CH—)		60
e	H	Ph	H	80
f	H	*t*-Bu	H	95

involves a reversible protonation of the quinone followed by loss of nitrogen concerted with carbon–carbon bond breakage. The resulting acylium ion cyclized on the hydroxyl oxygen giving the lactone (equation 25).

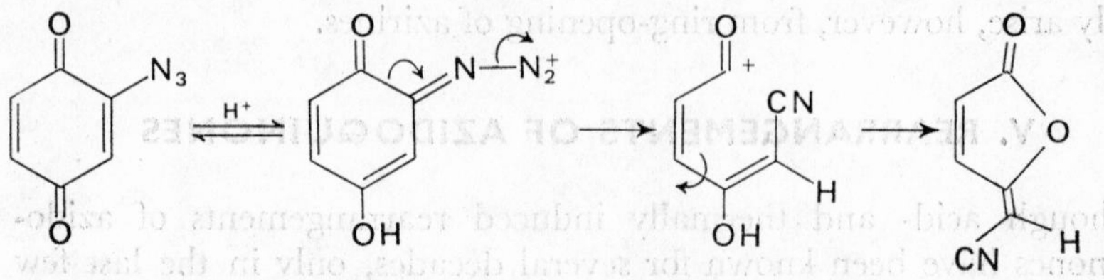

(25)

The azido-*o*-quinones, in contrast to the *para* isomers, do not undergo ring-contraction; instead, on treatment with strong aqueous acids they expand to azepine derivatives. For example, an 82% yield of the hydroxybenzazepinedione **138** was isolated following treatment of 4-azido-1,2-naphthaquinone (**135**) with sulphuric acid[48]. The mechanism proposed[48] for this reaction involves protonation of **135** to give iminodiazonium ion **136**. Loss of nitrogen concerted with phenyl migration leads to a vinyl carbonium ion **137** which reacts with water and tautomerizes to the amide **138**.

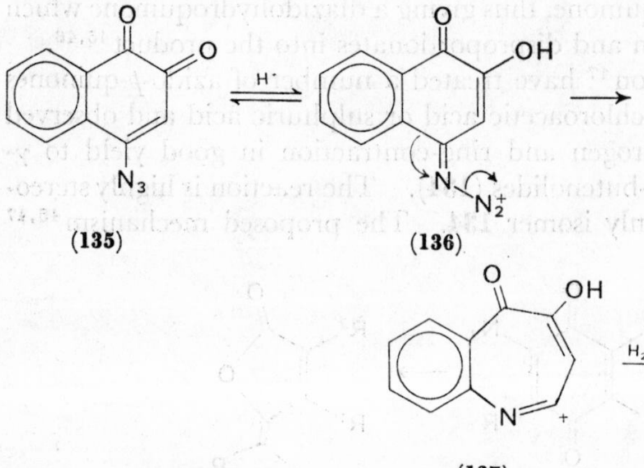

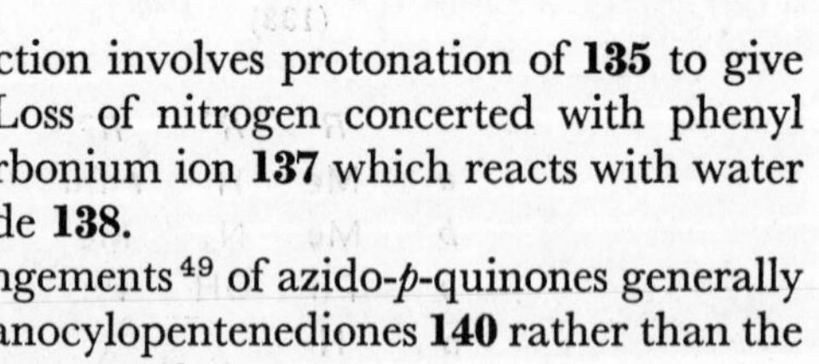

(135) (136) (137) (138)

Thermally induced rearrangements[49] of azido-*p*-quinones generally give the ring-contracted 2-cyanocylopentenediones **140** rather than the lactones **134**. It has been proposed[48] that this reaction involves loss of nitrogen concerted with carbon–carbon bond breakage to give zwitterion **139** which cyclizes to the pentenedione **140**.

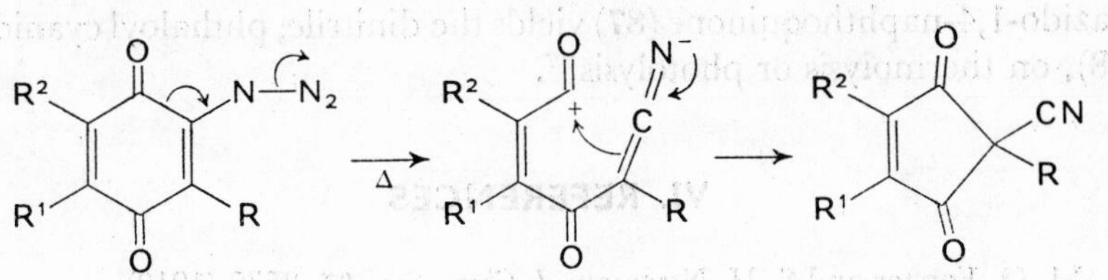

(139) (140)

	R	R¹	R²	% yield
a	Me			95
b	OMe			70
c	Me	NH	Me	96
d	Ph	H	Ph	55

In the case of the 3,6-diphenyl derivative **139*d*** a small amount of the lactone **141** was isolated from the reaction mixture, along with a 20% yield of the carbazole derivative **142**. This insertion product might

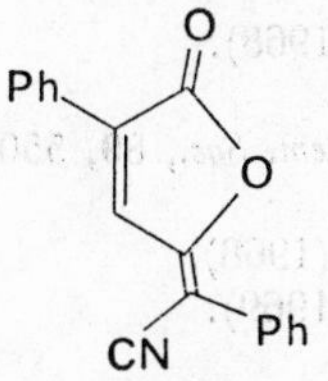

(141) 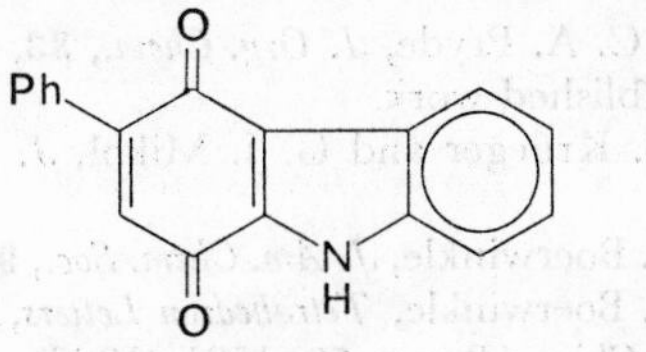 (142) 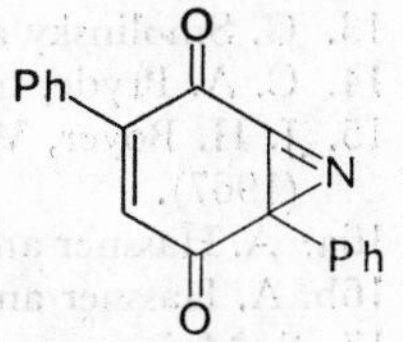 (143)

be pictured as arising via the strained azirine **143** in analogy to the mechanism proposed in section IV for the formation of indoles from β-phenyl vinyl azides.

A zwitterionic intermediate has been invoked[49] to account for the finding that on pyrolysis 2,5-diazido-3,6-diphenyl-1,4-benzoquinone (**144**) undergoes fragmentation into two molecules of phenylcyanoketene (**145**)[49]. As mentioned before, the vicinal diazide, 2,3-

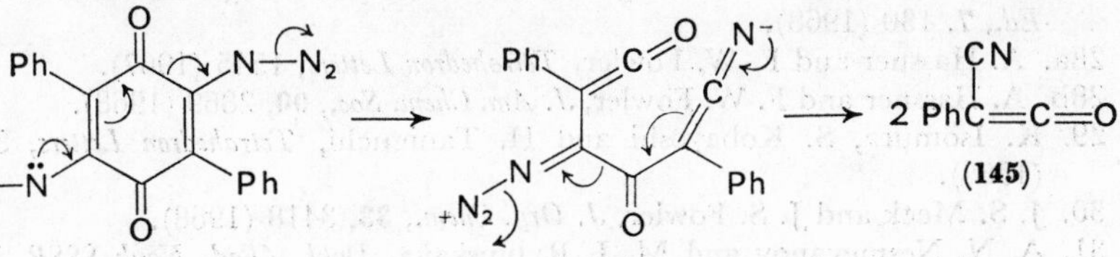

(144) (145)

diazido-1,4-naphthoquinone (**87**) yields the dinitrile, phthaloyl cyanide (**88**), on thermolysis or photolysis[37].

VI. REFERENCES

1. M. O. Forster and S. H. Newman, *J. Chem. Soc.*, **97**, 2570 (1910).
2. R. H. Wiley and J. Moffat, *J. Org. Chem.*, **22**, 995 (1957).
3. A. N. Nesmeyanov and M. I. Rybinskaya, *Izv. Akad. Nauk SSSR, Otd. Khim. Nauk*, 816 (1962); English translation: *ibid.*, 761 (1962).
4. V. G. Ostroverkhov and E. A. Shilov, *Ukrain. Khim. Zhar.*, **23**, 615 (1957).
5. A. Hassner, R. J. Isbister and A. Friederang, *Tetrahedron Letters*, 2939 (1969).
6. A. Hassner and R. J. Isbister, *J. Am. Chem. Soc.*, **91**, 6126 (1969).
7. A. Hassner, *J. Org. Chem.*, **33**, 2684 (1968).
8. J. H. Boyer and R. Selvarajan, *J. Am. Chem. Soc.*, **91**, 6122 (1969).
9. G. Smolinsky, *J. Org. Chem.*, **27**, 3557 (1962).
10. S. Hanessian and N. R. Plessas, *Chem. Commun.*, 706 (1968).
11. A. Hassner and F. W. Fowler, *J. Org. Chem.*, **33**, 2686 (1968).
12. F. W. Fowler, A. Hassner and L. A. Levy, *J. Am. Chem. Soc.*, **89**, 2077 (1967).
13. G. Smolinsky and C. A. Pryde, *J. Org. Chem.*, **33**, 2411 (1968).
14. C. A. Pryde, unpublished work.
15. J. H. Boyer, W. E. Krueger and G. J. Mikol, *J. Am. Chem. Soc.*, **89**, 5504 (1967).

16a. A. Hassner and F. Boerwinkle, *J. Am. Chem. Soc.*, **90**, 216 (1968).
16b. A. Hassner and F. Boerwinkle, *Tetrahedron Letters*, 3309 (1969).

17. S. Maiorana, *Ann. Chim. (Rome)*, **56**, 1531 (1966).
18. W. Bauer and K. Hafner, *Angew. Chem., Int. Eng. Ed.*, **8**, 772 (1969).
19. J. S. Meek and J. S. Fowler, *J. Am. Chem. Soc.*, **89**, 1967 (1967).
20. K. Friedrich, *Angew. Chem., Int. Eng. Ed.*, **6**, 959 (1967).
21. I. L. Knunyants and E. G. Bykhovskaya, *Dokl. Akad. Nauk SSSR*, **131**, 1338 (1960); English translation: *ibid.*, **131**, 411 (1960).
22. C. S. Cleaver and C. G. Krespan, *J. Am. Chem., Soc.*, **87**, 3716 (1965).
23. R. E. Banks and G. J. Moore, *J. Chem. Soc., C*, 2304 (1966).
24. G. R. Harvey and K. W. Ratts, *J. Org. Chem.*, **31**, 3907 (1966).
25. E. Zbiral and J. Stroh, *Monatsh. Chem.*, **100**, 1438 (1969).
26. A. Mustafa, S. M. A. D. Zayed and S. Khattov, *J. Am. Chem. Soc.*, **78**, 145 (1956).
27. F. P. Woerner, H. Reimlinger and D. R. Arnold, *Angew. Chem., Int. Eng. Ed.*, **7**, 130 (1968).

28a. A. Hassner and F. W. Fowler, *Tetrahedron Letters*, 1545 (1967).
28b. A. Hassner and F. W. Fowler, *J. Am. Chem. Soc.*, **90**, 2869 (1968).

29. K. Isomura, S. Kobayoshi and H. Taniguchi, *Tetrahedron Letters*, 3499 (1968).
30. J. S. Meek and J. S. Fowler, *J. Org. Chem.*, **33**, 3418 (1968).
31. A. N. Nesmeyanov and M. I. Rybinskaya, *Dokl. Akad. Nauk SSSR*, **166**, 1362 (1966).

32. S. Sato, *Bull. Chem. Soc. Japan*, **41**, 2524 (1968).
33a. B. Singh and E. F. Ullman, *J. Am. Chem. Soc.*, **89**, 6911 (1967).
33b. D. W. Kurtz and H. Shechter, *Chem. Commun.*, 689 (1966).
34. T. Nishiwaki, *Tetrahedron Letters*, 2049 (1969).
35. K. Isomura, M. Okada and H. Taniguchi, *Tetrahedron Letters*, 4073 (1969).
36. J. H. Hall and E. Patterson, *J. Am. Chem. Soc.*, **89**, 5856 (1967).
37. J. A. Van Allan, W. J. Priest, A. S. Marshall and G. A. Reynolds, *J. Org. Chem.*, **33**, 1100 (1968).
38. W. D. Emmons and J. P. Freeman, *J. Org. Chem.*, **22**, 456 (1957).
39. H. Lindemann and A. Mulhaus, *Ann. Chem.*, **446**, 1 (1925).
40. R. C. De Selms, *Tetrahedron Letters*, 1179 (1969).
41a. A. Hassner, R. J. Isbister, R. B. Greenwald, J. T. Klug and E. C. Taylor, *Tetrahedron*, **25**, 1637 (1969).
41b. A. K. Bose, B. Anjaneyulu, S. K. Bhatta-Charya and M. S. Manhas, *Tetrahedron*, **23**, 4769 (1967).
42a. J. H. Boyer in *Mechanisms of Molecular Migrations*, Vol. 2 (Ed. B. S. Thyagarajan) Wiley-Interscience, New York, 1969, p. 267.
42b. R. A. Abramovitch and B. A. Davis, *Chem. Rev.*, **64**, 149 (1964).
43. G. Smolinsky, E. Wasserman and W. A. Yager, *J. Am. Chem. Soc.*, **84**, 3220 (1962).
44. K. Fries and P. Ochwatt, *Ber.*, **56**, 1299 (1923).
45. H. W. Moore and H. R. Sheldon, *J. Org. Chem.*, **33**, 4019 (1968).
46. H. W. Moore, H. R. Sheldon and D. F. Shellhamer, *J. Org. Chem.*, **34**, 1999 (1969).
47. H. W. Moore and H. R. Sheldon, *Tetrahedron Letters*, 5431 (1968).
48. H. W. Moore, H. R. Sheldon and W. Weyler, Jr., *Tetrahedron Letters*, 1243 (1969).
49. H. W. Moore, W. Weyler, Jr. and H. R. Sheldon, *Tetrahedron Letters*, 3947 (1969).

Author Index

This author index is designed to enable the reader to locate an author's name and work with the aid of the reference numbers appearing in the text. The page numbers are printed in normal type in ascending numerical order, followed by the reference numbers in parentheses. The numbers in *italics* refer to the pages on which the references are actually listed.

Aaronson, A. M. 312 (254), *329*
Abramovitch, R. A. 59 (5), *178*, 222, 235 (3), 245 (78), 249 (89, 90), 251 (90), 252, 254 (89, 90), 255, 258 (3), 261 (78), 263 (3, 78), 264 (126), 265 (3, 78), 268 (148), 269 (152, 153, 155), 270 (78, 156, 157), 271 (78, 157), 272 (78), 274 (126), 279 (179), 281 (179, 188), 282 (188), 283 (148, 189), 284 (188, 189), 285 (148, 152, 153, 179, 188, 189), 286 (152, 155, 189), 287, 288, 294 (199), 299 (89, 90), 300 (78, 89, 90), 304 (90), 307 (89, 90), 312 (148, 256), 313, 314 (256), 315 (259), 317 (262, 263), 321 (155, 156, 269, 270), *322*, *324–327*, *329*, 366 (164, 165), *393*, 398, 401, 403 (8), 404 (64, 65), 422–426 (8), *435*, *436*, 442 (1), *498*, 525 (209, 210), *550*, 580 (42b), *585*
Acampora, M. 376 (205), *394*
Achiwa, K. 538 (329), *553*
Adam, W. 301 (234–236), *328*
Adams, J. B., Jr. 528 (231), *550*
Adams, K. A. H. 264, 274 (126), *325*, 366 (164, 165), *393*
Adams, R. 127, 128 (332, 334), 130 (332), *187*, 335 (11), *389*
Agami, C. 87 (172), *182*
Akimova, G. S. 357 (129), *392*, 416 (118), *437*, 531 (256, 257), *551*
Alder, K. 260 (111, 112), *325*, 373 (191, 192), *393*
Alemagna, A. 90 (203), *183*
Ali, Y. 76, 103 (98), 105 (98, 256), 106–108 (256), 109, 110 (98), 111, 112 (273), *180*, *185*, 335 (10), *389*
Allen, C. F. H. 62, 174 (25), *178*, 350 (92), *391*
Allenstein, E. 242 (68), *324*, 343 (62), *390*, 421 (141), *438*
Altshuller, A. P. 18 (51), *53*
Amble, E. 18 (52), *54*
Anastassiou, A. G. 452 (67, 68), 460 (68), *500*, 539 (335), *553*
Andersen, E. 262 (120), *325*
Anderson, G. W. 89, 90 (183), *182*
Andreades, S. 121 (314), *186*
Andrisano, R. 212 (34), *220*
Angeli, A. 431 (217), *440*
Angstrom, A. D. 195 (29), *201*
Anisuzzaman, A. K. M. 344 (67), *390*
Anjaneyulu, B. 200 (80), *202*, 574 (41b), *585*
Anschütz, A. 352 (103, 104), *391*
Anselme, J.-P. 61 (13), 168 (441–443), 170 (445), 177 (483), *178*, *189*, *190*, 192 (3), 193 (11), 196 (40), *200*, *201*, 229, 230 (29), 253, 254 (96), 292 (211, 212), 293 (213), *323*, *324*, *328*, 342 (55), 358 (135, 136), *390*, *392*, 449, 465, 486 (47), 490 (126), *499*, *501*, 506 (57), 522, (197), 543 (362), *546*, *549*, *554*
Antkowiak, W. 540 (337), *553*
Antoniades, E. P. 409–411, 414 (82), *436*
Appl, M. 257 (105, 107), 258 (108), *325*, 388 (251), *395*, 424 (172), 425 (173), *439*, 443 (15), 495 (15, 137, 138), 496 (138), *498*, *501*
ApSimon, J. W. 385 (244), *395*, 403 (49), *436*, 449, 455 (48), *499*, 506 (64), 538 (316, 317), 540 (316, 317, 337), *546*, *553*
Arcus, C. L. 78 (122), 79 (122, 127–129), *181*, 225 (15), 229 (23–27), 230 (32–35), 231, 232 (32), *323*, 344 (66), 385 (66, 245), 386 (66),

390, *395*, 410 (96), 416 (120), 417 (122, 123), 420 (138), 422 (153, 154), *437*, *438*
Arens, J. F. 62, 130 (27), 132 (343), *178*, *187*, 217 (47), *220*
Armstrong, A. T. 44 (101), *55*
Arnold, D. R. 428 (193), *439*, 477 (114), *501*, 564 (27), *584*
Arnold, Z. 353 (109), *391*
Ashby, B. 409 (81), *436*
Ashley, J. N. 287 (198), *327*
Assche, D. v. 525 (212), *550*
Atkinson, R. S. 541 (351), *554*
Audrieth, L. F. 62 (32), *178*
Aue, D. H. 369 (179), *393*, 538 (326), 543 (361), *553*, *554*
Aufdermarsh, C. A. 171 (446), *190*
Aures, D. 104 (249), *184*
Avramenko, L. F. 90, 91 (201), *183*
Awad, W. I. 83, 123 (149), 124 (321, 322), *182*, *186*, 194 (27), *201*
Aylward, J. B. 93, 176 (217), *183*, 433 (230), *440*, 517 (146, 147), *548*
Ayyangar, N. R. 336 (20), *389*
Azogu, C. I. 268 (148), 283 (148, 189), 284 (189), 285 (148, 189), 286 (189), 312 (148, 256), 313, 314 (256), 317 (262, 263), *326*, *327*, *329*

Bacchetti, T. 90 (203), *183*, 544 (364), *554*
Bader, A. 525 (212), *550*
Badger, G. M. 413 (112), *437*
Bailey, A. S. 114, 115 (281), *185*, 209, 210 (24, 25), 211 (24), *220*, 346 (75), *390*, 530 (247), *551*
Bak, T. A. 262 (119, 120), *325*, 407 (77), 422 (155), *436*, *438*, 509, 510 (104), *547*
Baker, B. R. 76, 103, 105 (95), 107 (259), *180*, *185*, 338 (32), *389*
Baker, J. W. 122 (316), *186*
Balabanov, G. P. 195, 198 (30), *201*, 279 (184), 280 (186), 283 (184), *327*
Baldwin, J. E. 350 (91), *391*, 542 (358), *554*
Ballhausen, C. J. 44 (100), *55*
Balli, H. 116 (296, 297), *186*, 353 (107, 108), 358 (134), *391*, *392*
Ballinger, P. 79 (133), *181*
Bamberger, E. 148 (378), 150 (389), *188*, 234 (52), 236 (52, 55–58), 237 (57), *323*, *324*, 338 (37), *389*
Bamford, W. R. 151 (396), *188*
Banks, R. F. 121 (312), *186*, 562, 571, 580, 581 (23), *584*
Banthorpe, D. V. 399, 401, 409 (15), *435*
Barash, L. 307 (246), *328*, 472 (107), *501*
Barger, G. 335 (15), *389*
Barker, M. F. 372 (187), *393*
Barnes, J. D. 76, 77 (102), *180*
Barrett, G. 230 (35), *323*
Barton, D. H. R. 98, 100 (230–232), 102 (242), *184*, 298 (224–226), *328*, 340 (43), 362 (43, 152), *390*, *392*, 426 (179, 180), *439*, 461 (83), 468 (83, 101, 102), 469 (103), 479 (83), *500*
Bassi, D. 22, 23 (65), *54*
Bauer, G. 277, 292 (175), *327*, 403 (46), *436*, 456, 463, 464, 467, 470, 489, 490 (75), *500*, 535, 536 (291), 538, 540, 544 (319), *552*, *553*
Bauer, V. J. 527 (218), *550*
Bauer, W. 561, 566, 579 (18), *584*
Baum, M. E. 336 (21), *389*
Beach, W. F. 387 (248), *395*, 542 (355), *554*
Beck, E. R. v. 505 (35), *545*
Beckman, A. O. 442 (6, 7), 447, 467 (6), *498*
Beckmann, S. 504 (18), *545*
Beckwith, A. L. J. 537, 539, 542 (310, 311), *552*
Beetham, D. 360 (143), *392*
Behr, F. E. 263 (123), *325*
Behringer, H. 130, 131 (339), *187* 511, 517 (111), *547*
Behringer, L. 160 (427), 161 (427, 428), *189*, 194 (21), *200*, 335 (13), *389*, 429 (206), *440*
Bell, A. 350 (92), *391*
Bell, F. 399 (20), *435*
Beltrame, P. 70 (61), 90 (203), *179*, *183*
Bender, M. L. 68 (57), *179*
Bennett, R. P. 274 (166), *326*, 349 (90), *391*
Benson, C. 518 (155), *548*
Benson, F. R. 115 (292), 127 (326), *185*, *187*, 398 (1, 2), 415 (1), 432 (1, 2), 433 (1), *434*, 531 (261), *551*
Benson, R. E. 505 (21), *545*
Berezovskii, V. M. 431 (214), *440*
Berger, A. 228 (21, 22), 229 (22), *323*, 421 (144), *438*

Berger, G. H. 527 (219), *550*
Berger, H. 174 (462), *190*
Bergmann, E. 207 (12), *219*, 294 (217), *328*
Bergmann, M. 176 (479), *190*
Berlin, K. D. 531 (258), *551*
Bernardi, L. 534 (279), *552*
Bernsmann, J. 76, 103, 110 (103), *180*
Berry, R. S. 460 (82), 483 (127), *500*, *501*, 538 (314, 315), *553*
Berry, W. L. 410, 413 (91), *437*
Berthier, G. 20, 32 (56), *54*
Bertho, A. 103 (248), 104 (248–250), *184*, 265, 267, 270 (138), *326*, 336, (26), 381 (226), *389*, *394*, 404 (56), *436*, 442 (5), 479, 482 (118), *498*, *501*, 506 (73), 525 (205), 527, 534 (222), *546*, *550*
Bertrans, M. 126 (325), *186*
Bestmann, H. 533 (274), *551*
Bettinetti, G. F. 88 (174), 167 (437, 438), *182*, *189*
Beyer, H. 518 (154), *548*
Bhalerao, U. T. 242 (67), *324*, 412 (102), *437*
Bhaskar, K. R. 197 (56), *201*
Bhatta-Charya, S. K. 574 (41b), *585*
Bianchetti, G. 290 (205, 206), *327*, 347, (79, 80), 374 (79), 375 (199), *390*, *393*, 434 (237), *440*, 529, 530 (245), *551*
Biemann, K. 420 (139), *438*
Bingham, T. E. 401, 403 (34), *435*
Birkhimer, E. A. 262 (119, 120), *325*, 422 (185), *438*
Birkofer, L. 62 (29), 134 (349), *178*, *187*, 413 (109), *437*
Blackburn, D. E. 264 (128), *325*, 364 (158), *392*, 422 (159), *438*
Blandamer, M. J. 3 (20), *53*
Blaschke, H. 369 (178), 371 (185), *393*, 450, 465 (52), 479, 481, 482 (128), *499*, *501*, 531 (264), 537, 539 (301), 543 (264, 301, 360), *551*, *552*, *554*
Bleiholder, R. F. 291 (208), *327*, 346 (76), *390*
Blomquist, A. T. 408 (80), *436*
Blomstrom, D.C. 127, 128 (334), *187*, 335 (11), *389*
Blumbergs, P. 108 (262, 264), *185*
Bock, H. 357 (133), *392*, 532 (269), *551*
Bockhorn, G. H. 369 (176), *393*, 527 (220), 540 (344), *550*, *553*
Bockmuhl, A. 156 (409), *189*
Boerwinkle, F. 140 (360–362), 142 (360), 144 (360, 361), 145 (360), *187*, 560 (16a, b), *584*
Bognar, R. 506 (67), *546*
Böhme, H. 77, 79 (115–117), *181*, 198 (64, 65), *202*
Boissonnas, R. A. 534 (280), *552*
Bolto, B. A. 66 (49), *179*
Bolton, R. 120, 130, 134, 139, 143 (309), *186*
Bonaccorsi, R. 45, 48, 50 (103), *55*
Bonnemay, A. 41 (96), *55*
Bonner, W. A. 409 (85), *436*
Bonnett, R. 32 (81a), *54*
Borsche, W. 243 (72, 73), 244 (73), *324*, 339 (40), *390*
Borsotti, G. 274 (167), *326*, 425 (175), *439*
Bose, A. K. 76, 95 (96), *180*, 200 (80), *202*, 338 (30, 31), *389*, 574 (41b), *585*
Bothner-By, A. A. 400 (26), *435*
Boulton, A. J. 262, 263 (121), *325*, 373 (190), *393*
Boutin, H. P. 16 (43), *53*
Bowes, G. 198 (71), *202*, 308 (249–251), *328*, *329*, 443 (25), 447 (37, 40), 455 (40), 458 (37, 40), 464 (25), *498*, *499*
Bown, D. W. 65, 67 (43d), *179*
Boyer, J. H. 2, 3 (8), *52*, 61 (15, 17), 76, 77 (83, 99, 100, 105), 79 (126), 80 (138), 90, 91 (197, 198), 94 (105), 119, 122–125 (17), 148 (377), *178*, *180*, *181*, *183*, *188*, 192 (4), 193 (17), 195 (34), 196 (43, 50), 198 (50), *200*, *201*, 204, 214 (2), *219*, 222 (1), 223 (9), 224 (9, 14), 232 (9, 38, 39), 233 (9, 40–42), 253 (93–95), 312 (255), *322–324*, *329*, 334 (6, 8), 339 (42), 342 (58), 344 (70), 348 (83), 366 (163), 367 (167), 370 (180), 371 (186), 389 (239, 240), *389–394*, 398 (3, 9), 415 (3), 419, 420 (134, 135), 421 (3), 424 (166), 426 (134), *434*, *435*, *438*, *439*, 518 (150), 531 (254), *548*, *551*, 557 (8), 560, 567 (15), 572 (8), 580 (15, 42a), *584*, *585*
Bradbury, S. 274, 275 (172), *327*
Bradshaw, J. S. 456 (77), *500*
Brass, K. 234 (49–51), *323*
Braun, J. von 62 (30), *178*, 383 (236), *394*, 402 (47), 415 (115), 432, 433

(226), *436*, *437*, *440*, 531 (260), *551*
Bravo, P. 376 (205, 206), *394*
Brenner, M. 336 (22), *389*
Bresler, H. W. 150 (387), *188*
Breslow, D. S. 90 (191), *183*, 265 (134–136), 279 (134, 180), 280 (134), 281 (134, 187), 282 (180), 283 (134), 284 (134, 191), 285, 286 (134), 288, 289 (180), *326*, *327*, 340 (49), 341 (51), *390*, 528 (234), 537 (297–300), 538 (297, 325), 539 (297), 544 (298), *550*, *552*, *553*
Bretschneider, H. 76, 77 (101), 151 (395), *180*, *188*, 334 (5), 357 (130), *389*, *392*
Brewster, P. 73 (69), *179*, 337 (27), *389*
Briggs, L. H. 407 (78), *436*
Britten, F. C. 514 (128), *548*
Britton, D. 372 (189), *393*
Brockhorn, G. H. 479, 482 (129), *501*
Brower, J. R. 401 (32), *435*
Brown, B. B. 157 (411, 412), 158 (411), *189*, 210 (29), *220*, 235, 237–239 (54), 263 (122), 269 (149), *324–326*, 339 (38), 365 (160), 371 (183), 372 (160), 385 (183), *389*, *392*, *393*, 462 (88), 468 (98), 491 (88), *500*
Brown, B. R. 432 (221), *440*
Brown, C. W. 21 (59), *54*
Brown, H. C. 135 (351), *187*, 206 (9), 215 (42), *219*, *220*
Brown, I. 360 (142), 363 (154), *392*, 427 (192), *439*, 506 (63), 530 (252), 539 (63), 542 (353), *546*, *551*, *554*
Brown, M. 505 (21), *545*
Brown, R. D. 117 (298a), *186*
Brown, R. F. C. 87 (171), *182*, 403 (50), *436*, 506 (66), 540 (340), *546*, *553*
Browne, A. W. 62 (31), *178*, 223, 237 (8), *322*, 342 (57), *390*, 419 (133), *438*, 507 (88), *547*
Brownlee, T. H. 92 (210), *183*
Brugman, F. W. 132 (343), *187*
Brun, J. 236 (55, 56, 58), *324*
Bruner, L. B. 151, 171 (398), *188*, 358 (137), *392*
Bruns, L. 374 (193), *393*
Bryant, J. I. 21, 22 (58), *54*
Bublitz, D. E. 228 (20), *323*
Buchanan, G. L. 287 (198), *327*
Büchi, G. 420 (139), *438*
Büchner, W. 335 (18), *389*
Buckley, G. D. 376 (202), *393*
Buell, G. R. 78, 129 (123), *181*
Buenker, R. J. 40, 50 (95), *55*, 445 (31), *499*
Buhler, A. 193 (15), *200*
Buijle, R. 378 (211), *394*
Bunce, S. C. 411 (98), *437*
Buncel, E. 217 (49), *220*
Bunnett, J. F. 212 (39), *220*
Bunton, C. A. 73 (68), *179*
Burak, I. 15, 26, (41), 27 (41, 73), 29, 47 (41), *53*, *54*, 445 (28), 447 (42), *499*
Burgess, J. M. 176 (480), *190*, 517 (148), *548*
Burkhardt, A. 521, 522, 524 (190), *549*
Burmester, S. 509, 510 (104), *547*
Burpitt, R. D. 530 (251), *551*
Burr, J. G. 419 (131), *438*
Buss, D. H. 111 (270), *185*
Butler, P. E. 244 (74), *324*
Butler, R. 527 (224), *550*
Butler, R. N. 512 (113, 114), *547*
Bykhovskaya, E. G. 121 (311), 122 (315), *186*, 424 (170, 171), *439*, 562, 571, 576 (21), *584*

Cadogan, J. I. G. 258 (109), 274 (168), *325*, *326*
Cais, M. 351 (95), *391*
Calderaro, E. 119, 127 (308), 133 (345), *186*, *187*
Cameron-Wood, M. 274 (168), *326*
Campagni, A. 69 (60), *179*
Campbell, A. 399, 407 (18), *435*
Campbell, C. D. 317 (265), *329*
Campbell, N. 410 (94), *437*
Campbell, T. W. 431 (216), *440*
Canter, F. C. 2, 3 (8), *52*, 61 (15), 76, 77 (105), 79 (126), 94 (105), *178*, *180*, *181*, 192 (4), *200*, 204, 214 (2) *219*, 222 (1), 223 (9), 224 (9, 14), 232, 233 (9), *322*, *323*, 342 (58), *390*, 398, 415 (3), 419, 420 (134, 135), 421 (3), 426 (134), *434*, *438*
Canuto, G. 171, 173 (447), *190*
Caprara, G. 274 (167), *326*, 425 (175), *439*
Carboni, R. A. 158 (415), *189*, 263 (124, 125), *325*, 381 (229), *394*
Cardwell, D. 84 (155), *182*

Caress, E. A. 254, 299, 302, 306 (99), *325*, 344 (69), *390*, 427 (151), *438*, 472 (109), *501*
Caronna, G. 75, 119 (82), *180*
Carpenter, W. R. 22, 23 (62), *54*, 194 (22), 197 (57), *200*, *201*, 382, 383 (232), *394*, 531 (263), *551*
Carpino, B. A. 89 (185), *182*, 527, 534 (225), *550*
Carpino, L. A. 89 (185), 90, 175 (186, 187), *182*, 527, 534 (225), *550*
Carrie, R. 367 (166), *393*
Carroll, D. G. 44 (101), *55*
Carter, J. V. 136 (355), *187*
Case, J. R. 209, 210 (25), *220*
Castellucci, N. C. 506 (68), 538 (321), *546*, *553*
Castle, J. E. 158 (415), *189*, 263 (124, 125), *325*, 381 (229), *394*
Castle, R. N. 174 (464), *190*
Castrantas, H. M. 527 (224), *550*
Catlette, W. H. 62, 76, 77 (28), *178*, 378 (215), *394*
Cattania, M. G. 70 (61), *179*
Cave, A. 76 (93), 96 (225), 98 (93, 233), *180*, *184*
Cavell, E. A. S. 93, 94 (219), *183*
Caveng, P. 66, 113 (48), *179*, 217 (48), *220*
Cecere, M. 145–147 (373), *188*
Cerny, V. 98 (234), *184*
Chambers, V. C. 215 (43), *220*
Chang, M. S. 92 (212), *183*, 513 (117), *547*
Chang, P. K. 374 (194), *393*
Chao, T. S. 3 (12), 23 (66), 28, 30 (12), *52*, *54*, 76, 77 (84), *180*, 195 (36), *201*, 380 (222), *394*, 445, 446 (33), *499*
Chapman, L. E. 405 (67), *436*, 506, 521, 542 (74), *546*
Chapman, N. B. 413 (105), *437*
Chatt, J. 134 (350), *187*
Chattaway, F. D. 148 (376), *188*, 416 (117), *437*
Chellathurai, T. 321 (269), *329*
Chemey, L. C. 505 (30), *545*
Cheney, L. C. 505 (45), *545*
Cheng, Y. M. 301 (235, 236), *328*
Cheronis, N. D. 195 (28), *201*
Chiang, M. 17 (47), *53*
Choi, C. S. 14, 15 (36), 15 (36, 43), *53*
Christensen, J. E. 112 (275, 276), *185*
Christmann, A. 59, 89, 90 (6), *178*, 245 (76), 279, 289 (181), 311 (253), 315, 316 (181), *324*, *327*, *329*, 355 (114), 356 (125), *391*, *392*, 403 (53), 426 (182), 427 (186), *439*, 442 (2), 449 (49), 467 (96), 468 (97), 481 (96), 490 (49), 492, 493 (97), *498–500*, 538, 544 (318), *553*
Christokletov, V. N. 357 (129), *392*, 416 (118), *437*, 531 (256, 257), *551*
Chua, J. 338 (36), *389*
Ciereszko, L. S. 419 (131), *438*
Ciganek, E. 96 (224), *183*
Clark, D. G. 174 (464), *190*
Clark, M. T. 409 (85), *436*
Clark, R. A. 261, 262 (114), *325*, 345, 360 (73), *390*
Clark, R. G. 36, 50 (93), *55*
Clarke, J. B. 411 (98), *437*
Cleaver, C. S. 121, 122 (313), *186*, 562, 571, 580 (22), *584*
Clegg, J. M. 158 (419), 173, 175 (468), *189*, *190*, 193 (10), *200*, 365, 383 (161), *392*, 462, 491 (89), *500*
Clement, R. A. 175 (473), *190*, 505 (44), 506 (70), *545*, *546*
Clementi, E. 36 (94a), 48 (105–107), 50 (106), *55*
Cleophax, J. 105 (255), 111 (274), *184*, *185*, 193 (12), 200 (79), *200*, *202*
Clifford, A. F. 144 (366), *188*
Closson, W. D. 28–30 (28), *53*, 197, 199 (63), *202*, 303–305 (243), *328*, 444, 445 (26), *499*, 506 (56), *546*
Clusius, K. 149 (382, 383), 152 (403), 159 (423, 424), 160 (425), 164 (433), 171 (449), 172 (450, 451), 173 (453), *188–190*, 428 (198, 199), 429 (200–202, 204), 430 (210, 213), *439*, *440*
Coburn, W. C., Jr. 519 (160, 169), *548*, *549*
Coda, L. 165, 166 (435), 167 (438), *189*
Coffin, B. 250, 269 (91), *324*, 427 (190), *439*
Cohen, J. 383 (238), *394*
Coleman, R. A. 402 (41), *435*, 535 (283), *552*
Collman, J. P. 349 (89), *391*, 534 (223), *550*
Conant, D. R. *25*
Coniglio, B. O. 66, 67, 71, 72 (53), *179*
Conley, R. T. 410 (87), *436*
Conrad, L. 504, 507 (7), *545*
Conrad, W. E. 414 (113), *437*

Cook, R. L. 14, 15 (37), *53*
Cookson, R. 102 (242), *184*
Coombs, M. M. 78 (122), 79 (122, 128), *181*, 193, 194 (14), *200*, 226, 227 (16), 229 (24, 26, 28), 230 (32, 33), 231, 232 (32), *323*, 344, 385, 386 (66), *390*, 410 (96), 416 (120), 419 (132), 422 (153), *437*, *438*
Cope, A. C. 96 (224), *183*
Coppola, A. 127 (327), *187*
Cornell, D. W. 460 (82), 483 (127), *500*, *501*, 538 (314, 315), *553*
Cothran, J. C. 62 (31), *178*
Cotter, R. J. 90 (190), *183*, 387 (248), *395*, 542 (355), *554*
Coulson, C. A. 8 (29), 25 (69), 36, 40, 52 (29), *53*, *54*
Cowan, D. O. 456 (77), *500*
Cram, D. J. 300 (232), *328*
Cramer, F. D. 103, 104 (245, 246), *184*
Crampton, M. R. 63 (38), *179*
Crast, L.-B. 505 (30), *545*
Craubner, H. 152 (403), *188*
Cremlyn, R. J. W. 99 (224), *183*, 196 (44), *201*
Croce, P. D. 290 (205), *327*, 375 (199), *393*
Cronin, D. A. 512 (114), 519 (173), *547*, *549*
Crow, W. D. 199 (76), *202*, 271 (158–160, 162, 163, 165), 272 (158), 273 (165), *326*, 425 (176, 177), *439*
Crowley, P. 90, 175 (186), *182*
Culmann, C. 209, 210 (19), *220*
Cuny, G. 175 (472), *190*, 506 (75), *456*
Curran, W. V. 89, 175 (177), *182*
Currie, C. L. 62 (23), *178*, 247, 253, 297 (97), *325*, 449, 450, 467, 468 (45), *499*
Curtice, J. S. 2, 4, 13, 17, 19 (7), *52*, 61, 91, 127 (9), *178*, 192, 195 (5), *200*, 207 (15), *220*, 245 (77), *324*, 398 (6), *435*
Curtius, T. 58, 62 (2), 63 (35), 75 (75), 152, 154 (401), 156 (409), 174 (459), *177*, *178*, *180*, *188–190*, 223 (5–7), 255 (101), 256 (102), 287 (193, 195, 196), 288 (200), 289 (196), 292 (196, 209), *322*, *325*, *327*, 335 (16), 342 (56), *389*, *390*, 402 (48), 404 (58), 416 (119), 421 (146), *436–438*, 442 (4), 479, 482 (117), *498*, *501*, 505 (23, 33, 36, 49), 506 (50, 51, 58, 71, 73), 507 (33, 62, 76), 521 (180–182, 190), 522 (190, 193), 524 (190, 193, 204), 525 (193, 204, 206–208), 526 (213–216), 527 (23), 531 (215), *545*, *546*, *549*, *550*

Dailey, B. P. 18 (52), *54*
Dalmonte-Casoni, D. 209 (22), 212 (22, 34), *220*
Dalton, C. 456 (77), *500*
Dalton, J. C. 302, 303 (241), *328*
Darapsky, A. 75 (75), 156 (409), 174 (454, 461–463), 177 (454, 482), *180*, *189*, *190*, 223 (5, 6), *322*, 416 (119), *437*
Darwent, B. de B. 62 (23), *178*, 246 (84), 247, 253, 297 (97), *324*, *325*, 449, 450, 467, 468 (45), *499*
Darzens, G. 511, 516 (142), *548*
Datta, P. K. 287 (197), *327*
Daudel, R. 41 (96), *55*
David, H. 75, 114 (81), *180*
Davies, A. J. 123, 124 (318), *186*, 343 (60), *390*, 406 (73), *436*
Davies, S. J. 149 (384), *188*, 374 (195), *393*
Davis, B. A. 59 (5), *178*, 222, 235, 255, 258, 263, 265 (3), *322*, 398, 401, 403, 422–426 (8), *435*, 442 (1), *498*, 580 (42b), *585*
Day, A. R. 379 (220), *394*
Day, B. F. 431 (216), *440*
Deb, S. K. 27 (72), *54*
Decius, J. C. *25*
Defeye, J. 338 (34), *389*
Dekker, M. 317 (264), *329*
DeMauriac, R. 315 (260), *329*, 342 (54), *390*, 399 (43), *436*, 449, 483 (50), *499*, 522 (198), 523 (199, 203), 527 (198), 540 (342), *549*, *550*, *553*
Demuth, E. 148 (378), 150 (389), *188*
Denivelle, L. 509 (96), *547*
Deorha, D. S. 114 (286), *185*, 212 (33), *220*
DePuy, C. H. 168 (439), *189*, 350 (93), *391*
Dergunov, Y. I. 279 (184), 280 (186), 283 (184), *327*
Dermer, D. C. 404 (62), *436*
De Selms, R. C. 574 (40), *585*
Determann, H. 89, 175 (176), *182*
Deters, D. W. 239, 241 (66), *324*
De Wolfe, R. H. 70, 83 (63), *179*
Dibeler, V. H. 17, 31 (48a), *53*
Dice, L. R. 410 (93), *437*

Dick, A. J. 110 (267), *185*
Dickenson, R. G. 442 (6, 7), 447, 467 (6), *498*
DiDomenico, J. 79 (124), *181*, 417 (126), *437*
Diels, O. 505 (43), *545*
Dietrich, M. W. 289 (202), *327*, 347 (81), 360 (141), *391*, *392*
Dietzsch, K. 420 (137), *438*
Dimaio, G. 412 (101), *437*
Dimroth, O. 75 (76), 119, 127, 130 (306), 148 (380), *180*, *186*, *188*, 357 (126–128), *392*
Dingle, H. E. 505 (19), *545*
Disselnkotter, H. 126 (324), *186*
Dixon, R. N. 457 (142), *501*
Djerassi, C. 32 (81), 33 (83), *54*, 199 (74), *202*
Doerges, J. 456, 463, 464, 467, 470, 489, 490 (75), *500*
Doering, W. von E. 168 (439), *189*, 257, 312 (106), *325*, 350 (93), 388 (250), *391*, *395*, 399 (14), 427 (187), *435*, *439*, 465, 493, 495, 496 (95), *500*
Donald, A. S. R. 77 (118), 123 (118, 318, 319), *181*, *186*, 343 (60), *390*
Dongowski, G. 374 (197), *393*
Dorges, J. 403 (46), *436*, 538, 540, 544 (319), *553*
Dornow, A. 175 (471), *190*, 379 (221), *394*
Dörr, W. 526 (213), *550*
Doub, L. 206 (10), *219*
Douglas, A. E. 14, 22 (35), *53*
Dows, D. A. 22, 23 (64), *54*
Drago, R. S. 34 (92), *55*
Drefahl, G. 145 (367), *188*
Dresdner, R. D. 401 (29), *435*
Drost, P. 209, 210 (23), *220*
Drozd, V. N. 118 (304), *186*
Due, M. 509 (101), *547*
Dufaux, R. 335 (18), *389*
Durden, J. A., Jr. 62, 76, 77 (28), *178*, 378 (215), *394*
Durham, L. 118 (302), *186*
Durrell, W. S. 401 (29), *435*
Dutcher, J. D. 103 (243), *184*
Dutt, P. V. 151 (392, 393), *188*
Dyall, L. K. 197 (55), *201*, 261 (117, 118), 262–264 (118), *325*, 422 (161), *438*

Eason, A. P. T. 287 (198), *327*
Ebert, R. 150 (388), *188*
Edmison, M. T. 285, 286 (192), *327*, 404 (62), *436*, 443 (19), *498*
Edwardes, W. E. 372 (188), *393*
Edwards, E. I. 265 (135), *326*, 537 (300), 538 (325), *552*, *553*
Edwards, J. O. *3*, 65 (44), *179*
Edwards, O. E. 96 (228), *184*, 342 (59), 360 (142), 363 (154), 385 (244), *390*, *392*, *395*, 403 (49), 427, (192), *436*, *439*, 449, 455 (48), *499*, *501*, 506 (63, 64), 530 (252), 538 (316, 317), 539 (63), 540 (316, 317, 337), 542 (353), *546*, *551*, *553*, *554*
Ege, S. N. 78, 79 (120), *181*
Eglinton, G. 126 (323), *186*
Ehrhart, G. 256 (102), *325*, 342 (56), *390*, 421 (146), *438*
Eichhorn, D. 145 (367), *188*
Einbeck, H. 505 (41), *545*
Elderfield, R. C. 410 (89), *437*
El Khadem, H. 381 (227), *394*
Ellzey, S. E. 334 (8), *389*
Eloy, F. 90–92 (199), *183*, 510, 512 (107), *547*
Emmons, W. D. 523 (200), *549*, 574 (38), *585*
Emoto, S. 216 (46), *220*, 335 (19), *389*
Endtinger, F. 164 (433), *189*
Engelman, F. 174 (463), *190*
Engels, W. 174 (463), *190*
Erickson, A. 87 (170), *182*
Erkstrom, B. 336 (24), *389*
Erlanger, B. F. 89, 175 (177), *182*
Escales, R. 127, 130, 131 (330), *187*
Esseny, J. M. 505 (30), *545*
Ether, R. M. 483 (131), *501*
Evans, B. L. 2, 14, 16–18, 21, 24, 26 (1), *52*, 448 (43), *499*
Evans, D. E. 96 (224), *183*
Evans, J. V. 229 (27), 230–232 (32), *323*, 417 (122), *437*
Ewins, A. J. 335 (15), *389*
Eyster, E. H. 197 (59), *201*

Fagley, T. F. 17 (45), *53*, 76 (106), *180*, 261 (115), *325*, 422 (156), *438*
Fahmy, A. F. M. 124 (321, 322), *186*
Fahr, E. 197 (53), *201*, 402 (42), *435*, 535 (285, 286), *552*
Fanshawe, W. J. 527 (218), *550*
Farber, L. 76, 95 (96), *180*, 338 (30, 31), *389*
Fargher, J. M. 266 (141), *326*, 462, 487 (87), *500*, 539 (334), *553*
Farkas, I. 506 (67), *546*

Farnum, D. G. 151 (397), *188*
Favini, G. 13 (31, 32), 20 (31), 30, 46 (32), 47 (31), *53*, 70 (61), *179* 207 (14), *220*
Feild, G. B. 528 (233), *550*
Ferber, E. 234 (49), *323*
Fester, G. 119, 127, 130 (306), *186*
Feuer, B. I. 90 (189), *183*, 267 (142, 143), 298 (142), *326*, 364, 389 (157), *392*, 462, 469 (86), *500*, 538 (327), *553*
Feyrenbach, K. 174 (465), *190*
Fiedler, H. 87 (169), *182*, 505 (46), *546*
Fierz, H. E. 76 (112), 84 (153), *181*, *182*, 209, 211, 212 (26), *220*, 504, 527 (1), *544*
Fierz-David, H. E. 193 (15), *200*
Fieser, L. F. 128 (333), *187*
Filler, R. 336 (20), *389*
Finnegan, W. G. 421, 433 (143), *438*, 505, 510 (24), 513 (125), 514 (125, 129), *545*, *548*
Finzi, P. V. 354 (111), *391*
Firestone, R. A. 529 (243), *551*
Fischer, E. 150, 152 (385), 171 (448), *188*, *190*
Fischer, H. J. 511, 517 (111), *547*
Fischer, W. 168 (441–443), *189*, 193 (11), 196 (40), *200*, *201*, 292 (211, 212), 293 (213), *328*, 358 (135, 136), *392*
Fleckenstein, L. J. 96 (224), *184*
Fleury, J. P. 525 (212), 534 (282), *550*, *552*
Folkers, K. 239 (61), *324*
Forman, R. A. 5 (23), 35 (23, 89), *53* *54*
Forsblad, I. 294, 295 (218), *328*, 357 (131), *392*
Forster, M. O. 62 (24), 75 (79, 80), 76 (108, 112), 80 (137), 84 (153, 155), 86 (163), 92 (215), 148 (381), *178*, *180–183*, *188*, 209, 211, 212 (26), *220*, 335 (12), 372 (187), *389*, *393*, 504 (1, 3), 512 (115), 527 (1), *544*, *547*, 555–557 (1), *584*
Fosblad, I. 532 (270), *551*
Fowler, F. W. 80 (136), 136 (136, 357b), 138 (357b), 140, 141 (136, 357b), 142 (357b), *181*, *187*, 200 (78), *202*, 217 (51, 52), 218 (51), *220*, 361 (149), *392*, 428 (194), *439*, 475 (111, 112), 476 (113), *501*, 558, 559 (11, 12), 565, 566 (28a, b), 573 (11), 579 (28b), *584*
Fowler, J. S. 81 (141), *181*, 423 (163), *438*, 561 (19), 567, 568 (30), 577 (19), *584*
Fox, M. F. 3 (20), *53*
Frank, W. 170 (444), *190*
Franke, W. K. 175 (472), *190*, 506 (75), *546*
Frankel, M. 336 (23), *389*, 536 (294), *552*
Frankevich, Ye. L. 18 (49), *53*
Franklin, J. L. 17, 31 (48a), *53*
Franz, J. E. 289 (201, 202), 296 (221), *327*, *328*, 347 (81), 357 (132), 360 (141), *391*, *392*
Fraser, V. 458 (80), *500*
Fraunberg, K. v. 266, 271 (140), 318 (140, 266), 319 (266), *326*, *329*, 389 (241, 242), *395*, 513 (121), 519 (121, 162, 163), *547*, *549*
Frazer, R. R. 505 (30), *545*
Freeman, J. P. 574 (38), *585*
Freiberg, L. A. 87 (167), 98, 100, 101 (235), *182*, *184*
Freund, M. 507 (78–81), *546*
Frey, A. J. 504, 505 (11), *545*
Friederang, A. 142 (364), *188*, 556, 557, 573, 574 (5), *584*
Friedlaender, P. 234 (47, 48), *323*
Friedman, L. 400 (26), *435*
Friedman, W. H. 150 (387), *188*
Friedrich, K. 344 (71), *390*, 561, 572, 576 (20), *584*
Fries, K. 83 (146), *181*, 581 (44), *585*
Fritz, J. S. 193 (13), *200*
Froitzheim-Kühlhorn, H. 374 (193), *393*
Frommeld, H.-D. 513 (118), *547*
Frosin, V. N. 424 (170), *439*
Frost, W. S. 62 (31), *178*
Fruton, J. S. 175 (477), *190*
Fujita, R. 280 (185), *327*
Fujita, S. 537, 539 (312), *553*
Fuks, R. 378 (211), *394*
Fuller, M. W. 214 (40), *220*
Furukawa, M. 507 (77), *546*
Fusco, R. 290 (206), *327*, 347 (79, 80), 374 (79), *390*, 410 (86), 434 (237), *436*, *440*, 518, 519 (156), 529 (245) 530 (156, 245), *548*, *551*

Gagneux, A. 84 (158), *182*, 430 (207), *440*
Galli, R. 145 (369–374), 146 (370, 372–374), 147 (371, 373, 374, *188*

Gal'perin, V. A. 195, 198 (30), *201*, 279, 283 (184), *327*
Galt, R. H. B. 388 (252), *395*, 414, 422 (114), *437*
Garbrecht, W. L. 514 (130), 515 (130, 135), *548*
Garcia-Munoz, G. 379 (219), *394*
Garner, R. 274 (170), *326*, 370 (181), *393*
Garrett, A. B. 401 (27), 402 (41), *435*, 535 (283), *552*
Garton, F. L. 148 (376), *188*
Gasiorowski, K. 209, 210 (19), *220*
Gaspar, P. P. 450 (58), *499*
Gatterman, L. 150 (388), *188*
Gaudemar, M. 126 (325), *186*
Gawargious, Y. A. 194 (27), *201*
Geiger, R. 505 (27), *545*
Geiseler, G. 247, 248, 253 (86), *324*
Gelderen, F. M. van 75 (79, 80), *180*
Geller, B. A. 166 (436), *189*
Genge, C. A. 90 (191), *183*, 265 (136), *326*, 340 (49), *390*, 528 (234), 537 (297, 299), 538, 539 (297), *550*, *552*
Geoghan, P. 135 (351), *187*
Germscheid, J. 174 (463), *190*
Gero, S. D. 105 (255), 111 (274), *184*, *185*, 193 (12), 200 (79), *200*, *202*
Gertner, D. 536 (294), *552*
Ghosh, P. B. 373 (190), *393*
Gibbs, C. F. 62 (32), *178*
Gibson, H. H., Jr. 276, 277 (174), *327*, 536 (293), *552*
Gibson, M. S. 176 (480), *190*, 517 (148), *548*
Gilchrist, T. L. 398, 401, 403, 404 (11), *435*
Gildenhorn, H. L. 402, 408 (37), *435*, 535 (284), *552*
Giles, D. E. 66, 67, 71, 72 (53), *179*
Gillam, A. E. 28 (74), *54*
Gillette, R. H. 197 (59), *201*
Gillilard, M. 417 (125), *437*
Givens, H. T. F. 383 (237), *394*
Giza, C. A. 89 (185), *182*, 527, 534 (225), *550*
Glasstone, S. 3 (16), *53*
Glazer, J. 210 (30), *220*
Glemser, D. 175 (470), *190*, 506, 521 (69), *546*
Gleu, K. 442 (8), *498*
Glinski, R. P. 108 (264), *185*
Godfrey, J. G. 505 (30), *545*
Goebel, N. 196 (43), *201*
Goerdeler, J. 296 (220), *328*, 356, (120), *391*
Goffredo, O. 534 (279), *552*
Gold, H. 126 (324), *186*, 377 (208), *394*
Goldstein, M. J. 399 (14), *435*
Golov, V. G. 280 (186), *327*
Gomes, E. 430 (212), *440*
Gomez-Revilla, A. 336 (24), *389*
Goncharova, I. N. 519 (164, 165), *549*
Goodlett, V. W. 530 (251), *551*
Goodman, L. 76, 103, 105 (95), 112 (275, 276), *180*, *185*, 338 (32), *389*
Gordy, W. 24 (68), *54*
Gotshal, Y. 66 (52), *179*, 261 (116), *325*, 422 (157), *438*
Gotzky, S. 76 (109), *180*, 383 (233), *394*, 434 (236), *440*
Goubeau, J. 242 (68), *324*, 343, (62), *390*, 421 (141), *438*
Gould, E. S. 67, 120 (54), *179*
Goutarel, R. 76 (93), 98 (93, 233), 100 (238), *180*, *184*
Graf, R. 87 (167), *182*
Grakauskas, V. A. 152 (405), 155 (408), *189*
Grammaticakis, P. 30 (79), *54*, 198 (66), *202*, 206 (11), *219*, 445 (35), *499*
Grandmougin, E. 212 (35), *220*
Grashey, R. 345, 373 (72), *390*, 529 (238), *550*
Grassmann, D. 517 (149), *548*
Grassmann, M. 544 (366), *554*
Gray, A. C. G. 262, 263 (121), *325*
Gray, H. B. 28–30 (28), 44 (100), *53* *55*, 197, 199 (63), *202*, 303–305 (243), *328*, 444, 445 (26), *499*, 506 (56), *546*
Gray, P. 2 (1, 2), 3 (2, 19), 5 (2), 14 (1, 2), 15 (19), 16 (1, 2), 17 (1), 18 (1, 2, 50), 21, 24 (1), 26 (1, 2), *52*, *53*, 448 (43), *499*
Green, M. 2, 8, 14, 16 (3), *52*, 65 (41), *179*
Greene, F. D. 301 (234), *328*
Greenwald, R. B. 196 (41), *201*, 348 (84), *391*, 574, 575 (41a), *585*
Grelecki, C. J. 246, 251 (83), *324*
Grieco, P. A. 114 (278, 279), 115 (279), *185*, 505 (32), *545*
Griesinger, A. 129 (336), *187*
Griess, P. 58, 148, 149 (1), 152 (404), *177*, *189*, 234 (44–46), *323*
Griot, R. 504, 505 (11), *545*

Groh, H. 96 (227), *184*
Gross, A. 76, 77 (107), *180*, 294, 295 (216), 311 (253), *328*, *329*, 355 (114), 356 (123), *391*, *392*, 400, 402 (25), 427 (186), *435*, *439*, 468 (97), 490 (119), 492, 493 (97), *500 501*
Grünanger, P. 354 (111), *391*, 432 (224), *440*
Grundmann, C. 61 (8), 62 (22), *178*, 333 (1), *389*, 513 (118), *547*
Grundmann, C. J. 76, 92 (113), *181*
Gruntuch-Jacobson, L. 76 (87), *180*
Guardino, G. 376 (205, 206), *394*
Gudmundsen, G. H. 227, 228 (19), *323*, 343 (61), *390*, 416, 417 (121), *437*
Guella, F. 70 (61), *179*
Guither, W. D. 174 (464), *190*
Gurvich, L. V. 18 (49), *53*
Gusarsky, E. 4 (11), *52*
Gustowski, W. 336 (20), *389*
Guthrie, R. D. 111 (268a, b, 270, 271), *185*, 337 (28), *389*
Gutmann, A. 194 (18–20), *200*
Gutmann, V. 505 (26), *545*

Haas, F. W. (194), *327*, 525 (208), *550*
Haberlein, H. 531 (258), *551*
Hafner, K. 360 (144), 369, 371 (175), 387 (247, 249), *392*, *393*, *395*, 405 (66, 69), *436*, 450 (54), 462 (90), 463 (90, 91), 465 (92, 94), 466, 467 (91), 479 (91, 123), 481 (123), 482 (54), 483 (130), 485 (92), 489 (54), *499–501*, 527 (221), 540 (343, 345, 347), 541 (350), 542 (347, 354), 543 (343, 359), *550*, *553*, *554*, 561, 566, 579 (18), *584*
Hage, S. M. 335 (17), *389*
Hahmann, O. 175 (471), *190*
Hahn, H. 243, 244 (73), *324*, 339 (40), *390*
Hai, S. M. A. 528 (236), *550*
Haines, A. H. 107 (259), *185*
Halans, E. 244 (75), *324*
Haldenwanger, H. 62 (22), *178*
Hall, J. H. 30 (80), *54*, 157 (413), 158 (413, 414, 419), *189*, 198 (69), *202*, 205–207, 209–212 (6), *219*, 258, 259 (110), 263 (123), 264 (127), 265 (139), 266 (141), 269 (110), *325*, *326*, 335 (9), 340 (45, 48), 348 (85), 365 (161), 368 (173), 381 (228), 383 (161), *389–394*, 422 (160), *438*, 443 (20), 447 (38), 462 (87, 89), 487 (87), 491 (89), *498–500*, 534 (276), 539 (332–334), *552 553*, 571 (36), *585*
Halvska, R. J. 427 (189), *439*
Ham, G. 215 (42), *220*
Hamer, J. 76, 77 (83, 105), 94 (105), *180*, 223, 224 (9), 232 (9, 38), 233 (9), *322*, *323*, 342 (58), 370, (180), *390*, *393*, 419, 420, 426 (134), *438*
Hammick, D. L. 372 (188), *393*, 432 (221), *440*
Hammond, G. S. 193 (13), *200*, 301 (233), *328*, 450 (58), 456 (77), *499*, *500*
Hampel, G. 505 (26), *545*
Hancock, J. 356 (116), *391*, 481, 490 (125), *501*
Hanessian, S. 104 (252), 106 (257), 108 (263), 111 (269), *184*, *185*, 215 (44), *220*, 558 (10), *584*
Hanisch, H. 174 (466), *190*
Hannerz, E. 204 (3), *219*
Hansen, O. R. 514 (134), *548*
Hantzsch, A. 132 (340), *187*, 197 (61, 62), *202*, 421, 433 (147), *438*, 513, 521 (124), *547*
Hanusch, F. 115 (295), *186*
Hardy, W. B. 274 (166), *326*, 349 (90), *391*
Harmon, R. E. 346 (78), *390*
Harrison, S. 338 (31), *389*
Hart, C. V. 115 (293), *185*, 192 (6), *200*, 516 (140), *548*
Hart, H. 252 (92), *324*
Hartenstein, A. 371 (184), *393*, 450, 466, 482 (53), *499*, 543, 544 (363), *554*
Hartwell, J. L. 128 (333), *187*
Hartzel, L. W. 115 (292), 127 (326), *185*, *187*, 518 (155), *548*
Harvey, G. R. 61, 126, 130 (16), *178*, 361 (148), 380 (224), *392*, *394*, 428 (195), *439*, 533 (273), *551*, 562, 565 (24), *584*
Harvey, S. H. 68, 73 (56), *179*
Harvill, E. K. 383 (237), *394*, 413 (104), *437*
Haskell, T. H. 111 (269), *185*
Hassan, S. S. M. 194 (27), *201*
Hassner, A. 80 (136), 136 (136, 356a, b, 357a, b), 138 (357a, b), 139 (358), 140 (136, 357a,b, 360–362), 141 (136, 356a,b, 357a,b), 142

(357a,b, 360, 364, 365), 144 (360, 361), 145 (360), *181*, *187*, *188*, 196 (41), 200 (78), *201*, *202*, 217 (51, 52), 218 (51), *220*, 348 (84), 361 (149), *391*, *392*, 428 (194), *439*, 475 (111, 112), 476 (113), *501*, 556 (5, 6), 557 (5–7), 558, 559 (11, 12), 560 (16a,b), 565, 566 (28a,b), 568 (7), 572 (6), 573 (5, 11, 12), 574 (5, 41a), 575 (41a), 579 (28b), *584*, *585*
Hauser, C. R. 401 (35), *435*, 540 (346), *553*
Hauser, E. 75 (77), *180*, 294, 295 (215), *328*, 356 (119), *391*, 532 (266), *551*
Haussmann, W. 515 (139), *548*
Haworth, R. D. 96 (224), *184*
Hayon, E. 22 (60), *54*
Heasley, V. L. 84 (159), *182*, 430 (208), *440*
Heathcock, C. H. 135 (353), 136, 141 (356a,b), *187*
Heathcock, J. F. 285, 286 (192), *327*, 443 (19), *498*
Hedaya, E. 271 (161), *326*
Hedayatullah, M. 509 (96), *547*
Hegarthy, A. F. 93, 176 (217), *183*, 433 (230), *440*, 517 (146, 147), *548*
Heidenreich, K. 505 (23), 506 (71), 521 (180, 181), 527 (23), *545*, *546*, *549*
Heisler, R. Y. 93–95 (218), *183*, 195 (35), *201*, 334, 337 (7), *389*
Helberg, J. 379 (221), *394*
Heller, G. 155 (407), *189*
Heller, H. E. 236 (59), *324*
Helmholz, L. 44 (99), *55*
Helwerth, E. 92 (213), *183*, 510, 517 (109), *547*
Hemery-Logan, K. R. 261, 262 (114), *325*, 345, 360 (73), *390*
Hempel, H. 507 (78), *546*
Hendrickson, D. N. 34, 47 (91), *55*
Hendrickson, J. B. 351, 352, 358 (100), *391*
Heneka, H. 512 (116), *547*
Henkel, K. 76 (90), *180*
Henke-Stark, F. 90, 132 (194), *183*
Henry, R. A. 383 (238), *394*, 421 (143), 432 (227), 433 (143), *438*, *440*, 505, 510 (24), 513 (125), 514 (125, 129), 516 (141), 519 (141, 174), *545*, *548*, *549*
Henseke, G. 174 (466), *190*
Henshall, A. E. 347 (81), *391*
Heppolette, R. L. 122 (317), *186*
Herbison-Evans, D. 34 (85), *54*
Herbst, R. M. 383 (237), *394*, 413 (104), *437*, 514 (130), 515 (130, 135, 136), *548*
Herring, D. L. 158 (416), *189*, 356 (122), *391*
Herzberg, G. 25, 38, 39 (71), *54*, 452, 460 (66), *499*
Heukelbach, E. 515 (138), *548*
Higbee, H. H. 209, 210 (20), *220*
Hilbert, H. E. 422 (152), *438*
Hildesheim, J. 105 (255), 111 (274), *184*, *185*, 193 (12), 200 (79), *200*, *202*
Hill, D. L. 66, 67 (47), *179*, 212 (37), *220*
Hill, J. 105 (253), 107 (253, 258), 108 (253), *184*, *185*
Hill, J. W. 265 (139), 266 (141), *326*, 340 (48), *390*, 462, 487 (87), *500*, 539 (333, 334), *553*
Hille, H. 506 (58), 507 (62), *546*
Hine, J. 142 (363), *188*, 402 (39), *435*
Hingardner, W. S. 505 (47), *546*
Hiron, F. 68 (55), 73 (55, 69), *179*, 337 (27), *389*
Hiskey, R. G. 528 (231), *550*
Hites, R. D. 133 (346, 348), *187*, 433 (233), *440*, 506 (84), 507 (84–86), *546*, *547*
Ho, K. C. 66 (47, 50), 67 (47), *179*, 212 (37), *220*
Hobson, J. D. 117 (299), 118 (300), 175 (467), *186*, *190*, 200, (77) *202*, 277 (177), 278 (178), *327*, 349 (88), *391*, 424 (168), *439*, 496 (139), *501*
Hoch, H. 194 (24), *200*, 516, 519 (143), *548*
Hocking, M. B. 399 (21), *435*
Hoegerle, K. 244 (74), *324*
Hoffman, C. W. W. 23 (66), *54*, 195 (36), 198 (70), *201*, *202*
Hoffmann, K. A. 194 (24), *200*, 516, 519 (143), *548*
Hoffmann, K. H. 410 (92), *437*
Hoffmann, W. 174 (457), *190*
Hofmann, A. 504, 505 (11), *545*
Hofmann, K. 175 (477), *190*
Hofmann, O. 402 (48), *436*, 505 (36), *545*
Hohenlohe-Oehringen, K. 382 (230, 231), *394*
Holcomb, W. D. 281, 282, 284, 285 (188), *327*

Holdrege, C. T. 505 (30), *545*
Holland, M. N. 519 (172), *549*
Holleman, A. F. 212 (31), *220*
Holly, S. 195, 198 (32), *201*
Holm, A. 509 (100–102), *547*
Holmes, R. E. 338 (35), *389*
Homeyer, A. H. 400 (22), *435*
Honda, I. 528 (229), *550*
Honsberg, U. 294, 295 (218), *328*, 357 (131), *392*, 532 (270), *551*
Honzl, J. 175 (474), *190*
Hoover, J. R. E. 379 (220), *394*, 530 (250), *551*
Hopkins, H. B. 122 (316), *186*
Hora, J. 98 (234), *184*, 528, 539 (230), *550*
Horino, H. 158 (417), *189*, 365 (162), *392*
Hormann, H. 76, 77 (101), *180*, 334 (5), *389*
Horne, R. J. 198 (71), *202*, 308 (249), *328*, 447, 458 (37), *499*
Horner, L. 59 (6), 76, 77 (107), 89, 90 (6), *178*, *180*, 245 (76), 277 (175), 279, 289 (181), 292 (175), 294, 295 (216), 311 (253), 315, 316 (181), *324*, *327–329*, 355 (114), 356 (123, 125), 378 (209), *391*, *392*, *394*, 400, 402 (25), 403 (46, 53), 426 (182), 427 (186), *435*, *436*, *439*, 442 (2), 449 (49), 456, 463, 464 (75), 467 (75, 96), 470 (75), 481 (96), 489 (75), 490 (49, 75, 119), *498–501*, 532 (268), 535, 536 (291), 538 (318, 319), 540 (319), 544 (318, 319), *551–553*
Horning, D. E. 87 (173), *182*, 504 (17), *545*
Horowitz, J. P. 408, 409 (83), *436*
Hortmann, A. G. 84 (154), *182*, 361 (146), *392*, 479 (116), *501*
Horton, D. 76, 103 (103), 105 (254), 110 (103, 266), *180*, *184*, *185*, 338 (33), *389*
Horwitz, J. P. 152 (405), 155 (408), *189*, 338 (36), *389*, 514, 515 (131), *548*
Hough, L. 105 (253), 107 (253, 258), 108 (253), 111 (270), *185*
Houpt, A. G. 223, 237 (8), *322*, 342 (57), *390*, 419 (133), *438*
Howe, R. 540 (337), *553*
Howland, L. 505 (38), *545*
Hoye, P. A. T. 73 (56), *179*
Huber, W. F. 336 (25), *389*
Hu-Chu Tsai 462, 487 (87), *500*
Hudson, R. F. 65 (40, 41), *179*
Hughes, E. D. 68 (55, 56), 73, (55, 56, 69), *179*, 210 (30), *220*, 236 (59), *324*, 337 (27), *389*
Huheey, J. E. *14*
Huisgen, R. 91 (209), 160 (425, 426), 164 (431), *183*, *189*, 257 (105, 107), 258 (108), 260 (113), 266, 271 (140), 290 (204), 291 (207), 318 (140, 266), 319 (266), *325–327*, *329*, 345 (72), 353 (110), 360 (140), 369 (178), 371 (185), 373 (72), 374 (110), 375 (200), 376 (204), 377 (200), 378 (210), 388 (251), 389 (241, 242), *390–395*, 401 (33), 410, 413 (90), 424 (172), 425 (173), 429 (203–205), 432 (222, 225), 434 (225), *435*, *437*, *439*, *440*, 443 (14, 15), 449 (47), 450 (52), 465 (47, 52), 479 (120, 128), 481, 482 (128), 486 (47), 495 (15, 136–138), 496 (138), *498*, *499*, *501*, 505 (22), 506 (57), 513 (121, 122), 519 (121, 162, 163), 529 (238–241, 244), 530 (239), 531 (255, 262, 264), 537, 539 (301), 543 (264, 301, 360, 362), *545–547*, *549–552*, *554*
Huneck, S. 403 (51), *436*, 540 (341), *553*
Hung, Y.-L. 105 (254), *184*
Hunger, A. 410 (92), *437*
Hunger, K. 90 (192), *183*, 528 (232), *550*
Huntress, E. H. 79 (125), *181*
Hurzeler, H. 149 (382), 159 (423), *188*, *189*, 429 (201), *439*
Huttel, R. 127 (328), *187*
Hyde, H. W. 90, 91 (198), *183*, 389 (240), *394*

Iglesias, J. 379 (219), *394*
Ikeler, T. J. 302, 310 (239, 240), *328*
Imoto, E. 507 (77), *546*
Imoto, M. 280 (185), *327*
Ingham, J. D. 93, 94 (220), *183*
Ingold, C. K. 68 (56), 73 (56, 69), 74, 76 (104), *179*, *180*, *204*, 205, 206 (5), 210 (30), 212, 214 (32), *219*, *220*, 235 (53), 236 (59), *323*, *324*, 337 (27), *389*
Inouye, K. 175 (478), *190*
Iqbal, Z. 21 (59), *54*

Isbister, R. J. 142 (364, 365), *188*, 196 (41), *201*, 348 (84), *391*, 556, 557 (5, 6), 572 (6), 573 (5), 574 (5, 41a), 575 (41a), *584*, *585*
Iselin, B. M. 89, 90 (182), 175 (476), *182*, *190*
Isomura, K. 362 (150), 365 (159), *392*, 566, 567 (29), 569 (35), 580 (29), *584*, *585*
Itai, T. 90, 115, 174 (204), *183*, 277 (60), *324*, 519 (168), *549*
Ito, G. 519 (168), *549*
Ito, S. 151, 171 (399), *188*

Jacox, M. E. 443 (10, 11), 468 (100), *498*, *500*
Jaffe, H. H. 206 (8), *219*
Jäger, G. 339 (41), *390*, 421 (142), *438*, 535 (287–289), *552*
Jager, J. 242 (69–71), 245 (70), *324*
James, A. T. 210 (30), *220*
Jansen, J. 506 (51), *546*
Jappy, J. 196 (48), *201*
Jarreau, F. X. 98 (233), 100 (238), *184*
Jary, J. 108 (265), *185*
Jencks, W. P. 409 (84), *436*
Jensen, H. 505 (38), *545*
Jensen, K. A. 509 (93, 100–102, 104), 510 (104), 514 (132, 134), *547*, *548*
Johnson, D. M. 378 (214), *394*
Johnson, M. D. 159, 160 (422), *189*
Johnson, R. L. 269 (154), *326*, 465, 481, 485 (93), *500*, 537, 542 (308), *552*
Johnson, T. B. 505 (47), *546*
Jones, D. N. 76, 98–100 (91), *180*
Jones, E. R. H. 126 (323), *186*
Jones, G. T. 210 (30), *220*
Jones, J. K. N. 110 (267), *185*
Jones, W. J. 14, 22 (35), *53*
Joshi, S. S. 114 (286), *185*, 212 (33), *220*
Joshua, W. P. 76 (112), *181*
Just, G. 538, 540 (322), *553*

Kahn, A. A. 320 (267, 268), *329*
Kahovec, L. 197 (58), *201*
Kainer, H. 153 (406), *189*
Kaiser, W. 134 (349), *187*, 360 (144), *392*, 405 (66), *436*, 465 (92, 94), 483 (130), 485 (92), *500*, *501*, 540 (345), 541 (350), 543 (359), *553*, *554*
Kamiya, S. 90, 115, 174 (204), *183*, 277 (60), *324*
Kamm, D. R. 368 (173), *393*
Kan, R. O. 30 (80), *54*, 61 (11), 157, 158 (413), *178*, *189*, 198 (69), *202*, 205–207, 209–212 (6), *219*, 335 (9), *389*, 447 (38), *499*, 534 (276), *552*
Kanayama, M. 175 (478), *190*
Kanda, T. 35 (88), *54*
Kang, H. H. 336 (20), *389*
Kantor, S. W. 401 (35), *435*, 540 (346), *553*
Kao, P.-L. 544 (365), *554*
Kappeler, H. 89 (180, 182), 90 (182), 175 (180), *182*, 534 (281), *552*
Karplus, M. 34, 35 (86), *54*
Kaska, W. C. 535 (290), *552*
Kato, H. 47 (104), *55*
Kato, M. 506 (68), 538 (321), *546*, *553*
Katritzky, A. R. 262, 263 (121), *325* 373 (190), *393*, 431 (219), *440*
Kauer, J. C. 263 (125), *325*, 381 (229), *394*, 508 (91), *547*
Kauffmann, T. 335 (17), *389*
Kawamura, K. 35 (88), *54*
Kawatani, H. 90 (188), *182*
Kay, P. S. 136 (355), *187*
Keana, J. F. W. 360 (143), *392*
Keana, S. B. 360 (143), *392*
Keating, J. 301 (237), *328*
Keller, W. 415 (115), *437*
Kemp, J. E. 197 (55), *201*, 261 (117, 118), 262–264 (118), *325*, 422 (161), *438*
Kendall, F. H. 72 (67), *179*
Kenny, D. H. 508 (92), *547*
Kent, M. E. 271 (161), *326*
Kenyon, J. 399 (16–18), 407 (17, 18), *435*
Kereszty, P. 434 (238), *440*
Kermack, W. O. 505 (34), *545*
Kern, C. W. 34, 35 (86), *54*
Kersting, F. 116 (296), *186*, 358 (134), *392*
Kesting, W. 506, 521 (72), *546*
Key, A. 151 (393), *188*
Khan, N. H. 175 (477), *190*
Khattab, S. 83 (150), *182*, 374 (196), *393*, 563 (26), *584*
Khuong-Huu-Qui 98 (233), 100 (238), *184*
King, G. J. 265, 280, 297 (133), *326*, 457 (143), *501*
Kikuchi, S. 198 (72), *202*
Kim, Y. K. 374, 377 (198), *393*

Kirchhoff, B. 374 (197), *393*
Kirenskaya, L. I. 86 (162), *182*, 510 (105), *547*
Kirk, J. C. 410 (88), *436*
Kirmreuther, H. 194 (24), *200*
Kirmse, W. 378 (209), *394*, 434 (229, 234), *440*, 510 (106), *547*
Kischa, K. 215 (45), *220*
Kistner, J. F. 76, 95 (96), *180*, 338 (30), *389*
Kitahanoki, K. 530 (249), *551*
Klages, F. 431 (218), *440*
Klavehn, W. 292 (209), *327*, 526, 531 (215), *550*
Kleinberg, J. 78, 129 (123), *181*, 228 (20–22), 229 (22), *323*, 421 (144), *438*
Kliegman, J. M. 307 (247, 248), 308 (248), *328*, 383 (234, 235), *394*, 428 (196, 197), *439*, 472 (105, 106, 108), *500*, *501*
Klimstra, P. D. 96 (226), *184*
Klingbeil, J. E. 515 (136), *548*
Klug, J. T. 196 (41), *201*, 348 (84), *391*, 574, 575 (41a), *585*
Knaggs, I. E. 14 (39), *53*
Knaus, E. E. 270, 321 (156), *326*
Knaus, G. N. 269, 286, 321 (155, 270), *326*, *329*, 525 (210), *550*
Knobler, Y. 336 (23), *389*
Knoll, A. G. 415 (116), *437*
Knorr, R. 290 (204), *327*, 378 (210), *394*, 531 (255), *551*
Knox, G. R. 317 (264), *329*
Knudsen, G. A., Jr. 301 (234), *328*
Knunyants, I. L. 121 (311), 122 (315), *186*, 424 (170, 171), *439*, 562, 571, 576 (21), *584*
Kobayashi, S. 365 (159), *392*, 566, 567, 580 (29), *584*
Koch, E. 297, 298 (223), *328*
Koczarski, A. 404 (57), *436*, 521, 522 (191), *549*
Koda, Y. 507 (77), *546*
Koenig, C. 479, 481 (123), *501*
Koga, G. 170 (445), 177 (483), *190*, 253, 254 (96), *324*
Koga, N. 168 (442), *189*, 196 (40), *201*, 229, 230 (29), 253, 254 (96), *323*, *324*, 342 (55), *390*, 522 (197), *549*
Kohlrausch, K. W. F. 197 (58), *201*
Kohnstam, G. 79 (133), *181*
Kokowsky, N. 89, 175 (177), *182*
Kolkoltseva, I. G. 357 (129), *392*
Komppa, B. 504 (18), *545*
Kondratyev, V. N. 18 (49), *53*
König, C. 387 (247), *395*, 542 (354), *554*
Konig, W. 247, 248, 253 (86), *324*
Konishi, H. 47 (104), *55*
Korczynski, A. 83 (147), *181*
Kovacic, A. 90, 115, 116 (205), *183*
Koyama, K. 525 (211), *550*
Krauch, K. 521 (185), *549*
Krauss, M. 17, 31 (48a), *53*
Krbechek, L. 264 (129), 269 (151), *325*, *326*, 340 (46), 368 (172), 369 (174), 385, 387 (46), *390*, *393*, 422 (158), 427 (188), *438*, *439*
Krebs, A. 378 (216), *394*
Kreevoy, M. M. 505 (25), *545*
Kreher, R. 242 (69–71), 245 (70), *324*, 339 (41), 369 (176, 177), *390*, *393*, 405 (68), 421 (142), *436*, *438*, 479, 482 (129), *501*, 527 (219, 220), 535 (287–289), 540 (344), *550*, *552*, *553*
Krespan, C. G. 121, 122 (313), *186*, 362 (151), *392*, 562, 571, 580 (22), *584*
Kreuter, C. 174 (463), *190*
Krishnamurthy, V. N. 19 (55), *54*
Kroto, H. W. 452, 460 (69), *500*
Krueger, W. E. 80 (138), *181*, 233 (42), *323*, 348 (83), *391*, 560, 567, 580 (15), *584*
Ksandr, Z. 108 (265), *185*
Kubo, M. 507 (77), *546*
Kubota, M. 349 (89), *391*, 534 (223), *550*
Kucera, J. 353 (109), *391*
Kuczynski, H. 79 (134), *181*
Kuhling, D. 369 (177), *393*, 405 (68), *436*
Kuhn, L. P. 79 (124), *181*, 417 (126), *437*
Kuhn, R. 153 (406), *189*
Kuhn, W. 32 (82), *54*
Kuna, M. 84, 85 (157), *182*
Kuntzen, H. 75 (78), *180*
Kunzle, F. 367 (169), *393*
Kurtz, D. W. 569 (33b), *585*
Kurtz, P. 126 (324), *186*
Kurz, P. 512 (116), *547*
Kussner, C. L. 90, 91 (200), *183*, 519 (170), *549*
Kuznesof, P. M. 34, 37 (91), *55*
Kuzuhara, H. 216 (46), *220*
Kwart, H. 320 (267, 268), *329*

Kyba, E. P. 249 (89, 90), 251 (90), 252, 254, 299, 300 (89, 90), 304 (90), 307 (89, 90), *324*

L'Abbe, G. 59 (7), *178*, 222, 235, 251, 260, 261, 265, 289, 317 (4), *322*, 333 (3, 4), 380 (225), *389*, *394*, 398, 401, 405, 414, 415 (7), *435*, 529, 531 (242), 533 (274), 544 (242), *551*
Labler, L. 98 (234), *184*
Ladik, J. 25, 41 (70), *54*, 534 (277), *552*
Lagowski, J. M. 431 (219), *440*
Lakritz, J. 173, 175 (468), *190*, 193 (10), *200*
Lamola, A. A. 456 (77), *500*
Langsdorf, W. B. 65, 67 (43c), *179*
Lansbury, P. T. 411 (99, 100), *437*
Lappert, M. F. 2, 4 (6), *52*
Larsen, J. W. 73 (73), *180*
Lawyer, C. B. 507 (89), *547*
Leadill, W. K. 410 (94), *437*
Leboeuf, M. 76 (93), 98 (93, 233), *180*, *184*
Lederer-Ponzer, E. 87 (167), *182*
Lee, C. C. 407, (76), *436*
Lee, S. H. 401 (27), *435*
Leermakers, J. A. 245 (81, 82), 246 (81), 247 (81, 82), 248, 251 (82), *324*
Leffler, J. E. 194 (26), *201*, 276, 277 (174), 279 (182, 183), 280 (183), 281, 291 (182), 294 (218), 295 (218, 219), 296 (182), *327*, *328*, 357 (131), *392*, 532 (265, 270, 272), 536 (293), *551*, *552*
Le Gras, J. 126 (325), *186*
Leitmann, O. 505 (26), *545*
Le Men, J. 96 (225), *184*
Lemke, H. 518 (154), *548*
Lendorff, P. 76 (87), *180*, 504 (16), *545*
Leon, E. 513 (120), *547*
Leone, R. 538 (325), *553*
Leusen, A. M. van 352 (105), *391*
Leutzow, A. E. 338 (33), *389*
Levene, P. A. 33 (83a), *54*, 84, 85 (157), *182*, 199 (73), *202*, 445 (41), *499*
Levering, D. R. 195 (33), *201*, 519 (171), *549*
Levinson, A. S. 362 (153), *392*, 506 (65), 540 (339), *546*, *553*
Levy, L. A. 136, 138, 140–142 (357a,b), *187*, 217 (52), *220*, 476 (113), *501*, 558, 559, 573 (12), *584*
Lewis, E. S. 159, 160 (422), *189*
Lewis, F. D. 174 (460), *190*, 254 (100). 299 (229), 302 (238, 241), 303, (229, 238, 241, 242), 307 (100), *325*, *328*, 426 (183–185), 430 (184), *439*, 449, 450 (46), 455 (74), 456 (76), 474 (74), *499*, *500*
Lewis, J. R. 528 (233), *550*
Lide, D. R. 5, 35 (23), *53*
Lieber, E. 2 (7), 3 (12), 4, 13, 17, 19 (7), 22 (63), 23 (63, 66), 28 (12, 75), 30 (12), *52*, *54*, 61 (9, 12), 76, 77 (84), 90 (12), 91, 127 (9), 132 (12), 133 (346–348), 176 (12), *178*, *180*, *187*, 192 (5), 195 (5, 31, 36, 37), 196 (45, 47), 197 (31, 45), 198 (37), *200*, *201*, 207 (15), *220*, 245 (77), *324*, 380 (222, 223), 383 (238), *394*, 398 (5, 6), 404, 415 (5), 421 (143), 433 (5, 143, 231, 233), *435*, *438*, *440*, 445 (33, 34), 446 (33), *499*, 505 (19), 506 (52, 84), 507 (52, 84–87, 89, 90), 513 (125), 514 (125, 129), 519 (171), 521 (52, 189), 533, 535 (275), *545–549*, *551*
Liedhegener, A. 351 (97, 99, 103), *391*
Lieser, T. 86 (165), *182*
Limpricht, H. 174 (455), *190*
Lin, C. C. 33 (84), *54*
Lin, L. S. 264 (128), *325*, 364 (158), *392*, 422 (159), *438*
Lindemann, H. 76 (111), 86 (161), *180*, *182*, 504 (5, 8, 15), *545*, 574 (39), *585*
Lindenmann, A. 175 (477), *190*
Lindsay, R. O. 62, 174 (25), *178*
Linke, S. 193 (16), *200*, 401 (36), *435*, 462 (84, 85), 487 (85), *500*, 505 (31), 507 (31, 61), 537 (296), 538 (61, 296), 539 (31), *545*, *546*, *552*
Lipatova, L. F. 519 (166), *549*
Litzius, W. 196 (46), *201*
Livingston, R. L. 14, 18 (38), *53*, 304 (244), *328*, 418 (130), *438*
Lloyd, D. 351 (96), *391*
Locatell, L. 84 (159), *182*
Löflund, F. 505 (43), *545*
Lofquist, R. 505, 510 (24), *545*
Logothetis, A. L. 76 (97), 79 (131), 92 (97), *180*, *181*, 256 (103), *325*, 363 (155), *392*, 423 (164), *439*
Lombardino, J. G. 61 (13), *178*, 192 (3), *200*
Loncrini, D. F. 336 (21), *389*
Longo, G. 173 (469), *190*

Lora-Tamayo, M. 379 (219), *394*
Lorber, M. 136, 141 (356b), *187*
Losin, E. T. 410 (89), *437*
Lossing, F. P. 271 (161), *326*
Loudon, J. D. 388 (252), *395*, 414, 422 (114), *437*
Löw, R. 353 (108), *391*
Lucas, H. J. 135 (354), *187*
Lucken, E. A. 79 (129), *181*, 229 (25), *323*, 417 (123), *437*
Luehn, F. 209, 210 (20), *220*
Luenow, L. 523 (201), *549*
Lundina, I. B. 90, 91 (202), 115 (289), *183*, *185*
Lunts, L. H. C. 96 (224), *184*
Lüthi, U. 428 (199), *439*
Lüttke, W. 430 (211), *440*
Lutz, N. 535 (286), *552*
Lwowski, W. 193 (16), *200*, 265, 266 (137), 269 (154), 315 (258, 260), 316 (258), *326*, *329*, 341 (50), 342 (54), 356 (117), 360 (50, 145), 371 (184), *390–393*, 398 (10), 399 (43), 401 (36), 403 (52), 404 (55, 60, 63), *435*, *436*, 449 (50, 51), 450 (53, 55), 451 (59–65), 455 (61), 460 (60, 82), 461 (61), 462 (84, 85), 463 (63), 464 (61), 465 (51, 55, 60, 62, 64, 93), 466 (53), 479 (55), 480 (55, 122), 481 (93, 124), 482 (53), 483 (50, 122, 127), 484 (65), 485 (93, 132), 487 (51, 62, 85, 122), 490 (124), *499–501*, 505 (31), 507 (31, 61), 517 (149), 522 (198), 523 (201–203), 525 (211), 527 (198), 528, 535 (235), 537 (296, 302, 303, 305–309, 313), 538 (61, 296, 309, 314, 315, 320, 323, 324), 539 (31, 235, 303, 309), 540 (302, 303, 342), 541 (235, 305–307, 313, 348, 349, 352), 542 (308, 356, 357), 543 (363), 544 (363, 365, 366, 368), *545*, *546*, *548–550*, *552–554*
Lyttleton, J. W. 407 (78), *436*

Ma, T. S. 195 (28), *201*
Mack, C. H. 196 (43), *201*
Maffei, S. 88 (174), 165, 166 (434, 435), 167 (437, 438), *182*, *189*
Magee, M. Z. 175 (477), *190*
Mahesh, V. K. 114 (286), *185*, 212 (33), *220*
Mahler, G. 223, 224, 228 (10), *322*, 419 (136), *438*
Mai, J. 150 (386, 387), *188*
Maier, D. P. 518 (153), *548*
Maimind, V. I. 430 (212), *440*
Maiorana, S. 80, 82 (140), *181*, 518, 519, 530 (156), *548*, 561, 566, 568, 576 (17), *584*
Malpass, J. R. 117 (299), 118 (300), 175 (467), *186*, *190*, 200 (77), *202*, 277 (177), 278 (178), *327*, 349 (88), *391*, 424 (168), *439*, 496 (139), *501*
Maltz, H. 532 (267), *551*
Mancuso, N. R. 411 (99, 100), *437*
Mandronero, R. 379 (219), *394*
Mangini, A. 209, 212 (22), *220*
Manhas, M. S. 200 (80), *202*, 574 (41b), *585*
Mansfield, G. H. 126 (323), *186*
Mansour, H. A. R. 381 (227), *394*
Marburg, S. 114 (278), *185*, 505 (32), *545*
Marcantonio, A. F. 90 (191), *183*, 265 (136), *326*, 340 (49), *390*, 528 (234), 537 (297–299), 538, 539 (297), 544 (298), *550*, *552*
Mare, P. B. D. de la 70 (62), 79 (133), 83 (62), 120, 130, 134, 139, 143 (309), *179*, *181*, *186*
Margozzi, A. P. 444 (27), *499*
Maricich, T. J. 296 (222), *328*, 356 (117), *391*, 451 (62, 62) 463 (63), 465 (62), 480 (122), 481 (124), 483 (122), 487 (62, 122), 490 (124), *499*, *501*, 537 (302, 303), 538 (324), 539 (303), 540 (302, 303), 542 (356), *552–554*
Marks, R. E. 77 (118), 78, 79 (122), 123 (118, 318, 319), 124 (318, 319), *181*, *186*, 225 (15), 229 (26), 230 (34), *323*, 343 (60), 344, 385, 386 (66), *390*, 406 (73), 420 (138), 422 (153), *436*, *438*
Marley, R. 247 (85), 306 (245), 308 (85, 245, 250), 309 (245), 311, 312 (85), *324*, *328*, 447 (40), 452 (70, 72), 455 (40), 456 (70), 458 (40), 460, 461, 464 (72), *499*, *500*
Marsh, F. D. 77 (119), *181*
Marshall, A. S. 348 (86), *391*, 424 (167), *439*, 497 (140), *501*, 571, 584 (37), *585*
Martin, D. 433 (232), *440*, 509 (94, 95, 97–99, 103), *547*
Martinelli, J. E. 84 (154), *182*, 361 (146), *392*, 479 (116), *501*
Masamune, S. 506 (68), 538 (321), 540 (338), *546*, *553*

Mason, J. P. 114, 115 (279), *185*
Mason, K. G. 2 (5), *52*
Matcha, R. 48 (107), *55*
Matsuda, K. 514 (127), *548*
Mattingly, T. W., Jr. 315 (260), *329*, 341 (50), 342 (54), 356 (117), 360 (50), *390*, *391*, 399 (43), *436*, 449 (50), 451 (61, 62), 455, 461, 464 (61), 465 (62), 481 (124), 483 (50), 487 (62), 490 (124), *499*, *501*, 522 (198), 527 (198, 227), 528, 535 (235), 537 (302), 538 (323), 539 (235), 540 (302, 342), 541 (235), 542 (356), *549*, *550*, *552–554*
Matuszko, A. J. 92 (212), *183*, 513 (117), *547*
McAllister, T. 271 (161), *326*
McCaffery, A. J. 3 (18a), *53*
McCarty, J. E. 300 (232), *328*
McClellan, A. L. 207 (16), *220*
McCombie, H. 413 (105), *437*
McConaghy, J. S. 360 (145), *392*, 451 (59, 60, 64), 460 (60), 465 (60, 64), *499*, 537, 541 (305–307), *552*
McDaniel, R. S. 346 (74), *390*
McDonald, J. R. 3, 12, 25–27, 30, 31, 45, 47 (9), *52*
McDonald, W. R. 66, 67, 71, 72 (53), *179*
McEwen, W. E. 78 (121, 123), 93–95 (218), 123, 125 (320), 129 (123), *181*, *183*, *186*, 195 (35), *201*, 227 (17–19), 228 (17–22), 229 (22), *323*, 334, 337 (7), 343 (61), *389*, *390*, 407 (79), 414 (113), 416 (121), 417 (79, 121, 124, 125, 127), 421 (144), *436–438*, 514 (126), *548*
McGlynn, S. P. 3, 12, 25–27, 30, 31 (9), 44 (101), 45, 47 (9), *52*, *55*
McGregor, A. C. 89, 90 (183), *182*
McIntosh, J. M. 363 (154), *392*, 542 (353), *554*
McKee, R. L. 519 (159), *548*
McKenna, J. 96 (224), *184*
McLean, A. D. 48, 50 (106), *55*
McMaster, I. T. 283–286 (189), *327*
McNeil, D. W. 271 (161), *326*
Medvedev, V. A. 18 (49), *53*
Meek, J. S. 81 (141), *181*, 423 (163), *438*, 561 (19), 567, 568 (30), 577 (19), *584*
Mehta, N. 417 (124), *437*
Meier, H. 524, 525 (204), *550*
Meier, R. 170 (444), *190*

Mainwald, J. 369 (179), *393*, 538 (326), 543 (361), *553*, *554*
Meisenheimer, J. 150 (390), *188*
Meisinger, M. A. P. 96 (224), *184*
Meldola, R. 75 (78), *180*
Mele, A. 17 (48), *53*
Menz, F. 352 (102), *391*
Merkle, M. 90 (195), *183*, 404 (59), *436*, 504 (14), 522 (195, 196), 524, 525 (195), 526 (217), *545*, *549*, *550*
Mertz, R. 534 (282), *552*
Merz, W. 129 (335), *187*
Mesch, W. 431 (218), *440*
Meshreki, M. H. 381 (227), *394*
Mesley, R. J. 79 (127), *181*, 229 (23), *323*, 385 (245), *395*, 422 (154), *438*
Messmer, A. 25, 41 (70), *54*, 194 (23), *200*, 356 (124), *392*, 534 (277), *552*
Metha, N. B. 227, 228 (17), *323*
Meth-Cohn, O. 158 (418), *189*, 267 (145), *326*, 368 (170), *393*
Meyer, H. 505 (35), *545*
Meyer, J. 294 (214), *328*, 356 (118), *391*, 532 (266), *551*
Meyer, W. L. 362 (153), *392*, 506 (65), 540 (339), *546*, *553*
Meyer zu Reckendorf, W. 103 (247a, b,c,d), 105 (247a), 111 (272), *184*, *185*
Michael, A. 209, 210 (20), *220*
Micheel, F. 104 (251), *184*
Michel, O. 152, 157 (400), *188*, 212 (35), *220*
Middleton, W. J. 362 (151), *392*
Miesenheimer, J. 432 (223), *440*
Mikol, G. J. 80 (138), *181*, 312 (255), *329*, 348 (83), *391*, 560, 567, 580 (15), *584*
Millar, I. T. 192 (1), *200*
Millar, J. H. 209, 210 (21), *220*
Miller, E. J. 90, 91 (197), *183*, 196, 198 (50), *201*, 389 (239), *394*, 518 (150), *548*
Miller, F. A. 22, 23 (65), *54*
Miller, J. 65 (39, 42), 66 (42, 46, 47, 49–51), 67 (46, 47), 70 (65), 71 (51), 72 (51, 67), 83 (152), 114 (277), 115 (152, 280), 117 (298b), 122 (317), 158, 159 (421), *179*, *182*, *185*, *186*, *189*, 207 (17), 208 (18), 209 (17), 212 (37, 38), 213 (17, 38), 214 (18, 40), *220*
Milligan, D. E. 443 (10, 11), 468 (99, 100), *498*, *500*
Minami, S. 412 (103), *437*

Minisci, F. 145 (368–375), 146 (368, 370, 372–375), 147 (371, 373), *188*
Minnis, R. L. 22, 23 (63), *54*, 61, 90, 132, 176 (12), *178*, 195 (31), 196 (47), 197 (31), *201*, 398, 404, 415, 433 (5), *435*, 506, 507 (52), 521 (52, 189), *546*, *549*
Minor, W. F. 505 (30), *545*
Miotti, U. 71, 72 (66), *179*
Miquel, A. M. 338 (34), *389*
Mishra, A. 485 (132), *501*, 537, 541 (313), *553*
Misiti, D. 239 (61), *324*
Misner, R. E. 531 (253), *551*
Mitra, S. S. 21 (59), *54*
Mlinko, S. 194 (23), *200*
Möbius, L. 260 (113), 290 (204), 291 (207), *325*, *327*, 353 (110), 360 (140), 374 (110), 375, 377 (200), 378 (210), *391–394*, 531 (255, 262), *551*
Modena, G. 69 (60), *179*
Modler, R. 233 (42), *323*
Moebius, L. 432 (222), *440*
Moffat, J. 62, 75, 80 (26), *178*, 378 (214), *394*, 556, 558, 576 (2), *584*
Moje, W. 127, 128, 130 (332), *187*
Mollberg, R. 336 (24), *389*
Moller, F. 398, 406 (13), *435*
Monneret, C. 100 (238), *184*
Montgomery, J. A. 90, 91 (200, 207), 92, 164 (207), *183*, 519 (157–160, 169, 170), *548*, *549*
Moore, G. J. 121 (312), *186*, 562, 571, 580, 581 (23), *584*
Moore, H. W. 83 (144, 145), 127, 130 (331), *181*, *187*, 200 (81, 82), *202*, 239 (61–66), 240 (62, 63), 241 (62, 65, 66), 277 (176, 176a), *324*, *327*, 388 (253), *395*, 420 (140), 421 (145), *438*, 581 (45–47), 582 (48, 49), 583 (49), *585*
Morf, D. 77, 79 (115–117), *181*, 198 (64, 65), *202*
Morgan, L. R. 98, 100 (230, 231), *184*, 196 (43), *201*, 232 (39), 233 (40, 41), 298 (224, 225), *323*, *328*, 339 (42), 340, 362 (43), *390*, 426 (179), *439*, 461 (83), 468 (83, 101), 479 (83), *500*
Moriarty, R. M. 265, 280, 297 (133), 298 (227), 299 (227, 228), 304 (228), 307 (228, 247, 248), 308 (248), *326*, *328*, 340, 343, 362 (44), 383 (234, 235), *390*, *394*, 426 (181), 428 (196, 197), *439*, 455 (73), 457 (143), 461, 463 (73), 470 (73, 104), 472 (105, 106, 108), 490 (73), *500*, *501*
Moriconi, E. J. 531 (253), *551*
Morin, L. T. 514 (127), *548*
Moritz, K. L. 387 (249), *395*, 463, 466, 467, 479 (91), *500*, 540, 542 (347), *553*
Moscowitz, A. 33 (83), *54*, 199 (74), *202*
Moseley, R. B. 65, 67 (43d), *179*
Motornyi, L. I. 510 (105), *547*
Motornyi, S. P. 86 (162), *182*
Moulin, F. 114 (284), *185*, 378 (213), *394*
Muchowski, J. M. 87 (173), *182*, 504 (17), *545*
Mucke, W. 433 (232), *440*, 509 (98), *547*
Mudretsova, I. I. 90, 91 (202), *183*
Muhlhaus, A. 76 (111), *180*, 574 (39), *585*
Muir, W. 505 (34), *545*
Muller, E. 174 (457), *190*
Muller, F. 134 (349), *187*
Müller, G. 260 (113), *325*, 360 (140), *392*, 530 (246), *551*
Müller, H. 50 (108), *55*
Muller, R. 76 (108), *180*
Muller, V. 353 (107), *391*
Mulliken, R. S. 44, 50 (102), *55*, 445 (30), *499*
Mullock, E. B. 274 (170, 171), *326*, *327*, 370 (181, 182), *393*
Mundlos, E. 77, 79 (117), *181*, 198 (64), *202*
Munekata, S. 198 (72), *202*
Munk, M. E. 374, 377 (198), *393*
Murphy, D. 111 (268a,b, 270, 271), *185*, 337 (28), *389*
Murray, R. A. 523 (202), *550*
Murray, R. W. 457 (78, 79, 144), 458 (79), *500*, *501*
Murrell, J. N. 447 (39), *499*
Mustafa, A. 83 (150), *182*, 374 (196), *393*, 563 (26), *584*
Myers, H. W. 17 (45), *53*, 76 (106), *180*

Nadkarni, S. 108 (260), *185*
Naegeli, C. 76 (87), 86 (164), 87 (168), *180*, *182*, 504 (4, 7, 16), 507 (7), *544*, *545*
Nagieb, F. 83, 123 (149), *182*

Nagy, H. K. 514, 515 (131), *548*
Nair, V. 156 (410), *189*
Nakanishi, K. 197 (54), *201*
Namba, K. 114 (282), *185*
Nambury, C. V. N. 22, 23 (63), *54*, 195, 197 (31), *201*, 506, 507, 521 (52), 533, 535 (275), *546*, *551*
Namyslowski, S. 83 (147), *181*
Neidlein, R. 356 (121), *391*, 515 (138, 139), 521 (186–188), *548*, *549*
Nelles, J. 76 (88), *180*, 505 (20), *545*
Nesbet, R. K. 36, 50 (94), *55*
Nesmeyanov, A. N. 81 (142, 143), 82 (143), 118 (304), *181*, *186*, 556, 561 (3), 568 (31), 576, 578 (3), *584*
Nestler, G. 215 (45), *220*
Netz, A. 92 (214), *183*, 510, 511, 517 (108), *547*
Neubert, M. E. 505 (30), *545*
Neumann, L. 197 (53), *201*, 402 (42), *435*, 535 (285), *552*
Neunhoeffer, H. 175 (472), *190*, 196 (46), *201*, 506 (75), *546*
Neunhoeffer, M. 196 (46), *201*
Newburg, N. R. 265, 279–281, 283–286 (134), *326*
Newman, M. S. 76 (89), 140 (359), *180*, *187*, 401 (27), 402 (37, 40, 41), 407 (40), 408 (37), *435*, 535 (283, 284), *552*
Newman, S. H. 80 (137), *181*, 555–557 (1), *584*
Newth, F. H. 94, 111 (221), *183*
Nichols, P. L., Jr. 93, 94 (220), *183*
Nickel, G. 234 (51), *323*
Nickon, A. 316, 321 (261), *329*, 536 (292), *552*
Nieland, H. 90 (195), *183*, 504 (14), 522 (195, 196), 524, 525 (195), *545*, *549*
Nielsen, D. R. 417 (127), *438*
Nishiwaki, T. 569 (34), *585*
Noakes, D. E. *14*
Noelting, E. 152, 157 (400), *188*, 212 (35), *220*
Noland, W. E. 372 (189), *393*
Nordstrom, D. K. 264 (127), *325*
Norman, R. O. C. 204, 205 (4), 206, 212 (7), *219*
Norris, A. R. 217 (49, 50), *220*
Norris, W. P. 432 (227), *440*, 516, 519 (141), *548*
Norup, B. 262 (119), *325*, 422 (155), *438*
Notation, A. D. 366 (165), *393*
Noto, F. 119 (307a), 132 (307a, 341), 133 (307a), *186*, *187*, 507 (82), 520 (82, 176), 521 (176), *546*, *549*
Novak, P. 108 (265), *185*
Noyori, R. 537, 539 (312), *553*
Nozaki, H. 537, 539 (312), *553*
Nozoe, T. 158 (417), *189*, 365 (162), *392*
Nussel, H. 103, 104 (248), *184*

Ochwatt, P. 83 (146), *181*, 581 (44), *585*
O'Dell, M. S., Jr. 246 (84), *324*
Odum, R. A. 257 (106), 312 (106, 254), 314 (257), *325*, *329*, 388 (250), *395*, 427 (187), *439*, 465, 493, 495, 496 (95), *500*
Oediger, H. 532 (268), *551*
Oehlschlager, A. C. 289 (203), *327*, 346 (74), *390*, 530 (248), *551*
Oesterlin, M. 63 (36), *178*
Oftedahl, E. 22, 23 (63), *54*, 133 (348), *187*, 195 (31), 196 (45), 197 (31, 45), *201*, 433 (231), *440*, 506 (52, 84), 507 (52, 84, 90), 521 (52), *546*, *547*
Oglukian, R. L. 261 (115), *325*, 422 (156), *438*
Ogrins, B. 3 (17), *53*
Ohrui, H. 216 (46), *220*, 335 (19), *389*
Ohse, E. 62, 115 (19), *178*
Okabe, H. 17 (48), *53*, 443, 444, 447, 448 (13), *498*
Okada, M. 362 (150), *392*, 569 (35), *585*
Okamoto, Y. 206 (9), *219*
Oliveri-Mandala, E. 75 (82), 119 (82, 305, 307a,b,c,d, 308), 120 (305), 127 (308, 327), 132 (307a, 341), 133 (307a,b,c,d, 345), 164 (432), *180*, *186*, *187*, *189*, 406 (75), *436*, 507 (82, 83), 520 (82, 176, 177), 521 (176), *546*, *549*
Olofson, R. A. 510 (110), *547*
Olsen, C. E. 377 (207), *394*
Omran, S. M. A. R. 83, 123 (149), *182*
Opitz, G. 129 (335, 336), *187*
Opitz, H. 505 (39), *545*
Orlando, C. M. 399 (44), *436*, 528 (237), *550*
Orville-Thomas, W. J. 5 (24–26), 16 (26), 24 (25, 26), 25 (25), *53*
Ostroverkhov, V. G. 130, 131 (338), *187*, 556 (4), *584*

Osuch, C. 289 (201, 202), 296 (221), *327*, *328*, 347 (81), 357 (132), 360 (141), *391*, *392*
Otsuji, Y. 507 (77), *546*
Otsuka, H. 175 (478), *190*
Ott, E. 62, 115 (19), *178*
Ott, H. 504, 505 (11), *545*
Otten, E. A. 115 (292), *185*, 518 (155), *548*
Otterbach, D. H. 108 (262), *185*
Otterbach, H. 103, 104 (245), *184*
Overberger, C. G. 61 (13), *178*, 192 (3), *200*

Pabst, A. 504 (8), *545*
Paetzad, P. 424 (169), *439*
Paine, R. H. 300, 302 (231), *328*
Palazzo, S. 410 (97), *437*
Pallini, U. 145 (369), *188*
Papa, A. J. 505 (28), *545*
Papazian, H. A. 21 (57), *54*, 444 (27), *499*
Paquette, L. A. 427 (189), *439*
Palzzo, G. 534 (278), *552*
Paradies, T. 507 (81), *546*
Parker, A. J. 66 (49, 51, 53, 64), 67 (53), 70 (64), 71, 72 (51, 53), *179*, 207, 209, 213 (17), *220*
Parker, R. E. 93, 94 (219), *183*
Parkes, G. D. 148 (376), *188*, 416 (117), *437*
Partyka, R. A. 505 (30), *545*
Patai, S. 66 (52), 80 (139), *179*, *181*, 261 (116), *325*, 422 (157), *438*
Patrick, J. E. 350 (91), *391*
Patterson, E. 348 (85), *391*, 571 (36), *585*
Patterson, L. J. 195 (33), *201*
Pauling, L. 5 (22), 17 (46), 43 (22), 51 (110), *53*, *55*
Paulsen, H. 199 (75), *202*
Paviak, S. C. 413, 414 (111), *437*
Pearson, R. G. 65 (44, 45), *179*
Pedersen, C. 377 (207), *394*, 509 (93), 514 (132), *547*, *548*
Pegolotti, J. A. 85 (160), *182*
Pense, W. 83 (146), *181*
Perkin, W. H. 3 (18), *53*
Perlinger, H. 160 (427), 161 (427, 428), *189*, 194 (21), *200*, 335 (13), *389*, 429 (206), *440*
Permutti, V. 412 (101), *437*
Perron, Y. G. 505 (30), *545*
Perveer, F. Ya. 93 (223), *183*
Petrikalns, A. 3 (17), *53*
Petrongolo, C. 45, 48, 50 (103), *55*
Petrov, A. A. 357 (129), *392*, 416 (118), *437*, 531 (256, 257), *551*
Petty, W. L. 93, 94 (220), *183*
Peyerimhoff, S. D. 40, 50 (95), *55*, 445 (31), *499*
Pfaff, I. 102 (240), *184*
Philip, J. C. 204 (1), *219*
Phillip, H. 301 (237), *328*
Phillips, C. S. G. 4 (10), *52*
Phillips, H. 399 (16), *435*
Piening, J. R. 505 (45), *545*
Pierce, J. H. 99 (237), *184*
Pillai, C. N. 133 (346, 348), *187*, 433 (233), *440*, 506 (84), 507 (84–86), *546*, *547*
Pimentel, G. C. 22, 23 (64), *54*, 443 (12), *498*
Pink, L. A. 422 (152), *438*
Pinter, I. 356 (124), *392*
Pittman, V. P. 399 (16), *435*
Plessas, N. R. 104 (252), *184*, 558 (10), *584*
Pocar, D. 290 (205, 206), *327*, 347 (79, 80), 374 (79), 375 (199), *390*, *393*, 434 (237), *440*, 529, 530 (245), *551*
Pochinok, V. Ya. 90, 91 (201), *183*
Pohl, G. 234 (43), *323*, 343 (64), *390*
Polzhofer, K. P. 527 (226), *550*
Ponsold, K. 33 (83), *54*, 76 (92), 96 (227), 98 (229), 102 (92, 239–241), 145 (367), *180*, *184*, *188*, 199 (74), *202*
Pontrelli, G. J. 452, 460 (68), *500*
Ponzio, G. 171, 173 (447), *190*
Porter, C. W. 401 (28), *435*
Porter, G. 458 (81), *500*
Portner, E. 505 (49), *546*
Postovskii, I. Ya. 90, 91 (202), 115 (289, 290), *183*, *185*, 519 (161, 164–166), *548*, *549*
Potier, A. 41 (97), *55*
Potier, P. 96 (225), *184*
Powell, G. 114 (283), *185*, 400 (23), *435*, 504 (9), *545*
Prabhakar, M. 174 (461), *190*
Preibsch, W. 98 (229), *184*
Presst, B. M. 79 (133), *181*
Preston, P. N. 196 (48), *201*
Priest, W. J. 348 (86), *391*, 424 (167), *439*, 497 (140), *501*, 571, 584 (37), *585*
Prince, W. M. 399 (44), *436*, 528 (237), *550*

Pringle, G. E. *14*
Pritchard, H. O. 13 (33), *53*
Pritzkow, W. 223, 224, 228 (10), 230 (36), 234 (43), 248, 298, 299 (88), *322–324*, 343 (64, 65), *390*, 413 (110), 419 (110, 136), 421 (149), *437*, *438*
Prosser, T. J. 90 (191), *183*, 265 (136), *326*, 340 (49), *390*, 528 (234), 537 (297–299), 538, 539 (297), 544 (298), *550*, *552*
Proudlock, W. 217 (49), *220*
Pryde, C. A. 83, 93 (151), *182*, 349, 365, 387 (87), *391*, 423 (162), *438*, 558 (13), 560 (13, 14), 561, 565, 567, 570, 574, 579 (13), *584*
Pschorr, R. 505 (41), *545*
Pulver, S. 75, 106 (74), *180*, 215 (41), *220*
Purgotti, A. 174 (456), *190*, 212 (36), *220*
Purushothaman, K. K. 96 (228), *184*, 342 (59), *390*
Puterbaugh, W. H. 140 (359), *187*
Putney, R. K. 76, 77, 94 (105), 157, 158 (411), *180*, *189*, 223, 224, 232, 233 (9), 263 (122), *322*, *325*, 342 (58), 371, 385 (183), *390*, *393*, 419, 420, 426 (134), *438*, 468 (98), *500*
Puttner, R. 360 (144), 369, 371 (175), *392*, *393*, 405 (66, 69), *436*, 450 (54), 462, 463 (90), 465 (92), 482 (54), 483 (130), 485 (92), 489 (54), *499–501*, 527 (221), 540 (343, 345), 541 (350), 543 (343), *550*, *553*, *554*
Pyszora, H. 2, 4 (6), *52*

Raaen, V. F. 409 (85), *436*, 497 (141), *501*
Rabalais, J. W. 3, 12, 25–27, 30, 31, 45, 47 (9), *52*
Radenhausen, R. 506 (50), *546*
Rager, H. 151 (395), *188*, 357 (130), *392*
Rahman, M. 265, 280, 297 (133), 298, 299 (227), *326*, *328*, 340, 343, 362 (44), *390*, 426 (181), *439*, 455 (73), 457 (143), 461, 463 (73), 470 (73, 104), 490 (73), *500*, *501*
Rajkumar, T. V. 380 (223), *394*
Ramachandran, J. 133 (347, 348), *187*, 433 (233), *440*, 507 (86), *547*
Ramsperger, H. C. 245–247 (80), *324*
Ranganathan, R. 531 (258), *551*
Rao, C. N. R. 2 (7), 3 (12), 4, 13 (7), 14 (38), 17 (7), 18 (38), 19 (7), 22 (63), 23 (63, 66), 28 (12), 30 (12, 77, 78), *52–54*, 61 (9, 12), 76, 77 (84), 90 (12), 91, 127 (9), 132 (12), 133 (348), 176 (12), *178*, *180*, *187*, 192 (5), 195 (5, 31, 36), 196 (47), 197 (31), 198 (68, 70), *200–202*, 207 (15), *220*, 245 (77), 304 (244), *324*, *328*, 380 (222, 223), *394*, 398 (5, 6), 404, 415 (5), 418 (130), 433 (5, 231, 233), *435*, *438*, *440*, 445, 446 (33), *499*, 505 (19), 506 (52), 507 (52, 86, 90), 521 (52), *545–547*
Rao, K. A. N. 335 (12), *389*
Rao, P. A. D. S. 73 (69), *179*, 337 (27), *389*
Rappoport, Z. 69, 70 (59), 80 (59, 139), 81 (59), *179*, *181*
Raschig, K. 255 (101), *325*, 526 (214), *550*
Rathsbung, H. 335 (14), *389*
Ratts, K. W. 61, 126, 130 (16), *178*, 361 (148), *392*, 428 (195), *439*, 562, 565 (24), *584*
Reagan, M. 316, 321 (261), *329*, 536 (292), *552*
Reardon, R. C. 299, 304, 307 (228), *328*
Rector, D. L. 346 (78), *390*
Redmont, J. W. 537, 539, 542 (310, 311), *552*
Rees, C. W. 274, 275 (172), 317 (265), *327*, *329*, 398, 401, 403, 404 (11), *435*, 541 (351), *554*
Reese, R. M. 17, 31 (48a), *53*
Regitz, M. 168 (440), *189*, 292 (210), *328*, 351 (94, 97–99), 352 (101–104, 106), 355 (112), *391*, 432 (220), *440*
Reich, P. 509 (97), *547*
Reichle, W. T. 283 (190), *327*, 406, 424, 425 (74), *436*
Reiher, M. 505 (27), *545*
Reilly, J. 404 (57), *436*, 514 (128), 521, 522 (191), *548*, *549*
Reimlinger, H. 477 (114), *501*, 564 (27), *584*
Reinisch, R. F. 157, 158 (411), *189*, 263 (122), *325*, 371, 385 (183), *393*, 468 (98), *500*
Reiser, A. 198 (71), *202*, 247 (85), 306 (245), 308 (85, 245, 249–251), 309 (245), 311, 312 (85), *324*, *328*, *329*,

443 (24, 25), 447 (37, 40), 452 (70, 72), 455 (40), 456 (70), 458 (37, 40, 80), 460, 461 (72), 464 (25, 72), *498–500*
Reist, E. J. 76, 103, 105 (95), *180*, 338 (32), *389*
Reitz, A. W. 197 (58), *201*
Rembarz, G. 374 (197), *393*
Remlinger, H. 428 (193), *439*
Renfrow, W. B. 265, 279, 280 (134), 281 (134, 187), 283–286 (134), *326*, *327*, 341 (51), *390*
Reutov, O. A. 135, 136 (352), *187*
Revesz, A. 104 (250), *184*
Reynolds, G. A. 90, 91 (196), *183*, 196 (51), *201*, 348 (86), 378 (217), *391*, *394*, 424 (167), *439*, 445 (36), 497 (140), *499*, *501*, 518 (152, 153), 532 (271), *548*, *551*, 571, 584 (37), *585*
Ribaldone, G. 274 (167), *326*, 425 (175), *439*
Rice, F. O. 246, 251 (83), *324*
Rice, S. N. 485 (132), *501*, 537, 541 (313), *553*
Richards, H. C. 96 (224), 99 (237), *183*, *184*
Richards, R. E. 34 (85), *54*
Richardson, A. C. 76, 103 (98), 105 (98, 253, 256), 106 (256), 107 (253, 256, 258), 108 (253, 256), 109, 110 (98), 111 (270, 273), 112 (273), *180*, *184*, *185*, 335 (10), *389*
Rickenbacher, H. R. 336 (22), *389*
Ried, W. 378 (218), *394*, 531 (259), *551*
Rigg, M. M. 514 (126), *548*
Rinkenbach, W. H. 62 (20), *178*
Rissom, J. 287 (193), 288 (200), *327*
Rittel, W. 89, 90 (182, 184), *182*
Ritter, A. 62 (29), *178*
Rivolta, A. M. 165, 166 (434), *189*
Robbins, R. F. 250, 269 (91), 405 (67), 427 (190), *436*, *439*, 506, 521, 542 (74), *546*
Roberts, C. W. 383 (237), *394*, 413 (104), *437*
Roberts, E. 210 (30), *220*
Roberts, J. D. 12, 13 (30), *53*, 91, 127, 163 (208), *183*, 215 (43), *220*, 430 (209), *440*, 511 (112), *547*
Robins, R. K. 338 (35), *389*
Robinson, C. A. 127, 130 (329), *187*, 378 (212), *394*
Robinson, R. 505 (48), *546*
Robson, P. 544 (367), *554*
Roesky, H. 175 (470), *190*, 506, 521 (69), *546*
Rondestvedt, C. S. 149 (384), *188*, 374 (194, 195), *393*
Ropp, G. A. 409 (85), *436*
Rosengren, K. 443 (12), *498*
Roser, O. 434 (235), *440*
Rossi, S. 410 (86), *436*, 518, 519, 530 (156), *548*
Roth, R. 516, 519 (143), *548*
Rothen, A. 33 (83a), *54*, 84, 85 (157), *182*, 199 (73), *202*, 445 (41), *499*
Rowe, C. D. 151, 171 (398), *188*, 358 (137), *392*
Roy, J. 269, 285, 286 (152), *326*, 404 (65), *436*
Rozum, Yi. S. 90, 91 (201), *183*
Rudinger, J. 175 (474), *190*
Rudolf, J. 432, 433 (226), *440*
Rudolph, W. 383 (236), *394*, 531 (260), *551*
Ruschig, H. 504 (10), *545*
Rüter, J. 352 (102), *391*
Rutherford, K G. 402, 407 (40) *435*
Rybinskaya, M. I. 81 (142, 143), 82 (143), *181*, 556, 561 (3), 568 (31), 576, 578 (3), *584*

Safir, R. S. 527 (218), *550*
Sah, P. T. 192 (7–9), *200*
Saikachi, H. 132 (344), *187*
Saito, Y. 35 (88), *54*
Sakakibara, S. 528 (229), *550*
Saltiel, J. 456 (77), *500*
Samek, Z. 108 (265), *185*
Sammour, A. M. A. 124 (321), *186*
Samosvat, L. S. 166 (436), *189*
Samuel, R. J. 4 (21), *53*
Sandorfy, C. 8, 14, 18 (27), *53*
Sato, S. 477 (115), *501*, 568, 569 (32), *585*
Sauer, H. 374 (193), *393*
Sauer, J. 345, 373 (72), *390*, 529 (238), *550*
Saunders, B. C. 413 (105) *437*
Saunders, K. H. 367 (168), *393*
Saunders, W. H. 79 (132), 174 (460), *181*, *190*, 254 (98–100), 299 (99, 229), 300 (231), 302 (99, 231, 238), 303 (229, 238, 242), 306 (99, 229), 307 (100), *325*, *328*, 344 (68, 69), *390*, 421 (150), 426 (183–185), 427 (150, 151), 430 (184), *438*, *439*, 449, 450 (46), 455 (74), 456 (76), 472 (109), 474 (74), *499–501*

Savell, W. L. 398, 432 (2), *434*
Sazonova, V. A. 118 (304), *186*
Scaplehorn, A. W. 93, 94 (219), *183*
Schaad, R. E. 120 (310), *186*
Schander, A. 507 (79), *546*
Schatz, P. N. 3 (18a), *53*
Schechter, H. 291 (208), *327*, 346 (76), *390*, 410 (88), *436*
Scheiffele, E. 315 (258, 260), 316 (258), *329*, 342 (54), *390*, 399 (43), 404 (60, 63), *436*, 449, 483 (50), *499*, 522, 527 (198), 540 (342), *549*, *553*
Scheiner, P. 346 (77), 360 (138, 139), 376 (201, 203), *390*, *392*, *393*
Schiff, H. 152 (402), *188*
Schleyer, P. v. R. 538 (325), *553*
Schmid, H. J. 410 (92), *437*
Schmid, K. H. 343 (63), *390*
Schmidt, A. 92 (211), *183*, 196 (49), *201*, 242 (68), *324*, 343 (62), *390*, 421 (141), *438*, 516 (144), 517 (145), *548*
Schmidt, F. 287 (195), *327*, 404 (58), *436*, 442 (4), *498*, 521 (182), 525 (207), *549*, *550*
Schmidt, K. F. 230 (30), *323*, 413 (107, 108), *437*
Schmitt, H. F. 87 (169), *182*, 505 (46), *546*
Schmutz, J. 367 (169), *393*
Schnabel, E. 89, 175 (181), *182*
Schnabel, W. J. 76, 92 (113), *181*
Schneider, J. 374 (193), *393*
Schoenecker, B. 102 (240), *184*
Schon, M. 378 (218), *394*, 531 (259), *551*
Schonberg, A. 61 (10), *178*, 347 (82), *391*, 402, 403 (45), *436*
Schrader, E. 114 (285), 115 (291), *185*
Schrecker, A. W. 399, 407 (19), *435*
Schreiner, E. C. 413 (104), *437*
Schroeter, G. 76 (110a, b), 86 (110b), *180*, 504 (2), *544*
Schubert, G. 102 (239), *184*
Schubert, H. W. 129 (336), *187*
Schuberth, A. 230 (36), *323*, 413, 419 (110), *437*
Schuerch, C. 79 (125), *181*
Schultheiss, W. 86 (161), *182*, 504 (5), *545*
Schumacher, U. 542 (357), *554*
Schutz, W. 207 (12), *219*
Schwall, H. 432 (220), *440*
Schwarz, H. P. 507 (80), *546*
Schwarzenbach, K. 172 (451), 173 (453), *190*, 429 (200), *439*
Schwyzer, R. 89 (180, 182, 184), 90 (182, 184), 175 (180), *182*, 534 (281), *552*
Scott, C. B. 65, 67 (43b), 68 (58), 159 (43b), *179*
Scott, F. L. 90 (193), 93, 176 (217), *183*, 196 (52), *201*, 404 (57, 61), 433 (230), *436*, *440*, 512 (113, 114), 514 (128), 517 (146, 147), 519 (172, 173), 521, 522 (191, 192), *547–549*
Scott, J. M. 19, 20 (54), *54*
Scott, M. T. 90 (193), *183*, 521, 522 (192), *549*
Scotti, C. 354 (111), *391*, 432 (224), *440*
Scriven, E. F. V. 270, 271 (157), *326*
Scrocco, E. 45, 48, 50 (103), *55*
Sedor, E. A. 252 (92), *324*
Seidler, J. H. 413 (112), *437*
Selman, L. 541 (348, 349), *553*, *554*
Selvarajan, R. 557, 572 (8), *584*
Senior, J. K. 245 (79), *324*, 421 (148), *438*
Senn, O. 150 (390), *188*, 432 (223), *440*
Senyavina, L. B. 22, 23 (61), *54*, 195 (38, 39), *201*, 506, 528 (55), *546*
Serban, N. 98 (233), *184*
Serre, J. 20, 32 (56), *54*
Serros, E. 535 (290), *552*
Seubold, F. H., Jr. 300 (230), *328*
Shakhnovskaya, F. B. 62 (34), *178*
Shapira, D. 27 (73), *54*
Shark, K. W. 419 (133), *438*
Sharon, N. 103 (244a, b), *184*
Sharp, T. M. 505 (42), *545*
Shechter, H. 569 (33b), *585*
Sheehan, J. C. 127, 130 (329), *187*, 378 (212), *394*
Sheinker, Yu. N. 22, 23 (61), 30 (76), *54*, 115 (289), *185*, 195 (38, 39), 197 (60), 198 (67), *201*, *202*, 445 (32), *499*, 506 (53–55), 528 (53,55), *546*
Sheldon, H. R. 83 (144, 145), 127, 130 (331), *181*, *187*, 200 (81, 82), *202*, 239 (62–66), 240 (62, 63), 241 (62, 65, 66), 277 (176), *324*, *327*, 388 (253), *395*, 420 (140), 421 (145), *438*, 581 (45–47), 582 (48, 49), 583 (49), *585*

Shellhamer, D. F. 127, 130 (331), *187*, 200 (82), *202*, 239 (64), *324*, 581 (46), *585*
Shemyakin, M. M. 430 (212), *440*
Sheppard, W. A. 508 (91), *547*
Sheradsky, T. 336 (23), *389*
Sherk, K. W. 78, 79 (120), *181*, 223, 237 (8), *322*, 342 (57), *390*
Sherman, E. 383 (238), *394*
Shilov, E. A. 130, 131 (338), *187*, 556 (4), *584*
Shimonishi, Y. 528 (229), *550*
Shine, H. J. 431 (215), *440*
Shingaki, T. 275, 291 (173), *327*, 544 (366), *554*
Shoppee, C. W. 96 (224), 99 (237), *183*, *184*
Shott-L'vova, E. A. 3, 19 (13), *52*
Shovlin, C. 307 (247), *328*, 383 (235), *394*, 428 (197), *439*, 472 (105, 106), *500*, *501*
Shozda, R. J. 86 (166), 88 (175), *182*
Sicher, J. 96 (224), *184*
Sidgwick, V. N. 17 (44), *53*
Sidhu, G. S. 242 (67), *324*
Sieber, P. 89, 175 (180), *182*, 534 (281), *552*
Sieber, W. 526 (216), *550*
Simmons, H. E. 263 (125), *325*, 381 (229), *394*
Simson, J. M. 537–539 (309), *552*
Singh, B. 569 (33a), *585*
Singh, K. 14, 41, 51 (34), *53*
Sinnema, Y. A. 62, 130 (27), *178*, 217 (47), *220*
Sipos, F. 96 (224), *184*
Sirokman, F. 108 (264), *185*
Sjoberg, B. 336 (24), *389*
Skell, P. S. 450 (56, 57), 483 (131), *499*, *501*
Skinner, H. A. 13 (33), *53*
Skovronek, H. S. 483 (131), *501*
Skraup, S. 151 (394), *188*
Slater, G. P. 407 (76), *436*
Sloan, A. B. 388 (252), *395*, 414, 422 (114), *437*
Sloan, M. F. 265, 279, 280 (134), 281 (134, 187), 283–286 (134), *326*, *327*, 341 (51), *390*
Slotta, K. H. 132 (342), *187*
Smalley, R. K. 158 (418), *189*, 267 (145), 268 (147), 274 (169), *326*, 341 (52), 368 (170, 171), *390*, *393*, 401, 403 (34), *435*, 504 (12), *545*
Smart, N. T. 89, 175 (178), *182*
Smets, G. 390 (225), *394*
Smick, R. L. 371 (184), *393*, 450, 466, 482 (53), *499*, 543, 544 (363), *554*
Smid, P. M. 352 (105), *391*
Smirnova, N. B. 90, 91 (202), 115 (290), *183*, *185*, 519 (161), *548*
Smith, G. B. L. 507 (88), *547*
Smith, N. R. 378 (214), *394*
Smith, P. A. S. 30 (80), *54*, 58 (3), 59 (4), 61 (3, 4), 76, 86 (85), 127, 130 (4), 151 (398), 157 (411–413), 158 (411, 413, 419), 171 (398), 173, 175 (468), *178*, *180*, *188–190*, 192 (2), 193 (10, 17), 194 (25), 198 (69), *200–202*, 205–207, 209 (6), 210 (6, 28, 29), 211, 212 (6), *219*, *220*, 222, 228 (2), 230 (37), 235, 237–239 (54), 258, 259 (110), 263 (122), 269 (110, 149), *322–326*, 333 (2), 335 (9), 339 (38), 340 (45), 341 (53), 358 (137), 365 (160, 161), 366 (163), 371 (183, 186), 372 (160), 383 (161), 385 (183), *389*, *390*, *392*, *393*, 398 (4, 12), 401, 402 (4), 406 (4, 72), 407 (4), 408 (83), 409 (81–83), 410 (82, 91, 93, 95), 411 (4, 82), 412 (4), 413 (72, 91, 95, 106), 414 (82), 415 (72), 417–419, 421 (4), 422 (160), 433, 434 (228), *435–438*, *440*, 443 (20), 447 (38), 462 (88, 89), 468 (98), 491 (88, 89), *498–500*, 504 (6), 505 (6, 37), 506 (6), 508 (92), 513 (119, 120), 534 (276), *545*, *547*, *552*
Smith, R. A. 542 (358), *554*
Smolinsky, G. 76 (94), 77 (94, 114), 80 (114), 83 (151), 90 (189), 93 (151), 158 (420), *180–183*, *189*, 196 (42), *201*, 265 (130–132), 267 (142–144, 146), 268 (146), 269 (150), 297 (130–132), 298 (132, 142), 307 (132), 308, 310 (130), *325*, *326*, 340 (47), 349 (87), 355 (47), 361 (147), 364 (156, 157), 365 (87), 385 (243), 387 (87), 389 (157), *390–392*, *395*, 423 (162, 165), 426 (178), 428 (165), *438*, *439*, 443 (16–18, 22, 23), 456 (22, 23), 457 (22, 23, 78, 79, 144), 458 (79), 462 (17, 18, 86), 467 (18), 469 (23, 86), 475 (110), 496 (18), *498*, *500*, *501*, 538 (327, 328), *553*, 558 (9, 13), 560, 561 (13), 564 (9), 565, 567, 570, 574, (13), 579 (9, 13), 580 (43), *584*, *585*

Smyth, C. P. 3, 19 (14), *52*
Sneen, R. A. 73 (70–73), 74 (72), 136 (355), *180, 187*
Snyder, L. C. 265, 297 (131), *325*
Sokolov, V. I. 135, 136 (352), *187*
Sommers, A. H. 76, 77 (102), *180*
Sonnenberg, J. 79 (135), *181*
Sorm, F. 83 (148), *181*
Soundarajan, S. 19 (55), *54*
Spangenberg, O. 505 (41), *545*
Sparr, B. I. 417 (125), *437*
Spauschus, H. O. 19, 20 (54), *54*
Speakman, P. H. R. 544 (367), *554*
Spencer, R. D. 408 (80), *436*
Spencer, R. R. 76, 103, 105 (95), *180*, 338 (32), *389*
Spietschka, E. 400, 402 (25), *435*, 490 (119), *501*
Spinks, J. W. T. 407 (76), *436*
Spoerri, P. E. 87 (170), *182*
Springall. H. D. 192 (1), *200*
Springmann, H. 103, 104 (245), *184*
Stadler, D. 355 (112), *391*
Stadler, J. 234 (49, 50), *323*
Stadler, P. A. 504, 505 (11), *545*
Stange, O. 521 (178, 179), *549*
Stangl, H. 260 (113), *325*, 360 (140), *392*
Stanovnik, B. 90 (205, 206), 115, 116 (205, 288), *183, 185*, 520 (175), *549*
Stansbury, H. A. 62, 76, 77 (28), *178*, 378 (215), *394*
Starratt, A. N. 298 (226), *328*, 362 (152), *392*, 426 (180), *439*, 469 (103), *500*
Staudinger, H. 75 (77), *180*, 294 (214, 215), 295 (215), *328*, 356 (118, 119), *391*, 532 (266), *551*
Stedman, R. J. 530 (250), *551*
Stefanovitch, G. 86 (164), *182*, 504 (4), *544*
Steglich, W. 175 (475), *190*
Stein, G. 260 (111, 112), *325*, 373 (191, 192), *393*
Steiner, E. R. 372 (188), *393*
Steiner, G. 33 (83), *54*, 199 (74), *202*
Steinkopf, W. 87 (169), *182*, 505 (46), *546*
Steinruck, K. 151 (394), *188*
Stephanie, J. G. 264 (127), *325*
Stephenson, K. J. 99 (237), *184*
Stern, E. S. 28 (74), *54*
Stern, R. 62 (33), *178*
Stevens, C. L. 108 (262, 264), *185*
Stevens, T. S. 151 (396), *188*
Stewart, E. T. 8 (29), 36 (29, 93), 40 (29), 50 (93), 52 (29), *53, 55*
Stolle, R. 90 (194, 195), 92 (213, 214), 115 (294, 295), 132 (194), 174 (465), *183, 186, 190*, 404 (59), 434 (235), *436, 440*, 504 (13, 14), 510 (108, 109), 511 (108), 517 (108, 109), 518 (151), 521 (184, 185), 522 (194–196), 524 (194, 195), 525 (195), 526 (217), *545, 547–550*
Storch, H. 115 (294), *186*, 518 (151), *548*
Storch, I. 413 (109), *437*
Storr, R. C. 274, 275 (172), *327*
Stowe, R. W. 321 (270), *329*
Strachan, E. 458 (81), *500*
Strani, G. 534 (278), *552*
Strating, J. 352 (105), *391*
Straw, D. 76, 77 (99, 100), *180*, 253 (93–95), *324*, 344 (70), 367 (167), *390, 393*, 424 (166), *439*
Strecker, E. 336 (26), *389*
Streitwieser, A. 75, 106 (74), *180*, 215 (41), *220*
Stroh, J. 387 (246), *395*, 562, 563 (25), *584*
Strum, H. J. 389 (241), *395*
Struve, G. 506 (50), *546*
Sturm, H. J. 519 (162), *549*
Sudarsanum, V. 200 (80), *202*
Sugihara, J. M. 108 (261), *185*
Summers, G. H. R. 96 (224), 99 (237), *183, 184*
Sumuleanu, C. 150 (391), *188*
Sun, J. Y. 349 (89), *391*, 534 (223), *550*
Sundberg, R. J. 264 (128), *325*, 364 (158), *392*, 422 (159), *438*
Suschitzky, H. 158 (418), *189*, 267 (145), 268 (147), 274 (169–171), *326, 327*, 341 (52), 368 (170), 370 (181, 182), *390, 393*, 504 (12), *545*
Sutherland, R. G. 268 (148), 279, 281 (179), 283 (148), 285 (148, 179), 312 (148, 256), 313, 314 (256), 317 (262, 263), *326, 327, 329*
Sutter, J. R. 261 (115), *325*, 422 (156), *438*
Sutton, C. 535 (290), *552*
Sutton, D. A. 84 (156), *182*
Sutton, L. E. 207 (13), *220*
Swain, C. G. 65, 67 (43a,b,c,d), 68 (58), 159 (43b), *179*, 505 (25), *545*
Swenton, J. S. 302 (239, 240), 310 (239, 240, 252), *328, 329*, 355 (115),

391, 427 (191), *439*, 452, 455 (71), *500*
Swern, D. 337 (29), *389*
Swift, A. C. 530 (250), *551*
Swift, G. 337 (29), *389*
Syrkin, Ya. K. 3, 19 (13), *52*, 197 (60), *202*, 506, 528 (53), *546*
Szabó, I. F. 506 (67), *546*
Szathmáry, J. 195, 198 (32), *201*
Szego, F. 356 (124), *392*
Szeimies, G. 260 (113), 290 (204), 291 (207), *325*, *327*, 353 (110), 360 (140), 374 (110), 375 (200), 376 (204), 377 (200), 378 (210), *391*, *393*, *394*, 432 (222), *440*, 531 (255), *551*

Takai, K. 132 (344), *187*
Takano, Y. 530 (249), *551*
Takaya H. 537, 539 (312), *553*
Takaya, T. 287, 288, 294 (199), *327*
Takemato, K. 280 (185), *327*
Takimoto, H. 264 (129), 269 (151), *325*, *326*, 340 (46), 368 (172), 369 (174), 385, 387 (46), *390*, *393*, 422 (158), 427 (188), *438*, *439*
Tan, L. 76, 98 (93), *180*
Tanaka, K. 79 (130), *181*, 224 (11–13), *322*, *323*, 418 (128, 129), 420 (129), *438*
Tanida, H. 530 (249), *551*
Taniguchi, H. 362 (150), 365 (159), *392*, 566, 567 (29), 569 (35), 590 (29), *584*, *585*
Tappi, G. 514 (133), *548*
Taylor, C. 62 (20), *178*
Taylor, E. C. 196 (41), *201*, 348 (84), *391*, 574, 575 (41a), *585*
Taylor, K. G. 108 (262, 264), *185*
Taylor, R. 204, 205 (4), 206, 212 (7), *219*
Tedder, J. M. 355 (113), *391*
Teerlink, N. J. 108 (261), *185*
Teetsov, J. 505 (19), *545*
Temple, C., Jr. 90, 91 (200, 207), 92, 164 (207), *183*, 519 (157–160, 169, 170), *548*, *549*
Temple, R. D. 295 (219), *328*, 532 (265), *551*
Terashima, S. 403 (54), *436*, 484 (133, 134), *501*, 538 (329–331), *553*
Terry, G. C. 247, 308, 311, 312 (85), *324*, 443 (24), 452, 460, 461, 464 (72), *498*, *500*
Texier, F. 367 (166), *393*
Thayer, J. S. 23 (67), *54*
Thelin, H. 336 (24), *389*
Thiele, J. 76 (86), 173 (452), *180*, *190*, 513, 516, 519 (123), 521 (178, 179, 183), *547*, *549*
Thiemann, F. 442 (3), *498*
Thomas, A. E. 22, 23 (63), 28 (75), *54*, 195 (31, 37), 197 (31), 198 (37), *201*, 445 (34), *499*, 506, 507, 521 (52), *546*
Thompson, A. 110 (266), *185*
Thorpe, M. C. 519 (160, 169), *548*, *549*
Thrush, B. A. 443 (9), *498*
Thyagarajan, G. 242 (67), *324*, 412 (102), *437*
Thyssen, H. 507 (76), *546*
Tichy, M. 96 (224), *184*
Ticozzi, C. 376 (206), *394*
Tiemann, F. 536 (295), *552*
Tietz, R. F. 78 (121), *181*, 227, 228 (18), *323*, 407, 417 (79), *436*
Tilden, W. A. 209, 210 (21), *220*
Tillman, P. 530 (248), *551*
Timm, D. 248, 298, 299 (88), *324*, 343 (65), *390*, 421 (149), *438*
Tinker, J. F. 90, 91 (196), *183*, 196 (51), *201*, 445 (36), *499*, 518 (152), *548*
Tisler, M. 90 (205, 206), 115, 116 (205, 288), *183*, *185*, 520 (175), *549*
Tisue, G. T. 193 (16), *200*, 401 (36), 403 (52), *435*, *436*, 449 (51), 462 (84, 85), 465 (51), 487 (51, 85), *499*, *500*, 507 (59, 61), 537 (296), 538 (61, 296, 320), *546*, *552*, *553*
Tiurekova, G. N. 519 (166), *549*
Tobin, J. C. 512 (113), *547*
Toda, T. 158 (417), *189*, 365 (162), *392*
Todd, M. J. 258 (109), *325*
Todd, W. M. 505 (48), *546*
Todesco, P. E. 69 (60), *179*
Tomasi, J. 45, 48, 50 (103), *55*
Tomita, M. 412 (103), *437*
Tomson, A. J. 338 (36), *389*, 514, 515 (131), *548*
Tori, K. 530 (249), *551*
Trattner, R. B. 314 (257), *329*
Travis, D. N. 452, 460 (66), *499*
Treinin, A. 4 (11), 15 (41), 22 (60), 26 (41), 27 (41, 73), 29, 47 (41), *52–54*, 445 (28), 447 (42), *499*
Trinius, W. 174 (463), *190*
Trivedi, J. P. 507 (89), *547*

Trozzolo, A. M. 457 (78, 79, 144), 458 (79), *500*, *501*
Tsai, H. C. 265 (139), *326*, 340 (48), *390*, 539 (333), *553*
Tschesche, R. 132 (342), *187*
Tsuji, T. 530 (249), *551*
Tsuno, Y. 194 (26), *201*, 279 (182, 183), 280 (183), 281, 291 (182), 294, 295 (218), 296 (182), *327*, *328*, 357 (131), *392*, 401 (30, 31), 402 (38), *435*, 507 (60), 532 (270, 272), *546*, *551*
Tuerck, U. 130, 131 (339), *187*
Tulchinskaya, L. S. 431 (214), *440*
Tureck, O. 61 (18), *178*
Turro, N. J. 456 (77), *500*
Tyabji, A. 87 (168), *182*, 504, 507 (7), *545*

Ugi, I. 129 (337), 160 (425–427), 161 (427, 428), 162 (429, 430), *187*, *189*, 194 (21), *200*, 335 (13), *389*, 429 (203–206), *439*, *440*, 505 (22), *545*
Ugo, R. 290 (206), *327*, 347 (80), *390*
Uhlfelder, E. 521 (183), *549*
Ullman, E. F. 569 (33a), *585*
Ullmann, H. 296 (220), *328*, 356 (120), *391*
Uma, V. 269 (152, 153, 155), 285 (152, 153), 286 (152, 155), 315 (259), 321 (155), *326*, *329*, 404 (64, 65), *436*, 525 (209, 210), *550*
Umani-Ronchi, A. 376 (205, 206), *394*
Urbach, H. 339 (39), *390*
Urban, W. 347 (82), *391*
Urbanski, J. A. 338 (36), *389*
Urbanski, T. 62 (21), *178*
Uyeo, P. 406 (71), *436*
Uyeo, S. 412 (103), *437*

Vagt, A. 132 (340), *187*, 513, 521 (124), *547*
Van Allan, J. A. 90, 91 (196), *183*, 196 (51), *201*, 348 (86), *391*, 424 (167), *439*, 445 (36), 497 (140), *499*, *501*, 518 (152, 153), 532 (271), *548*, *551*, 571, 584 (37), *585*
Vandenbelt, J. M. 206 (10), *219*
Van der Werf, C. A. 84 (159), 93–95 (218), *182*, *183*, 195 (35), *201*, 334, 337 (7), *389*, 414 (113), 430 (208), *437*, *440*
Varsányi, Gy. 195, 198 (32), *201*
Vasil'eva, A. S. 86 (162), *182*
Vastine, F. 349 (89), *391*, 534 (223), *550*
Vaughan, J. 433, 434 (228), *440*
Vecchi, M. 149 (383), 159 (424), 160 (425), *188*, *189*, 429 (202, 204), *439*
Vedeneyev, V. I. 18 (49), *53*
Veillard, A. 48 (107), *55*
Veldstra, H. 515 (137), *548*
Vereshchagina, N. N. 90, 91 (202), *183*
Vernon, J. A. 88 (175), *182*
Vernon, J. M. 260 (113), *325*, 360 (140), *392*
Vetterlein, R. 225 (15), *323*, 420 (138), *438*
Viehe, H. G. 378 (211), *394*
Vita, C. de 371 (184), *393*, 450, 466, 482 (53), *499*, 543, 544 (363), *554*
Vitafinzi, P. 432 (224), *440*
Vocelle, D. 363 (154), *392*, 542 (353), *554*
Vogel, A. I. 3 (15), *53*
Vogt, A. 421, 433 (147), *438*
Vogt, V. 456 (77), *500*
Vogt, W. 505 (40), *545*
Voitov, A. P. 93 (223), *183*
Volpp, G. P. 527 (224), *550*
Vorlander, D. 75 (81), 114 (81, 287), *180*, *185*
Vossius, D. 257 (105), *325*, 388 (251), *395*, 424 (172), *439*, 495 (137), *501*

Wacker, L. 148 (379), *188*
Waddington, T. C. 3, 15 (19), 18 (50), *53*
Wadsworth, W. S. 523 (200), *549*
Wagner, D. 536 (294), *552*
Wagner, E. L. 41, 43, 44 (98), *55*
Wagner, H. M. 308 (250, 251), *328*, *329*, 443 (25), 447, 455, 458 (40), 464 (25), *498*, *499*
Wahl, W. H. 3, 28, 30 (12), *52*, 445, 446 (33), *499*
Walborsky, H. M. 336 (21), *389*
Walker, B. H. 420 (139), *438*
Walker, P. 259, 260, 265 (87), *324*, 425 (174), *439*, 443 (21), *498*
Walkowicz, C. 79 (134), *181*
Wallis, E. S. 400 (24), *435*
Walsh, A. D. 15, 38 (40), *53*, 445 (29), *499*

Wanger, J. 197 (58), *201*
Ward, J. J. 118 (303), *186*
Ware, J. C. 79 (132), *181*, 254 (98), *325*, 344 (68), *390*, 421, 427 (150), *438*
Wasserman, E. 265, 297 (130–132), 298 (132), 307 (132, 246), 308, 310 (130), *325*, *328*, 443, 456 (22, 23), 457 (22, 23, 78, 79, 144), 458 (79), 469 (23), 472 (107), *498*, *500*, *501*, 537 (304), *552*, 580 (43), *585*
Wasson, F. I. 351 (96), *391*
Waters, W. A. 259, 260, 265 (87), *324*, 425 (174), *439*, 443 (21), *498*
Webster, B. 355 (113), *391*
Weckherlin, S. 430 (211), *440*
Wedgwood, J. J. 346 (75), *390*
Weil, T. 351 (95), *391*
Weinstock, J. 89 (179), *182*, 505 (29), *545*
Weisser, H. R. 171 (449), *190*, 428 (198), 430 (210, 213), *439*, *440*
Welge, K. H. 448 (44), *499*
Wendler, N. L. 98 (236), *184*
Wen-Hou Yin 192 (9), *200*
Wentrup, C. 199 (76), *202*, 271 (158–160, 162–165), 272 (158, 164), *273* (165), *326*, 425 (176, 177), *439*, 509 (102), *547*
Wessel, W. 504 (15), *545*
West, R. 23 (67), *54*
Westland, R. D. 123, 125 (320), *186*
Weygand, F. 76 (90), 90 (192), 175 (475), *180*, *183*, *190*, 505 (27), 528 (232), *545*, *550*
Weyler, W., Jr. 83 (145), *181*, 239, 241 (65), 277, (176, 176a), *324*, *327*, 388 (253), *395*, 420 (140), *438*, 582 (48, 49), *585*
Wheeler, R. 17 (47), *53*
Wheland, G. W. 51 (110), *55*
Whistler, R. L. 344 (67), *390*
White, E. H. 171 (446), *190*
White, J. E. 209–211 (24), *220*, 530 (247), *551*
White, W. D. 34 (92), *55*
Whitehead, H. R. 151 (392), *188*
Whiting, M. C. 126 (323), *186*
Whitmore, F. C. 400 (22), *435*
Wiardi, P. W. 515 (137), *548*
Wiberg, N. 343 (63), *390*
Wiegrabe, W. 357 (133), *392*, 532 (269), *551*
Wieland, H. 92 (216), *183*, 512 (116), *547*
Wieland, T. 89, 175 (176), *182*, 339 (39), *390*
Wiener, H. 73 (70–72), 74 (72), *180*
Wikholm, R. J. 239, 241 (66), *324*
Wilcoxon, F. 507 (88), *547*
Wiley, R. H. 62, 75, 80 (26), *178*, 378 (214), *394*, 556, 558, 576 (2), *584*
Wilkerson, C. 301 (236), *328*
Willets, F. W. 247, 308, 311, 312 (85), *324*, 443 (24), 452, 460, 461, 464 (72), *498*, *500*
Williams, B. H. 302, 310 (239, 240), *328*
Williams, M. W. 89, 175 (178), *182*
Williams, N. R. 108 (260), *185*
Williams, R. E. 105 (255), *184*, 200 (79), *202*
Williams, R. J. P. 4 (10), *52*
Williams, V. 247, 308, 311, 312 (85), *324*, 452, 460, 461, 464 (72), *500*
Wiltshire, J. F. K. 410 (94), *437*
Windaus, A. 505 (39, 40), *545*
Winitz, M. 175 (477), *190*
Winnerwisser, M. 14, 15 (37), *53*
Winstein, S. 79 (135), 84 (158), 135 (354), *181*, *182*, *187*, 430 (207), *440*
Wislicenus, W. 75 (76), *180*
Witanowski, M. 34, 35 (87), *54*
Withers, J. C. 62 (24), *178*
Wittig, G. 378 (216), *394*
Woerner, F. P. 428 (193), *439*, 451 (65), 477 (114), 484 (65), *499*, *501*, 541 (352), *554*, 564 (27), *584*
Wohl, A. 152 (402), *188*
Wolf, W. A. 351, 352, 358 (100), *391*
Wolfe, H. 230 (31), *323*, 406 (70), *436*
Wolff, A. 294 (217), *328*
Wolff, A. P. L. 434 (238), *440*
Wolff, L. 257, 312 (104), *325*, 495 (135), *501*
Wolfrom, M. L. 76, 103 (103), 105 (254), 110 (103, 266), *180*, *184*, *185*
Wolfsberg, M. 44 (99), *55*
Wolter, G. 518 (154), *548*
Wong, K. W. 66 (46, 50), 67 (46), *179*
Woodward, R. B. 510 (110), *547*
Woodworth, R. C. 450 (56, 57), *499*
Wormall, A. 151 (392), *188*
Wright, C. M. 176 (481), *190*
Wu, T. K. 34 (90), *55*
Wulff, H. 104 (251), *184*
Wulfman, C. E. 50 (109), *55*, 118 (301, 302), *186*
Wulfman, D. S. 118 (301–303), *186*

Yager, W. A. 265, 297 (130, 132), 298 (132), 307 (132, 246), 308, 310 (130), *325, 328,* 443, 456 (22, 23), 457 (22, 23, 78, 79, 144), 458 (79), 469 (23), 472 (107), *498, 500, 501,* 580 (43), *585*
Yajima, H. 90 (188), *182*
Yamada, S. I. 403 (54), *436,* 484 (133, 134), *501,* 538 (329–331), *553*
Yamashita, T. 114 (282), *185*
Yarnell, C. F. 118 (301), *186*
Yarovenko, N. N. 86 (162), *182,* 510 (105), *547*
Ykman, P. 380 (225), *394*
Yoffe, A. D. 2 (1, 4), 14, 16 (1), 17 (1, 4), 18, 21, 24, 26 (1), *52,* 448 (43), *499*
Yonezawa, T. 47 (104), *55*
Young, D. P. 399, 407 (17), *435*
Young, G. T. 89, 175 (178), *182*
Young, L. 401 (28), *435*
Young, W. G. 70, 83 (63), 84 (158), 85 (160), *179, 182,* 430 (207), *440*
Yukawa, Y. 79 (130), *181,* 224 (11–13), *322, 323,* 401 (29–31), 402 (38), 418 (128, 129), 420 (129), *435, 438*
Zahler, R. E. 212 (39), *220*
Zaidi, W. A. 301 (236), *328*
Zakharova, T. A. 90, 91 (201), *183*
Zalkow, L. H. 289 (203), *327,* 530 (248), *551*
Zayed, S. M. A. D. 83 (150), *182,* 374 (196), *393,* 563 (26), *584*
Zbiral, E. 215 (45), *220,* 387 (246), *395,* 562, 563 (25), *584*
Zehetner, W. 538, 540 (322), *553*
Zeitlin, M. 234 (47), *323*
Zenda, H. 506 (68), 538 (321), *546, 553*
Zervas, L. 176 (479), *190*
Zheltova, V. N. 22, 23 (61), *54,* 195 (39), *201,* 506, 528 (55), *546*
Ziegler, K. 374 (193), *393*
Zilkha, A. 536 (294), *552*
Zimmerman, P. 150 (390), *188,* 432 (223), *440*
Zinser, D. 387 (249), *395,* 463, 466, 467, 479 (91), *500,* 540, 542 (347), *553*
Zollinger, H. 66, 113 (48), *179,* 217 (48), *220*
Zook, H. D. 413, 414 (111), *437*
Zutarein, P. 413 (108), *437*

Subject Index

Acidity function, for reaction of ketones with HN_3 407
Acidolysis, of azides, leading to amines 338, 339
leading to azomethines 342, 343
Acids, reaction with HN_3 408
α,β-unsaturated, addition of HN_3 123
Acyl azides, addition of, to multiple bonds 529–531
to nitriles 531
adducts of 531–534
decomposition, radical-induced 535, 536
electronic spectra 445
infrared spectra 195, 507
in peptide synthesis 88
nitrene formation from 536–544
nucleophilic displacement of azido group 534
photolysis of 479–491
pyrolysis of 398–402
quantitative analysis of 194
reaction with acids 535
rearrangement of 88, 176, 194, 341, 342, 350, 398–405, 535
synthesis of 86–89, 175, 504–507
Acylium ion, in reaction of acids with HN_3 408
Addition—*see*—Electrophilic addition and Nucleophilic addition
free radical—*see*—Free radicals, addition
Addition–elimination reaction, by azide ion 113, 561, 562
Alcohols, by azide solvolysis 223
Aldehydes, aromatic, reaction with azides 339
reaction with azidoalcohols 370
reaction with HN_3 405–412
of α,β-unsaturated 123
reaction with silyl azides 134
Alicyclic azides—*see*—Azides, alicyclic
Aliphatic azides—*see*—Azides, aliphatic
Alkenes, addition to, of acyl azides 529–531
of azides 260, 373–377, 416
of carbonyl nitrenes 541, 542
of halogen azides 136–142
of mercuric azide 135, 136
of nitrenes 360, 464, 465
of sulphonyl azides 289–291
conjugated, addition of halogen azides 141, 142
addition of HN_3 122–130
with contiguous bonds, addition of HN_3 132, 133
Alkoxycarbonyl azides—*see*—Azidoformates
Alkyl azides, decomposition, thermally induced 245–256, 467
transition metal catalysed 317
with trivalent phosphorus compounds 294
dipole moments 19, 207
electronic spectra 27–30, 445
optical rotatory dispersion measurements 199
photolysis of 297–308, 426–428, 449, 450, 456, 467–474
quantitative analysis of 194
reaction with Lewis acids 242, 243
reaction with protonic acids 223–234
rearrangement, acid-catalysed 416–421, 535
photochemically induced 426–428
thermally induced 421–425
synthesis of, by nucleophilic substitution 75–78
from hydrazine derivatives 175
from sulphonyl azides 171
unsaturated, conversion to heterocycles 363
Alkynes, addition to, of azides 377–379, 531
of azidoformates 369
of halogen azides 142–144
of hydrazoic acid 127, 130–132
of nitrenes 465, 466, 543
conversion to vinyl azides 556, 557

Allyl azides, photolysis of 479
rearrangement of 84, 85
secondary and tertiary 135
synthesis of, by elimination 219
by nucleophilic substitution 83–85
from azirines 156
from hydrazoic acid 164
Allyl compounds, adduct with azides 346
nucleophilic substitution of 70
reaction with azide ion 126
Amides, by acyl nitrene insertions 538
Amidines, from triazoles 434
Amines, from azides, by acidolysis 338, 339
by reduction 223, 333–338
via nitrene intermediates 339–342
Aminosugars, from azidocarbohydrate intermediates 103, 110–113
Ammonia, in azide synthesis 148–152
Analysis, of azides, qualitative 192, 193
quantitative 193–195
Anchimeric assistance—*see*—Neighbouring group effect
Aroyl azides, photolysis of 456, 464, 489–491
rearrangement 400
ultraviolet spectra 195
Arrhenius activation parameters, for alkyl azide pyrolysis 246, 248
Aryl azides, decomposition of, thermally induced 256–278, 467
transition metal catalysed 317–319
with trivalent phosphorus compounds 295
dipole moments 19, 207
electrophilic substitution of 209–212
infrared spectra 197
mass spectra 199
photolysis of 308–314, 388, 426–428, 452, 456, 465, 491–498
potential energy of RN—N_2 bond 453
quantitative analysis of 194
reaction with Lewis acids 243, 244
reaction with protonic acids 234–242
rearrangement of, acid-catalysed 416–421
photochemically induced 426–428
thermally induced 421–425
synthesis of, by diazotization 148–155, 158, 429
by nucleophilic substitution 113–115
from hydrazine derivatives 171–173, 177
from nitroso compounds 165–168, 428
from pentazoles 161
from semicarbazides 177
from sulphonyl azides 169–171
Azepines, synthesis of, from azides 311, 312, 388, 405, 420, 424, 427, 478, 481
from nitrenes 387, 481, 485, 542
Azide ion, displacement of 66
force constant 25
nuclear magnetic resonance spectra 34, 35
nucleophilic substitution by 63–119
for azide synthesis 75–119
mechanism of 63–75
quadrupole coupling 34, 35
theoretical calculations for 34–52
vibration spectra 21, 22
Azide radical, from azidoferric ion 146, 147
from chlorine azide 140
Azides—*see also*—Azido group and Azido compounds
acyl—*see*—Acyl azides
alicyclic, synthesis of 75–80
aliphatic, synthesis of 75–80
alkenyl—*see*—Allyl azides and Vinyl azides
alkoxycarbonyl—*see*—Azidoformates
alkyl—*see*—Alkyl azides
aromatic, synthesis of 113–119, 157
aryl—*see*—Aryl azides
as synthetic starting materials 332–388
for heterocycle synthesis 358–388
transformation of azido to other functional groups 332–358
carbamoyl—*see*—Carbamoyl azides
characterization and determination 191–200
decomposition of 222–322
haloalkyl—*see*—Azidohaloalkanes
halogeno—*see*—Halogen azides
heteroaromatic, synthesis of 115–117, 157, 173

Azides (*cont.*)
hydrazidic—*see*—Hydrazidic azides
ionic—*see*—Ionic azides
photochemistry of 442–498
rearrangements of 398–434
silyl—*see*—Silyl azides
structure of 14–34
sulphonato—*see*—Azidosulphonates
synthesis of 63–177
by addition reactions 119–147
by diazotization 147–176
by nucleophilic substitution by azide ion 63–119
toxicity and explosiveness of 61–63
Azidinium salts—*see*—*N*-Diazonium compounds
Azidoalcohols, decomposition, acid-catalysed 232
dipole moments 19
infrared spectra 195
reaction with aldehydes 370
synthesis from epoxides 93–95, 102
Azidoaldehydes, synthesis of, by diazotization 150
from hydrazoic acid 123
thermolysis of 253, 424
Azidoalkenes, dipole moments 19
synthesis of 167, 218
thermolysis of 256
Azidoanilines, dissociation constants of 205
Azidoazomethines, synthesis of 92
tautomerism of 86, 90, 91, 116, 383
Azidobenzalazines 517
Azidocarbazoles, by azide photolysis 312
Azidocarbohydrates, circular dichroism of 199
in azide analysis 193
leading to aminosugars 103, 110
reduction of 335
synthesis of 103–113
Azidocarbonyl group, effect on nucleophilic substitution 214
Azidocarboxylic acids, dissociation constants 204, 205
infrared spectra 196
photolysis of 470, 471
reduction of 336
synthesis of 123, 146
Azidodehalogenation 115
α-Azidoesters, dipole moment 19
photolysis of 472
synthesis of 123, 141, 146
Azidoethers, infrared spectra 195
ultraviolet spectra 198
Azidoferrocenes, decomposition, photolytically induced 312–314
thermally induced 268, 312–314
synthesis of 118
Azidofluorenes, synthesis of 79, 158
Azidoformamidines—*see*—Guanyl azides
Azidoformates, conversion to nitrenes 89
photolysis of 450, 451, 455, 462, 464, 466, 479–486
protecting agents for amino groups 89, 90
reaction with alkynes 369
rearrangement of 342, 405, 535
synthesis of 89, 90, 527, 528
thermolysis of 451, 462
ultraviolet and infrared spectra 528
Azido group, as leaving group 212, 213
directing and activating effect of 203–219
in addition reactions 216, 217
in electrophilic substitution 209–212
in elimination reactions 217–219
in nucleophilic substitution 212–216
Azido halides—*see*—Halogen azides
Azidohaloalkanes, conversion to vinyl azides 80, 136
formation from iodine azide 136, 137
Azidohalogenation, of alkenes 147
Azidohydrins, hydrogen bonding in 233
Azidoimines, infrared spectra 196
Azidoketones, acidolysis of 342, 343
conversion to nitriles 348
conversion to pyrazines 387
dipole moments 19
infrared spectra 195
reduction of 334
synthesis of 123, 141, 146
thermolysis of 253, 371, 424
Azidolactam, mass spectra of 200
Azidomercuration 135
Azidonium ions 106, 215
Azidooximes 512, 513
synthesis of 92
tautomerism of 90
Azidopeptides, synthesis of 89

Azidoquinones, decomposition of, acid-catalysed 239–241
thermally induced 277, 278
mass spectra 200
rearrangement of 420, 421, 581–584
synthesis of, from aminoquinones 150
from haloquinones 83, 581
from quinones 127
Azidosteroids, synthesis of 95–103, 145
Azidosulphonates, for azide synthesis 88
Azidosulphones, ultraviolet spectra 198
Azidotropones, conversion to nitriles 349, 497
mass spectra of 200
photolysis of 496, 497
thermolysis of 277, 278
Aziridines, conversion to aryl azides 149
formation of, from azides 248, 359–361, 451, 485, 487, 526
from nitrenes 464, 541
from triazolines 530
Azirines, conversion to allyl azides 156
synthesis from azides 121, 361, 362, 423, 427, 475, 563–565
Azo compounds, synthesis of 354–356
by diazo transfer 354, 355
by nitrene/azide reaction 467
by nitrene dimerization 355, 356, 461
Azomethines, synthesis of, by azide acidolysis 342, 343
by decomposition of azide adducts 345–348
by nitrene rearrangement 343–345

Benzofuroxans, synthesis from azides 371, 372
Bond angles, of azides 14–16
of hydrazoic acid 14–16
Bond lengths, of azide ion 14
of azides 14–16
of hydrazoic acid 14–16
Bond order, of azides 5
σ Bonds, in azides 5, 6
π Bonds, in azides 6, 7
Bromonium ion 140

Carbamoyl azides 521–527
conversion to nitrenes 90
dissociation of 521
infrared spectra 195, 196
photolysis of 483, 522
rearrangement of 342, 404, 522–524
synthesis of 90, 132, 133, 176, 520, 521
Carbazoles, synthesis from azides 365, 492
Carbolines, synthesis from azides 365
Circular dichroism, of azides 32, 33, 199
Configuration, retention in azide rearrangement 399, 403
Conformation effect, on migratory aptitude of groups 301, 304
Conjugation, of azides 4, 208, 213, 214
Conjugative effect, of azido group 206, 207, 216
Correlation diagram, for the azide ion 38–40
Curtius rearrangement, of acyl azides 88, 176, 194, 341, 342, 350, 398–405, 504
Cyanides—*see*—Nitriles
Cyclization, intramolecular in azide pyrolysis 261–265, 267

Decomposition, of azide adducts 345–348
of azides 222–322
acid-catalysed 222–245
photolytic 297–317
thermally induced 245–294
transition metal catalysed 317–322
with trivalent phosphorus compounds 294–296
Diazides, conversion to dinitriles 348
conversion to tetrazoles 383, 472
photolysis of 307, 472
synthesis of 76, 130, 146, 563
Diazidoalcohols, synthesis from epichlorhydrin 93, 94
Diazidoquinones, photolysis of 497
synthesis from quinones 127
Diaziridines, synthesis from azides 362
Diazo compounds, synthesis of 292, 350–354, 428, 533
Diazonium compounds, reaction with ammonia and derivatives 148–152
reaction with hydrazoic acid 429

N-Diazonium compounds 116, 352, 357
Diazotization, leading to azides 147–176
mechanism of reaction 158
Diazo transfer reaction 168–171, 292, 350–355
Diazo transfer reagent 170
Dinitrenes, electronic spectra of 458
Dipole interactions, in azide ion substitutions 105
Dipole moments, of azides 18–20, 207
of hydrazoic acid 18, 20
Displacement, of azide ion 66
Dissociation, of acyl nitrenes 538
Dissociation constant, of azidocarboxylic acids 204, 205
of hydrazoic acid 3

Electrolytic reaction, leading to azides 176, 177
Electronegativity, of azide radical and halogens 141
Electronic spectra, of azide ion 25–27
of azides 26
aliphatic 27–30, 445, 446
aromatic 30, 447
of hydrazoic acid 26, 27, 444
of nitrenes 308, 456–461
Electronic structure, of covalent azides 4–8
of nitrenes 456–461
Electron spin resonance spectra, in sulphonyl azides thermolysis 280
of nitrenes 265, 456
Electrophilic addition, of halogen azides 136–144
of hydrazoic acid 128–130, 132, 133, 217
of vinyl azides 216
Electrophilic substitution, effect of azido group 207
of azides 209–212
Elimination reactions, of azides 217–219
with concerted migration 306
Enamines, conversion to triazolines 374
Epimines, conversion to azidosugars 103, 110, 111
Episulphides, conversion to azidosugars 103, 110
Episulphonium intermediate, in azidosugars synthesis 112

Epoxides, conversion to azidoalcohols 93–95
conversion to azidosugars 103, 110–113
Esters, α,β-unsaturated, addition of azides 141
addition of hydrazoic acid 123

Force constants, of azide ion 25
of azides 24
of hydrazoic acid 24
Free radicals, addition of, leading to azides 144–147
effect of, on acyl azides 535, 536
on organic azide pyrolysis 275–277
on sulphonyl azide pyrolysis 291, 292
on vinyl azides 576

Guanyl azides 513–516

Haloalkyl azides—*see*—Azidohaloalkanes
Halogenation, of azides 210
Halogen azides, addition to alkynes 142–144, 556, 557
addition to alkenes 80, 120, 136–142, 217
as azide radical source 140
elimination reactions 217
free radical additions 144, 145
infrared spectra 196
Hammett relationship, for aryl azide rearrangement 417
Heats of formation, of azide ion 18
of azides 16
Heterocycles—*see also*—specific classes of compounds
synthesis from azides 358–388, 422, 423
Hofmann rearrangement, of azo compounds 177
Hückel-type calculations, for azide ion 8, 9, 40–45
for azido group 9–12
for hydrazoic acid 40–45
sp Hybridization, in azides 5
Hydrazides, conversion to azides 154, 505
Hydrazidic azides, synthesis of 93, 517
tautomerism of 90
Hydrazidines, conversion to azides 176

Hydrazines, conversion to azides, by diazotization 152–156
by nitrosation 171–176
preparation, by azide photolysis 312
from carbamoyl azides 342
Hydrazoic acid, additions of 120–133
1,3-dipolar 120
electrophilic 128, 129
to alkynes 130–132, 556
to compounds with contiguous double bonds 132, 133
to cyanides 415
to olefins 120–130
bond angles and lengths 14–16
bond energy 17
diazotization leading to azides 156–165, 429
dipole moment 18, 20
dissociation constant 3
electronic spectra 26, 27, 444
force constants 24
ionization potentials 31
n.m.r. spectra 34, 35
photolysis of 297, 442, 447–449, 467
quadrupole coupling 33–35
reaction with carbonyl compounds 405–412
reaction with nitroso compounds 165–168
Rydberg transitions 30, 31
theoretical calculations for 34–52
vibration spectra 23
Hydrogen azide—*see*—Hydrazoic acid
Hydrogen bonding, in azidohydrins 233
Hydroxy azides—*see*—Azidoalcohols
Hydroxylamine, conversion to aryl azide 150

Imidazoles, synthesis from azides 367, 368
Imides, aromatic, conversion to azides 128
Imidoyl azides 510
conversion to phosphazides 532
intermediate in acyl addition to nitriles 531
tautomerism 510, 513, 514
with C=N function in ring 518, 519
with open chain C=NX 510–517
Imines, from azides 223, 468–470, 474
Imino azides—*see*—Azidoimines
Iminophosphoranes—*see*—Phosphinimines
Imino radical, in NH_3 photolysis 442, 443
singlet state of 460
Iminosulphuranes, synthesis from azides 356
Indazoles, synthesis from azides 368, 369
Indoles, synthesis from azides 364, 563
Indolines, synthesis from azides 364
Inductive effect, of azide radicals 3
of azido group 205, 212, 216, 217
Infrared spectra, of azides 195–197, 207, 507, 528
Iodonium ion, ring opening by azide ion 139
Ionic azides—*see also*—Azide ion and Mercuric azide
"activated" for organic azide synthesis 76, 86
explosive properties 62
Ionization constants—*see*—Dissociation constants
Ionization potential, of HN_3 and azides 31
Isocyanates, conversion to azides 132, 133
synthesis from azides 87, 349, 350, 449
Isotopic labelling, in azide/aniline reaction 257
in azide structure study 171
in diazonium compound/NH_3 reaction 149
in diazonium compound/hydrazines reaction 152
in diazonium compound/HN_3 reaction 159, 165
in nitroso compound/HN_3 reaction 166
Isoxazoles, synthesis from azides 371, 563

Keto azides—*see*—Azidoketones
Ketones, conversion to diketones 342
reaction with hydrazoic acid 405–412
reaction with silyl azides 134
α,β-unsaturated, addition of azides 141
addition of hydrazoic acid 123

Lactams, from acyl azides 539
LCAO SCF molecular orbitals 48–50
Lewis acids, catalysts for azide rearrangement 402, 420
  for azide decomposition 242–245, 420
Lone pair electrons, in azides 5, 7, 8

Mass spectra, of azides 199, 200
Mercuric azide, for organic azide preparation 134–136
Migration, in azide decomposition, acid-catalysed 223, 224, 228
    photolytic 299–302, 304, 426, 471
    thermally induced 248, 299–301, 399
  in Schmidt reaction 408–410

Neighbouring group effect, in nucleophilic substitution reactions 74 75, 105, 215
  in pyrolysis of aryl azides 261, 262
Nitration, of azides 209, 210
Nitrenes, addition to alkenes 360, 464, 465, 541, 542
  addition to alkynes 465, 466, 543
  addition to isonitriles 544
  addition to nitriles 543, 544
  addition to sulphoxides 544
  conversion to carbene 272
  dimerization to azo compounds 355, 356, 461
  dissociation 538
  electronic spectra 308, 456–461
  e.s.r. experiments on 265, 266, 456
  electrophilic attack on non-bonding pairs 466–468
  hydrogen abstraction by 463, 464
  insertion of, into C—H bonds 265, 426, 461, 462, 487, 525, 538–540
    into N—H bonds 462, 463, 540, 541
    into O—H bonds 462, 527, 540
    intramolecular leading to heterocycles 362–368, 384, 385
  intermediate in azide photolysis 426, 442, 449
  intermediate in azide reduction 339–342
  preparation 537
    by azide decomposition 17, 245, 249, 258, 265, 339, 425
    from azidoformates 89
    from carbamoyl azides 90
  reaction with aromatic systems 542, 543
  rearrangement of 343–345
  singlet 537, 541, 542, 544
    addition to double bonds 483, 485, 486
    for azepine formation 481, 485
    in azide photolysis 303, 310, 427, 481, 484
    in azide pyrolysis 247, 265, 266, 278, 286
  triplet 537, 541
    addition to double bonds 483, 485, 486
    in azide photolysis 302, 310, 403, 427
    in azide pyrolysis 247, 265–267, 286
Nitrenium ion, intermediated in azide decomposition 235, 242
Nitriles, addition of azides 374, 382, 383, 531
  addition of hydrazoic acid 123
  addition of nitrenes 543
  conversion to oxadiazoles 371, 543
  conversion to tetrazoles 415
  synthesis from azides 348, 349
Nitro compounds, α,β-unsaturated, addition of HN_3 123
Nitrosation, of hydrazine derivatives 171–176
Nitrose compounds, conversion to azides 165–168, 428
Nuclear magnetic resonance spectra, of azides and HN_3 34, 35
Nucleophiles, azide ion as 65–70, 106, 143
  effect on organic azide pyrolysis 274, 275, 388
  effect on sulphonamide pyrolysis 287–289
Nucleophilic addition, leading to azides 119–147
  of azide ion to organic azide 217
Nucleophilic substitution, by azide ion 63–119
  for azide synthesis 75–119
    mechanism of 63–75
  effect of azidocarbonyl group 214
  of azides 207, 212–216, 534

Olefins—*see*—Alkenes
Optical rotatory dispersion, of azides 31–33, 199

Organometallic azides, infrared spectra 196
thermal rearrangement of 424
ortho–para Orientation, by azido group 4, 207, 209
Oxadiazoles, synthesis from azides 371–373, 486, 543
Oxazoles, synthesis from azides 369–371, 543
synthesis from ketone/HN_3 reaction 413
Oxidation, of arylhydrazines 177
Oximino azides—*see*—Azidooximes

Pan-activating substituents 205
Pentazene 162, 163, 167
Pentazoles 160
decomposition of 161, 163–166
formation of 163, 165
intermediates 162
Peptide synthesis, azidoformates as protecting agents 89
using acyl azides 88, 175, 534
Phosphinimines, synthesis from azides 356, 357, 532
Phosphoranes, reaction with azides 533
Phosphorus compounds, for azide decompositions 294–296, 356
for vinyl azide synthesis 562, 563
Photolysis, flash 311, 443
of azides 297–317, 361, 388, 402–404, 426–428, 442–498
of triazolines 359
of vinyl azides 564–575
sensitized 455, 456
Piperidones, from azide photolysis 484
Polar effect, on azide ion substitutions 108, 109
Polarization effect, of azido group 204–209
Protecting agents, acyl azides as 175
Protonic acids, for azide decomposition 223–242
Pseudohalides, azides as 2–4, 215
Pyrazines, synthesis from azides 386, 387, 476, 563
Pyrazoles, synthesis from azides 368
Pyrolidines, synthesis from azides 366, 468, 469
Pyrolysis, of azides 245–294, 361, 371, 388, 398–402, 421–425
of vinyl azides 564–575

Quadrupole coupling, of azides and HN_3 33–35
Quantum yield, of alkyl azide photolysis 449, 450
of aromatic azide photolysis 453, 454
Quinolines, by azide photolysis 469
Quinones, addition of hydrazoic acid 127

Radicals—*see*—Free radicals
Raman spectra, of azides 197
Rearrangement—*see also*—named rearrangements
leading to azides 428–430
of allyl azides 84, 85
of azides 398–434
acid-catalysed 402, 416–421
in unusual substrates 404, 405
photochemically induced 402–404, 426–428
thermally induced 398–402, 421–425
of triazenes 430–435
Redox systems, azide formation in the presence of 145–147
Reduction, in azide decomposition 223, 333–338
Regioselectivity 139, 140, 145
Regiospecificity 96, 139, 140, 144, 217, 218
Resonance effect, of azides 4
Resonance energy, of the azido group 17
Ring contraction, in sulphonates substitution by azide ion 108, 109
Rotation barrier, in azides 25
Rydberg transitions, in HN_3 spectrum 30, 31

Schmidt reaction 405–421
of azides 224, 227
of carbonyl compounds 230, 405–412
of diarylethylenes 227, 228
of quinones 239
Schmidt-type reactions, for decomposition of protonated azide 124
Self-consistent field methods (SCF) 45–47
Semicarbazide, conversion to aryl azides 177
from hydrazidic azides 517

Silyl azides, addition to carbonyl compounds 120, 134
Singlet state, of azides 303
of nitrenes 247, 265, 266, 272, 286, 303, 310, 427, 451, 481
Solvent effect, in nucleophilic substitution reactions 70–72, 82, 106
Stereochemistry, in azide substitution reactions 69, 72–75
Stereospecificity, in elimination reactions of azidohaloalkanes 219
of azide reduction to amines 337
of azido group introduction 95, 136, 138
of nitrenes addition 360, 451, 483
Steric effect, on substitution by azide ion 105, 108
Steroids, reaction with azide ion 95–103
Structure, of azido group 12–34
Substituent constants, of azido group 30, 204–206, 210, 213, 214
Substitution—*see*—Electrophilic substitution and Nucleophilic substitution
Sulphenyl azides 296
Sulphinyl azides, decomposition of 296
Sulphonyl azides, conversion to azides 168–171
conversion to triazenes 151
decomposition, acid-catalysed 242
radical-induced 536
thermally induced 279–294, 404
transition metal catalysed 320–322
with trivalent phosphorus compounds 296
infrared spectra 196
quantitative analysis of 194
photolysis of 314–317
Sultams, obtained from sulphonyl azides 285
Swain–Scott equation 159

Tautomerism, of azidooximes 90
of hydrazidic azides 90
of tetrazoles 86, 91, 116, 383, 384, 510, 513, 514, 518, 519
Tetrazene, formation of 152–154
intermediate in azide formation 152, 177
intermediate in triazene formation 358
Tetrazoles, synthesis of 125, 382–384, 412, 413, 415, 432–434, 472, 510, 513, 531
tautomerism of 86, 91, 116, 383, 384, 510, 513, 518, 519
Thermodynamic properties, of azides 16–18
Thermolysis—*see*—Pyrolysis
Thiatriazoles, photolysis of 510
synthesis 133, 507–509
thermolysis of 509
Thioacyl azides 507–510
Toxicity, of azides 61–63
Transition metals, catalysts in azide decomposition 317–322
Triazenes, cyclic—*see*—Triazoles
product of azo rearrangement 177
rearrangement of 430, 431
synthesis from azides 357, 532, 534
Triazoles, rearrangement of 431–434
synthesis of 377–382
by alkyne/HN_3 reaction 127, 130 131
by azide photolysis 427
by organic azide reaction, with alkenes 377–379
with alkynes 130, 377, 531
with ketones 377
Triazolines, conversion to aziridine 530
photolysis of 359
synthesis from azides 373–377, 530
Triplet state, of azides 302, 310, 313
of nitrenes 247, 265–267, 286, 302, 310, 403, 427, 451
Tropones, conversion to azides 117

Ultraviolet spectra, of azides 197, 198, 528
Urethanes, by azide photolysis 489, 490

Vibration spectra, of azide ion 21, 22
of azides 22, 23
of hydrazoic acid 18, 20
Vinyl azides, addition reactions of 216
as intermediates 573–575
conversion to 1-azirines 361, 423, 427, 475, 563
conversion to nitriles 348
decomposition, acid-catalysed 576, 578
free radical induced 576
thermally induced 564–575

Vinyl azides (*cont.*)
infrared spectra 195, 196
internal 559, 565
mechanism of reactions 577–581
photolysis of 475–479, 564–575
synthesis of 556–563
by addition–elimination reactions 561, 562
by nucleophilic substitution 80–83
from alkynes 556, 557
from azides 557–561
from azidohaloalkanes 136, 138, 218
using phosophoroorganic compounds 562, 563
terminal 559, 560, 563
Vinyl compounds, nucleophilic substitution of 69
Waldern inversion 72, 73, 75, 95, 105